ANNUAL REVIEW OF PLANT PHYSIOLOGY AND PLANT MOLECULAR BIOLOGY

ANNUAL REVIEW OF PLANT PHYSIOLOGY AND PLANT MOLECULAR BIOLOGY

VOLUME 50, 1999

RUSSELL L. JONES, *Editor*
University of California, Berkeley

HANS J. BOHNERT, *Associate Editor*
University of Arizona

VIRGINIA WALBOT, *Associate Editor*
Stanford University

http://www.AnnualReviews.org   science@annurev.org   650-493-4400

ANNUAL REVIEWS   4139 EL CAMINO WAY   P.O. BOX 10139   PALO ALTO, CALIFORNIA 94303-0139

ANNUAL REVIEWS
Palo Alto, California, USA

International Standard Serial Number: 1040-2519
International Standard Book Number: 0-8243-0650-3
Library of Congress Catalog Card Number: 50-13143

⊗ The paper used in this publication meets the minimum requirements of American National Standards for Information Sciences—Permanence of Paper for Printed Library Materials, ANZI Z39.48-1992.

TYPESET BY TECHBOOKS, FAIRFAX, VA
PRINTED AND BOUND IN THE UNITED STATES OF AMERICA

Annual Review of Plant Physiology and Plant Molecular Biology
Volume 50 (1999)

CONTENTS

(*continued*)   v

vi   CONTENTS (*continued*)

RELATED ARTICLES IN OTHER *ANNUAL REVIEWS*

From the *Annual Review of Biochemistry*, Volume 68, 1999:

Initiation of Base Excision Repair: Glycosylase Mechanisms and Structures,
Amanda K. McCullough, M. L. Dodson, and R. Stephen Lloyd

Sterols and Isoprenoids: Signaling Molecules Derived from the Cholesterol Biosynthetic Pathway, Peter A. Edwards and Johan Ericsson

CREB: A Stimulus-Induced Transcription Factor Activated by a Diverse Array of Extracellular Signals, Adam J. Shaywitz and Michael E. Greenberg

The 26S Proteasome: A Molecular Machine Designed for Controlled Proteolysis,
D. Voges, P. Zwickl, and W. Baumeister

In Vitro Selection of Functional Nucleic Acids, David S. Wilson and Jack W. Szostak

Structural Motifs in RNA, P. B. Moore

De Novo Design and Structural Characterization of Proteins and Metalloproteins,
William F. DeGrado, Christopher M. Summa, Vincenzo Pavone, Flavia Nastri, and Angela Lombardi

Mutagenesis of Glycosidases, Hoa D. Ly and Stephen G. Withers

Catalysis by Metal-Activated Hydroxide in Zinc and Manganese Metalloenzymes,
David W. Christianson and J. David Cox

Nuclear-Receptor Ligands and Ligand-Binding Domains, Ross V. Weatherman, Robert J. Fletterick, and Thomas S. Scanlan

Cellular and Molecular Biology of the Aquaporin Water Channels, Mario Borgnia, Søren Nielsen, Andreas Engel, and Peter Agre

Regulation of the Cytoskeleton and Cell Adhesion by the Rho Family GTPases in Mammalian Cells, K. Kaibuchi, S. Kuroda, and M. Amano

Inorganic Polyphosphate: A Molecule of Many Functions, Arthur Kornberg, Narayana N. Rao, and Dana Ault-Riché

Membrane Fusion and Exocytosis, Reinhard Jahn and Thomas C. Südhof

The Anaphase-Promoting Complex: New Subunits and Regulators,
A. M. Page and P. Hieter

Control of Carpel and Fruit Development in Arabidopsis, Cristina Ferrándiz, Soraya Pelaz, and Martin F. Yanofsky

From the *Annual Review of Biophysics and Biomolecular Structure*, Volume 27, 1998:

Signaling Complexes: Biophysical Constraints on Intracellular Communication,
Dennis Bray

(*continued*)

From the *Annual Review of Cell & Developmental Biology,* Volume 14, 1998:

(*continued*)

Martin Gibbs

Annu. Rev. Plant Physiol. Plant Mol. Biol. 1999. 50:1–25

EDUCATOR AND EDITOR

Martin Gibbs

Department of Biology, Brandeis University, Waltham, Massachusetts 02454-9110;
e-mail: cabral@binah.cc.brandeis.edu

KEY WORDS: photosynthesis, isotopes, intermediary metabolism, *Plant Physiology*, journal
management

CONTENTS

The First Years

I was born on Armistice Day, 1922, in the city of Brotherly Love. My years
in the Philadelphia public school system encompassed the stock market sell
down of 1929 and the Great Depression of the 1930s. Although my brother
and I grew up in modest circumstances, we were able to holiday each August in
Atlantic City, then a family-oriented beach community with games of chance
limited to a penny arcade and the promise of a jackpot of a candy bar. Pocket
money was sufficient to afford a little chemistry lab in the basement of our
home until a burst of hydrogen sulfide gas convinced my parents to conclude
the tinkering below. The neighborhood elementary (grades 1–6) schoolhouse
was a nineteenth-century sandstone building subdivided entirely into traditional
classrooms, while lunch was taken at home and some classes (wood shop)
were accommodated in other schools. Class began each day with inspirational
readings from the Bible. Another event that I recall vividly was the halting of
instruction with the class standing in silence for one minute on the eleventh
hour of November 11 to commemorate the signing of the treaty ending the
1914–18 War. Junior High School experience differed considerably from that
of the neighborhood school. The four-mile distance between our home and
Junior High was accomplished by walking inasmuch as school busing was not
in effect and public transportation was unavailable. I do not remember a single

1

occasion when snowstorms, and there were many, delayed the instructional schedule. In contrast to the neighborhood school, the student body was more ethnically diverse. I entered the all-male (students and teachers) Central High School established in 1836 for students who had earned above average grades in the previous nine years. I took courses in biology, chemistry, physics, math, Latin, and German and on graduation in February 1940, buttressed by winning the monetary chemistry award, I knew that I wanted a college education with a major in chemistry.

Inasmuch as entry into college was restricted to the fall, I made good use of the interval to earn my keep and tuition. I was the gofer in a bookbinding shop, home-delivered circulars of the local supermarket, and in the summer worked on the Pennsylvania Railroad line between Camden and Atlantic City. Fortunately, railroad employment continued throughout the undergraduate years, providing more than adequate financial support but precious little time for extracurricular activities.

College Years, Undergraduate and Graduate

I matriculated into the Philadelphia College of Pharmacy and Science in the fall of 1940 as a concentrator in chemistry and my brother, following military service, into the Temple College of Pharmacy. The College was primarily a specialized undergraduate-level institution where the curriculum and requirements in the sciences dictated much of the annual schedule. A senior honors or independent study was not offered but the chemistry staff did not discourage lab initiative. Only two courses, I recall, associated with a traditional liberal arts curriculum were listed in the instructional book. One, to which I took a yearlong liking because Herman Wittmeyer was a gifted teacher, focused on the major writers of classic and contemporary literature. The other was a one-semester introduction to economics taught by an adjunct professor.

The baccalaureate under an accelerated wartime program was scheduled for completion in the fall of 1943. In the third and final year, I moved into botany almost by accident. Of the science electives offered, I opted for pharmacognosy, my first full course in plant science. Edmund MacLaughlin, dedicated and knowledgeable, emphasized the history of drugs and, making use of herbarium specimens, identification of the important medicinal plants. Theodor P. Haas, curator at the Munich-Nymphenburg Botanical Institute until he had to flee the Nazis in 1938, traveling to the United States by way of Russia, Siberia, and San Francisco, joined the College as an instructor in botany and medicinal plants. On weekends, Haas, an educator, serious and wise, with strong will and character, conducted field trips accompanied by a selected group of chemistry concentrators into the parks of Philadelphia where he described the principles of plant systematics and explained plant parts and their function. Thus my interest in biology and particularly plant biology took hold.

Having decided to win a graduate degree (with the approval of the draft board), and acting on the advice of Arthur Osol, the professor of chemistry, four classmates and I motored in a gas-guzzler supported by precious ration coupons to Pittsburgh to attend the 1943 meeting of the American Chemical Society. I registered with the placement service for available positions and of the two offered, I accepted willingly a teaching fellowship for $750 per year in the Department of Chemistry at the University of Illinois in Urbana. My weekly assignments would be 12 hours of assisting in the general and analytical labs and an additional eight hours of exam grading and tutoring. During the next four years I earned a doctoral degree in botany, with minors in chemistry and agronomy, and met an undergraduate, Karen Kvale, who after a career as an international flight attendant, became my wife.

The Illinois faculty was outstanding. I took organic chemistry with Reynold C. Fuson, biochemistry (Richard Byerrum was a lab assistant) with William C. Rose, and physical chemistry with Frederick T. Wall. They were great teachers and recognized investigators who linked together the two functions of teaching and research. As the year progressed, I decided to major in the application of chemistry to botany but the biochemistry group, then a division of the chemistry department, had little interest in plants. Therefore I transferred into the Department of Botany as a research assistant funded by a grant administered by F. Lyle Wynd, who became my nominal advisor. The Botany Annex, his lab, was spacious and well equipped but its location of one mile from campus isolated his students from contact with other graduate students and from Wynd himself who was housed in the centrally located botany building. The other Annex occupant in his final year was G. Ray Noggle. Arthur Galston had departed the previous year. My obligation to the donor was fulfilled by ashing endless samples of grasses followed by mineral content analyses using techniques developed by Noggle. Clearly Noggle was not an ardent New Dealer since I was admonished for tuning into the Franklin D. Roosevelt funeral procession described by Arthur Godfrey. I spent the academic year completing the required course work and was preparing a thesis outline when Wynd resigned. His sudden departure in the summer of 1945 for Michigan State University, where he would chair the Department of Botany and Plant Pathology, left me in an awkward position. I remained behind and Harry Fuller replaced Wynd as advisor. Noggle, who earned the doctoral degree with Wynd that summer, stayed on until the spring of 1946 before moving to the University of Virginia. I was the sole occupant of the Annex but without a game plan or fellowship support.

An unexpected opportunity solved the financial questions. D.D. DeTurk, a recognized authority on soil fertility, had secured two years of funding (again $750 per year) to support an assistant in the Department of Agronomy. The research project, the influence of soil type on the chemical composition of common grasses, was conducted in the Dixon Springs Experiment Station near Vienna, IL, about 200 miles south of Urbana. Gathering samples involved periodic trips, by car, between the Station and the Annex, where the analyses were performed.

With that dilemma in hand, I attended to the thesis. Here I took advantage of Noggle's research, which dealt with ploidy in plants, and of John T. Buchholz's lectures presented in his morphology course. Buchholz, President of the Botanical Society, was a summer investigator in the Department of Genetics in Cold Spring Harbor, where he published with the distinguished geneticist, Albert F. Blakeslee, on the genetics of *Datura*.

My PhD topic was the chemical changes occurring during the growth of diploid and tetraploid *Datura stramonium*. Blakeslee supplied the seeds and Buchholz advised the plan of research. Polyploid plants have a slower rate of physiological development as compared to that of diploids. With respect to *Datura*, the diploids took 35 days and the tetraploids needed an additional 15 days to complete the vegetative phase of growth. Clearly, at equal chronological ages the tetraploids were physiologically younger than the diploid plants. The results of the analyses (mineral content, weight, sugars) were plotted against age in days (chronological age) or leaf number (physiological age). The slower rate of development explained most of the differences in the chemical composition of the two tissues of equal chronological age. I deposited the thesis and final report to the Agronomy Department in May 1947.

In my haste to conclude the two responsibilities, I had neglected a job search. To this end, I registered with a placement service and scanned each issue of *Science*. Nary an invitation or hint of a position. At the advice of Robert Emerson, a new arrival in the Department of Botany, I directed inquiries to David Goddard, Ray Dawson, and Kenneth Thimann. None had postdoctoral funds but Thimann did urge submission of a resume to the Brookhaven National Laboratory, scheduled to be activated in July. I returned to my parent's home in June with the advanced degree but without employment.

Following a month of restless inactivity, I dusted off my three-speed Raleigh bicycle and set out solo on a tour of New England and eastern Canada armed with an American Youth Hostel pass. I had made good use in 1939 and 1941 of their overnight facilities billed at 25 cents per evening. On the first trip lasting two weeks, I visited Montreal, Sherbrooke, and the Lake Memphremegog area, reaching St. Albans, VT, by train from Philadelphia. The second began in Rochester, NY, and ended in North Bay, the birthing city of the Dionne quintuplets. The immediate trip starting in Northfield, MA, and reaching Canada with a route along the Connecticut River, scheduled for at least six weeks, was halved when I collected mail addressed to General Delivery, Boston Post Office on August 12. Within the packet, in addition to an encouraging note from

Oswald Tippo, a member of my doctoral committee and later, Chancellor of the University of Massachusetts-Amherst, was an interview invitation from the Brookhaven National Laboratory in response to the resume I had forwarded in April. The week following, I sat opposite the Director of Personnel, a physicist. I returned September 2 with the title of Junior Scientist in the Department of Biology.

Brookhaven National Laboratory

Brookhaven National Laboratory, formerly Camp Upton named after Brigadier General Emery Upton, Civil War hero and military tactician, is situated on Long Island roughly of equal distance between New York City and Montauk Point. Upton was a US Army induction center in the great wars of 1917 and 1941. Possibly its best-known inductee was Sergeant Irving Berlin who composed and produced musicals such as Yip Yip Yaphank about army life in the camp in the US War Army Theater. The camp, inactivated in 1946, was transferred to the US Atomic Energy Commission who in turn contracted it to Associated Universities, Inc., a consortium of Ivy League institutions.

The Laboratory was divided into four units: physics, chemistry, medicine, and biology. When I arrived, laboratories and library were in the planning stages.

Most employees, myself included, were billeted on the site in officers' quarters. The old mess hall fortified with a more fitting name concluded service with the Friday lunch. In good time, the Laboratory operated a Friday bus to nearby Patchogue, the link to New York City via the Long Island Railroad, with a return pick-up on Sunday evening. Since the big city was prohibitively out-of-reach for the stipend of a junior scientist, I remained on site with provisions stored in the departmental refrigerator.

At my first meeting with the departmental chairman, Leslie F. Nims, a mammalian physiologist from Yale, I was informed that one function of Brookhaven personnel was to supply ^{14}C-labeled compounds to the research community of Associated Universities, Inc. He assigned me the responsibility of synthesizing radiocarbon-labeled simple sugars. Inasmuch as barium carbonate was the only radioactive compound available, photosynthesis in leafy material seemed a practical approach.

The barrack adjacent to the departmental office was converted to a laboratory in the spring of 1948 and greenhouse space was rented off-site. Competent machinists and glassblowers fabricated an apparatus for exposure of leaves of *Canna indica*, an accumulator of sucrose, to ^{14}CO$_2$ and by conventional methodology, sucrose, and on hydrolysis, the monosaccharides were available for gratis distribution. I realized that my tenure at Brookhaven would be restricted to a routine until the sugars were commercially accessible. To this end,

I convinced a company of the ready market for the radiotagged reagents and when listed in their catalog, my production line was shut down.

One immediate problem was how to localize the ^{14}C in the sugars. Isotopic carbon distribution patterns in sugar were determined at that time by a procedure introduced in 1945 by the noted microbial biochemist, Harland G. Wood, the discoverer of CO_2 fixation by nonphotosynthetic organisms. Here, the two lactic acids formed from glucose or fructose in the *Lactobacillus casei* glycolytic homolactic fermentation were degraded chemically. The method yielded tracer not in individual but in pairs of carbon atoms.

There was need for a method to determine radioactivity in the individual carbons. The solution came when Ralph DeMoss, my first postdoctoral student, and his professor I.C. (Gunny) Gunsalus brought a novel microorganism to Brookhaven. They and R.C. Bard had reported that *Leuconostoc mesenteroides* produced one molecule each of CO_2, ethanol, and lactic acid per glucose fermented. Using labeled glucoses prepared biosynthetically and chemically, we demonstrated that *Leuconostoc* ferments glucose via a new reaction sequence: Carbon dioxide arises from carbon 1; the methyl and carbinol carbons of ethanol from carbons 2 and 3, respectively; and the carboxyl, alpha- and beta-carbons of lactic acid from carbons 4, 5, and 6. To account for the novel data, we proposed that a portion of the oxidative pentose phosphate pathway (hexosemonophosphate shunt) was involved. More important, for a determination of isotopic carbon patterns in carbohydrates, this degradation of glucose yields each carbon separately—a decided advantage over *L. casei*.

The findings with *Leuconostoc* established glucose dissimilation by a pathway other than classical glycolysis. With L.M. Paege, Vincent W. Cochrane, Martin Busse, Jack Sokatch, and W.A. Wood, fermentative sequences of glucose metabolism were surveyed in a number of microorganisms. The anaerobe *Pseudomonas lindneri* was selected because of its high yields of ethanol and CO_2 arising from glucose and fructose. The conversion of the simple sugars to end products involved a mechanism differing, in part, from both classical glycolysis and the *Leuconostoc* pathway. The results indicated a close resemblance to the aerobic dissimilatory pathway of Entner and Doudoroff proposed for *Pseudomonas*. In a relatively short time, glucose breakdown pathways were determined in the fungus *Rhizopus*, cell-free extracts of yeast, homofermentative and heterofermentative lactic acid formers, and pentose fermentative patterns in *Fusarium* and *Escherichia coli*.

Shortly after the *Leuconostoc* procedure was published, it was used in many investigations. Among the early studies was a collaborative effort carried out with Bernard Horecker, National Institutes of Health, to reveal the mechanism for the conversion of pentose phosphates formed in the hexosemonophosphate pathway to hexose phosphate. In the collaborative work, the course of the reaction was followed in extracts of rat liver with ribose-5-P-1-^{14}C and -2,3-^{14}C as substrates. The isotopic distribution pattern in the glucose determined with *Leuconostoc* confirmed a transketolase-transaldolase series of reactions involving sedoheptulose-7-phosphate as intermediate. This work assisted Horecker and Ephraim Racker to elucidate the cyclic nature of the direct oxidation pathway and provided a mechanism for the total oxidation of glucose without inclusion of the Krebs Cycle.

The summer of 1953 brought Horecker to Brookhaven to repeat the experiments with pea leaf and root preparations. We found that, while the root extract showed a labeling pattern essentially the same as that of liver, the leaf preparation differed. The difference between the root and leaf data was eventually accounted for by the photosynthetic carbon reduction cycle of Calvin and Benson.

It was good fortune that Howard Gest, Sidney Udenfriend, Jerome Schiff, G. Robert Greenberg, and Harry Beevers were additional summer guests. For a more complete accounting, see the "Boys of Summer" in ASPP Newsletter, Jan/Feb 1996.

Severo Ochoa and Feodor Lynen were analyzing the mechanism of the condensing enzyme, citrate synthase, and were in need of acetyl coenzyme A labeled with sulfur of high specific radioactivity. In response to a phone call from Ochoa, I procured radioactive sulfur of questionable purity from the on-site nuclear reactor staff. The Nobel laureates came to Brookhaven where we proceeded to purify the starting material. The two were urged, but without success, to follow my lead and wear protective gloves. After the first purification step, they urgently scrubbed until the monitoring instrument detected roughly 20,000 cpm and 50,000 cpm on the hands of Lynen and Ochoa, respectively, whereupon the puckish Lynen with a winking of the eye suggested the difference was due to the baseline level of "cold" Germanic and "hot" Spanish blood. Following dinner, the small vial containing the purified material, later found to be in curie amounts, was packed in a 25-gallon trash can filled with Kieselgur to eliminate radioactivity between car trunk and passengers on their return to New York City.

The large trash can was a familiar tool in the lab. NADP prepared from 100 kilograms of beef liver procured from immediately slain animals in the slaughterhouses of Manhattan was transported frozen in the can wherein the tissue was inactivated by high-pressured steam. Since the yield was roughly 300 milligrams, the slaughterhouse was seen not infrequently. The coenzyme was required in our investigation of the phosphorylating and nonphosphorylating glyceraldehyde-3-phosphate dehydrogenases localized in leaf tissue.

In 1954, I was invited to address the International Botanical Congress in Paris. Brookhaven, after approval of the itinerary and briefing by security, provided generous travel arrangements. I disembarked in Plymouth after crossing the Atlantic on the Ile de France first class, and the return was by propeller aircraft, which owing to deep fog overflew Gander and refueled in a Canadian military airfield in Labrador. I began the trip in England staying in the home of Frank Dickens whose lectures at Brookhaven on the direct oxidation pathway in liver influenced me to plan a similar study with plant tissue. He, in turn, introduced me to Robin Hill and Harold Davenport whose laboratory lifestyles have been chronicled repeatedly. Dan Arnon presented a seminar that day in Cambridge where for the first time, I heard his stimulating accounting of the photo-assimilation of CO_2 and incorporation of inorganic phosphate by isolated chloroplasts. The England schedule concluded with a "reverse Harry Beevers pathway" by paying my respects to W.O. James (postdoctoral advisor), pioneer of the glycolytic pathway in plants, and to Meirion Thomas (PhD mentor) and Stanley Ranson, noted for publications on crassulacean acid metabolism and authorship of a classic plant physiology text. Attending the Congress led to a personal friendship with A. Moyse, organizer of the photosynthesis section.

After the Congress, I presented a seminar in the biochemical institute of Feodor Lynen in Munich. Following the lecture delivered in German, a request came from the audience that discussion be in English. I was pleased to oblige. I met Otto Kandler that day and at a lunch arranged by his wife, Traudl, in their little apartment in the Nymphenburg Botanical Institute, their extended stay at Brookhaven was planned. They came the following year to apply the *Leuconostoc* fermentation to sugars isolated from algae and higher plants that had photoassimilated $^{14}CO_2$ for brief periods. The isotopic distribution patterns were asymmetrical in contrast to a symmetrical one predicted by the photosynthetic cycle. To account for the asymmetry, we proposed a reduction followed by cleavage of the addition product of ribulose-1-5-diphosphate and CO_2 to glyceraldehyde-3-phosphate and glycerate-3-phosphate. Termed the Gibbs effect and quoted as an argument against the cycle, the asymmetrical labeling was later shown to result from a lack of isotopic equilibrium between the two three-carbon intermediates that combine to yield the six-carbon sugar and is not an indicator of modification.

In the summer of 1950, I was privileged to organize the first Brookhaven conference, CO_2 Assimilation Reactions in Biological Systems, concerned with the application of isotopic tracers in biological systems. Heady stuff, three years after winning a doctoral degree, to forward invitational letters to the speakers: Harland G. Wood, Severo Ochoa, Merton F. Utter, Stanley Carson, Eric Conn and Birgit Vennesland, Robert Stutz, Sam Aronoff and Leo P. Vernon, and A.A. Benson. Other attendees were Chris B. Anfinsen, Otto Meyerhof, Carl Neuberg, G. Ray Noggle, Jack Myers, Robert Emerson, Hubert B. Vickery, and Ray Dawson. I edited the proceedings, which were distributed without charge by the US Atomic Energy Commission.

The second conference, organized by my colleague and good friend Robert Steele, dealt with nitrogen metabolism. Camp Upton, isolated and without public transportation, offered few extra-evening activities. Nobelist Chris Anfinsen came loaded with dice. After a few rolls, a security officer confiscated the cubes and demanded name identification of the participants. The ever-alert owner of the dice admitted to Louis Pasteur, Philip Handler, later President of the National Academy of Sciences, confessed to Ludwig van Beethoven, R.C. Fuller responded with Gregor Mendel, and I, unfortunately known, had no option. The following morning in the conference hall, the officer requested that Pasteur, Beethoven, and Mendel report to the security office to complete clearance forms.

Run-ins with security did happen. I concluded that Western Reserve University was a hotbed of political controversy since I was questioned repeatedly by an officer prior to and after each visit from the Cleveland school. I was called in once for an extended formal hearing complete with stenotypist who recorded the proceedings. I was asked to respond to the accusation that an uncle, a pensioned wounded veteran of WWI, whom I saw when I attended the 1947 Rose Bowl (Illinois 47-UCLA14), was alleged to subscribe to the Communist outlet, The Daily Worker, and to have frequent encounters with undesirables. After one hour of interrogation, I was dismissed. Apparently my statements satisfied my inquisitors since the follow-up scheduled for New York City was canceled. Traveling abroad brought in security for briefing and debriefing. Considering the times and fear of the junior senator from Wisconsin, Brookhaven security agents were, on the whole, cooperative, understanding, and respectful of the research staff.

Nearing the end of a seven-year contract and informed by the newly installed chairman that an extension would not be forthcoming, a stroke of accidental good fortune rescued my career and family. Harold Williams, head of the Department of Biochemistry at Cornell University, on a sabbatical semester with Robert Steele, offered an Associate Professor's position in his department. In 1956, after many good and productive years in which I edited a conference book and was in the authorship mix of 33 full-length manuscripts, my family departed for Ithaca. The laboratory, carved out of the bowling alley in the old post exchange building that I vacated, has been well used by Clint Fuller, Robert Smillie, William Siegelman, and presently Geoffrey Hind.

The College of Agriculture at Cornell University

Whereas my Brookhaven responsibility was directed to research all day and every day, a faculty appointee, I was to learn, juggles his or her schedule to accommodate undergraduate and graduate advising and instruction, departmental, college, and university assignments, as well as endless grant proposals to federal and private sector sources. My teaching experience was limited to cellular physiology when David Goddard, Editor-in-Chief of *Plant Physiology*, spent a sabbatical semester at the University of Washington in 1954. My family became the caretakers of his home in a quiet suburb of Philadelphia. I enjoyed the companionship of Alex Shrift, Ed Cantino, and Jerome Schiff whose doctoral thesis topic was sulfate reduction and incorporation into plant protein, a subject of deep interest of Goddard since a postdoctoral fellowship with Leonor Michaelis at the Rockefeller Institute. I was enthusiastic to team-teach the elementary biochemistry course with a new arrival, George P. Hess, a student of Vincent DeVigneaud. In the eight years of teaching this course, the highlight occurred on a typical upstate chilly February morning, 8:30 a.m., when the departmental secretary shouted from the rear of the lecture hall, "It's a boy." After applause and completion of the lecture, I sped to the hospital.

The College of Agriculture funded by the State of New York was established in 1907 on the campus of a university founded about 50 years earlier. College faculty were state not Cornell employees. State regulations interpreted rigorously by the College administration did cause discomforting results. Morris Cynkin, holder of a Cornell University doctorate in microbiology supported by an NIH fellowship with an annual stipend of $4000 paid directly to him by the federal agency, joined my laboratory in 1957 to study the distribution of radioisotope in glucose formed photosynthetically from CO_2 in the isolated intact chloroplast. Since the stipend did not commingle with college funds, I was summoned to the Dean's office to sort out Cynkin's listing in the appointment letter and directory. The Dean's quick thumbing of the state salary schedule book brought forth the unappealable decision that he be appointed Technician Level 2. And so he was for two years.

Most of my students in class were from the departments in Agriculture, and I served on many doctoral examination committees of graduate students in Botany, Microbiology, Agronomy, and Horticulture. One notable was a doctoral candidate in Microbiology, Sam Conti, who after a distinguished academic career became Provost of the University of Massachusetts-Amherst. I enjoyed friendships with Adrian Srb, Walter Bonner, George Hess, and F.C. Steward. We ate in the faculty club where the outspoken Steward held forth on his topic of the day until, prearranged, we agreed not to disagree with him.

Cornell University attracted able graduate students. Paul Kindel, Evelyn Havir, Nona Calo, Robert Togasaki, and Elchanan Bamberger dealt with photosynthesis and respiration in cells, organelles, and reconstituted systems, while Louise Anderson, supported by a grant from the American Coffee Institute, studied the biosynthesis of caffeine. W.Y. Cheung continued the work initiated at Brookhaven on the oxidative pentose phosphate cycle in the intact cell of the blue green alga, *Tolypothrix*, which oxidized glucose but with an impaired citric acid cycle. James Willard and Marvin Schulman came later and characterized aldolases and glyceraldehyde-3-phosphate dehydrogenases in higher plants and algae. The postdoctoral fellows were Martin Busse, who followed glucose fermentative patterns in lactic acid bacteria; Jean Galmiche, who teamed with Herb Marsh to demonstrate that light did not effect the turnover of the citric acid cycle in the alga, *Scenedesmus*; George Cheniae, who concentrated on acetate metabolism; George Russell, who showed the involvement of the Calvin-Benson cycle in photoreduction and the oxyhydrogen reaction in algae adapted to a hydrogen metabolism; and Charles Fewson and Clanton Black, who measured action spectra and quantum requirement for the photochemical production of NADPH and ATP in chloroplasts. Their research problem was begun in collaboration with David Krogmann, John F. Turner, CSIRO, Plant Physiology Unit in Sydney, who joined the lab in 1960, and Solon Gordon, Argonne National Laboratory, where the study was conducted using the Argonne biological grating spectrograph as the radiation source.

In a college community, there is considerable social activity in which my wife and I participated. It was not uncommon for her to arrange dinners in our home for seminar speakers and visiting faculty. We also took satisfaction from the holidays mixers that she organized for the undergraduates I advised.

In the summer of 1957, accompanied by my family, I had the great pleasure of bringing the *Leuconostoc* technique to the laboratory of Gleb Krotkov in Kingston, Ontario. My visit to Queen's College had its beginning ten years earlier when Krotkov, en route to Berkeley for a sabbatical year with W.Z. Hassid, visited the Botany Annex in Urbana. The summer stay was additionally unforgettable by the presence of two exuberant Australian students, Robert Smillie and Kenneth Scott. Jerome Schiff, then at Brandeis, stayed with us in the Krotkov home, where we completed a chapter for Steward's *Treatise of Plant Physiology*.

Considering the political climate in 1958 between the two superpowers, a weeklong visit by a Soviet plant physiologist N.G. Doman, Academy of Sciences of the USSR, was unexpected but most welcome. His research focused on the primary product of the ribulose-1,5-bisphosphate carboxylase reaction. With a paper chromatography apparatus stationed on the rooftop of the Academy lab in the midst of a Moscow winter, he claimed to have evidence for an intermediate prior to glycerate-3-phosphate. I cannot recall a full report in a reviewed journal.

The 1959 Montreal Botanical Congress was a second opportunity to see A. Moyse who with Marie Louise Champigny came to Ithaca. Without doubt, an enduring friendship got me elected into the French Academy of Sciences. I became acquainted with A.A. Krasnovsky and Alexandrovich A. Nichiporovich; they announced that a section on photosynthesis would be included in the 1961

Biochemical Congress in Moscow. Many American participants arrived by a chartered Soviet jet departing Brussels. Cameras were confiscated on entry and returned on exiting. My roommate, Bessel Kok, misplaced the one irreplaceable room key but with absolute hotel control, it was unnecessary. I toured the Timiriazev Botanical Institute and dined in the home of Natasha Vosskrensenskaya. Her husband was editor of the *Soviet Botanical Journal*. A worthwhile event was visiting in Leningrad the laboratory of Oleg Zalensky whose research on photosynthesis and ecology was internationally recognized. My first visit to the USSR was an eye-opening experience, a visit that has been repeated many times. I exited the Soviet Union by train to Helsinki and journeyed to Lund to attend the meeting of the Scandinavian Society of Plant Physiologists, with a stopover in Stockholm where Lawrence Bogorad was on a sabbatical leave. I met Hans Burstrom and Anders Kylin, who would succeed Burstrom as editor of *Physiologia Plantarum*. Also present were James Smith and Stacey French.

Editor-in-Chief of Plant Physiology

When I was running what seemed to be endless ashings of leaf tissue and mineral determinations, analysis of the data indicated a relationship worthy of publication. At Ray Noggle's urging, the paper was communicated to *Plant Physiology* and receipt acknowledged by Editor Walter F. Loehwing on March 19, 1947. My first and only submission ever accepted without revision appeared promptly in the JuneJuly book.

Immediate approval was satisfying but approval without change more so since I was the two-fingered typist using a borrowed instrument. The Brookhaven papers were published in *Plant Physiology*, *Nature*, *Journal of Biological Chemistry*, *Archives of Biochemistry and Biophysics*, *Journal of Bacteriology*, *Journal of the American Chemical Society*, *Zeitschrift fur Naturforschung* and *Proceedings of the National Academy*. Inasmuch as I was not a dedicated *Plant Physiology* author, I was totally unprepared in March 1962 for a telephonic message from Harry Beevers and Aubrey Naylor, President and immediate past President of the American Society of Plant Physiologists (ASPP), offering me the editorship of the Journal. I postponed my decision until the opportunity to discuss the position with Editor Allan Brown. He was supportive and assured a smooth, overlapping transition of three to four months; he would continue to receive manuscripts through August 31. Manuscripts received September 1 or later would be processed in Ithaca and I would get out Volume 38, starting with the January 1963 issue. Allan's editorial assistant was scheduled to come to the Cornell campus in October to familiarize my staff (half-time copy editor and third-time clerk-typist) with editorial and office routines. After acceptance in a letter to Harry Beevers dated March 26, 1962, I turned my attention to the College of Agriculture administration. I had two essential needs: space to accommodate the staff and a budget line in the business office funded by ASPP to pay for postage, telephone, and office supplies. In my first annual report to the Executive Committee of ASPP, August 1963, I expressed appreciation to Richard H. Barnes, Dean of the School of Nutrition, whose faculty shared Savage Hall with the Department of Biochemistry, for moral and financial support. He provided a typewriter, file cabinets, and space for the clerk-typist, the wife of a graduate student (who later became Dean of Agriculture, University of Arizona) in the School. The copy editor, wife of a Cornell faculty physicist, marked-up manuscripts at home. To handle funds, I opened a personal account in a local bank. Mail was transported to the post office by me personally or by the staff members. Telephone charges were billed to my home until the remarkably efficient and conscientious Executive-Secretary Treasurer of ASPP, William (Bill) Klein, forwarded a credit card. I looked forward to my first meeting with Bill and his wife, Winifred, benefactors with sound advice throughout my editorship. In time, I came to appreciate and recognize the distinct privilege to be associated with the Kleins, two warm and good friends of outstanding ability and professionalism.

Akin to my predecessors, children of the Great Depression, office and editorial expenses were truly minimal. I cajoled reviewers, made decisions on all manuscripts, coordinated author's alterations with my reading of galley proofs, checked the annual subject index, and dealt with the printer, Craftsman, Inc. in Kutztown, PA. Management of the Journal was sandwiched among my duties as a teaching faculty member and a mentor for doctoral and postdoctoral students.

In the 1963 report to the Editorial Board, I highlighted a record number of 243 contributions, a 25% increase over the preceding year. My policy of having Editorial Board members critique all manuscripts unless there was an exceptional need for a specialist not on the board resulted in a doubling of the names listed on the masthead to 26. The unexpected increase in submissions necessitated hiring an additional copy editor who also worked at home. When it became clear that the mechanics of the Journal office were threatened, relocation was prudent. Abram Sachar, President of Brandeis University, and Jerome Schiff provided a haven and a warm welcome. An editorial office adjacent to my lab was penciled into the plan of a building under construction for the Department of Biology. President Sachar orchestrated an overhead-free budget line and the fiscal officer of the university directed financial statements to the Executive Secretary-Treasurer, while mail was delivered to the editorial office in a separate pouch.

The first years on the Waltham campus were not without dramatic changes. By 1967, my fifth year as editor, submissions had more than doubled. Journal pages per volume concomitantly increased from 779 to 1806. Between 1926 when Editor Charles A. Shull (1926-45) got out Volume 1, No. 1, and 1963 the Journal was issued bimonthly, three years later, monthly except for July and August, and finally in 1967, monthly. In the tradition of Shull and continued by his successors Walter F. Loehwing (1945–1953), David R. Goddard (1953–1958), and Allan H. Brown (1958–1963), the editorial handling of manuscripts through review, decision, and revision was dealt with solely by the chief editor. Inasmuch as I carried a teaching load and directed a research lab, coping with 413 contributions in 1967 meant no interludes for much else. To work down the weighty backlog, the family (now five children) summered in Woods Hole, Cape Cod. After auditing the 9 a.m. lecture in the Physiology course organized by Brandeis colleague Andrew Szent-Gyorgi, which included Rod Clayton and George Hoch as instructors, I withdrew to a rented (those were the days before ASPP expense accounts) desk hidden in the stacks of the Marine Biological Laboratory library, which was open 24 hours daily including July 4. Eugene I. Rabinowitch, I was told by Hans Gaffron, wrote his monumental treatise on *Photosynthesis* at that desk with a view of the Eel Pond and the crafts moored therein. Scrutiny of a nearby shelf turned up a box of material labeled Rabinowitch. The editorial office was not neglected. The clerk typist and I shuttled between Waltham and Woods Hole on alternate weeks.

Brown in his final report to the Editorial Board predicted that the Journal would soon grow to unmanageable size. He could not foresee a twofold increase in every criterion for growth in five years. Yet papers, on the average, appeared in print about five to six months after receipt. The heavy volume of work made large inroads into my research time. Even though I benefited from the experience, challenge, and recognition, I did not wish to leave research for full-time Journal management. Solo control of an expanding monthly periodical was a crapshoot not only for contributors but also for my family. At the 1967 annual meeting, I introduced a plan to the Executive Committee of ASPP for reducing the increased workload of the Editor-in-Chief by appointment of two and perhaps three Associate Editors. Each Associate Editor and I would be responsible for decisions on manuscripts distributed by the Editor-in-Chief in a particular field of specialization. All correspondence with the Press including galley and page proof would remain in Waltham. Appointment of three Associate Editors (Leon Bernstein, Johannes von Overbeek, Joe Varner) was approved and money was budgeted for the stipends and expenses. Jerome Schiff whose principal duty was to assign manuscripts in my absence was already serving. The editorial review board selected by and partnered with the 5 decision editors expanded to 42. With this arrangement, I hoped that we had achieved a workable number of reliable and timely reviewers.

The printer, often overlooked, is an essential component in the issuance of a publication. Between 1964–69 my transactions with printing shops should have been, but were not, monotonous and continued throughout apace. Allan Brown prized Craftsman, Inc., principally the plant superintendent, Harry (Nick) Wylie, who took pride in and gave deference to the Journal. Located in Kutztown, the firm had never been seen by a representative of ASPP. I coaxed Bill and Winifred Klein into our convening there to assess the status of the enterprise and be introduced to Nick Wylie. Scrutiny of the plant convinced me that closure was imminent. At the suggestion of the parent company (Printing Corporation of America owned both Craftsman and Business Press), the operation was moved to Business Press in nearby Lancaster. The plant superintendent's father supervised *Plant Physiology* when Shull was editor. After two years in Lancaster, the Journal returned to Kutztown; reunification with Nick Wylie under the new ownership of DH Conover and, more important, a redactory service, albeit only one, thus reducing mailing time and work load in Waltham.

Distrustful of small firms with limiting and aging presses, I was determined to search for a printer well-positioned with state-of-the-art equipment, a network of professional redactors, and nearby the Business Office. The latter reservation was made since I envisioned the day when much of the tediousness of the editorial office could be transferred to a restricted bureaucracy located in a permanent and substantial home office.

Bill Klein and I investigated a number of printing houses and after serious consideration with input from Aubrey Naylor, concluded a fair contract with Waverly Press in Baltimore and Easton, MD. And none too soon. Shortly after this event, our friendly mole telephoned that not only had Conover employees been tendered bad checks for wages, but also FICA, and perhaps other taxes were in arrears. We were warned that IRS was poised to padlock the building and auction off the contents, including our galley plates. Immediately, Bill and I rendezvoused in Kutztown early on the morning of December 24, 1969, and with the assistance of Nick Wylie, loaded the galley plates of the November and December issues onto a Waverly truck dispatched by Sales Representative Constance Kiley (later President) less than an hour ahead of IRS. The crates were transported to Baltimore, corrected there, and under supervision by Waverly Press, Volume 44 was finished off and in the mail by mid-January.

I persuaded the executive committee of ASPP that the next development for the Journal should be a one-day winter meeting where the Associate Editors and I, sitting with Waverly Press personnel and on occasion the Executive Director, could survey the entire operation. Van Overbeek hosted the first meeting in a secluded room in the main library at Texas A&M University. We neglected an alarm bell at 4:45 p.m. and a second louder peal at 5 p.m. Within 15 minutes a guard strode into the room stating that we would remain in the closed building until cleared by a duty officer. Van Overbeek's exhortations notwithstanding, exiting was denied. Our frustrated host telephoned his wife, our dinner hostess, a publicly renowned lawyer and judge in College Station, who sprang us forthwith.

In 1978, academic institutions in New England were unwilling or unable to pay the increase in the price of heating fuel resulting from the Middle East oil embargo. Classes at Brandeis were dismissed from mid-December until early February. I was invited by Mack Dugger, Dean of the College of Natural and Agricultural Sciences, and Charles Coggins, Chair Department of Botany and Plant Sciences, to become a visiting professor at the University of California, Riverside. My duties other than delivering lectures to undergraduate and graduate students of plant physiology were to interact with John Thomson, R.T. Wedding, Irwin Ting, and Robert Heath and their research groups. I enjoyed the friendship of other faculty members in various departments especially the field specialists, who kept my family happy with a cornucopia of freshly collected fruits and vegetables. The appointment would be renewed for ten additional winters. When on-campus housing was no longer available, we lived in a motel where, one January, our neighbors were the Langs, Anton, the editor of *Planta*, and Lydia, also visiting for a brief stay. Celebrating Anton's birthday with champagne (his favorite beverage), it seemed natural for the conversation to drift into a comparison of managerial styles in editing the two prestigious plant science journals. Karen, a proofreader, was a salaried employee of ASPP whereas Lydia, the office manager of *Planta*, blurted out that she worked gratis. A disturbed Lydia challenged the birthday boy, and it is my understanding that thereafter, she collected compensation from Konrad Springer Verlag.

On the retirement of Van Overbeek from the board, the editorial winter meeting shifted to Riverside. A spin-off suggested by Mack Dugger and Robert Leonard was a symposium built around the editors and local faculty members. Held in January 1978, with Jack Dainty as leadoff speaker, the one-day event developed into the annual University of California Riverside Symposium in Botany. The proceedings of the 1982 symposium, edited by Irwin Ting and Martin Gibbs and published by ASPP, were issued promptly and economically by reproducing the manuscripts from camera-ready copy. Jack Shannon organized an equivalent service on the Penn State campus. These were the first books sponsored and marketed by ASPP since the well-sold defunct monographs edited 30 years earlier by W.F. Loomis.

Three times yearly the Kleins and I reviewed extensively Journal procedures with Waverly Press personnel assigned to the publication. Frequently, I took advantage of trips to Washington to visit the Kleins in Bethesda the day before and enjoy the warm hospitality of their home. On a few occasions, travel was accomplished by rail since the meeting site was located a short walk from the Pennsylvania station. The rail trip from Boston was achieved either in a sleeper car, involving a lengthy in-station stay awaiting the onset of a Waverly Press day, or by spending the previous evening with family in Philadelphia, followed by an early morning departure. Using the latter approach, I was stranded in the Philadelphia station when a sudden cold snap caused a total shutdown of the New York–Washington route. While taking my options into account, announcement of a southbound train propelled me into the trainmaster's office where I became aware that the "special" parked in a heated shed was scheduled for immediate departure to the Bowie racetrack. However, the unavailability in Philadelphia of the daily racing forum—must reading for horse racing touts— necessitated an unscheduled stop in Baltimore. Identifying myself as an old railroad hand, we bonded and armed with a pass he issued, I boarded with the most determined passengers I had ever encountered. I arrived at the appointed hour.

Each Journal printed by Waverly Press is assigned a company symbol similar to a procedure common in the stock exchange. Hence *Journal of Biological Chemistry* was BC. In the case of *Plant Physiology*, PP was abandoned and replaced by PLANT.

Shull announced to the fledgling ASPP in 1925 that he "accepted the position of managing editor, or editor-in-chief of the Journal." Clearly, he, I, and the intermediary editors were both. Transfer of redactory service from the editorial office to the press was the first substantive decision to disaffiliate editing from management. When manuscripts received per year soared beyond 1000, it became increasingly clear that my successor, most likely a competitive academic scholar balancing teaching and research with service to a Journal, might question an editorial position coupled to a considerable residuum of managerial responsibilities. The opportunity to separate the remaining operations such as production schedule, format, reading of galley and page proofs came when ASPP established the Gude Plant Science Center, hired Melvin Josephs in 1986 as Executive Officer, who then appointed Melinda (Jody) Carlson as senior production editor. Finally, after 60 years, modification of the Shull dictum was under way.

The constitution of ASPP directs the Editor-in-Chief of *Plant Physiology* to prepare and present an annual report to the Editorial Board. The place of presentation is left to the Editor's discretion. I opted for a posh Society-sponsored editorial dinner that would include spouses to thank associate and review editors for the issuance of a plant journal with the highest impact rating (*Current Contents*) and the lowest cost per page. Intensive discussion of the report was reserved for a location listed in the program of the national meeting. The first joint meeting with the Canadian Society of Plant Physiologists was held in Calgary in 1973. The local representative had selected a site for the banquet without forewarning of an ordinance restricting an open-bar in a public restaurant. The proprietor came up with a clever solution. Bill Klein and I rushed to city hall, ASPP purchased a vendor's license, we backed our car into the loading platform of a provincial liquor store, and finally placed the cases of refreshments in the good keeping of the bartender.

I made it a practice not to be absent from the editorial office for an extended period. The one exception, as recipient of an Alexander von Humboldt award (August–November 1988), was a sabbatical stay in the labs of the two Erwins, Latzko and Beck. Clanton Black filled in as the distributing editor.

In my tenure as Editor, there were substantial increases in published issues (6 vs 12), volumes (1 vs 3), number of pages (1000 vs 4800), submissions (200 vs 1300), and manuscripts rejected (20% vs 35%). My office logged in about 25,000 manuscripts. Inspection of Figure 1 indicates that the annual submissions curve divides into two linear phases, with a break between 1971 and 1975. For the final five years, foreign papers accounted for the linearity. In 1977, the ratio of US to non-US papers was 2.7, a value that approached parity by 1991. I was always disturbed by any limitations on submission of manuscripts. For instance, requirement of membership discriminated against plant physiologists in the developing countries. Resumption of foreign subscriptions in 1985 was probably related to the removal of this restriction that had been in effect since 1982. Perhaps also coupled to the linear increase in

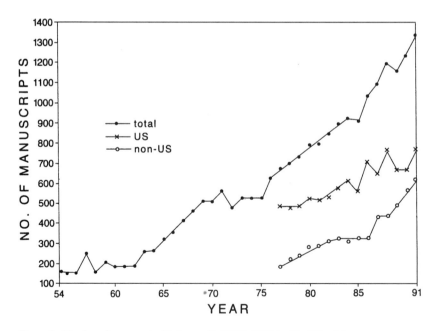

Figure 1   Number of papers submitted annually (1954–1991) to *Plant Physiology*.

foreign submissions was a turnabout in foreign library subscriptions, which historically accounted for more than 50% of the total. Clearly, *Plant Physiology* had become an international journal with a diverse authorship.

Two important and valuable innovations were introduced. Mini-reviews, suggested by Maarten Chrispeels, first appeared in 1986. The Plant Gene Register (PGR), conceived by Hans Bohnert from a need to publish plant gene sequences in the form of a brief communication, was included in the October 1990 book.

Since the establishment of the Journal, it was customary to publish under News and Notes the annual report of the Editor-in-Chief, minutes including the budget of the annual meeting, obituaries, etc. To restrict the Journal to research papers, starting in 1964, my annual report was distributed to the membership at the national meeting. That trend continued with the publication of a newsletter by the business office in 1967. By 1968, the News and Notes section was concluded.

In 1988, with the assistance of Charles Smith and Wilbert McNamara, Waverly Press personnel, the appearance of the Journal was made over. The cover, which had been designed some 20 years earlier and printed black on brown in Times Roman, was changed to a green sheet with white and black print in Helvetica typeface. The new cover of Carolina high gloss coating replaced the lighter-weighted St. Regis Royalite. Unbeknownst at least to me, Waverly Press entered *Plant Physiology* in the National Composition Association's Annual Typographic Awards competition. The award: *Plant Physiology* has been selected as one of the best publications in existence today for typographic design, layout, clarity, and readability.

After three decades of a happy and eventful association with *Plant Physiology*, this chapter in my career was over.

Brandeis Universitiy

We moved in the fall of 1964 into a newly constructed wing of the Science Complex on the Brandeis University campus. My neighbors were Attila Klein, Jerome Schiff, and Albert Kelner.

Work on the spinach chloroplast continued with the regulation of the CO_2 assimilatory pathway by pH and O_2 (J. Michael Robinson, Zvi Plaut), osmotic stress (Gerald Berkowitz), temperature (Chee Fook Fu), and inhibitors (R.G. Everson, P.W. Ellyard), and preillumination (M. Avron). Comparative research on chloroplasts from Crassulacean acid metabolism plants (George Piazza, C.A. Levi), and corn (Denny O'Neal, C.S. Hew) was initiated. When Uwe Klein (a postdoctoral student from the University of Bonn) and Changguo Chen joined the lab, we broadened the comparative study to include the first isolation of functional chloroplasts from an alga, *Chlamydomonas reinhardtii*.

Gentle lysis with digitonin of protoplasts produced by autolysine treatment of synchronized cell cultures yielded chloroplasts that photoassimilated CO_2 at roughly half the cellular rate. With respect to dependency upon light intensity, pH, and other parameters, the results were similar to those reported for higher plant plastids. Chen extended this investigation and showed that chloroplasts of Chlamydomonas F-60, a mutant deficient in phosphoribulokinase, a component of the Calvin-Benson cycle, contained some enzymes and properties of the reductive carboxylic acid cycle. In an anaerobic environment rich with acetate (from cellular starch) and CO_2, documented in the raw sewage lagoons in the Antelope Valley of the California Mojave Desert where the rapid growth of nearly unialgal cultures of Chlamydomonas has been reported by R.W. Eppley, the reductive carboxylic acid cycle first described by Dan Arnon and Bob B. Buchanan in the photosynthetic green bacteria may play an important role in the ecological physiology and survival of the green algae.

It was my very good fortune that two gifted enzymologists, Erwin Latzko and Grahame Kelly (student of John F. Turner), came to the lab. Kelly clarified the status of glyceraldehyde-3-phosphate dehydrogenases in the plant cell by characterizing a cytoplasmic irreversible NADP-linked enzyme, therefore not coupled to a substrate phosphorylation. One potential function of the enzyme, as shown by Elchanan Bamberger, was the indirect transfer of photosynthetically reduced NADP from chloroplast to cytoplasm. In addition to authoring many papers with Erwin Latzko on the regulatory enzymes of photosynthesis, we edited Photosynthesis II, Encyclopedia of Plant Physiology. We and our families have enjoyed a durable and lasting friendship.

When Kelly completed his postdoctoral years (including marriage to a Brandeis undergraduate), he became a member of Latzko's lab in the Chemical Institute, Weihenstephan, and moved with him to the Department of Botany, University of Munster. We three scheduled many auto trips throughout western and central Europe. One of the more memorable was the professionally guided tour of Prague by Z. Sestak coupled to his arranging a quiet (secretive) meeting in a Trebon restaurant with his colleague, I. Setlik, whose lab we desired to visit. On gate arrival and after a highly emotional argument between our host and the administrator of the complex, we were denied entry to the lab. However, from the vantage of a flooring with knotholes in Setlik's apartment, we had an ample view of the lab below, and his staff with data books briefed us in the apartment. Crossing, entry and exit, of the Czechoslovakian border in a car with three occupants holding diverse passports was not without incident. Each inspection of car and passengers lasted at least one half hour.

Under conditions where unicellular green algae are adapted to a hydrogen metabolism, Hans Gaffron in the 1940s documented photoevolution of H_2 stimulated by acetate, photoreduction—the reduction of CO_2 coupled to the uptake of H_2—and the oxyhydrogen reaction—the simultaneous consumption of CO_2 and H_2 in the dark, in both reductive events CO_2 is processed by the Calvin-Benson cycle. I was intrigued by and had not forgotten Gaffron's papers, which I first read as a graduate student. With the elucidation of chloroplastic and mitochondrial electron transport sequences, the citric acid and glyoxylate cycles, and the isolation of *Chlamydomonas* chloroplasts capable of fixing CO_2, mechanisms could be proposed and tested to account for these processes. Using inhibitors in intact cells and chloroplasts of *Chlamydomas*, postdoctoral fellows T.E. Maione and D.L. Erbes revealed the role of the thylakoidal electron transport pathway in photoreduction and in the passage of reducing equivalents from carbon substrates to the transport chains in the fermentative dark and photoevolution of H_2. In follow-up experiments, Rene Gfeller and Chen found convincing evidence that gas (CO_2 and H_2) release during anaerobic photometabolism of a carbon substrate (acetate, glucose) required the cooperation of the plastid electron transport chain with the mitochondrial citric acid cycle and the microbody glyoxylate cycle. Key to unraveling the oxyhydrogen reaction is the site of ATP synthesis, the limiting step for the reduction of CO_2 in the darkened chloroplast. In a 1963 review, I proposed the mitochondria as the ATP generator because of the requirement of O_2. This suggestion may be valid since the addition of an external supply of ATP to *Chlamydomonas* chloroplasts stimulated generously the baseline rate of CO_2 uptake. Investigation of isolated chloroplasts and mitochondria for H_2-dependent ATP synthesis will be needed to understand fully the organization of this complex reaction.

In my final study, prior to retirement, I returned to pathways of carbohydrate breakdown, but in spinach, corn mesophyll, and *Chlamydomonas* chloroplasts. In a collaborative effort with K.O. Willeford, K.K. Singh, and K.J.K. Ahluwalia, our experimental approach was to monitor the release of CO_2 in darkened chloroplast preparations externally supplied with specifically labeled monosaccharides. Inasmuch as chloroplasts have a partial glycolytic pathway, gluconate-6-phosphate dehydrogenase is most likely solely responsible for CO_2 release. From the isotopic flow patterns, it was clear that the oxidative pentose phosphate cycle, a sequence for the complete conversion of glucose to 6 CO_2 first proposed by Horecker and Racker in mammalian cells, was functioning in the plastids and clearly without the mitochondrial citric acid cycle. Chloroplast versatility with respect to electron acceptors was documented by recording CO_2 release when nitrate or oxaloacetate were supplied anaerobically. We defined in our reports the release of CO_2 from sugars by the organelle as chloroplast respiration. This designation may be questioned since O_2 uptake and ATP synthesis, two additional basic components of true respiration, were not measured.

Woods Hole and the Marine Biological Lab became an annual summer event. Jerry Schiff shared our home. We were fortunate to have Hans Gaffron,

Carl Price, and David Mauzerall as neighbors. In 1970, I joined with Harold Siegelman (senior instructor), Anthony San Pietro, Frank Loewus, and Bob Wilce to constitute the faculty of the Marine Botany course. When Siegelman retired from the course, Loewus was chosen director. Trevor Goodwin and Synove and Amy Jensen were visiting faculty, and a few of the notable students were Sam Beale, Ralph Quatrano, Joe Remus, and James Saunders. The students associated with me surveyed freshly collected algae for H_2 photoproduction. The ferry dock (Woods Hole to Martha's Vineyard) area furnished the most active specimens that we analyzed. Financial support for the course (lab and lectures) was provided by the NSF, Metabolic Biology Panel, Program Director Eli Romanoff, who for his beneficial service to and betterment of plant science was given the Gude Award of ASPP.

The 1972 symposium Genetic Aspects of Photosynthesis organized by Yusif Nasyrov, Rector of the Tajik Agricultural Institute, was held in Dushanbe and was the occasion of my second trip to the Soviet Union. This conference afforded another opportunity to meet with Nichiporovich, his wife Sophia Tageeva, Natasha Vosskrensenskaya, A.T. Mokronosov, Oleg Zalensky, and A.A. Shylk, who had lectured at Brandeis University in 1964. The meeting, broad in scope and international in attendance, was summarized in a book edited by Nasyrov and Z. Sestak. A highlight was the excursion to view the Vavilov Institute established in 1935 in the foothills of the Pamir mountain range for inducing mutant plants (primarily *Arabidopsis*) in an elevated environment. The gravel-covered road from Dushanbe to the Institute, still occasionally blocked by sheep, was claimed to have been trod by Alexander the Great en route to India. Three years after the Dushanbe symposium, I was greatly honored to be a vice president of the Leningrad International Botanical Congress.

One of my most challenging off-campus assignments was being a member of the Eastern European panel of the Council for the International Exchange of Scholars, the agency that administers the Fulbright Program. The panel evaluated applications for the countries in the former Soviet Bloc. It was my pleasure to serve on this diverse committee for two consecutive five-year terms from 1973–1983, the second as Chair. I traveled quite widely throughout this large area, usually accompanied by the very able panel director, Georgene Lovecky, and local US embassy personnel. On occasion, a Washington-based State Department representative attended. An overview of the program was held with the ambassador prior to university visitations negotiated by embassy staff and the local Ministry of Education. In the countries I dealt with, ministerial authority determined access to all academic institutions and it was their stationed deputies who set the itinerary and the agenda. A typical tour of the former USSR started with Moscow State University, an overnight train to Kiev, a flight to Novosibirsk, and concluded with a stay in Tbilisi. In one extended sweep, we also took in Hungary, Bulgaria, Romania, Poland, and Yugoslavia. Despite the numerous drawn-out, frustrating, emotional, and often unproductive negotiations with governmental and academic apparatuses, the few successes were more than compensatory. I enjoyed the opportunity to give seminars in each country, meet with students and faculty, and to broaden my experiences and contacts in Eastern Europe. A pleasing aspect of the Fulbright experience was interaction with colleagues and their families when I dined in their homes, a privilege denied my companions by domestic protocol. Another benefit was my arranging Fulbright Teaching Awards at Brandeis University for Nichiporovich and Nasyrov. Lifetime employees of the Academy of Sciences, neither had teaching experience, but they did their homework and their lectures were warmly received.

I was invited by Debra Wince, Division of International Affairs, NSF, to organize a Hungaria-USA binational symposium on photosynthesis. After a preliminary trip to Hungary and discussion with Agnes Faludi-Daniel, the two sides met in Szeged, May 1982, and again in Salve Regina College, August 1986, as a satellite to the 7th International Photosynthesis Congress in Providence. Sadly, Faludi-Daniel who played a pivotal role in the jointly administered meetings and organized the Szeged conference, died shortly before the second meeting. The proceedings dedicated to her, which I edited, were published with NSF and industrial support.

My final endeavor was as the American organizer of a Russian-US workshop on photosynthesis under the sponsorship of the Academy of Sciences of the USSR and the National Academy of Sciences of the United States. My Soviet counterpart was V.V. Shuvalov. The NSF approved a proposal to fund ten young investigators to double the number of American participants. In May 1992, the workshop convened in Pushchino in the Institute of Soil Science and Photosynthesis, where V.I. Kefeli, Director of the Institute, presented the opening lecture. The 42 contributions covered a broad spectrum of topics. The banquet dinner held in an Institute bomb shelter, resplendent with the traditional communist slogans and portraits of Soviet political leaders, was claimed by the featured speaker to have been constructed against Cruise missiles. I was accompanied by a grandson who during the conference was taken care of by Raisa Demidova and her family. Two years later, the workshop limited to senior investigators reconvened in Woods Hole. Prior to Woods Hole, the Soviet delegation was divided into two touring groups, one guided by I. Zelitch, and the other by William L. Ogren and Robert Blankenship.

The Gordon Conference on Photosynthesis in the 1960s allotted to carbon metabolism one three-hour evening session. I petitioned the governing board to reorder the program, with carbon metabolism split out as a separate conference. With I. Zelitch as co-chair and with supportive letters from Melvin Calvin,

Dan Arnon, and Martin Kamen, a turndown in the fall of 1974 was unexpected and disappointing. Reversal was the result of a chance meeting in October 1975 with Ernie Jaworski at the International Conference on Crop Productivity—Research Imperatives organized by Sylvan Wittwer, Stanley Ries, and Marvin Lamborg. Jaworski, a senior Monsanto Company employee and chair of a Gordon advisory board dealing with corporate sponsorship, came upon the proposal in the dead-letter file. Within two weeks of our meeting, approval was in hand and the conference was held in 1976. The 1996 meeting, Photosynthetic CO_2 Fixation and Metabolism in Green Plants, organized by Steve Huber and Hans Bohnert, emphasized the new trend of research on the regulation of photosynthesis resulting from the recent development of molecular biology methodology. Marie Louise Champigny and I published a conference report in *Compte Rendu*.

Mary Bunting, President of Radcliffe College, invited me to be a Sigma Xi national lecturer, with a challenging itinerary and a schedule of ten lectures within a fortnight. In March 1967, with the initial lecture in Greeley and the last in Calgary, presentations were given in Fort Collins, Laramie, Bozeman, Missoula, Spokane, Pullman, Twin Falls, and Bellingham. The lectures were based on our research on glycolate formation, regulation of CO_2 assimilation in the chloroplast, and carbohydrate dissimilatory pathways in microorganisms and plants. Accomplished by plane, train, and automobile, the tour was progressing as programmed until the unforeseen occurred in the Calgary Airport. My homebound route was Calgary to Boston with a connecting flight in Toronto. Of the two airlines linking Calgary with Toronto, it was my misfortune to be scheduled on the one grounded by a strike. Relating my woes to a coach of a professional hockey team readying for an immediate departure, I "joined" the group (disguised with a team shirt) and arrived into Toronto with ample time to connect with the Boston flight.

Participation in the New England sectional meeting of ASPP afforded an opportunity to visit with colleagues unseen since the summer meeting of the national society. During the difficult 1930s, this was the only meeting for a large majority of New Englanders. According to one source, the banquet was a black tie event. The section formed in 1933 convened annually on university campuses. In order to conserve expenses, the proceedings commenced after Friday lunch and terminated in the early afternoon of the following day. Preference was given to undergraduate and graduate presenters, with allowance for nonacademic institutions. My introduction to the section was the 1952 meeting at Smith College. The collection of dues was a not-to-be-forgotten event. The secretary, an elderly University of Massachussetts professor, sequestered the treasury of less than $200 in a well-worn sugar sackcloth identified with XXX and secured by a tightly drawn knot. During the business meeting, the empty bag was passed row by row while he urged "donations" of not less than 25 cents. Reknotted, the contents were shaken and estimated by the clinking of the coinage. Tea was poured from silver-plated pots and at some meetings served in porcelain china cups. In 1955, I petitioned the sectional executive committee to expand and include the states of New York, Pennsylvania, and New Jersey. The proposal was accepted but not without a lengthy and heated discussion. The Brookhaven meeting, the first held outside the six states encompassing New England, resulted in another drawn-out discussion by the executive committee, when a preprandial cocktail mixer was fit into the program. An Illinois summit lunch with Art Galston, Ray Noggle, Oswald Tippo, and spouses was the highlight of the 1997 Massachusetts meeting for me. If merely to provide a small yet critical forum for fledgling plant scientists who do not participate in the national meeting, the sectional meeting is worthwhile to preserve. To this retiree, the little meeting recalls the past when the parent society met on campuses in a relaxed and comfortable setting.

Two years after retiring from the editorship of *Plant Physiology*, I stopped classroom instruction, gave up my lab, and moved into a campus office.

ACKNOWLEDGMENT

I wish to acknowledge the many contributions of undergraduate (honors program at Brandeis), graduate, and postdoctoral students and visiting faculty not cited in the text. I am grateful to Lisa Gibbs for preparing the manuscript for publication, to Cheryl Clemens for assistance with the figure, and to Marcia Cabral for help in communicating with contributors.

Annu. Rev. Plant Physiol. Plant Mol. Biol 1999. 50:27–45

PHOSPHATE TRANSLOCATORS IN PLASTIDS

Ulf-Ingo Flügge

Botanisches Institut der Universität zu Köln, Gyrhofstrasse 15, D-50931 Köln, Germany; e-mail: uiflue@biolan.uni-koeln.de

KEY WORDS: carbon transport, heterologous expression, shikimate pathway, starch biosynthesis

ABSTRACT

During photosynthesis, energy from solar radiation is used to convert atmospheric carbon dioxide into intermediates that are used within and outside the chloroplast for a multitude of metabolic pathways. The daily fixed carbon is exported from the chloroplasts as triose phosphates and 3-phosphoglycerate. In contrast, nongreen plastids rely on the import of carbon, mainly hexose phosphates. Most organelles require the import of phosphoenolpyruvate as an immediate substrate for carbon to enter the shikimate pathway, leading to a variety of important secondary compounds. The envelope membrane of plastids contains specific translocators that are involved in these transport processes. Elucidation of the molecular structure of some of these translocators during the past few years has provided new insights in the functioning of particular translocators. This review focuses on the characterization of different classes of phosphate translocators in plastids that mediate the transport of the phosphorylated compounds in exchange with inorganic phosphate.

CONTENTS

27

1040-2519/99/0601-0027$08.00

INTRODUCTION

Communication between plastids and the surrounding cytosol occurs via the plastid envelope membrane. The inner envelope membrane contains a variety of transporters that mediate the exchange of metabolites between both compartments (19). Carbon fixed during the day can be exported from the chloroplasts into the cytosol for the synthesis of sucrose, which is subsequently allocated to heterotrophic organs of the plant such as roots, seeds, fruits, or tubers. Export of the newly fixed carbon in the form of triose phosphates (and 3-phosphoglycerate) proceeds via the triose phosphate/phosphate translocator (TPT). During the night, the breakdown products of transitory starch are exported in the form of glucose via a glucose translocator (58). Plastids are also involved in nitrogen assimilation, the synthesis of amino acids and fatty acids, and in the synthesis of a series of plant secondary products that are formed via the shikimic acid pathway. This pathway requires the provision of the plastids with phosphoenolpyruvate (PEP) as an immediate precursor. The nature of a corresponding transporter has remained elusive until the recent discovery of the plastidic PEP/phosphate translocator (PPT) (11). Plastids of nonphotosynthetic tissues have to import carbon as a source of energy and for driving biosynthetic pathways, for example, leading to fatty acids, amino acids, or starch. Carbon import into nongreen plastids can proceed in the form of hexose phosphates via a recently discovered glucose 6-phosphate/phosphate translocator (GPT) (37). This review focuses on the characterization of the different classes of phosphates translocators that are present in plastids of various plant tissues.

MEASUREMENTS OF PLASTIDIC PHOSPHATE TRANSPORT ACTIVITIES

Background

A basic technique for measurements of metabolite transport in plastids is the silicone oil filtering centrifugation method (29). Microfuge tubes are filled with a bottom layer of a denaturing solution (e.g. perchloric acid), followed by a layer of silicone oil. The plastids are pipetted on top of the oil, incubated with a radioactively labeled substrate, and subsequently centrifuged through the silicone oil layer, whereby the transport reaction is terminated. A more advanced technique is the double layer silicone oil centrifugation system, which consists of an aqueous substrate layer sandwiched between two silicone oil layers. This allows measurements of uptake rates in the range of 1 to two seconds (24, 34). Alternatively, transporters, especially antiporters, can be measured by reconstitution into artificial membranes (liposomes) that have been prepared by sonification of phospholipids in a buffered solution containing an exchangeable substrate. Incorporation of detergent-solubilized proteins into the liposomal membrane can be achieved by a freeze/thaw step (40) and the external substrate can subsequently be removed by gel filtration of the proteoliposomes. The reconstituted transport activity can be assessed by measuring the amount of radioactively labeled substrate that is transported into the liposomes. Using various metabolites for preloading of the liposomes, the transport characteristics of an antiporter can be determined, i.e. which countersubstrate can be exchanged by the reconstituted translocator. The reconstitution method can also be used to follow a particular transport activity during purification (11, 18, 37, 46). Not only can highly purified translocator proteins be functionally reconstituted into liposomes but also whole membranes, plastids, or even crude homogenates from different plant tissues (20). The reconstitution system thus allows direct access to antiporters of different plant tissues without the necessity of isolating intact organelles.

The Triose Phosphate/Phosphate Translocator

The triose phosphate/phosphate translocator (TPT) of chloroplasts was the first phosphate translocator to be described in terms of substrate specificities and kinetic constants (12). It mediates the export of fixed carbon in the form of triose phosphates and 3-phosphoglycerate from the chloroplasts into the cytosol. The exported photosynthates are then used for the biosynthesis of sucrose and amino acids and the released phosphate is shuttled back into the chloroplasts via the TPT for the formation of ATP (see Figure 1).

In its functional form, the TPT is a dimer composed of two identical subunits (13, 74). As substrates, the TPT accepts either inorganic phosphate or a phosphate molecule attached to the end of a three-carbon chain, such as triose phosphates or 3-phosphoglycerate. C3-compounds with the phosphate molecule at C-2 position, for example, phosphoenolpyruvate, 2-phosphoglycerate, are only poorly transported (12). Under physiological conditions, the substrates are transported via a strict 1:1 exchange. Transport proceeds via a ping-pong type of reaction mechanism, i.e. the first substrate is transported across the membrane and then leaves the transport site before the second substrate can be bound and transported. The transport site thus alternatively faces either membrane side, thereby transporting substrates in opposite directions (15). Furthermore, only the transport site facing the cytosol is accessible to inhibitors of the TPT, namely

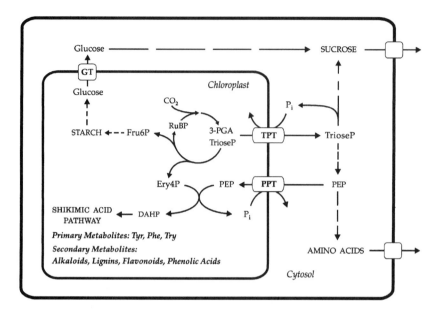

Figure 1 Transport processes mediated by the TPT protein and the PPT protein. 3-PGA, 3-phosphoglycerate; DAHP, 3-deoxy-D-arabino-heptulosonate 7-phosphate; Ery4P, erythrose 4-phosphate; Fru6P, fructose 6-phosphate; Fru1,6P$_2$, fructose 1,6-bisphosphate; GT, glucose translocator; P$_i$, inorganic phosphate; PEP, phosphoenolpyruvate; Phe, phenylalanine; PPT, phosphoenolpyruvate/phosphate translocator; RuBP, ribulose 1,5-bisphosphate; TPT, triose phosphate/phosphate translocator; TrioseP, triose phosphates; Tyr, tyrosine; Try, tryptophane. For details, see text.

pyridoxal 5′-phosphate and 4,4′-diisothiocyanostilbene-2,2′-disulfonate. This asymmetry in the structure of the transport site is linked to different transport affinities on both membrane sides; these are about five times lower on the stromal side of chloroplasts (15).

In intact chloroplasts, unidirectional transport of phosphate can be observed but with a V_{max} that is two to three orders of magnitude lower than that of the antiport mode (12, 49). Using the reconstituted system in which the concentrations of phosphate in both the internal and the external compartments are accessible to experimental variations, the transport activity of the reconstituted TPT can reach values that exceed those measured for an antiport mode by at least one order of magnitude. It is suggested that transport under these conditions proceeds by a mechanism different from the antiport mode, probably by a (channel-like) uniport mechanism. Evidence for ion channel properties of the TPT is provided (*a*) by decreasing the activation energy for phosphate transport from 46 kJ/mol (antiport mode) to 18 kJ/mol (uniport mode), a value that is in the range observed for ion channels and (*b*) by measuring the TPT-mediated unidirectional transport by the patch-clamp technique (64). It can be concluded from these electrophysiological experiments that the TPT can behave as a voltage-dependent ion channel, preferentially permeable to anions, as well as an antiporter. As suggested by Saier (57), different classes of transporters might share common structural motifs and may have arisen from a common ancestor. A small structural change within the translocation pore might then allow transport via an ion channel mode or might result in strong coupling of substrate binding linked to conformational changes, as observed for transporters operating in the antiport mode.

The Phosphoenolpyruvate/Phosphate Translocator

Mesophyll chloroplasts of C_4-plants possess a TPT-like translocator that mediates the export of phosphoenolpyruvate (PEP) from the chloroplasts as substrate for the PEP carboxylase in the cytosol. The resulting inorganic phosphate is shuttled back into the chloroplasts via this translocator. It has been suggested that the phosphate translocator from C_4-mesophyll chloroplasts transports, in addition to the substrates of the TPT, also PEP (7, 10, 24, 35, 56). In view of the recently discovered phosphoenolpyruvate/phosphate translocator (PPT) that transports only PEP and inorganic phosphate but accepts triose phosphates and 3-phosphoglycerate only very poorly, it can now be suggested that mesophyll chloroplasts contain two different phosphate translocators, a TPT that is involved in the triose phosphate/3-phosphoglycerate shuttle and a PPT that transports PEP in exchange with inorganic phosphate.

PEP is only very poorly transported by the TPT of C_3-chloroplasts (11, 12, 24). However, a PEP transport activity is also present in chloroplasts or, at least, in a subtype of plastids that is present in preparations of mesophyll chloroplasts. Reconstitution of chloroplasts or chloroplast envelope membranes (21, 44) always shows a low, but significant transport activity of PEP that cannot be attributed to TPT. A PEP/phosphate exchange activity was also detected in nongreen plastids, including plastids from pea roots (2), tomato fruit plastids (62), cauliflower bud plastids (20), maize kernels (16), and sweet pepper plastids (70).

In all types of plastids, PEP serves different functions, e.g. as precursor for the biosynthesis of fatty acids or of aromatic amino acids. PEP is an immediate substrate for the plastid-localized shikimate pathway leading, via the synthesis of aromatic amino acids, to a large number of secondary metabolites, e.g. alkaloids, flavonoids, and lignins (see Figure 1). Apart from plastids of lipid-storing tissues, most chloroplasts, e.g. from pea, spinach, or *Arabidopsis* (1, 61, 65, 73), and nonphotosynthetic plastids, e.g. from pea roots (2) or cauliflower buds (36), are unable to convert 3-phosphoglycerate into PEP via the glycolytic pathway. Due to the absence or low activities of phosphoglucomutase and/or enolase, glycolysis cannot proceed further than to 3-phosphoglycerate (47). Therefore, these plastids rely on the supply of PEP from the cytosol.

The Hexose Phosphate/Phosphate Translocator

Nongreen plastids of heterotrophic tissues are carbohydrate-importing organelles and, in the case of amyloplasts in storage tissues, the site of starch and/or fatty acid synthesis. Since these plastids are normally unable to generate hexose phosphates from C3-compounds owing to the absence of fructose 1,6-bisphosphatase activity (9), they rely on the import of cytosolically generated hexose phosphates that are formed from sucrose delivered from source tissues. Sucrose is unloaded from the phloem either via symplasmic connections or via the apoplast and is cleaved by either invertase or sucrose synthase. The resulting hexoses are then converted into hexose phosphates and imported into the plastids as the source of carbon for starch and fatty acid biosynthesis and, in addition, as a substrate for the oxidative pentose phosphate pathway (see Figure 2). This pathway can deliver reductants for nitrogen metabolism and fatty acid biosynthesis (2, 3, 22, 39).

The results of transport measurements with intact organelles or reconstituted tissues from different plants (e.g. pea roots, pea embryos, cauliflower inflorescences, maize endosperm, potato tubers, pepper fruits) suggest that the hexose phosphate transport is mediated by a phosphate translocator that imports hexose phosphates in exchange with inorganic phosphate or C3-sugar phosphates (2, 16, 20, 32, 33, 50, 53, 59, 62). In sink tissues from most plants studied to date, glucose 6-phosphate (Glc6P) is the preferred hexose phosphate taken up

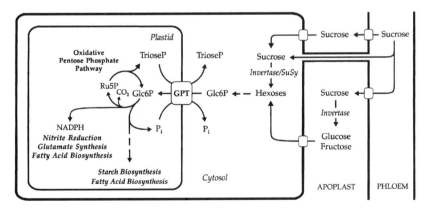

Figure 2 Proposed function of the GPT protein in heterotrophic tissues. GPT, glucose 6-phosphate/phosphate translocator; Glc6P, glucose 6-phosphate; Ru5P, ribulose 5-phosphate; P_i, inorganic phosphate; SuSy, sucrose synthase; TrioseP, triose phosphates. For details, see text.

by nongreen plastids. However, in amyloplasts from wheat endosperm, glucose 1-phosphate (Glc1P) rather than Glc6P is the precursor for starch biosynthesis (68, 69, 72).

In conclusion, nongreen plastids appear to possess a phosphate translocator that is specific for the transport of phosphate, phosphorylated C3-compounds and hexose phosphates, which suggests that these transport processes are mediated by a TPT-like phosphate translocator with an extended substrate specificity (31).

A Glc6P transport activity is also present in chloroplasts from guard-cells (51). Like nongreen plastids, these chloroplasts are devoid of fructose 1,6-bisphosphatase activity (27), the key enzyme for the conversion of triose phosphates into hexose phosphates, and, therefore, rely on the provision of hexose phosphates for starch biosynthesis. Starch is mobilized during stomatal opening and converted to malate that is then used as a counterion for potassium.

A hexose phosphate/phosphate transport activity can also be detected in spinach chloroplasts after feeding of detached leaves with glucose (54). The chloroplasts of these leaves are photosynthetically active but contain unusually large quantities of starch. It could be shown that the precursor for starch biosynthesis is imported into the chloroplasts in the form of Glc6P by a hexose phosphate/phosphate translocator, as is the case for heterotrophic plastids in sink tissues. Glucose-feeding has obviously induced a switch in the function of the chloroplasts from carbon-exporting source organelles to carbon-importing sink-organelles and has led to the induction of a sink-linked plastidic hexose phosphate/phosphate transport activity.

MOLECULAR IDENTIFICATION AND CHARACTERIZATION OF PLASTIDIC PHOSPHATE TRANSLOCATORS

We used a biochemical approach to clone cDNAs coding for plastidic transport systems. The particular membrane proteins were purified to homogeneity, cleaved by endoproteases, and the resulting peptides were then used to design oligonucleotides and to generate PCR-fragments for screening of cDNA libraries.

The spinach TPT was the first plant membrane transport system for which the primary sequence could be determined (17). Meanwhile, TPT-sequences are available from various plants, e.g. those from *Arabidopsis*, pea, potato, maize, *Flaveria*, and tobacco. All TPT-sequences have a high similarity to each other (10, 11, 41, 77).

More recently, we isolated cDNAs coding for two other plastidic phosphate translocators from heterotrophic tissues: the phosphoenolpyruvate/phosphate

translocator (PPT) from maize endosperm, maize roots, cauliflower buds, tobacco leaves, and *Arabidopsis* leaves, and the Glc6P/phosphate translocator
(GPT) from maize endosperm, pea roots, and potato tubers (11, 37). As is the
case for the TPTs, the PPT and the GPT proteins share a high degree of identity
with each other (mature proteins about 75–95% identity). A comparison of the
phosphate translocator cDNA sequences with entries in the sequence databases
revealed no significant homologies with known proteins. This is an indication
for the unique transport specificity of the plastidic phosphate translocator proteins. In contrast, the plastidic 2-oxoglutarate/malate translocator as well as
the recently discovered plastidic adenylate translocator share some similarities
with transporters of bacterial origin (38, 76).

In contrast to the high homologies among the translocators of one class, the
overall similarities between the members of the TPT, PPT, and GPT families
are about 35% and are restricted in all translocator proteins to five regions,
each 15 to 30 amino acid residues in length. A phylogenetic tree constructed
by using the distance matrix method confirmed the existence of three different
classes of plastidic phosphate translocators and showed that members of each
translocator family cluster together but that the three classes of transporters
cluster at approximately equal distances from each other (37).

All members of the different classes of phosphate translocators are nuclear-
encoded and possess N-terminal transit peptides (about 80 amino acid residues)
that direct the adjacent protein correctly to the plastids (4, 11, 37, 42). Import of the translocators into plastids is driven by ATP and depends on the
translocation machinery of the envelope membrane. The mature parts of the
phosphate translocators consist of approximately 330 amino acid residues per
monomer, are highly hydrophobic, and contain information (envelope insertion signals) for the integration of the proteins into the inner envelope membrane. Each monomer is predicted to consist of 5–7 hydrophobic segments,
which are assumed to form α-helices that traverse the membrane in zig-zag
fashion connected by hydrophilic loops. The phosphate translocators thus
belong to the group of translocators with a 6+6 helix folding pattern, as
is the case for mitochondrial carrier proteins (45). In contrast, the plastidic
2-oxoglutarate/malate translocator (76) and the ATP/ADP translocator (38)
have transmembrane topologies with a 12-helix motif, which resembles that
of other plasma membrane transporters from prokaryotes and eukaryotes that
presumably all function as monomers.

Based on a tentative model for the arrangement of the TPT in the membrane,
it is probable that all 12 α-helices of the phosphate translocator dimer take part
in forming a hydrophilic translocation channel through which the substrates
could be transported across the membrane (75). Interestingly, two successive
charged residues in helix V (Lysine, Arginine) that have been proposed to

be involved in substrate binding (10) are conserved in all phosphate transport proteins. Site-directed mutagenesis of the lysine residue in helix V (Lys273Gln, spinach TPT) led to a complete loss of transport activity (B Kammerer & UI Flügge, unpublished observations), a result suggesting that this residue is indeed essential for the transport reaction.

FUNCTIONAL STUDIES OF PLASTIDIC PHOSPHATE TRANSLOCATORS

Heterologous Expression of Plastidic Phosphate Translocators in Yeast Cells

The final proof for the identity of an isolated transporter cDNA is the expression of the corresponding coding region to produce the functional protein in heterologous systems, for example, yeast cells, bacteria, or oocytes. *Escherichia coli*, which is commonly used as a host for the expression of foreign proteins, failed to express plastidic phosphate translocator proteins, because of the toxicity of the gene product. We have demonstrated that yeast (*Schizosaccharomyces pombe*) can be successfully used for the expression of functional plastidic translocators. The recombinant plastidic translocator proteins, representing about 1% of the total protein from transformed cells, were associated with yeast internal membranes, either mitochondrial membranes or membranes of the rough endoplasmic reticulum (44). For yeast transformation, an expression vector containing cDNAs coding for either the whole precursor protein or only the mature translocator protein was used (44, 76). For the subsequent measurements of the transport activities, either total membrane fractions or the transport proteins isolated therefrom were reconstituted into artificial membranes. To facilitate the isolation of the phosphate translocators from the transformed cells, recombinant proteins were engineered to contain a C-terminal tag of six consecutive histidine residues (His$_6$-tag). This allows the purification of the tagged transporter proteins to apparent homogeneity by a single chromatography step on metal-affinity columns (11, 37, 44).

Figure 3 shows the substrate specificities and Table 1 the kinetic constants of the TPT, PPT, and GPT proteins reconstituted into liposomes that had been preloaded with different phosphorylated metabolites that function as exchangeable countersubstrates. It is evident that the purified TPT only transports triose phosphates and 3-phosphoglycerate in exchange with inorganic phosphate, but not PEP and hexose phosphates. In contrast, the PPT transports inorganic phosphate preferentially in exchange with PEP. Triose phosphates and 3-phosphoglycerate, the only substrates of the TPT, are only poorly transported by the PPT, with apparent inhibition constants 1–2 orders of magnitude

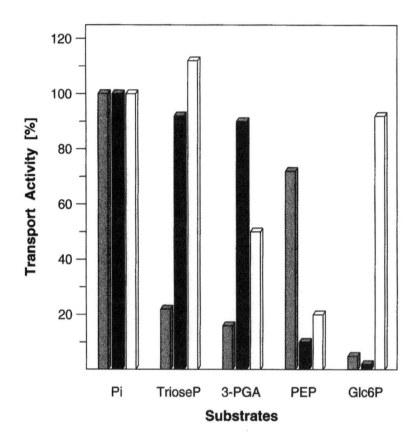

Figure 3   Substrate specificities of the TPT, PPT, and GPT. The recombinant and histidine-tagged phosphate transport proteins were isolated from yeast cells by metal affinity chromatography and were reconstituted into liposomes that had been preloaded with the indicated substrates. The phosphate transport activities of the translocators [PPT, phosphoenolpyruvate/phosphate translocator (light gray), left; TPT, triose phosphate/phosphate translocator (dark gray), middle; GPT, glucose 6-phosphate/phosphate translocator (white), right] are given as a percentage of the activity measured for proteoliposomes preloaded with inorganic phosphate. 3-PGA, 3-phosphoglycerate; PEP, phosphoenolpyruvate; Pi, inorganic phosphate; TrioseP, dihydroxyacetone phosphate; Glc6P, glucose 6-phosphate. Data from Reference 37.

Table 1   Apparent K_m (phosphate) and K_i values of recombinant plastidic phosphate translocators for various phosphorylated metabolites

		TPT*	PPT	GPT
		(mM)		
Phosphate	K_m (app)	1.0	0.8	1.1
Triose phosphate	K_i (app)	1.0	8.0	0.6
3-Phosphoglycerate	K_i (app)	1.0	4.6	1.8
2-Phosphoglycerate	K_i (app)	12.6	5.7	—
Phosphoenolpyruvate	K_i (app)	3.3	0.3	2.9
Glucose 6-phosphate	K_i (app)	>50	>50	1.1

*The [^{32}P]phosphate transport activities of the recombinant translocators (TPT, triose phosphate/phosphate translocator from spinach chloroplasts; PPT, phosphoenolpyruvate/phosphate translocator from cauliflower bud plastids; GPT, glucose 6-phosphate/phosphate translocator from pea root plastids), purified from yeast cells, were measured in a reconstituted system using proteoliposomes that had been preloaded with inorganic phosphate. Data from References 11, 37.

higher than that for PEP. The PPT protein thus functions as a PEP/phosphate transporter that is able to provide the plastids with PEP even in the presence of other phosphorylated intermediates.

Unlike the TPT and PPT proteins, the GPT accepts inorganic phosphate, triose phosphates, and Glc6P about equally well as countersubstrates, whereas the affinity of the GPT toward 3-phosphoglycerate is lower. PEP only serves as a poor substrate with an apparent inhibition constant that is three to ten times higher that the apparent K_m values for the transport of phosphate and Glc6P, respectively (37) (Table 1). Glc1P or fructose phosphates are virtually not transported by any of the phosphate transport proteins. The GPT thus links the cytosolically located conversion of sucrose and hexoses to Glc6P with metabolic reactions within the plastid, i.e. the biosynthesis of starch, fatty acids, and the oxidative pentose phosphate pathway that delivers reduction equivalents for nitrogen metabolism and fatty acid biosynthesis. Inorganic phosphate and triose phosphate that are formed during these processes can both be used as counter substrates by the GPT in exchange with Glc6P (see Figure 2).

Until recently, it was accepted that the transport of phosphate, phosphorylated C3-compounds, and hexose phosphates, observed in nongreen plastids, are mediated by a TPT-like phosphate translocator. Our findings clearly show that these metabolites are not transported by a single transport system, but rather by different members of the phosphate translocator family with partially overlapping substrate specificities. Such a system enables the efficient uptake of individual phosphorylated substrates even in the presence of high concentrations of other phosphorylated metabolites, which would otherwise compete for the binding site of a single transport system.

The identity of the hexose phosphate transporter that is specific for the uptake of Glc1P is unknown. The observation that mutants of *Arabidopsis* with a defect in the plastidic phosphoglucomutase are unable to synthesize starch (5, 6) clearly indicates that this plant does not possess a functional Glc1P translocator. Since these mutants possess a cytosolic phosphoglucomutase isoenzyme mediating the conversion of Glc6P into Glc1P, starch biosynthesis should be driven from Glc1P imported via the Glc1P translocator. Furthermore, transgenic potato plants with reduced activity of the plastidic phosphoglucomutase are also defective in starch accumulation and show a phenotype comparable to that found in ADP-glucose pyrophosphorylase antisense lines (R Trethewey, personal communication). This observation indicates that (*a*) Glc6P is the preferred substrate taken up by plastids and (*b*) the conversion of Glc6P to Glc1P inside the plastids, catalyzed by phosphoglucomutase, is a prerequisite for starch formation. The dependence of starch formation on the plastidic phosphoglucomutase would also argue against the observation that starch synthesis in potato tubers is driven by cytosolic Glc1P but not Glc6P (48).

Gene Expression Studies

The different physiological functions of the phosphate translocator families are linked to differential patterns of expression. The TPT activity is associated with photosynthetic carbon metabolism. Consequently, expression of the TPT gene is observed only in photosynthetically active tissues, but not in unambiguously heterotrophic tissues such as roots or potato tubers (11, 16, 60). As shown by in situ hybridization studies, the TPT gene is present in both mesophyll cells and bundle sheath cells of C_4-plants (P Nicolay & UI Flügge, unpublished observations). The expression level of PPT-specific transcripts could be detected in both photosynthetic and heterotrophic tissues, although transcripts were more abundant in nongreen tissues (11). The level of PPT steady-state RNA in photosynthetic tissues is lower by at least one order of magnitude than is the level of the TPT mRNA, both in C_3- and in C_4-plants. However, we have recently isolated a cDNA from maize that is homologous to the PPT cDNA. The corresponding gene is highly expressed in mesophyll cells of C_4-plants and, presumably, codes for the C_4-PEP/phosphate translocator that exports PEP from the chloroplasts as substrate for the PEP carboxylase in the cytosol (K Fischer & UI Flügge, unpublished observations).

Transcripts of the GPT gene were almost lacking in photosynthetic tissues but are abundant in heterotrophic tissues such as roots, developing maize kernels, potato tubers, or reproductive organs (37). This is in line with the proposed function of the GPT protein in these tissues that utilize Glc6P as a precursor for starch biosynthesis. The slight expression of the GPT observed in photosynthetic tissues might be due to the presence of the GPT protein in chloroplasts of guard cells (see above).

It has been shown recently that in the seed endosperm of some cereals, the key enzyme for starch synthesis, ADP-glucose pyrophosphorylase, is mainly present in the cytosol and not in the plastids (8, 71). The ADP-glucose formed in the cytosol is presumably transported into the plastids for starch biosynthesis via an ADP-glucose/adenylate antiporter that is functionally and structurally distinct from the recently identified ADP/ATP translocator (38, 63). It is assumed, but not yet proven, that the Brittle-1 protein serves as an ADP-glucose/adenylate transporter, which would thus represent an alternative route to provide the plastids with a precursor for starch biosynthesis (66, 67). In maize, the Brittle-1 protein is expressed during the later stages of endosperm development (66), whereas the level of GPT mRNA reaches a plateau shortly after pollination that was subsequently maintained through 20 days (37). It remains to be established how the activities of both proteins are coordinated in seed development.

Transgenic Plants with Altered Activities of Plastidic Phosphate Translocators

The TPT is an important link between metabolism in the chloroplast and the cytosol. Only about 10% of the total transport activity of the TPT can be used for (productive) net triose phosphate export to provide the carbon skeleton for further biosynthetic processes. Since both the cytosol and the stroma contain triose phosphates, 3-phosphoglycerate, and inorganic phosphate competing for transport in either direction, it appears feasible that much of the TPT activity is used for catalyzing (nonproductive) homologous exchanges (14).

From the observation of subcellular metabolite concentrations in intact spinach leaves, it has been proposed that the TPT can exert a kinetic limitation during sucrose biosynthesis in vivo (23). We have assessed the role of the TPT on photosynthetic metabolism by creating transgenic antisense potato plants in which both the amount and the activity of the TPT were reduced to 70% of the controls (55). In ambient CO_2 and intermediate light, there was no significant effect on photosynthetic rates, growth, and tuber development in the transformants. However, moderate reduction of the TPT activity resulted in a marked perturbation of leaf metabolism. Most remarkably, the content of stromal 3-phosphoglycerate was greatly increased compared to the corresponding value in wild-type plants. This should result in a large decrease of the stromal content of inorganic phosphate since the TPT mediates a strict counterexchange of substrates. The increased stromal 3-phosphoglycerate/phosphate ratio should lead to an increase of starch synthesis due to the allosteric activation of the

ADP-glucose pyrophosphorylase (52) and reflect the situation of chloroplasts in which an increased starch synthesis was observed due to a decreased availability of phosphate within the cytosol (30). Indeed, the starch content in the leaves of the transformants was much higher than in wild-type plants, suggesting that the daily assimilated carbon is mainly maintained within the plastids and directed into the accumulation of starch. Since there was no obvious effect on plant growth, the transformants were obviously able to efficiently compensate for their deficiency in TPT activity.

The transgenic plants mobilize and export the major part of the daily accumulated carbon during the following night, in contrast to wild-type plants that generally export the major part of the fixed carbon during ongoing photosynthesis (28). The altered day-night rhythm of carbon allocation to sink tissues also leads to a change in the diurnal growth pattern of the TPT antisense plants: The growth rate during the night is considerably increased as compared to the growth rate during the day period (J Fisahn & L Willmitzer, personal communication). The transformants likely circumvent the reduced TPT activity by mobilizing the daily accumulated starch via amylolytic starch breakdown. This results in the formation of hexoses that are subsequently exported via a glucose translocator (58; see Figure 1). Interestingly, transgenic antisense TPT plants from tobacco accumulate starch as potato plants do, but start to mobilize the accumulated starch during ongoing photosynthesis. These plants showed increased rates of amylolytic starch mobilization and a higher transport capacity for glucose across the envelope membrane (26).

In further studies, transgenic tobacco plants with gradually decreased or increased TPT activities were utilized to study the control the TPT exerts on the fluxes of carbon into starch and sucrose as well as on the rate of CO_2 assimilation (RE Häusler, NH Schlieben, P Nicolay & UI Flügge, unpublished observations). The data indicate that the TPT exerts a considerable control on the rate of both CO_2 assimilation and sucrose biosynthesis under saturating CO_2. These studies also revealed that the rate of sucrose biosynthesis from glucose (deriving from starch degradation) could account for up to 60–70% of the wild-type rate in the absence of the TPT.

From the experiments with the antisense TPT plants, it can be concluded that the transformants can efficiently compensate for their deficiency in TPT activity provided that a carbon sink (i.e. starch) can be generated during photosynthesis that can subsequently be mobilized. Transgenic potato plants with a reduced ability to synthesize assimilatory starch (e.g. by antisense repression of the ADP-glucose pyrophosphorylase) show also no effect on growth and productivity. Export of the daily fixed carbon via the TPT is obviously so efficient in these plants that heterotrophic tissues can be adequately supplied with reduced carbon. However, if both starch formation and the activity of the TPT are reduced, the corresponding transformants show a severe phenotype (25). These transformants are unable to export sufficient amounts of fixed carbon during the day, nor do they have a carbon store that they could use during the dark period.

The in planta role of genes can also be studied by analysis of mutants that have been created, for example, via insertion mutagenesis. Recently, *Arabidopsis* mutants with a reduced expression of the chlorophyll a/b binding protein (cab) in responses to pulses of light were isolated [CAB underexpressed, *CUE* mutants (43)]. The phenotype of the null mutations in *Arabidopsis* is quite severe. The plants underexpress genes for chloroplast components, both in the light and in response to a light pulse. The seedlings are not able to establish photoautotrophic growth and die unless they are germinated on sucrose. The paraveinal regions of the mutant leaves are still green but the interveinal regions are pale green, resulting in a reticulate pattern. Antisense PPT plants from tobacco showed a comparable, but transient, visible phenotype (RE Häusler, A Weber, P Nicolay, L Voll & UI Flügge, unpublished observations). Different alleles of *cue1* with reduced light-responsive gene expression were isolated from *Arabidopsis* (*cue1-1* to *cue1-8*) and the corresponding gene was identified in a T-DNA-tagged mutant population. Surprisingly, the *cue1* gene corresponds to the PPT (J Chory & S Streatfield, personal communication). Future work will elucidate how this severe phenotype is linked to the role of the PPT in plant metabolism and development.

CONCLUDING REMARKS

There has been considerable progress during the past few years in studying plastidic translocators at the biochemical and, more recently, molecular levels. In the near future, the *Arabidopsis* and rice genome sequencing programs will provide the sequences of complete higher plant genomes. To date, 35,000 *Arabidopsis* ESTs and about 11,000 rice ESTs (Expressed Sequence Tags) are already available. Future work will likely concentrate on identifying the functions of genes coding for putative envelope translocators, for example, and on elucidating the specific role of a particular gene in plant metabolism. This will require combined efforts on genetic, molecular, biochemical, and physiological levels.

ACKNOWLEDGMENTS

Work in the author's laboratory was funded by the Deutsche Forschungsgemeinschaft, the Fonds der Chemischen Industrie, the Bundesministerium für Bildung und Forschung, and by the European Communities' BIOTECH Programme, as part of the Project of Technological Priority 1993–1996.

Literature Cited

1. Bagge P, Larsson C. 1986. Biosynthesis of aromatic amino acids by highly purified spinach chloroplasts. Compartmentation and regulation of the reactions. *Physiol. Plant.* 68:641–47

2. Borchert S, Harborth J, Schünemann D, Hoferichter P, Heldt HW. 1993. Studies of the enzymatic capacities and transport properties of pea root plastids. *Plant Physiol.* 101:303–12

3. Bowsher CG, Boulton EL, Rose J, Nayagam S, Emes MJ. 1992. Reductant for glutamate synthase is generated by the oxidative pentose phosphate pathway in non-photosynthetic root plastids. *Plant J.* 2:893–98

4. Brink S, Fischer K, Klösgen RB, Flügge UI. 1995. Sorting of nuclear-encoded chloroplast membrane proteins to the envelope and the thylakoid membrane. *J. Biol. Chem.* 270:20808–15

5. Caspar T, Huber SC, Somerville C. 1985. Alterations in growth, photosynthesis, and respiration in a starchless mutant of *Arabidopsis thaliana* (L.) deficient in chloroplast phosphoglucomutase activity. *Plant Physiol.* 79:11–17

6. Caspar T, Lin T-S, Kakefuda G, Benbow L, Preiss J, Somerville C. 1991. Mutants of *Arabidopsis* with altered regulation of starch metabolism. *Plant Physiol.* 95:1181–88

7. Day DA, Hatch MD. 1981. Transport of 3-phosphoglyceric acid, phosphoenolpyruvate, and inorganic phosphate in maize mesophyll chloroplasts, and the effect of 3-phosphoglyceric acid on malate and phosphoenolpyruvate production. *Arch. Biochem. Biophys.* 211:743–49

8. Denyer K, Dunlap F, Thorbjørnsen T, Keeling P, Smith AM. 1996. The major form of ADP-glucose pyrophosphorylase in maize endosperm is extra-plastidial. *Plant Physiol.* 112:779–85

9. Entwistle G, ap Rees T. 1990. Lack of fructose-1,6-bisphosphatase in a range of higher plants that store starch. *Biochem. J.* 271:467–72

10. Fischer K, Arbinger B, Kammerer K, Busch C, Brink S, et al. 1994. Cloning and *in vivo* expression of functional triose phosphate/phosphate translocators from C₃- and C₄-plants: evidence for the putative

participation of specific amino acids residues in the recognition of phosphoenolpyruvate. *Plant J.* 5:215–26

11. Fischer K, Kammerer B, Gutensohn M, Arbinger B, Weber A, et al. 1997. A new class of plastidic phosphate translocators: a putative link between primary and secondary metabolism by the phosphoenolpyruvate/phosphate antiporter. *Plant Cell* 9:453–62

12. Fliege R, Flügge UI, Werdan K, Heldt HW. 1978. Specific transport of inorganic phosphate, 3-phosphoglycerate and triosephosphates across the inner membrane of the envelope in spinach chloroplasts. *Biochim. Biophys. Acta* 502:232–47

13. Flügge UI. 1985. Hydrodynamic properties of the Triton X-100 solubilized chloroplast phosphate translocator. *Biochim. Biophys. Acta* 815:299–305

14. Flügge UI. 1987. Physiological function and physical characteristics of the chloroplast phosphate translocator. In *Progress in Photosynthesis Research*, ed. J Biggins, 3:739–46. The Hague: Nijhoff

15. Flügge UI. 1992. Reaction mechanism and asymmetric orientation of the reconstituted chloroplast phosphate translocator. *Biochim. Biophys. Acta* 1110:112–18

16. Flügge UI. 1995. Phosphate translocation in the regulation of photosynthesis. *J. Exp. Bot.* 46:1317–23

17. Flügge UI, Fischer K, Gross A, Sebald W, Lottspeich F, et al. 1989. The triose phosphate-3-phosphoglycerate-phosphate translocator from spinach chloroplasts: nucleotide sequence of a full-length cDNA clone and import of the *in vitro* synthesized precursor protein into chloroplasts. *EMBO J.* 8:39–46

18. Flügge UI, Heldt HW. 1981. The phosphate translocator of the chloroplast envelope. Isolation of the carrier protein and reconstitution of transport. *Biochim. Biophys. Acta* 638:296–304

19. Flügge UI, Heldt HW. 1991. Metabolite translocators of the chloroplast envelope. *Annu. Rev. Plant Physiol. Plant Mol. Biol.* 42:129–44

20. Flügge UI, Weber A. 1994. A rapid method for measuring organelle-specific substrate transport in homogenates from plant tissues. *Planta* 194:181–85

21. Flügge UI, Weber A, Fischer K, Loddenkötter B, Wallmeier H. 1992. Structure and function of the chloroplast triose phosphate/phosphate translocator. In *Research in Photosynthesis*, ed. N Murata, 3:667–74. Dordrecht: Kluwer

22. Foster JM, Smith AM. 1993. Metabolism of glucose-6-phosphate by plastids from developing pea embryos. *Planta* 190:17–24

23. Gerhardt R, Stitt M, Heldt HW. 1987. Subcellular metabolite levels in spinach leaves. Regulation of sucrose synthesis during diurnal alterations in photosynthetic partitioning. *Plant Physiol.* 83:399–407

24. Gross A, Brückner G, Heldt HW, Flügge UI. 1990. Comparison of the kinetic properties, inhibition and labelling of the phosphate translocators from maize and spinach mesophyll chloroplasts. *Planta* 180:262–71

25. Hattenbach B, Müller-Röber B, Nast G, Heineke D. 1997. Antisense repression of both ADP-glucose pyrophosphorylase and triose phosphate translocator modifies carbohydrate partitioning in potato leaves. *Plant Physiol.* 115:471–75

26. Häusler RE, Schlieben NH, Schulz B, Flügge UI. 1998. Compensation of decreased triose phosphate/phosphate transport activity by accelerated starch turnover and glucose transport in transgenic tobacco. *Planta* 204:366–76

27. Hedrich R, Raschke K, Stitt M. 1985. A role for fructose-2,6-bisphosphate in regulating carbohydrate metabolism in guard cells. *Plant Physiol.* 79:977–82

28. Heineke D, Kruse A, Flügge UI, Frommer WB, Riesmeier JW, et al. 1994. Effect of antisense repression of the chloroplast triose-phosphate translocator on photosynthetic metabolism in transgenic potato plants. *Planta* 193:174–80

29. Heldt HW. 1980. Measurement of metabolite movement across the envelope and of the pH in the stroma and the thylakoid space in intact chloroplasts. *Methods Enzymol.* 69:604–13

30. Heldt HW, Chon CJ, Maronde D, Herold A, Stancovic ZS, et al. 1977. Role of orthophosphate and other factors in the regulation of starch formation in leaves and isolated chloroplasts. *Plant Physiol.* 59:1146–55

31. Heldt HW, Flügge UI, Borchert S. 1991. Diversity of specificity and function of phosphate translocators in various plastids. *Plant Physiol.* 95:341–43

32. Hill LM, Smith AM. 1991. Evidence that glucose 6-phosphate is imported as the substrate for starch biosynthesis by the plastids of developing pea embryos. *Planta* 185:91–96

33. Hill LM, Smith AM. 1995. Coupled movements of glucose 6-phosphate and triose phosphate through the envelopes of plastids from developing embryos of pea (*Pisum sativum* L.). *J. Plant Physiol.* 146:411–17

34. Howitz KT, McCarty RE. 1985. Kinetic characteristics of the chloroplast envelope glycolate transporter. *Biochemistry* 24:2645–52

35. Huber SC, Edwards GE. 1977. Transport in C_4 mesophyll chloroplasts. Evidence for an exchange of inorganic phosphate and phosphoenolpyruvate. *Biochim. Biophys. Acta* 462:603–12

36. Journet EP, Douce R. 1985. Enzymic capacities of purified cauliflower bud plastids for lipid synthesis and carbohydrate metabolism. *Plant Physiol.* 79:458–67

37. Kammerer B, Fischer K, Hilpert B, Schubert S, Gutensohn M, et al. 1998. Molecular characterization of a carbon transporter in plastids from heterotrophic tissues: the glucose 6-phosphate/phosphate antiporter. *Plant Cell* 10:105–17

38. Kampfenkel K, Möhlmann T, Batz O, Van Montagu M, Inzé D, et al. 1995. Molecular characterization of an *Arabidopsis thaliana* cDNA encoding a novel putative adenylate translocator of higher plants. *FEBS Lett.* 74:351–55

39. Kang F, Rawsthorne S. 1996. Metabolism of glucose-6-phosphate and utilization of multiple metabolites for fatty acid synthesis by plastids from developing oilseed rape embryos. *Planta* 199:321–27

40. Kasahara M, Hinkle PC. 1997. Reconstitution and purification of the D-glucose transporter from human erythrocytes. *J. Biol. Chem.* 252:7384–90

41. Knight JS, Gray JC. 1994. Expression of genes encoding the tobacco chloroplast phosphate translocator is not light-regulated and is repressed by sucrose. *Mol. Gen. Genet.* 242:586–94

42. Knight JS, Gray JC. 1995. The N-terminal hydrophobic region of the mature phosphate translocator protein is sufficient for targeting to the chloroplast inner envelope membrane. *Plant Cell* 7:1421–32

43. Li H-m, Culligan K, Dixon RA, Chory J. 1995. CUE1: a mesophyll cell-specific positive regulator of light-controlled gene expression in Arabidopsis. *Plant Cell* 7:1599–610

44. Loddenkötter B, Kammerer B, Fischer K, Flügge UI. 1993. Expression of the functional mature chloroplast triose phosphate

translocator in yeast internal membranes and purification of the histidine-tagged protein by a single metal-affinity chromatography step. *Proc. Natl. Acad. Sci. USA* 90:2155–59

45. Maloney PC. 1990. A consensus structure for membrane transport. *Res. Microbiol.* 141:374–83

46. Menzlaff E, Flügge UI. 1993. Purification and functional reconstitution of the 2-oxoglutarate/malate translocator from spinach chloroplasts. *Biochim. Biophys. Acta* 1147:13–18

47. Miernyk JA, Dennis DT. 1992. A developmental analysis of the enolase isoenzymes from *Ricinus communis. Plant Physiol.* 99: 748–50

48. Naeem M, Tetlow IJ, Emes MJ. 1997. Starch synthesis in amyloplasts purified from developing potato tubers. *Plant J.* 11:1095–103

49. Neuhaus HE, Maass U. 1996. Unidirectional transport of orthophosphate across the envelope of isolated cauliflower-bud amyloplasts. *Planta* 198:542–48

50. Neuhaus HE, Thom E, Batz O, Scheibe R. 1993. Purification of highly intact plastids from various heterotrophic plant tissues. Analysis of enzyme equipment and precursor dependency for starch biosynthesis. *Biochem. J.* 296:395–401

51. Overlach S, Diekmann W, Raschke K. 1993. Phosphate translocator of isolated guard-cell chloroplasts from *Pisum sativum* L. transports glucose-6-phosphate. *Plant Physiol.* 101:1201–7

52. Preiss J, Levi C. 1980. Starch synthesis and degradation. In *The Biochemistry of Plants*, ed. PK Stumpf, EE Conn, 3:371–423. New York: Academic

53. Quick WP, Neuhaus HE. 1996. Evidence for two types of phosphate translocators in sweet-pepper (*Capsicum annuum* L.) fruit chromoplasts. *Biochem. J.* 320:7–10

54. Quick WP, Scheibe R, Neuhaus HE. 1995. Induction of a hexose-phosphate translocator activity in spinach chloroplasts. *Plant Physiol.* 109:113–21

55. Riesmeier JW, Flügge UI, Schulz B, Heineke D, Heldt HW, et al. 1993. Antisense repression of the chloroplast triose phosphate translocator affects carbon partitioning in transgenic potato plants. *Proc. Natl. Acad. Sci. USA* 90:6160–64

56. Rumpho ME, Edwards GE. 1985. Characterization of 4,4′-diisothiocyano-2,2′-disulfonic acid stilbene inhibition of 3-phosphoglycerate-dependent O_2-evolution in isolated chloroplasts. Evidence for a common binding site of the C_4 phosphate translocator for 3-phosphoglycerate, phos-

phoenolpyruvate and inorganic phosphate. *Plant Physiol.* 78:537–44

57. Saier MH. 1990. Coupling of energy to transmembrane solute translocation in bacteria. *Res. Microbiol.* 141:282–86

58. Schäfer G, Heber U, Heldt HW. 1977. Glucose transport into spinach chloroplasts. *Plant Physiol.* 60:286–89

59. Schott K, Borchert S, Müller-Röber B, Heldt HW. 1995. Transport of inorganic phosphate and C_3- and C_6-sugar phosphates across the envelope of potato tuber amyloplasts. *Planta* 196:647–52

60. Schulz B, Frommer WB, Flügge UI, Hummel S, Fischer K, et al. 1993. Expression of the triose phosphate translocator gene from potato is light dependent and restricted to green tissues. *Mol. Gen. Genet.* 238:357–61

61. Schulze-Siebert D, Heineke D, Scharf H, Schulz G. 1984. Pyruvate-derived amino acids in spinach chloroplasts: synthesis and regulation during photosynthetic carbon metabolism. *Plant Physiol.* 76:465–71

62. Schünemann D, Borchert S. 1994. Specific transport of inorganic phosphate and C_3- and C_6-sugar-phosphates across the envelope membranes of tomato (*Lycopersicon esculentum*) leaf-chloroplasts, tomato fruit-chloroplasts and fruit-chromoplasts. *Bot. Acta* 107:461–67

63. Schünemann D, Borchert S, Flügge UI, Heldt HW. 1993. ADP/ATP translocator from pea root plastids: comparison with translocators from spinach chloroplasts and pea leaf mitochondria. *Plant Physiol.* 103:131–37

64. Schwarz M, Gross A, Steinkamp T, Flügge UI, Wagner R. 1994. Ion channel properties of the reconstituted chloroplast triose phosphate/phosphate translocator. *J. Biol. Chem.* 269:29481–89

65. Stitt M, ap Rees T. 1979. Capacities of pea chloroplasts to catalyse the oxidative pentose phosphate pathway and glycolysis. *Phytochemistry* 18:1905–11

66. Sullivan T, Kaneko Y. 1995. The maize *brittle1* gene encodes amyloplast membrane polypeptides. *Planta* 196:477–84

67. Sullivan TD, Strelow LI, Illingworth CA, Phillips CA, Nelson OE. 1991. Analysis of the maize *brittle-1* alleles and a defective *Suppressor-mutator*-induced mutable allele. *Plant Cell* 3:1337–48

68. Tetlow IJ, Blisset KJ, Emes MJ. 1994. Starch synthesis and carbohydrate oxidation in amyloplasts from developing wheat endosperm. *Planta* 194:454–60

69. Tetlow IJ, Bowsher CG, Emes MJ. 1996. Reconstitution of the hexose phosphate

translocator from the envelope membranes of wheat endosperm amyloplasts. *Biochem. J.* 319:717–23

70. Thom E, Möhlmann T, Quick WP, Camara B, Neuhaus HE. 1998. Sweet pepper plastids: enzymic equipment, characterisation of the plastidic oxidative pentose-phosphate pathway, and transport of phosphorylated intermediates across the envelope membrane. *Planta* 204:226–33

71. Thorbjørnsen T, Villand P, Denyer K, Olsen O-A, Smith AM. 1996. Distinct isoforms of ADPglucose pyrophosphorylase occur inside and outside the amyloplasts in barley endosperm. *Plant J.* 10:243–50

72. Tyson RH, ap Rees T. 1988. Starch synthesis by isolated amyloplasts from wheat endosperm. *Planta* 175:33–38

73. Van der Straeten D, Rodrigues-Pousada RA, Goodman HM, Van Montagu M. 1991. Plant enolase: gene structure, expression and evolution. *Plant Cell* 3:719–35

74. Wagner R, Apley EC, Gross A, Flügge UI. 1989. The rotational diffusion of chloroplast phosphate translocator and of lipid molecules in bilayer membranes. *Eur. J. Biochem.* 182:165–73

75. Wallmeier H, Weber A, Gross A, Flügge UI. 1992. Insights into the structure of the chloroplast phosphate translocator protein. In *Transport and Receptor Proteins of Plant Membranes*, ed. DT Cooke, DT Clarkson, pp. 77–89. New York: Plenum

76. Weber A, Menzlaff E, Arbinger B, Gutensohn M, Eckerskorn C, Flügge UI. 1995. The 2-oxoglutarate/malate translocator of chloroplast envelope membranes: molecular cloning of a transporter protein containing a 12-helix motif and expression of the functional protein in yeast cells. *Biochemistry* 34:2621–27

77. Willey DL, Fischer K, Wachter E, Link TA, Flügge UI. 1991. Molecular cloning and structural analysis of the phosphate translocator from pea chloroplasts and its comparison to the spinach phosphate translocator. *Planta* 183:451–61

Annu. Rev. Plant Physiol. Plant Mol. Biol. 1999. 50:47–65

THE 1-DEOXY-D-XYLULOSE-5-PHOSPHATE PATHWAY OF ISOPRENOID BIOSYNTHESIS IN PLANTS

Hartmut K. Lichtenthaler

Botanisches Institut (Plant Physiology and Biochemistry), University of Karlsruhe, D-76128 Karlsruhe, Germany;
e-mail: Hartmut.Lichtenthaler@bio-geo.uni-karlsruhe.de

KEY WORDS: carotenoid biosynthesis, chloroplast metabolism, isopentenyl diphosphate, isoprene formation, non-mevalonate IPP formation

---

ABSTRACT

In plants the biosynthesis of prenyllipids and isoprenoids proceeds via two independent pathways: (*a*) the cytosolic classical acetate/mevalonate pathway for the biosynthesis of sterols, sesquiterpenes, triterpenoids; and (*b*) the alternative, non-mevalonate 1-deoxy-D-xylulose-5-phosphate (DOXP) pathway for the biosynthesis of plastidic isoprenoids, such as carotenoids, phytol (a side-chain of chlorophylls), plastoquinone-9, isoprene, mono-, and diterpenes. Both pathways form the active C_5-unit isopentenyl diphosphate (IPP) as the precursor from which all other isoprenoids are formed via head-to-tail addition. This review summarizes current knowledge of the novel 1-deoxy-D-xylulose-5-phosphate (DOXP) pathway for isopentenyl diphosphate biosynthesis, apparently located in plastids. The DOXP pathway of IPP formation starts from D-glyceraldehyde-3-phosphate (GA-3-P) and pyruvate, with DOXP-synthase as the starting enzyme. This pathway provides new insight into the regulation of chloroplast metabolism.

CONTENTS

47

1040-2519/99/0601-0047$08.00

INTRODUCTION

The biosynthesis of plant isoprenoids, carotenoids, phytol, sterols, plasto-quinone-9 as well as monoterpenes, sesquiterpenes, diterpenes, or polyterpenes seemed to have been well understood since the late 1950s (36). Labeling experiments with ^{14}C-labeled substrates indicated that the photosynthetic plants and algae form their isoprenic C_5-unit (IPP) and all isoprenoids—as in animal systems and fungi—via the acetate/mevalonate (MVA) pathway (23–25, 34, 35, 70), although some observations were not in agreement with the MVA pathway. For example, photosynthetically fixed $^{14}CO_2$ was rapidly incorporated into the plastidic isoprenoids (carotenoids, phytol, plastoquinone-9), whereas ^{14}C-labeled acetate and MVA were readily incorporated into the cytosolic sterols, but only at low rates into the plastidic isoprenoids (8, 9, 21, 22, 27, 34, 37). Moreover, mevinolin, a highly specific inhibitor of the HMG-CoA reductase, efficiently inhibited the cytosolic sterol and ubiquinone accumulation, but did not affect the accumulation of phytol, carotenoids, and plastoquinone-9 in plastids (4–6, 13, 58). In addition, isolated plastids could not make IPP from MVA (44). The discovery that the isoprenoid hopanoids (sterol surrogates) of certain eubacteria are formed via a non-MVA pathway (53, 55) was the starting point in 1993 for the author's group to re-investigate the biosynthesis of plastidic isoprenoids, in cooperation with Michael Rohmer (Strasbourg), and Frieder W. Lichtenthaler (Darmstadt). Applying ^{13}C- and ^{2}H-labeling techniques, NMR spectroscopy, and GC-MS analyses, it was shown that green algae (chlorophyta), higher plants, and other algal groups synthesize their plastidic isoprenoids including isoprene via the novel 1-deoxy-D-xylulose-5-phosphate (DOXP) pathway (3, 38, 39, 41, 42, 63–68, 75). This pathway is also involved in the biosynthesis of various other terpenoids (16, 17, 30, 46).

^{13}C-LABELING OF PLASTIDIC ISOPRENOIDS FROM [1-^{13}C]GLUCOSE

The ^{13}C-labeling of ß-carotene, lutein, phytol, and the nona-prenyl chain of plastoquinone-9 in green algae and higher plants, grown photoheterotrophically on [1-^{13}C]glucose, provided a ^{13}C-labeling pattern (Figure 1) that was not in agreement with the formation of the IPP precursor unit from acetate and MVA (39–42, 63–68). The IPP C_5-units of these plastidic isoprenoids did not exhibit the expected ^{13}C-enrichment in the three C-atoms C-2, C-4, and C-5, but rather showed labeling in the two C-atoms C-1 and C-5 (Figure 1). This finding clearly indicated the existence of a completely different IPP biosynthesis pathway in green algae and higher plants for the biosynthesis of plastidic isoprenoids. Examination of the ^{13}C-labeling pattern of cytosolic sterols from [1-^{13}C]glucose revealed that green algae (*Scenedesmus, Chlorella, Chlamydomonas*) exhibited the same non-MVA labeling pattern for sterols as for plastidic isoprenoids

A. ^{13}C-labeling pattern of ß-carotene, phytol and plastoquinone-9 in higher plants and green algae

B. ^{13}C-labeling of IPP from [1-^{13}C] glucose

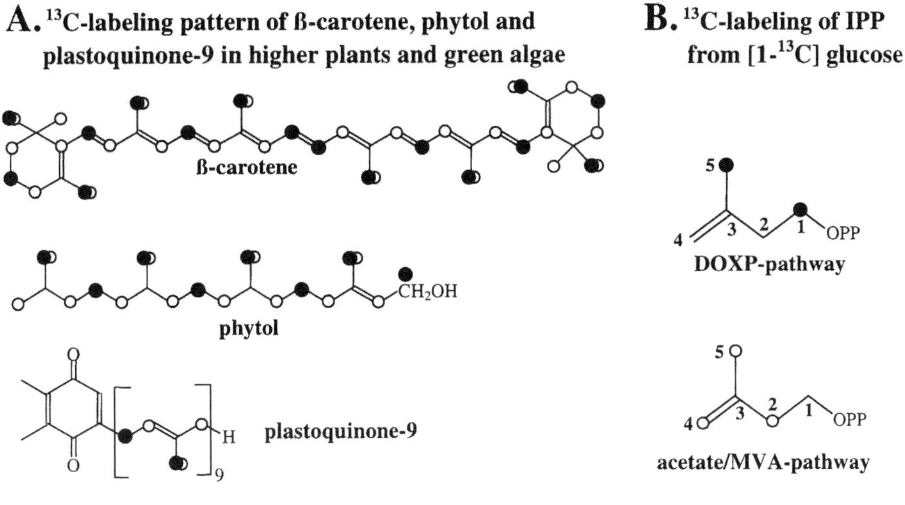

(●) *Labeling via [1-^{13}C] glucose:* **DOXP-pathway.**

(○) *Expected labeling pattern via the acetate/MVA-pathway.*

Figure 1 Labeling patterns in (*A*) plastidic isoprenoids of higher plants and various algae when supplied with [1-^{13}C]glucose. (*B*) Two labeling patterns in isopentenyl diphosphate (IPP) resulting from the DOXP pathway (*upper*) and the MVA-pathway (*lower*). The labeling of the plastidic isoprenoids and IPP proceeded only via the DOXP pathway; the expected labeling via the MVA pathway could not be detected. *Black circles*: Labeling of C-atoms from [1-^{13}C]glucose via the DOXP pathway; *white circles*: expected labeling of C-atoms via the acetate/MVA pathway of IPP formation.

(12, 65, 67), whereas in higher plants (barley, carrot, duckweed) sitosterol was labeled via the classical acetate/MVA pathway (38, 41). Thus, unicellular green algae only have the DOXP pathway for IPP formation, whereas higher plants possess two different IPP biosynthesis pathways (36, 38).

THE DOXP PATHWAY OF IPP BIOSYNTHESIS

The 1-Deoxy-D-Xylulose-5-Phosphate Synthase, First Enzyme of the DOXP Pathway

The starting substrates of the DOXP pathway are glyceraldehyde-3-phosphate (GA-3-P) and pyruvate. In a thiamin-dependent transketolase-type reaction, a C_2-unit derived from pyruvate (hydroxyethyl-thiamine) is transferred to GA-3-P, whereby DOXP is formed (Figure 2). This step is catalyzed by the enzyme DOXP-synthase, or DXS. For photosynthetic organisms this enzymic step was first verified in green algae by extensive ^{13}C-NMR studies using various glucoses, ^{13}C-labeled at different C-atoms, and uniformly labeled [U-^{13}C$_6$]glucose (63–68). It was subsequently also demonstrated in higher plants applying labeling from [1-^{13}C]glucose (39, 41). The labeling pattern of the C_5-units of IPP (Figure 1) is identical to that found in eubacteria (53–55). Further evidence for this initial step was the efficient incorporation of 1-deoxy-D-xylulose (DOX) into plastidic isoprenoids (3, 68, 75) (see below). As final proof for

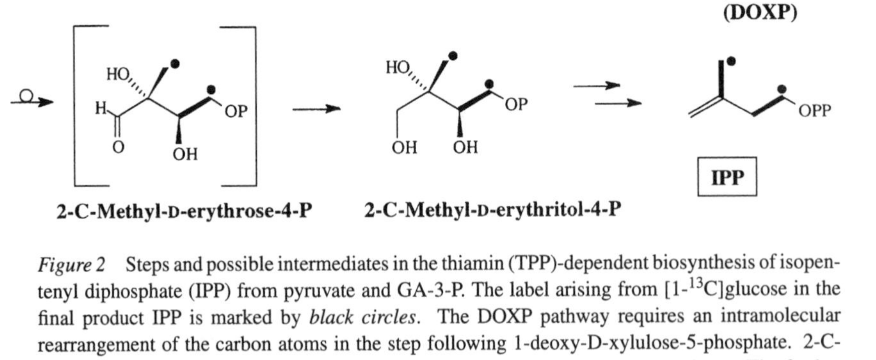

Figure 2   Steps and possible intermediates in the thiamin (TPP)-dependent biosynthesis of isopentenyl diphosphate (IPP) from pyruvate and GA-3-P. The label arising from [1-^{13}C]glucose in the final product IPP is marked by *black circles*. The DOXP pathway requires an intramolecular rearrangement of the carbon atoms in the step following 1-deoxy-D-xylulose-5-phosphate. 2-C-methyl-D-erythrose-4-P and 2-C-methyl-D-erythritol-4-P are possible intermediates. The further enzymatic steps and intermediates are not yet known.

DOXP-synthase as the starting step, it was demonstrated that a plant DOXP-synthase of *Mentha* (33) and a bacterial DOXP-synthase of *Escherichia coli* (43, 69), both overexpressed in *E. coli*, form DOXP from GA-3-P and pyruvate.

C-Skeleton Rearrangement

In further steps that are not yet fully clarified DOXP is transformed into IPP, possibly via 2-C-methyl-D-erythrose-4-phosphate and 2-C-methyl-D-erythritol-4-phosphate (Figure 2). These steps from DOXP to IPP require several reductases, dehydratases, and a kinase, and as co-factors possibly 3 NADPH and one ATP. This transformation of DOXP to IPP is based on an intramolecular C-skeleton rearrangement, whereby the C_2-unit of DOXP, originating from pyruvate, is inserted between the C-atoms C-1 and C-2 of GA-3-P (Figure 2). The incorporation of the complete C_2- and C_3-units from glucose into IPP was shown by the $^{13}C/^{13}C$ coupling constants of the NMR spectra seen after growing *Scenedesmus* on uniformly labeled [U-^{13}C]glucose (63, 67). Whether IPP is formed as the first isoprenoid C_5-unit or its isomer DMAPP in the DOXP pathway is unresolved.

1-Deoxy-D-Xylulose as Precursor Substrate

Evidence for DOXP as the first intermediate in the alternative IPP biosynthesis pathway came from the specific incorporation of deuterium (d)-labeled [1-$^2H_1$]deoxy-D-xylulose (d-DOX) and its methyl-glycoside (methyl-d-DOX) into the plastidic isoprenoid phytol in green algae (*Scenedesmus, Chlamydomonas*), a red alga (*Cyanidium*), and a higher plant (*Lemna*) (68), as well as into isoprene (*Populus, Chelidonium, Salix*) (68, 75, 76) as analyzed by NMR and/or GC-MS spectra. ^{13}C-MVA, when applied at a high concentration to a leaf, can be incorporated into isoprene and phytol, albeit to a lower extent (68). Plants and most algae apparently have the capacity to readily hydrolyze the applied xyluloside methyl-d-DOX to the free pentulose d-DOX, and to phosphorylate it to DOXP as the endogenous intermediate that is incorporated into the final isoprenoid. The transfer of methyl-d-DOX and d-DOX via IPP into isoprene and phytol (68) is additional evidence for the C-skeleton rearrangement in the DOXP pathway occurring in one of the steps after the formation of DOXP. The specific incorporation of a ^{13}C-labeled DOX into ß-carotene of *Catharanthus* (3), and of double ^{13}C-labeled DOX into ubiquinone of *E. coli* (52) provides additional corroboration.

2-C-Methyl-D-Erythritol-4-Phosphate as a Possible Intermediate

The further enzymic steps in the biosynthesis of IPP from DOXP have not yet been established in plants. One highly probable candidate is 2-C-methyl-D-erythritol-4-phosphate, which, after further reduction, dehydration and phosphorylation steps should yield IPP or DMAPP (Figure 2). When applied to plants or green algae, however, deuterium-labeled methyl-erythritol (14) was not incorporated into isoprenoids or isoprene (76), possibly due to the lack of a kinase that could convert 2-C-methyl-D-erythritol to its phosphate being the putative endogenous intermediate. In bacteria, by contrast, this deuterium-labeled methyl-erythritol is incorporated into the prenyl side-chain of menaquinone and ubiquinone at a low rate (15), and DOXP is transformed in *E. coli* to 2-C-methyl-D-erythritol-4-phosphate in the presence of NADPH by a reducto-isomerase (32). In *Corynebacterium*, methyl-D-erythritol-2,4-cyclodiphosphate is accumulated and marked from ^{13}C-glucose according to the DOXP pathway (14). In *Liriodendron*, a ^{13}C-labeled 1-DOX was converted into 2-C-methyl-D-erythritol (56): A 2-C-methyl-erythronolactone has been detected in higher plants (18, 31, 60). DOXP could possibly yield this lactone after oxidation and benzilic acid rearrangement. Although the intermediates following 2-C-methyl-D-erythritol-4-phosphate in the DOXP pathway of IPP-formation have not yet been identified, their structure can be presumed to be two reduction and dehydration steps, with one phosphorylation step being involved in these final steps of IPP formation.

COMPARTMENTATION OF IPP AND ISOPRENOID BIOSYNTHESIS IN HIGHER PLANTS

In their IPP and isoprenoid biosynthesis, there is a dichotomy in higher plants, one related to the plastid and the other to the cytosol (36, 41). The acetate/MVA pathway, producing IPP for sterol biosynthesis (11, 38, 41), proceeds in the cytosol and can be inhibited by mevinolin (4, 5, 13, 58). Sesquiterpenes are formed in the cytosol (2, 7), as well as polyterpenes by a consecutive chain elongation (Figure 3) (19). Given the existence of the DOXP pathway of IPP formation, polyterpene biosynthesis requires investigation to determine if it is solely based on the MVA pathway or is partly dependent on the DOXP pathway.

LOCALIZATION OF THE DOXP PATHWAY IN PLASTIDS

The plastid, in turn, is the site of the DOXP pathway of IPP formation (see below). This IPP biosynthesis starts from GA-3-P, an intermediate in the photosynthetic carbon reduction cycle, and pyruvate, which can be formed within the plastid from 3-phosphoglyceric acid. The DOXP pathway delivers isoprene (68, 75, 76), carotenoids (3, 38, 39, 41), phytol, and the nona-prenyl chain of plastoquinone-9 (3, 11, 38, 39, 41) as well as mono- (2, 17) and diterpenoids (7, 16, 30, 62), as indicated in Figure 3. This IPP and isoprenoid pathway can

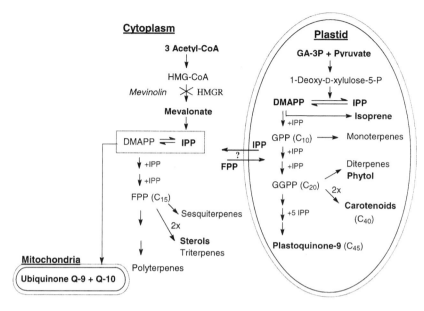

Figure 3 Suggested compartmentation of IPP and isoprenoid biosynthesis in higher plants between cytosol (acetate/MVA pathway) and plastids (DOXP pathway). The specific inhibition of the cytosolic HMG-CoA reductase (HMGR) by the antibiotic mevinolin (4–6, 13) is indicated. The nona- and deca-prenyl chain formation of ubiquinones Q-9 and Q-10 apparently proceeds in mitochondria from cytosolic IPP. Abbreviations used: DMAPP, dimethylallyl diphosphate; GPP, geranyl diphosophate; FPP, farnesyl diphosphate; GGPP, geranylgeranyl diphosphate.

easily be labeled from photosynthetically fixed $^{14}CO_2$, as Goodwin et al observed some 40 years ago (8, 21, 22).

At least the final steps of the biosynthesis of the plastidic isoprenoids proceed in chloroplasts (35). Supporting evidence for the plastid localization of the DOXP pathway is the observation that the light-dependent emission of isoprene is formed from DMAPP (74) within the chloroplast (68, 75, 76). Furthermore, the biosynthesis of thiamine and pyridoxal occurs in chloroplasts (28, 29). The DOXP pathway is also present in the cyanobacterium *Synechocystis* for biosynthesis of phytol and ß-carotene (Table 1) (12, 38, 49). If cyanobacteria are progenitors of chloroplasts, they could have conserved their originally bacterial DOXP pathway of IPP biosynthesis during co-evolution with the eukaryotic host cells. Moreover, the fact that the genes for DOXP synthase in *Arabidopsis* and *Chlamydomonas* possess a plastid transit peptide sequence (40, 47) is strong evidence for the localization of the DOXP pathway in plastids.

Table 1 Formation of isoprenoids in plants and photosynthetic organisms via the acetate/ mevalonate (MVA) or the new 1-deoxy-D-xylulose-5-phosphate (DOXP) pathway of IPP formation. The data were obtained by determining the ^{13}C-labeling pattern of the isoprenoids from ^{13}C-glucoses or from deuterium- or ^{13}C-labeled 1-deoxy-D-xylulose via ^{13}C-NMR or mass spectrometry. PQ-9 = plastoquinone-9.

Organism	Isoprenoid	IPP pathway	References
Cyanobacteria			
Synechocystis PCC 6714	Phytol, ß-carotene	DOXP	12, 38
Green algae			
Scenedesmus obliquus	Phytol, ß-carotene, lutein,	DOXP	42
	Plastoquinone-9, Chondrilla-	DOXP	63, 65
	sterol, ergost-7-enol,	DOXP	66, 67
	Ubiquinone-10	DOXP	12
Chlorella fusca	Phytol, ß-carotene	DOXP	38
	Chondrillasterol	DOXP	12
Chlamydomonas	Phytol, ß-carotene	DOXP	38, 39
reinhardtii	Chondrillasterol	DOXP	12
Red algae			
Cyanidium caldarium	Phytol	DOXP	68
	Ergosterol	MVA	12
Heterokontophyta			
Ochromonas danica	Phytol	DOXP	68
	Ergosterol	MVA	12
Euglenophyta			
Euglena gracilis	Phytol	MVA	68
	Ergosterol	MVA	12
Liverworts			
Riccocarpus natans	Ricciocarpin A (sesquiterpene)	MVA	72
	Phytol	DOXP	1
Conocephalum conicum	Bornylacetate (monoterpene)	DOXP	72
	Phytol	DOXP	1
	Cubebanol (sesquiterpene)	MVA	1
Higher plants			
Carotenoids, phytol,			
isoprene, sterols			
Lemna gibba	Phytol, ß-carotene, PQ-9,	DOXP	38–41
	Sitosterol, stigmasterol,	MVA	38–41
	campesterol	MVA	38–41
Daucus carota	Phytol,	DOXP	38–41
	Sitosterol, stigmasterol,	MVA	38–41
	campesterol	MVA	38–41
Hordeum vulgare	Phytol	DOXP	38–41
	Sitosterol	MVA	38–41

(Continued)

Table 1   *(Continued)*

Organism	Isoprenoid	IPP pathway	References
Populus nigra	Isoprene (hemiterpene)	DOXP	75, 76
Chelidonium maius	Isoprene	DOXP	75, 76
Salix viminalis	Isoprene	DOXP	75, 76
Catharanthus roseus	Phytol, carotene	DOXP	3
	Sitosterol	MVA	3
Lycopersicon esculentum	Lycopene	DOXP	64
Nicotiana tabacum	Plastoquinone-9	DOXP	11
	Sitosterol, stigmasterol	MVA	11
	Ubiquinone-10	MVA	11
Mono-, sesqui- or diterpenoids			
Ginkgo biloba	Ginkgolide A (diterpene)	DOXP	62
Taxus chinensis	Taxol (diterpene)	DOXP	16
Marrubium vulgare	Marrubiin (diterpene)	DOXP	30
Mentha x piperita	Menthone (monoterpene)	DOXP	17
Mentha pulegium	Pulegone (monoterpene)	DOXP	17
Pelargonium graveolens	Geraniol (monoterpene)	DOXP	17
Thymus vulgaris	Thymol (monoterpene)	DOXP	17
Matricaria recutita	Sesquiterpenes	DOXP[a]	2
Hordeum vulgare	Sesquiterpenoid derivative	DOXP	46
Salvia officinalis	Kauren (diterpene)	DOXP	b
Eucalyptus globulus, ⎫	Volatile mono-, sesqui-	DOXP	b
Clematis vitisalba ⎭	and diterpenes	DOXP	b

[a]Primarily DOXP pathway, third C_5-unit also via MVA pathway.
[b]J Piel & W Boland, personal communication.

COOPERATION BETWEEN THE TWO IPP PATHWAYS OF HIGHER PLANTS

Whether the two cellular IPP pools cooperate and exchange IPP or other prenyl diphosphates, such as GPP, FPP or GGPP, is unresolved at present (Figure 3). Several observations suggest at least some exchange. One example is the low labeling rates of plastidic isoprenoids from applied ^{14}C-MVA. In ^{13}C-labeling of the diterpene ginkgolide from ^{13}C-glucose, three isoprene units were found to be labeled via the MVA pathway, and the fourth isoprene unit via the DOXP pathway (62). In the liverwort *Heteroscyphus*, the first three isoprenic units of phytol showed some label from applied ^{13}C-MVA, whereas the fourth unit was not labeled (50, 51). Both observations point to the transfer of a cytosolic FPP into the plastid where FPP was condensed with a DOXP-derived IPP. In our ^{13}C-labeling studies of phytol and carotenoids from ^{13}C-glucoses, we detected no such import of FPP into the plastid.

Some export of IPP or GPP from plastids into the cytosol may occur, yet such a transfer cannot proceed to a large extent, as deduced from inhibitor studies with mevinolin. When cytosolic MVA and sterol biosynthesis were blocked by the inhibitor mevinolin (4–6, 58), transfer of IPP or higher prenyl homologues from the chloroplast was insufficient for cytosolic sterol biosynthesis although labeling experiments with ^{13}C-MVA and deuterium-labeled DOXP in algae demonstrated some export of IPP or other prenyl diphosphates from the plastids (68). Also, recent studies in chamomile indicated in sesquiterpenes the first two C_5-units were derived from ^{13}C-glucose via the DOXP pathway, and the third C_5-unit was labeled by either the DOXP or the MVA pathway (1). Future research must define at what physiological conditions and developmental stages the plastidic DOXP-dependent biosynthesis of IPP, isoprene, monoterpene, diterpene (phytol), and tetraterpenes (carotenoids) is fully autonomous or partially dependent on the cytosolic IPP pathway, and vice versa.

BIOSYNTHESIS OF THE PRENYL SIDE-CHAINS OF UBIQUINONES

Mitochondria, which contain ubiquinones with prenyl side-chains (34, 58, 60), apparently do not possess their own IPP biosynthesis pathway. Their prenyl chain biosynthesis is dependent on cytosolic IPP formation (11) (see below). Plant mitochondria contain ubiquinone-9 (Q-9) and ubiquinone-10 (Q-10) (57, 59). The final steps of ubiquinone biosynthesis, the prenylation of the benzoquinone nucleus, apparently proceed in the mitochondria. The accumulation of sterols and ubiquinones was strongly mevinolin inhibited (4, 5, 13), which suggests that formation of the prenyl side-chains of ubiquinones is dependent on cytosolic IPP biosynthesis. Moreover, labeled MVA-5-P was not incorporated by mitochondria isolated from higher plants, whereas IPP was (45). In higher plants the mitochondrial ubiquinone biosynthesis is dependent on the cytosolic IPP formation. It has recently been shown in non-green tobacco cell cultures that sterols and the prenyl side-chain of Q-10 came from the same IPP pool synthesized via MVA (11) (Figure 3).

In green algae, however, not only the plastidic isoprenoids are formed via the DOXP pathway, but so too are the cytosolic sterols (63, 65–68). With ^{13}C-labeled glucose it was demonstrated that the deca-prenyl chain of ubiquinone Q-10 in *Scenedesmus* is also synthesized via the DOXP pathway (12).

DISTRIBUTION OF THE DOXP-PATHWAY IN ALGAE

The DOXP pathway for IPP biosynthesis is widely distributed in photosynthetic organisms, such as algae and higher plants, and is required for the synthesis of plastidic isoprenoids (Table 1). This pathway also occurs in cyanobacteria (12), in several green algae (36, 63, 65, 66), the red alga *Cyanidium* (12, 68), and in the chrysophyte *Ochromonas* (12, 68). In *Cyanidium* and *Ochromonas*, the cytoplasmic sterols are formed via the classical MVA pathway as in higher plants (12, 49, 68). In contrast, the unicellular green algae tested synthesize not only their plastidic isoprenoids, but also their sterols via the DOXP pathway (38, 42, 63, 65, 66).

In *Euglena*, the situation is inverse; both the plastidic phytol and the cytoplasmic ergosterol are ^{13}C-labeled from glucose via the MVA pathway (12, 49, 68). When [2-^{13}C]MVA is supplied to *Euglena*, a large amount of the label shows up in ergosterol, and to a lesser degree in plastidic phytol (49, 68). These recent results confirm the very early labeling studies of *Euglena* ß-carotene via the MVA pathway (71). *Euglena* may have lost the DOXP pathway during the genetic rearrangement after the second endosymbiotic event (Figure 4). In contrast, in *Ochromonas*, which is believed to represent a secondary endosymbiotic event (73), the plastidic DOXP pathway was conserved. Green algae, in turn, seem to have lost their cytosolic MVA pathway of IPP formation. This suggests that during the evolution of various extant algal groups different strategies of genetic and metabolic organization took place.

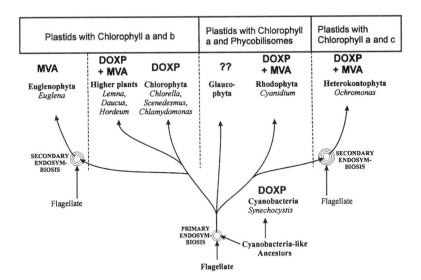

Figure 4   Putative evolution of some photosynthetic algae and higher plants with indication of the presence of one or both types of isopentenyl diphosphate (IPP) biosynthesis: MVA and/or DOXP pathway. Primary and secondary endosymbiotic events (73) leading to chloroplasts with an envelope consisting of 2, 3, or 4 biomembranes are indicated.

Low labeling of phytol with [2-^{13}C]MVA was observed in *Cyanidium* and *Ochromonas* (49, 68), but not in the green alga *Scenedesmus*. When applying intermediates of the DOXP pathways, such as [1-^{2}H]DOX to *Cyanidium* and *Ochromonas*, the deuterium label showed up not only in phytol, but also in ergosterol (49, 68), indicating that in both algae some exchange may exist between the two IPP pools of different biosynthetic origin. Incorporation of minor label of methyl[1-^{2}H$_1$]DOX into phytol and ergosterol of *Euglena* (49, 68) is thought to be caused by a breakdown of d-DOX.

THE DOXP PATHWAY AND BIOSYNTHESIS OF TERPENOIDS

The new DOXP pathway for IPP biosynthesis has now been established unequivocally in eubacteria (53–55), cyanobacteria (12, 49), various algal groups (42, 63, 65, 66), and higher plants (3, 16, 17, 30, 41, 68). In higher plants, it is further responsible for the formation of the volatile hemiterpene isoprene (68, 75, 76), diterpenes, such as ginkgolides (62), taxol (16), marrubiin (30), the monoterpenes menthone (17), and borneol (72); for the secondary carotenoid lycopene in tomato fruits (36, 64); for sesquiterpenoid biosynthesis in chamomile (2) and in mycorrhizal barley roots (46) and the volatile mono-, sesqui- and diterpenoids of several flowers (Table 1). Although the DOXP pathway for IPP and terpenoid biosynthesis is widely distributed in higher plants (Table 1), it has yet to be determined whether the basic carbon skeleton of the numerous other plant terpenoids is derived from the MVA or plastidic DOXP pathway, or by a cooperation of both.

GENES OF 1-DEOXY-D-XYLULOSE-5-PHOSPHATE SYNTHASE (DXS)

1-Deoxy-d-xylulose-5-phosphate synthase (dxs) is the first enzyme of the DOXP pathway to be characterized at the enzymatic and molecular level. DXS of *E. coli* is a transketolase-like enzyme with a molecular weight of about 65 KDa (43, 69). It is one of a distinct family of DXS-like protein sequences that have been found in several bacteria and plants (See Figure 5). The DXS are highly conserved and share sequences of a special class of transketolases (Figure 5). The sequence motif (between "VILNDN" and "VGAL") allows the bacterial and plant DXS sequences to be distinguished from each other. One of these DXS-like genes is *CLA1* (47), a single copy gene that is positively regulated by light. The protein sequence includes a predicted chloroplast transit peptide. Thus, *CLA1* is likely a plastidic enzyme with a key function in pigment biosynthesis (47). A

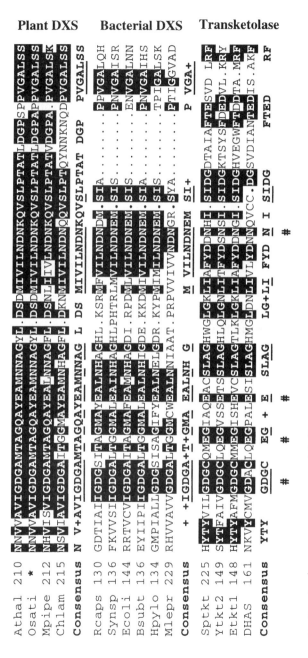

Figure 5 Alignments of transketolase-type DXS sequences of plants and bacteria (or DXS homologous open reading frames) and of transketolases in the range of the thiamine binding site. The putative thiamine binding-site includes $GDGX_{7-8}EX_{3-4}AX_{11-13}N$ (26). # denotes residues involved in TPP-binding in yeast transketolase (48); *dots* denote the absence of amino acid residues; the *boldface* and *shaded* characters denote amino acid residues that are highly conserved. Conserved residues are indicated in each group with the symbol +. If a residue of the consensus occurs in more than one group, it is underlined. Athal, *Arabidopsis thaliana CLA1* (accession number U27099) (33); Osati, *Oryza sativa* (af024512); Mpipe, *Mentha x piperita* (af019383) (33); Chlamy, *Chlamydomonas reinhardtii* (aj007559); Rcaps, *Rhodobacter capsulatus* (z11165); Synsp, *Synechocystis sp.* (d90903); Ecoli, *Escherichia coli* (u82664); Bsubt, *Bacillus subtilis* (d84432); Hpylo, *Helicobacter pylori* (ae000552); Mlepr, *Mycobacterium leprae* (u15181); Sptkt, *Spinacia oleracea* chloroplast transketolase (L76554); Ytkt2, Yeast transketolase 2 (p33315); Etkt1, *E. coli* transketolase 1 (p27302); DHAS, *Hansenula polymorpha* formaldehyde transketolase (p06834).

highly similar gene was recently cloned from *Mentha piperita* and the expressed protein was shown to be active in DOXP synthesis (33). The *Mentha* gene sequence predicts 68% identical amino acid residues with those of *CLA1*.

In our laboratory a cDNA clone was isolated from *Chlamydomonas*, which is highly similar to *CLA1* (85% homologous amino acid residues) and with a predicted plastid transit peptide. We also detected a second DXS-homologous sequence in *Chlamydomonas*. Thus, *Chlamydomonas* appears to have two different DXS genes, possibly for cytosolic and plastidic IPP formation. Whether the plant DXS has a regulatory role in IPP and isoprenoid biosynthesis is not yet clear. One could expect several isoforms of DXS, assuming that biosynthesis of photosynthetic isoprenoids, of essential oils in oil glands, and of terpenoid phytoalexins are dependent on particular DXS activities.

BRANCH POINTS WITH OTHER CHLOROPLAST BIOSYNTHETIC PATHWAYS

The early observations that ^{14}C-labeled CO_2, GA-3-P, and pyruvate are better precursors of plastidic isoprenoids than ^{14}C-acetate or ^{14}C-MVA (8, 9, 21, 22, 27) are now being clarified with the operation of the DOXP pathway of IPP formation, in which case GA-3-P and pyruvate are direct substrates of the DOXP synthase. ^{14}CO_2$ is rapidly transferred into 3-phosphoglyceric acid (3-PGA) and GA-3-P via photosynthetic CO_2 assimilation (Figure 6). The quick formation of IPP and DMAPP from CO_2 via GA-3-P and pyruvate also explains the rapid isoprene emission under heat stress conditions (74) and the fast labeling and emission of isoprene from photosynthetically fixed ^{13}CO_2$ (10). Exogenously applied ^{14}C-acetate is quickly incorporated into fatty acids via the plant's plastidic de novo fatty acid synthetase (e.g. 20), but not into carotenoids and other plastid isoprenoids, since acetate is not a substrate of the DOXP pathway. DOXP, in turn, is an intermediate not only in the plastidic IPP biosynthesis, but also in the synthesis of thiamine and pyridoxal (28, 29). Pyruvate, in turn, is an essential branch point of the plastid metabolism; it serves as substrate of the DOXP pathway, of acetyl-CoA formation, and de novo fatty acid biosynthesis, and is also required for the biosynthesis of valin, leucin, and isoleucin (61) (Figure 6). Whether pyruvate is made in plastids from 3-PGA or arises as a byproduct of the ribulosebisphosphate carboxylase activity (2a) or may be delivered, in part, from the cytosol, has yet to be determined. Moreover, phosphoenol pyruvate (PEP) is a substrate of the shikimic acid pathway that, in plants, also occurs in plastids. Thus, there are many branch points in the use and metabolite flow of the primary photosynthetic products 3-PGA and GA-3-P to the various end products that require a fine regulation of chloroplast metabolism.

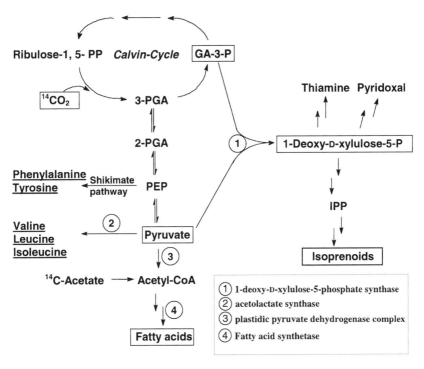

Figure 6   Metabolic pathways and branch points in plastids. The flow of metabolites from the photosynthetic reductive pentosephosphate cycle (Calvin cycle) into different end products, such as IPP, plastidic isoprenoids, isoprene, fatty acids, amino acids as well as thiamine and pyridoxal, is indicated. The central role of <u>GA-3-P</u> and <u>pyruvate</u> in the formation of 1-deoxy-D-xylulose-5-phosphate, IPP, and plastidic isoprenoids is emphasized.

CONCLUSION AND OUTLOOK

The incorporation studies over the past four years demonstrated that the DOXP pathway of IPP and isoprenoid biosynthesis is widely distributed in photosynthetic organisms. Future research should be directed to (*a*) elucidating the individual enzymatic steps between DOXP and IPP, (*b*) characterizing the corresponding genes and enzymes, and (*c*) evaluating the regulation of the DOXP pathway with respect to other metabolic pathways in chloroplasts. Finally, the possibility must also be examined of a partial cooperation of the two IPP yielding cellular pathways, the MVA and DOXP routes, in the biosynthesis of plant terpenoids. Enzymes of the DOXP pathway represent targets for new inhibitors. We may therefore anticipate the development of novel herbicides against plants and algae as well as antibacterial substances against pathogenic bacteria possessing the DOXP pathway. In fact, fosmidomycin has now been described as the first herbicide blocking the DOXP pathway (77).

ACKNOWLEDGMENTS

A major part of the work described here was supported by a grant from the German Research Council, DFG Bonn, which is gratefully acknowledged. Part of the work was performed in cooperation with Frieder W. Lichtenthaler, Darmstadt/Germany, and Michel Rohmer, Strasbourg/France. I thank my PhD students Jörg Schwender, Johannes Zeidler, and Christian Müller for assistance; Antonella Barelli-Kummer for typing the manuscript; and Gabrielle Johnson for language assistance.

> **Visit the *Annual Reviews* home page at**
> **http://www.AnnualReviews.org**

Literature Cited

1. Adam KP, Thiel R, Zapp J, Becker H. 1998. Involvement of the mevalonic acid pathway and the glyceraldehyde-pyruvate pathway in terpenoid biosynthesis of the liverworts *Riccio carpus natans* and *Conocephalum conicum. Arch. Biochem. Biophys.* 354:181–87

2. Adam KP, Zapp J. 1998. Biosynthesis of the isoprene units of chamomile sesquiterpenes. *Phytochemistry* 48:653–59

2a. Andrews TJ, Kane HJ. 1991. Pyruvate is a by-product of catalysis by ribulosebisphoshate carboxylase/oxygenase. *J. Biol. Chem.* 266:9447–52

3. Arigoni D, Sagner S, Latzel C, Eisenreich W, Bacher A, Zenk MH. 1997. Terpenoid biosynthesis from 1-deoxy-D-xylulose in higher plants by intramolecular skeletal rearrangement. *Proc. Natl. Acad. Sci. USA* 94:10600–5

4. Bach TJ, Lichtenthaler HK. 1982. Mevinolin, a highly specific inhibitor of microsomal 3-hydroxy-3-methyl-glutaryl-coenzyme A reductase of radish plants. *Z. Naturforsch. Teil C* 37:46–50

5. Bach TJ, Lichtenthaler HK. 1982. Inhibition of mevalonate biosynthesis and plant growth by the fungal metabolite mevinolin. See Ref. 74b, pp. 515–21

6. Bach TJ, Lichtenthaler HK. 1983. Inhibition by mevinolin of plant growth, sterol formation and pigment accumulation. *Physiol. Plant.* 59:50–60

7. Bohlmann J, Meyer-Gauen G, Croteau R. 1998. Plant terpenoid synthases: molecular biology and phytogenetic analysis.

Proc. Natl. Acad. Sci. USA 95:4126–33

8. Braithwaite GD, Goodwin TW. 1960. Studies on carotenogenesis. 27. Incorporation of [2-^{14}C]acetate, DL-[2-^{14}C] mevalonate and $^{14}CO_2$ into carrot-root preparations. *Biochem. J.* 76:194–97

9. Braithwaite GD, Goodwin TW. 1960. Studies on carotenogenesis. 25. The incorporation of [1-^{14}C]acetate, [2-^{14}C]acetate and $^{14}CO_2$ into lycopene by tomato slices. *Biochem. J.* 76:1–5

10. Delwiche CF, Sharkey TD. 1993. Rapid appearance of ^{13}C in biogenic isoprene when $^{13}CO_2$ is fed to intact leaves. *Plant Cell Environ.* 16:587–91

11. Disch A, Hemmerlin A, Bach TJ, Rohmer M. 1998. Mevalonate-derived isopentenyl diphoshate is the biosynthetic precursor of ubiquinone prenyl side-chain in tobacco BY-2 cells. *Biochem. J.* 331:615–21

12. Disch A, Schwender J, Müller C, Lichtenthaler HK, Rohmer M. 1998. Mevalonate versus glyceraldehyde phosphate pathway for isoprenoid biosynthesis in unicellular algae and the cyanobacterium *Synechocystis. Biochem. J.* 333:381–88

13. Döll M, Schindler S, Lichtenthaler HK, Bach TJ. 1984. Differential inhibition by mevinolin of prenyllipid accumulation in cell suspension cultures of *Silybum marianum* L. See Ref. 68a, pp. 277–80

14. Duvold T, Bravo JM, Pale-Grosdemange C, Rohmer M. 1997. Biosynthesis of 2-C-methyl-D-erythritol, a putative C_5 intermediate in the mevalonate-independent

pathway for isoprenoid biosynthesis. *Tetrahedron Lett.* 38:4769–72

15. Duvold T, Cali P, Bravo JM, Rohmer M. 1997. Incorporation of 2-C-methyl-D-erythritol, a putative isoprenoid precursor in the mevalonate-independent pathway, into ubiquinone and menaquinone of *Escherichia coli. Tetrahedron Lett.* 38:6181–84

16. Eisenreich W, Menhard B, Hylands PJ, Zenk MH, Bacher A. 1996. Studies on the biosynthesis of taxol: The taxane carbon skeleton is not of mevalonoid origin. *Proc. Natl. Acad. Sci. USA* 93:6431–36

17. Eisenreich W, Sagner S, Zenk MH, Bacher A. 1997. Monoterpenoid essential oils are not of mevalonate origin. *Tetrahedron Lett.* 38:3889–92

18. Ford CW. 1981. A new lactone from water-stressed chickpea. *Phytochemistry* 20:2019–20

19. Gershenzon J, Croteau RB. 1993. Terpenoid biosynthesis: the basic pathway and formation of monoterpenes, sesquiterpenes and diterpenes. See Ref. 48a, pp. 339–88

20. Golz A, Focke M, Lichtenthaler HK. 1994. Inhibitors of *de novo* fatty acid biosynthesis in higher plants. *J. Plant Physiol.* 143:426–33

21. Goodwin TW. 1958. Incorporation of $^{14}CO_2$, [2-^{14}C]acetate and [2-^{14}C]mevalonic acid into ß-carotene in etiolated maize seedlings. *Biochem. J.* 68:26–27

22. Goodwin TW. 1958. Studies in carotenogenesis 25. The incorporation of $^{14}CO_2$, [2-^{14}C]acetate and [2-^{14}C]mevalonic acid into ß-carotene by illuminated etiolated maize seedlings. *Biochem. J.* 70:612–17

23. Goodwin TW. 1965. Regulation of terpenoid biosynthesis in higher plants. In *Biosynthetic Pathways in Higher Plants*, ed. JB Pridham, T Swain, pp. 57–71. London: Academic

24. Goodwin TW. 1977. The prenyllipids of the membranes of higher plants. In *Lipids and Lipid Polymers in Higher Plants*, ed. M Tevini, HK Lichtenthaler, pp. 29–47. Berlin: Springer-Verlag

25. Goodwin TW. 1981. Biosynthesis of plant sterols and other triterpenoids. See Ref. 70, pp. 444–80

26. Hawkins CF, Borges A, Perham RN. 1989. A common structural motif in thiamin pyrophosphate-binding enzymes. *FEBS Lett.* 255:77–82

27. Heintze A, Görlach J, Leuschner C, Hoppe P, Hagelstein P, et al. 1990. Plastidic isoprenoid synthesis during chloroplast development. Change from metabolic autonomy to division-of-labor stage. *Plant Physiol.* 93:1121–22

28. Julliard JH. 1992. Biosynthesis of the pyridoxal ring (vitamin B6) in higher plant chloroplasts and its relationship with the biosynthesis of the thiazol ring (vitamin B1). *C.R. Acad. Sci. Ser.* 314:285–90

29. Julliard JH, Douce R. 1991. Biosynthesis of the thiazole moiety of thiamin (vitamin B1) in higher plant chloroplasts. *Proc. Natl. Acad. Sci. USA* 88:2041–45

30. Knöss W, Reuter B, Zapp J. 1997. Biosynthesis of the labdane diterpene marrubiin in *Marubium vulgare* via a non-mevalonate pathway. *Biochem. J.* 326:449–54

31. Kringstad R, Singsaas AO, Rusten G, Baekkemoen G, Paulsen BS, Nordal A. 1980. 2-C-methylaldotetronic acid present in plants. *Phytochemistry* 19:543–45

32. Kuzuyama T, Takahashi S, Watanabe H, Seto H. 1998. Direct formation of 2-methyl-D-erythritol 4-phosphate from 1-deoxy-D-xylulose 5-phosphate by 1-deoxy-D-xylulose 5-phosphate reductoisomerase, a new enzyme in the non-mevalonate pathway to isopentenyl diphosphate. *Tetrahedron Lett.* 39:4509–12

33. Lange B, Wildung M, McCaskill R, Croteau R. 1998. A family of transketolases that directs isoprenoid biosynthesis via a mevalonate-independent pathway. *Proc. Natl. Acad. Sci. USA* 95:2100–4

34. Lichtenthaler HK. 1987. Functional organization of carotenoids and prenylquinones in the photosynthetic membrane. In *The Metabolism, Structure and Function of Plant Lipids*, ed. P Stumpf, JB Mudd, WD Nes, pp. 63–73. New York: Plenum

35. Lichtenthaler HK. 1993. The plant prenyllipids including carotenoids, chlorophylls and prenylquinones. See Ref. 48a, pp. 427–70

36. Lichtenthaler HK. 1998. The plant's 1-deoxy-D-xylulose-5-phospate pathway for biosynthesis of isoprenoids. *Fett/Lipid.* 100:128–38

37. Lichtenthaler HK, Bach TJ, Wellburn AR. 1982. Cytoplasmic and plastidic isoprenoid compounds of oat seedlings and their distinct labeling from ^{14}C-mevalonate. See Ref. 74b, pp. 489–500

38. Lichtenthaler HK, Rohmer M, Schwender J. 1997. Two independent biochemical pathways for isopentenyl diphosphate and isoprenoid biosynthesis in higher plants. *Physiol. Plant.* 101:643–52

39. Lichtenthaler HK, Rohmer M, Schwender

J, Disch A, Seemann M. 1997. A novel mevalonate-independent pathway for the biosynthesis of carotenoids, phytol and prenyl chain of plastoquinone-9 in green algae and higher plants. See Ref. 74a, pp. 177–79

40. Lichtenthaler HK, Schwender J. 1998. The 1-deoxy-D-xylulose-5-phosphate pathway for biosynthesis of carotenoids and plastidic isoprenoids. See Ref. 56a, pp. 419–24

41. Lichtenthaler HK, Schwender J, Disch A, Rohmer M. 1997. Biosynthesis of isoprenoids in higher plant chloroplasts proceeds via a mevalonate independent pathway. *FEBS Lett.* 400:271–74

42. Lichtenthaler HK, Schwender J, Seemann M, Rohmer M. 1995. Carotenoid biosynthesis in green algae proceeds via a novel biosynthetic pathway. See Ref. 47a, pp. 115–18

43. Lois LM, Campos N, Putra SR, Danielsen K, Rohmer M, Boronat A. 1998. Cloning and characterization of a gene from *Escherichia coli* encoding a transketolase-like enzyme that catalyzes the synthesis of D-1-deoxyxylulose 5-phosphate, a common precursor for isoprenoid, thiamin, and pyridoxol biosynthesis. *Proc. Natl. Acad. Sci. USA* 95:2105–10

44. Lütke-Brinkhaus F, Kleinig H. 1987. Formation of isopentenyl diphosphate via mevalonate does not occur within etioplasts or etiochloroplasts of mustard (*Sinapis alba* L.) seedlings. *Planta* 171:401–11

45. Lütke-Brinkhaus F, Liedvogel B, Kleinig H. 1984. On the biosynthesis of ubiquinones in plant mitochondria. *Eur. J. Biochem.* 141:537–41

46. Maier W, Schneider B, Strack D. 1998. Biosynthesis of sesquiterpenoid cyclohexane derivatives in mycorrhizal barley roots proceeds via the glyceraldehyde-3-phosphate/pyruvate pathway. *Tetrahedron Lett.* 39:521–24

47. Mandel MA, Feldmann KA, Herrera-Estrella L, Rocha-Sosa M, Leon P. 1996. *CLA1*, a novel gene required for chloroplast development, is highly conserved in evolution. *Plant J.* 9:649–58

47a. Mathis P, ed. 1995. *Photosynthesis: From Light to Biosphere.* Amsterdam: Kluwer

48. Meshalkina L, Nilsson U, Wikner C, Kostikowa T, Schneider G. 1997. Examination of thiamin diphosphate binding site in yeast transketolase by site-directed mutagenesis. *Eur. J. Biochem.* 244:646–52

48a. Moore TS, ed. 1993. *Lipid Metabolism in Plants.* Boca Raton, FL: CRC

49. Müller C, Schwender J, Disch A, Rohmer M, Lichtenthaler FW, Lichtenthaler HK. 1998. Occurrence of the 1-deoxy-D-xylulose-5-phosphate pathway of isopentenyl diphosphate biosynthesis in different algae groups. See Ref. 56a, pp. 425–28

50. Nabeta K, Ishikawa T, Okuyama H. 1995. Sesqui- and diterpene biosynthesis from ^{13}C labeled acetate and mevalonate in cultured cells of *Heteroscyphus planus. J. Chem. Soc. Perkin. Trans.* 1:3111–15

51. Nabeta K, Kawae T, Saitoh T, Kikuchi T. 1997. Synthesis of chlorophyll α and ß-carotene from ^{2}H and ^{13}C-labeled mevalonates and ^{13}C-labeled glycin in cultured cells of liverworts *Heteroscyphus planus* and *Lophocolea heterophylla. J. Chem. Soc. Perkin Trans.* 1:261–67

52. Putra SR, Lois LM, Campos N, Boronat A, Rohmer M. 1998. Incorporation of $[2,3-^{13}C_2]$- and $[2,4-^{13}C_2]$-D-deoxyxylulose into ubiquinone of *Escherichia coli* via the mevalonate-independent pathway for isoprenoid biosynthesis. *Tetrahedron Lett.* 39:23–26

53. Rohmer M, Knani M, Simonin P, Sutter B, Sahm H. 1993. Isoprenoid biosynthesis in bacteria: a novel pathway for early steps leading to isopentenyl diphosphate. *Biochem. J.* 295:517–24

54. Rohmer M, Seemann M, Horbach S, Bringer-Meyer S, Sahm H. 1996. Glyceraldehyde 3-phosphate and pyruvate as precursors of isoprenic units in an alternative non-mevalonate pathway for terpenoid biosynthesis. *J. Am. Chem. Soc.* 118:2564–66

55. Rohmer M, Sutter B, Sahm H. 1989. Bacterial sterol surrogates. Biosynthesis of the side chain of bacteriohopanetetrol and of a carbocyclic pseudopentose from ^{13}C-labeled glucose in *Zymomonas mobilis. J. Chem. Soc. Chem. Commun.,* pp. 1471–72

56. Sagner S, Eisenreich W, Fellermeier M, Latzel C, Bacher A, Zenk MH. 1998. Biosynthesis of 2-C-methyl-D-erythritol in plants by rearrangement of the terpenoid precursor 1-deoxy-D-xylulose 5-phosphate. *Tetrahedron Lett.* 39:2091–94

56a. Sanchez J, Cerda-Olmedo E, Martinez-Force E, eds. 1998. *Advances in Plant Lipid Research.* Univ. Sevilla: Secretariado Publ.

57. Schindler S. 1984. Verbreitung und Konzentration von Ubichinon-Homologen in Pflanzen. *Karlsr. Contrib. Plant Physiol.* 12:1–240

58. Schindler S, Bach TJ, Lichtenthaler HK.

1985. Differential inhibition by mevinolin of prenyllipid accumulation in radish seedlings. *Z. Naturforsch. Teil C* 40:208–14

59. Schindler S, Lichtenthaler HK. 1984. Comparison of the ubiquinone homologue pattern in plant mitochondria and their possible prokaryotic ancestors. See Ref. 68a, pp. 273–76

60. Schramm RW, Tomaszewska B, Petersson G. 1979. Sugar-related hydroxy acids from *Phaseolus* and *Trifolium* species. *Phytochemistry* 18:1393–94

61. Schulze-Siebert D, Heinecke D, Scharf H, Schultz G. 1984. Pyruvate derived amino acids in spinach chloroplasts. *Plant Physiol.* 76:465–71

62. Schwarz MK. 1994. *Terpenbiosynthese in Ginkgo biloba.* PhD thesis. Eidgenoss. Tech. Hochsch., Zürich

63. Schwender J. 1995. Untersuchungen zur Biosynthese der Isoprenoide bei der Grünalge *Scenedesmus obliquus* mittels ^{13}C-Isotopenmarkierung. *Karlsr. Contrib. Plant Physiol.* 31:1–85

64. Schwender J, Lichtenthaler HK. 1998. Biosynthesis of lycopene in tomato fruits proceeds via the non-mevalonate isoprenoid pathway. See Ref. 56a, pp. 429–32

65. Schwender J, Lichtenthaler HK, Disch A, Rohmer M. 1997. Biosynthesis of sterols in green algae (*Scenedesmus, Chlorella*) according to a novel, mevalonate-independent pathway. See Ref. 74a, pp. 180–82

66. Schwender J, Lichtenthaler HK, Seemann M, Rohmer M. 1995. Biosynthesis of isoprenoid chains of chlorophylls and plastoquinone in *Scenedesmus* by a novel pathway. See Ref. 47a, pp. 1001–4

67. Schwender J, Seemann M, Lichtenthaler HK, Rohmer M. 1996. Biosynthesis of isoprenoids (carotenoids, sterols, prenyl side-chains of chlorophyll and plastoquinone) via a novel pyruvate/glyceraldehyde-3-phosphate non-mevalonate pathway in the green alga *Scenedesmus. Biochem. J.* 316:73–80

68. Schwender J, Zeidler J, Gröner R, Müller C, Lichtenthaler HK, et al. 1997. Incorporation of 1-deoxy-D-xylulose into isoprene and phytol by higher plants and algae. *FEBS Lett.* 414:129–34

68a. Siegenthaler PA, Eichenberger W, eds. 1984. *Structure, Function and Metabolism of Plant Lipids.* Amsterdam: Elsevier

69. Sprenger GA, Schörken U, Wiegert T, Grolle S, de Graaf AA, et al. 1997. Identification of a thiamin-dependent synthase in *Escherichia coli* required for the formation of 1-deoxy-D-xylulose-5-phosphate precursor to isoprenoids, thiamin, and pyridoxol. *Proc. Natl. Acad. Sci. USA* 94:12857–62

70. Spurgeon SL, Porter JW. 1981. Introduction. In *Biosynthesis of Isoprenoid Compounds*, ed. JW Porter, SL Spurgeon, 1:1–46. New York: Wiley

71. Steele JW, Gurin S. 1960. Biosynthesis of ß-carotene in *Euglena gracilis. J. Biol. Chem.* 235:2778–85

72. Thiel R, Adam KP, Zapp J, Becker H. 1997. Isopentenyl diphosphate biosynthesis in liverworts. *Pharm. Pharmacol. Lett.* 7:103–5

73. Van den Hoek C, Mann DG, Jahns HM. 1995. *Algae, An Introduction to Phycology.* Cambridge: Cambridge Univ. Press

74. Wildermuth MC, Fall R. 1996. Light-dependent isoprene emission. *Plant Physiol.* 112:171–82

74a. Williams JP, Khan MU, Lem NW, eds. 1997. *Physiology, Biochemistry and Molecular Biology of Plant Lipids.* Dordrecht: Kluwer

74b. Wintermans JFGM, Kuiper PJC, eds. 1982. *Biochemistry and Metabolism of Plant Lipids.* Amsterdam: Elsevier

75. Zeidler JG, Lichtenthaler HK, May HU, Lichtenthaler FW. 1997. Is isoprene emitted by plants synthesized via the novel isopentenylpyrophosphate pathway? *Z. Naturforsch. Teil C* 52:15–23

76. Zeidler JG, May HU, Lichtenthaler FW, Lichtenthaler HK. 1998. Isoprene emitted by plants is formed via the 1-deoxy-D-xylulose phosphate pathway of isopentenyl diphosphate biosynthesis. See Ref. 56a, pp. 446–49

77. Zeidler JG, Schwender J, Müller C, Wiesner J, Lichtenthaler HK, et al. 1998. Inhibition of the non-mevalonate 1-deoxy-D-xylulose-5-phosphate pathway of plant isoprenoid biosynthesis by fosmidomycin. *Z. Naturforsch.* 53C:980–86

Annu. Rev. Plant Physiol. Plant Mol. Biol. 1999. 50:67–95

CHLOROPHYLL DEGRADATION

Philippe Matile, Stefan Hörtensteiner
University of Zürich, Institute of Plant Biology, Zollikerstrasse 107, CH-8008 Zürich, Switzerland; e-mail: phibus@botinst.unizh.ch

Howard Thomas
Cell Biology Department, Institute of Grassland and Environmental Research, Plas Gogerddan, Aberystwyth, Ceredigion, SY23 3EB, United Kingdom; e-mail: sid.thomas@bbsrc.ac.uk

KEY WORDS: catabolites, catabolic pathway, regulation, genetics, evolution

---

ABSTRACT
Although the loss of green color in senescent leaves and ripening fruits is a spectacular natural phenomenon, research on chlorophyll breakdown has been largely neglected until recently. This review summarizes knowledge about the fate of chlorophyll in degreening tissues that has been gained during the past few years. Structures of end- and intermediary products of degradation as well as the biochemistry of the porphyrin-cleaving reaction have been elucidated. The intracellular localization of the catabolic pathway is particularly important in the regulation of chlorophyll breakdown. None of the genes encoding the related catabolic enzymes has so far been isolated, which makes chlorophyll degradation an area of opportunity for future research.

CONTENTS

67

1040-2519/99/0601-0067$08.00

INTRODUCTION

A few months before the start of the twentieth century, Albert F. Woods addressed the American Association for the Advancement of Science in Columbus, Ohio, on the subject, "The destruction of chlorophyll by oxidizing enzymes" (185). His survey includes some spectacularly erroneous assertions (e.g. "It has long been known that chlorophyll could be readily converted by oxidation, into a yellow coloring matter, xanthophyll..."); but some of the enzymology he describes, and the questions he raises, are still keeping researchers busy, 100 years later. By 1912, the second major player in the tale of chlorophyll (Chl) breakdown, chlorophyllase, had been discovered by Arthur Stoll (155). And for most of this century, in spite of the appearance of some repetitive research reports and a few reviews, there was little more to be said about the biochemical basis of Chl breakdown. The publication in 1987 of Hendry et al's review entitled "The degradation of chlorophyll—a biological enigma" (47) marks a revival of interest in the subject. By drawing on information from satellite imaging, tourism, petrochemistry, and archaeology, as well as mainstream biology and chemistry, the authors revealed a rich field of research possibilities. Aspects of the subject have been reviewed subsequently (15, 37, 38, 85, 89, 90, 133, 163), telling a story of growing understanding, as well as of continuing frustrations. The present account of Chl breakdown not only brings this story up-to-date but also tries to anticipate how current and imminent developments, such as genomics, will affect the study of Chl catabolism, just as they will revolutionize most other areas of experimental plant biology.

PLASTID DEVELOPMENT

Chloroplast to Gerontoplast Transition

Chlorophyll degradation is a symptom of transition of chloroplasts to gerontoplasts. A distinctive term for the senescence-specific form of plastids (148) is justified because the metabolism of gerontoplasts, in contrast to that of all other forms of a plastid, is solely catabolic. The yellowing of senescent leaves is due to unmasking and partial retention of carotenoids rather than to the new biosynthesis of yellow pigments such as occurs when chloroplasts differentiate into chromoplasts, for example. Developing gerontoplasts persist and remain intact throughout leaf senescence (84, 117). They lose volume and density as a consequence of extensive losses of stromal components and thylakoids, and the number and size of lipophilic plastoglobules increase. Thus, fully developed gerontoplasts consist of a still intact envelope surrounding a number of large plastoglobules. It is well established that gerontoplast development is under nuclear control and that the expression of plastid-encoded genes plays only a minor part, if at all (see 84). In rice coleoptiles and leaves, ctDNA is extensively degraded even before Chl breakdown marks the onset of differentiation of gerontoplasts (149, 150). On the other hand, leaf senescence in many species is reversible (31, 95, 104, 176), which suggests that fully developed gerontoplasts retain enough genetic information to support regreening and chloroplast reassembly.

Chloroplast to Chromoplast Transition

Chromoplasts are the typical form of Chl-less plastids in fruits and petals (16). Under conditions of light stress, chromoplast-like organelles may also develop in green algae (6). Transitions of chloroplasts into chromoplasts or gerontoplasts are comparable as far as Chl breakdown and the disappearance of thylakoids are concerned. A distinctive feature of chromoplast development is the incorporation of a new set of proteins, encoded in both nuclear and plastidic genomes, which functions in the synthesis of secondary carotenoids and their incorporation in fibrillar and globular structures (16).

DEGRADATION WITHOUT NET LOSS

Chlorophyll Turnover

When cells accumulate a particular metabolite, or maintain it at a steady level, there is usually flux through the related pool. Chl is probably no exception, but there are very few reliable reports in the literature on its rates of turnover. This section gives a picture of a highly incomplete and poorly understood area of plant metabolism, but at least it may point out fruitful research possibilities.

The estimation of rates of Chl synthesis and degradation during the greening of etiolated cereal leaves by following pigment accumulation in the presence and absence of an inhibitor of aminolevulinate (ALA) synthesis resulted in calculated half-lives of as little as 6 to 8 h for leaves in the early stages of de-etiolation, increasing to more than 50 h as greening became complete (48, 154). Considerable incorporation of radiolabeled acetate into the Chl of mature dicot leaves was observed, but little into the pigment of monocots (122). Some Chl turnover during photosynthesis and photoinhibition is likely associated with the short half-life of the D1 polypeptide, part of the heterodimeric reaction center of photosystem II (28, 124). Appreciable rates of Chl turnover were also observed in algae (39, 126).

The Chl biosynthetic pathway is detectable until very late in the lifespan of photosynthetic cells (56). The capacity to bleach the pigment appears also to be built into such cells. Thus, the potential for the simultaneous anabolism and catabolism of Chl is present; but mostly we do not know if this potential is realized. It seems likely also that degradation of Chl during net accumulation or at the steady state is only one aspect of a general requirement to regulate the family of macrocyclic tetrapyrrole intermediates in the biosynthetic pathway en route to photosynthetically functional pigment complexes (180).

Degradation During Chlorophyll Biosynthesis

The pathway of Chl synthesis comprises the action of more than a dozen enzymes to convert glutamate into Chl (180). All intermediates from uroporphyrin III onwards are potentially phototoxic. Hence, "traffic" restrictions along the biosynthetic route that result in the accumulation of such intermediates, e.g. as provoked artificially by antisense transformation at the level of uroporphyrin III decarboxylase (102), cause susceptibility of plants to photodamage. It would make sense if diversionary routes were provided to shunt such intermediates away harmlessly in the event of traffic restrictions farther along the main pathway leading to Chl. In one such detour, phototoxic precursors are exported from the plastid and further processed in the cytosol or at the plasma membrane (62). The *tigrina* series of barley mutants provide genetic evidence of at least four loci regulating metabolic flux between the C5 precursor ALA and protochlorophyllide (Pchlide), the intermediate immediately preceding the light-requiring step at Pchlide oxidoreductase (180).

It is unlikely that the catabolic system degrading Chl during senescence is responsible for turning over Chl or its derivatives in pre-senescent tissues. End-product catabolites that accumulate in senescent cells have never been found in cells in which the pigment is undergoing net synthesis or steady-state turnover (P Matile's group, unpublished data). Moreover, non-yellowing senescence mutants display neither abnormal accumulation of Chl nor extended pigment synthetic capacity (4, 110). Turnover of the D1 protein during photosynthesis is also perfectly normal in the *Festuca* senescence mutant Bf993 (49).

There are a few reports of enzyme activities that may function in fine-tuning levels of macrocyclic intermediates in the Chl biosynthetic pathway. The first Mg-containing intermediates are Mg protoporphyrin IX and its monomethyl ester (MgProtoMe); etiolated *Phaseolus* leaves were shown to contain heat-labile and oxygen-dependent activity able to bleach MgProtoMe (55). A similar enzyme from barley was purified (33); developmental changes in its activity suggested that it may be responsible for pigment bleaching during greening as well as during senescence-associated degreening. These interesting observations appear not to have been pursued further. A somewhat similar activity was demonstrated in preparations from cucumber cotyledons (184). The enzyme was shown to utilize MgProtoMe, to be heat-labile and oxygen-dependent, to correlate in a reciprocal fashion with MgProtoMe cyclase, and to decrease in activity as plastids differentiated. The authors presented evidence that this kind of bleaching process in plastid development is distinct from the peroxidative and fatty acid–dependent pigment oxidizing activities reported in connection with ripening, senescence, and pathological deterioration (see below).

Heme and Phytochrome

A branch of the C5 pathway that may be significant for the regulation of Chl biosynthesis leads from protoporphyrin IX via ferrochelatase to heme (181). The precursor of phytochrome chromophore synthesis is biliverdin IX, the product of opening the heme macrocycle (8). Mutants of pea, tomato, and *Arabidopsis* (159) that lack the ring-opening ferredoxin-dependent heme oxygenase (8) are yellow-green in color, but this is more likely to be a consequence of defective photocontrol of Chl biosynthesis than because Chl precursors have been rerouted. Before a specific Chl-degrading oxygenase was discovered, heme oxygenase was an object of fascination for those seeking to understand plant pigment breakdown. We now know that the reaction mechanisms of the two enzymes are quite distinct.

DEGRADATION ASSOCIATED WITH NET LOSS

Phenology of Net Loss

The loss of Chl is the preferred and most easily measured parameter for describing the yellowing of senescing leaves or color changes in ripening fruits. Degreening is clearly part of developmental processes taking place in fully viable cells. This can easily be appreciated, for example, by observing petal development, which in many species is associated with the rapid loss of Chl shortly before anthesis. The capacity of metabolic degreening is ubiquitous in the plant kingdom and is manifest not only in spectacular natural phenomena such as autumnal color changes in the foliage of deciduous trees but also in seeming evergreens, ferns, and algae.

Products of Chlorophyll Degradation

IDENTIFICATION   Unambiguous identification of degradation products required the specific radiolabeling of Chl in the porphyrin moiety. Feeding of [4-^{14}C]ALA during greening of etiolated plantlets of barley (121) or to expanding and greening canola cotyledons (35) resulted in the specific radiolabeling of Chl for the study of catabolism. As Chl disappeared during subsequent senescence, the radiolabel was progressively recovered in the water-soluble fraction of leaf extracts, indicating that Chl degradation is associated with hydrolysis into the phytol and porphyrin moieties. In barley plantlets developing under natural conditions, the label incorporated into the pyrrole units of Chl was retained in the senescent leaves (119); there was no indication of export of radiolabel from the senescent primary leaf to other parts of the plant.

PHYTOL   As Chl is degraded, substantial proportions of phytol remain esterified (83, 118, 120) with fatty acids (118) and with acetic acid (12). In senescent barley leaves, free and esterified forms of phytol are located in the plastoglobules of developing gerontoplasts (12). Losses of total phytol during leaf senescence have been attributed to photooxidative conversion into various isoprenoid compounds (132).

GREEN PRODUCTS OF DEGRADATION   Several derivatives of Chl that have an intact macrocycle and are likely to represent intermediary products of degradation have been detected in one form or another: pheophytin, chlorophyllides (Chlide), and pheophorbide (Pheide) *a* (e.g. 3, 64, 127, 164, 186). During breakdown of Chl these compounds are detectable only in trace amounts (if at all), indicating that removal of phytol and the central Mg-atom is quickly followed by processes that cause the loss of green color. Dephytylation is commonly considered to be the first catabolic reaction, but it may be preceded by Mg-removal, as suggested by the occurrence of pheophytin in degreening leaves (3). Whether 13^2-hydroxy Chl *a*, which has occasionally been observed (e.g. 94, 186), belongs to the group of natural intermediary breakdown products is unclear.

TETRAPYRROLIC CATABOLITES   Until recently, the fate of the porphyrin part of Chl in senescent leaves remained cryptic. For as long as the search for products of degradation that were expected to accumulate concomitant with the loss of Chl remained unsuccessful, it was speculated that a continuous, rapid breakdown to simple molecules such as CO_2 and NH_3 might explain the apparently traceless disappearance (47). In retrospect, the difficulties with the discovery of Chl catabolites are easily explained. Like bile pigments, all catabolites identified so far in senescent leaves have a linear tetrapyrrolic structure, but unlike the products of heme degradation, they are colorless and, hence, escaped detection.

In every instance where the chemical structure of such a terminal Chl catabolite has been determined, it has been shown to be formally derived from pheophorbide a through the oxygenolytic cleavage of the macrocycle at the C4/C5 mesoposition (Figure 1). The methine bridge C5 carbon is preserved as a formyl group (a clear difference from heme catabolism, where this carbon atom is lost as CO), and a second O atom appears as a lactam group at C4. As well as the ring-opening mechanism, all known final catabolites share the complete eradication of the conjugated system between the pyrroles. Hydroxylation of the ethyl side chain in pyrrole B is another common structural feature.

These final products have been termed NCCs (nonfluorescent Chl catabolites) and thereby distinguished from blue-fluorescing intermediary catabolites (FCCs, fluorescent Chl catabolites) and RCCs, the red-colored type of catabolite discovered in *Chlorella protothecoides* (26). Because the basic structure of NCCs is modified in some plant species, the following convenient nomenclature has been proposed (35). A prefix indicates the plant species (e.g. Bn = *Brassica napus*) and the individual compounds are numbered according to decreasing polarity as judged by retention times during reverse phase HPLC; e.g. Bn-NCC-3 represents the most apolar catabolite accumulated in senescent leaves of oilseed rape.

The structures of NCCs elucidated so far are illustrated in Figure 1. In the fully senescent leaves of *Cercidiphyllum japonicum* (18), *Liquidambar styraciflua*, and *L. orientalis* (61), the full complement of Chl-porphyrin is converted mole-for-mole into the basic structure of NCCs. In Bn-NCC-3 from oilseed rape this structure is modified as the 13^2-carboxymethyl group is demethylated (105), but in the major NCC of this species, Bn-NCC-1, the 13^2-hydroxyl group is esterified with malonic acid (106) and in Bn-NCC-2 it is glucosylated (105). The only structure of an NCC known so far from senescent barley leaves, Hv-NCC-1, has an intact 13^2-carboxymethyl group but is hydroxylated in the vinyl group of pyrrole A (69, 70). It may be anticipated that some or even many further structural variants in the family of NCCs remain to be discovered. Radiolabeling in barley has revealed the occurrence of about a dozen different NCCs (119, 121). It would not be surprising if, in addition to esterification with malonic acid, other conjugations reminiscent of secondary metabolites, e.g. with acetic or cinnamic acid, were to be discovered in various plant species.

Hydroxylation and conjugation confer increased water solubility on NCCs, just as they do on secondary metabolites such as phenolics, which are dissolved in the vacuolar sap. Not unexpectedly, therefore, Chl catabolites are accumulated in the vacuoles of senescent mesophyll cells (13, 24, 50, 88).

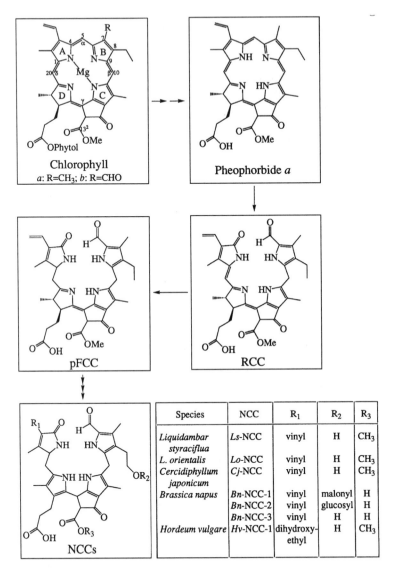

Figure 1 Structures of intermediary and final Chl catabolites arranged according to the "pheophorbide *a* oxygenase" (PaO) pathway of chlorophyll degradation. Explanations, abbreviations, and references are presented in the text.

During periods of rapid Chl breakdown, trace amounts of colorless blue-fluorescing compounds are detectable in senescent leaves (23, 24). These FCCs were originally discovered in acidic extracts from leaves of *Festuca pratensis*: In an acidic milieu FCCs are readily oxidized to pink-colored compounds (85). Such "pink pigments" were identified as catabolites of Chl because they occurred only in senescent leaves of the wild-type cultivar Rossa but not in those of a mutant genotype, Bf 993, which is unable to degrade Chl (23, 87). Trace amounts of FCCs were found to occur only temporarily and to be positively correlated with rates of Chl breakdown. Such kinetics would be expected of intermediary products of a catabolic pathway, eventually leading to the progressively accumulating end products, NCCs.

FCCs have a fluorescence spectrum (excitation at 320 nm, emission at 450 nm) typical for Schiff-base structures $-N=C-C=C-N-$. The linear tetrapyrrolic structure of the canola FCC depicted in Figure 1 (107) suggests that the fluorescence is due to the unsaturated γ methine bridge linking pyrroles C and D. Such structures also occur in lipofuscin-like fluorescent compounds observed in senescent tissues (e.g. 96, 97) and suggested to be products of lipid peroxidation. The absence of lipofuscin-like compounds in senescent leaves of the stay-green meadow fescue (23) suggests, however, that they originate from Chl rather than from malondialdehyde. Incidentally, NCCs and FCCs react with acidic ninhydrin and may mimic proline (23): The accumulation of proline observed in senescent leaves (e.g. 125) in fact was probably due to the degradation of Chl.

Chlorella protothecoides excretes red pigments into the medium when cells are induced to bleach under appropriate conditions (115). The structure of such a pigment (Figure 1) not only reveals its origin from Chl but also the same regio-specific cleavage of the macrocycle (26) as in the catabolites from angiosperms. In contrast to Chl catabolism in senescent leaves, which seems to yield derivatives of Chl *a* exclusively, a red bilin (RCC) derived from Chl *b* has also been isolated from Chl-degrading cultures of *C. protothecoides* (60).

A special type of linear tetrapyrrolic catabolite, apparently derived from Chl *b* and ring opening at the C20/C1 mesoposition, has been identified as the light emitter in the bioluminescent dinoflagellate, *Euphausia pacifica* (108).

Catabolic Enzymes and Catabolic Pathway

CHLOROPHYLLASE   In the course of the many decades since its discovery (155), chlorophyllase (Chlase, E.C. 3.1.1.14) has acquired a vast literature. Commonly, the hydrolysis of Chl into Chlide and phytol is regarded as the initial step of breakdown (Figure 2); and yet, despite detailed knowledge about catalytic properties of the enzyme as studied in vitro, the action of Chlase in vivo has remained mysterious.

Chlase is a hydrophobic protein of plastid membranes (e.g. 14, 32, 51, 157, 172) that is distinguished by its functional latency: In preparations of chloroplast membranes Chl is not dephytylated unless the membranes are solublized in the presence of detergent (e.g. 2, 93) or acetone at high concentrations (e.g. 32). Even during Chl breakdown in senescent leaves Chlase remains latent. Hence, all the properties of Chlase determined under highly unphysiological conditions in vitro, such as kinetic parameters, dependencies on pH and temperature, and so forth (e.g. 29, 66, 81, 100, 144, 173) are likely irrelevant for the understanding of dephytylation in vivo. The most intriguing problem of regulation of Chl breakdown at the level of Chlase concerns the mechanism by which the interaction between Chlase and its substrates is achieved. Latency of the enzyme may be explained simply by the spatial separation between Chl in the thylakoid pigment-protein complexes and Chlase, which appears to be located in the plastid envelope (93).

Chlases have been purified repeatedly but despite the availability of N-terminal amino acid sequences (172, 173) and a specific antibody (172), the corresponding gene(s) has so far been recalcitrant to molecular cloning. Such unexpected difficulty is not easily explained[1]. Still another puzzling feature of Chlase is its apparent glycoprotein nature as inferred from binding to concanavalin A (158, 173), indeed an unusual property of a component of chloroplast envelope membranes.

Under certain conditions Chlase can act as a transesterase and, therefore, it has occasionally been considered to have a function in the phytylation of Chlide (e.g. 30). After the elucidation of the last step of Chl biosynthesis (134), such a function of Chlase can now be disregarded.

Mg-DECHELATASE   The enzymic release of Mg^{2+} from Chlide in exchange for $2H^+$ has been demonstrated in preparations from algae (116, 189) as well as from higher plants (145), but detailed knowledge about the properties of Mg-dechelatase is scarce. Under conditions preventing further catabolism of the reaction product, Pheide, the activity can be demonstrated by assessing Pheide accumulation in vivo as well as in isolated chloroplasts and chloroplast membranes (73). Apparent dechelatase activity can be assayed with chlorophyllin as substrate (63, 179). This activity is associated with chloroplast membranes and like Chlase seems to be constitutive. In detergent-solubilized membranes of barley chloroplasts, dechelatase activity (substrate: chlorophyllin) was localized in a distinct complex (135). Attempts to purify Mg-dechelatase have yielded unexpected results: The activity was heat-stable and associated with a

---

[1]Note added in proof: The gene for chlorophyllase has now been cloned from ethylene-treated orange fruit and expressed in *E. coli* (D Jacob-Wilk, Y Eyal, D Holland, J Riov & EE Goldschmidt, personal communication).

low-molecular-weight compound rather than with a protein (146, 177). Whether this "Mg-dechelating substance" (146) is identical with the activity responsible for the release of Mg that occurs in the catabolic pathway of Chl remains to be demonstrated.

PHEOPHORBIDE a OXYGENASE, RCC REDUCTASE   The ring-opening step of the catabolic pathway is decisive for the loss of green color. The first identifiable colorless product, pFCC (Figure 1) is formally derived from Pheide a by the addition of two atoms of O and four atoms of H. The enzymic conversion of Pheide a into pFCC in vitro requires two protein components from gerontoplast membranes and stroma, respectively (36, 53, 137). Oxygenolysis of Pheide a is catalyzed by the membrane component, Pheide a oxygenase (PaO), and yields the red bilin RCC that in a channeled reaction is reduced by RCC reductase to yield pFCC (128) (Figures 1, 2). Chl breakdown in C. protothecoides is terminated by the action of the oxygenase and RCC is released into the culture medium (26).

In Chlorella (20) as well as in Brassica napus (54), incorporation studies in the presence of $^{18,18}O_2$ showed that only the formyl oxygen originates from dioxygen, whereas the lactam oxygen at C4 is probably derived from water. The mechanism of the monooxygenase-catalyzed ring opening has not yet been elucidated in detail. A hypothesis has been proposed in which the initial step is regioselective C4/C5 epoxide formation, followed by hydrolysis and rear-rangement of double bonds (20). Stereoselective final reduction of the C9/C10 double bond has been demonstrated (19).

Inhibition by appropriate chelators (35) as well as regeneration studies (53) suggest that PaO is an Fe-containing monooxygenase. Its redox cycle is driven by reduced ferredoxin (Figure 2); in the light, this reductant is generated by photosystem I (128), whereas in the dark, NADPH and a corresponding sys-tem involving the stromal oxidative pentose-phosphate-cycle and glucose-6-phosphate as ultimate e$^-$ donor are required (36, 53, 128). In senescent leaves, all components of such a reducing system (177), including ferredoxin (128), are retained or even newly synthesized as long as Chl breakdown continues.

Among the properties of PaO, the absolute specificity for Pheide a as substrate (53) is noteworthy because it explains the exclusive occurrence of Chl a-derived final catabolites (Figure 1). Pheide b is a competitive inhibitor of PaO but is no more effective as a substrate than pyroPheide, protoporphyrin IX, and Chlide a (S Hörtensteiner, unpublished data). The oxygenase of C. protothecoides seems to be less specific, as suggested by the production of an RCC derived from Pheide b (60).

So far, the activity of PaO has been detected only in senescent leaves of several species (36, 53, 137, 167, 178) as well as in ripening fruits (1, 103). Indeed, PaO

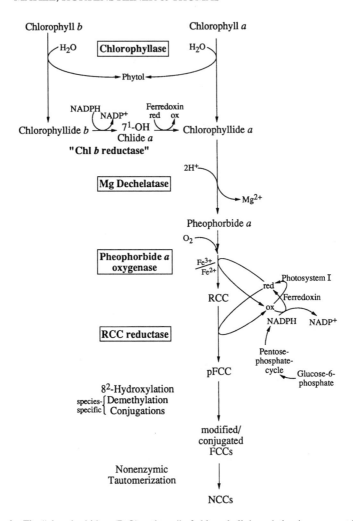

Figure 2 The "pheophorbide *a* (PaO) pathway" of chlorophyll degradation in senescent leaves. See text for discussion.

appears to be the only enzyme that plays a pace-making role in the breakdown of Chl, and the catabolic pathway (Figure 2) may, therefore, be designated as the "pheophorbide *a* pathway."

RCC is released from PaO only upon its reduction by RCC reductase in a reaction where reduced ferredoxin is again the immediate e⁻ donor (Figure 2). If, in the absence of PaO, RCC is employed as substrate of RCC reductase, pFCC is produced only under anaerobic conditions, suggesting that in situ a channeled reaction takes place in an oxygen-free microenvironment at the site of PaO (128). Depending on the plant species, the cleavage of Pheide *a* yields one or the other of two C20-stereoisomers of primary FCCs. The species-specific type of pFCC is determined by RCC reductase, not by PaO (130, 177). RCC reductase is a soluble constitutive component of plastids that occurs not only in green tissues but also in etiolated leaves and even in roots (130).

METHYLESTERASE   The structures of NCCs (Figure 1) demonstrate that the 13^2-carboxymethyl group may be demethylated. This modification is likely due to the action of a methylesterase sismilar to that discovered in *Chenopodium album* (147), which accepts Pheide as substrate. This constitutive "pheophorbidase" is absent from barley, the NCC of which (Figure 1) has an intact 13^2-carboxymethyl group (147).

HYDROXYLATION AND CONJUGATION   All NCCs are hydroxylated in the ethyl side chain of pyrrole B. A cytochrome P_{450}-type enzyme may be responsible for the corresponding modification of FCCs. In senescent cotyledons of canola, the predominant final catabolite, *Bn*-NCC-1, is esterified with malonic acid (Figure 1). A constitutive cytosolic transferase catalyzes the esterification of *Bn*-NCC-3 with malonyl-SCoA as cosubstrate (52). A corresponding glucosyltransferase responsible for the formation of *Bn*-NCC-2 has not yet been identified.

TAUTOMERIZATION   The conversion of FCCs into NCCs is due to a rearrangement of double bonds in pyrrole D and the adjacent γ methine bridge. This tautomerization takes place nonenzymically under slightly acidic conditions (S Hörtensteiner, unpublished data) and, therefore, is likely to occur upon the deposition of FCCs in the vacuole.

Chlorophyll b to Chl a Conversion

All final catabolites described so far not only are derived from Chl *a* but represent the breakdown products of Chl *b* as well (18, 35). In other words, degreening must be associated with the funneling of Chl *b* into the pool of *a*-forms. The corresponding reduction of the 7^1-formyl into the 7^1-methyl group was originally discovered in etioplasts (57) and is thought to be part of a Chl cycle by which the two forms of Chl are balanced in the photosynthetic apparatus (113). Reduction of Pheide *b* to 7^1-hydroxy Chlide *a*, Chlide *a*, and eventually to Chl *a*, has been demonstrated by infiltration of the Zn-analogue of Chlide *b* (138) or pyroChlide *b* (59) into etiolated leaves. "Chl *b* reductase" appears to consist of two components (Figure 2), an NADPH-dependent enzyme (58) producing the 7^1-hydroxy intermediate and a second reductase that is dependent on reduced ferredoxin (140).

In barley leaves the induction of senescence is accompanied by a marked increase in Chl b reductase activity (V Scheumann, unpublished data). Thus, Chl b reductase appears to play a role in the breakdown of Chl. The substrate specificity as determined in preparations of etioplasts (139) suggests that reduction may take place at the level of the dephytylated pigments, as indicated in Figure 2. Chlide b recycling must be very efficient in senescent leaves since inhibition of PaO activity leads to accumulation of only pheophorbide a, but never b (73, 178).

REGULATION OF BREAKDOWN

Subcellular Organization of the PaO Pathway

The PaO pathway of Chl breakdown extends over several intracellular compartments, starting in the thylakoids and inner-envelope membrane, and ending in the vacuole. As illustrated in Figure 3 it comprises not only stepwise enzymic breakdown but also transport processes within the senescent chloroplasts and across membranes. At least two enzymes of the pathway, Chlase (93) and PaO (91), have been localized to the plastid envelope. The location of the second enzyme of the pathway, Mg dechelatase, has not yet been established but it is not too speculative to assume that it is also located in the envelope. Since the ring-opening reaction occurs by the joint action of PaO and stromal RCC reductase, and since both enzymes depend on reduced ferredoxin (128) (Figure 2), the exact location can only be in the inner membrane of the envelope. Indeed, FCCs, including pFCCs, are present within gerontoplasts (36, 136, 137). Upon modification, presumably involving side chain hydroxylation(s), FCCs are eventually released into the cytosol. In isolated barley gerontoplasts, the export of Hv-FCC-2 was demonstrated to require the hydrolysis of ATP supplied in the medium, i.e. in the cytosol (92), suggesting that the inner membrane of the envelope is equipped with a corresponding carrier. Further metabolism of FCCs in the cytosol comprises species-specific modifications such as possibly demethylation of the 13^2-carboxymethyl group and conjugations such as the malonylation of Bn-NCC-3 in the case of canola (52).

The final deposition of Chl catabolites in the vacuole is associated with the transport of these catabolites, presumably FCCs, across the tonoplast. Studies with isolated vacuoles from barley mesophyll cells have revealed the existence of a specific carrier that is directly energized by ATP hydrolysis (50). It may belong to the family of ATP-binding-cassette (ABC) transporters, as do multidrug resistance–associated proteins (MRP2, MRP3) from *Arabidopsis thaliana* that transport glutathione conjugates as well as an NCC (75, 171). Such a possible connection between detoxification of xenobiotics and the final disposal of Chl catabolites is anything but coincidental: It implies that Chl degradation might resemble Chl detoxification (see below in Significance of Degradation).

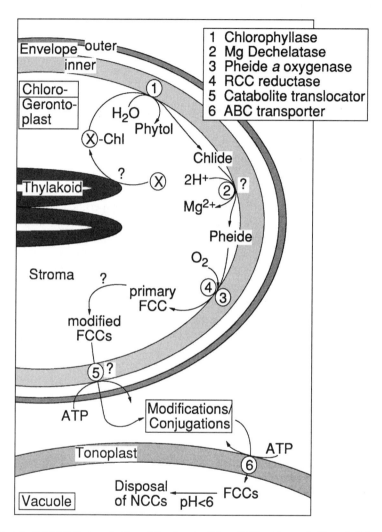

Figure 3   Intracellular organization of the pheophorbide *a* (PaO) pathway of chlorophyll degradation in senescent mesophyll cells. The hypothetical part of the catabolic system marked "X" is thought to be responsible for dismantling the thylakoid pigment-protein complexes and transport of the resulting Chl molecules to the inner-envelope membrane, site of the first catabolic enzyme, Chlase. Further explanations and details in the text and in Figure 2.

In isolated intact gerontoplasts, breakdown of Chl to FCCs depends on the supply of ATP in the medium (36, 136). However, none of the reactions of the PaO pathway required ATP. A similar discrepancy was observed in the case of Chl *b* to Chl *a* transformation, which in intact, but not in lysed etioplasts, depends on ATP (140).

Regulation of PaO Pathway

With the exception of PaO itself, all known components of the PaO pathway are constitutive enzymes. Activities may be modulated during development, as widely documented in the case of Chlase (e.g. 27, 64, 78, 80, 81, 98, 172), but PaO is detectable exclusively in senescent leaves (36, 53, 178), where its activity is positively correlated with rates of Chl loss (129).

It is not immediately clear why the overall regulation of breakdown should be at the third enzymic step of the pathway, rather than at the initial reaction. That there is some degree of feedback control is clear from experiments in which the action of PaO in senescent leaves is blocked by appropriate treatments. Under these conditions, Chlides and Pheide *a* accumulate (73, 120). This phenomenon can also be observed in a PaO-deficient mutant genotype of *Festuca pratensis* (178); but the accumulation of dephytylated Chls does not take place when senescent leaves of the mutant are treated with cycloheximide (164), which suggests that cytoplasmic protein synthesis is required for providing Chlase with its substrates. The respective protein(s) may be responsible for dismantling the thylakoid pigment-protein complexes and/or the transport of Chl to the site of Chlase.

Chlorophyll-Bleaching Enzymes of Unknown Functions

Chl is readily bleached in vitro by the action of oxidative enzymes such as peroxidases in the presence of H_2O_2 and a phenolic compound (e.g. 65, 71, 143, 188) or lipoxygenase in the presence of linolenic acid (e.g. 68). Chl is also bleached when chloroplast membranes (11, 76, 82) or isolated photosystem preparations (77) are incubated in the presence of unsaturated fatty acids. The significance of these reactions for Chl breakdown in senescent leaves is dubious. In any case, it is not justified to refer to "peroxidase" or "Chl oxidase" pathways (63, 187) as long as products of bleaching have not been identified as natural catabolites.

Genetics of Degradation

The number of enzyme activities known or suspected to open the ring of Chl or its derivatives, and to process the products of ring-opening, may approach 20. Unless it has been already identified for some other, apparently unrelated, metabolic function, not one of the corresponding 20 genes or gene families has been described in molecular terms. Either these genes remain to be isolated or they are among the anonymous ESTs and open reading frames emerging from the various plant genomics projects. Functional analysis by, for example, transposon mutagenesis, should be a rewarding approach to assigning gene identities to those enzymes that bleach Chl, since mutant phenotypes will be easily screened. It is also likely that such mutations are tolerated and would not threaten the viability of the whole plant. Many naturally occurring mutations and genetic variants deficient in pigment-degrading capacity (referred to by the general term "stay-green") have been described (111, 168).

Despite the lack of molecular detail, by studying the characteristics of the different kinds of stay-green, some inferences can be made about the genetic mechanisms regulating expression of Chl degradation. Three classes of stay-greens are recognized. In the first group, the Chl degradation machinery is intact but is activated late. Alternatively, degradation may start on time but proceed abnormally slowly. In a third type of variation, initiation and rate occur normally but one or more steps in the degradation pathway may be deficient. These variations on the stay-green theme may be referred to as types A, B, and C, respectively (168). Stay-greens in a number of species have been classified, but only in the case of a few type Cs has Chl degradation been analyzed in terms other than simple pigment measurements.

The most intensively studied stay-green variant was originally identified in the pasture grass *Festuca pratensis* (169). Green tissues of the mutant retain Chl more or less indefinitely. The mutation is inherited as a single recessive nuclear gene (162). It is an example of a type C stay-green, since most of the senescence syndrome, including a decrease in photosynthetic capacity and degradation of soluble leaf proteins, proceeds normally (46, 160, 162, 169). Similar phenotypes have been observed in *Phaseolus vulgaris* (131) and *Pisum sativum* (the latter corresponding to the green cotyledon character originally described by Mendel), and in all three cases, pheophorbide *a* oxygenase activity is not detectable (4, 167, 178). The genetic lesion is likely to be situated either in the structural gene for PaO itself, or in a locus that regulates its induction at the onset of senescence. It would be a major step forward for the study of Chl degradation if the corresponding locus could be isolated.

The occurrence of this mutation in *Festuca* has some distinct advantages. Recombinant chromosomes resulting from introgression of *F. pratensis* genes into species of the related genus *Lolium* are readily identified by genomic in situ hybridization. This has allowed the stay-green locus to be physically mapped in the *Festuca-Lolium* genome (165, 170). Repeated backcrossing has reduced the size of the recombinant segment until it now has become feasible to identify the mutant locus, or at least genes tightly linked to it, by searching for *Festuca*-specific polymorphisms in genomic or cDNA clones. Because the stay-green phenotype is so easily screened, it is a very convenient character for testing the combined introgression/expression mapping approach to genetic analysis.

Not only deficiency regarding PaO but also lesions in Chlase, Mg dechelatase, or RCC reductase should result in a type C phenotype. But no corresponding angiosperm mutant has yet been described. Either it is necessary to increase the range of plant material to be screened, or else the constitutive nature of these enzymes means that knock-out mutations have as yet unforeseen lethal consequences for higher plants. We may need to isolate the corresponding genes and subject them to analysis by reverse genetics before we obtain any significant insight.

Legumes seem to be especially tolerant of mutations affecting Chl degradation, perhaps because nitrogen fixation compensates for the lower availability of internal nitrogen from which stay-greens tend to suffer. At least nine separate genetic loci have been associated with the stay-green phenotype and other senescence traits in soybean (168). One of particular interest is a cytoplasmic gene, *cytG*. The corresponding phenotype is characterized by near-normal loss of Chl *a* but comparative stability of Chl *b* during senescence (40). Chl *b* is likely converted to Chl *a* prior to degradation by the ring-opening oxygenase route (Figure 2). The product of the wild-type allele of *cytG* could well be one of the enzymes or controlling factors in the *b* to *a* conversion sequence. It would be interesting to examine the ctDNA of *cytG* soybean for a sequence anomaly that may identify the specific gene.

Stay-greens of types A and B are clearly not chlorophyll degradation mutants per se, although the variant genes are critical for the coordination of pigment loss with the other components of the senescence syndrome. Because of the agronomic significance of extended photosynthetic duration, type A/B traits have been studied physiologically and by linkage analysis. Quantitative trait loci (QTLs) related to foliar senescence are currently the subject of inheritance studies and linkage mapping in a number of species, including sorghum and millet (74, 156, 175, 182). As genes for components of the Chl degradation pathway and associated senescence processes are gathered by other approaches, it will be instructive to establish whether they map anywhere near the QTLs for leaf color and lifespan. Further developments on the molecular genetics of Chl degradation may also be anticipated following the isolation of three *Arabidopsis* stay-greens (112). Once the corresponding genes are classified and mapped, the powerful tools of *Arabidopsis* genetic analysis can at last be applied to the problem of Chl degradation, and rapid progress will surely follow.

SIGNIFICANCE OF DEGRADATION

Senescence Benefits from Chlorophyll Degradation

The amount of mobilizable material (principally Mg and nitrogen) invested in the Chl molecule is small relative to that available from other salvaged cell constituents. For example, the 240-kDa photosystem II core complex is associated with about 36 Chl *a* molecules (44). If all the nitrogen in this complex were mobilized, less than 6% would be contributed by Chl. Moreover, isolated Chl is not particularly stable. And yet, at the end of its lifespan, the green cell takes special and metabolically expensive measures to degrade Chl because active catabolism of Chl is beneficial to the plant.

Both in vivo and in vitro, proper assembly of thylakoid pigment-protein complexes has an absolute requirement for Chls in precisely stoichiometric amounts (163). Thus, Chl has a constructional and stabilizing role in addition to its light-harvesting and photosynthetic functions. Building of Chl into complexes with other components, particularly carotenoids, is also important for controlling its photodynamism. But removing Chl from such complexes to recover protein nitrogen and lipid carbon is delicate and must be strictly coupled with the macrocycle opening by PaO and RCC reductase rendering Chl photodynamically harmless and preserving the viability of the senescing cells. The obligate relationship between pigment removal and nutrient recycling is convincingly demonstrated in certain type C stay-green plants, where a lesion in ring opening preserves Chl and significantly reduces the lability of associated proteins, lipids, and carotenoids (5, 41, 45, 46). In the stay-green mutant of *Festuca pratensis* the persistence of photostable Chl (166) must be accompanied by maintenance of processes that dissipate absorbed quanta other than through photosynthetic CO_2 fixation. Enhanced levels of photorespiration and carotenoids (10, 41) suggest that this mutant is attuned to the excitation state of Chl and adjusts its antioxidant status according to the need to channel excess energy away harmlessly.

Chl is degraded not because its products are reusable but primarily because otherwise it would block access to more valuable materials. The nitrogen and carbon from which Chl is constructed remain in the cell for good. This is the price the plant pays for access to thylakoid proteins and lipids. The notion of Chl catabolism as being essential for salvage while not itself being a salvage process leads to a conclusion that may make sense for other aspects of the pathway. Not only is the Chl macrocycle converted into non-photodynamic linear forms, these are irreversibly transported to the vacuole and sequestered as oxidized and conjugated by-products (50, 75). In a sense, Chl is not so much catabolized as detoxified.

Chlorophyll Degradation May Be Beneficial During Cell Death

Discoloration, necrotic and chlorotic lesions, and other visible disfigurations are diagnostic of the pathological or terminal state of cells and tissues. During the hypersensitive (or a similar) injury response, rapidly propagated oxidation is important to cauterize the traumatized zone (72, 123). When cells collapse there is neither the need nor the metabolic integrity to induce the oxygenase pathway of Chl catabolism, and rapid pigment destruction is likely to be due to the action of constitutive peroxidases, lipoxygenase, and fatty acid-dependent with Chl-bleaching activities (e.g. 76, 82, 141, 161).

EVOLUTION OF DEGRADATION

Chlorophyll-Degrading Organisms

When during evolution did autotrophic organisms gain the ability to degrade Chl? This question deserves attention in relation to the evolution of photosynthesis, coinciding with the appearance of the potentially toxic and photodynamic Chl molecule.

Chl-degrading processes have been observed in diverse species across the entire taxonomic range: in red algae (79), diatoms (152), green algae (22, 25), and prokaryotes (42, 99), and the leaves of many bryophytes, pteridophytes, and gymnosperms lose Chl during a process that closely resembles foliar senescence in angiosperms (7, 9, 21, 43, 86, 109, 114). In most of these instances of Chl bleaching, only Chlide, pheo-pigments (Pheide and Phein) and their pyroforms have been identified as degradation products so far (34, 79, 151, 189). Accordingly, data on enzyme activities are limited to Chlase (e.g. 144, 158) and Mg-dechelatase (116, 189). Thus, for many Chl-degrading systems the fate of Chl remains unclear, and we only can speculate as to whether non-angiosperms might be capable of generating Chl catabolites similar to the NCCs of higher plants.

Only in three cases have linear tetrapyrroles been identified as Chl breakdown products in lower plants (22, 26, 101). The structure of the main red bilin (RCC) excreted in *Chlorella prothecoides* (25) is identical to that of the intermediary product of Pheide *a* to pFCC transformation in higher plants (128). In addition, the mechanism of chlorin macrocycle opening in *Chlorella* and oilseed rape is identical with respect to oxygen incorporation (20, 54). The Chl breakdown mechanism of higher plants might have evolved from green algae, which are generally accepted as their phylogenetic ancestors (174). Thus, Chl catabolism appears to be an early innovation of photosynthetic organisms.

In contrast to *Chlorella*, where RCCs as the final catabolites are excreted into the surrounding medium, plants living in nonaqueous habitats have to degrade further the red, and thus still potentially photodynamic, RCC. RCC reductase, catalyzing the stereospecific reduction of the C20/C1 double bond of RCC to the primary FCC, is present in angiosperms (130) as well as in more ancient species such as *Selaginella* (S Rodoni & P Matile, unpublished data). It may be argued, therefore, that a crucial step in the evolution of the Chl catabolic pathway might have been the appearance of RCC reductase.

Plastid Transition and Evolution

Gerontoplasts and chromoplasts are plastids characteristic of the terminal, differentiated state of the above-ground organs of angiosperms. Mosses, ferns, and conifers are also likely to be genetically competent to carry out the conversion from chloroplasts to gerontoplasts. Ultrastructural modifications of thylakoid membranes during Chl breakdown have been described for unicellular algae (142), and even in *Synechococcus lividus* bleaching is accompanied by the formation of "vesicularized thylakoids" (99). Thus, the chloroplast-to-gerontoplast transition appears to be a common feature of the degreening process, which either predates the appearance of chromoplasts in evolution, or is at least derived from some common ancestral plastid form, perhaps of the kind observed in stressed cells of *Chlorella zofingiensis* (6). Color changes during leaf senescence can act as a visual signal or "fruit flag" (153, 183). We, therefore, speculate that chromoplast differentiation (16), a key process in entomophily, fruit dispersal, and angiosperm evolution, is a variation on the more ancient theme of plastid transition leading to gerontoplasts, and that the Chl catabolism pathway is a common, defining biochemical feature. The corolla was an innovation of mid-Cretaceous rosids (17). Before this, the stimulus that attracted insects to flowers was probably not visual. It may be, therefore, that the diversification of plastid differentiation pathways can be dated to somewhere around the late Cretaceous period—a time of rapid angiosperm diversification. It is tempting to link the two events, but fossils preserve neither color nor organelle structure, so we may never know whether this really happened.

CONCLUSIONS

A century after Albert F. Woods (185), we have made considerable progress in understanding some aspects of Chl degradation, though as recently as 20 years ago we could not have made this claim. We can now look forward to a period of accelerating progress, particularly if the goal of cloning the genes encoding PaO and the other catabolic enzymes is realized. The developmental and tissue specificities of gene expression, as well as complementation of PaO-deficient stay-greens, will tell us more about Chl catabolism by the PaO route and its regulation. Moreover, the PaO pathway is far from fully described; and as for alternative systems of Chl degradation, all we have at present are tantalizing glimpses of unknown physiological significance. One of the most important of the areas of present ignorance—important because it relates to N-recycling associated with leaf senescence—is the mechanism responsible for dismantling the pigment-protein complexes in the thylakoid and directing the resulting Chl molecules and apoproteins into catabolism. There is more than enough work here to keep researchers busy for another 100 years.

ACKNOWLEDGMENTS

The authors are indebted to numerous students and collaborators who have contributed substantially to the elucidation of Chl degradation. They are also obliged to several colleagues for communicating valuable information, and they acknowledge the generous financial support by the Swiss National Science Foundation and by the UK Biotechnology and Biological Sciences Research Council.

> Visit the *Annual Reviews home page* at
> http://www.AnnualReviews.org

Literature Cited

1. Akhtar MS, Goldschmidt EE, John I, Rodoni S, Matile P, Grierson D. 1999. Altered patterns of senescence and ripening in *gf*, a stay-green mutant of tomato. *J. Exp. Bot.* Submitted

2. Amir-Shapira D, Goldschmidt EE, Altman A. 1986. Autolysis of chlorophyll in aqueous and detergent suspensions of chloroplast fragments. *Plant Sci.* 43: 201–6

3. Amir-Shapira D, Goldschmidt EE, Altman A. 1987. Chlorophyll catabolism in senescing plant tissue: *In vivo* breakdown intermediates suggest different degradative pathways for *Citrus* fruit and parsley leaves. *Proc. Natl. Acad. Sci. USA* 84:1901–5

4. Bachmann A, Fernandez-Lopez J, Ginsburg S, Thomas H, Bowkamp JC, et al. 1994. Stay-green genotypes of *Phaseolus vulgaris* L-chloroplast proteins and chlorophyll catabolites during foliar senescence. *New Phytol.* 126:593–600

5. Bakken AK, Macduff J, Humphreys M, Raistrick N. 1997. A stay-green mutation of *Lolium perenne* affects NO_3^- uptake and translocation of N during prolonged N starvation. *New Phytol.* 135:41–50

6. Bar E, Rise M, Vishkautsan M, Arad S. 1995. Pigment and structural changes in *Chlorella zofingiensis* upon light and nitrogen stress. *J. Plant Physiol.* 146:527–34

7. Bauer H, Gallmetzer C, Sato T. 1991. Phenology and photosynthetic activity in sterile and fertile sporophytes of *Dryopteris filix-mas* (L) Schott. *Oecologia* 86:159–62

8. Beale SI. 1994. Biosynthesis of open-chain tetrapyrroles in plants, algae, and cyanobacteria. *CIBA Found. Symp.* 180:156–68

9. Behera YN, Biswal B. 1990. Differential response of fern leaves to senescence modulating agents of angiospermic plants. *J. Plant Physiol.* 136:480–83

10. Biswal B, Rogers LJ, Smith AJ, Thomas H. 1994. Carotenoid composition and its relationship to chlorophyll and D1 protein during leaf development in a normally senescing cultivar and a stay-green mutant of *Festuca pratensis*. *Phytochemistry* 37:1257–62

11. Blackbourn HD, Jeger MJ, John P. 1990. Inhibition of degreening in the peel of bananas ripened at tropical temperatures. V. Chlorophyll bleaching activity measured *in vitro*. *Ann. Appl. Biol.* 117:175–86

12. Bortlik K. 1990. *Chlorophyllabbau: Charakterisierung von Kataboliten in seneszenten Gerstenblättern.* PhD thesis. Univ. Zürich

13. Bortlik K, Peisker C, Matile P. 1990. A novel type of chlorophyll catabolite in senescent barley leaves. *J. Plant Physiol.* 136:161–65

14. Brandis A, Vainstein A, Goldschmidt EE. 1996. Distribution of chlorophyllase among components of chloroplast membranes in orange (*Citrus sinensis*) leaves. *Plant Physiol. Biochem.* 34:49–54

15. Brown SB, Houghton JD, Hendry GAF. 1991. Chlorophyll breakdown. See Ref. 134a, pp. 465–89

16. Camara B, Hugueney P, Bouvier F, Kuntz M, Moneger R. 1995. Biochemistry and molecular biology of chromoplast development. *Int. Rev. Cytol.* 163:175–247

17. Crepet WL, Friis EM, Nixon KC. 1991.

Fossil evidence for the evolution of biotic pollination. *Philos. Trans. R. Soc. London Ser. B* 333:187–95

18. Curty C, Engel N. 1996. Detection, isolation and structure elucidation of a chlorophyll *a* catabolite from autumnal senescent leaves of *Cercidiphyllum japonicum. Phytochemistry* 42:1531–36

19. Curty C, Engel N. 1997. Chlorophyll catabolism: high stereoselectivity in the last step of the ring cleaving process. *Plant Physiol. Biochem.* 35:707–11

20. Curty C, Engel N, Gossauer A. 1995. Evidence for a monooxygenase-catalyzed primary process in the catabolism of chlorophyll. *FEBS Lett.* 364:41–44

21. De Greef JA, Butler WL, Roth TF, Fredericq H. 1971. Control of senescence in *Marchantia* by phytochrome. *Plant Physiol.* 48:407–12

22. Doi M, Shima S, Egashira T, Nakamura K, Okayama S. 1997. New bile pigment excreted by a *Chlamydomonas reinhardtii* mutant: a possible breakdown catabolite of chlorophyll *a. J. Plant Physiol.* 150:504–8

23. Düggelin T, Bortlik K, Gut H, Matile P, Thomas H. 1988. Leaf senescence in a non-yellowing mutant of *Festuca pratensis*: accumulation of lipofuscinlike compounds. *Physiol. Plant.* 74:131–36

24. Düggelin T, Schellenberg M, Bortlik K, Matile P. 1988. Vacuolar location of lipofuscin- and proline-like compounds in senescent barley leaves. *J. Plant Physiol.* 133:492–97

25. Engel N, Curty C, Gossauer A. 1996. Chlorophyll catabolism in *Chlorella* protothecoides. 8. Facts and artifacts. *Plant Physiol. Biochem.* 34:77–83

26. Engel N, Jenny TA, Mooser V, Gossauer A. 1991. Chlorophyll catabolism in *Chlorella protothecoides.* Isolation and structural elucidation of a red bilin derivative. *FEBS Lett.* 293:131–33

27. Fang Z, Bouwkamp JC, Solomos T. 1998. Chlorophyllase activities and chlorophyll degradation during leaf senescence in non-yellowing mutant and wild type of *Phaseolus vulgaris* L. *J. Exp. Bot.* 49:503–10

28. Feierabend J, Dehne S. 1996. Fate of the porphyrin cofactors during the light-dependent turnover of catalase and of photosystem II reaction-center D1 in mature rye leaves. *Planta* 198:413–22

29. Fernandez-Lopez JA, Almela L, Soledad-Almansa M, Lopez-Roca JM. 1992. Partial purification and properties of chlorophyllase from chlorotic *Citrus limon* leaves. *Phytochemistry* 31:447–49

30. Fiedor L, Rosenbach-Belkin V, Scherz A. 1992. The stereospecific interaction between chlorophylls and chlorophyllase. Possible implication for chlorophyll biosynthesis and degradation. *J. Biol. Chem.* 267:22043–47

31. Fischer A, Feller U. 1994. Seasonal changes in the pattern of assimilatory enzymes and of proteolytic activities of juvenile ivy. *Ann. Bot.* 74:389–96

32. Garcia AL, Galindo L. 1991. Chlorophyllase in citrus leaves: localization and partial purification of the enzyme. *Photosynthetica* 25:105–11

33. Gassman M, Ramanujam P. 1986. Relation between enzymatic destruction of magnesium porphyrins and chloroplast development. In *Regulation of Chloroplast Differentiation*, ed. G Akoyunoglou, H Senger, pp. 115–23. New York: Liss

34. Gerdol R, Bonora A, Poli F. 1994. The vertical pattern of pigment concentrations in chloroplasts of *Sphagnum capillifolium. Bryologist* 97:158–61

35. Ginsburg S, Matile P. 1993. Identification of catabolites of chlorophyll-porphyrin in senescent rape cotyledons. *Plant Physiol.* 102:521–27

36. Ginsburg S, Schellenberg M, Matile P. 1994. Cleavage of chlorophyll-porphyrin. Requirement for reduced ferredoxin and oxygen. *Plant Physiol.* 105:545–54

37. Gossauer A. 1994. Catabolism of tetrapyrroles. *Chimia* 48:352–61

38. Gossauer A, Engel N. 1996. Chlorophyll catabolism—structures, mechanisms, conversions. *J. Photochem. Photobiol. B* 32:141–51

39. Grombach KH, Lichtenthaler HV, Erismann KH. 1978. Incorporation of $^{14}CO_2$ in photosynthetic pigments of *Chlorella pyrenoidosa. Planta* 140:37–43

40. Guiamét JJ, Schwartz E, Pichersky E, Noodén LD. 1991. Characterization of cytoplasmic and nuclear mutations affecting chlorophyll and chlorophyll-binding proteins during senescence in soybean. *Plant Physiol.* 96:227–31

41. Gut H, Rutz C, Matile P, Thomas H. 1987. Leaf senescence in a non-yellowing mutant of *Festuca pratensis*: degradation of carotenoids. *Physiol. Plant.* 70:659–63

42. Haidl H, Knodlmayr K, Rüdiger W, Scheer H, Schoch S, Ullrich J. 1985. Degradation of bacteriochlorophyll *a* in

Rhodopseudomonas sphaeroides R26. *Z. Naturforsch. Teil C* 40:685–92

43. Hakala K, Sewon P. 1992. Reserve lipid-accumulation and translocation of C^{14} in the photosynthetically active and senescent shoot parts of *Dicranum elongatum*. *Physiol. Plant.* 85:111–19

44. Hankamer B, Nield J, Zheleva D, Boekema E, Jansson S, Barber J. 1996. Isolation and biochemical characterization of monomeric and dimeric PSII complexes from spinach and their relevance to the organisation of photosystem II *in vivo*. *Eur. J. Biochem.* 243:422–29

45. Harwood JL, Jones AVHM, Thomas H. 1982. Leaf senescence in a non-yellowing mutant of *Festuca pratensis*. III. Total lipids of leaf tissue during senescence. *Planta* 156:152–56

46. Hauck B, Gay AP, Macduff J, Griffiths CM, Thomas H. 1997. Leaf senescence in a non-yellowing mutant of *Festuca pratensis*: implications of the stay-green mutation for photosynthesis, growth and nitrogen nutrition. *Plant Cell Environ.* 20:1007–18

47. Hendry GAF, Houghton JD, Brown SB. 1987. The degradation of chlorophyll—a biological enigma. *New Phytol.* 107:255–302

48. Hendry GAF, Stobart AK. 1986. Chlorophyll turnover in greening barley. *Phytochemistry* 25:2735–37

49. Hilditch P, Thomas H, Rogers LJ. 1986. Two processes for the breakdown of the Q_B protein of chloroplasts. *FEBS Lett.* 208:313–16

50. Hinder B, Schellenberg M, Rodoni S, Ginsburg S, Vogt E, et al. 1996. How plants dispose of chlorophyll catabolites. Directly energized uptake of tetrapyrrolic breakdown products into isolated vacuoles. *J. Biol. Chem.* 271:27233–36

51. Hirschfeld KR, Goldschmidt EE. 1983. Chlorophyllase activity in chlorophyll-free citrus chromoplasts. *Plant Cell Rep.* 2:117–18

52. Hörtensteiner S. 1998. NCC malonyl-transferase catalyzes the final step of chlorophyll breakdown in rape (*Brassica napus*). *Phytochemistry* 49:953–56

53. Hörtensteiner S, Vicentini F, Matile P. 1995. Chlorophyll breakdown in senescent leaves: enzymic cleavage of pheophorbide *a in vitro*. *New Phytol.* 129:237–46

54. Hörtensteiner S, Wüthrich KL, Matile P, Ongania K-H, Kräutler B. 1998. The key step in chlorophyll breakdown in

higher plants: cleavage of pheophorbide a macrocycle by a monooxygenase. *J. Biol. Chem.* 273:15335–39

55. Hougen CL, Meller E, Gassman ML. 1982. Magnesium protoporphyrin monoester destruction by extracts of etiolated red kidney bean leaves. *Plant Sci. Lett.* 24:289–94

56. Hukmani P, Tripathy BC. 1994. Chlorophyll biosynthetic reactions during senescence of excised barley (*Hordeum vulgare* L. cv IB65) leaves. *Plant Physiol.* 105:1295–300

57. Ito H, Tanaka Y, Tsuji H, Tanaka A. 1993. Conversion of chlorophyll *b* to chlorophyll *a* by isolated cucumber etioplasts. *Arch. Biochem. Biophys.* 306:148–51

58. Ito H, Takaichi S, Tsuji H, Tanaka A. 1994. Properties of synthesis of chlorophyll *a* from chlorophyll *b* in cucumber etioplasts. *J. Biol. Chem.* 269:22034–38

59. Ito H, Tanaka A. 1996. Determination of the activity of chlorophyll *b* to chlorophyll *a* conversion during greening of etiolated cucumber cotyledons by using pyrochlorophyll *b*. *Plant Physiol. Biochem.* 34:35–40

60. Iturraspe J, Engel N, Gossauer A. 1994. Chlorophyll catabolism. Isolation and structure elucidation of chlorophyll *b* catabolites in *Chlorella protothecoides*. *Phytochemistry* 35:1387–90

61. Iturraspe J, Moyano N. 1995. A new 5-formylbilinone as the major chlorophyll *a* catabolite in tree senescent leaves. *J. Org. Chem.* 60:6664–65

62. Jacobs JM, Jacobs NJ. 1993. Porphyrin accumulation and export by isolated barley (*Hordeum vulgare*) plastids. Effect of diphenyl ether herbicides. *Plant Physiol.* 101:1181–87

63. Janave MT. 1997. Enzymic degradation of chlorophyll in cavendish bananas: in vitro evidence for two independent degradative pathways. *Plant Physiol. Biochem.* 35:837–46

64. Johnson-Flanagan AM, Spencer MS. 1996. Chlorophyllase and peroxidase activity during degreening of maturing canola (*Brassica napus*) and mustard (*Brassica juncea*) seed. *Physiol. Plant.* 97:353–59

65. Kato M, Shimizu S. 1987. Chlorophyll metabolism in higher plants. VII. Chlorophyll degradation in senescing tobacco leaves; phenolic-dependent peroxidative degradation. *Can. J. Bot.* 65:729–35

66. Khalyfa A, Kermasha S, Marsot P, Goetghebeur M. 1995. Purification and

characterization of chlorophyllase from alga *Phaeodactylum tricornutum* by preparative native electrophoresis. *Appl. Biochem. Biotechnol.* 53:11–27

67. Kingston-Smith AH, Thomas H, Foyer CH. 1997. Chlorophyll *a* fluorescence, enzyme and antioxidant analyses provide evidence for the operation of alternative electron sinks during leaf senescence in a stay-green mutant of *Festuca pratensis. Plant Cell Environ.* 20:1323–37

68. Köckritz A, Schewe T, Hieke B, Hass W. 1985. The effects of soybean lipoxygenase-1 on chloroplasts from wheat. *Phytochemistry* 24:381–84

69. Kräutler B, Jaun B, Amrein W, Bortlik K, Schellenberg M, Matile P. 1992. Breakdown of chlorophyll: constitution of a secoporphinoid chlorophyll catabolite isolated from senescent barley leaves. *Plant Physiol. Biochem.* 30:333–46

70. Kräutler B, Jaun B, Bortlik K, Schellenberg M, Matile P. 1991. On the enigma of chlorophyll degradation: the constitution of a secoporphinoid catabolite. *Angew. Chem. Int. Ed. Engl.* 30:1315–18

71. Kuroda M, Ozawa T, Imagawa H. 1990. Changes in chloroplast peroxidase activities in relation to chlorophyll loss in barley leaf segments. *Physiol. Plant.* 80:555–60

72. Lamb C, Dixon RA. 1997. The oxidative burst in plant disease resistance. *Annu. Rev. Plant Physiol. Plant Mol. Biol.* 48:251–75

73. Langmeier M, Ginsburg S, Matile P. 1993. Chlorophyll breakdown in senescent leaves: demonstration of Mg-dechelatase activity. *Physiol. Plant.* 89:347–53

74. Liu CJ, Witcombe JR, Pittaway TS, Nash M, Hash CT, et al. 1994. An RFLP-based genetic map of pearl millet (*Pennisetum glaucum*). *Theor. Appl. Genet.* 89:481–87

75. Lu Y-P, Li Z-S, Drozdowicz YM, Hörtensteiner S, Martinoia E, Rea PA. 1998. AtMRP2, an *Arabidopsis* ATP binding cassette transporter able to transport glutathione-S-conjugates and chlorophyll catabolites: functional comparisons with AtMRP1. *Plant Cell* 10: 267–82

76. Lüthy B, Martinoia E, Matile P, Thomas H. 1984. Thylakoid-associated "chlorophyll oxidase": distinction from lipoxygenase. *Z. Pflanzenphysiol.* 113:423–34

77. Lüthy B, Thomas H, Matile P. 1986.

Linolenic acid-dependent "chlorophyll oxidase"-activity: a property of photosystems I and II. *J. Plant Physiol.* 123: 203–9

78. Majumdar S, Ghosh S, Glick BR, Dumbroff EB. 1991. Activities of chlorophyllase, phosphoenolpyruvate carboxylase and ribulose-1,5-bisphosphate carboxylase in the primary leaves of soybean during senescence and drought. *Physiol. Plant.* 81:473–80

79. Marquardt J. 1998. Effects of carotenoid depletion on the photosynthetic apparatus of a *Galdieria sulphururia* (Rhodophyta) strain that retains its photosynthetic apparatus in the dark. *J. Plant Physiol.* 152:372–80

80. Martinez GA, Chaves AR, Anon MC. 1996. Effects of exogenous application of gibberellic acid on color-change and phenylalanine ammonia lyase, chlorophyllase, and peroxidase activities during ripening of strawberry fruit (*Fragaria x ananassa* Duch.). *J. Plant Growth Regul.* 15:139–46

81. Martinez GA, Civello PM, Chaves AR, Anon MC. 1995. Partial characterization of chlorophyllase from strawberry fruit (*Fragaria ananassa* Duch.). *J. Food Biochem.* 18:213–26

82. Martinoia E, Dalling MJ, Matile P. 1982. Catabolism of chlorophyll: demonstration of chloroplast-localized peroxidative and oxidative activities. *Z. Pflanzenphysiol.* 107:269–79

83. Matile P. 1986. Blattseneszenz. *Biol. Rundsch.* 24:349–65

84. Matile P. 1992. Chloroplast senescence. In *Crop Photosynthesis: Spatial and Temporal Determinants*, ed. NR Baker, H Thomas, pp. 413–40. Amsterdam: Elsevier

85. Matile P, Düggelin T, Schellenberg M, Rentsch D, Bortlik K, et al. 1989. How and why is chlorophyll broken down in senescent leaves? *Plant Physiol. Biochem.* 27:595–604

86. Matile P, Flach B, Eller B. 1992. Spectral optical properties, pigments and optical brighteners in autumn leaves of *Ginkgo biloba* L. *Bot. Acta* 105:13–17

87. Matile P, Ginsburg S, Schellenberg M, Thomas H. 1987. Catabolites of chlorophyll in senescent leaves. *J. Plant Physiol.* 129:219–28

88. Matile P, Ginsburg S, Schellenberg M, Thomas H. 1988. Catabolites of chlorophyll in senescing leaves are localized in the vacuoles of mesophyll cells. *Proc. Natl. Acad. Sci. USA* 85:9529–32

89. Matile P, Hörtensteiner S, Thomas H,

Kräutler B. 1996. Chlorophyll breakdown in senescent leaves. *Plant Physiol.* 112:1403–9

90. Matile P, Kräutler B. 1995. Wie und warum bauen Pflanzen das Chlorophyll ab? *Chem. Unserer Zeit* 29:298–306

91. Matile P, Schellenberg M. 1996. The cleavage of phaeophorbide *a* is located in the envelope of barley gerontoplasts. *Plant Physiol. Biochem.* 34:55–59

92. Matile P, Schellenberg M, Peisker C. 1992. Production and release of a chlorophyll catabolite in isolated senescent chloroplasts. *Planta* 187:230–35

93. Matile P, Schellenberg M, Vicentini F. 1997. Localization of chlorophyllase in the chloroplast envelope. *Planta* 201:96–99

94. Maunders MJ, Brown SB, Woolhouse HW. 1983. The appearance of chlorophyll derivatives in senescing tissue. *Phytochemistry* 22:2443–46

95. McLaughlin JC, Smith SM. 1995. Glyoxylate cycle enzyme-synthesis during the irreversible phase of senescence in cucumber cotyledons. *J. Plant Physiol.* 146:133–38

96. Meir S, Philosoph-Hadas S, Aharoni N. 1992. Ethylene-increased accumulation of fluorescent lipid-peroxidation products detected during senescence of parsley by a newly developed method. *J. Am. Soc. Hortic. Sci.* 117:128–32

97. Merzlyak MN, Rumyantseva VB, Shevyryova VV, Gusev MV. 1983. Further investigations of liposoluble fluorescent compounds in senescing plant cells. *J. Exp. Bot.* 34:604–9

98. Mihailovic N, Lazarevic M, Dzeletovic M, Yuckov M, Durdevic M. 1997. Chlorophyllase activity in wheat, *Triticum aestivum* L. leaves during drought and its dependence on the nitrogen form applied. *Plant Sci.* 129:141–46

99. Miller LS, Holt SC. 1977. Effect of carbon dioxide on pigment and membrane content in *Synechococcus lividus. Arch. Microbiol.* 115:185–98

100. Minguez-Mosquera MI, Gandulrojas B, Gallardo-Guerrero LG. 1994. Measurement of chlorophyllase activity in olive fruit (*Olea europaea*). *J. Biochem.* 116: 263–68

101. Miyake K, Ohtomi M, Yoshizawa H, Sakamoto Y, Nakayama K, Okada M. 1995. Water soluble pigments containing xylose and glucose in gametangia of the green alga, *Bryopsis maxima. Plant Cell Physiol.* 36:109–13

102. Mock H-P, Trainotti L, Kruse E, Grimm B. 1995. Isolation, sequencing and expression of cDNA sequences encoding uroporphyrinogen decarboxylase from tobacco and barley. *Plant Mol. Biol.* 28: 245–56

103. Moser D, Matile P. 1997. Chlorophyll breakdown in ripening fruits of *Capsicum annuum. J. Plant Physiol.* 150:759–61

104. Mothes K. 1960. Über das Altern der Blätter und die Möglichkeit ihrer Wiederverjüngung. *Naturwissenschaften* 47:337–51

105. Mühlecker W, Kräutler B. 1996. Breakdown of chlorophyll: constitution of nonfluorescing chlorophyll-catabolites from senescent cotyledons of the dicot rape. *Plant Physiol. Biochem.* 34:61–75

106. Mühlecker W, Kräutler B, Ginsburg S, Matile P. 1993. Breakdown of chlorophyll: a tetrapyrrolic chlorophyll catabolite from senescent rape leaves. *Helv. Chim. Acta* 76:2976–80

107. Mühlecker W, Ongania K-H, Kräutler B, Matile P, Hörtensteiner S. 1997. Tracking down chlorophyll breakdown in plants: elucidation of the constitution of a fluorescent chlorophyll catabolite. *Angew. Chem. Int. Ed. Engl.* 36:401–4

108. Nakamura H, Musicki B, Kishi Y. 1988. Structure of the light emitter in krill (*Euphausia pacific*) bioluminescence. *J. Am. Chem. Soc.* 110:2683–85

109. Nebel B, Matile P. 1992. Longevity and senescence in needles of *Pinus cembra. Trees* 6:156–61

110. Nock LP, Rogers LJ, Thomas H. 1992. Metabolism of protein and chlorophyll in leaf tissue of *Festuca pratensis* during chloroplast assembly and senescence. *Phytochemistry* 31:1465–70

111. Noodén LD, Guiamét JJ. 1996. Genetic control of senescence and aging in plants. In *Handbook of the Biology of Aging*, ed. EL Schneider, JW Rowe, pp. 94–117. New York: Academic. 4th ed.

112. Oh SA, Park J-H, Lee GI, Paek KH, Park SK, Nam HG. 1997. Identification of three genetic loci controlling leaf senescence in *Arabidopsis thaliana. Plant J.* 12:527–35

113. Ohtsuka T, Ito H, Tanaka A. 1997. Conversion of chlorophyll *b* to chlorophyll *a* and the assembly of chlorophyll with apoproteins by isolated chloroplasts. *Plant Physiol.* 113:137–47

114. Ong BL, Ng SL, Wee YC. 1995. Photosynthesis in *Platycerium coronarium* (Koenig ex Mueller) Desv. *Photosynthetica* 31:59–69

115. Oshio Y, Hase E. 1969. Studies on red pigments excreted by cells of *Chlorella protothecoides* during the process of bleaching induced by glucose or acetate. I. Chemical properties of the red pigments. *Plant Cell Physiol.* 10:51–59

116. Owens TG, Falkowski PG. 1982. Enzymatic degradation of chlorophyll *a* by marine phytoplankton in vitro. *Phytochemistry* 21:979–84

117. Parthier B. 1988. Gerontoplasts—the yellow end in the ontogenesis of chloroplasts. *Endocytobiosis Cell Res.* 5:163–90

118. Patterson GW, Hugly S, Harrison D. 1993. Sterols and phytyl esters of *Arabidopsis thaliana* under normal and chilling temperatures. *Phytochemistry* 33:1381–83

119. Peisker C. 1991. *Chlorophyllabbau in vergilbenden Gerstenblättern: Entwicklung und Anwendung einer Methode zur* ^{14}*C-Markierung von Chlorophyll*. PhD thesis. Univ. Zürich

120. Peisker C, Düggelin T, Rentsch D, Matile P. 1989. Phytol and the breakdown of chlorophyll in senescent leaves. *J. Plant Physiol.* 135:428–32

121. Peisker C, Thomas H, Keller F, Matile P. 1990. Radiolabelling of chlorophyll for studies on catabolism. *J. Plant Physiol.* 136:544–49

122. Perkins HJ, Roberts DWA. 1963. Chlorophyll turnover in monocotyledons and dicotyledons. *Can. J. Bot.* 41:221–26

123. Prasad TK, Anderson MD, Martin BA, Stewart CR. 1994. Evidence for a chilling-induced oxidative stress in maize seedlings and a regulatory role for hydrogen peroxide. *Plant Cell* 6:65–74

124. Raskin VI, Fleminger D, Marder JB. 1995. Integration and turnover of photosystem II pigment. In *Photosynthesis: From Light to Biosphere*, ed. P Mathis, 3:945–48. Amsterdam: Kluwer

125. Reddy PS, Veeranjaneyulu K. 1991. Proline metabolism in senescing leaves of horsegram. *J. Plant Physiol.* 137:381–83

126. Riper DM, Owens TG, Falkowski PG. 1979. Chlorophyll turnover in *Skeletonema costatum*, a marine plankton diatom. *Plant Physiol.* 64:49–54

127. Rise M, Goldschmidt EE. 1990. Occurrence of chlorophyllides in developing, light grown leaves of several plant species. *Plant Sci.* 71:147–51

128. Rodoni S, Mühlecker W, Anderl M, Kräutler B, Moser D, et al. 1997. Chloro-

phyll breakdown in senescent chloroplasts. Cleavage of pheophorbide *a* in two enzymatic steps. *Plant Physiol.* 115:669–76

129. Rodoni S, Schellenberg M, Matile P. 1998. Chloroplast breakdown in senescing barley leaves as correlated with phaeophorbide a oxygenase activity. *J. Plant Physiol.* 152:139–44

130. Rodoni S, Vicentini F, Schellenberg M, Matile P, Hörtensteiner S. 1997. Partial purification and characterization of RCC reductase, a stroma protein involved in chlorophyll breakdown. *Plant Physiol.* 115:677–82

131. Ronning CM, Bouwkamp JC, Solomos T. 1991. Observations on the senescence of a mutant non-yellowing genotype of *Phaseolus vulgaris* L. *J. Exp. Bot.* 42:235–41

132. Rontani J-F, Cuny P, Grossi V. 1996. Photodegradation of chlorophyll phytol chain in senescent leaves of higher plants. *Phytochemistry* 42:347–51

133. Rüdiger W, Schoch S. 1989. Abbau von Chlorophyll. *Naturwissenschaften* 76:453–57

134. Rüdiger W, Schoch S. 1991. The last step of chlorophyll biosynthesis. In *Chlorophylls*, ed. H Scheer, pp. 451–64. Boca Raton: CRC Press

135. Schellenberg M, Matile P. 1995. Association of components of the chlorophyll catabolic system with pigment-protein complexes from solubilized chloroplast membranes. *J. Plant Physiol.* 146:604–8

136. Schellenberg M, Matile P, Thomas H. 1990. Breakdown of chlorophyll in chloroplasts of senescent barley leaves depends on ATP. *J. Plant Physiol.* 136:564–68

137. Schellenberg M, Matile P, Thomas H. 1993. Production of a presumptive chlorophyll catabolite in vitro: requirement of reduced ferredoxin. *Planta* 191:417–20

138. Scheumann V, Helfrich M, Schoch S, Rüdiger W. 1996. Reduction of the formyl group of zinc pheophorbide *b* in vivo: a model for the chlorophyll *b* to *a* transformation. *Z. Naturforsch. Teil C* 51:185–94

139. Scheumann V, Ito H, Tanaka A, Schoch S, Rüdiger W. 1996. Substrate specificity of chlorophyll(ide) *b* reductase in etioplasts of barley (*Hordeum vulgare* L.). *Eur. J. Biochem.* 242:163–70

140. Scheumann V, Schoch S, Rüdiger W. 1998. Chlorophyll *a* formation in the chlorophyll *b* reductase reaction requires

reduced ferredoxin. *J. Biol. Chem.* In press

141. Shibata H, Kono Y, Yamashita S, Sawa Y, Ochiai H, Tanaka K. 1995. Degradation of chlorophyll by nitrogen-dioxide generated from nitrite by peroxidase reaction. *Biochim. Biophys. Acta* 1230:45–50

142. Shihira-Ishikawa I, Hase E. 1964. Nutritional control of cell pigmentation in *Chlorella protothecoides* with special reference to the degeneration of chloroplasts induced by glucose. *Plant Cell Physiol.* 5:227–40

143. Shimokawa K, Uchida Y. 1992. A chlorophyllide *a* degrading enzyme (hydrogen peroxide-2,4 dichlorophenol requiring) of *Citrus unshiu* fruits. *J. Jpn. Soc. Hortic. Sci.* 61:175–81

144. Shioi Y, Sasa T. 1986. Purification of solubilized chlorophyllase from *Chlorella protothecoides*. *Methods Enzymol.* 123:421–27

145. Shioi Y, Tatsumi Y, Shimokawa K. 1991. Enzymatic degradation of chlorophyll in *Chenopodium album*. *Plant Cell Physiol.* 32:87–93

146. Shioi Y, Tomita N, Tsuchiya T, Takamiya K. 1996. Conversion of chlorophyllide to pheophorbide by Mg-dechelating substance in extracts of *Chenopodium album*. *Plant Physiol. Biochem.* 34:41–47

147. Shioi Y, Watanabe K, Takamiya K. 1996. Enzymatic conversion of pheophorbide *a* to the precursor of pyropheophorbide *a* in leaves of *Chenopodium album*. *Plant Cell Physiol.* 37:1143–49

148. Sitte P, Falk H, Liedvogel B. 1980. Chromoplasts. In *Pigments in Plants*, ed. F-C Czygan, pp. 117–48. Stuttgart: Gustav Fischer. 2nd ed.

149. Sodmergen S, Kawano S, Tano S, Kuroiwa T. 1989. Preferential digestion of chloroplast nuclei (nucleoids) during senescence of the coleoptile of *Oryza sativa*. *Protoplasma* 152:65–68

150. Sodmergen S, Kawano S, Tano S, Kuriowa T. 1991. Degradation of chloroplast DNA in second leaves of rice (*Oryza sativa*) before leaf yellowing. *Protoplasma* 160:89–98

151. Spooner N, Harvey HR, Pearce GES, Eckardt CB, Maxwell JR. 1994. Biological defunctionalization of chlorophyll in the aquatic environment. 2. Action of endogenous algal enzymes and aerobic bacteria. *Org. Geochem.* 22:773–80

152. Spooner N, Keely BJ, Maxwell JR. 1994. Biologically mediated defunction-alization of chlorophyll in the aquatic environment. 1. Senescence decay of the diatom *Phaeodactylum tricornutum*. *Org. Geochem.* 21:509–16

153. Stiles EW. 1982. Fruit flags—two hypotheses. *Am. Nat.* 120:500–9

154. Stobart AK, Hendry GAF. 1984. Chlorophyll turnover in greening wheat leaves. *Phytochemistry* 23:27–30

155. Stoll A. 1912. *Über Chlorophyllase und die Chlorophyllide*. PhD thesis. Nr. 49. ETH Zürich

156. Tao YZ, Jordan DR, Henzell RG, McIntyre CL. 1998. Construction of a genetic map in sorghum recombinant inbred line using probes from different sources and its comparison with other sorghum maps. *Aust. J. Agric. Res.* 49:729–36

157. Tarasenko LG, Khodasevich EV, Orlovskaya KI. 1986. Localization of chlorophyllase in chloroplast membranes. *Photobiochem. Photobiophys.* 12:119–21

158. Terpstra W. 1981. Identification of chlorophyllase as a glycoprotein. *FEBS Lett.* 126:231–35

159. Terry MJ. 1997. Phytochrome chromophore-deficient mutants. *Plant Cell Environ.* 20:740–45

160. Thomas H. 1982. Leaf senescence in a non-yellowing mutant of *Festuca pratensis*. I. Chloroplast membrane polypeptides. *Planta* 154:212–18

161. Thomas H. 1986. The role of polyunsaturated fatty acids in senescence. *J. Plant Physiol.* 123:97–105

162. Thomas H. 1987. *Sid*: a Mendelian locus controlling thylakoid membrane disassembly in senescing leaves of *Festuca pratensis*. *Theor. Appl. Genet.* 73:551–55

163. Thomas H. 1997. Chlorophyll, a symptom and regulator of plastid development. *New Phytol.* 136:163–81

164. Thomas H, Bortlik K, Rentsch D, Schellenberg M, Matile P. 1989. Catabolism of chlorophyll in vivo: significance of polar chlorophyll catabolites in a non-yellowing senescence mutant of *Festuca pratensis* Huds. *New Phytol.* 111:3–8

165. Thomas H, Evans C, Thomas HM, Humphreys MW, Morgan WG, et al. 1997. Introgression, tagging and expression of a leaf senescence gene in *Festulolium*. *New Phytol.* 137:29–34

166. Thomas H, Matile P. 1988. Photobleaching of chloroplast pigments in leaves of a non-yellowing mutant genotype of *Festuca pratensis*. *Phytochemistry* 27:345–48

167. Thomas H, Schellenberg M, Vicentini F, Matile P. 1996. Gregor Mendel's green and yellow pea seeds. *Bot. Acta* 109:3–4

168. Thomas H, Smart C. 1993. Crops that stay green. *Ann. Appl. Biol.* 123:193–219

169. Thomas H, Stoddart JL. 1975. Separation of chlorophyll degradation from other senescence processes in leaves of a mutant genotype of meadow fescue (*Festuca pratensis*). *Plant Physiol.* 56:438–41

170. Thomas HM, Morgan WG, Meredith MR, Humphreys MW, Thomas H, Legget JM. 1994. Identification of parental recombined chromosomes of *Lolium multiflorum x Festuca pratensis* by genomic *in situ* hybridization. *Theor. Appl. Genet.* 88:909–13

171. Tommasini R, Vogt E, Fromenteau M, Hörtensteiner S, Matile P, et al. 1998. An ABC-transporter of *Arabidopsis thaliana* has both glutathione-conjugate and chlorophyll catabolite transport activity. *Plant J.* 13:773–80

172. Trebitsch T, Goldschmidt EE, Riov J. 1993. Ethylene induces *de novo* synthesis of a chlorophyll degrading enzyme in *Citrus* fruit peel. *Proc. Natl. Acad. Sci. USA* 90:9441–45

173. Tsuchiya T, Ohta H, Masuda T, Mikami B, Kita N, et al. 1997. Purification and characterization of two isozymes of chlorophyllase from mature leaves of *Chenopodium album*. *Plant Cell Physiol.* 38:1026–31

174. Van den Hoek C, Mann DG, Jahns HM. 1995. *Algae. An Introduction to Phycology.* Cambridge: Cambridge Univ. Press

175. Van Oosterom EJ, Jayachandran R, Bidinger FR. 1996. Diallel analysis of the stay-green trait and its components in sorghum. *Crop Sci.* 36:549–55

176. Venkatrayappa T, Fletcher RA, Thompson JE. 1984. Retardation and reversal of senescence in bean leaves by benzyladenine and decapitation. *Plant Cell Physiol.* 25:407–18

177. Vicentini F. 1997. *Seneszenz und Blattgrün: Enzyme des Chlorophyllabbaus.* PhD thesis. Univ. Zürich

178. Vicentini F, Hörtensteiner S, Schellenberg M, Thomas H, Matile P. 1995. Chlorophyll breakdown in senescent leaves: identification of the lesion in a stay-green genotype of *Festuca pratensis. New Phytol.* 129:247–52

179. Vicentini F, Iten F, Matile P. 1995. Development of an assay for Mg-dechelatase of oilseed rape cotyledons, using chlorophyllin as the substrate. *Physiol. Plant.* 94:57–63

180. Von Wettstein D, Gough S, Kannangara CG. 1995. Chlorophyll biosynthesis. *Plant Cell* 7:1039–57

181. Walker CJ, Yu GH, Weinstein JD. 1997. Comparative study of heme and Mg-protoporphyrin (monomethyl ester) biosynthesis in isolated pea chloroplasts: effects of ATP and metal ions. *Plant Physiol. Biochem.* 35:213–21

182. Walulu RS, Rosenow DT, Wester DB, Nguyen HT. 1994. Inheritance of the stay green trait in sorghum. *Crop Sci.* 34:970–72

183. Webb CJ. 1985. Fruit flags in a temperate evergreen shrub *Corokia cotoneaster* (Escalloniaceae). *NZ J. Bot.* 23:343–44

184. Whyte BJ, Castelfranco PA. 1993. Breakdown of thylakoid pigments by soluble proteins of developing chloroplasts. *Biochem. J.* 290:361–67

185. Woods AF. 1899. The destruction of chlorophyll by oxidizing enzymes. *Zentralbl. Bakteriol. Parasitenkd. Infektionskr.* 5:745–54

186. Yamauchi N, Akiyama Y, Kako S, Hashinaga F. 1997. Chlorophyll degradation in Wasa satsuma mandarin (*Citrus unshiu* Marc.) fruit with on-tree maturation and ethylene treatment. *Sci. Hortic.* 71:35–42

187. Yamauchi N, Watada AE. 1991. Regulated chlorophyll degradation in spinach leaves during storage. *J. Am. Soc. Hortic. Sci.* 116:58–62

188. Yamauchi N, Xia XM, Hashinaga F. 1997. Involvement of flavonoid oxidation with chlorophyll degradation by peroxidase in Wase satsuma mandarin fruits. *J. Jpn. Soc. Hortic. Sci.* 66:283–88

189. Ziegler R, Blaheta A, Guha N, Schönegge B. 1988. Enzymatic formation of pheophorbide and pyropheophorbide during chlorophyll degradation in a mutant of *Chlorella fusca* Shihira et Kraus. *J. Plant Physiol.* 132: 327–33

Annu. Rev. Plant Physiol. Plant Mol. Biol. 1999. 50:97–131

PLANT PROTEIN SERINE/THREONINE KINASES: Classification and Functions

D. G. Hardie

Biochemistry Department, Dundee University, Dundee, DD1 5EH, Scotland, United Kingdom; e-mail: d.g.hardie@dundee.ac.uk

KEY WORDS: protein kinases, higher plants, sequence families, signal transduction, cellular signaling

ABSTRACT

The first plant protein kinase sequences were reported as recently as 1989, but by mid-1998 there were more than 500, including 175 in *Arabidopsis thaliana* alone. Despite this impressive pace of discovery, progress in understanding the detailed functions of protein kinases in plants has been slower. Protein serine/threonine kinases from *A. thaliana* can be divided into around a dozen major groups based on their sequence relationships. For each of these groups, studies on animal and fungal homologs are briefly reviewed, and direct studies of their physiological functions in plants are then discussed in more detail. The network of protein-serine/threonine kinases in plant cells appears to act as a "central processor unit" (cpu), accepting input information from receptors that sense environmental conditions, phytohormones, and other external factors, and converting it into appropriate outputs such as changes in metabolism, gene expression, and cell growth and division.

CONTENTS

97

1040-2519/99/0601-0097$08.00

INTRODUCTION

The past decade has proved something of a golden age in plant protein kinase research. Although protein kinases had been previously detected and analyzed via biochemical approaches (e.g. 50), the first higher plant protein kinase cDNA sequences were reported by Lawton et al in 1989 (97). A search conducted in August 1998 revealed that the EMBL database contained 549 higher plant protein kinase sequences, of which 175 were from *Arabidopsis thaliana* alone. While a small proportion of these represent multiple entries of the same gene sequence, these numbers give some insight into the recent increase in knowledge. Figure 1 shows that since 1989 the number of higher plant kinase DNA sequences in the database has increased inexorably, with the rate of deposition still accelerating. The golden age is therefore still under way, although it could be argued that the field is maturing. While such predictions are always risky, one might imagine that representatives of most of the major subfamilies of plant protein kinases have already been discovered, and that future discoveries will usually fall into one of the existing groups. This statement must be qualified by stressing that, for the purposes of this review, the term protein kinase is restricted to members of the classical "eukaryotic" protein kinase superfamily (46) ("eukaryotic" is placed within quotes because members of this family have recently also been found in some prokaryotes). It is quite possible that novel plant protein kinases not related to this family remain to be found. Intriguing precedents for this are genes involved in the response to ethylene and cytokinins in *A. thaliana*, which are members of the "prokaryotic" two-component histidine/aspartate kinase family (13, 84) and are not related to the "eukaryotic" kinase family.

DNAs encoding higher plant protein kinases have generally been cloned by one of four routes:

1. *Cloning by sequence homology*, either utilizing PCR with degenerate primers, based on the sequence motifs known to be conserved in mammalian

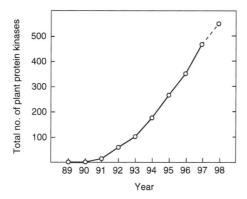

Figure 1 Total number of higher plant protein kinase entries in the EMBL database plotted against the year of initial deposition. On August 5th, 1998, the EMBL/EMBLNEW databases were searched with the term "kinase" in the "description" field, "viridiplantae" in the "organism" field, and the year of first entry in the "date" field. Proteins other than protein kinases (usually around 10% to 20% of the total hit by the "kinase" search term) were removed by manual inspection. This search may miss a few protein kinases if the word "kinase" does not appear in the "description" field [e.g. the two-kinase sequences deposited by Lawton et al (97) in 1989, which have been added]. No attempt was made to delete multiple entries of the same sequence. The 1998 data are shown with a dashed line as the calendar year was not complete at the time of analysis.

and yeast protein kinases, or by screening cDNA libraries with protein kinase DNAs derived from non-plant species. The first approach, pioneered by Hanks (45), was used in the cloning of the first four plant kinase DNAs reported (35, 97, 162), and is largely responsible for the explosion of sequences evident in Figure 1. However, although this approach provides the predicted amino acid sequence, it gives no information about the function of the protein kinase other than by virtue of sequence similarity with kinases of known function, usually of animal or fungal origin.

2. *Cloning by functional homology*, e.g. selection of cDNAs by their ability to complement mutations in other species, especially in the yeast *Saccharomyces cerevisiae*. Although many plant DNAs have been subsequently shown to complement yeast mutations, surprisingly this approach does not appear to have been widely used for the initial cloning. However, it was used to clone DNA encoding the *LAMMER* kinase AFC1 (6) and a CDK-activating kinase (158).

3. *Cloning from a partial amino acid sequence of a purified protein* Although this is the "classical" approach for cloning of a known protein, in

practice it has been rarely used in the case of plant protein kinases, probably because very few have been purified to homogeneity from plant sources. A rare example was the original cloning of a calmodulin-like domain protein kinase (CDPK) by Harper et al (53).

4. *Cloning of mutants*   This approach involves selection of mutants defective in some physiological process, and then cloning of the mutant gene, usually by genetic mapping and chromosome walking. The latter approach is laborious, but if successful the rewards can be great, because the mutant phenotype immediately gives direct clues as to the function of the gene. A significant short cut is possible if the mutants are generated by insertional mutagenesis using *Agrobacterium tumifaciens* T-DNA, since it is then much easier to locate the mutated gene. Although this approach is not specifically aimed at cloning of protein kinases, given their important role in cellular regulation, it is not surprising that many have emerged from these genetic screens. Interesting examples of protein kinase genes cloned by this approach include the disease-resistance genes *Xa21* from rice (147) and *Pto* from tomato (108), and the genes giving rise to the developmental mutants *CLAVATA1* (15) and *TOUSLED* (136), and the brassinosteroid-resistant mutant *BRI1* (100), all from *A. thaliana*.

FUNCTIONS OF PROTEIN KINASES: MODELS IN ANIMALS AND YEAST

Despite the recent advances in higher plants, the functions of protein kinases are much better understood for animals and for the budding yeast *Saccharomyces cerevisiae*. Since many plant protein kinases have homologs in animals and yeast, these systems provide useful models to guide studies of their functions in plants. In an influential review in 1987 (70), Hunter pointed out that a protein kinase-phosphatase cycle was analogous to a transistor, and like the latter can act either as a switch, changing the function of the target protein on or off, and/or as an amplifier, increasing the amplitude of a signal. I have extended this analogy [see (9)] to point out that the network of protein kinases and phosphatases within the cell can be likened to the assemblage of transistors in the central processor unit (CPU) of a computer. This cellular "CPU" receives input information about the external environment from sensor and receptor proteins, processes it, and then triggers appropriate output responses, such as changes in metabolism, ion fluxes, gene expression, or cell growth and division. To make this concept more specific, some of the major inputs and outputs for which protein kinases serve as information processors in animal and fungal cells can be summarized as follows:

1. *The response to soluble extracellular messenger molecules* such as hormones, growth factors, and neurotransmitters in animals, or mating factors in yeast. The receptors for these messengers may themselves be protein kinases, particularly transmembrane proteins with protein kinase domains on the intracellular face, or they may interact with cascades of protein kinases via protein-protein interactions. Classic examples include growth factor receptors of the protein-tyrosine kinase class, the MAP kinase cascades that lie downstream of many receptors, and protein kinases that respond to second messengers such as cyclic AMP (protein kinase A), phosphatidylinositol-3,4,5-trisphosphate (protein kinase B), diacylglycerol (protein kinase C), and calcium (calmodulin-dependent protein kinases).

2. *The response to messenger molecules anchored in the the surface of neighboring cells*   In these cases protein kinases play important roles in development and differentiation by conveying information about the identity of neighboring cells. A good example is the Eph family of protein-tyrosine kinase receptors, and their protein ligands, the ephrins (179). These are involved, for example, in guiding axons of nerve cells to grow towards appropriate targets by inhibiting inappropriate cell-cell interactions. It could be argued that the responses to soluble and anchored messengers are not intrinsically different. However, I mention anchored messengers separately because perception of positional information would appear to be crucial to plant development, and receptor-linked kinases may play an important role in this process.

3. *Defense responses*   Although the antibodies on B cells and the receptors on T cells that recognize "foreign" molecules are not themselves protein kinases, they do activate cascades of downstream kinases to initiate immune responses.

4. *The timing of events that occur discontinuously in the cell cycle*   This is the particular province of the cyclin-dependent protein kinases (CDKs) that, in association with different cyclins, control entry into S phase and mitosis in response to input cues such as mitogens, the availability of nutrients, and the integrity or state of replication of DNA.

5. *The response to stressful environmental conditions, including starvation for key nutrients*   Examples here include the Hog1 kinase, involved in the response to osmotic stress (106), and the Snf1 kinase (49), involved in the response to glucose deprivation, both in *S. cerevisiae*.

These animal and fungal systems provide useful models for the functions of plant protein kinases, particularly because when a plant kinase is first defined there are often no clues to its function other than its sequence similarity to animal and fungal homologs. The approach is justified because the use of protein kinases as cellular control devices appears to have arisen very early during eukaryotic evolution, and the basic mechanisms have often been conserved. While the animal and fungal models can be very useful for framing experiments, it should be remembered that plants have evolved independently of animals and fungi for perhaps a billion years, and the systems should not be expected to be identical. In the presentations of protein kinase subfamilies that follow, I generally introduce the subfamily by discussing what is known about the role of their relatives in animals and yeast, and then consider to what extent these models have been helpful in our understanding of the function of the higher plant homologs.

CLASSIFICATION OF PROTEIN KINASES

Protein kinases can be studied by biochemical assays, utilizing their ability to phosphorylate particular substrate proteins. This approach has been used widely to study plant protein kinases that phosphorylate metabolic enzymes such as pyruvate dehydrogenase, phosphoenolpyruvate (PEP) carboxylase, nitrate reductase, pyruvate, Pi dikinase, or sucrose phosphate synthase. However, an inherent problem with a purely biochemical approach is the difficulty in comparing results obtained by different laboratories, especially when different species or tissues are used, as it is often difficult to know whether the protein kinases under study correspond to the same molecular entities. A protein kinase is only rigorously defined when its DNA and/or amino acid sequence is established. Similarities in amino acid sequence would therefore seem the most logical method to classify protein kinases. This is also why I have omitted from this review some interesting cases such as PEP carboxylase kinase (54, 160) and the double-stranded RNA-dependent protein kinase (96), which do not yet appear to have been cloned.

Figure 2 shows a phylogenetic tree illustrating the relationships between the sequences of the kinase domains of 89 protein kinases from *A. thaliana* that were in the EMBL database in August 1998. Table 1, which may be viewed in the Supplementary Materials Section at http://www.annualreviews.org, provides a listing enabling individual kinases in Figure 2 to be identified. I restricted this analysis to *A. thaliana* partly to make the task more manageable but also because it avoided a difficult problem inherent in comparing sequences from different species, i.e. deciding whether two sequences are true interspecies homologs (*orthologs*) or merely closely related genes (*paralogs*). The number of sequences to be considered was restricted to 89 by eliminating identical or nearly identical sequences, and by only considering cases in which the full length sequence was available (thus omitting, among others, expressed sequence tags).

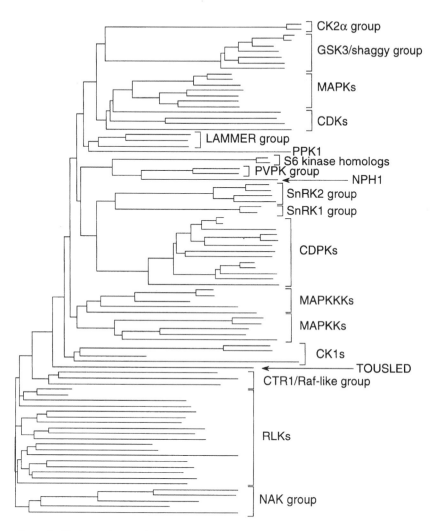

Figure 2 Phylogenetic tree of cloned *A. thaliana* protein kinases. Unique, full-length protein kinase sequences were extracted from the EMBL database. The sequences were aligned using PILEUP (28) and the kinase catalytic domains [as defined by (46)] were then used to construct a phylogenetic tree using CLUSTREE (60). The length of the branches between two sequences is a measure of the degree of sequence divergence between them. To identify individual protein kinases, see Supplementary Materials Section, http://www.annualreviews.org.

Fortunately, the analysis shows (Figure 2) that most of the 89 protein kinase sequences cluster together into about 12 major subfamilies. Although the tree was constructed on the basis of the sequences of the kinase catalytic domains alone, members of a subfamily often also exhibit sequence similarities outside of this region, particularly within regulatory domains. Because of this, they tend to exhibit functional as well as structural similarity.

The remainder of this review provides an overview of the structure and function of plant protein kinases, utilizing these subfamilies as the basis. They are considered in a somewhat arbitrary order, commencing with those for which the most direct biochemical information in plant systems is available. Figure 2 was constructed using *A. thaliana* sequences only, which inevitably introduces a bias in coverage towards that species, but where appropriate I also cover examples from other species. Regretfully, length restrictions require that the coverage is selective.

THE CALCIUM-DEPENDENT PROTEIN KINASE (CDPK) SUBFAMILY

See (141) for another view of this family. Calcium is a ubiquitous intracellular messenger in animal cells, and many of its effects are mediated by binding to the ubiquitous Ca^{2+}-binding protein calmodulin. The Ca^{2+}-calmodulin complex in turn activates calmodulin-binding proteins, an important subset of which are the calmodulin-dependent protein kinases (CaMKs). The CaMKs have a large variety of different functions, with some being specific for individual targets (e.g. myosin light chain kinases, phosphorylase kinase), whereas others have multiple targets (e.g. CaMKI, CaMKII, CaMKIV). In general, they are maintained in an inactive state in the absence of calmodulin via autoinhibitory regions, just C-terminal to the kinase domain, that associate with the active site cleft and thus block access to exogenous substrates (40, 65). In some CaMKs (e.g. CaMKII), the autoinhibitory region contains an autophosphorylation site that is phosphorylated upon binding of Ca^{2+}-calmodulin. In most other cases, the autoinhibitory region resembles a phosphorylation site but does not have a phosphorylatable amino acid, when it is referred to as a pseudosubstrate (48).

Calcium is also a ubiquitous intracellular messenger in plant cells, increasing in response to a wide variety of stimuli, such as touch, wind, gravity, light, cold, auxin, abscisic acid, giberellic acid, salt stress, and fungal elicitors (12, 129). These observations triggered a hunt for calmodulin-dependent protein kinases in plants, but when the first Ca^{2+}-dependent protein kinase was extensively purified from soybean, it was found to be activated by Ca^{2+} in the absence of exogenous calmodulin (50). This behavior was explained when the DNA was cloned and found to encode a protein kinase catalytic domain followed by a C-terminal domain with 39% identity to calmodulin, with the four "EF-hand" Ca^{2+}-binding motifs of calmodulin being conserved (53). The plant Ca^{2+}-dependent protein kinases can also therefore be termed "calmodulin-like domain protein kinases," and the acronym (CDPK) remains the same. This subfamily does not appear to be present in animals and fungi, although members have been found in protists such as *Paramecium tetraurelia* and *Plasmodium falciparum* (146, 177).

At least 13 members of the CDPK family have now been sequenced from *A. thaliana* (Figure 2), and multiple isoforms have also been characterized in many other plant species. They form the largest well-defined protein kinase subfamily in plants (the receptor-like kinases are currently more numerous, but are also much more diverse in sequence and structure). Why such a large number of CDPK isoforms should exist in a single species is unclear, but this may provide both specificity and flexibility in the response to different Ca^{2+}-elevating stimuli. Consistent with this, different CDPK isoforms in *A. thaliana* are expressed in a cell type– and developmental stage–specific manner (64), whereas different isoforms from soybean have been reported to differ in Ca^{2+} sensitivity and specificity for peptide substrates (99).

When compared with mammalian protein kinases, the sequences of plant CDPKs are most similar to the CaMKs (46), and they could conceivably have arisen by fusion of a gene encoding an ancestral CaMK with a calmodulin-like gene (Figure 3). There is now strong evidence that the CDPKs and CaMKs are regulated by similar mechanisms. CDPKs have a junction region between

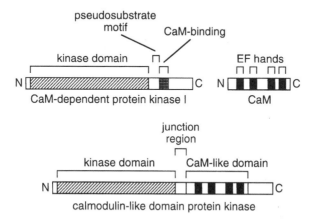

Figure 3   Comparison of domain structure of mammalian calmodulin- (CaM-) dependent protein kinase I and plant CDPKs. The four "EF hand" Ca^{2+}-binding motifs in calmodulin, and in the calmodulin-like domain of CDPK, are indicated by black boxes.

the kinase domain and the calmodulin-like domain (CLD), in the same position as the autoinhibitory region of mammalian CaMKs. There is now extensive evidence that the junction region is an autoinhibitory sequence whose effect is relieved by Ca^{2+}-dependent binding of the CLD within the same polypeptide (51, 52, 68, 171). Conceivably, having the Ca^{2+}-binding sites on the same polypeptide, rather than on a separate calmodulin molecule which would need to find its target by diffusion, makes the response more rapid. This arrangement represents one case where plants have adopted a different strategy than animals and fungi, although it is a variation on a theme rather than a radical departure. Plants do also contain Ca^{2+}-dependent protein kinases lacking a CLD (see below).

Although CDPKs are usually purified using mammalian proteins or synthetic peptides as substrates, a wide range of plant proteins have been shown to be substrates in cell-free assays. In most cases it remains unclear whether these are substrates in vivo. Targets for CDPKs where an effect of phosphorylation on function has been reported include the following:

1. Nodulin 26, an aquaporin (water channel) that is incorporated into the symbiosome membrane (SM) surrounding the N_2-fixing bacteria in root nodules of legumes, is phosphorylated at Ser-262 by an SM-associated CDPK and a recombinant CDPK. This converts the aquaporin from a fully open, high-conductance state to a voltage-sensitive form with a higher preponderance of low conductance states (98).

2. Tonoplast intrinsic protein-α(α-TIP), another aquaporin found in the membrane of protein storage vacuoles in *Phaseolus vulgaris* seeds, is phosphorylated by a Ca^{2+}-dependent kinase (presumably a CDPK) present in the same membranes. Using expression of α-TIP in *Xenopus laevis* oocytes, evidence was obtained that phosphorylation at the same serine residues can increase the channel's water permeability (78).

3. Ca^{2+}-dependent kinases from spinach leaves were identified by immunological criteria (1) or partial amino acid sequencing (32) as members of the CDPK family. One of these reversibly inactivated nitrate reductase by phosphorylation at Ser-543, as did the most closely related isoform from *A. thaliana* (CDPK6) after expression in bacteria (32).

4. Application of purified recombinant CDPK, in the presence of MgATP, to membrane patches from vacuolar membranes of *Vicia faba* guard cells resulted in activation of a Cl^- channel (128). This may be involved in the increased Cl^- uptake that occurs during stomatal opening.

5. Overexpression of constitutively active, truncated CDPK mutants in maize leaf protoplasts resulted in increased expression of reporter genes coupled to the promoter of *HVA1*, a stress-inducible gene from barley (144). This effect was obtained using only two out of eight *A. thaliana* CDPKs tested, indicating a degree of isoform specificity. Although the immediate downstream targets in this system were not determined, these results indicate that some CDPK isoforms may be involved in induction of gene expression by stress.

6. Ca^{2+}-dependent protein kinases from maize leaf (69) and soybean root nodules (176) phosphorylated sucrose synthase from the same sources. Huber et al (69) reported that phosphorylation reduced the K_m values for UDP and sucrose by fivefold, suggesting that this phosphorylation could activate sucrose catabolism.

Although most Ca^{2+}-dependent protein kinases in plants appear to be members of the immediate CDPK subfamily (characterized by an intrinsic CLD), other types have been reported. A protein kinase has been cloned from apple (166) with sequence similarity to mammalian CaMKII, including a putative calmodulin-binding region but lacking a CLD. Poovaiah and coworkers (126) cloned a cDNA from developing anthers of *Lilium longiflorum* encoding a protein kinase (CCaMK) that had a kinase domain followed by a calmodulin-binding region closely related to that of CaMKII, then a domain with three EF hand motifs related to the neural Ca^{2+}-binding protein visinin. When expressed in bacteria, this kinase was almost completely calmodulin-dependent for phosphorylation of an exogenous substrate. However, the visinin-like domain did bind Ca^{2+}, and appeared to be responsible for a Ca^{2+}-activated autophosphorylation (154). Finally, a CDPK-related sequence from carrot (102) has a CLD that is more distantly related to calmodulin than conventional CDPKs. When maize homologs of this CDPK-related kinase (CRK) were expressed in bacteria, they were found to be Ca^{2+}-independent (39).

THE SNF1-RELATED PROTEIN KINASE (SNRK) SUBFAMILY

The yeast (*S. cerevisiae*) SNF1 protein kinase is involved in the response to glucose starvation, whereas its mammalian homolog AMP-activated protein kinase (AMPK) is involved in the response to cellular stresses which cause ATP depletion (44, 49). Both the yeast and animal kinases exist as heterotrimeric complexes with a catalytic α subunit (encoded by the *SNF1* gene in yeast), and noncatalytic β and γ subunits, which appear to have roles in regulation and targeting of the complex. The yeast *SNF1* gene is required for derepression of genes that are repressed by glucose, i.e. the reversal of glucose repression (also known as catabolite repression) (137). The SNF1 complex is rapidly activated on glucose removal by a mechanism involving phosphorylation by an upstream kinase kinase (168), and it appears to relieve glucose repression, at least in part, by phosphorylating the repressor protein Mig1 (125). In contrast to yeast SNF1, where most of the information has been gleaned from genetic studies, mammalian AMPK is well characterized at the biochemical level (49). The complex is activated by elevation of 5′-AMP, both directly via allosteric activation, and indirectly by promoting phosphorylation by an upstream kinase kinase (AMPKK). Elevation of AMP occurs in intact cells under any stress conditions where ATP is depleted, including heat shock (19) and, in some cell types, starvation for glucose (140). Activation of AMPK conserves ATP by phosphorylating and inactivating certain biosynthetic enzymes, and at the same time switches on alternative catabolic pathways to generate more ATP. There are many parallels between the mammalian AMPK and yeast SNF1 systems, although surprisingly, the SNF1 complex is not activated by AMP.

DNA sequencing has revealed two subfamilies of SNF1/AMPK-related protein kinases in plants (44), i.e. the SnRK1 subfamily (currently two in *A. thaliana*), which are related to SNF1 and AMPK throughout their catalytic subunits, and the SnRK2 family (described below, currently four in *A. thaliana*), which are only related within the kinase domains (Figure 4). Independently

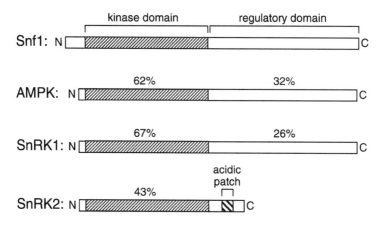

Figure 4 Comparison of domain structures of the Snf1-like subfamily, including Snf1 from *S. cerevisiae*, AMP-activated protein kinase (AMPK, α subunit) from mammals, and the higher plant Snf1-related kinases-1 and -2 (SnRK1 and SnRK2). Numbers above the kinase and regulatory domains represent % sequence identity with the equivalent region in Snf1, assessed using the program GAP (Genetics Computer Group), utilizing the rat AMPK-2 and *A. thaliana* AKIN10 and PROKINA (GENBANK:L05561) sequences as representative examples. The C-terminal domains of the SnRK2 family are unrelated to Snf1.

of these molecular biological studies, biochemical approaches identified plant protein kinases very similar to mammalian AMPK (3, 29, 104). Although these were initially termed HMG-CoA reductase kinases, it is now clear that they represent the products of *SnRK1* genes (2, 5, 29). Like their yeast and animal counterparts, they exist as large multimeric complexes, although regulatory β and/or γ subunits have not yet been identified in plants. Unlike AMPK (but like SNF1), they do not appear to be activated by AMP, but they are regulated by phosphorylation by an upstream kinase kinase. The latter is currently poorly characterized, but its effect can be mimicked using mammalian AMPKK (29, 104). This latter observation is consistent with the observation that the specific threonine residue, which is phosphorylated by AMPKK on AMPK [Thr-172 (55)], is conserved in all plant SnRK1 sequences, as is the sequence immediately surrounding it. By analogy with the yeast and animal systems, the plant SnRK1 kinases will likely be activated by stresses such as starvation for a carbon source, although this has not yet been directly demonstrated. The yeast, animal, and plant SNF1-related kinases have very similar substrate specificity in vitro (3, 22). Although no in vivo substrates have yet been identified with certainty for the plant kinases, in cell-free assays they phosphorylate and inactivate HMG-CoA reductase, sucrose phosphate synthase and nitrate reductase (21, 29, 156). These are key regulatory enzymes of isoprenoid and sucrose synthesis, and nitrogen assimilation, respectively, so that activation of SnRK1 complexes would be expected to inhibit these anabolic pathways. This is related to the function of mammalian AMPK although, with the exception of HMG-CoA reductase, the actual protein targets differ. Activation of SnRK1 kinases is also likely to upregulate catabolic pathways by regulating gene expression. The phenomenon of catabolite repression has been reported to occur in plant cells [reviewed in (75)] and it is interesting that glyoxylate cycle enzymes are repressed by glucose in both plants and yeast (42). Recently, transgenic potato plants have been established that express SnRK1 DNA in antisense orientation. Results obtained with these plants suggest that the SnRK1 kinases are involved in the regulation of expression of the sucrose synthase gene in response to the availability of carbohydrate (131). Despite its name, sucrose synthase is involved in the breakdown of sucrose in plant "sink" tissues, and occupies an analogous metabolic position to the invertase encoded by the *SUC2* gene, a classical target for regulation by SNF1 in yeast.

Members of the SnRK2 group (Figure 4) contain kinase domains related to those of Snf1 and AMPK, but the C-terminal domains are unrelated and are usually characterized by stretches of acidic residues, either poly-Glu or poly-Asp. The functions and biochemical properties of these kinases remain unknown. Using the synthetic peptides used to assay AMPK and SnRK1s, kinase activities have been isolated from cauliflower inflorescence (3) and spinach leaves (29) that are small, monomeric proteins of the size predicted for the SnRK2 subfamily. They have substrate specificities similar to the SnRK1 kinases and, like the latter, are regulated by phosphorylation (3). It is tempting to suggest that they are products of SnRK2 genes, although direct evidence is currently lacking.

THE RECEPTOR-LIKE KINASE SUBFAMILY

The receptor-like kinases (RLKs) are characterized by an N-terminal hydrophobic signal peptide, and an internal hydrophobic sequence followed by a basic "stop-transfer" sequence, suggesting that they are Type I membrane proteins with a single transmembrane helix. The C-terminal, intracellular regions are the locations of the protein kinase domains. There are at least 18 RLKs in *A. thaliana*: Their kinase domain sequences cluster into a discrete subfamily (Figure 2), although they are more divergent than those of the other subfamilies discussed in this review. Another review of the RLK family has recently been published (141).

The models for this group are the receptor-linked kinases in animal cells, which are also Type I membrane proteins with an intracellular kinase domain. The majority of animal receptor-linked kinases are protein-tyrosine kinases (a subfamily not yet found in plants) but there are also a small number of protein-serine/threonine kinases. The mechanisms of signal transduction downstream of mammalian receptor-linked kinases may be summarized as follows:

1. The receptor-linked protein-tyrosine kinases usually exist as monomers in the absence of ligand. Ligand binding causes the receptors to form homo- or hetero-oligomers, usually dimers. The cytoplasmic domains phosphorylate each other, and this creates docking sites for proteins with domains (e.g. SH2 domains) that form high-affinity binding sites for specific sequence motifs containing phosphotyrosine. By this means, large signaling complexes assemble at the membrane in response to ligand binding. Components of these complexes include enzymes that generate second messengers such as inositol-1,4,5-trisphosphate (IP3) and phosphatidylinositol-3,4,5-trisphosphate (PIP3), factors that promote conversion of Ras family proteins between their active (GTP-bound) and inactive (GDP-bound) states, protein-tyrosine phosphatases, and additional protein-tyrosine kinases (see 2 below).

2. Some protein-tyrosine kinases, e.g. those in the Src family, are not transmembrane proteins. However, they associate with the inner surface of the plasma membrane via myristoylated N termini, and contain SH2 domains that cause them to associate with receptor-linked kinases after the latter are activated.

3. Mammalian serine/threonine-kinase receptors include those for transforming growth factor-, activins, and bone morphogenetic proteins. Each ligand has distinct type I and type II receptors (both of which are protein-serine/threonine kinases) and ligand binding causes formation of heterodimers. The type II receptor phosphorylates the type I receptor, activating the latter which then phosphorylates proteins of the SMAD family. The phosphorylated SMADs associate with other SMADs and translocate to the nucleus, where they trigger changes in gene expression [reviewed in (95)].

The plant receptor-like kinases (RLKs) are thought to be receptors mainly by virtue of their structural similarities to the mammalian receptor kinases, although a ligand has not yet been identified for any of them. However, the idea that they are receptors was strengthened with the recent cloning of the *BRI1* gene from *A. thaliana* (100). *Bri1* mutants are resistant to the brassinosteroid brassinolide, and map-based cloning of the *BRI1* gene showed that it encoded an RLK. This is consistent with the idea that BRI1 is a brassinosteroid receptor, although the true ligand may be a complex between the steroid and a binding protein.

Many of the 18 *A. thaliana* RLKs analyzed for this review fall into one of four subclasses when their extracellular domain sequences are analyzed (Figure 5: note that they do not cluster in exactly the same manner if the kinase domains are analyzed). These subclasses include:

1. The leucine-rich repeat (LRR) group [reviewed in (11, 161)], currently the largest group in *A. thaliana*. LRRs occur in numerous eukaryotic proteins (93), and are thought to be involved in mediating protein-protein interactions. Intriguingly, LRRs are present in several mammalian receptors for protein/peptide messengers, including some receptor-linked kinases such as the nerve growth factor receptor. LRR proteins contain multiple tandem copies of a circa 25-residue repeat with leucines at characteristic spacings. The crystal structure of ribonuclease inhibitor protein shows that its 16 LRRs form a horseshoe-shaped structure with a 16-strand parallel β-sheet on the inside and 16 short helices on the outside. It is proposed that the exposed surface of the β-sheet forms the site for protein-protein interactions (92).

2. The S-domain group, members of which have extracellular domains related to the S-locus glycoprotein (SLG) of *Brassica* species, involved in the self-incompatibility (SI) response (122). This response is discussed in more detail below, but it is worth noting that the three S-domain RLKs found in *A. thaliana* to date (Figure 5) cannot be involved in self-incompatibility because they are expressed in inappropriate locations, and this species does not display self-incompatibility. An S-domain is characterized by a characteristic array of cysteine residues, and other conserved motifs (161).

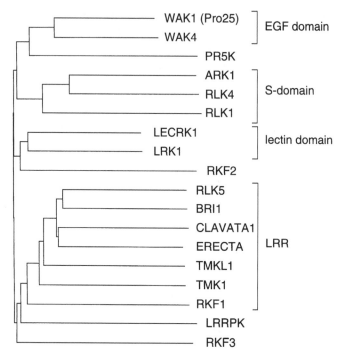

Figure 5 Phylogenetic tree of extracellular domains of RLKs, constructed using the same methods as for Figure 2. In each case, the extracellular domains were obtained by removing from the amino acid sequence the transmembrane domain and all regions C-terminal to it.

3. The lectin-like domain group, currently represented in *A. thaliana* by LECRK1 and LRK1. The extracellular domain of LECRK1 is related to legume lectins and to the Lec2 protein of *A. thaliana* (59). This relationship raises the intriguing possibility that these receptors bind oligosaccharides, such as the elicitors derived from breakdown of the cell wall (either of the host or of the pathogen) during fungal infection.

4. The EGF repeat receptors, represented in *A. thaliana* by WAK1 [wall-associated protein kinase-1 (57), formerly called Pro25 (94)], and WAK4. These have extracellular domains containing sequence repeats related to mammalian epidermal growth factor. The significance of this remains unclear.

Other *A. thaliana* RLKs that cannot readily be placed into any of these groups include PR5K and LRRPK (Figure 5). PR5K (165) has an extracellular domain related to the *A. thaliana* PR5 protein, which accumulates in the extracellular space in response to infection by various pathogens. Unlike PR5, PR5K is not induced upon infection, but their sequence relationship raises the possibility that PR5K might be a receptor involved in the response to infection. LRRPK (26) has an extracellular domain with a novel leucine zipper motif. It has the interesting property that levels of its mRNA are repressed by light, but its function remains unclear.

The extracellular domains of many RLKs are related to other proteins and these relationships can, as in a few cases discussed above, give tantalizing clues as to their possible function. However, much more direct evidence regarding function is available for those cases where the DNA was obtained by cloning of mutant genes. The *BRI1* gene (100), encoding an LRR kinase that may represent a brassinosteroid receptor, has already been mentioned. Two other members of the LRR subfamily in *A. thaliana*, i.e. *ERECTA* and *CLAVATA1 (CLV1)*, give rise to developmental mutants. *ERECTA* is expressed at and around apical meristems, and mutations lead to plants with compact inflorescences with short, thick peduncles and fruits. It is proposed that the Erecta protein may be involved in cell-cell signaling during meristem development (155). *CLV1* is expressed in a patch of cells in the center of the apical meristem, and mutations lead to greatly enlarged shoot and floral meristems with large clusters of undifferentiated cells at the center. These mutations therefore appear to upset the balance between cell proliferation and differentiation. Signaling through CLAVATA1 may inhibit proliferation of stem cells, promote their differentiation into organ primordia, or both (15).

A particularly interesting group of plant kinases has been defined by cloning genes responsible for disease resistance. The *Xa21* gene in rice is responsible for resistance to some strains of the bacterial pathogen *Xanthomonas oryzae*, and encodes an RLK with an LRR-type extracellular domain (147). This suggests that it is a receptor for some product present during infection, although the actual ligand is not known. In tomato, the *Pto* gene confers resistance to strains of *Pseudomonas syringae* carrying the corresponding virulence gene, *avrPTO*, triggering a localized *hypersensitive response* involving programmed cell death at the site of infection. The closely related (and tightly linked) *Fen* gene confers a hypersensitive reponse to the insecticide fenthion, which may perhaps be mimicking a product of another pathogen. *Pto* and *Fen* encode protein kinases that are 80% identical in sequence (103, 108). Outside of their kinase domains they only have short N-terminal extensions (with potential myristoylation sites), and have no potential extracellular or transmembrane domains. Nevertheless, the sequences of their kinase domains are related to those of RLKs. To confer pathogen resistance and fenthion sensitivity, respectively, *Pto* and *Fen* require another gene present in the same gene cluster, *Prf*, that encodes a protein containing leucine-rich repeats (139). It seems likely that complexes between

Pto and Prf, and Fen and Prf, form the receptors for products of infection (7). However, Prf does not have a signal sequence, and if this model is correct the receptors may be entirely intracellular. DNA encoding another protein kinase (Pti1) related to Pto has been cloned from tomato by yeast two-hybrid analysis, using Pto as the "bait." Pto phosphorylates Pti1 in vitro (but not vice versa), and overexpression of Pti1 enhances the hypersensitive response to bacterial strains carrying *avrPTO* (178). This suggests that Pti1 may be part of a protein kinase cascade acting downstream of Pto.

A final instance where functions of plant RLKs have been studied by genetic approaches is the phenomenon of self-incompatibility (SI) in *Brassicaceae*. SI prevents pollen grains from genetically similar plants from germinating when they fall on the stigma. In the type found in *Brassicaceae*, SI is controlled by the *S* locus complex, a highly polymorphic gene cluster that occurs in a number of distinct haplotypes (122). Two of the genes encoded at the *S* locus are the *S* locus glycoprotein (SLG) and the S-locus receptor-like kinase (SRK), which has an extracellular domain closely related to the SLG present in the same haplotype (149). SRK was the original representative of the S-domain RLK subfamily (see above). Both SLG and SRK are coordinately expressed in the stigma (148). A current hypothesis (148) is that SRK (possibly in combination with SLG) forms a receptor for some component of the pollen grain specified by the S-locus. A match between the pollen component and SLG/SRK causes a response that prevents the pollen grain from germinating.

At present, little is known about the signal transduction pathways downstream of RLKs. However, the kinase domains of most RLKs autophosphorylate (161), and one possibility is that this creates docking sites for components of signaling complexes, by analogy with the mechanism of signaling from mammalian protein-tyrosine kinase-linked receptors (see above). Stone et al (150) screened an *A. thaliana* expression library for proteins that interacted with the autophosphorylated kinase domain of RLK5, a member of the LRR subfamily of RLKs. This resulted in cloning of DNA encoding a protein called KAPP (kinase-associated protein phosphatase), which contains a type I membrane anchoring domain, a kinase interaction domain (responsible for the interaction with RLK5), and a domain related to the protein-serine/threonine phosphatase, PP2C. KAPP also appears to interact with the phosphorylated kinase domains of some other RLKs (10), and evidence from transgenic plants expressing KAPP suggests that it acts as a negative regulator of the CLAVATA1 signal transduction pathway (151, 167). A protein unrelated to KAPP (ARC1) that interacts with the phosphorylated kinase domain of the *Brassica* S-locus receptor-like kinase was recently cloned in the yeast two-hybrid system (43). It was expressed specifically in the stigma, and was of interest because it contained "arm" repeats of the type found in *Drosophila melanogaster* armadillo and vertebratecatenins. These latter proteins are involved in the Wnt/Wingless signaling pathway (71).

THE MAP KINASE (MAPK), MAP KINASE KINASE (MAPKK), AND MAP KINASE KINASE KINASE (MAPKKK) SUBFAMILIES

Although these fall into different subfamilies when their kinase domain sequences are compared (Figure 2), it is convenient to consider them together. In yeast and mammalian cells they form cascades or "modules" operating in the direction MAPKKK → MAPKK → MAPK. The *S. cerevisiae* genome encodes six MAPKs, with the two-best characterized MAPK modules acting downstream of mating factor receptors (Ste11 → Ste7 → Fus3) and sensors of high osmolarity (Ssk2/22 → Pbs2 → Hog1) (105). In mammalian cells the classical MAPK cascade (Raf1 → MEK1/2 → ERK1/2) lies downstream of growth factor receptors that act via the GTP-binding protein Ras. At least two other MAPK cascades (MEKK1/2 → JNKK1 → JNK, ASK1 → MKK3/6 → p38) are also activated by cytokines and various cellular stresses (112). DNAs encoding at least seven MAPKs (114, 116), five MAPKKs [(121) and unpublished sequences in the database], and five MAPKKKs (118, 124) have been cloned from *A. thaliana* (Figure 2; See Supplementary Materials Section, http://www.annualreviews.org). Representatives have also been found in other plant species including alfalfa, tobacco, oat, and petunia [reviewed in (63)]. All of the MAPKs contain the characteristic MAPK motif Thr-X-Tyr in the "T-loop" or "activation segment" (79) between conserved kinase motifs VII and VIII. In the yeast and mammalian systems, phosphorylation at both the threonine and the tyrosine, catalyzed by a single MAPKK, is necessary to cause activation. The serine or threonine residues on MAPKKs, which are phosphorylated during activation by MAPKKKs, are also conserved in the plant MAPKKs (121, 145). There have been few direct biochemical studies of whether these kinases really exist as cascades in plants, although two *A. thaliana* MAPKs are phosphorylated and activated by a *Xenopus laevis* MAPKK in vitro (114). Overexpression of plant DNAs has also been shown to suppress mutations in specific components of MAP kinase cascades in *S. cerevisiae* (4, 82, 118, 124, 130). These results indicate that these plant kinases can take part in specific MAP kinase cascades at the appropriate level when expressed in yeast.

Although no plant MAPK cascade has yet been mapped out in fine detail, there is abundant evidence that they are rapidly activated by plant hormones and by environmental stimuli such as touch, cold shock, wounding, and pathogen infection. The almost universal method by which MAPK activation in plants has been addressed is the "in-gel" kinase assay, in which an artificial MAPK substrate (myelin basic protein, MBP) is cast into a gel, extracts are separated on the gel by electrophoresis in SDS, and after renaturation the gel is incubated with $[\gamma^{32}P]ATP/Mg$ to locate active MBP kinases. MAPKs are identified as MBP kinases migrating at around 45 kDa. Added refinements are to show that activation is associated with tyrosine phosphorylation of the polypeptide using an anti-phosphotyrosine antibody, and that activation is abolished by prior incubation with protein-tyrosine phosphatases. Ideally, the specific MAPK involved should also be identified by immunoprecipitation with isoform-specific antibodies; as yet this has only been done in a few cases.

Using this approach, a MAPK (and an activity which phosphorylated a recombinant MAPK, i.e. a MAPKK) were shown to be activated by auxin within five minutes of addition to cultured tobacco cells (114), and a MAPK was activated by abscisic acid (ABA) within one minute of addition to barley aleurone protoplasts (91). In alfalfa leaves, a MAPK (identified as MMK4 by immunoprecipitation) was activated by cold shock, water stress, and wounding; these effects were not mediated by ABA (8, 81). MAPKs were also activated (and induced at the transcriptional level) within one minute of wounding of tobacco leaves (142, 159). Surprisingly, overexpression of DNA encoding a wound-induced MAPK (*WIPK*) reduced the basal kinase activity and abolished its activation by wounding, apparently a "gene silencing" effect. In these transgenic plants, wounding did not lead to expression of wound-induced genes or accumulation of jasmonic acid (a phytohormone normally associated with the wounding response) but led instead to accumulation of salicylic acid (a messenger normally associated with the response to pathogen attack) (142). These results indicate that WIPK is involved in the activation of jasmonic acid synthesis by wounding. A MAPK called SIP kinase was shown to be activated by salicylic acid in tobacco (175). In tomato leaves, a MAPK was activated by wounding, during grazing by insect larvae, by fungal elicitors of defense responses, or by treatment with systemin, a peptide messenger believed to be a primary signal in the wounding response (152). In parsley cells, peptides derived from the fungal pathogen *Phytophthora sojae* act as elicitors of defense responses, inducing activation of defense-related genes and production of reactive oxygen species. These peptides activated a MAPK (ERMK) related to alfalfa MMK4, and caused its translocation to the nucleus, a phenomenon observed after activation of mammalian MAPKs. The use of pharmacological probes suggested that the MAPK activation was downstream of changes in ion flux produced by the elicitors, but upstream of the oxidative burst that is part of the defense response (101). In summary, MAPKs appear to be involved at various levels in plant responses to hormonal and environmental stimuli. The challenge now is determine which MAPK, MAPKK and MAPKKKs are involved in each response, and to see how they link to the putative receptor or sensor protein lying upstream that monitors the presence of the stimuli.

In most of the cases listed above, MAPK activation was transient, with the kinase being rapidly inactivated despite the continuing presence of the stimulus. Addition of protein synthesis inhibitors did not affect activation, but did block subsequent inactivation [e.g. (8)]. Similar behavior of MAPK cascades in mammals can be explained by induction of transcription of specific dual-specificity MAPK phosphatases (86). Hirt's group recently cloned, by suppression of the mating factor-induced cell cycle arrest in *S. cerevisiae*, an alfalfa DNA (*MP2C*) encoding a protein phosphatase (110). Although this enzyme was a member of the protein phosphatase-2C family rather than the dual-specificity phosphatase family, mRNA encoding it was induced after wounding of alfalfa leaves, and the bacterially expressed protein could inactivate in vitro the MAPK which was activated by wounding. Its induction upon wounding may therefore be responsible for the transient nature of the MAPK response.

THE CYCLIN-DEPENDENT KINASE (CDK) SUBFAMILY

The yeast and mammalian cyclin-dependent kinases (CDKs), exemplified by the *cdc2* and *CDC28* gene products from *Schizosaccharomyces pombe* and *S. cerevisiae*, respectively, have catalytic subunits of about 34 kDa that are minimal kinase domains. The CDKs are characterized by being inactive except when bound to a *cyclin* protein by interactions involving the conserved PSTAIRE motif on the CDK (76). Some CDKs are critical regulators of entry into key phases of the cell cycle, such as S phase and mitosis. Other, such as Pho85 in *S. cerevisae*, appear to have roles not connected with the cell cycle (83). In general, there are more cyclins than there are CDKs, with a single CDK often having multiple roles, depending on the cyclins with which it associates. The activity of CDKs during the cell cycle is regulated by association with cyclins that are synthesized and degraded at different times, by association with CDK inhibitor proteins whose content and/or activity varies during the cycle, and by phosphorylation and dephosphorylation by upstream kinases and phosphatases. An activating phosphorylation event occurs at a conserved threonine within the "T loop," catalyzed by CDK-activating kinases, and an inactivating phosphorylation occurs at a tyrosine residue within the sequence GXGXYG (conserved kinase motif I), catalyzed in *S. pombe* by the Wee1 kinase.

Despite differences in the mechanics of cell division between fungi, animals, and plants, the basic cell cycle control machinery appears to be conserved. In 1990, Feiler & Jacobs (35) described cloning of a *cdc2*-like gene, and they also found a Cdc2-like protein kinase activity, in pea. Two *cdc2*-like genes, termed

CDC2a and *CDC2b*, have been cloned from *A. thaliana* (36, 61). *CDC2a*, which has an exact match to the PSTAIRE motif, partially complements *cdc2-* mutations in *S. pombe*, whereas *CDC2b*, which has a PPTALRE motif instead, does not. In general, PSTAIRE variants are broadly expressed throughout the plant cell cycle, whereas PPTALRE variants are expressed in a narrower window from S phase through to M phase (38, 107). It has been proposed that the PPTALRE variants might be involved in formation of the pre-prophase band (30). Recently, a CDK-activating kinase has also been cloned from *A. thaliana* by complementation of an *S. cerevisiae cak1* mutant. When purified by immunoprecipitation, it phosphorylated and activated a human CDK (158). A fourth CDK-like kinase from *A. thaliana* is MAK homologous kinase (MHK), related to the mammalian male germ cell-associated kinase MAK (120). The latter is expressed specifically in testicular germ cells and may have a role in meiosis (77), but the functions of MHK and MAK are not well understood.

DNAs encoding at least ten A/B-type mitotic cyclins, plus three D-type G1 cyclins, have been cloned from *A. thaliana* (23, 132). Homologs of several important regulatory proteins that interact with CDKs have also been reported in plants. These include a CDK inhibitor related to p27^{Kip1} (163), a CDK-binding protein related to *S. pombe* Suc1 (24), and a homolog of a key target for phosphorylation by CDKs in G1 phase, the retinoblastoma protein (41, 170). For a detailed discussion of the plant cell cycle, readers are referred to previous reviews (30, 143). However, while there are many indications that CDKs are involved in regulating the cell cycle in plants, the evidence is often only correlative in nature. One exception is a paper by Hemerly et al (58) examining the effects of expression of a dominant negative mutant of *A. thaliana CDC2a*. When expressed in *S. pombe*, it produced abnormally large cells similar to those of a *cdc2*-mutant. When expressed in *A. thaliana* plants this mutant appeared to be lethal, but when expressed in tobacco it gave rise to transgenic lines with smaller numbers of abnormally large cells when compared with controls, although overall development was in many respects normal. Leaf mesophyll protoplasts derived from these plants did not undergo any cell divisions in vitro, unlike cells from control plants (58). These results provide direct evidence that a normal level of Cdc2 kinase is necessary to drive plant cells through the cell cycle. Another interesting paper (109) showed that expression of *S. pombe cdc25* (encoding the phosphotyrosine phosphatase that reverses the inactivating phosphorylation on CDKs) gave rise to transgenic tobacco plants with more frequent, smaller lateral roots that had smaller cells at mitosis. This suggests that the inactivation of CDKs by tyrosine phosphorylation can cause them to become rate-limiting for mitosis. These results are consistent with observations that *Nicotiana plumbaginifola* cell cultures deprived of cytokinins are arrested in late G2 phase with an inactive, tyrosine phosphorylated, Cdc2-like histone H1 kinase (173). The kinase could be reactivated by *S. pombe* Cdc25 in vitro.

THE CASEIN KINASE I (CK1) AND CASEIN KINASE II (CK2) SUBFAMILIES

These two subfamilies are considered together, although they are not closely related (Figure 2). Casein kinases I and II are so named because they were originally detected using the milk protein casein as substrate. Both phosphorylate serine or threonine residues in the context of nearby acidic side chains. An interesting feature of this specificity is that phosphoamino acids can substitute for acidic residues, so that other protein kinases can sometimes "prime" phosphorylation by CK1 and CK2. For CK1 the acidic residues are preferred on the N-terminal side of the serine/threonine, with the P-3 position being particularly important, whereas for CK2 they are preferred on the C-terminal side, with the P+3 position being important [P-3/P+3 represents the position 3 residues N-terminal/C-terminal with respect to the phosphorylated amino acid]. A perplexing feature of both CK1 and CK2, which means that their functions remain somewhat enigmatic, is that they do not appear to be highly regulated in vivo. Conceivably, they provide a constitutive, basal kinase activity against which protein phosphatases (and kinases phosphorylating neighboring priming sites) could provide regulation of target protein function.

Mammalian CK1s exist as monomers that are widely expressed and found in both the cytoplasm and nucleus (16). They phosphorylate many substrate proteins in vitro; often this does not produce a change in function of the target, although an interesting exception is glycogen synthase, where phosphorylation at Ser-7 by cyclic AMP-dependent protein kinase primes the phosphorylation and inactivation at Ser-10 by CK1 (37). In *S. cerevisiae*, deletion of a redundant pair of CK1 genes (*YCK1 and YCK2*, also known as *CK11/CK12*) is lethal (134, 164), while disruption of a third (*HRR25*) leads to hypersensitivity to DNA-damaging agents, suggesting a function in DNA repair (27). DNAs encoding five CK1s have been cloned from *A. thaliana* [(90, 113) and GENBANK:U97568]. A CK1 has also been purified to homogeneity from broccoli (88). The biochemical properties of this, and of recombinant *A. thaliana CKI1* expressed in bacteria (90), are similar to those of mammalian CK1s. As yet there are few insights into the functions of CK1s in plant cells.

An unusual feature of mammalian CK2s is that they will utilize GTP as well as ATP and are relatively insensitive to the general kinase inhibitor staurosporine. They normally exists as $\alpha_2\beta_2$ tetramers, where the α subunits are catalytic. The β subunits are not essential for activity, but modify activity towards certain substrates. CK2s phosphorylate a bewildering array of substrate proteins in vitro, and in some cases, these targets are phosphorylated in vivo. Particularly interesting targets include DNA-binding proteins such as p53 (85) and DNA topoisomerase II (73). Disruption of both of the CK2 genes (*CKA1/2*) in *S. cerevisiae* is lethal, but experiments with temperature-sensitive mutants suggest a role for *CKA1* in the maintenance of cell polarity (133), and for *CKA2* in cell cycle progression (47). DNAs encoding two isoforms of the CK2 catalytic (α) subunit (*CKA1* and *CKA2*) have been cloned from *A. thaliana* (119), as have two forms of the β subunit (*CKB1* and *CKB2*) (17). Active $\alpha_2\beta_2$ tetramers can be reconstituted from recombinant CKA1 and CKB1 (89). CK2 has also been purified from *A. thaliana* in two forms, i.e. a monomeric catalytic subunit and a tetrameric $\alpha_2\beta_2$ form (34). Plant CK2s have been shown to phosphorylate a wide variety of plant proteins in vitro. In most cases, it is not clear whether this affects the function of the target protein, but phosphorylation of the transcription factor GBF1 stimulates its binding to DNA containing a "G box" (89). The catalytic subunit of CK2 from maize occupies a special place in that it recently became the first plant protein kinase for which a crystal structure is available (123). The structure confirmed the location of basic residues within the active site that bind the acidic residues around the target serine, and also provided clues as to why CK2s can utilize GTP as well as ATP.

THE GSK3/SHAGGY SUBFAMILY

Mammalian glycogen synthase kinase-3 (GSK3) was originally described as one of several protein kinases phosphorylating and inactivating rabbit muscle glycogen synthase. When DNA encoding it was cloned, it turned out to be closely related to the *shaggy/zeste-white 3* gene product from *Drosophila melanogaster* (169). In mammals, GSK3 occurs in a signal transduction pathway downstream of the insulin and insulin-like growth factor-1 receptors. It is unusual in that it is inactivated (rather than activated) by stimulation of the pathway, via phosphorylation by protein kinase B (also known as Akt) at a serine residue near the N terminus (20). Since GSK3 inactivates glycogen synthase this accounts, at least in part, for the observed activation of glycogen synthesis by insulin. In *Drosophila* the *shaggy/zeste-white 3* gene is required for the establishment of segment polarity. It is now known that the GSK3/shaggy protein kinase lies on the pathway downstream of Wingless, a secreted protein messenger involved in the control of development (71). A related pathway is found in mammalian cells, where the equivalent of Wingless is the Wnt1 protein. Stimulation of mammalian fibroblasts with *Drosophila* Wingless results in inactivation of GSK3 due to phosphorylation; the signal transduction pathway is not known but it is clearly distinct from that utilized by insulin (18).

Figure 2 reveals seven GSK3/*shaggy* homologs in *A. thaliana*, although if partial sequences are included there are at least ten (31). There are also multiple representatives in rice, tobacco, petunia, and alfalfa [reviewed in (31)]. Because of their sequence similarity with *Drosophila shaggy*, they have been assumed to play roles during plant development. However, although different isoforms show tissue-specific expression patterns (25, 33, 80, 127), there is currently little evidence available that provides any real insight into the functions of the plant GSK3 family.

OTHER SUBFAMILIES AND MISCELLANEOUS KINASES

In this section I review the remaining cloned kinases from *A. thaliana* that either form only small subfamilies or do not readily fit into any of the other groupings.

The CTR1/Raf-Like Subfamily

The *CTR1* (*constitutive triple response-1*) gene was identified via loss-of-function mutations that displayed a "triple response" in the absence of the hormone ethylene. Cloning of the mutant gene revealed that it encoded a protein kinase related to the mammalian Raf subfamily (87). The archetypal Raf family member, i.e. Raf-1, couples the GTP-binding protein Ras to the MAP kinase (ERK1/ERK2) cascade, and acts at the level of a MAPKKK (although not closely related to other MAPKKKs). It was therefore proposed (87) that CTR1 lies upstream of a MAPK cascade, although direct evidence for this is still lacking. The proposed receptors for ethylene, the products of the genes *ETR1* (13), *ETR2* (138), and *ERS* (66), all of which give rise to ethylene-insensitive mutants, are members of the prokaryotic two-component histidine/aspartate kinase family. Since loss of *CTR1* function gives rise to a constitutive ethylene-like response, the CTR1 pathway must normally switch off ethylene responses, and be inactivated by binding of ethylene to its receptor(s). The N-terminal regulatory domain of CTR1 has recently been shown to interact with the histidine kinase domains of ETR1 and ERS by two-hybrid analysis (14). Two protein kinases of unknown function that appear to cluster into the same group as CTR1 according to their kinase domain sequences are ATN1 (157) and ATMRK1 (GENBANK:AB006810).

The LAMMER Subfamily

Bender & Fink (6) cloned *A. thaliana* DNA (*AFC1*) by virtue of its ability to complement a double mutation (*fus3 kss1*) in genes encoding MAPKs that transmit the effect of mating factors to expression of mating-specific genes in yeast. AFC1 was not a MAPK but a member of the CDK-related *LAMMER* family (172), members of which have the sequence *EHLAMMERIL* at kinase conserved motif XI. They also cloned two other *LAMMER* genes, *AFC2* and *AFC3*; unlike *AFC1, AFC2* did not complement the *fus3 kss1* mutant. The functions of the *LAMMER* subfamily remain unclear.

S6 Kinase Homologs

DNAs encoding two kinases related to mammalian p70 ribosomal protein S6 kinase have been cloned (117, 174). These were induced at the transcriptional level by cold or salt stress (117). Zhang et al (174) found that one of the kinases (PK1) was active after expression in insect (Sf9) cells but not in bacteria, probably because it requires posttranslational modifications, including phosphorylation. It phosphorylated two proteins in *A. thaliana* ribosomes in vitro, but its physiological role remains unknown.

The PVPK-1 Subfamily

Both PK64 and ATPK5 (56, 115) appear to be homologs of PVPK-1 from *Phaseolus vulgaris* and G11A from rice. Although the latter two were the first plant protein kinases to be cloned (97), the functions of the members of this subfamily remain unknown.

NAK Subfamily

The original representative of this subfamily was NAK (120), but it also includes APK1 (62), APK2a (74), ARSK1 (72), and GENBANK:U53501. Interesting features of the NAK and APK1 sequences are that they bear some resemblance to mammalian protein-tyrosine as well as protein-serine/threonine kinases. When APK1 was expressed in bacteria it caused bacterial proteins to become reactive with anti-phosphotyrosine antibodies. However, the expressed kinase autophosphorylated, and phosphorylated exogenous proteins such as casein and myosin light chain, on serine and threonine and not tyrosine (62). It therefore does not appear to be a true protein-tyrosine kinase.

Miscellaneous Kinases

Mutations in the *Tousled* gene give rise to plants that exhibit random loss of floral organs, and organ development is impaired. The gene encodes a protein kinase (136) with an N-terminal domain that contains a coiled-coil region responsible for oligomerization, and a localization signal that directs the protein to the nucleus (135). Plants with mutations in the *NPH1* gene, which appears to encode a soluble protein kinase, lack all phototropic responses. The N-terminal domain contains two repeated sequences that are related to sequences found in bacterial proteins regulated by redox status (67). Finally, the PPK1 kinase

(GENBANK:Y11930) from *A. thaliana* is most closely related to the human TTK (threonine/tyrosine kinase) (111). Little is known about its function.

CONCLUSIONS AND PERSPECTIVES

Although the past decade has indeed been a golden age for the discovery of plant protein serine/threonine kinases, the elucidation of their functions is only just beginning. The advent of cloning by homology has meant that the rate of accumulation of knowledge about the number and sequence of plant kinase genes has greatly outstripped the rate at which we have learned about their physiological roles. Even for the plant kinases identified by cloning of mutant genes, there is still a great deal to be understood. Knowledge of the phenotype of a mutation does not mean that one fully understands the function of the encoded protein; the phenotype only reveals one defect produced by the mutation under the particular circumstances of the genetic screen. In most cases, we also know about only one component in any plant signal transduction chain, and we do not understand the full sequence of events. In the next decade, we will need to explore upstream and downstream of the known kinases in order to identify their regulators and target proteins respectively. In my view, this will require greater emphasis on biochemical analysis than has occurred in the past, although the yeast two-hybrid screen and interaction cloning (150) are useful approaches to obtain initial clues as to upstream and downstream partners. Whatever approach is adopted, this field of plant research will remain dynamic and exciting for the foreseeable future, and many surprises and new twists undoubtedly await us.

ACKNOWLEDGMENTS

Studies on plant protein kinases in this laboratory have been supported by the UK Biotechnology and Biological Sciences Research Council (BBSRC). This work benefited from the use of the SEQNET facility of the BBSRC, Daresbury, UK.

Visit the *Annual Reviews home page* at
http://www.AnnualReviews.org

Literature Cited

1. Bachmann M, Huber JL, Athwal GS, Wu K, Ferl RJ, Huber SC. 1996. 14-3-3 proteins associate with the regulatory phosphorylation site of spinach leaf nitrate reductase in an isoform-specific manner and reduce dephosphorylation of ser-543 by endogenous protein phosphatases. *FEBS Lett.* 398:26–30

2. Ball KL, Barker J, Halford NG, Hardie DG. 1995. Immunological evidence that HMG-CoA reductase kinase-A is the cauliflower homologue of the Rkin1 subfamily of plant protein kinases. *FEBS Lett.* 377:189–92

3. Ball KL, Dale S, Weekes J, Hardie DG. 1994. Biochemical characterization

of two forms of 3-hydroxy-3-methyl-glutaryl-CoA reductase kinase from cauliflower (*Brassica oleracea*). *Eur. J. Biochem.* 219:743–50

4. Banno H, Hirano K, Nakamura T, Irie K, Nomoto S, et al. 1993. *NPK1*, a tobacco gene that encodes a protein with a domain homologous to yeast BCK1, STE11, and Byr2 protein kinases. *Mol. Cell. Biol.* 13:4745–52

5. Barker JHA, Slocombe SP, Ball KL, Hardie DG, Shewry PR, Halford NG. 1996. Evidence that barley HMG-CoA reductase kinase is a member of the SNF1-related protein kinase family. *Plant Physiol.* 112:1141–49

6. Bender J, Fink GR. 1994. AFC1, a LAMMER kinase from *Arabidopsis thaliana*, activates *STE12*–dependent processes in yeast. *Proc. Natl. Acad. Sci. USA* 91: 12105–9

7. Bent AF. 1996. Plant disease resistance genes—function meets structure. *Plant Cell* 8:1757–71

8. Bogre L, Ligterink W, Meskiene I, Barker PJ, Heberlebors E, et al. 1997. Wounding induces the rapid and transient activation of a specific MAP kinase pathway. *Plant Cell* 9:75–83

9. Braun DM, Garcia XU, Stone JM. 1996. Protein phosphorylation: examining the plant CPU. *Trends Plant Sci.* 1:289–91

10. Braun DM, Stone JM, Walker JC. 1997. Interaction of the maize and Arabidopsis kinase interaction domains with a subset of receptor-like protein kinases: implications for transmembrane signaling in plants. *Plant J.* 12:83–95

11. Braun DM, Walker JC. 1996. Plant transmembrane receptors—new pieces in the signaling puzzle. *Trends Biochem. Sci.* 21:70–73

12. Bush DS. 1995. Calcium regulation in plant cells and its role in signaling. *Annu. Rev. Plant Physiol. Plant Mol. Biol.* 46:95–122

13. Chang C, Meyerowitz EM. 1995. The ethylene hormone response in *Arabidopsis*—a eukaryotic 2-component signaling system. *Proc. Natl. Acad. Sci. USA* 92: 4129–33

14. Clark KL, Larsen PB, Wang XX, Chang C. 1998. Association of the Arabidopsis CTR1 Raf-like kinase with the ETR1 and ERS ethylene receptors. *Proc. Natl. Acad. Sci. USA* 95:5401–6

15. Clark SE, Williams RW, Meyerowitz EM. 1997. The *CLAVATA1* gene encodes a putative receptor kinase that controls shoot and floral meristem size in *Arabidopsis*. *Cell* 89:575–85

16. Cobb MH. 1995. Casein kinase 1 (vertebrates). In *The Protein Kinase Factsbook*, ed. DG Hardie, SK Hanks, pp. 1:347–49. London: Academic

17. Collinge MA, Walker JC. 1994. Isolation of an *Arabidopsis thaliana* casein kinase II β subunit by complementation in *Saccharomyces cerevisiae*. *Plant Mol. Biol.* 25:649–58

18. Cook D, Fry MJ, Hughes K, Sumathipala R, Woodgett JR, Dale TC. 1996. Wingless inactivates glycogen synthase kinase-3 via an intracellular signalling pathway which involves a protein kinase C. *EMBO J.* 15:4526–36

19. Corton JM, Gillespie JG, Hardie DG. 1994. Role of the AMP-activated protein kinase in the cellular stress response. *Curr. Biol.* 4:315–24

20. Cross DA, Alessi DR, Cohen P, Andjelkovich M, Hemmings BA. 1995. Inhibition of glycogen synthase kinase-3 by insulin mediated by protein kinase B. *Nature* 378:785–89

21. Dale S, Arró M, Becerra B, Morrice NG, Boronat A, et al. 1995. Bacterial expression of the catalytic domain of 3-hydroxy-3-methylglutaryl CoA reductase (isoform HMGR1) from *Arabidopsis thaliana*, and its inactivation by phosphorylation at serine-577 by *Brassica oleracea* 3-hydroxy-3-methylglutaryl CoA reductase kinase. *Eur. J. Biochem.* 233:506–13

22. Dale S, Wilson WA, Edelman AM, Hardie DG. 1995. Similar substrate recognition motifs for mammalian AMP-activated protein kinase, higher plant HMG-CoA reductase kinase-A, yeast SNF1, and mammalian calmodulin-dependent protein kinase I. *FEBS Lett.* 361:191–95

23. Day IS, Reddy ASN. 1998. Isolation and characterization of two cyclin-like cDNAs from Arabidopsis. *Plant Mol. Biol.* 36:451–61

24. de Veylder L, Segers G, Glab N, Casteels P, Van Montagu M, Inze D. 1997. The Arabidopsis Cks1At protein binds the cyclin-dependent kinases Cdc2aAt and Cdc2bAt. *FEBS Lett.* 412:446–52

25. Decroocqferrant V, Vanwent J, Bianchi MW, Devries SC, Kreis M. 1995. *Petunia hybrida* homologs of shaggy/zeste-white-3 expressed in female and male reproductive organs. *Plant J.* 7:897–911

26. Deeken R, Kaldenhoff R. 1997. Light-repressible receptor protein kinase: a novel photo-regulated gene from *Arabidopsis thaliana*. *Planta* 202:479–86

27. DeMaggio AJ, Lindberg RA, Hunter T, Hoekstra MF. 1992. The budding yeast

HRR25 gene product is a casein kinase I isoform. *Proc. Natl. Acad. Sci. USA* 89:7008–12

28. Devereux J, Haeberli P, Smithies O. 1984. A comprehensive set of sequence analysis programs for the VAX. *Nucleic Acids Res.* 12:387–95

29. Deleted in proof

30. Doonan J, Fobert P. 1997. Conserved and novel regulators of the plant cell cycle. *Curr. Opin. Cell Biol.* 9:824–30

31. Dornelas MC, Lejeune B, Dron M, Kreis M. 1998. The *Arabidopsis* SHAGGY-related protein kinase (ASK) gene family: structure, organization and evolution. *Gene* 212:249–57

32. Douglas P, Moorhead G, Hong Y, Morrice N, MacKintosh C. 1998. Purification of a nitrate reductase kinase from *Spinacia oleracea* leaves, and its identification as a calmodulin-domain protein kinase. *Planta* 206:435–42

33. Einzenberger E, Eller N, Heberlebors E, Vicente O. 1995. Isolation and expression during pollen development of a tobacco cDNA clone encoding a protein kinase homologous to shaggy glycogen synthase kinase-3. *Biochim. Biophys. Acta* 1260:315–19

34. Espunya MC, Martinez MC. 1997. Identification of two different molecular forms of *Arabidopsis thaliana* casein kinase II. *Plant Sci.* 124:131–42

35. Feiler HS, Jacobs TW. 1990. Cell division in higher plants: a cdc2 gene, its 34 kDa product, and histone H1 kinase activity in pea. *Proc. Natl. Acad. Sci. USA* 87:5397–401

36. Ferreira PCG, Hemerly AS, Villarroel R, Vanmontagu M, Inze D. 1991. The *Arabidopsis* functional homolog of the p34cdc2 protein kinase. *Plant Cell* 3:531–40

37. Flotow H, Roach PJ. 1989. Synergistic phosphorylation of rabbit muscle glycogen synthase by cyclic AMP-dependent protein kinase and casein kinase I. Implications for hormonal regulation of glycogen synthase. *J. Biol. Chem.* 264:9126–9128

38. Fobert PR, Gaudin V, Lunness P, Coen ES, Doonan JH. 1996. Distinct classes of cdc2-related genes are differentially expressed during the cell-division cycle in plants. *Plant Cell* 8:1465–76

39. Furumoto T, Ogawa N, Hata S, Izui K. 1996. Plant calcium-dependent protein kinase-related kinases (CRKs) do not require calcium for their activities. *FEBS Lett.* 396:147–51

40. Goldberg J, Nairn AC, Kuriyan J. 1996.

Structural basis for the autoinhibition of calcium/calmodulin-dependent protein kinase I. *Cell* 84:875–87

41. Grafi G, Burnett RJ, Helentjaris T, Larkins BA, Decaprio JA, et al. 1996. A maize cDNA-encoding a member of the retinoblastoma protein family—involvement in endoreduplication. *Proc. Natl. Acad. Sci. USA* 93:8962–67

42. Graham IA, Denby KJ, Leaver CJ. 1994. Carbon catabolite repression regulates glyoxylase cycle gene expression in cucumber. *Plant Cell* 6:761–72

43. Gu TS, Mazzurco M, Sulaman W, Matias DD, Goring DR. 1998. Binding of an arm repeat protein to the kinase domain of the S-locus receptor kinase. *Proc. Natl. Acad. Sci. USA* 95:382–87

44. Halford NG, Hardie DG. 1998. SNF1-related protein kinases: global regulators of carbon metabolism in plants? *Plant Mol. Biol.* 37:735–48

45. Hanks SK. 1987. Homology probing: identification of cDNA clones encoding members of the protein-serine kinase family. *Proc. Natl. Acad. Sci. USA* 84:388–92

46. Hanks SK, Hunter T. 1995. The eukaryotic protein kinase superfamily: kinase (catalytic) domain structure and classification. *FASEB J.* 9:576–96

47. Hanna DE, Rethinaswamy A, Glover CV. 1995. Casein kinase II is required for cell cycle progression during G1 and G2/M in *Saccharomyces cerevisiae*. *J. Biol. Chem.* 270:25905–14

48. Hardie DG. 1988. Pseudosubstrates turn off protein kinases. *Nature* 335:592–93

49. Hardie DG, Carling D, Carlson M. 1998. The AMP-activated/SNF1 protein kinase subfamily: metabolic sensors of the eukaryotic cell? *Annu. Rev. Biochem.* 67:821–55

50. Harmon AC, Putnam-Evans C, Cormier MJ. 1987. A calcium-dependent but calmodulin-independent protein kinase from soybean. *Plant Physiol.* 83:830–837

51. Harmon AC, Yoo BC, McCaffery C. 1994. Pseudosubstrate inhibition of CDPK, a protein kinase with a calmodulin-like domain. *Biochemistry* 33:7278–87

52. Harper JF, Huang JF, Lloyd SJ. 1994. Genetic identification of an autoinhibitor in CDPK, a protein kinase with a calmodulin-like domain. *Biochemistry* 33:7267–77

53. Harper JF, Sussman MR, Schaller GE, Putnam-Evans C, Charbonneau H, Harmon AC. 1991. A calcium-dependent protein kinase with a regulatory domain

similar to calmodulin. *Science* 252:951–954

54. Hartwell J, Smith LH, Wilkins MB, Jenkins GI, Nimmo HG. 1996. Higher plant phosphoenolpyruvate carboxylase kinase is regulated at the level of translatable mRNA in response to light or a circadian rhythm. *Plant J.* 10:1071–78

55. Hawley SA, Davison M, Woods A, Davies SP, Beri RK, et al. 1996. Characterization of the AMP-activated protein kinase kinase from rat liver, and identification of threonine-172 as the major site at which it phosphorylates and activates AMP-activated protein kinase. *J. Biol. Chem.* 271:27879–87

56. Hayashida N, Mizoguchi T, Yamaguchi-Shinozaki K, Shinozaki K. 1992. Characterization of a gene that encodes a homolog of protein kinase in *Arabidopsis thaliana*. *Gene* 121:325–30

57. He ZH, Fujiki M, Kohorn BD. 1996. A cell wall-associated, receptor-like protein-kinase. *J. Biol. Chem.* 271:19789–93

58. Hemerly A, Engler JD, Bergounioux C, Vanmontagu M, Engler G, et al. 1995. Dominant negative mutants of the cdc2 kinase uncouple cell division from iterative plant development. *EMBO J.* 14:3925–36

59. Herve C, Dabos P, Galaud JP, Rouge P, Lescure B. 1996. Characterization of an *Arabidopsis thaliana* gene that defines a new class of putative plant receptor kinases with an extracellular lectin-like domain. *J. Mol. Biol.* 258:778–88

60. Higgins DG, Sharp PM. 1988. CLUSTAL: a package for performing multiple sequence alignment on a microcomputer. *Gene* 73:237–44

61. Hirayama T, Imajuku Y, Anai T, Matsui M, Oka A. 1991. Identification of 2 cell cycle controlling *cdc2* gene homologs in *Arabidopsis thaliana*. *Gene* 105:159–65

62. Hirayama T, Oka A. 1992. Novel protein kinase of *Arabidopsis thaliana* (APK1) that phosphorylates tyrosine, serine and threonine. *Plant Mol. Biol.* 20:653–62

63. Hirt H. 1997. Multiple roles of MAP kinases in plant signal transduction. *Trends Plant Sci.* 2:11–15

64. Hong Y, Takano M, Liu CM, Gasch A, Chye ML, Chua NH. 1996. Expression of 3 members of the calcium-dependent protein kinase gene family in *Arabidopsis thaliana*. *Plant Mol. Biol.* 30:1259–75

65. Hu SH, Parker MW, Lei JY, Wilce MCJ, Benian GM, Kemp BE. 1994. Insights into autoregulation from the crystal structure of twitchin kinase. *Nature* 369:581–84

66. Hua J, Chang C, Sun Q, Meyerowitz EM. 1995. Ethylene insensitivity conferred by Arabidopsis *ers* gene. *Science* 269:1712–14

67. Huala E, Oeller PW, Liscum E, Han IS, Larsen E, Briggs WR. 1997. Arabidopsis NPH1: a protein kinase with a putative redox-sensing domain. *Science* 278:2120–23

68. Huang JF, Teyton L, Harper JF. 1996. Activation of a Ca^{2+}-dependent protein kinase involves intramolecular binding of a calmodulin-like regulatory domain. *Biochemistry* 35:13222–30

69. Huber C, Huber JL, Liao PC, Gage DA, McMichael RW, et al. 1996. Phosphorylation of serine-15 of maize leaf sucrose synthase—occurrence in vivo and possible regulatory significance. *Plant Physiol.* 112:793–802

70. Hunter T. 1987. A thousand and one protein kinases. *Cell* 50:823–29

71. Hunter T. 1997. Oncoprotein networks. *Cell* 88:333–46

72. Hwang IW, Goodman HM. 1995. An *Arabidopsis thaliana* root-specific kinase homolog is induced by dehydration, ABA, and NaCl. *Plant J.* 8:37–43

73. Ishida R, Iwai M, Marsh KL, Austin CA, Yano T, et al. 1996. Threonine 1342 in human topoisomerase IIalpha is phosphorylated throughout the cell cycle. *J. Biol. Chem.* 271:30077–82

74. Ito T, Takahashi N, Shimura Y, Okada K. 1997. A serine/threonine protein kinase gene isolated by an *in vivo* binding procedure using the *Arabidopsis* floral homeotic gene product, AGAMOUS. *Plant Cell Physiol* 38:248–58

75. Jang JC, Sheen J. 1994. Sugar sensing in higher plants. *Plant Cell* 6:1665–79

76. Jeffrey PD, Russo AA, Polyak K, Gibbs E, Hurwitz J, et al. 1995. Mechanism of CDK activation revealed by the structure of a cyclinA-CDK2 complex [see comments]. *Nature* 376:313–20

77. Jinno A, Tanaka K, Matsushime H, Haneji T, Shibuya M. 1993. Testis-specific MAK protein kinase is expressed specifically in the meiotic phase in spermatogenesis and is associated with a 210-kilodalton cellular phosphoprotein. *Mol. Cell. Biol.* 13:4146–56

78. Johnson KD, Chrispeels MJ. 1992. Tonoplast-bound protein kinase phosphorylates tonoplast intrinsic protein. *Plant Physiol.* 100:1787–95

79. Johnson LN, Noble MEM, Owen DJ. 1996. Active and inactive protein kinases: structural basis for regulation. *Cell* 85:149–58

80. Jonak C, Heberlebors E, Hirt H. 1995.

Inflorescence-specific expression of ATK-1, a novel *Arabidopsis thaliana* homolog of shaggy/glycogen synthase kinase-3. *Plant Mol. Biol.* 27:217–21

81. Jonak C, Kiegerl S, Ligterink W, Barker PJ, Huskisson NS, Hirt H. 1996. Stress signaling in plants—a mitogen-activated protein kinase pathway is activated by cold and drought. *Proc. Natl. Acad. Sci. USA* 93:11274–79

82. Jonak C, Kiegerl S, Lloyd C, Chan J, Hirt H. 1995. Mmk2, a novel alfalfa MAP kinase, specifically complements the yeast mpk1 function. *Mol. Gen. Genet.* 248:686–94

83. Kaffman A, Herskowitz I, Tjian R, O'Shea EK. 1994. Phosphorylation of the transcription factor PHO4 by a cyclin-CDK complex, PHO80–PHO85. *Science* 263:1153–56

84. Kakimoto T. 1996. Cki1, a histidine kinase homolog implicated in cytokinin signal-transduction. *Science* 274:982–85

85. Kapoor M, Lozano G. 1998. Functional activation of p53 via phosphorylation following DNA damage by UV but not gamma radiation. *Proc. Natl. Acad. Sci. USA* 95:2834–37

86. Keyse SM. 1998. Protein phosphatases and the regulation of MAP kinase activity. *Semin. Cell Dev. Biol.* 9:143–52

87. Kieber JJ, Rothenberg M, Roman G, Feldmann KA, Ecker JR. 1993. CTR1, a negative regulator of the ethylene response pathway in *Arabidopsis*, encodes a member of the raf family of protein kinases. *Cell* 72:427–41

88. Klimczak LJ, Cashmore AR. 1993. Purification and characterization of casein kinase I from broccoli. *Biochem. J.* 293:283–88

89. Klimczak LJ, Collinge MA, Farini D, Giuliano G, Walker JC, Cashmore AR. 1995. Reconstitution of *Arabidopsis* casein kinase II from recombinant subunits and phosphorylation of transcription factor Gbf1. *Plant Cell* 7:105–15

90. Klimczak LJ, Farini D, Lin C, Ponti D, Cashmore AR, Giuliano G. 1995. Multiple isoforms of Arabidopsis casein kinase I combine conserved catalytic domains with variable carboxyl-terminal extensions. *Plant Physiol.* 109:687–96

91. Knetsch MLW, Wang M, Snaarjagalska BE, Heimovaaradijkstra S. 1996. Abscisic-acid induces mitogen-activated protein-kinase activation in barley aleurone protoplasts. *Plant Cell* 8:1061–67

92. Kobe B, Deisenhofer J. 1993. Crystal structure of porcine ribonuclease inhibitor, a protein with leucine-rich repeats. *Nature* 366:751–56

93. Kobe B, Deisenhofer J. 1994. The leucine-rich repeat: a versatile binding motif. *Trends Biochem. Sci.* 19:415–21

94. Kohorn BD, Lane S, Smith TA. 1992. An *Arabidopsis* serine/threonine kinase homologue with an epidermal growth factor repeat selected in yeast for its specificity for a thylakoid membrane protein. *Proc. Natl. Acad. Sci. USA* 89:10989–92

95. Kretzschmar M, Massague J. 1998. SMADs: mediators and regulators of TGF-beta signaling. *Curr. Opin. Genet. Dev.* 8:103–11

96. Langland JO, Langland LA, Browning KS, Roth DA. 1996. Phosphorylation of plant eukaryotic initiation factor-2 by the plant-encoded double-stranded RNA-dependent protein kinase, pPKR, and inhibition of protein synthesis in vitro. *J. Biol. Chem.* 271:4539–44

97. Lawton MA, Yamamoto RT, Hanks SK, Lamb CJ. 1989. Molecular cloning of plant transcripts encoding protein kinase homologs. *Proc. Natl. Acad. Sci. USA* 86:3140–44

98. Lee JW, Zhang Y, Weaver CD, Shomer NH, Louis CF, Roberts DM. 1995. Phosphorylation of nodulin 26 on serine 262 affects its voltage-sensitive channel activity in planar lipid bilayers. *J. Biol. Chem.* 270:27051–57

99. Lee JY, Yoo BC, Harmon AC. 1998. Kinetic and calcium-binding properties of three calcium-dependent protein kinase isoenzymes from soybean. *Biochemistry* 37:6801–9

100. Li JM, Chory J. 1997. A putative leucine-rich repeat receptor kinase involved in brassinosteroid signal transduction. *Cell* 90:929–38

101. Ligterink W, Kroj T, zurNieden U, Hirt H, Scheel D. 1997. Receptor-mediated activation of a MAP kinase in pathogen defense of plants. *Science* 276:2054–57

102. Lindzen E, Choi JH. 1995. A carrot cDNA encoding an atypical protein kinase homologous to plant calcium-dependent protein kinases. *Plant Mol. Biol.* 28:785–97

103. Loh YT, Martin GB. 1995. The disease resistance gene *Pto* and the fenthion-sensitivity gene *Fen* encode closely related functional protein kinases. *Proc. Natl. Acad. Sci. USA* 92:4181–84

104. MacKintosh RW, Davies SP, Clarke PR, Weekes J, Gillespie JG, et al. 1992. Evidence for a protein kinase cascade in higher plants: 3-hydroxy-3-methyl-

glutaryl-CoA reductase kinase. *Eur. J. Biochem.* 209:923–31

105. Madhani HD, Fink GR. 1998. The riddle of MAP kinase signaling specificity. *Trends Genet.* 14:151–55

106. Maeda T, Wurglermurphy SM, Saito H. 1994. A two-component system that regulates an osmosensing MAP kinase cascade in yeast. *Nature* 369:242–45

107. Magyar Z, Meszaros T, Miskolczi P, Deak M, Feher A, et al. 1997. Cell cycle phase specificity of putative cyclin-dependent kinase variants in synchronized alfalfa cells. *Plant Cell* 9:223–35

108. Martin GB, Brommonschenkel SH, Chunwongse J, Frary A, Ganal MW, et al. 1993. Map-based cloning of a protein kinase gene conferring disease resistance in tomato. *Science* 262:1432–36

109. McKibbin RS, Halford NG, Francis D. 1998. Expression of fission yeast cdc25 alters the frequency of lateral root formation in transgenic tobacco. *Plant Mol. Biol.* 36:601–12

110. Meskiene I, Bogre L, Glaser W, Balog J, Brandstotter M, et al. 1998. MP2C, a plant protein phosphatase 2C, functions as a negative regulator of mitogen-activated protein kinase pathways in yeast and plants. *Proc. Natl. Acad. Sci. USA* 95:1938–43

111. Mills GB, Schmandt R, McGill M, Amendola A, Hill M, et al. 1992. Expression of TTK, a novel human protein kinase, is associated with cell proliferation. *J. Biol. Chem.* 267:16000–6

112. Minden A, Karin M. 1997. Regulation and function of the JNK subgroup of MAP kinases. *Biochim. Biophys. Acta* 1333:F85–104

113. Mindrinos M, Katagiri F, Yu GL, Ausubel FM. 1994. The *A. thaliana* disease resistance gene RPS2 encodes a protein containing a nucleotide-binding site and leucine-rich repeats. *Cell* 78:1089–99

114. Mizoguchi T, Gotoh Y, Nishida E, Yamaguchishinozaki K, Hayashida N, et al. 1994. Characterization of 2 cDNAs that encode MAP kinase homologs in *Arabidopsis thaliana* and analysis of the possible role of auxin in activating such kinase activities in cultured cells. *Plant J.* 5:111–22

115. Mizoguchi T, Hayashida N, Yamaguchishinozaki K, Harada H, Shinozaki K. 1992. Nucleotide sequence of a cDNA encoding a protein kinase homolog in *Arabidopsis thaliana. Plant Mol. Biol.* 18:809–12

116. Mizoguchi T, Hayashida N, Yamaguchishinozaki K, Kamada H, Shinozaki K. 1993. AtMPKs—a gene family of plant MAP kinases in *Arabidopsis thaliana. FEBS Lett.* 336:440–44

117. Mizoguchi T, Hayashida N, Yamaguchishinozaki K, Kamada H, Shinozaki K. 1995. 2 genes that encode ribosomal protein S6 kinase homologs are induced by cold or salinity stress in *Arabidopsis thaliana. FEBS Lett.* 358:199–204

118. Mizoguchi T, Irie K, Hirayama T, Hayashida N, Yamaguchishinozaki K, et al. 1996. A gene encoding a mitogen-activated protein kinase kinase kinase is induced simultaneously with genes for a mitogen-activated protein kinase and an S6 ribosomal protein kinase by touch, cold, and water stress in *Arabidopsis thaliana. Proc. Natl. Acad. Sci. USA* 93:765–69

119. Mizoguchi T, Yamaguchi-Shinozaki K, Hayashida N, Kamada H, Shinozaki K. 1993. Cloning and characterization of two cDNAs encoding casein kinase II catalytic subunits in *Arabidopsis thaliana. Plant Mol. Biol.* 21:279–89

120. Moran TV, Walker JC. 1993. Molecular cloning of two novel protein kinase genes from *Arabidopsis thaliana. Biochim. Biophys. Acta* 1216:9–14

121. Morris PC, Guerrier D, Leung J, Giraudat J. 1997. Cloning and characterisation of MEK1, an Arabidopsis gene encoding a homologue of MAP kinase kinase. *Plant Mol. Biol.* 35:1057–64

122. Nasrallah JB, Stein JC, Kandasamy MK, Nasrallah ME. 1994. Signaling the arrest of pollen-tube development in self-incompatible plants. *Science* 266:1505–1508

123. Niefind K, Guerra B, Pinna LA, Issinger OG, Schomburg D. 1998. Crystal structure of the catalytic subunit of protein kinase CK2 from *Zea mays* at 2.1 Å resolution. *EMBO J.* 17:2451–62

124. Nishihama R, Banno H, Kawahara E, Irie K, Machida Y. 1997. Possible involvement of differential splicing in regulation of the activity of *Arabidopsis* ANP1 that is related to mitogen-activated protein kinase kinases (MAPKKKs). *Plant J.* 12:39–48

125. Ostling J, Ronne H. 1998. Negative control of the Mig1p repressor by Snf1p-dependent phosphorylation in the absence of glucose. *Eur. J. Biochem.* 252:162–68

126. Patil S, Takezawa D, Poovaiah BW. 1995. Chimeric plant calcium/calmodulin-dependent protein-kinase gene with a neural visinin-like calcium-binding

domain. *Proc. Natl. Acad. Sci. USA* 92: 4897–901

127. Pay A, Jonak C, Bogre L, Meskiene I, Mairinger T, et al. 1993. The MSK family of alfalfa protein kinase genes encodes homologs of shaggy/glycogen-synthase kinase-3 and shows differential expression patterns in plant organs and development. *Plant J.* 3:847–56

128. Pei ZM, Ward JM, Harper JF, Schroeder JI. 1996. A novel chloride channel in *Vicia faba* guard cell vacuoles activated by the serine/threonine kinase, CDPK. *EMBO J.* 15:6564–74

129. Poovaiah BW, Reddy ASN. 1993. Calcium and signal transduction in plants. *Crit. Rev. Plant Sci.* 12:185–211

130. Popping B, Gibbons T, Watson MD. 1996. The *Pisum sativum* MAP kinase homologue (PsMAPK) rescues the *Saccharomyces cerevisiae hog1* deletion mutant under conditions of high osmotic stress. *Plant Mol. Biol.* 31:355–63

131. Purcell PC, Smith AM, Halford NG. 1998. Antisense expression of a sucrose non fermenting-1–related protein kinase sequence in potato results in decreased expression of sucrose synthase in tubers and loss of sucrose inducibility of sucrose synthase transcripts in leaves. *Plant J.* 14:195–202

132. Renaudin JP, Doonan JH, Freeman D, Hashimoto J, Hirt H, et al. 1996. Plant cyclins: a unified nomenclature for plant A-, B- and D-type cyclins based on sequence organization. *Plant Mol. Biol.* 32:1003–18

133. Rethinaswamy A, Birnbaum MJ, Glover CV. 1998. Temperature-sensitive mutations of the *CKA1* gene reveal a role for casein kinase II in maintenance of cell polarity in *Saccharomyces cerevisiae*. *J. Biol. Chem.* 273:5869–77

134. Robinson LC, Hubbard EJ, Graves PR, DePaoli Roach AA, Roach PJ, et al. 1992. Yeast casein kinase I homologues: an essential gene pair. *Proc. Natl. Acad. Sci. USA* 89:28–32

135. Roe JL, Durfee T, Zupan JR, Repetti PP, McLean BG, Zambryski PC. 1997. TOUSLED is a nuclear serine/threonine protein kinase that requires a coiled-coil region for oligomerization and catalytic activity. *J. Biol. Chem.* 272:5838–45

136. Roe JL, Rivin CJ, Sessions A, Feldmann KA, Zambryski PC. 1993. The *Tousled* gene in *A. thaliana* encodes a protein kinase homolog that is required for leaf and flower development. *Cell* 75:939–50

137. Ronne H. 1995. Glucose repression in fungi. *Trends Genet.* 11:12–17

138. Sakai H, Hua J, Chen QHG, Chang CR, Medrano LJ, et al. 1998. *ETR2* is an *ETR1*-like gene involved in ethylene signaling in *Arabidopsis*. *Proc. Natl. Acad. Sci. USA* 95:5812–17

139. Salmeron JM, Oldroyd GED, Rommens CMT, Scofield SR, Kim HS, et al. 1996. Tomato *Prf* is a member of the leucine-rich repeat class of plant disease resistance genes and lies embedded within the *Pto* kinase gene cluster. *Cell* 86:123–33

140. Salt IP, Johnson G, Ashcroft SJH, Hardie DG. 1998. AMP-activated protein kinase is activated by low glucose in cell lines derived from pancreatic β cells, and may regulate insulin release. *Biochem. J.* 335:533–59

141. Satterlee JS, Sussman MR. 1998. Unusual membrane-associated protein kinases in higher plants. *J. Membr. Biol.* 164:205–13

142. Seo S, Okamoto N, Seto H, Ishizuka K, Sano H, Ohashi Y. 1995. Tobacco MAP kinase—a possible mediator in wound signal transduction pathways. *Science* 270:1988–92

143. Shaul O, Vanmontagu M, Inze D. 1996. Regulation of cell-division in Arabidopsis. *Crit. Rev. Plant Sci.* 15:97–112

144. Sheen J. 1996. Ca^{2+}-dependent protein kinases and stress signal transduction in plants. *Science* 274:1900–2

145. Shibata W, Banno H, Ito Y, Hirano K, Irie K, et al. 1995. A tobacco protein-kinase, NPK2, has a domain homologous to a domain found in activators of mitogen-activated protein kinases (MAPKKs). *Mol. Gen. Genet.* 246:401–10

146. Son M, Gundersen RE, Nelson DL. 1993. A second member of the novel Ca^{2+}-dependent protein kinase family from *Paramecium tetraurelia*. Purification and characterization. *J. Biol. Chem.* 268:5940–48

147. Song WY, Wang GL, Chen LL, Kim HS, Pi LY, et al. 1995. A receptor kinase-like protein encoded by the rice disease resistance gene, xa21. *Science* 270:1804–6

148. Stein JC, Dixit R, Nasrallah ME, Nasrallah JB. 1996. SRK, the stigma-specific S-locus receptor kinase of *Brassica*, is targeted to the plasma membrane in transgenic tobacco. *Plant Cell* 8:429–45

149. Stein JC, Howlett B, Boyes DC, Nasrallah ME, Nasrallah JB. 1991. Molecular cloning of a putative receptor protein kinase gene encoded at the self-incompatibility locus of *Brassica*

oleracea. Proc. Natl. Acad. Sci. USA 88: 8816–20

150. Stone JM, Collinge MA, Smith RD, Horn MA, Walker JC. 1994. Interaction of a protein phosphatase with an *Arabidopsis* serine-threonine receptor kinase. *Science* 266:793–95

151. Stone JM, Trotochaud AE, Walker JC, Clark SE. 1998. Control of meristem development by CLAVATA1 receptor kinase and kinase-associated protein phosphatase interactions. *Plant Physiol.* 117:1217–25

152. Stratmann JW, Ryan CA. 1997. Myelin basic protein kinase activity in tomato leaves is induced systemically by wounding and increases in response to systemin and oligosaccharide elicitors. *Proc. Natl. Acad. Sci. USA* 94:11085–89

153. Sugden C, Donaghy P, Halford NG, Hardie DG. 1999. Two SNF1–related protein kinases from spinach leaf phosphorylate and inactivate HMG-CoA reductase, nitrate reductase and sucrose phosphate synthase. *Plant Physiol.* In press

154. Takezawa D, Ramachandiran S, Paranjape V, Poovaiah BW. 1996. Dual regulation of a chimeric plant serine threonine kinase by calcium and calcium-calmodulin. *J. Biol. Chem.* 271:8126–32

155. Torii KU, Mitsukawa N, Oosumi T, Matsuura Y, Yokoyama R, et al. 1996. The Arabidopsis *erecta* gene encodes a putative receptor protein kinase with extracellular leucine-rich repeats. *Plant Cell* 8:735–46

156. Toroser D, Huber SC. 1998. 3-Hydroxy-3-methylglutaryl-coenzyme A reductase kinase and sucrose-phosphate synthase activities in cauliflower florets: Ca^{2+} dependence and substrate specificities. *Arch. Biochem. Biophys.* 355:291–300

157. Tregear JW, Jouannic S, Schwebeldugue N, Kreis M. 1996. An unusual protein kinase displaying characteristics of both the serine/threonine and tyrosine families is encoded by the *Arabidopsis thaliana* gene *ATN1. Plant Sci.* 117:107–19

158. Umeda M, Bhalerao RP, Schell J, Uchimiya H, Koncz C. 1998. A distinct cyclin-dependent kinase-activating kinase of *Arabidopsis thaliana. Proc. Natl. Acad. Sci. USA* 95:5021–26

159. Usami S, Banno H, Ito Y, Nishihama R, Machida Y. 1995. Cutting activates a 46–kilodalton protein kinase in plants. *Proc. Natl. Acad. Sci. USA* 92:8660–64

160. Vidal J, Chollet R. 1997. Regulatory phosphorylation of C-4 PEP carboxylase. *Trends Plant Sci.* 2:230–37

161. Walker JC. 1994. Structure and function of the receptor-like protein kinases of higher plants. *Plant Mol. Biol.* 26:1599–609

162. Walker JC, Zhang R. 1990. Relationship of a putative receptor protein kinase from maize to the S-locus glycoproteins of Brassica. *Nature* 345:743–46

163. Wang H, Fowke LC, Crosby WL. 1997. A plant cyclin-dependent kinase inhibitor gene. *Nature* 386:451–52

164. Wang PC, Vancura A, Mitcheson TG, Kuret J. 1992. Two genes in *Saccharomyces cerevisiae* encode a membrane-bound form of casein kinase-1. *Mol. Biol. Cell* 3:275–86

165. Wang XQ, Zafian P, Choudhary M, Lawton M. 1996. The PR5K receptor protein kinase from *Arabidopsis thaliana* is structurally related to a family of plant defense proteins. *Proc. Natl. Acad. Sci. USA* 93:2598–602

166. Watillon B, Kettmann R, Boxus P, Burny A. 1993. A calcium/calmodulin-binding serine/threonine protein kinase homologous to the mammalian Type II calcium/calmodulin-dependent protein-kinase is expressed in plant cells. *Plant Physiol.* 101:1381–84

167. Williams RW, Wilson JM, Meyerowitz EM. 1997. A possible role for kinase-associated protein phosphatase in the *Arabidopsis CLAVATA1* signaling pathway. *Proc. Natl. Acad. Sci. USA* 94:10467–72

168. Wilson WA, Hawley SA, Hardie DG. 1996. The mechanism of glucose repression/derepression in yeast: SNF1 protein kinase is activated by phosphorylation under derepressing conditions, and this correlates with a high AMP:ATP ratio. *Curr. Biol.* 6:1426–34

169. Woodgett JR. 1991. A common denominator linking glycogen metabolism, nuclear oncogenes and development. *Trends Biochem. Sci.* 16:177–81

170. Xie Q, Sanz-Burgos AP, Hannon GJ, Gutierrez C. 1996. Plant cells contain a novel member of the retinoblastoma family of growth regulatory proteins. *EMBO J.* 15:4900–8

171. Yoo BC, Harmon AC. 1996. Intramolecular binding contributes to the activation of CDPK, a protein kinase with a calmodulin-like domain. *Biochemistry* 35:12029–37

172. Yun B, Farkas R, Lee K, Rabinow L. 1994. The *Doa* locus encodes a member of a new protein kinase family and is essential for eye and embryonic development in *Drosophila melanogaster. Genes Dev.* 8:1160–73

173. Zhang K, Letham DS, John PCL. 1996. Cytokinin controls the cell cycle at mitosis by stimulating the tyrosine dephosphorylation and activation of p34(cdc2)-like h1 histone kinase. *Planta* 200:2–12

174. Zhang SH, Lawton MA, Hunter T, Lamb CJ. 1994. Atpk1, a novel ribosomal-protein kinase gene from Arabidopsis. 1. Isolation, characterization, and expression. *J. Biol. Chem.* 269:17586–92

175. Zhang SQ, Klessig DF. 1997. Salicylic acid activates a 48-kD MAP kinase in tobacco. *Plant Cell* 9:809–24

176. Zhang XQ, Chollet R. 1997. Seryl-phosphorylation of soybean nodule sucrose synthase (nodulin-100) by a Ca^{2+}-dependent protein kinase. *FEBS Lett.* 410:126–30

177. Zhao Y, Kappes B, Franklin RM. 1993. Gene structure and expression of an unusual protein kinase from *Plasmodium falciparum* homologous at its carboxyl terminus with the EF hand calcium-binding proteins. *J. Biol. Chem.* 268:4347–54

178. Zhou JM, Loh YT, Bressan RA, Martin GB. 1995. The tomato gene *Pti1* encodes a serine/threonine kinase that is phosphorylated by Pto and is involved in the hypersensitive response. *Cell* 83:925–35

179. Zhou R. 1998. The Eph family receptors and ligands. *Pharmacol. Ther.* 77:151–81

Annu. Rev. Plant Physiol. Plant Mol. Biol. 1999. 50:133–61

IMPROVING THE NUTRIENT COMPOSITION OF PLANTS TO ENHANCE HUMAN NUTRITION AND HEALTH[1]

Michael A. Grusak
USDA/ARS Children's Nutrition Research Center, Department of Pediatrics, Baylor College of Medicine, Houston, Texas 77030; e-mail: mgrusak@bcm.tmc.edu

Dean DellaPenna
Department of Biochemistry, University of Nevada, Reno, Nevada 89557; e-mail: della_d@med.unr.edu

KEY WORDS: minerals, vitamins, iron, alpha-tocopherol, phytochemicals

ABSTRACT

Plant foods contain almost all of the mineral and organic nutrients established as essential for human nutrition, as well as a number of unique organic phytochemicals that have been linked to the promotion of good health. Because the concentrations of many of these dietary constituents are often low in edible plant sources, research is under way to understand the physiological, biochemical, and molecular mechanisms that contribute to their transport, synthesis and accumulation in plants. This knowledge can be used to develop strategies with which to manipulate crop plants, and thereby improve their nutritional quality. Improvement strategies will differ between various nutrients, but generalizations can be made for mineral or organic nutrients. This review focuses on the plant nutritional physiology and biochemistry of two essential human nutrients, iron and vitamin E, to provide examples of the type of information that is needed, and the strategies that can be used, to improve the mineral or organic nutrient composition of plants.

---

[1] The US Government has the right to retain a nonexclusive, royalty-free license in and to any copyright covering this paper.

CONTENTS

INTRODUCTION

The nutritional health and well-being of humans are entirely dependent on plant foods. Plants are critical components of the dietary food chain in that they provide almost all essential mineral and organic nutrients to humans either directly, or indirectly when plants are consumed by animals, which are then consumed by humans. Because plants are autotrophic, they can acquire elemental compounds and convert these into the building blocks (e.g. amino acids, fatty acids, nucleic acids, secondary metabolites) needed to make all complex macromolecules necessary to support plant growth and reproduction. Humans, on the other hand, require many of the same mineral nutrients as plants, but have additional requirements for various complex organic molecules. With the exception of vitamins B_{12} and D, a plant-based food supply can ensure the adequate nutrition of humans at all stages of life.

Not all plant foods, however, contain all the essential nutrients needed for human health, nor do they usually contain given nutrients in sufficiently concentrated amounts to meet daily dietary requirements in a single serving. For instance, seed foods are good sources of carbohydrates, proteins, lipids, and lipid-soluble vitamins, but tend to have low concentrations of Fe and Ca. Leafy vegetables are good sources of most minerals and vitamins, but are less nutrient dense with respect to protein and carbohydrates. Fruits provide carbohydrates, water-soluble vitamins, and various carotenoids, but generally are minor sources of protein and certain minerals. Thus, not only is a diverse, complex diet necessary to fully support human growth and health, but the concentration of various nutrients within the dietary mix also is an important determinant of whether nutritional requirements are being met.

Unfortunately, many people do not consume a sufficiently diverse diet. In the developing world, many low-income families exist on a simple diet composed primarily of staple foods (e.g. rice, wheat, maize) that are poor sources of some macronutrients and many micronutrients (22). As a result, recent estimates are that 250 million children are at risk for vitamin A deficiency (of which 250,000–500,000 will suffer irreversible blindness every year), 2 billion people (33% of the world's population) are at risk for iron deficiency (infants, children, and women of reproductive age are particularly vulnerable), and 1.5 billion people are at risk for iodine deficiency (38). Furthermore, even in the United States, the average intake of fruits and vegetables was recently assessed to be only 3.4 servings per day, well below the five-per-day recommendation of the US National Research Council (110, 145). This recommendation is derived from numerous epidemiological studies, which suggest that the daily consumption of five or more servings of fruits and vegetables is associated with reduced risk for several types of cancer as well as other degenerative diseases (12). Many of these studies have implicated antioxidants, such as carotenoids (e.g. β-carotene and lycopene), tocopherols, ascorbic acid, and selenium, as contributors to the healthful properties of such a diet. Furthermore, fruits and vegetables also contain a vast and complex array of bioactive secondary compounds (nonessential for humans), collectively referred to as phytochemicals, that show promise in contributing to the promotion of good health and protection against human diseases (139, 165).

To ensure an adequate dietary intake of all essential nutrients and to increase the consumption of various health-promoting compounds, researchers have been interested in improving the nutritional quality of plants, with respect to both nutrient composition and concentration (13, 47, 161). In this article, we discuss the role that plant foods play in human nutrition and health, and review some of the efforts pertinent to plant nutrient composition. We then review our current knowledge regarding the physiological, biochemical, and molecular mechanisms that are involved in the eventual deposition of iron, a mineral nutrient, and vitamin E, an organic nutrient. Because it would be impossible to discuss all plant-derived nutrients in this article, we focus on just these two nutrients to provide examples of the type of information that is needed, and the strategies that can be used, to improve the mineral or organic nutrient composition of plants.

ROLE OF PLANTS IN HUMAN HEALTH

Essential Human Nutrients

Humans require an energy supply (in addition to water and oxygen) that can be provided by a mixture of carbohydrates, lipids, and protein (amino acids), as well as 17 mineral nutrients and 13 vitamins. Among the lipids, linoleic acid and linolenic acid cannot be synthesized by humans (65), and thus must be obtained from dietary sources (plants synthesize both of these fatty acids). Nine essential amino acids must be obtained in various amounts from ingested protein (90), which also provides dietary sulfur and nitrogen. The 15 remaining essential mineral nutrients and 13 vitamins are required in varying amounts, as indicated in Table 1. To provide examples of some of the higher daily nutrient requirements that would need to be met from dietary sources, Recommended Dietary Allowances (RDAs) are presented for adults (Table 1). RDAs are the daily levels of intake of essential nutrients judged to be adequate to meet the known nutrient needs of practically all healthy persons (110). RDAs vary with age, sex, and physiological status (i.e. pregnancy, lactation), and generally are lowest for newborn infants and highest for adult males and pregnant or lactating women.

Dietary standards have been around for over 100 years, and initially were developed as recommendations to alleviate starvation and associated nutrient-deficiency diseases during economic and wartime crises (57, 58). The US RDAs were first published in 1941, with the most recent, the 10th edition, published in 1989 (110). This soon will be replaced with a new set of values called the Dietary Reference Intakes (DRIs) (1) that are scheduled to be published in the year 2000. DRIs are being developed to encompass three existing values: RDAs, the Estimated Average Requirements, and the Tolerable Upper Intake Levels (66). The need for the new DRI values reflects the growing knowledge base regarding the roles of nutrients in human health. Vitamins, minerals, and other dietary constituents at varying levels are now known to be significant contributors to the reduction of risk of chronic diseases such as cancers, cardiovascular diseases, and degenerative diseases associated with aging (11, 20, 77, 139, 148, 165), in addition to alleviating the classical nutritional deficiency diseases. Thus, new guidelines are needed, especially with regard to upper limits of safe intake.

Plant foods can contribute significantly to human nutrition and health, because they contain almost all essential human nutrients. However, as noted in Table 1, nutrient composition varies among different plant foods, and nutrient content in a single serving rarely fulfills the RDA for any given vitamin or mineral. Note also that the content values are for whole, unprocessed and uncooked foods. Many vitamins and minerals are lost or greatly reduced in concentration during cooking, storage, or processing (127). Thus, enhancing the nutrient concentration of many plant food products would contribute significantly to human nutrition and health.

Bioavailability

The total content, or absolute concentration, of a given nutrient in a food is not always a good indicator of its useful nutritional quality, because not all of the nutrients in food are absorbed. Human nutritionists use the term bioavailability to describe the proportion of an ingested nutrient that is digested, absorbed, and ultimately utilized (17). Digestion involves various physical, chemical, enzymatic, and secretory processes that combine to break down the food matrix as much as possible (keep in mind that cellulose cannot be digested because humans do not express a cellulase). These digestive processes serve to release and solubilize nutrients so they can diffuse out of the bulk food matrix to the enterocytes of the intestine. Absorption then depends on whether the nutrient is in an appropriate form (e.g. charged, complexed), whether the necessary absorptive transport systems are in place in the gut (this depends in part on the individual's nutritional status), and whether inhibitory or promotive substances are present in the food matrix (90). Because of these issues, the bioavailability of many minerals and organic micronutrients varies greatly (110), and can be as low as 5% in the case of most plant sources of Fe (23) or in the case of Ca when it is present in foods as crystalline Ca-oxalate (40, 158). This raises the possibility that one strategy for improving plant nutritional quality could be merely to alter the composition of the food (e.g. changing the form of the stored nutrient, removing inhibitory compounds), in order to enhance the bioavailability of existing nutrient levels.

Phytochemicals

Besides the well-established minerals and vitamins, plants can provide an array of interesting, nonessential phytochemicals in the diet. Phytochemicals can be separated into several groups, based on their biosynthetic origin, with some groups containing several thousand chemically distinct compounds. These include flavonoids, flavones, phytosterols, phenols and polyphenols, phytoestrogens, glucosinolates, and indoles, to name but a few (11, 27, 33, 37, 69, 82, 152). With respect to human health, estimates of potential therapeutic value have been made for numerous compounds produced by plants. Unfortunately, most of our information in this area stems from epidemiological evidence that only can provide associations between health benefits and particular foods or classes of food components. Thus, the current research focus in this area is to identify the exact chemical(s) responsible for a given beneficial effect and to delineate the underlying biochemical and molecular mechanisms involved. For many groups of phytochemicals, we know little about their bioavailability, whether

Table 1 Essential human nutrients, daily requirements, and plant food sources

Nutrient	Maximum adult RDA[a]	Safe upper intake limits (relative to RDA)[b]	Predominant plant source	Mean nutrient content (source)[c]	Human health reference
Minerals					
Potassium	2000 mg	9X	Various	447 mg (kale)	94
Calcium	1200 mg	2X	Vegetables	48 mg (broccoli)	4
Phosphorus	1200 mg	2X	Various	376 mg (peanut)	4
Chloride	750 mg	5X	Various	60 mg (tomato)	102
Sodium	500 mg	5X	Various	79 mg (spinach)	102
Magnesium	350 mg	1X	Seeds, leafy vegetables	138 mg (wheat flour)	32
Iron	15 mg	5X	Seeds, leafy vegetables	1.47 mg (pea)	164
Zinc	15 mg	1X	Seeds	2.93 mg (wheat flour)	2
Manganese[d]	2–5 mg	1X	Seeds	4.9 mg (oat)	76
Fluoride[d]	1.5–4 mg	1X	Aerial tissues	0.11 mg (spinach)	79
Copper[d]	1.5–3 mg	1X	Seeds	1.14 mg (peanut)	91
Molybdenum[d]	75–250 μg	1X	Seeds	70 μg (oat)	119
Chromium[d]	50–200 μg	1X	Various	33 μg (potato)	105
Iodine	150 μg	13X	Various	12 μg (kale)	28
Selenium	70 μg	13X	Seeds	7.2 μg (peanut)	89
Water-soluble vitamins					
Vitamin C	60 mg	16X	Fruits, vegetables	42.2 mg (cantaloupe)	129
Niacin (vitamin B$_3$)	19 mg NE[e]	150X	Seeds, leafy vegetables	9.9 mg NE (wheat flour)	68
Pantothenic acid[d]	4–7 mg	150X	Seeds	1.35 mg (oat)	146

Vitamin B_6	2 mg	125X	Seeds	0.35 mg (peanut)	88
Riboflavin (vitamin B_2)	1.7 mg		Cereal grains, leafy vegetables	0.22 mg (wheat flour)	103
Thiamin (vitamin B_1)	1.5 mg	67X	Seeds	0.64 mg (peanut)	121
Folate	200 μg	50X	Legume seeds, leafy vegetables	240 μg (peanut)	21
Biotin[d]	30–100 μg	300X	Seeds	13 μg (oat)	106
Vitamin B_{12}	2.0 μg	500X	Not found in plants	0 (all plants)	60
Fat-soluble vitamins					
Vitamin E	10 mg α-TE[f]	100X	Seeds, leafy vegetables	1.23 mg α-TE (wheat flour)	154
Vitamin A[g]	1000 μg RE[h]	5X (retinol) 100X (β-carotene)	Colored fruits and vegetables	2813 μg RE (carrot)	124
Vitamin K	80 μg	375X	Leafy vegetables	350 μg (spinach)	155
Vitamin D	10 μg	4X	Not found in plants	0 (all plants)	29

[a]Recommended dietary allowances (RDA) are the daily levels of intake of essential nutrients judged to be adequate to meet the known nutrient needs of practically all healthy persons. Values presented are the highest RDA either for male or female adults, excluding pregnant or lactating women. Values are from Reference 110.

[b]The concept assumes that there is individual variation in both requirement for the nutrient and tolerance for high intake. The safe upper intake limit is associated with a low probability of adverse side effect. The authors are not advocating intake of supplements at these levels. Values were derived from References 83, 110, and 168.

[c]Concentrations are for 100 g raw, edible portion of food; values are representative for predominant source and are not the highest known example. All seed sources represent whole-seed or whole-grain values. Note that water content varies among sources. Values are from References 138 and 154a.

[d]Recommended values for these nutrients are provided as estimated safe and adequate daily dietary intake ranges, because less information is available on which to base the RDAs. Values are from Reference 110.

[e]One mg NE (niacin equivalent) is equal to 1 mg of niacin or 60 mg of dietary tryptophan.

[f]One mg α-TE (α-tocopherol equivalent) is equal to 1 mg (R,R,R)-α-tocopherol.

[g]Preformed vitamin A is not found in plant foods. However, plants contain a number of provitamin A carotenoids (e.g. β-carotene), which can be metabolized to vitamin A.

[h]Vitamin A activity is expressed in retinol equivalents (RE). One mg RE is equal to 1 μg all-*trans* retinol, 6 μg all-*trans* β-carotene, or 12 μg of other provitamin A carotenoids.

the absorbed compounds are metabolized into active or inactive forms in the body, or their modes of action at the cellular, biochemical, or molecular level.

PLANT IMPROVEMENT THROUGH CULTIVAR SELECTION

The primary objective of modern agriculture and breeding programs over the past 50 years has been to increase productivity by increasing yields. This has been achieved primarily by selecting for resistance to diseases, increased fruit set and size, specific plant growth forms, increased grain fill, etc. The quest for increased yield is without a doubt a primary concern for a growing world population, and should be continued with vigor. However, equally as important, but often overlooked, are the nutrient composition and density of crops, especially with regard to micronutrients.

When nutrient content has been assessed in various plants, significant genotypic variation has been observed for minerals, vitamins and phytochemicals. For example, pod Ca concentration varied nearly twofold among snap bean genotypes (118), β-carotene concentrations varied fourfold among broccoli cultivars (132), and folate concentrations varied fourfold among red beet cultivars (157). These few examples highlight the fact that significant variation in nutrient content already exists within the germplasm of these species, and probably exists in many others. Thus, classical breeding approaches can and are being used to provide nutritionally improved cultivars (135, 151). In addition, genotypes with dissimilar nutrient concentration are helping to identify the genetic (144) or physiological (54) basis for nutrient variation. Clearly, more advances will be possible if attention is given to the evaluation of currently available germplasm resources (149).

IMPROVEMENT STRATEGIES FOR MINERAL NUTRIENTS: IRON AS AN EXAMPLE

General Considerations

Minerals can be grouped into three categories: the macronutrient minerals (N, S, P, Ca, K, Mg) that are needed in highest concentration by plants (mg/g dry weight range), the micronutrient minerals (Fe, Mn, B, Cl, Zn, Cu, Mo, Ni) that are needed in lesser amounts (μg/g dry weight range), and various generally nonessential minerals (e.g. Na, F, Se, Cr, I) that are found in plant tissues in varying concentrations (71). A plant's ability to increase the total content of a mineral from any one of these categories always depends on soil composition and the availability of that mineral in the plant's environment. Different improvement strategies may be needed for essential and nonessential minerals, depending on the existence of specific or nonspecific transport systems and the capacity for safe bioavailable storage in the edible tissues. Additionally, because energy costs to the plant may dictate that content changes are more feasible at the μg level than the mg level, percentage changes in mineral content may be more dramatic for the micronutrient and nonessential minerals.

The mineral composition of each plant organ is determined by a sequence of events that begins with membrane transport in the roots, proceeds to the xylem system for transit to the vegetative organs, may involve temporary storage in stem or leafy tissues, in some cases utilizes mobilization via the phloem pathway, and concludes with deposition in one or more cellular compartments (see Figure 1). All of these processes are integrated and regulated by the plant to

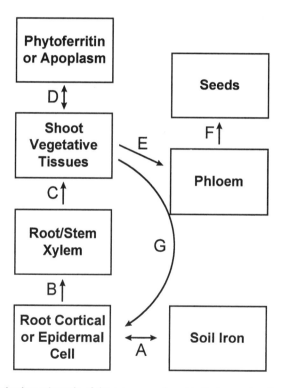

Figure 1 Whole plant schematic of the processes relevant to Fe transport and accumulation in higher plant tissues. Potential control points that would influence the movement of Fe from one compartment to the next include: (*A*) Fe acquisition/uptake phenomena, including the release of compounds by roots to chelate or solubilize soil Fe; (*B*) intracellular/intercellular transport, including the involvement of xylem parenchyma; (*C*) transpiration rates of vegetative tissues; (*D*) storage and remobilization phenomena; (*E*) Fe-chelate expression and capacity for phloem Fe loading; (*F*) phloem transport capacity of photoassimilates from a given source region; (*G*) communication of shoot Fe status via phloem-mobile signal molecules to regulate root processes.

ensure that adequate, but not toxic, quantities of mineral nutrients are available for the plant's growth and development. For any given mineral, a holistic understanding of the relevant transport and partitioning mechanisms, and the molecular to whole-plant factors that regulate them, is critical to enable improvement strategies to be developed. Strategically, one must determine the process or processes that rate limit the eventual deposition of each mineral in a given structure, such that these can be targeted for genetic or molecular modification. Our current understanding of these processes in iron nutrition is given as an example.

Iron Processes in Plants

Iron is an essential nutrient found in leaves and other plant tissues, typically at low concentrations (<200 $\mu g/g$ dry weight). As a component of various redox and iron-sulfur enzymes (100), iron plays an important role in general plant metabolism. Additionally, it is essential for chlorophyll formation (117). However, iron is an extremely reactive transition metal that can catalyze the formation of free radical species (59). Because these secondary species can damage lipids, proteins, or nucleic acids (77), Fe uptake is tightly regulated, and most of the Fe that does enter the plant must be chelated or sequestered in nonreactive forms. Thus, to improve plant Fe composition, not only do acquisition processes need to be upregulated, but attention must also be paid to how any excess Fe is handled as it moves through or is sequestered within the plant.

ROOT IRON ACQUISITION   Higher plants utilize one of two strategies for Fe acquisition (101). Strategy I involves an obligatory reduction of ferric iron (usually as an Fe[III]-compound) prior to membrane influx of Fe^{2+}; this strategy is used by all dicotyledonous plants and the non-grass monocots. Strategy II (used by grasses) employs ferric chelators, called phytosiderophores, that are released by roots and chelate ferric iron in the rhizosphere. The Fe(III)-phytosiderophore is absorbed intact via a plasmalemma transport protein. When plants of either strategy are challenged with Fe-deficiency stress, the processes associated with one or the other strategy are upregulated in the plant's root system.

Mechanistically, Strategy I plants utilize a plasmalemma reductase that transfers an electron from an internal reductant to an external Fe(III)-chelate (63). Reduced Fe^{2+} is released from the chelate and is transported across the root-cell plasmalemma via a ferrous transport protein (34, 39). At present, root Fe(III) reductase activity has been characterized in a number of species (107), but no plant Fe(III) reductase has been isolated and sequenced. However, four candidate genes (*frohA*, *frohB*, *frohC*, and *frohD*), suggested to encode Fe(III) reductases, have been identified in *Arabidopsis* by PCR, using degenerate primers based on conserved motifs found in yeast Fe(III) reductases (122, 123).

More information is known about the Fe^{2+} transporter. The *Arabidopsis IRT1* gene (for *I*ron *R*egulated *T*ransporter) was identified by functional complementation of a yeast mutant defective in iron uptake (34). Expression of *IRT1* is localized to roots in *Arabidopsis*, and is induced when plants are challenged with Fe deficiency.

Strategy II plants release one of a group of closely related phytosiderophores belonging to the mugineic acid family (147). Phytosiderophores are low-molecular-weight peptides derived from methionine (109, 134); their biosynthetic pathway is fairly well established (97, 108). A few of the critical biosynthetic enzymes have recently been identified (62, 73, 74), as a result of differences observed between Fe-sufficient and Fe-deficient plants. Strategy II plants require at least two membrane transport systems for phytosiderophores: one to facilitate the efflux of uncomplexed phytosiderophore and the other the influx of Fe(III)-phytosiderophore. Release follows a diurnal periodicity in most species (167) and may involve vesicular targeting to the plasmalemma (112). Fe(III)-phytosiderophore influx is generally increased in Fe-deficient plants and the uptake system appears capable of transporting most of the identified Fe(III)-phytosiderophore species (96). Both high-affinity and low-affinity phytosiderophore uptake systems have been shown to function in maize (156). Neither of the transporters has been identified at the protein or gene level, but a transport-defective maize mutant (*ys1*) may prove useful in this regard (156).

Although much has been learned about the mechanisms of iron acquisition in higher plants, including the upregulation of root biosynthetic and transport systems in response to Fe deficiency, we only have limited understanding of the molecular regulation of these processes. Recent evidence suggests that the control may lie in the shoot tissues. Studies with two Fe-hyperaccumulating pea mutants demonstrated that shoot-to-root transmission of a phloem-mobile signal was responsible for the elevated rates of root Fe(III) reduction (53). Root reductase activity in wild-type, Fe-sufficient pea also was shown to be dynamically modulated throughout the plant's life cycle in response to whole-plant Fe demand (49). These and other studies (84, 125) indicate that whole-plant Fe status is somehow monitored and assessed by the shoot tissues, with Fe need or Fe demand subsequently communicated to the roots, probably through an intracellular network (93). Because Fe(III) reduction is thought to be the rate-limiting Fe acquisition process in Strategy I plants (55, 166) and the entire phytosiderophore synthesis and transport cascade is critical for Strategy II plants (25, 156), it is no wonder that these processes are tightly regulated by the shoot to prevent toxic accumulation of Fe. The phloem-mobile signal compound has yet to be identified, but possible candidates include hormones, Fe-binding compounds, and even re-translocated Fe (10, 85, 98, 125, 141). Identification of the signal and a characterization of its molecular interaction with Fe-status-responsive genes (e.g. *IRT1*) is crucial if we wish to control/ enhance root acquisition of Fe.

TRANSPORT TO AND STORAGE IN EDIBLE ORGANS   Once absorbed by root epidermal or cortical cells, Fe is transported radially to the root cortex for loading into the xylem pathway. It has been suggested that the Fe(II)-chelating peptide, nicotianamine (141), may act to stabilize free Fe^{2+} and assist in its intracellular trafficking to the xylem parenchyma (81, 131, 160). However, because Fe moves within the xylem pathway as Fe(III)-citrate (100, 163), the Fe^{2+} absorbed by Strategy I plants would appear to be oxidized to Fe^{3+} at some location within the root, presumably in a controlled manner. Oxidation and subsequent citrate chelation may occur at the site of Fe uptake, because the nicotianamine-less tomato mutant, *chloronerva*, is able to transport Fe to its shoots (115). In Strategy II plants, symplasmic citrate chelation could similarly occur in the root periphery where Fe is absorbed as the ferric form. The processes of Fe movement through roots and within the xylem system may not be rate-limiting, because mutants that exhibit elevated Fe uptake capacity demonstrate excessive Fe hyperaccumulation in leaves (56, 80).

Transport in the xylem pathway carries Fe to all transpiring organs and delivers it initially to the organ's apoplasmic compartment (100). A combination of redox reactions, pH equilibria, and transport processes will determine the eventual fate of Fe within the tissue, and thereby influence its bioavailability as a human food source. Fe(III)-citrate can be reduced by light energy (18), a plasmalemma-localized reductase (19), or possibly apoplasmic ascorbate (95) to generate free Fe^{2+}, which can be absorbed by leaf cells through an Fe^{2+} transport protein (19). Alternatively, the Fe^{2+} can reoxidize and precipitate in the cell wall space, possibly as Fe-hydroxide or Fe-phosphate species. Fe^{2+} absorbed by leaf cells can be used in various enzymes, assist in chlorophyll biosynthesis, or be stored within the chloroplastic iron storage protein, phytoferritin (14, 15, 116). How Fe is handled within the symplasm and transported between cytoplasm and various organelles is poorly understood, although nicotianamine may play a role (130). An Fe-citrate transporter may aid the transport of Fe into chloroplasts (87), where ascorbate is thought to facilitate Fe partitioning into phytoferritin (86). Iron can be stored in phytoferritin for future use; phytoferritin also can be used to sequester excess Fe (92).

Some of the shoot Fe is exported to developing sinks and growing roots via the phloem pathway. Studies with the *brz* and *dgl* Fe-hyperaccumulating pea mutants demonstrated that Fe must be chelated prior to phloem loading (48), and that elevated rates of Fe loading are possible (99), presumably due to the overexpression of a chelator. In normal plants, Fe mobilization to developing seeds is rate-limited by the level of synthesis of the phloem-mobile chelator (48); this possibly serves to prevent Fe overload in the reproductive propagules. Little is known about the nature of the chelator, the compartment in which Fe chelation occurs, or the mechanism of phloem Fe loading (52, 140, 159).

Strategies for Iron Improvement

Plant sources of Fe include both xylem-fed leafy vegetables and phloem-fed seeds. Increasing the Fe content of either type usually necessitates increases in total Fe input to the plant, and may require modifications to whole-plant partitioning. Unfortunately, the homeostatic processes that control Fe influx and movement throughout the plant (see Figure 1) appear tightly matched to minimize Fe toxicity at all points within the system (52). Thus, if one step is altered to allow a higher flux of Fe, the next step may not necessarily conduct this enhanced flux. For instance, when two Fe(III) reductase genes from yeast (*FRE1* and *FRE2*) were constitutively expressed in tobacco (128), total root reductase activity was enhanced fourfold relative to Fe-grown controls, but leaf Fe content was increased only 50% in the transformed plants. This suggests that the activity and/or spatial localization of the Fe^{2+} uptake system was not enhanced significantly to take advantage of the excess Fe^{2+} generated by the yeast reductases. More significant improvements in total Fe uptake may require a holistic approach involving overexpression of multiple components of the Fe acquisition system, overexpression of the Fe-status signal molecule, or identification and expression of transcriptional regulators that mediate iron deficiency responses, in order to activate all necessary root processes (9, 53).

In cases of plant organs, such as seeds, whose predominant supply of nutrients is provided via the phloem pathway, improvements in Fe content will require modifications to the phloem loading system. Overexpression of a phloem-mobile Fe-chelator can enable increased phloem Fe transport (99), but in species like pea, in which 75% of the shoot Fe content already is localized to the seeds (49), an increased uptake and delivery of Fe to the loading regions will also be required, preferably to coincide with the period of seed fill to prevent toxic accumulation in the leaves. Alternatively, in cereals such as rice, whose seeds import only about 4% of total shoot Fe (MA Grusak, unpublished), targeting increased phloem-mobile chelate expression to source regions that contain available Fe could help to increase seed Fe content. The Fe transported to seeds must then be sequestered in a nonreactive form, although phytoferritin expression may increase automatically in response to the elevated Fe load (92). Modifying seeds to store the excess Fe in heme-containing enzymes, or chelated to peptides, might further enhance the seed's Fe nutritional quality by improving bioavailability (7, 45).

IMPROVEMENT STRATEGIES FOR ORGANIC NUTRIENTS: VITAMIN E AS AN EXAMPLE

General Considerations

Plants contain and elaborate many unique, interconnected biochemical pathways that produce an astonishing array of organic compounds that not only perform vital functions in plant cells, but also are essential or beneficial for human nutrition. One such class of compounds is the tocopherols (collectively known as vitamin E), a class of lipid soluble antioxidants that is synthesized only by plants and other photosynthetic organisms. The essential nutritional value of tocopherols was recognized more than 70 years ago and the compound responsible for the highest vitamin E activity was first identified specifically as α-tocopherol in 1922 (36). Despite the well-documented benefits of tocopherols in human diets (20, 148), only recently has significant progress been made at the molecular level regarding the synthesis and accumulation of tocopherols in plant tissues.

The low levels of activity and the membrane-bound nature of plant tocopherol biosynthetic enzymes have historically made the isolation of the corresponding genes, via protein purification, a daunting task. In this regard, tocopherols and other important lipid- and water-soluble plant-derived organic nutrients (e.g. carotenoids, folate, biotin, thiamin) have much in common. While the classical biochemical approaches of purification, enzyme assays, and radiolabeled tracer studies have been and will continue to be invaluable in furthering our understanding of the biosynthesis and transport of organic nutritional components, these approaches are limited in many ways. Often, protein stability and enzyme activity levels are low, substrates for biochemical assays are often not commercially available, and multiple components of a single biosynthetic step cannot be biochemically resolved.

Biochemical approaches can be complemented by genetic and molecular approaches in which one uses the organism to identify which steps are important in a biochemical pathway. Over the past several years, new genomic technologies, combined with the increasing ease of integrating molecular, genetic, and biochemical approaches in the field of plant biochemistry (16, 126, 153), have allowed researchers to make significant progress in dissecting the biosynthetic pathways for several classes of organic plant nutrients. The most successful in this regard has been the carotenoid pathway, in which almost all the biosynthetic enzymes have been cloned during the past several years (reviewed in 26). More recently, enzymes involved in folate and thiamin synthesis (8, 111, 120), biotin synthesis (5), and tocopherol synthesis (44, 78, 113, 114, 133a) have also been cloned. These genes undoubtedly will be used in the future to study and manipulate the synthesis of these nutrients at the molecular and biochemical levels,

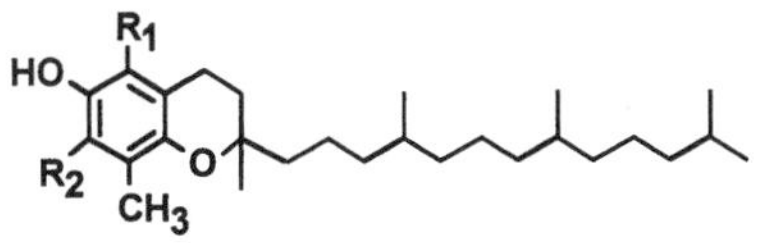

Type	R_1	R_2
α-tocopherol	CH_3	CH_3
γ-tocopherol	H	CH_3
β-tocopherol	CH_3	H
δ-tocopherol	H	H

Figure 2 Tocopherol structures. The number and position of ring methyls in α, β, γ, and δ are indicated in the accompanying table.

with the ultimate goal of modifying the levels of these nutrients in agronomically important plants. The remainder of this section focuses on recent progress in the molecular dissection and manipulation of the tocopherol biosynthetic pathway in plants. α-Tocopherol manipulation is presented here as a specific example of the potential of such integrative approaches for manipulation of plant-derived phytochemicals.

Structures and Functions of Tocopherols

The general structures of tocopherols are shown in Figure 2. All tocopherols are amphipathic molecules in which the hydrophobic tail associates with membrane lipids and the polar head groups remain at the membrane surface. α-, β-, γ-, and δ-tocopherols differ only in the number and position of methyl substituents on the aromatic ring, with α- having three, β- and γ- having two, and δ-tocopherol having only one substituent. Tocotrienols also occur in plants, differing from the tocopherols only in the degree of saturation of their hydrophobic tails. Tocopherols have saturated side chains and are the most common in plants. Tocotrienols are thought to be either biosynthetic intermediates that accumulate when specific steps in tail synthesis are blocked, or end products that specifically accumulate in some tissues, often to high levels (136).

The best-characterized and arguably most important function of tocopherols in biological membranes is to act as recyclable chain-reaction terminators of polyunsaturated fatty acid (PUFA) free radicals generated by lipid oxidation (41, 72). The in vivo antioxidant activities of tocopherols against lipid oxidation are $\alpha > \beta \cong \gamma > \delta$ with one molecule of each tocopherol protecting up to 220, 120, 100, and 30 molecules of PUFA, respectively, before being consumed (42). However, the relative antioxidant activity of tocopherols when tested in fats and oils in vitro is reversed to $\delta > \beta \cong \gamma > \alpha$.

Of the different tocopherol species present in foods, α-tocopherol is the most important to human health and has the highest vitamin E activity (100, 50, 10, and 3 percent relative activity for α-, β-, γ- and δ-tocopherols, respectively) (72). Naturally synthesized α-tocopherol occurs as a single (R,R,R)-α-tocopherol isomer. Chemically synthesized α-tocopherol, the most common tocopherol in vitamin E supplements, is a racemic mixture of eight different stereoisomers that range from 21% to 100% activity, relative to (R,R,R)-α-tocopherol (35). Although α-, β-, γ- and δ-tocopherols are absorbed equally during digestion, (R,R,R)-α-tocopherol is preferentially retained and distributed throughout the body (154). This retention is mediated by a hepatic tocopherol binding protein that shows a marked preference for α-tocopherol over β-, γ- and δ-tocopherols and non-(R,R,R)-α-tocopherol species (64). Over the past 20 years, a large and convincing body of epidemiological evidence has indicated that vitamin E supplementation at therapeutic doses (400 International Units, or approximately 250 mg of $[R,R,R]$-α-tocopherol daily) results in decreased risk for cardiovascular disease and cancer, aids in immune function, and prevents or slows a number of degenerative disease processes in humans (20, 148, 154). Note that this intake is much higher than the current adult RDAs for α-tocopherol (8 mg adult women, 10 mg adult men), levels intended merely to prevent a deficiency of this vitamin.

Plant Oils: The Major Dietary Source of Tocopherols

Plant tissues vary enormously in their total tocopherol content and tocopherol composition (Table 2), with concentrations ranging from extremely low levels in potato to very high levels in oil palm leaves and oil seeds (104). In green leafy tissues, α-tocopherol is often the most abundant tocopherol; however, such tissues contain relatively low concentrations of total tocopherols (i.e. between 10 and 50 μg/g fresh weight) (61, 104). This predominance of α-tocopherol in photosynthetic tissues presumably reflects a critical and highly conserved structural or functional role.

Unlike photosynthetic tissues, seeds often are more concentrated in total tocopherols, with their corresponding oils generally containing from 500 to 2000 μg/g tocopherols (61, 104). However, in most seed crops, including those from which the major edible oils are derived, α-tocopherol is present only as a minor component (150; Table 2). Nonetheless, seed oils still represent the major source of naturally derived dietary α-tocopherol due to the large amount of vegetable oils in the average American diet.

The Tocopherol Biosynthetic Pathway

The biosynthetic pathway for tocopherol synthesis in higher plants and algae (Figure 3) was elucidated in the early 1970s from precursor/product studies

Table 2   Tocopherol levels and composition in selected crops and plant oils[a]

Plant and organ	Total tocopherol (μg/g fresh weight)	Percent α-tocopherol	Percent others and major types
Potato tuber	0.7	90	10% γ,β-tocopherols
Lettuce leaf	7.5	55	45% γ-tocopherol
Cabbage leaf	17	100	—
Spinach leaf	30	63	5% γ-tocopherol, 33% δ-tocopherol
Synechocystis sp. PCC6803[b]	10	95	5% γ-tocopherol
Arabidopsis leaf	40	90	10% γ-tocopherol
Arabidopsis seed	350	1	95% γ-tocopherol, 4% δ-tocopherol
Oil palm leaf	300–500	100	—
Palm seed oil	500	25	30% α-tocotrienol, 40% γ-tocotrienol
Rapeseed oil	500–700	28	73% γ-tocopherol
Sunflower seed oil	700	96	4% γ,β-tocopherols
Corn seed oil	1000	20	70% γ-tocopherol, 7% δ-tocopherol
Soybean seed oil	1200	7	70% γ-tocopherol, 22% δ-tocopherol

[a]From References 61, 150.
[b]This is a photosynthetic, unicellular cyanobacterium.

using radiolabeled intermediates. Through these efforts it was shown that photosynthetic organisms synthesize α-tocopherol using a common set of enzymatic reactions (162). The first step in the pathway is the formation of homogentisic acid (HGA), the aromatic precursor common to both tocopherols and plastoquinones (162), by the enzyme p-hydroxyphenylpyruvate dioxygenase (HPPDase). The HPPDase enzyme locus (*PDS1*) has been identified by mutant analysis in *Arabidopsis* and shown to be essential for tocopherol and plastoquinone biosynthesis in plants (113). Several laboratories have now isolated cDNAs encoding HPPDase from various plant species and, surprisingly, have shown it to be a cytosolic enzyme (44, 114). HPPDase represents the first enzyme of the tocopherol biosynthetic pathway to be cloned from any photosynthetic organism.

HGA is subject to phytylation or prenylation (phytyl-PP and solanyl-PP, C_{20} and C_{45}, respectively) to form the first true tocopherol and plastoquinone intermediates, 2-methyl-6-phytylplastoquinol and 2-methyl-6-solanylplastoquinol-9, respectively. Genetic studies have identified a single locus in *Arabidopsis* (the *PDS2* locus) whose mutation disrupts both phytylation and prenylation, suggesting both reactions are mediated by a single enzyme (113). This activity is

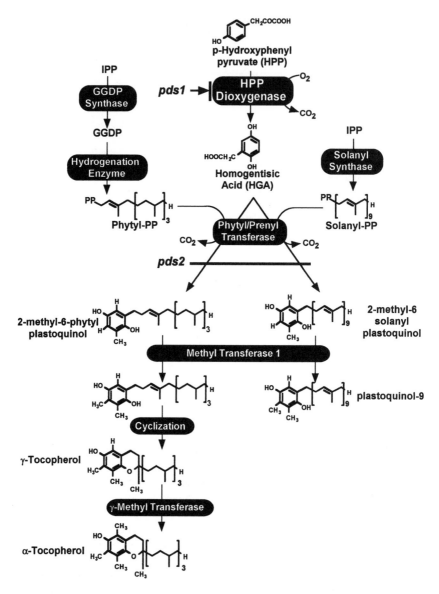

Figure 3 Tocopherol biosynthetic pathway. The pathway shown is present in all photosynthetic organisms. Enzymatic activities are labeled in black boxes. The sequence of steps to α-tocopherol after addition of the phytyl tail to HGA is the most widely accepted of many possible sequences proposed from biochemical studies. HPPDase is generally accepted as having a cytosolic localization; all other enzymes are presumably localized to plastids.

the branch point for the tocopherol and plastoquinone biosynthetic branches of the pathway, and represents a second potentially key enzymatic step regulating flux through the pathway. In addition to the level and activity of the enzyme encoded by the *PDS2* locus, the availability of various substrates for the reaction (HGA, solanyl-PP, GGDP, and phytol-PP) also are potentially important in determining the total amount and ratios of tocopherol, tocotrienol, or PQ intermediates made in a tissue. For tocopherols, production of the hydrophobic tail would minimally require a (possibly specific) GGDP synthase and a reductase for saturating GGDP to generate phytol-PP. A GGDP reductase has recently been cloned that is active toward both free GGDP and geranylated chlorophyll derivatives (78). In the coming years, overexpression of this enzyme should allow determination of its involvement in tocopherol synthesis.

2-Methyl-6-phytylplastoquinol is the common intermediate in the synthesis of all tocopherols. The next steps in α-tocopherol synthesis are ring methylations and ring cyclization. The preferred reaction sequence for α-tocopherol synthesis in isolated spinach chloroplasts is thought to be: (*a*) ring methylation at position 3 to yield 2,3-dimethyl-6-phytylplastoquinol, (*b*) cyclization to yield d-7,8 dimethyltocol (γ-tocopherol), and finally (*c*) a second ring methylation at position 5 to yield α-tocopherol (137). The first ring methylation reaction is common to both tocopherol and plastoquinone synthesis and is thought to be carried out by a single enzyme that is specific for the site of methylation on the ring but has broad substrate specificity and accommodates both classes of compounds (24, 137). The second ring methylation enzyme (γ-tocopherol methyltransferase) has an enzymatic activity distinct from the first and has been purified from both higher plants and algae (31, 67, 133). The tocopherol cyclization enzyme has been purified to homogeneity and biochemically characterized from *Anabena variabilis* (142, 143).

Although α- and γ-tocopherol biosynthesis proceeds as described above, it is not entirely clear how δ- and β-tocopherols are produced. Although not commonly found in all plant tissues, δ- and β-tocopherols can nonetheless be present at relatively high levels in certain tissues, most notably seeds. It is most likely that δ- and β-tocopherols are synthesized by a subset of the same complement of enzymes involved in α-tocopherol synthesis and only accumulate under conditions when one or both methylation enzymes are rate limiting. The overall tocopherol composition is therefore determined by the combined activities and substrate specificities of the tocopherol cyclase and two methylation enzymes present in a given tissue.

Strategies and Gene Targets for α-Tocopherol Improvement

The tocopherol biosynthetic enzymes described above can be classified into two general groups: those predominantly affecting quantitative aspects of the pathway (flux through the pathway) and those predominantly affecting qualitative aspects of the pathway (the relative amounts of α-, β-, γ, and δ-tocopherols produced). Current evidence suggests that steps involved in the formation and phytylation of HGA (HPPDase and the prenyl/phytyl transferase) and production of the phytol tail (GGDP synthase and GGDP reductase) function in a quantitative manner to regulate flux through the pathway (43). The subsequent cyclization and methylation reactions function primarily in a qualitative manner to regulate the tocopherol composition of a given plant tissue (30).

Quantitative manipulation of tocopherol levels requires metabolic engineering of what are likely to be multiple enzymatic activities in order to increase carbon flux through the pathway. HPPDase and GGDP reductase have already been cloned, and as other relevant pathway enzymes are cloned, quantitative manipulation of tocopherol levels in plant tissues may indeed become a reality. Qualitative manipulation of the existing tocopherol composition in a tissue would require positive or negative alteration of one or both of the methyltransferases. Negative alteration would cause accumulation of specific biosynthetic intermediates (β-, γ-, and δ-tocopherols), while positive manipulation would result in the conversion of any biosynthetic intermediates to the pathway end product, α-tocopherol. A tocopherol biosynthetic enzyme that is likely to have the greatest impact on the levels of α-tocopherol accumulated in a tissue is the final enzyme of the pathway, γ-tocopherol methyltransferase.

IMPROVING DIETARY α-TOCOPHEROL LEVEL BY MANIPULATING γ-TOCOPHEROL METHYLTRANSFERASE (γ-TMT) ACTIVITY   Although the most highly consumed vegetable oils in American diets (i.e. soybean, corn, and rapeseed oils) (3) contain very high levels of total tocopherol, these oils are relatively poor sources of α-tocopherol (the form with the highest vitamin E activity), because γ-tocopherol predominates. As described earlier, γ-tocopherol is methylated to form α-tocopherol in a reaction catalyzed by the enzyme γ-tocopherol methyltransferase (γ-TMT). These observations suggest that γ-TMT activity is likely limiting in the seeds of most agriculturally important oil crops and may be responsible for the low proportion of α-tocopherol synthesized and accumulated. As such, γ-TMT is a prime molecular target for manipulation of α-tocopherol levels in crops.

Given the levels and types of tocopherols in most of the important oil seed crops (Table 2), the following hypothetical scenario is an example of the impact that altering γ-TMT activity in seeds could have on the average daily intake of vitamin E. It has been estimated that between 20% and 25% of the calories consumed in American diets are derived from seed oils, with soybean oil accounting for ~70% of the edible oil consumed (28–38 g daily). However, though 1 g of soybean oil contains 1.2 mg total tocopherols, only 7% is α-tocopherol (6). One would need to consume 190–380 g of soybean oil daily

(1800–3600 calories) in order to obtain the recently recommended daily intake of 15–30 mg α-tocopherol (154). If all the γ-tocopherol in soybean oil could be converted to α-tocopherol by overexpressing γ-TMT activity in seeds, 28–38 g of such an oil would provide 26–36 mg of α-tocopherol per day, a >tenfold increase over existing soybean oil. Note that this could be achieved without needing to alter the level of total tocopherols in the oil.

ISOLATION OF GENES ENCODING γ-TMT FROM SYNECHOCYSTIS PCC6803 AND ARABIDOPSIS   In order to isolate a cDNA encoding γ-TMT, complementary molecular genetic approaches were pursued concurrently in *Arabidopsis* and the photosynthetic bacteria *Synechocystis* PCC6803 (133a). These two model organisms were selected because both synthesize α-tocopherol of identical stereochemistry by presumably identical pathways (162), and both are highly tractable genetic, molecular, and biochemical systems. The ease with which gene disruption (46) can be used to test gene function in *Synechocystis*, combined with the recent report of the complete *Synechocystis* genome sequence (75), provided a unique opportunity for taking a genomics-based approach toward identifying genes encoding tocopherol biosynthetic enzymes.

By searching the *Synechocystis* genomic database with the *Arabidopsis* HPP-Dase protein sequence, a single open reading frame (ORF) with high homology was identified. The *Synechocystis* HPPDase gene was located within a 10-ORF operon. Because bacteria often organize enzymes of biosynthetic pathways into operons to ensure their coordinate regulation, it was hypothesized that the 10-ORF operon might also contain one or more genes that encode for additional enzymes involved in tocopherol synthesis in *Synechocystis*.

Examination of this operon identified one ORF (SLR0089) that shared a high degree of similarity to Δ-(24)-sterol-C-methyltransferases. SLR0089 also contained a predicted leader peptide that would target the protein to the bacterial plasma membrane, the site of tocopherol synthesis in this organism. To test the hypothesis that SLR0089 encodes a *Synechocystis* tocopherol methyltransferase, gene replacement experiments were performed to create a SLR0089 null mutant that could be analyzed for alterations in the normal *Synechocystis* tocopherol profile. Wild-type *Synechocystis* synthesizes greater than 95% of its total tocopherols as α-tocopherol (Table 2). The SLR0089 null mutant was unable to synthesize α-tocopherol and instead accumulated the immediate precursor γ-tocopherol as its sole tocopherol, a phenotype consistent with a disruption of γ-TMT activity. Expression of the SLR0089 open reading frame in *Escherichia coli*, followed by activity assays, directly demonstrated that the expressed enzyme was able to convert γ-tocopherol to α-tocopherol in vitro. Having conclusively defined SLR0089 as a γ-TMT, its protein sequence was used to identify an ortholog from the *Arabidopsis* database. The *Arabidopsis* protein also demonstrated γ-TMT activity when expressed in *E. coli* (133a).

OVEREXPRESSION OF γ-TMT IN ARABIDOPSIS SEEDS INCREASES α-TOCOPHEROL CONTENT   Having cloned a higher plant γ-TMT, the hypothesis could be tested that γ-TMT is a key and limiting enzyme regulating the α-tocopherol composition of seeds. For this experiment, *Arabidopsis* was chosen as the model system because its seeds are composed of >95% γ-tocopherol and ~1% α-tocopherol. In such a system, even small increases in α-tocopherol levels could be easily detected. The *Arabidopsis* γ-TMT cDNA was overexpressed on a seed specific promoter; pooled segregating T2 seeds from primary transformants were analyzed for changes in tocopherol content and composition. Several independent γ-TMT overexpressing lines contained 85–95% of their total tocopherol pool as α-tocopherol (133a), a >80-fold increase over wild-type controls in α-tocopherol levels. Importantly, total seed tocopherol levels were not altered in these plants. Given the differing vitamin E potency of α-, β-, γ- and δ-tocopherols, the total vitamin E activity of 50 g of wild-type *Arabidopsis* seed oil would be 7.5 IU, while that of transgenic lines would be 67.5 IU, a ninefold increase in total vitamin E activity of the oil without increasing total tocopherols! This represents the first example of increasing a vitamin level in a plant tissue by molecular manipulation in plants. Similar increases in vitamin E activity as a result of γ-TMT overexpression can be envisioned for commercially important oils.

ISSUES AND PROSPECTS

Thomas Jefferson wrote: "The greatest service which can be rendered any country is, to add an useful plant to its culture" (70). Clearly, the development of nutritionally improved crops has immense significance to humankind, especially as our world population is expanding to over 6 billion people (38). As researchers focus more attention on the molecular mechanisms of plant nutritional physiology and biochemistry, as well as the variation that currently exists in our germplasm reserves, we in the plant science community will be in position to contribute significantly to the improvement of our plant-based food supply. However, decisions will have to be made regarding which nutrients to target and which crops to modify, such that the greatest nutritional impact is achieved. Because these decisions will require an understanding of human physiology and food chemistry, strong interdisciplinary collaborations will be needed among plant scientists, human nutritionists, and food scientists.

Improvement strategies can and should be developed now for the established, essential nutrients, as long as attention is paid to the upper safe limit of intake for each nutrient (see Table 1). However, regarding many of the nonessential phytochemicals with putative health benefits, more information is needed on their bioavailability and dose dependency, and an identification of specific molecular compounds having health efficacy is required, before plant improvement strategies should be pursued (51). Plant scientists can assist human nutritionists in this arena by providing stable isotope-labeled plant material to determine the bioavailability and subsequent metabolism of phytochemicals from whole foods (50). Additionally, efforts to identify cultivars or mutants with varied phytochemical composition can provide unique materials to be used in clinical investigations, such that the health-promoting activity of a single compound, or a class of compounds, can be deciphered.

ACKNOWLEDGMENTS

The writing of this review was supported in part by the US Department of Agriculture, Agricultural Research Service under Cooperative Agreement number 58-6250-6-001 to MAG. Work of the authors' discussed in this review was supported in part by the US Department of Agriculture, NRI-CGP through grant Numbers 94-37100-0823 to MAG and 98-01445 to DDP, as well as National Science Foundation grant number IBN-9630341 to MAG. The contents of this publication do not necessarily reflect the views or policies of the US Department of Agriculture, nor does mention of trade names, commercial products, or organizations imply endorsement by the US Government.

> Visit the *Annual Reviews home page* at
> http://www.AnnualReviews.org

Literature Cited

1. Aggett PJ, Bresson J, Haschke F, Hernell O, Koletzko B, et al. 1997. Recommended Dietary Allowances (RDAs), Recommended Dietary Intakes (RDIs), Recommended Nutrient Intakes (RNIs), and Population Reference Intakes (PRIs) are not "recommended intakes". *J. Pediatr. Gastroenter. Nutr.* 25:236–41

2. Aggett PJ, Comerford JG. 1995. Zinc and human health. *Nutr. Rev.* 53:S(II)16–22

3. American Soybean Association. 1997. *Soy Stats: A Reference Guide to Important Soybean Facts and Figures.* http://www.ag.uiuc.edu/~stratsoy/97/soystats/

4. Arnaud CD, Sanchez SD. 1996. Calcium and phosphorus. See Ref. 168, pp. 245–55

5. Baldet P, Alban C, Douce R. 1997. Biotin synthesis in higher plants: purification and characterization of bioB gene product equivalent from *Arabidopsis thaliana* overexpressed in *Escherichia*

coli and its subcellular localization in pea leaf cells. *FEBS Lett.* 419:206–10

6. Ball G. 1988. *Fat-Soluble Vitamin Assays in Food Analysis. A Comprehensive Review.* London: Elsevier. 317 pp.

7. Beard JL, Dawson H, Pinero DJ. 1996. Iron metabolism: a comprehensive review. *Nutr. Rev.* 54:295–317

8. Belanger FC, Leustek T, Chu B, Kriz AL. 1995. Evidence for the thiamin biosynthetic pathway in higher plants and its developmental regulation. *Plant Mol. Biol.* 29:809–21

9. Bienfait HF. 1988. Proteins under the control of the gene for Fe efficiency in tomato. *Plant Physiol.* 88:785–87

10. Bienfait HF. 1989. Prevention of stress in iron metabolism of plants. *Acta Bot. Neerl.* 38:105–29

11. Bingham SA, Atkinson C, Liggins J, Bluck L, Coward A. 1998. Phytooestrogens: Where are we now? *Br. J. Nutr.* 79:393–406

12. Block G, Patterson B, Subar A. 1992.

Fruit, vegetables, and cancer prevention: a review of the epidemiological evidence. *Nutr. Cancer* 18:1–29

13. Bouis H. 1996. Enrichment of food staples through plant breeding: a new strategy for fighting micronutrient malnutrition. *Nutr. Rev.* 54:131–37

14. Briat J-F, Fobis-Loisy I, Grignon N, Lobréaux S, Pascal N, et al. 1995. Cellular and molecular aspects of iron metabolism in plants. *Biol. Cell* 84:69–81

15. Briat J-F, Lobréaux S. 1997. Iron transport and storage in plants. *Trends Plant Sci.* 2:187–93

16. Briggs SP. 1998. Plant genomics: more than food for thought. *Proc. Natl. Acac. Sci. USA* 95:1986–88

17. Bronner F. 1993. Nutrient bioavailability, with special reference to calcium. *J. Nutr.* 123:797–802

18. Brown JC, Cathey HM, Bennett JH, Thimijan RW. 1979. Effect of light quality and temperature on Fe^{3+} reduction, and chlorophyll concentration in plants. *Agron. J.* 71:1015–21

19. Brüggemann W, Maas-Kantel K, Moog PR. 1993. Iron uptake by leaf mesophyll cells: the role of the plasma membrane-bound ferric-chelate reductase. *Planta* 190:151–55

20. Buring JE, Hennekens CH. 1997. Antioxidant vitamins and cardiovascular disease. *Nutr. Rev.* 55:S53–60

21. Butterworth CE Jr, Bendich A. 1996. Folic acid and the prevention of birth defects. *Annu. Rev. Nutr.* 16:73–97

22. Calloway DH. 1995. *Human Nutrition: Food and Micronutrient Relationships.* Washington, DC: Int. Food Policy Res. Inst. 23 pp.

23. Consaul JR, Lee K. 1983. Extrinsic tagging in iron bioavailability research: a critical review. *J. Agric. Food Chem.* 31:684–89

24. Cook WB, Miles D. 1992. Nuclear mutations affecting plastoquinone accumulation in maize. *Photosyn. Res.* 31:99–111

25. Crowley DE, Wang YC, Reid CPP, Szaniszlo PJ. 1991. Mechanisms of iron acquisition from siderophores by microorganisms and plants. In *Iron Nutrition and Interactions in Plants*, ed. Y Chen, Y Hadar, pp. 213–32. Dordrecht: Kluwer

26. Cunningham FX Jr, Gantt E. 1998. Genes and enzymes of carotenoid biosynthesis in plants. *Annu. Rev. Plant Physiol. Plant Mol. Biol.* 49:557–83

27. Decker EA. 1997. Phenolics: prooxi-dants or antioxidants? *Nutr. Rev.* 55:396–98

28. Delange F. 1994. The disorders induced by iodine deficiency. *Thyroid* 4:107–28

29. DeLuca HF, Zierold C. 1998. Mechanisms and functions of vitamin D. *Nutr. Rev.* 56:S4–10

30. Demurin Y. 1993. Genetic variability of tocopherol composition in sunflower seeds. *Helia* 16:59–62

31. D'Harlingue A, Camara B. 1985. Plastid enzymes of terpenoid biosynthesis: purification and characterization of a gamma-tocopherol methyltransferase. *J. Biol. Chem.* 260:15200–3

32. Dreosti IE. 1995. Magnesium status and health. *Nutr. Rev.* 53:S(II)23–27

33. Dreosti IE. 1996. Bioactive ingredients: antioxidants and polyphenols in tea. *Nutr. Rev.* 54:S51–58

34. Eide D, Broderius M, Fett J, Guerinot ML. 1996. A novel iron-regulated metal transporter from plants identified by functional expression in yeast. *Proc. Natl. Acad. Sci. USA* 93:5624–28

35. Eitenmiller RR. 1997. Vitamin E content of fats and oils: nutritional implications. *Food Technol.* 51:78–81

36. Evans HM, Bishop KS. 1922. On the existence of a hitherto unrecognized dietary factor essential for reproduction. *Science* 56:650–51

37. Fahey JW, Zhang Y, Talalay P. 1997. Broccoli sprouts: an exceptionally rich source of inducers of enzymes that protect against chemical carcinogens. *Proc. Natl. Acad. Sci. USA* 94:10367–72

38. Food and Agriculture Organization of the United Nations, Int. Life Sci. Inst. 1997. *Preventing Micronutrient Malnutrition: A Guide to Food-Based Approaches.* Washington, DC: Int. Life Sci. Inst. 105 pp.

39. Fox TC, Guerinot ML. 1998. Molecular biology of cation transport in plants. *Ann. Rev. Plant Physiol. Plant Mol. Biol.* 49:669–96

40. Franceschi VR, Horner HT Jr. 1980. Calcium oxalate crystals in plants. *Bot. Rev.* 46:361–427

41. Fryer MJ. 1992. The antioxidant effects of thylakoid vitamin E (α-tocopherol) *Plant Cell Environ.* 15:381–92

42. Fukuzawa K, Tokumura A, Ouchi S, Tsukatani H. 1982. Antioxidant activities of tocopherols on Fe^{2+}-ascorbate-induced lipid peroxidation in lectithin liposomes. *Lipid* 17:511–13

43. Furuya T, Yoshikawa T, Kimura T, Kaneko H. 1987. Production of toco-

pherols by cell cultures of safflower. *Phytochemistry* 26:2741–47

44. Garcia I, Rodgers M, Lenne C, Rolland A, Sailland A, Matringe M. 1997. Subcellular localization and purification of a p-hydroxyphenylpyruvate dioxygenase from cultured carrot cells and characterization of the corresponding cDNA. *Biochem. J.* 325:761–69

45. Glahn RP, Van Campen DR. 1997. Iron uptake is enhanced in Caco-2 cell monolayers by cysteine and reduced cysteinyl glycine. *J. Nutr.* 127:642–47

46. Golden S. 1988. Mutagenesis of cyanobacteria by classical and gene-transfer-based methods. *Meth. Enzymol.* 167:714–27

47. Graham RD, Welch RM. 1996. *Breeding For Staple Food Crops With High Micronutrient Density.* Washington, DC: Int. Food Policy Res. Inst. 79 pp.

48. Grusak MA. 1994. Iron transport to developing ovules of *Pisum sativum.* I. Seed import characteristics and phloem iron-loading capacity of source regions. *Plant Physiol.* 104:649–55

49. Grusak MA. 1995. Whole-root iron(III)-reductase activity throughout the life cycle of iron-grown *Pisum sativum* L. (Fabaceae): relevance to the iron nutrition of developing seeds. *Planta* 197:111–17

50. Grusak MA. 1997. Intrinsic stable isotope labeling of plants for nutritional investigations in humans. *Nutr. Biochem.* 8:164–71

51. Grusak MA, DellaPenna D, Welch RM. 1999. Physiological processes affecting the content and distribution of phytonutrients in plants. *Nutr. Rev.* In press

52. Grusak MA, Pearson N, Marentes E. 1999. The physiology of micronutrient homeostasis in field crops. *Field Crops Res.* 60:41–56

53. Grusak MA, Pezeshgi S. 1996. Shoot-to-root signal transmission regulates root Fe(III) reductase activity in the *dgl* mutant of pea. *Plant Physiol.* 110:329–34

54. Grusak MA, Stephens BW, Merhaut DJ. 1996. Influence of whole-plant net calcium influx and partitioning on calcium concentration in snap bean pods. *J. Am. Soc. Hortic. Sci.* 121:656–59

55. Grusak MA, Welch RM, Kochian LV. 1990. Does iron deficiency in *Pisum sativum* enhance the activity of the root plasmalemma iron transport protein? *Plant Physiol.* 94:1353–57

56. Guerinot ML, Yi Y. 1994. Iron: nutritious, noxious, and not readily available. *Plant Physiol.* 104:815–20

57. Harper AE. 1985. Origin of Recommended Dietary Allowances—an historic overview. *Am. J. Clin. Nutr.* 41:140–48

58. Harper AE. 1987. Evolution of Recommended Dietary Allowances—new directions? *Annu. Rev. Nutr.* 7:509–37

59. Henle ES, Luo Y, Linn S. 1996. Fe^{2+}, Fe^{3+}, and oxygen react with DNA-derived radicals formed during iron-mediated Fenton reactions. *Biochemistry* 35:12212–19

60. Herbert V. 1996. Vitamin B-12. See Ref. 168, pp. 191–205

61. Hess JL. 1993. Vitamin E, α-tocopherol. In *Antioxidants in Higher Plants*, ed. R Alscher, J Hess, pp. 111–134. Boca Raton, FL: CRC

62. Higuchi K, Kanazawa K, Nishizawa N-K, Chino M, Mori S. 1994. Purification and characterization of nicotianamine synthase from Fe-deficient barley roots. *Plant Soil* 165:173–79

63. Holden MJ, Luster DG, Chaney RL, Buckhout TJ, Robinson C. 1991. Fe^{3+}-chelate reductase activity of plasma membranes isolated from tomato (*Lycopersicon esculentum* Mill.) roots. Comparison of enzymes from Fe-deficient and Fe-sufficient roots. *Plant Physiol.* 97:537–44

64. Hosomi A, Arita M, Sato Y, Kiyose C, Ueda T, Igarashi O, Arai H, Inoue K. 1997. Affinity for α-tocopherol transfer protein as a determinant of the biological activities of vitamin E analogs. *FEBS Lett.* 409:105–8

65. Innis SM. 1996. Essential dietary lipids. See Ref. 168, pp. 58–66

66. Institute of Medicine (U.S.), Food and Nutrition Board. 1994. *How Should the Recommended Dietary Allowances be Revised?* Washington, DC: Natl. Acad. 36 pp.

67. Ishiko H, Shigeoka S, Nakano Y, Mitsunaga T. 1992. Some properties of gamma-tocopherol methyltransferase solubilized from spinach chloroplast. *Phytochemistry* 31:1499–500

68. Jacob RA, Swendseid ME. 1996. Niacin. See Ref. 168, pp. 184–90

69. Jang M, Cai L, Udeani GO, Slowing KV, Thomas CF, et al. 1997. Cancer chemopreventive activity of resveratrol, a natural product derived from grapes. *Science* 275:218–20

70. Jefferson T. 1984. A memorandum (Services to my country), circa 1800. In *Writings*, ed. MD Peterson, pp. 702–4. New York: Library of America. 1600 pp.

71. Kabata-Pendias A, Pendias H. 1992. *Trace Elements in Soils and Plants*. Boca Raton, FL: CRC. 365 pp. 2nd ed.

72. Kamal-Eldin A, Appelqvist LA. 1996. The chemistry and antioxidant properties of tocopherols and tocotrienols. *Lipids* 31:671–701

73. Kanazawa K, Higuchi K, Nishizawa N-K, Fushiya S, Chino M, et al. 1994. Nicotianamine aminotransferase activities are correlated to the phytosiderophore secretions under Fe-deficient conditions in Gramineae. *J. Exp. Bot.* 45:1903–6

74. Kanazawa K, Higuchi K, Nishizawa N-K, Fushiya S, Mori S. 1998. Detection of two distinct isozymes of nicotianamine aminotransferase in Fe-deficient barley roots. *J. Exp. Bot.* 46: 1241–44

75. Kaneko T, Sato S, Kotani H, Tanaka A, Asamizu E, et al. 1996. Sequence analysis of the genome of the unicellular Cyanobacterium *Synechocystis sp.* strain PCC6803. II. Sequence determination of the entire genome and assignment of potential protein-coding regions. *DNA Res.* 3:109–36

76. Keen CL, Zidenberg-Cherr S. 1996. Manganese. See Ref. 168, pp. 334–43

77. Kehrer JP, Smith CV. 1994. Free radicals in biology: sources, reactivities, and roles in the etiology of human diseases. In *Natural Antioxidants in Human Health and Diseases*, ed. B Frei, pp. 2:25-62. San Diego: Academic. 588 pp.

78. Keller Y, Bouvier F, d'Harlingue A, Camara B. 1998. Metabolic compartmentation of plastid prenyllipid biosynthesis—evidence for the involvement of a multifunctional geranylgeranyl reductase. *Eur. J. Biochem.* 251:413–17

79. Kleerekoper M, Balena R. 1991. Fluorides and osteoporosis. *Annu. Rev. Nutr.* 11:309–24

80. Kneen BE, LaRue TA, Welch RM, Weeden NF. 1990. Pleiotropic effects of *brz*. A mutation in *Pisum sativum* (L.) cv 'Sparkle' conditioning decreased nodulation and increased iron uptake and leaf necrosis. *Plant Physiol.* 93:717–22

81. Kochian LV. 1991. Mechanisms of micronutrient uptake and translocation in plants. In *Micronutrients in Agriculture*, ed. JJ Mordvedt, FR Cox, LM Shuman, RM Welch, pp. 229–96. Madison, WI: Soil Sci. Soc. Am. 2nd ed.

82. Kurzer MS, Xu X. 1997. Dietary phytoestrogens. *Annu. Rev. Nutr.* 17:353–81

83. Lachance PA. 1998. Overview of key nutrients: micronutrient aspects. *Nutr. Rev.* 56:S34–39

84. Landsberg E-C. 1984. Regulation of iron-stress-response by whole-plant activity. *J. Plant Nutr.* 7:609–21

85. Landsberg E-C. 1996. Hormonal regulation of iron-stress response in sunflower roots: a morphological and cytological investigation. *Protoplasma* 194:69–80

86. Laulhère J-P, Barcelò F, Fontecave M. 1995. Dynamic equilibria in iron uptake and release by ferritin. *BioMetals* 9:303–9

87. Laulhère J-P, Briat J-F. 1993. Iron release and uptake by plant ferritin: effects of pH, reduction and chelation. *Biochem. J.* 290:693–99

88. Leklem JE. 1996. Vitamin B-6. See Ref. 168, pp. 174–83

89. Levander OA, Burk RF. 1996. Selenium. See Ref. 168, pp. 320–28

90. Linder MC. 1991. *Nutritional Biochemistry and Metabolism: With Clinical Applications*. New York: Elsevier. 603 pp. 2nd ed.

91. Linder MC. 1996. Copper. See Ref. 168, pp. 307–19

92. Lobréaux S, Thoiron S, Briat J-F. 1995. Induction of ferritin synthesis in maize leaves by an iron-mediated oxidative stress. *Plant J.* 8:443–49

93. Lucas WJ, Ding B, Van der Schoot C. 1993. Plasmodesmata and the supracellular nature of plants. *New Phytol.* 125:435–76

94. Luft FC. 1996. Potassium and its regulation. See Ref. 168, pp. 272–76

95. Luwe MWF, Takahama U, Heber U. 1993. Role of ascorbate in detoxifying ozone in the apoplast of spinach (*Spinacia oleracea* L.) leaves. *Plant Physiol.* 101:969–76

96. Ma J-F, Kusano G, Kimura S, Nomoto K. 1993. Specific recognition of mugineic acid-ferric complex by barley roots. *Phytochemistry* 34:599–603

97. Ma J-F, Shinada T, Matsuda C, Nomoto K. 1995. Biosynthesis of phytosiderophores, mugineic acids associated with methionine recycling. *J. Biol. Chem.* 270:16549–54

98. Maas FM, van de Wetering DAM, van Buesichem ML, Bienfait HF. 1988. Characterization of phloem iron and its possible role in the regulation of Fe-efficiency reactions. *Plant Physiol.* 87:167–71

99. Marentes E, Grusak MA. 1998. Iron transport and storage within the seed coat and embryo of developing seeds of

pea (*Pisum sativum* L.). *Seed Sci. Res.* 8:367–75

100. Marschner H. 1995. *Mineral Nutrition of Higher Plants.* San Diego: Academic. 889 pp.

101. Marschner H, Römheld V. 1994. Strategies of plants for acquisition of iron. *Plant Soil* 165:261–74

102. Maxwell MH, Kleeman CR, Narins RG. 1987. *Clinical Disorders of Fluid and Electrolyte Metabolism.* New York: McGraw-Hill

103. McCormick DB. 1994. Riboflavin. In *Modern Nutrition in Health and Disease*, ed. ME Shils, JA Olson, M Shike, pp. 366–75. Philadelphia: Lea & Febiger. 8th ed.

104. McLaughlin P, Weihrauch JC. 1979. Vitamin E content of foods. *J. Am. Diet. Assoc.* 75:647–65

105. Mertz W. 1993. Chromium in human nutrition: a review. *J. Nutr.* 123:626–33

106. Mock DM. 1996. Biotin. See Ref. 168, pp. 220–35

107. Moog PR, Brüggemann W. 1994. Iron reductase systems on the plant plasma membrane—a review. *Plant Soil* 165:241–60

108. Mori S. 1994. Mechanisms of iron acquisition by graminaceous (Strategy II) plants. In *Biochemistry of Metal Micronutrients in the Rhizosphere*, ed. JA Manthey, DE Crowley, DG Luster, 15:225–249. Boca Raton, FL: Lewis. 372 pp.

109. Mori S, Nishizawa N. 1987. Methionine as a dominant precursor of phytosiderophores in *Graminaceae* plants. *Plant Cell Physiol.* 28:1081–92

110. National Research Council (U.S.), Food and Nutrition Board. 1989. *Recommended Dietary Allowances.* Washington, DC: Natl. Acad. 284 pp. 10th ed.

111. Neuburger M, Rebeille F, Jourdain A, Nakamura S, Douce R. 1996. Mitochondria are a major site for folate and thymidylate synthesis in plants. *J. Biol. Chem.* 271:9466–72

112. Nishizawa N, Mori S. 1987. The particular vesicle appearing in the barley root cells and its relation to mugineic acid secretion. *J. Plant Nutr.* 11:915–24

113. Norris SR, Barrette TR, DellaPenna D. 1995. Genetic dissection of carotenoid synthesis in Arabidopsis defines plastoquinone as an essential component of phytoene desaturation. *Plant Cell* 7:2139–48

114. Norris SR, Shen X, DellaPenna D. 1998. Complementation of the Arabidopsis *pds1* mutation with the gene encoding p-hydroxyphenylpyruvate dioxygenase. *Plant Physiol.* 117:1317–23

115. Pich A, Scholz G, Stephan UW. 1994. Iron-dependent changes of heavy metals, nicotianamine, and citrate in different organs and in the xylem exudate of two tomato genotypes. Nicotianamine as possible copper translocator. *Plant Soil* 165:189–96

116. Proudhon D, Wei J, Briat J-F, Theil EC. 1996. Ferritin gene organization: Differences between plants and animals suggest possible kingdom-specific selective constraints. *J. Mol. Evol.* 42:325–36

117. Pushnik JC, Miller GW. 1989. Iron regulation of chloroplast photosynthetic function: mediation of PSI development. *J. Plant Nutr.* 12:407–21

118. Quintana JM, Harrison HC, Nienhuis J, Palta JP, Grusak MA. 1996. Variation in calcium concentration among sixty S_1 families and four cultivars of snap bean (*Phaseolus vulgaris* L.). *J. Am. Soc. Hortic. Sci.* 121:789–93

119. Rajagopalan KV. 1988. Molybdenum: an essential trace element in human nutrition. *Annu. Rev. Nutr.* 8:401–27

120. Rebeille F, Macherel D, Mouillon JM, Garin J, Douce R. 1997. Folate biosynthesis in higher plants: purification and molecular cloning of a bifunctional 6-hydroxymethyl-7,8-dihydropterin pyrophosphokinase/7,8-dihydropteroate synthase localized in mitochondria. *EMBO J.* 16:947–57

121. Rindi G. 1996. Thiamin. See Ref. 168, pp. 160–66

122. Robinson NJ, Sadjuga, Groom QJ. 1997. The *froh* gene family from *Arabidopsis thaliana*: putative iron-chelate reductases. *Plant Soil* 196:245–48

123. Robinson NJ, Sadjuga MR, Groom QJ. 1997. The *froh* gene family from *Arabidopsis thaliana*: putative iron-chelate reductases. In *Plant Nutrition for Sustainable Food Production and Environment*, ed. T Ando, pp. 191–94. Dordrecht: Kluwer. 982 pp.

124. Rock CL. 1997. Carotenoids: biology and treatment. *Pharmacol. Ther.* 75:185–97

125. Romera FJ, Alcántara E, de la Guardia MD. 1992. Role of roots and shoots in the regulation of the Fe efficiency responses in sunflower and cucumber. *Physiol. Plant.* 85:141–46

126. Saier MH Jr. 1998. Genome sequencing and informatics: new tools for biochemical discoveries. *Plant Physiol.* 117:1129–33

127. Salunkhe DK, Desai BB. 1987. Effects of agricultural practices, handling, processing, and storage on vegetables. In *Nutritional Evaluation of Food Processing*, ed. E Karmas, RS Harris, 3:23–71. New York: AVI. 786 pp. 3rd ed.

128. Samuelsen AI, Martin RC, Mok DWS, Mok MC. 1998. Expression of the yeast *FRE* genes in transgenic tobacco. *Plant Physiol.* 118:51–58

129. Sauberlich HE. 1994. Pharmacology of vitamin C. *Annu. Rev. Nutr.* 14:371–91

130. Scholz G, Becker R, Pich A, Stephan UW. 1992. Nicotianamine—a common constituent of strategies I and II of iron acquisition by plants: a review. *J. Plant Nutr.* 15:1647–65

131. Scholz G, Becker R, Stephan UW, Rudolph A, Pich A. 1988. The regulation of iron uptake and possible functions of nicotianamine in higher plants. *Biochem. Physiol. Pflanzen* 183:257–69

132. Schonhof I, Krumbein A. 1996. Gehalt an wertgebenden Inhaltsstoffen verschiedener Brokkolitypen (Brassica oleracea var. italica Plenck). *Gartenbauwissenschaft* 61:281–88

133. Shigeoka S, Ishiko H, Nakano Y, Mitsunaga T. 1992. Isolation and properties of α-tocopherol methyltransferase in *Euglena gracilis. Biochim. Biophys. Acta* 1128:220-26

133a. Shintani D, DellaPenna D. 1998. Elevating the vitamin E content of plants through metabolic engineering. *Science.* 282:2098–100

134. Shojima S, Nishizawa N-K, Fushiya S, Nozoe S, Irifune T, Mori S. 1990. Biosynthesis of phytosiderophores. *In vitro* biosynthesis of 2′-deoxymugineic acid from L-methionine and nicotianamine. *Plant Physiol.* 93:1497–503

135. Simon PW, Wolff XY, Peterson CE, Kammerlohr DS. 1989. High carotene mass carrot population. *HortScience* 24:174

136. Soll J, Schultz G. 1979. Comparison of geranylgeranyl and phytyl substituted methylquinols in the tocopherol synthesis of spinach chloroplast. *Biochem. Biophys. Res. Commun.* 91:715–20

137. Soll J, Schultz G. 1980. 2-methyl-6-phytyquinol and 2,3-dimethyl-5-phytylquinol as precursors of tocopherol synthesis in spinach chloroplasts. *Phytochemistry* 19:215–18

138. Souci SW, Fachmann W, Kraut H. 1989. *Food Composition and Nutrition Tables 1989/90*. Stuttgart: Wiss. Verl.ges. 1028 pp. 4th ed.

139. Steinmetz KA, Potter JD. 1996. Vegetables, fruit, and cancer prevention: a review. *J. Am. Diet. Assoc.* 96:1027–39

140. Stephan UW, Schmidke I, Pich A. 1994. Phloem translocation of Fe, Cu, Mn, and Zn in *Ricinus* seedlings in relation to the concentrations of nicotianamine, an endogenous chelator of divalent metal ions, in different seedling parts. *Plant Soil* 165:181–88

141. Stephan UW, Scholz G. 1993. Nicotianamine: mediator of transport of iron and heavy metals in the phloem? *Physiol. Plant.* 88:522–29

142. Stocker A, Fretz H, Frick H, Ruttimann A, Woggon WD. 1996. The substrate specificity of tocopherol cyclase. *Bioorg. Med. Chem.* 4:1129–34

143. Stocker A, Netscher T, Ruttimann A, Muller RK, Schneider H, et al. 1994. The reaction mechanism of chromanol-ring formation catalyzed by tocopherol cyclase from *Anabaena variabilis* Kutzing (Cyanobacteria). *Helv. Chim. Acta* 77:1721–37

144. Stommel JR. 1994. Inheritance of beta carotene content in the wild tomato species *Lycopersicon cheesmanii. J. Hered.* 85:401–4

145. Subar A, Heimendinger J, Krebs-Smith S, Patterson B, Kessler R, Pivonka E. 1992. *5 A Day for Better Health: A Baseline Study of Americans' Fruit and Vegetable Consumption.* Washington, DC: Natl. Cancer Inst.

146. Tahiliani AG, Beinlich CJ. 1991. Pantothenic acid in health and disease. *Vitam. Horm.* 46:165–228

147. Takagi S-I. 1993. Production of phytosiderophores. In *Iron Chelation in Plants and Soil Microorganisms*, ed. LL Barton, BC Hemming, 4:111–31. San Diego: Academic. 490 pp.

148. Tangney CC. 1997. Vitamin E and cardiovascular disease. *Nutr. Today* 32:13–22

149. Tanksley SD, McCouch SR. 1997. Seed banks and molecular maps: unlocking genetic potential from the wild. *Science* 277:1063–66

150. Taylor P, Barnes P. 1981. Analysis of vitamin E in edible oils by high performance liquid chromatography. *Chem. Ind.* (Oct. 17):722–26

151. Tigchelaar EC, Tomes ML. 1974. Carorich tomato. *HortScience* 9:82

152. Tijburg LB, Mattern T, Folts JD, Weisgerber UM, Katan MB. 1997. Tea flavonoids and cardiovascular disease: a review. *Crit. Rev. Food Sci. Nutr.* 37:771–85

153. Timberlake WE. 1998. Agricultural genomics comes of age. *Nature Biotech.* 16:116–17

154. Traber MG, Sies H. 1996. Vitamin E in humans: demand and delivery. *Annu. Rev. Nutr.* 16:321–347

154a. US Department of Agriculture, Agricultural Research Service. 1998. *USDA Nutrient Database for Standard Reference, Release 12.* Nutrient Data Laboratory Home Page, http://www.nal.usda.gov/fnic/foodcomp

155. Vermeer C, Jie K-SG, Knapen MHJ. 1995. Role of vitamin K in bone metabolism. *Annu. Rev. Nutr.* 15:1–22

156. von Wirén N, Marschner H, Römheld V. 1995. Uptake kinetics of iron-phytosiderophores in two maize genotypes differing in iron deficiency. *Physiol. Plant.* 93:611–16

157. Wang M, Goldman IL. 1996. Phenotypic variation in free folic acid content among F_1 hybrids and open-pollinated cultivars of red beet. *J. Am. Soc. Hortic. Sci.* 121:1040–42

158. Weaver CM, Heaney RP. 1991. Isotopic exchange of ingested calcium between labeled sources. Evidence that ingested calcium does not form a common absorptive pool. *Calcif. Tissue Int.* 49:244–47

159. Welch RM. 1986. Effects of nutrient deficiencies on seed production and quality. *Adv. Plant Nutr.* 2:205–47

160. Welch RM. 1995. Micronutrient nutrition of plants. *Crit. Rev. Plant Sci.* 14:49–82

161. Welch RM, Combs GF Jr, Duxbury JM. 1997. Toward a 'greener' revolution. *Issues Sci. Technol.* (Fall):55–63

162. Whistance GR, Threlfall DR. 1970. Biosynthesis of phytoquinones; homogentisic acid: a precursor of plastoquinones, tocopherols and alpha-tocopherolquinone in higher plants, green algae and blue-green algae. *Biochem. J.* 117:593–600

163. White MC, Decker AM, Chaney RL. 1981. Metal complexation in xylem fluid. I. Chemical composition of tomato and soybean stem exudate. *Plant Physiol.* 67:292–300

164. Winzerling JJ, Law LH. 1997. Comparative nutrition of iron and copper. *Annu. Rev. Nutr.* 17:501–26

165. World Cancer Research Fund, American Institute for Cancer Research. 1997. *Food, Nutrition and the Prevention of Cancer: a Global Perspective.* Washington, DC: Am. Inst. Cancer Res.

166. Yi Y, Guerinot ML. 1996. Genetic evidence that induction of root Fe(III) chelate reductase activity is necessary for iron uptake under iron deficiency. *Plant J.* 10:835–44

167. Zhang F-U, Römheld V, Marschner H. 1991. Diurnal rhythm of release of phytosiderophores and uptake rate of zinc in iron-deficient wheat. *Soil Sci. Plant Nutr.* 37:671–78

168. Ziegler EE, Filer LJ Jr, eds. 1996. *Present Knowledge in Nutrition.* Washington, DC: Int. Life Sci. Inst. 684 pp. 7th ed.

Annu. Rev. Plant Physiol. Plant Mol. Biol 1999. 50:163–86

GAMETOPHYTE DEVELOPMENT IN FERNS

Jo Ann Banks
Department of Botany and Plant Pathology, Purdue University, West Lafayette, Indiana 47907-1153; e-mail: banks@btny.purdue.edu

KEY WORDS: alternation of generations, fertilization, mutants, photomorphogenesis, sex determination, spermatogenesis

---

ABSTRACT

The fern gametophyte has interested plant biologists for the past century because its structure and development is simple and amenable to investigation. Past studies have described many aspects of its development, including germination of the spore, patterns of cell division and differentiation, photomorphogenic or light-regulated responses, sex determination and differentiation of gametangia, hormone and pheromone responses, and fertilization. Several genes that are predicted to regulate some of these processes have been recently cloned, making it possible to analyze how these processes are controlled at a molecular level. The emergence of the fern *Ceratopteris richardii* as a model organism for readily identifying and characterizing mutations that affect key developmental processes in gametophytes makes it a powerful tool for dissecting the molecular mechanisms underlying these processes. If advances in gene cloning techniques and transformation are forthcoming in *Ceratopteris*, it is likely that the study of developmental processes in ferns will significantly contribute to our understanding of plant development and evolution beyond that which can be learned solely from studying angiosperms.

CONTENTS

163

Introduction

This review describes recent experimental research that has explored the growth and development of the gametophytes of ferns, the second largest group of modern vascular plants in terms of species diversity (26). Although ferns are more primitive than seed plants and are of little economic value, their primitive features have been exploited for studying many fundamental aspects of plant development. These studies, conducted over the past century, have resulted in a large and diverse body of literature that was, in many cases, pivotal in giving other biologists important insights into angiosperm development. Most of the older literature describing these studies has been concisely reviewed in two books: *The Experimental Biology of Ferns*, edited by AF Dyer (17), and *The Developmental Biology of Fern Gametophytes* by V Raghavan (53). The former book discusses both the gametophyte and the sporophyte generations of ferns, while the latter discusses only the gametophyte generation. These books are invaluable resources for citations to literature that is too old to be in electronic reference libraries. This review examines what has been learned of gametophyte development since the publication of these books, particularly in fern species that have been developed as model systems.

What are Ferns, and Why Study Them?

The group of plants discussed in this review includes those in the order Filicales (Class Filicopsida, Division Tracheophyta, Kingdom Plantae), following the classification of Stewart & Rothwell (59). Members of the Filicales are leptosporangiate (with a sporangial wall that is one cell layer thick), homosporous (producing only one type of spore), and have other well-known, fern-like traits. These traits include sporophytes with adventitious roots, large fronds (leaves) that unfurl from crosiers or fiddleheads, and sporangia on the abaxial surface of leaves. Their gametophytes are small, exosporic, and free-living. Recent molecular analyses indicate that the Filicales are a monophyletic group (27) that first appeared in the fossil record about 250 million years ago (59). Theories of the evolutionary origins of ferns have been recently discussed (49, 59).

   Although ferns are not important crops and have little impact on the human species, it is useful to understand what can be learned about plant development using fern gametophytes as experimental systems. Like all vascular plants, the fern life cycle alternates between two distinct phases or generations: a diploid sporophytic phase and a haploid gametophytic phase, the former representing the asexual, spore-producing phase and the latter the sexual, gamete-producing phase of the life cycle. In ferns and other more primitive plants, the haploid gametophytic phase of the life cycle, from spore germination to fertilization of the gametes, is completely independent of the diploid sporophytic plant. The mature fern gametophyte is a multicellular, autotrophic yet small (~1–2 mm) structure. As an experimental organism, the fern gametophyte is ideal for study given that it can be cultured to maturity in a petri dish and all aspects of its growth and development observed and manipulated in a nondestructive way. How environmental and genetic factors influence spore formation and germination, how the haploid spore initiates and develops the basic body plan of the gametophyte, how the gametophyte grows and differentiates sex cells, how fertilization occurs, and how the gametophyte patterns its development to nurture the developing sporophytic zygote and embryo are among the important questions unique to the gametophyte that can be addressed most effectively using ferns. The most recent and significant advance in addressing these questions has been the development and use of the fern *Ceratopteris richardii* as a model genetic system for study (35, 36). Like *Arabidopsis thaliana*, *Ceratopteris* can be used to identify and study mutants, particularly those that affect the fern gametophyte. Using this plant, biologists can now begin to identify and dissect the complex genetic regulatory pathways underlying the developmental processes unique to the gametophyte.

Ferns occupy a middle branch of the vascular plant evolutionary tree. Like angiosperms, ferns have true leaves, which is a reflection of their comparatively advanced vascular systems. Because ferns have characteristics typical of both primitive and advanced land plants, ferns are key to understanding how genetic changes in the basic regulatory machinery underlying plant development and morphology have evolved. The similarities and differences in the development of ferns and angiosperms are highlighted in this review in an effort to define the major evolutionary changes that occurred in land plant evolution as the lineages that gave rise to the modern ferns and angiosperms diverged from their common ancestor.

Following a brief description of *Ceratopteris*, the organization of this review follows the chronological development of the gametophyte, beginning with the acquisition and breaking of spore dormancy and ending with the fertilization of the egg. The different phases of the *Ceratopteris* life cycle, illustrating both the gametophyte and sporophyte generations, are shown in Figure 1 (see the color section at the end of the volume).

Ceratopteris Richardi, A Model Fern Species

A description of the biology of the fern *Ceratopteris* and the unique features of *Ceratopteris* that have made it a model system have been recently reviewed (12, 19, 36, 37, 79). Commonly referred to as the "C-Fern," *Ceratopteris* is an annual, semitropical genus that grows in semiaquatic environments, including rice paddies. Its life cycle is relatively short, about three to four months from spore to spore. The diploid sporophyte plant forms fronds that produce ~10^6

haploid spores per month. The same fronds also form hundreds of vegetative, adventitious buds in the axils of the leaves so that each plant can be propagated vegetatively. The ability to generate vast quantities of genetically identical spores from a sporophyte derived from an intra-gametophytic cross is important for genetic studies, as it permits the isolation of revertants or second-site suppressors of known mutations in a very short period of time. High spore production is also an asset for molecular and biochemical studies of early gametophyte development, because obtaining enough spores for study is not a limitation.

A variety of mutations that affect gametophyte development in *Ceratopteris* are listed in Table 1. Single-celled spores can be treated with EMS or irradiated

Table 1   Mutants of *Ceratopteris*

Type	Allele designation	Reference
Photomorphogenic		
Dark-germinating	*dkg1*	12, 56
De-etiolated	*germ3, 4*	11, 12
Reduced light-mediated inhibition of germination and/or elongation	*germ1, 2*	11, 12
Potential cytoskeletal defects		
Clumped chloroplasts	*cp1, 2*	67
Defective sperm flagella	*230x*	16
Sex determination and differentiation		
Hermaphoroditic, antheridiogen insensitive	*her* (50)[a]	2, 55, 77
Transformer	*tra* (6)[a]	4
Feminization	*fem* (15)[a]	2
Many antheridia	*man1*	4
Disorganized meristem	*dim1, 2*	R Smith & JA Banks (unpublished observations)
Irregular meristem	HαTUBE1[b]	36
Resistant to		
Abscisic acid	*abr48, 104*	30
Paraquat	*pq2, pq45, pqa*	9, 31–33
Aciflrorfen	6	34
Glyphosate	*blt1*	62
NaCl	*glt1, 2*	34, 68, 75, 78, 79
Hydroxy-L-proline	*stl1, 2*	58
Azetidine-2-carboxylate	HαYn[b]	34
2-aminoethyl-L-cysteine	HαAzn[b]	34
5-flurodeoxyuridine	HαCYn[b]	34
Al$_2$(SO$_4$)$_3$ at pH 4.4	HαFn[b]	80
	HαAT3, 7, 29[b]	

[a]The number of independent alleles is indicated in parenthesis.
[b]The strain number is designated.

to generate mutations that are easily scored in the gametophyte once the muta-genized spores are plated and grown on selective medium. Because the game-tophytes are haploid, there is no need to screen the progeny of self-fertilized, mutagenized gametophytes to select gametophytic mutations. The gameto-phyte develops rapidly, reaching sexual maturity only two weeks after spores are added to water or culture medium. *Ceratopteris* gametophytes develop as males or hermaphrodites that can either be intra-gametophytically self-fertilized or inter-gametophytically out-crossed. By self-fertilizing a hermaphrodite, one can obtain in only two weeks a diploid sporophyte plant that is homozygous at all loci. The large size of the *Ceratopteris* genome may require the mutagenesis and screening of a large number of spores in order to generate a population of ga-metophytes that collectively harbor random mutations at all loci required for sat-uration mutagenesis. Because the gametophyte is only between 1–2 mm in size at sexual maturity, many gametophytes can be grown in a small area. In a typi-cal mutagenesis screen, 2×10^6 gametophytes can be grown and scored in ten 100-mm-diameter petri dishes. Furthermore, because the spore is single-celled at the time of mutagenesis, the resulting gametophyte is not a mosaic of mutant and wild-type cells as would occur when an embryo in the seed is mutagenized.

Although the haploid nature of the gametophyte is useful for identifying mutants, it is difficult to assess whether a wild-type gene can complement a gametophytic mutation. This problem can be overcome by generating geneti-cally diploid gametophytes that are heterozygous for the mutation in question. DeYoung et al (14) produced such *Ceratopteris* gametophytes by incorporat-ing into the life cycle a process referred to as apospory. Apospory refers to the generation of gametophytes from diploid sporophyte leaves without an in-tervening meiotic division. This spontaneously occurs when pieces of young, diploid sporophyte leaves are placed on medium lacking hormones and sucrose. By self-fertilizing aposporously derived gametophytes, a tetraploid sporophyte can be produced. The segregation of phenotypes in the diploid gametophyte progeny of the tetraploid parent can be used to determine whether gametophytic mutations are recessive, loss-of-function alleles, or dominant gain-of-function alleles. The ploidy levels of such progeny are easily confirmed by counting chro-mosome numbers of spore mother cells and by measuring the DNA content in the sperm of diploid gametophytes by cell flow cytometry (14).

The ability to generate diploid gametophytes from diploid sporophyte tissues (apospory), as well as the ability to form haploid sporophytes from haploid gametophytes (apogamy) in ferns indicate that ploidy level is insufficient to account for the dramatic differences that exist between the gametophyte and sporophyte generations of the plant life cycle. The identification of other factors that are involved in maintaining the differences between the two generations re-mains one of the great mysteries in plant biology. The ability to genetically and morphologically manipulate the gametophytes and sporophytes of *Ceratopteris* makes it a useful system for investigating this problem.

Although *Ceratopteris* is useful for identifying mutations and dissecting complex developmental pathways in the gametophyte (an example is described in the section on sex determination), it has some serious disadvantages as a model system. As yet, there are no published reports of transformation yielding transgenic plants. This makes it difficult to manipulate then test the expression of genes that have been cloned from *Ceratopteris* to study their functions in transgenic gametophytes or sporophytes. A second disadvantage is the large genome size, which will make it difficult to clone genes by strictly physical methods such as chromosome walking, which have been identified by mutation.

The Spore—Preparation for Dormancy and Germination

The single-celled, haploid spore, which is the product of meiosis in all land plants, represents the dormant phase of the fern life cycle. Produced by and shed from the adult sporophyte plant by the millions, dormant fern spores can remain viable for decades. Spores range in size from 60 to 100 μm, depending on the species. A thick protective coat, usually consisting of two layers, surrounds each spore. The inner layer of the spore coat, or intine, is made of cellulose and the outer, decorated layer, or exine, is made of sporopollenin. Sporopollenin is the same substance found in pollen walls. Petitt (51) has reviewed studies that describe the ultrastructure and composition of fern spore walls.

In addition to forming a protective seed coat or fruit, the developing seeds of angiosperms and gymnosperms prepare for dormancy by synthesizing storage globulins to be used during germination (39). Among the seed-storage proteins produced are vicilin and legumin, which are structurally similar to one another (1), suggesting that they evolved from a common ancestor via a duplication event (25). Vicilin-related genes are expressed in the developing endosperm and cotyledons of angiosperms, and in the haploid endosperm of gymnosperms. A vicilin-like gene of the fern *Matteuccia struthiopteris* was recently cloned and shown to be expressed only during the late stages of fern spore development (57). The cloning of the vicilin-like gene of *Matteuccia* has served well in understanding the evolution of seed-storage proteins in plants, as it appears to represent the ancestral gene that was duplicated in the seed plant lineage (57). The results of this study also suggest that dormancy in seeds and fern spores requires the synthesis and accumulation of similar proteins, even though the two structures represent very different phases of the plant life cycle.

After dormancy is broken and the spore begins to germinate, it becomes metabolically active, using up its reserves of stored protein, sugars, and oils; respiration and nucleic acid and protein synthesis are stimulated as it undergoes numerous cytological changes (reviewed in 53). The metabolic events that accompany spore germination have been likened to the germination of the angiosperm seed (53). In germinating seeds, for example, oils are converted to carbohydrate via the glyoxylate cycle (7). The enzymatic activities of two key enzymes in this pathway, isocitrate lyase (ICL) and malate syntase (MS), increase immediately following fern spore germination, and then decline as the lipid content in the spore decreases (13, 24). Although the genes encoding these enzymes have not been cloned from ferns, one gene encoding a putative protein similar in sequence to aconitase has been cloned from both germinating *Ceratopteris* spores (C Wen & JA Banks, unpublished observations) and germinating *Arabidopsis* seeds (52). This enzyme converts citrate to isocitrate, the substrate for ICL that feeds into the glyoxylate cycle. In *Arabidopsis*, the expression of the aconitase gene increases dramatically during seed germination as well as during seed and pollen maturation (52). In *Ceratopteris*, the aconitase message accumulates to high levels about three days after spores are placed in water (C Wen & JA Banks, unpublished observations).

Although data are limited, similarities in the physiology and molecular biology of fern spore and angiosperm seed dormancy and germination suggest that these processes require the synthesis of similar enzymes and proteins. These processes are potentially regulated by similar mechanisms. If true, it raises the interesting question of how the seed habit evolved from plants that, like ferns, lacked a dormant embryonic phase of development. Did the acquisition of the seed habit in the seed plant lineage involve a heterochronic switch of a dormancy program from a pre-embryonic (spore) phase of the plant life cycle to a postembryonic phase? Although many genes are likely to be involved in dormancy and germination, a switch in the timing of their expression may require relatively few changes in major regulatory genes. Comparative studies on the genes controlling the acquisition and breaking of dormancy in the fern spore and angiosperm seed should provide answers to this important and interesting evolutionary question.

Early Growth and Development of the Gametophyte

Fern spores are spherical but have a distinct apical/basal polarity that is defined by monolete or trilete markings at the apex of the spore. The spore ruptures along these markings as it germinates. As it ruptures, the nucleus undergoes its first division. In all species of ferns examined, the first division of the spore is asymmetric, leading to one large cell contained within the body of the spore, and one smaller cell. Prior to the first mitotic division, the spore nucleus migrates within the cytoplasm of the cell in a stereotypic fashion. First described in *Onoclea sensibilis* (69), the typical migratory pathway followed by the spore nucleus is first to the proximal pole of the spore, then toward the equatorial face of the spore. This migration is thought to position the nucleus for the first asymmetric division of the spore. *Onoclea* spores that are treated with colchicine (a microtubule inhibitor) remain uninucleate or undergo a single symmetric division leading to spores with two cells of approximately equal sizes, neither of which differentiates further (70). It is not clear from these studies whether the lack of asymmetry prevents differentiation or colchicine itself inhibits differentiation. Bassel & Miller (6) produced two symmetric cells in *Onoclea* by centrifugation of spores, timed to coincide with their first mitotic division. In these spores, neither cell differentiated a rhizoid, indicating that the first asymmetric division is a necessary condition for rhizoid differentiation.

The first asymmetric division of the fern spore is reminiscent of the first asymmetric division of the angiosperm microspore (reviewed in 66), which leads to the formation of a large vegetative cell and a small generative cell within the developing pollen grain. The generative cell divides once to produce the two sperm cells, while the vegetative cell later forms the pollen tube, a structure that superficially resembles the fern rhizoid. As with fern spores, treatment of angiosperm microspores with microtubule inhibitors or centrifugation results in a symmetrical first division (64, 65). The two resulting cells express a gene that is normally expressed only in the vegetative cell, indicating that an asymmetric division is essential for differentiation of generative cells (18).

Although it appears that the migration of the spore nucleus to one pole is necessary to establish the first asymmetric division in angiosperm microspores and fern spores, what controls the direction of nuclear movement is poorly understood. Recent studies in *Ceratopteris* suggest that the direction of nuclear migration is controlled by gravity. In germinating *Ceratopteris* spores, the first rhizoids emerge and grow downward with respect to gravity. Shortly after adding *Ceratopteris* spores to water, the nucleus moves randomly within a region of the spore that is centered behind the trilete marking (21). This period corresponds to the time that the developmental polarity of the future rhizoid is oriented by gravity. It is possible that the nucleus is playing a role in gravity detection like a statolith (20). Furthermore, when spores are grown on a clinostat, the direction of nuclear migration and subsequent primary rhizoid growth is randomized (21). In other ferns, light also influences the polarity of rhizoid growth (53); however, light plays a much lesser role than gravity on the direction of rhizoid growth (21). Because the thick spore coat is an impediment to analysis of cellular structure, Edwards & Roux (22) examined the effects of gravity on protoplasts (isolated from *Ceratopteris* gametophytes) and found them to mimic the effects of gravity on the migration of the nucleus and the direction of rhizoid growth in intact spores.

Gravity positions the nucleus for the first asymmetric division in fern spores, but what causes the two cells to adopt different cell fates is unknown. Nine different patterns of cell division and cellular differentiation beyond the first asymmetric division have been observed in ferns thus far (53). In some species of ferns, the smaller cell divides once to produce one daughter cell that will differentiate as a rhizoid, and another that will differentiate as the protonemal initial. In *Ceratopteris*, the smaller apical cell differentiates as the protonemal initial, while the larger basal cell divides asymmetrically to form one smaller cell that differentiates as the rhizoid initial. The rhizoid is a single-celled, elongated, nonphotosynthetic cell that is thought to function in anchoring and absorption of nutrients. The protonema initial eventually gives rise to the photosynthetic prothallus of the fern gametophyte. The evolutionary or ecological significance of the diversity in patterns of these early cell divisions and the developmental fates of cells observed among fern gametophytes is unclear. Similarly, species differences in division patterns among angiosperm embryos are regular but enigmatic in meaning. Although mutations that affect cell fate decisions have not been screened, they should be simple to obtain in *Ceratopteris*. Because the germination and growth of *Ceratopteris* spores naturally occurs outside of the sporophyte parent, it is an excellent model for understanding how genetic factors effect the earliest phases of spore and gametophyte development and cell fate determination.

Unlike the rhizoid that never divides, the fern protonema initial divides in a plane parallel to the surface of the spore face, producing a filamentous protonema. The duration of the protonemal or filamentous phase of development, and the number of cells in the filament varies depending on the species, ranging from 3 cells in *Ceratopteris* (5) to 20 cells in some species of the Grammitidaceae (60). The protonema eventually undergoes a transition from one- to two-dimensional growth, which marks the beginning of the prothallial phase of differentiation. Unlike the three-dimensional sporophyte, the gametophyte prothallus generally divides only in two dimensions, resulting in a prothallus that is only one-cell layer thick at maturity. In many species of ferns, the transition from one- to two-dimensional growth is regulated by light. Because light affects so many processes in the gametophyte, recent studies focusing on light-regulated development in the fern gametophyte are discussed.

Photomorphogeneis in the Gametophyte

The quantity and quality of light that developing fern gametophytes are exposed to effects their growth in much the same way as it affects the growth of any flowering plant. One example of a photomorphogenic response is spore germination, which, like the germination of many angiosperm seeds, is dependent on the exposure to red light (650–670 nm) and is inhibited by far-red light (733–750 nm). Spores exposed to red light germinate, whereas spores exposed to far-red light, regardless of their prior exposure to red light or alternating exposures to red and far-red light, do not germinate. The reversibility of the red and

Table 2   The effects of red/far-red and blue light on gametophyte development in *Adiantum*

Response	Red/far-red light	Blue light	Reference
Germination	Promotes	Inhibits	72
First mitosis	Promotes	Inhibits	23
Cell cycle	G1 maintenance	Enters S	72
Filament tip growth	Toward red	None	61
2-D transition	Does not promote	Promotes	71
Apical tip swelling	None	Promotes	44
Chloroplast distribution in cell			
Low fluence: toward wall	Promotes	Promotes	73
High fluence: toward center	Promotes	Promotes	41
Nuclear position in cell	?[a]	?[a]	40

[a]In light-grown gametophytes, the nuclei are centered in the cell, whereas in dark-grown gametophytes, nuclei abut the cell wall. The photoreceptor responsible for differences in nuclear positioning is under investigation (40).

far-red light effects on spore germination indicates that this response is mediated by the pigment phytochrome. While red light promotes spore germination, blue light (380–440 nm) and UV radiation (260 nm) inhibit red-light-potentiated germination of dormant spores, indicating that germination is also mediated by flavin-type photoreceptors, or cryptochromes. The processes that are regulated by red/far-red and blue light in *Adiantum capillus verenis* gametophytes are listed in Table 2. Because the physiology of photomorphogenetic responses in ferns has been recently reviewed (72), discussion here focuses on recent studies of the molecular and cell biology of photomorphogenic responses in gametophytes.

The phytochromes of *Adiantum capillus verenis* and *Anemia phyllitidis* are encoded by gene families consisting of three or four members, and are named *Adiantum PHY1-3* and *Anemia PHY1-4*, respectively (43, 50, 74). The *Adiantum PHY1* and *PHY2* genes and proteins are similar in structure to the *PHYA-E* phytochrome genes of *Arabidopsis* in intron position, putative chromophore attachment site, and presence of a putative bacterial two-component protein kinase domain. The *Adiantum* PHY3 gene is unusual in that the carboxy-terminal half of the deduced protein is replaced by a eukaryotic protein kinase domain that has yet to be observed in angiosperm phytochromes. The function of this domain is unknown, but its presence suggests that some phytochromes may have unique functions in the fern gametophyte or sporophyte that are not found in angiosperms. In *Anemia*, *PHY1* and *4* mRNA levels increase in dark-grown imbibed spores and decrease when spores are transferred to red light. The *PHY3* gene is not expressed in dark-grown spores but is expressed when spores are exposed to red light. Differences in expression patterns suggest that the different phytochrome genes function during different stages of gametophyte development.

Three different cryptochrome-related cDNA sequences, named *AdiCRY1–3*, have recently been isolated from *Adiantum* (42). Like the *CRY1* and *CRY2* cryptochrome genes and proteins of *Arabidopsis*, all three fern cryptochromes contain an amino-terminal photolyase-homologous domain, a more variable carboxy-terminal extension, and residues that interact with FAD (flavin adenine dinucleotide), a catalytic co-factor of microbial type I photolyases. Although all three genes are expressed in gametophytic and sporophytic tissues of *Adiantum*, it is not yet known if or how these cryptochromes are involved in blue light-regulated processes.

Although none of the photoperception genes has been isolated from *Ceratopteris*, specific functions of cryptochrome and phytochrome genes can be correlated with a mutant photomorphogenic phenotype with a defect in a specific *PHY* or *CRY* gene in this species. Five *photomorphogenic* mutants, representing three phenotypic classes, have already been characterized in *Ceratopteris* (10–12). One class of mutants displays a de-etiolated phenotype in the dark, similar to the *det*, *fus*, and *cop* mutants of *Arabidopsis*. Although these mutants are not likely to be defective in pigment biosynthesis, many of these genes have been cloned from *Arabidopsis* and could be used to isolate similar genes from wild-type and mutant *Ceratopteris* plants. A second class of *Ceratopteris* mutants include those that display reduced light-mediated inhibition of germination and/or cell elongation and are similar in phenotype to the *hy* mutants of *Arabidopsis*. These genes are likely to be involved in pigment biosynthesis or signal transduction. Again, the cloning of wild-type and mutant *CRY* or *PHY* genes from *Ceratopteris* is one approach to assigning specific functions to these genes. A third class of mutants obtained in *Ceratopteris* includes one mutant that germinates only in darkness and displays reversed photoregulation. Although similar mutants in *Arabidopsis* have not been observed, this mutant phenotype suggests that this gene may couple the phytochrome signal transduction pathway and the light-sensitive steps in spore germination (11).

As illustrated in Figure 2, a dark-grown *Ceratopteris* prothallus is etiolated, whereas a light-grown prothallus is de-etiolated. When dark-grown prothalli are transferred to blue light, but not to red or far-red light, cell elongation in the subapical region of the prothallus is inhibited (45). Thus, blue light appears to repress the etiolated response (or promote a de-etiolated response) in these plants. To understand if blue light induces a reorientation of cortical microtubules (MTs) that would result in an inhibition of cell elongation, Murata et al (45) studied the arrangement of MTs in *Ceratopteris* prothalli grown under varying light conditions. Because the prothallus consists of a single layer of cells, individual cells of the prothallus could be irradiated and the cytological

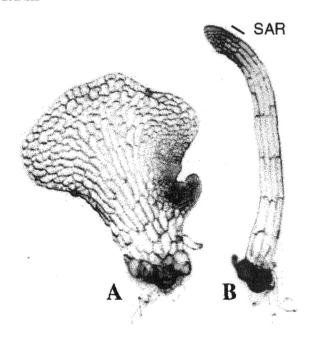

Figure 2   Light- and dark-grown *Ceratopteris* gametophytes. The gametophyte in (*A*) was grown in continuous light for 12 days, while the gametophyte in (*B*) was grown in the light for two days then transferred to the dark for 10 days. The subapical region (SAR) of the gametophyte that is sensitive to blue light is indicated.

and morphological effects easily examined. They observed that blue light reorients MTs from transverse to oblique or parallel to the growing axis within 3 h, but only in individual irradiated cells located within the subapical region of the prothallus (shown in Figure 2). Non-irradiated neighboring cells did not reorient their MTs and continued to elongate. This study indicates that each cell within the subapical region of the prothallus perceives blue light, reorients its cortical MTs from transverse to longitudinal, and ceases cell elongation independent of its surrounding cells.

This study also illustrates the utility of the fern gametophyte to study plant cell biology. Unlike other commonly used systems, such as *Tradescantia* stamen hairs and onion epidermis that are terminally differentiated, dynamic changes in cell ultrastructure associated with cell growth and differentiation can be easily studied in the fern gametophyte. In addition to the simplicity of the gametophyte, mutations that are likely to affect the cytoskeleton have been isolated in *Ceratopteris*. In the recessive *clumped chloroplast* mutant of *Ceratopteris*, for example, the chloroplasts are aggregated to one corner of the cell rather than evenly distributed throughout the cell (67). This mutant phenotype likely results from a cytoskeletal defect, although this mutant has not yet been characterized beyond the light microscopic level.

Sex Determination and Differentiation in the Gametophyte

Many species of ferns produce sexually dimorphic gametophytes that are male or hermaphroditic/female. Because most ferns are homosporous and each spore is capable of developing as either sex type, the sex of the gametophyte is not genetically determined. Rather, it is determined after spore germination but before the sex organs differentiate on the gametophyte, a period of only two days in *Ceratopteris* (5). The determinate of sexual phenotype in many homosporous ferns is a pheromone, referred to as antheridiogen. First identified by Döpp (15), antheridiogens are synthesized and secreted by female or hermaphroditic members of a population, and induce male development in younger, sexually undetermined gametophytes of the same population (reviewed in 4, 46–48). Because sex determination in the fern gametophyte is best understood in *Ceratopteris*, discussion of this topic focuses on this plant. Studies of sex determination in other species of ferns have been reviewed (46–48, 53).

A *Ceratopteris* spore always develops as a hermaphrodite when grown in isolation. Shortly after the transition from one- to two-dimensional growth, the hermaphrodite prothallus initiates a lateral meristem. Based on histological examination, the lateral meristem is derived from the middle of the three cells of the protonema (Figure 3*B*, *C*). This middle cell undergoes periclinal divisions to produce three cells of similar sizes. One of the marginal cells then undergoes one periclinal division and one anticlinal division to produce four cells (Figure 3*E*). Each of these then undergoes a number of anticlinal and periclinal divisions to form what will become the upper and lower halves of the lateral meristem (Figure 3*F*). The rates of cell division in cells within the two halves of the meristem are uneven, such that the meristem produces two bulges that eventually give the hermaphrodite a mitten shape, then later a heart shape. As the meristem is initiated, the most apical cell of the protomena divides at an angle of about 45°, which produces a pie-shaped, or lenticular, terminal cell (Figure 3*E*). Repeated divisions of the apical cell and its derivatives continue, but for only a short period of time. Cells derived from the apical cell do not contribute to the meristem of the hermaphrodite, but rather to the part of the prothallus that is occupied by the "fingers" of the mitten-shaped gametophyte (Figure 3*C*). At the stage of development when the apical cell and its derivatives cease dividing, the gametophyte consists of three zones: the basal part of the gametophyte closest to the spore from which rhizoids are continuously initiated; the meristem region, where all subsequent divisions of the prothallus occur; and the terminal tip of the gametophyte. Although the lineages of all of the cells in each zone

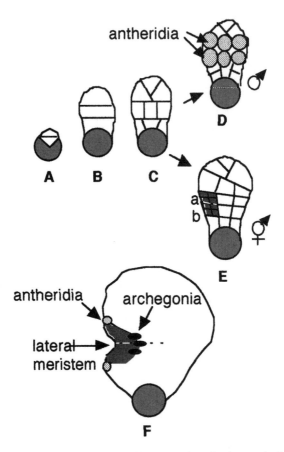

Figure 3 A diagram of male and hermaphrodite gametophyte development in *Ceratopteris*. *A*. As the spore cracks, the spore cell divides asymmetrically, producing one larger basal cell and one smaller apical cell. *B*. The apical cell divides in one plane, forming a three-celled, one-dimensional protonema. *C*. The most apical cell divides at ~45° angles, while the two more basal cells of the protonema undergo periclinal divisions, forming a two-dimensional prothallus. *D*. If exposed to antheridiogen, most cells of the prothallus, other than the most apical and most basal cells, differentiate as antheridia. *E*. If not exposed to antheridiogen, a marginal cell undergoes an anticlinal division to produce the sectors *A* and *B*. *F*. Continued anticlinal and periclinal divisions of the *A* and *B* sectors ultimately produce the lateral meristem and meristem notch. Antheridia and archegonia initiate at the perimeter of the meristem, shown as a gray-shaded area.

can be traced back histologically to one of the three cells of the protonema, there is no experimental evidence to confirm this. This pattern of growth in the hermaphrodite suggests that the fates of cells in the young germinating gametophyte become fixed rapidly.

A cell within the lateral meristem of the hermaphrodite has one of four possible cell fates. It may remain within the meristem and continue to divide (a condition necessary to maintain the meristem); it may get pushed out of the meristematic region and simply enlarge, thus contributing to the growing sheet of cells in the prothallus; or it may differentiate into an archegonia or antheridia (the gametangia). The archegonium is a flask-shaped structure that houses the egg, while the antheridium is the organ where sperm cells are made. Archegonia always initiate just below the meristem notch, while antheridia always initiate just above the meristem notch and at the periphery of the meristem (Figure 3F). Neither sex organ initiates outside of the meristematic region. Two weeks after spore inoculation, the hermaphrodite is heart-shaped, has many rhizoids, and many gametangia at various stages of development. Antheridia located in the "wings" of the gametophyte are the most mature and ready to release sperm, whereas the most mature archegonia (those furthest away from the meristem notch) are ready to be fertilized. Because the sperm require water to swim to the egg, fertilization cannot occur in the absence of water. If fertilization does not occur, the meristem continues to divide indefinitely, producing more antheridia, archegonia, and photosynthetic cells such that at any given point in time, there are always sperm and an egg ready to be fertilized. Because each archegonium produces one egg that is viable for less than two days and sperm are viable for only ~20 min, it is essential that the hermaphrodite continue to grow and produce gametes until fertilization occurs.

A *Ceratopteris* spore develops as a male gametophyte only if it is exposed to antheridiogen early in its development, no later than ~4–5 days after spore inoculation (5). By an unknown mechanism, the gametophyte subsequently loses the competence to respond to antheridiogen and will develop as a hermaphrodite even if exposed to antheridiogen after ~5 days. One of the first morphological differences observed between a hermaphrodite and a male gametophyte is the absence of divisions in the middle cell of the male protonema, which, in the hermaphrodite, leads to the development of the lateral meristem. Antheridiogen thus functions early in gametophyte development to suppress these divisions. For this reason, the males are said to be ameristic, or lacking in a meristem. The cells of the male protonema do divide, however, and produce a multicellular male prothallus. These divisions occur only within the lenticular apical cell, which at each division produces another apical cell and one daughter cell. The daughter quickly begins to differentiate an antheridium while the apical cell continues to be the source of all new cells of the male (Figure 3D). A second function of antheridiogen is to promote the rapid differentiation of antheridia in all cells of the gametophyte other than the most apical cell.

The antheridiogens of many ferns have been characterized as gibberellins, or GAs (63, 81–85). The antheridiogen of *Ceratopteris* has not been identified, but indirect evidence suggests that it may also be a GA. The presence of GA biosynthetic inhibitors AMO 1618 and ancymidol in culture media results in many more hermaphrodite gametophytes than in their absence, indicating that by inhibiting GA biosynthesis, both antheridiogen production and male induction are also inhibited (76). Abscisic acid (ABA), a known inhibitor of GA responses in flowering plants, also inhibits antheridiogen-induced male development (29, 30). If the antheridiogen of *Ceratopteris* is a GA, it is likely to be a novel GA, because the antheridiogens of other species of ferns, as well as GA1 and GA3, do not induce male development in *Ceratopteris* gametophytes (JA Banks, unpublished observations). Media conditioned with the growth of *Ceratopteris* gametophytes contain at least two compounds that are separable by chromatography and display biological activity (E Strain & JA Banks, unpublished observations).

To understand how antheridiogen controls the sex of the *Ceratopteris* gametophyte, mutations that alter the normal sex of the gametophyte have been identified (reviewed in 3). The identification of sex-determining mutants serves several purposes. First, the number of genes that are involved in sex determination can be estimated. Second, by studying the epistatic interactions between these genes, the genes can be ordered in a hierarchical pathway. Although genetics alone is insufficient to assign a biochemical function to a gene, this analysis highlights important genes and aids in choosing which should be cloned, and predicts in a general sense what the functions of these genes might be.

Four different phenotypic classes of sex-determining mutants have been identified in *Ceratopteris* thus far (2, 4). Those that are always hermaphroditic, even in the presence of antheridiogen, are referred to as *her* (for *hermaphroditic*) mutants. Those that are always male, even in the absence of antheridiogen, are called the *tra* (for *transformer*) mutants. Mutants that are always female are referred to as the *fem* (for *feminization*) mutants. The fourth category includes one that is male in the presence of antheridiogen, but in the absence of antheridiogen is a hermaphrodite that produces ten times more antheridia than normal, and is called *man* (for *many antheridia*). Other mutants that form too few antheridia or form an abnormal meristem have also been identified (E Smith & JA Banks, unpublished observations). By comparing the phenotypes of double-mutant gametophytes to single-mutant phenotypes, a model that describes the hypothetical interactions among the sex-determining genes has been proposed (reviewed in 3). The effects of the absence or presence of antheridiogen on the activities of these genes are illustrated in Figure 4.

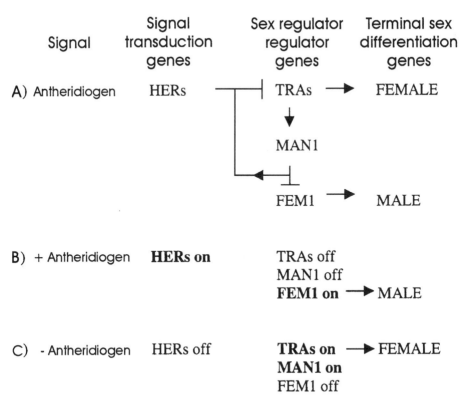

Figure 4 A model of the sex-determining pathway in *Ceratopteris*. The potential interactions among the sex-determining genes are illustrated in (*A*). The activities of the sex-determining genes in the presence (*B*) or absence (*C*) of antheridiogen are shown. Lines ending in arrows indicate activating interactions and lines ending in bars indicate repressing interactions.

One key feature of this model is that it proposes two major classes of sex-determining genes, one that promotes maleness (the *FEM* gene) and another that promotes femaleness (the *TRA* genes). Because each also represses, or mutually excludes the expression of the other, only one class of gene (*FEM* or *TRA*) can be expressed in an individual gametophyte. What determines which of these two predominates in the gametophyte depends on antheridiogen, which activates the *HER* genes. When the *HER* genes are active they repress the *TRA* genes, allowing the *FEM* gene to predominate. In the absence of antheridiogen, the *TRA* gene is not repressed, which leads to the repression of the *FEM* gene. Thus, only male traits are expressed in the presence of antheridiogen (*FEM* gene "on"), and female traits are expressed in the absence of antheridiogen

(*TRA* genes "on"). This flexible mechanism of sex determination allows the gametophyte to determine its sex by sensing its environment rather than by genetic pre-determination, as is the case in most animal species.

The meristem is a female trait of the gametophyte that requires the expression of the TRA genes. This meristem is typical of other plant meristems in that the processes of cell division and differentiation must be regulated to maintain a meristem size over time. If cells differentiate prematurely with a constant rate of cell division, the meristem will eventually become a terminally differentiated structure. The *man1* mutant of *Ceratopteris* can be considered a defect in the integration of these two processes. At sexual maturity, the *man1* gametophyte produces more antheridia than wild-type hermaphrodites. If left unfertilized, all cells of the meristem of the *man1* gametophyte eventually differentiate as antheridia, indicating that the rate of differentiation of cells in the meristem exceeds the rate of cell division. When *man1* gametophytes are self-fertilized, the resulting sporophyte is abnormal, producing several aberrant, small leaves before the meristem terminally differentiates as a leaf-like knob (4). Sporophytes homozygous for the *tra7* allele (a gene that is required for meristem development in the gametophyte) are similar to *man1* homozygous sporophytes (4). The defects in meristem development noted in the meristems of *man1* and *tra7* gametophytes and sporophytes indicate that both meristems use the same genes to pattern their development, even though the structures they give rise to are vastly different from one another. The expression of many of the *KNOTTED*-like meristem genes (M Kato, JA Banks & M Hasebe, unpublished observations) and MADS-box-like genes of *Ceratopteris* (28) in both the gametophytes and sporophytes of this fern also suggest that the two phases of the fern life cycle are regulated by the very same genes. Although most of these genes are expressed in the gametophyte and sporophyte, if and how their expression might be effected by posttranscription mechanisms (such as alternative splicing) is unknown.

One unexpected outcome of the genetic analyses of sex determination in *Ceratopteris* was discovery of the large number of *HER* loci. The *her* mutants are insensitive to antheridiogen and are thought to define either the receptor of antheridiogen, or define genes that make up the antheridiogen signal transduction pathway. The 12 independent *her* alleles characterized represent at least 5 different loci (2), indicating that the antheridiogen signaling pathway may be defined by more than 5 genes, as saturation of this phenotype has not yet been achieved. This contrasts with the small number of genes in *Arabidopsis* that are thought to define the GA signaling pathway. A technical reason could explain that why more genes are likely to be identified in *Ceratopteris* may be due to the response itself. Whereas antheridiogen is required only for promoting male development and is not essential to fern development or successful reproduction, certain GA responses in plants may be essential. Mutations in essential genes defining a GA signal transduction pathway may be lethal and hence difficult to identify. The sex-determination response in *Ceratopteris* thus appears to be a unique and useful phenotype for identifying genes necessary for hormone signaling in plants. Another interesting feature of sex determination in *Ceratopteris* is that ABA blocks the antheridiogen response as previously mentioned. Several mutants that are insensitive to ABA have been identified in *Ceratopteris* (30). By studying the interactions among the ABA and antheridiogen insensitive mutants, it may be possible to understand if, how, and at what steps the two signaling pathways interact with one another.

Gamete Formation, Fertilization and Death of the Gametophyte

The ultimate function of the gametophyte is to produce gametes. Oogenesis and spermatogenesis have been extensively studied and described in several species of ferns (reviewed in 53). Many of these studies have focused on sperm development, because the cytoskeleton of these cells is among the most complex in plants. In *Ceratopteris* (37), each antheridium contains a spermatogenous cell that divides 5 times to produce 32 isodiametric spermatocytes that soon differentiate into mature, multiflagellated, coiled sperm cells. The last two divisions are particularly remarkable. Prior to the fourth division, the blepharoplast (a microtubule organizing center, or MTOC) arises de novo and organizes a spindle microtubule array for the final mitotic division. During the transition from metaphase to anaphase of the final division, the blepharoplast appears as two procentrioles. Once these separate toward opposite poles of the cell, the blepharoplast reorganizes to form a multilayer strip, or MLS, which in turn organizes the spline and amorphous zone. The spline is a ribbon of microtubules that wraps around the nucleus as the spline elongates and gives the sperm, its nucleus, and mitochondrion a coiled shape. The amorphous zone appears as a layer that lies between the spline and plasmalemma. The basal bodies of the flagella appear to originate in this zone. While this brief description of sperm development in *Ceratopteris* only highlights the interesting and complex cellular events that occur during the production of the 32 motile sperm cells, very few studies have focused on fern sperm development over the past two decades. This is surprising considering that fern sperm are unique among plant cells because they have structurally recognizable MTOCs. To address whether the MTOCs of fern sperm are biochemically similar to animal MTOCs, Hoffman et al (38) probed sections of developing *Ceratopteris* antheridia with antibodies to several mammalian MTOC proteins. Antibodies that recognize mammalian centrosomes, as MPM-2 and C-9, label the blepharoplast prior to the final mitotic division, indicating that it is a centrosome analog. Antibodies to centrin, another MTOC protein, labels the MLS and amorphous zone. While

MTOCs clearly exist in ferns during this very brief period in the fern life cycle, what causes them to suddenly appear during spermatogenesis then disappear after fertilization is unknown.

The most fascinating event to observe in fern development is fertilization. Although this event has been described in several species of ferns, few studies have been conducted on this subject since the early 1980s. In preparation for fertilization, the archegonium, which is a multicellular, flask-shaped structure, must form a channel within its neck to allow sperm direct access to the egg. This occurs by an unknown mechanism that causes the central file of cells in the neck of the archegonium (or ventral canal cells) to disintegrate. Associated with the disintegration of these cells is the appearance of a mucilaginous drop of liquid at the tip of the archegonia. When water is added to mature gametophytes, sperm are released and, within minutes, swarm to the open neck of the archegonia. There they become trapped in the mucilaginous substance secreted by archegonia. While is it obvious that sperm are chemotactically drawn to the egg, the identity of the chemoattractant is unknown. In *Pteridium aquilinum* (8, 54) and *Ceratopteris* (JA Banks, unpublished observations), malic and other organic acids act as a sperm chemoattractant. Which of these acids, if any, are endogenous attractants is unknown.

Once the egg is fertilized, the zygote begins its sporophytic program of expression. Shortly after fertilization, cells within the meristem of the gametophyte cease dividing and no new gametangia are initiated. Cells that once made up the meristem enlarge, giving the once heart-shaped gametophyte an oval appearance. By the time the sporophyte develops its first root, the gametophyte is clearly dead, indicating that fertilization triggers some type of signal that is, in essence, poisonous to the gametophyte. While the nature of this signal is unknown, mutations in *Ceratopteris* that prevent the production or perception of the signal could be easily identified.

Concluding Remarks

At first glance, the fern gametophyte appears to be a very simple structure, generally consisting of a single sheet of cells dotted with gametangia. Unlike the sporophyte, it lacks organs (leaves, stems, and root) and complex tissues (such as vascular and epidermal tissue). While the functions of the gametophyte (to produce gametes) and sporophyte (to produce spores) are also different, recent studies reviewed here indicate that the gametophyte responds to the environment using mechanisms similar to those used by the more complex sporophytes of angiosperms. Given that fern gametophytes are haploid, simple in structure, and independent of the sporophyte plant, the fern gametophyte has great potential to yield important and significant insights in three areas of plant development research. The first area includes studies aimed at understanding functions unique to the gametophyte, including sex determination, gamete production, and fertilization. Similar processes in angiosperms are more difficult to study because the gametophytes are reduced in size, surrounded by maternal tissues of the sporophyte, and difficult to observe in a nondestructive way. A second area includes studies aimed at understanding how the gametophyte responds to its environment (e.g. by light and pheromones). Although studies of model angiosperms, such as *Arabidopsis*, have proven valuable in identifying genes that regulate plant responses to the environment, fern gametophytes may provide even more insights in this area. For example, the equivalent of one type of photomorphogenic mutant of the fern *Ceratopteris* (the *dkg* mutant) has not been observed in angiosperms. Whether this reflects fundamental differences between the biology of ferns and angiosperms, or reflects the ability to screen novel mutant phenotypes in ferns, remains to be seen. The third area of research is the evolution of complex traits. To understand the genetic and molecular changes that have occurred during land plant evolution, it will be important to understand how regulatory genes have evolved. This can be accomplished by studying the structure and regulation of genes in ferns (and other lower vascular plants) that are known to regulate key developmental processes in angiosperms. The greatest obstacles to addressing these questions are the lack of transformation techniques and inability to efficiently clone genes that have been identified by mutation, particularly in *Ceratopteris*. Should these obstacles be overcome, our understanding of plant development and evolution will undoubtedly advance using the tiny fern gametophyte.

Visit the *Annual Reviews* home page at
http://www.AnnualReviews.org

Literature Cited

1. Argos P, Narayana S, Nielsen N. 1985. Structural similarities between legumin and vicilin storage proteins from legumes. *EMBO J.* 2:1111–17
2. Banks J. 1994. Sex-determining genes in the homosporous fern *Ceratopteris*. *Development* 120:1949–58
3. Banks J. 1997. Sex determination in the fern *Ceratopteris*. *Trends Plant Sci.* 2:175–80
4. Banks J. 1997. The *TRANSFORMER* genes of the fern *Ceratopteris* simultaneously promote meristem and archegonia development and repress antheridia development in the developing gametophyte. *Genetics* 147:1885–97
5. Banks J, Hickok L, Webb M. 1993. The programming of sexual phenotype in the homosporous fern, *Ceratopteris richardii*. *Int. J. Plant Sci.* 154:522–34
6. Bassel A, Miller J. 1982. The effects of centrifugation on asymmetric cell division and differentiation of fern spores. *Ann. Bot.* 50:185–98
7. Bonner J, Varner J. 1965. *Plant Biochemistry*, pp. 226–27. New York: Academic. 1054 pp.
8. Brokaw C. 1958. Chemotaxis of bracken spermatozoids. Implications of electrochemical orientation. *J. Exp. Biol.* 35:197–212
9. Carroll EW, Schwarz GJ, Hickok LG. 1988. Biochemical studies of paraquat-tolerant mutants of the fern *Ceratopteris richardii*. *Plant Physiol.* 87:651–54
10. Cooke T, Racusen R, Hickok L, Warne T.

1987. The photocontrol of spore germination in the fern *Ceratopteris richardii*. *Plant Cell Physiol.* 28:753–59

11. Cooke T, Hickok L, Vanderwoude W, Banks J, Scott R. 1993. Photobiological characterization of a spore germination mutant with reversed photoregulation in the fern *Ceratopteris richardii*. *Photochem. Photobiol.* 57:1032–41

12. Cooke T, Hickok L, Sugai M. 1995. *Ceratopteris richardii* as a model system for studying plant photobiology. *Int. J. Plant Sci.* 156:367–73

13. DeMaggio, Agreen C, Stetler D. 1980. Biochemistry of fern spore germination. Glyoxylate and glycolate cycle activity in *Onoclea sensibilis* L. *Plant Physiol.* 66:922–24

14. DeYoung B, Weber T, Hass B, Banks J. 1997. Generating autotetraploid sporophytes and their use in analyzing mutations affecting gametophyte development in the fern *Ceratopteris*. *Genetics* 147:809–14

15. Döpp W. 1950. Eine die Antheridienbildung bei farnen fördernde Substanz in den Prothallien von *Pteridium aquilinum* L., Kuhn. *Ber. Dtsch. Bot. Ges.* 63:139–47

16. Duckett J, Klekowski E, Hickok L. 1979. Ultrastructural studies of mutant spermatozoids in ferns. I. The mature nonmotile spermatozoid of mutation 230X in *Ceratopteris thalictroides* (L.) Brongn. *Gamete Res.* 2:317–43

17. Dyer A. 1979. *The Experimental Biology of Ferns*. London: Academic. 657 pp.

18. Eady C, Lindsey K, Twell D. 1995. The significance of microspore division and division symmetry for vegetative cell-specific transcription and generative cell differentiation. *Plant Cell* 7:65–74

19. Eberle J, Hasebe M, Nemacheck J, Wen C, Banks J. 1995. *Ceratopteris*: a model system for studying sex-determining mechanisms in plants. *Int. J. Plant Sci.* 156:359–66

20. Edwards E, Roux S. 1994. Limited period of graviresponsiveness in germinating spores of *Ceratopteris richardii*. *Planta* 195:150–52

21. Edwards E, Roux S. 1998. Influence of gravity and light on the developmental polarity of *Ceratopteris richardii* fern spores. *Planta* 205:553–60

22. Edwards E, Roux S. 1998. Gravity and light control of the developmental polarity of regenerating protoplast isolated form prothallial cells of the fern *Ceratopteris richardii*. *Plant Cell Rep.* 17:711–16

23. Furuya M, Kanno M, Okamoto M, Fukuda S, Wada M. 1997. Control of mitosis by phytochrome and a blue-light receptor in fern spores. *Plant Physiol.* 113:677–83

24. Gemmrich A. 1979. Isocitrate lyase in germinating spore of the fern *Anemia phyllitidis*. *Phytochemistry* 18:1143–46

25. Gibbs P, Strongin K, McPherson A. 1989. Evolution of legume seed storage proteins—a domain common to legumins and vicilins is duplicated in vicilins. *Mol. Biol. Evol.* 6:614–23

26. Gifford E, Foster A. 1989. *Morphology and Evolution of Vascular Plants*. New York: Freeman. 626 pp.

27. Hasebe M, Omori T, Nakazawa M, Sano T, Kato M, Isatsuki K. 1994. rbcL gene sequences provide evidence for the evolutionary lineages of leptosporangiate ferns. *Proc. Natl. Acad. Sci. USA* 91:5730–34

28. Hasebe M, Wen C, Kato M, Banks J. 1998. Characterization of MADS homeotic genes in the fern *Ceratopteris richardii*. *Proc. Natl. Acad. Sci. USA* 95:6222–27

29. Hickok L. 1983. Abscisic acid blocks antheridiogen-induced antheridium formation in gametophytes of the fern *Ceratopteris*. *Can. J. Bot.* 61:888–92

30. Hickok L. 1985. Abscisic acid resistant mutants in the fern *Ceratopteris*: characterization and genetic analysis. *Can. J. Bot.* 63:1582–85

31. Hickok L, Schwarz O. 1986. An *in vitro* whole plant selection system: paraquat tolerant mutants in the fern *Ceratopteris*. *Theor. Appl. Genet.* 72:302–6

32. Hickok L, Schwarz O. 1986. Paraquat tolerant mutants in *Ceratopteris*: genetic characterization and reselection for enhanced tolerance. *Plant Sci.* 47:153–58

33. Hickok L, Schwarz O. 1989. Genetic characterization of a mutation that enhances paraquat tolerance in the fern *Ceratopteris richardii*. *Theor. Appl. Genet.* 77:200–4

34. Hickok L, Vogelien D, Warne T. 1991. Selection of a mutation conferring high NaCl tolerance to gametophytes of *Ceratopteris*. *Theor. Appl. Genet.* 81:293–300

35. Hickok L, Warne T, Slocum M. 1987. *Ceratopteris richardii*: applications for experimental plant biology. *Am. J. Bot.* 74:1304–16

36. Hickok L, Warne T, Fribourg R. 1995. The biology of the fern *Ceratopteris* and its use as a model system. *Int. J. Plant Sci.* 156:332–45

37. Hoffman J, Vaughn K. 1995. Using the developing spermatogenous cells of *Ceratopteris* to unlock the mysteries of the plant cytoskeleton. *Int. J. Plant Sci.* 156:346–58

38. Hoffman J, Vaughn K, Joshi H. 1994. Structural and immunocytochemical characterization of microtubule organizing

centers in pteridophyte spermatogenous cells. *Protoplasma* 179:46–60

39. Ensen U, Berthold H. 1989. Legumin-like proteins in gymnosperms. *Phytochemistry* 28:143–46

40. Kagawa T, Wada M. 1993. Light-dependent nuclear positioning in prothallial cells of *Adiantum capillus-veneris*. *Protoplasma* 177:82–85

41. Kagawa T, Kadota A, Wada M. 1994. Phytochrome-mediated photoorientation of chloroplasts in protonemal cells of the fern *Adiantum* can be induced by brief irradiation with red light. *Plant Cell Physiol.* 35:371–77

42. Kanegae T, Wada M. 1998. Isolation and characterization of the plant blue light photoreceptor (cryptochrome) homologous genes of the fern *Adiantum capillus-veneris*. In press

43. Maucher H, Scheuerlein R, Schraudolf F. 1992. Detection and partial sequence of phytochrome genes in the ferns *Anemia phyllitidis* (L.) Sw (Schizaceaeae) and *Dryopteris filix-mas* L. (Polypodiaceae) by using polymerase-chain reaction technology. *Photochem. Photobiol.* 56:759–63

44. Murata T, Wada M. 1989. Organization of cortical micortubules and microfibril deposition in repsonse to blue-light-induced apical swelling in a trip-growing *Adiantum* protonemal cell. *Planta* 178:334–41

45. Murata T, Kadota A, Wada M. 1997. Effects of blue light on cell elongation and microtubule orientation in dark-grown gametophytes of *Ceratopteris richardii*. *Plant Cell Physiol.* 38:201–209

46. Näf U. 1959. Control of antheridium formation in the fern species *Anemia phyllitidis*. *Nature* 184:798–800

47. Näf U. Antheridiogens and antheridial development. See Ref. 17, pp. 436–70

48. Näf U, Kakanishi K, Endo M. 1975. On the physiology and chemistry of fern antheridiogens. *Bot. Rev.* 41:315–59

49. Nicklas K. 1997. *The Evolutionary Biology of Plants*. Chicago: Univ. Chicago Press. 449 pp.

50. Okamoto H, Hirano Y, Abe H, Tomizawa K, Furuya M, Wada M. 1993. The deduced amino acid sequence of phytochrome from *Adiantum* includes consensus motifs present in phytochrome B from seed pants. *Plant Cell Physiol.* 34:1329–34

51. Pettitt J. 1979. Ultrastructure and cytochemistry of spore wall morphogenesis. See Ref. 17, pp. 213–52

52. Peyret P, Perez P, Alric M. 1995. Structure, genomic organization and expression of the *Arabidopsis thaliana* aconitase gene. Plant aconitase shows significant homology with mammalian iron-responsive element-binding protein. *J. Biol. Chem.* 270:8131–37

53. Raghavan V. 1989. *Developmental Biology of Ferns*. New York: Cambridge Univ. Press. 361 pp.

54. Rothschild N. 1956. Sperm-egg interacting substances, II. In *Fertilization*, pp. 39–50. New York: Wiley. 170 pp.

55. Scott R, Hickok L. 1987. Genetic analysis of antheridiogen sensitivity in *Ceratopteris richardii*. *Am. J. Bot.* 74:1872–77

56. Scott R, Hickok L. 1991. Inheritance and characterization of a dark-germinating/light-sensitive mutant in the fern *Ceratopteris*. *Can. J. Bot.* 69:2616–19

57. Shutov A, Braun H, Chesnokov Y, Baumlein H. 1998. A gene encoding a vicilin-like protein is specifically expressed in fern spores. Evolutionary pathway of seed storage globulins. *Eur. J. Biochem.* 252:79–89

58. Singh M. 1990. *Characterization of a hydroxyproline tolerant mutant in the fern* Ceratopteris. MS thesis. Univ. Tenn., Knoxville

59. Stewart W, Rothwell G. 1993. *Paleobotany and the Evolution of Plants*. Cambridge: Cambridge Univ. Press. 521 pp. 2nd ed.

60. Stokey A, Atkinson L. 1958. The gametophyte of the Grammitidaceae. *Phytomorphology* 8:391–403

61. Sugai M, Furuya M. 1985. Action spectrum in ultraviolet and blue light region for the inhibition of red-light induced spore germination in *Adiantum capillus veneris* L. *Plant Cell Physiol.* 26:953–56

62. Tai Chun P, Hickok L. 1992. Inheritance of two mutations conferring glyphosate tolerance in the fern *Ceratopteris richardii*. *Can. J. Bot.* 70:1097–99

63. Takano K, Yamane H, Yamauchi T, Takahashi N, Furber M, Mander L. 1989. Biological activities of the methyl ester of gibberellin a73, a novel and principal antheridiogen in *Lygodium japonicum*. *Plant Cell Physiol.* 30:201–5

64. Tanaka I, Ito M. 1981. Control of division patterns in explanted microspores of *Tulipa gesneriana*. *Protoplasma* 108:329–40

65. Terasaka O, Niitsu T. 1987. Unequal cell division and chromatin differentiation in pollen grain cells. I. Centrifugal, cold and caffeine treatments. *Bot. Mag. Tokyo* 100:205–16

66. Twell D, Park S, Lalanne E. 1998. Asymmetric division and cell-fate determination in developing pollen. *Trends Plant Sci.* 3:305–10

67. Vaughn K, Hickok L, Scott R. 1989. Structural analysis and inheritance of a clumped

chloroplast mutant in the fern *Ceratopteris*. *J. Hered.* 81:146–51

68. Vogelien D, Hickok L, Warne T. 1996. Differential effects of Na⁺, Mg2⁺, K⁺, Ca2⁺ and osmotic stress on the wild type and the NaCl-tolerant mutants *stl1* and *stl2* of *Ceratopteris richardii*. *Plant Cell Environ.* 19:17–23

69. Ogelmann T, Miller J. 1980. Nuclear migration in germinating spores of *Onoclea sensibilis*: the path and kinetics of movement. *Am. J. Bot.* 67:648–52

70. Vogelmann T, Bassel A, Miller J. 1981. Effects of microtubule-inhibitors on nuclear migration and rhizoid differentiation in germinating fern spores (*Onoclea sensibilis*). *Protoplasma* 109:295–316

71. Wada M, Murata T. 1988. Photocontrol of the orientation of cell division in *Adiantum*. IV. Light-induced cell flattening preceding two-dimensional growth. *Bot. Mag. Tokyo* 101:111–20

72. Wada M, Sugai M. 1994. Photobiology of ferns. In *Photomorphogenesis in Plants*, ed. R Kendrick,G Kronenberg, pp. 783–802. Dordrecht, Netherlands: Kluwer. 2nd ed.

73. Wada M, Grolig F, Haupt W. 1993. Light-oriented chloroplast positioning. Contribution to progress in photobiology. *J. Phytochem. Photobiol.* 17:3–25

74. Wada M, Kanegai T, Nozue K, Fukuda S. 1997. Cryptogam phytochromes. *Plant Cell Environ.* 20:695–90

75. Warne T, Hickok L. 1987. Single gene mutants tolerant to NaCl in the fern *Ceratopteris*: characterization and genetic analysis. *Plant Sci.* 52:49–55

76. Warne T, Hickok L. 1989. Evidence for a gibberellin biosynthetic origin of *Ceratopteris* antheridiogen. *Plant Physiol.* 89:535–38

77. Warne T, Hickok L, Scott R. 1988. Characterization and genetic analysis of antheridiogen-insensitive mutants in the fern *Ceratopteris*. *Bot. J. Linn. Soc.* 96:371–79

78. Warne T, Hickok L, Kinraide T, Vogelien D. 1996. High salinity tolerance in the *stl2* mutation of *Ceratopteris richardii* is associated with enhanced K⁺ influx and loss. *Plant Cell Environ.* 19:24–32

79. Warne T, Vogelien D, Hickok L. 1995. The analysis of genetically and physiologically complex traits using *Ceratopteris*: a case study of NaCl-tolerant mutants. *Int. J. Plant Sci.* 156:374–84

80. Wright S, Hickok L, Warne T. 1990. Characterization of mutants of *Ceratopteris richardii* selected on aluminum (Al₂(SO₄)₃-Na₂EDTA). *Plant Sci.* 68:257–62

81. Wynne F, Mander L, Oyama N, MurofushiN, Yamane H. 1998. An antheridiogen, 13–hydroxy-GA(73) methyl ester(GA(109)), from the fern *Lygodium circinnatum*. *Phytochemistry* 47:1177–82

82. Yamane H, Nohara K, Takahashi H, Schraudolf H. 1987. Identification of antheridic acid as an antheridiogen in *Anemia rotundifolia* and *Anemia flexuosa*. *Plant Cell Physiol.* 28:1203–7

83. Yamane H, Sato K, Nohara K, Nakayama M, Murofushi N, et al. 1988. The methyl ester of a new gibberellin GA73: the principal antheridiogen *in Lygodium japonicum*. *Tetrahedron Lett.* 29:3959–62

84. Yamauchi T, Oymama N, Yamane H, Murofushi N, Schraudolf H, et al. 1995. 3–EPI-GA 63, antheridiogen in *Anemia phyllitidis*. *Phytochemistry* 38:1345–48

85. Yamauchi T, Oymama N, Yamane H, Murofushi N, Schraudolf H, et al. 1996. Identification of antheridiogens in *Lygodium circinnatum* and *Lygodium flexuosum*. *Plant Physiol.* 111:741–45

Annu. Rev. Plant Physiol. Plant Mol. Biol. 1999. 50:187–217

C₄ GENE EXPRESSION

Jen Sheen

Department of Molecular Biology, Massachusetts General Hospital, Department
of Genetics, Harvard Medical School, Boston, Massachusetts 02114;
e-mail: Sheen@molbio.mgh.harvard.edu

KEY WORDS: photosynthesis, bundle sheath, mesophyll, transcription, signal transduction

---

ABSTRACT

C_4 plants, including maize, *Flaveria*, amaranth, sorghum, and an amphibious
sedge *Eleocharis vivipara*, have been employed to elucidate the molecular mechanisms and signaling pathways that control C_4 photosynthesis gene expression.
Current evidence suggests that pre-existing genes were recruited for the C_4
pathway after acquiring potent and surprisingly diverse regulatory elements.
This review emphasizes recent advances in our understanding of the creation
of C_4 genes, the activities of the C_4 gene promoters consisting of synergistic and combinatorial enhancers and silencers, the use of $5'$ and $3'$ untrans-
lated regions for transcriptional and posttranscriptional regulations, and the function of novel transcription factors. The research has also revealed new insights
into unique or universal mechanisms underlying cell-type specificity, coordinate nuclear-chloroplast actions, hormonal, metabolic, stress and light responses,
and the control of enzymatic activities by phosphorylation and reductive
processes.

CONTENTS

1040-2519/99/0601-0187$08.00

INTRODUCTION

The majority of plants assimilate atmospheric CO_2 through the C_3 pathway of photosynthesis by using ribulose-1,5-bisphosphate carboxylase/oxygenase (RUBISCO or RBC). Under high light intensity, high temperature, and arid environmental conditions, RBC's low CO_2 affinity and inability to distinguish O_2 from CO_2 result in low photosynthetic capacity and significant energy waste through photorespiration in C_3 plants (35, 36, 43, 48, 49, 67) (Figure 1). The so-called C_4 plants have evolved the C_4 cycle pathway that serves as a CO_2 pump. By integrating the two CO_2 assimilation pathways consecutively in the spatially cooperative mesophyll (MC) and bundle sheath cells (BSC), C_4 plants can achieve high photosynthetic efficiency, especially under conditions that cripple C_3 plants (35, 36, 43, 48, 49, 67). Studies of the molecular basis of C_4 photosynthesis have enhanced our understanding of fundamental and complex biological processes and provided information that can be exploited for potential agricultural applications.

Research in molecular biology of C_4 photosynthesis has been initiated from the study of chloroplast genes in maize 20 years ago (78, 152). Significant progress has been made in understanding the molecular basis of C_4 photosynthesis by using multiple model systems. Because each system evolved independently (27, 28, 35–37, 48, 67, 128), these studies offer a wealth of information about how plants use diverse and creative molecular mechanisms to

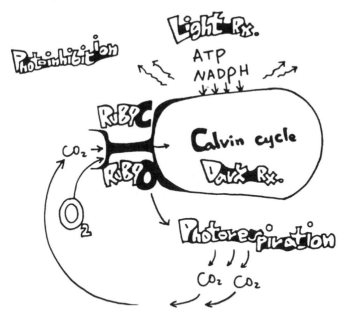

Figure 1 Rubisco is the bottleneck of C$_3$ photosynthesis. Ribulose-1,5-biphosphate carboxylase (RuBPC) and oxygenase (RuBPO).

acquire and regulate new genes. The elucidation of biochemistry, physiology, and leaf anatomy underlying the operation of C$_4$ photosynthesis constitutes a great accomplishment in plant biology (35, 36, 43, 48, 49, 67). The link between leaf development and gene expression has been extensively investigated (28, 29, 73, 98). Excellent reviews on some aspects of C$_4$ gene evolution and regulation are available (29, 43, 67, 73, 86, 88, 98, 134). More general coverage of C$_4$ photosynthesis is presented by a special issue of the *Australian Journal of Plant Physiology* (Volume 24, No 4, 1997), and a new book, *The Biology of C$_4$ Plants* (edited by R Sage & RK Monson, 1998). This review emphasizes the unique contributions in the understanding of molecular mechanisms underlying C$_4$ gene evolution and regulation. The most recent studies using transgenic plants, transient expression, and genetic approaches are summarized and discussed. Other aspects include nuclear-chloroplast coordination, stress and UV-B responses, and the integrated views of signal transduction from hormonal and metabolic regulations to enzymatic controls by protein phosphorylation that were first discovered in C$_4$ plants, but may turn out to be universal regulatory mechanisms in higher plants.

DIFFERENTIAL GENE EXPRESSION
AND C$_4$ PHOTOSYNTHESIS

Genes Involved in C$_4$ Photosynthesis

Despite their diverse origins, all C$_4$ plants exhibit characteristic features of two morphologically and biochemically distinct photosynthetic cell types, MC and BSC, in leaves or culms (28, 29, 35, 36, 43, 48, 49, 67, 73, 98, 134, 148). The assimilation of CO$_2$ is first carried out in MC through the C$_4$ cycle pathway. Carbonic anhydrase (CA) and phosphoenolpyruvate carboxylase (PEPC) are responsible for the hydration and fixation of CO$_2$ to oxaloacetate (OAA). This process is highly efficient and insensitive to O$_2$. The CO$_2$ acceptor PEP is generated from pyruvate by pyruvate orthophosphodikinase (PPDK). OAA is then converted to malate by NADP malate dehydrogenase (MDH) or to aspartate by aspartate aminotransferase (AST). Either malate or aspartate is transferred to BSC where CO$_2$ is released by NADP- or NAD-malic enzyme (ME) or PEP carboxykinase (PCK), and reassimilated by RBC through the Calvin cycle in a CO$_2$-enriched environment, thus avoiding photorespiration (Figure 2) (43, 48, 50, 49, 67, 98).

Many nuclear genes encoding the enzymes in the C$_4$ cycle pathway have been isolated from maize, amaranth, *Flaveria*, sorghum, and *Eleocharis vivipara* (1, 9, 14, 16, 45, 51, 54, 64, 76, 79, 80, 86, 93, 94, 97, 105, 120, 125, 126, 134, 141, 157). By using purified MC and BSC or in situ hybridization and immunolocalization, it has been demonstrated that these C$_4$ genes display MC- or BSC-specific expression patterns at both the protein and transcript levels in the mature leaves of all C$_4$ plants examined (Table 1). In addition, nuclear genes encoding the conserved photosynthetic enzymes or proteins involved in the Calvin cycle, photosystem II, and oxygen-evolving complex have been identified and shown to be differentially or preferentially expressed in MC or BSC (Table 1) (14, 118, 121, 159). Nuclear run-on experiments have indicated that MC-specific expression is mainly regulated at the transcriptional level, whereas BSC-specific expression is likely controlled at both transcriptional and posttranscriptional levels in maize leaves (109; J Sheen, unpublished data).

Cell-Type Specific Regulation of Diverse Genes

Besides photosynthesis genes, the expression of diverse genes involved in nitrogen and sulfate assimilation, amino acid metabolism, metabolite transport, and the biosynthesis of starch and sucrose also exhibits cell-type specific patterns (15, 20, 35, 36, 48, 66, 82, 99, 134, 138, 139, 159). The most striking example is the MC-specific expression of genes encoding nitrate reductase (NR), nitrite reductase (NiR), and other related functions for nitrate assimilation (134).

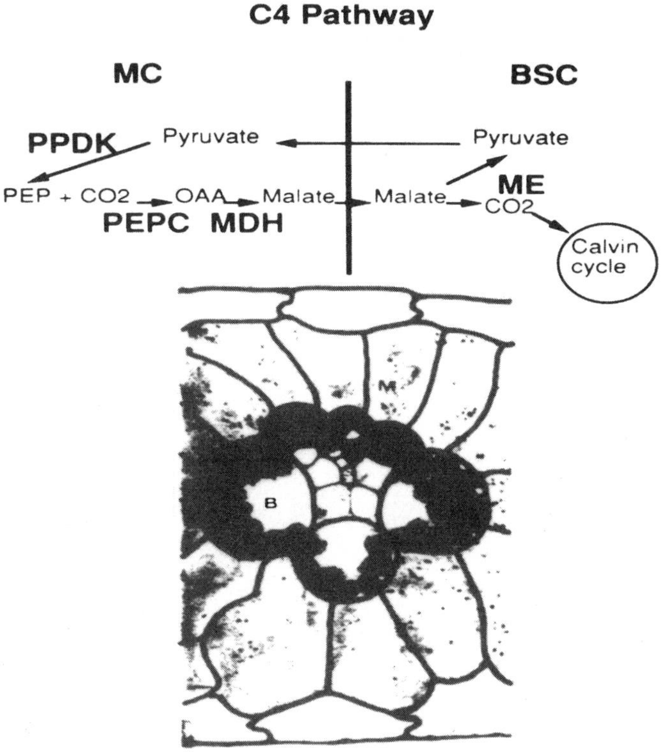

Figure 2 A simplified NADP-ME-type C$_4$ cycle pathway in maize leaves. The bundle sheath cells (BSC or B) accumulate starch and are darkly stained by iodine. Mesophyll cells (MC or M) are located between BSC and epidermal layers. Pyruvate orthophosphodiknase (PPDK), phospho-enolpyruvate carboxylase (PEPC), and NADP-malate dehydrogenase (MDH) are present in MC, whereas NAPD-malic enzyme (ME) is expressed in BSC.

Moreover, the genes encoding ferredoxin-dependent glutamate synthase (Fd-GOGAT) and glutamine synthase (GS2) for ammonia metabolism are preferentially induced by nitrate signals in MC but not in BSC. The differential induction of these genes in MC may accommodate the physiological specialization of each cell type for nitrogen metabolism (134). Recently, extensive differential screening has been performed in sorghum and maize to identify large number of genes that are specifically expressed in MC or BSC (Table 1). Sequence analysis of these genes has uncovered new candidates (e.g. BSC-specific *Pck* in maize) that will likely broaden our understanding of the physiological functions and gene regulation in MC and BSC (153, 159; T Furumoto & K Izui, personal communication).

Table 1   Regulation of nuclear genes involved in C_4 photosynthesis

Genes[a,b]	Cell type	Induction	Repression	Species
C4Pdk	M[c]	L[c], N, UV-A UV-B	S[c], A, gly ABA	Maize, *F. t.*[d], *A. h.*, Sorghum, *E. v.*
C4Ppc	M	L, N	S, A	Maize, *F. t.* *A. h.*, sorghum
C4Mdh	M	L	—	Maize, *F. t.*, sorghum
C4Cah	M	L, N	—	Maize, sorghum
Cab	M	L	S, A	Maize, sorghum
PsbO	M	L	—	Maize, sorghum
PsbP	M	L	—	Maize, sorghum
PsbQ	M	L	—	Maize, sorghum
PsbR	M	—	—	Sorghum
PsbS	M	—	—	Sorghum
PsbT	M	—	—	Sorghum
PsbW	M	—	—	Sorghum
GapB	M	—	—	Sorghum
PetF	M	—	—	Sorghum
PetH	M	—	—	Sorghum
Tpi	M	—	—	Sorghum
RbcS	B	L	S, A, ABA	Maize, *F. t.*, *A. h.*
Pck	B	—	—	Maize
NADPMe	B	L, UV-B	S, A	Maize, *F. t.*, sorghum
NADMe	B	L	—	*A. h.*
Rca	B	—	—	Sorghum
Prk	B	—	—	Sorghum
FbaC	B	—	—	Sorghum
TklC	B	—	—	Sorghum
Rpe	B	—	—	Sorghum
Omt	B	—	—	Sorghum

[a]References: 1, 9, 14, 17, 86, 97, 112, 116, 118, 119, 120, 121, 126, 134, 141, 157.

[b]Genes not found in the text: *Psb*: photosystem II; *GapB*: NADP-glyceraldehyde-3-phosphate dehydrogenase; *PetF*: ferredoxin; *PetH*: ferredoxin-NADP-oxidoreductase; *Tpi*: triosephosphate isomerase; *Rca*: rubisco activase; *Prk*: phosphoribulokinase; *Fba*: fructose-1,6-bisphosphatase aldolase; *Tkl*: transketolase; *Rpe*: ribulose-5-phosphate 3-epimerase; *Omt*: 2-oxoglutarate/malate translocator.

[c]M: mesophyll; B: bundle sheath; L: red and/or blue light; N: nitrogens; S: sugars; A: acetate; gly: glycerol; ABA: abscisic acid.

[d]*F. t.*: *Flaveria trinervia*; *A. h.*: *Amaranthus hypochondriacus*; *E. v.*: *Eleocharis vivipara*.

Leaf Development and Differential Gene Expression

Most C_4 genes examined show strict expression in leaves or leaf-like structures but not in non-photosynthetic roots and stems. Antibodies and cDNA probes for photosynthetic enzymes and genes were used as cell-specific markers by in situ hybridization and immunolocalization to follow leaf development from primodia or regenerating calli to mature leaves in maize (2, 28, 29, 71, 72, 73, 74, 97, 98). The expression of C_4 genes shows temporal and spatial regulation patterns that match the developmental stage and age gradients of leaves. Clonal analysis of photosynthetic BSC and MC in maize leaves suggests that MC development is dependent on position rather than lineage (71). The photosynthetic competence of MC and BSC is tightly coupled to vein development and is superimposed by light regulation. Light signaling not only enhances C$_4$ gene expression but also reduces the expression of *RbcS* and *RbcL* in MC (75, 119, 120). A current model points to unknown regulatory signals generated from veins for the control BSC and MC differentiation and C$_4$ pattern gene expression (28, 73, 98).

In the NAD-ME type of C$_4$ dicot *Amaranthus hypochondriacus*, it has been shown that posttranscriptional regulation determines initial C$_4$ gene expression patterns in developing cotyledons and leaves (9). Transcripts for *RbcS, RbcL, Ppc,* and *Pdk* accumulate in the shoot meristems and leaf primordia. RBCS and RBCL proteins, but not the C$_4$ enzymes, are detected in early tissues. Protein accumulation of the C$_4$ enzymes occurs only when the leaf vascular system begins to differentiate (101). Distinct from maize, light is not required for the cell-type-specific expression of *RbcS* and *RbcL* genes and other C$_4$ genes in amaranth (154). In the three-color leaves of *A. tricolor*, both transcription and translation are responsible for the lack of photosynthetic capacity in the red and yellow regions of leaves (92). There is a tight coordination between the basipetal C$_3$ to C$_4$ transition in *Rbc* gene expression and the basipetal sink-to-source transition in amaranth (155). In the developing cotyledons of C$_4$ dicot *Flaveria trinervia*, the expression of C$_4$ genes is light-dependent and more similar to that in monocot maize but not dicot amaranth, which suggests independent evolution of different C$_4$ plants (125).

C$_4$ GENE EVOLUTION AND REGULATION

Most C$_4$ genes have closely related homologs displaying low ubiquitous expression in C$_4$ and C$_3$ plants, suggesting that C$_4$ genes are recent products of gene duplications. Drastic modifications in regulatory sequences could have accommodated the changes in abundance and localization for C$_4$ photosynthesis. The lack of consensus regulatory sequences among C$_4$ genes documents the remarkable co-evolution of diverse genetic modifications. The molecular mechanisms for C$_4$ gene evolution have been deduced from comparisons of the C$_4$ genes with their non-photosynthetic counterparts in the same C$_4$ plant or with their orthologs in closely related C$_3$ plants.

Pdk Genes

The maize *C$_4$Pdk* gene shares almost all of its coding sequence with the ancestral gene (*cyPdk1*) encoding a cytosolic PPDK (Figure 3). The first exon

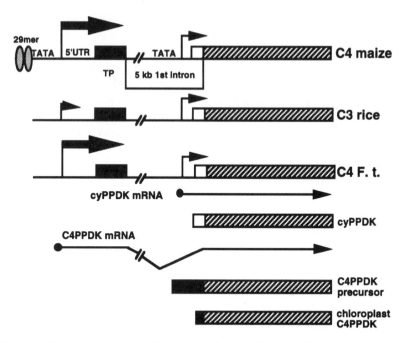

Figure 3 Dual promoters regulate *Pdk* expression in C₄ and C₃ plants. The *black box* represents the chloroplast transit peptide (TP), the *white box* is the unique N terminus of the cytosolic PPDK (cyPPDK), and the *hatched box* covers the shared coding sequence by C₄PPDK and cyPPDK. *Arrows* indicate transcription.

of *C₄Pdk* encoding a chloroplast transit peptide is separated from the first exon of *cyPdk1* and the rest of the shared coding region by a 5-kb intron, which includes repetitive sequences. The promoter of *cyPdk1* is located in the first intron of *C₄Pdk* and has ubiquitous activity. In contrast, the promoter of *C₄Pdk* directs leaf-specific and light-inducible transcription in MC. Another family member, *cyPdk2*, in maize shares similarity with *cyPdk1* up to the TATA box of the promoter. It was proposed that the creation of *C₄Pdk* could be mediated through repetitive sequences by unequal recombination that resulted in a new intron and brought a transit peptide and a new promoter to fulfill the need of a C₄ gene (45, 113). However, recent characterization of the *Pdk* gene in the C₃ monocot rice revealed similar dual promoters, gene structure, and expression patterns (56, 86). Furthermore, the single-copy *Pdk* gene in the C₄ dicot *F. trinervia* has strikingly similar structure (Figure 3) and shows high sequence similarity to the *Pdk* gene in the C₃ dicot *F. pringlei* (103). Therefore, the major difference between the C₄ and C₃ *Pdk* is at the strength of the first promoter as the transcripts for the chloroplast PPDK is expressed at very low level in C₃ rice and *F. pringlei* (56, 103).

Ppc Genes

At least three classes of *Ppc* genes have been found in maize and sorghum. The single copy C_4Ppc gene shows a unique and high level of expression in MC (32, 54, 64, 76, 110, 145). The comparison of the more divergent sequences upstream of the coding regions among three maize *Ppc* genes revealed that $C_4PpcZm1$ and *PpcZm3B* are very similar but their 5′ flanking sequences diverge before the TATA box. The observation leads to the suggestion that the C_4Ppc gene could be generated from an ancestral *Ppc* gene after the unequal recombination near the TATA box bringing regulatory elements required for the new expression patterns and high activity (110). Note that the two maize *cyPdk* genes also share similar sequences up to the TATA box but show distinct expression levels. Perhaps the regions near the transcription initiation site are hot spots for DNA recombination responsible for the generation of distinct expression patterns and levels among homologous genes in maize. The most informative sequence comparison between the C_4 and C_3 *Ppc* orthologs was performed in dicot *F. trinervia* (C_4) and *F. pringlei* (C_3). There are four classes of *Ppc* genes in each plant. The *PpcA* genes from these two species are 96% identical but expressed differently. Thus the major events during the evolution of the C_4PpcA gene occurred at the promoter level (38, 51, 52, 130, 157).

Me Genes

Two closely related NADP-ME genes have been isolated from *F. bidentis* and maize (84, 105). Surprisingly, both *Me* genes encode chloroplast transit peptides and similar sequences, but show different expression patterns. Although both *Me* genes are found in C_3 *F. pringlei*, the expression of its C_4 *Me* ortholog (*Me1*) is not observed (84). In the C_4 dicot amaranth, NAD-ME but not NADP-ME is used for decarboxylation in the mitochondria of BSC. The amaranth C_4 NAD-ME shares more similarity with human NADP-ME than with maize and thus may represent another origin and process for C_4 gene evolution (9, 79).

Mdh Genes

Although two distinct NADP-MDH genes have been found, the expression of only one *Mdh* gene is MC-specific and light-inducible, and serves in C_4 photosynthesis in maize and sorghum (80, 94). In various *Flaveria* species, however, there appears to be only a single *Mdh* gene, suggesting that a pre-existing gene has been re-regulated without gene duplication during the evolution from C_3 to C_4 plants (93).

Cah Genes

The cytosolic and chloroplast forms of CA have been characterized in plants (3, 16, 50, 67, 81). The cytosolic CA is MC-specific and important for C_4 photosynthesis. Two *Cah* cDNAs have been isolated in maize, sorghum, and

F. bidentis, but only one *Cah* gene shows enhanced and MC-specific expression in illuminated leaves and likely encodes the C_4 gene (16, 81, 159). As all *Cah* genes are closely related in C_4 and C_3 plants, simple genetic alternation is likely required for the C_4 function.

Other Photosynthesis Genes

The evolution of C_4 plants also included the altered cell-type specificity for the existing genes involved in the Calvin cycle and photosystem II. As in C_3 plants, RBCS and chlorophyll *a/b* binding proteins (CAB) are encoded by multi-gene families in maize, amaranth, and *Flaveria* (9, 39, 118, 119, 126). The six members of the maize *Cab* genes show complex light and cell-type expression patterns (118). However, unlike the C_4 genes whose expression pattern is distinct from their closely related homologs, all *RbcS* genes exhibit a similar BSC-specific expression pattern in C_4 plants (9, 39, 119, 126).

C_4 GENE EXPRESSION IN TRANSGENIC PLANTS

To identify, characterize, and definitively prove that any putative regulatory elements associated with C_4 genes are functional and can account for the unique expression patterns and levels in C_4 plants, it is essential to develop reliable transformation methods (21, 24, 63, 83, 102, 131, 141, 157). Despite technical difficulties, successful studies of C_4 gene regulation in transgenic *Flaveria* species and maize have recently been achieved (21, 24, 63, 83, 102, 131, 141, 157).

C_4 Dicot Flaveria

The most extensive transgenic analyses have been carried out in the C_4 dicot *F. bidentis* based on a facile transformation procedure and the use of the ß-glucuronidase reporter gene (GUS) (21). The transgenic plants were analyzed by histochemical and cell separation techniques (21, 141, 157). The 5′ flanking region (2185 bp) of the *F. trinervia* C_4PpcA1 gene is sufficient to reproduce MC-specific, developmental, and basipetal expression patterns. The promoter region (2583 bp) of the orthologous *PpcA1* gene from *F. pringlei* (C_3) directs reporter gene expression mainly in vascular tissues of leaves and stems and at low levels in MC. These experiments are the first to demonstrate that *cis*-acting elements are responsible for the C_4 expression patterns that require synergistic cooperation between distal (-2074 to -1501) and proximal (-570 to $+1$) regions (131, 157). Similarly, the 5′ region (1491 bp) of the C_4Pdk gene directs MC specificity and light induction in seedlings and mature leaves (102, 141).

Complex regulatory mechanisms have been unveiled from the study of the C_4Me1 gene (83, 141). Transgenic plants were obtained with a set of chimeric constructs using the 5′ (up to 2361 bp) and 3′ (up to 5900 bp) regions of the *F. bidentis* *C$_4$Me1* gene fused to GUS. Although the 5′ region determines BSC specificity, the 3′ region contains enhancer-like elements and confers high-level expression in leaves. This interaction of the 5′ and 3′ sequences appears to be specific to C$_4$ *F. bidentis* because the same construct does not direct significant expression in transgenic C$_3$ tobacco. Although the 3′ region (900 bp) of the *C$_4$Me1* gene can serve as a transcriptional terminator with a heterologous promoter in tobacco protoplasts, it does not enhance the promoter activity (83). It remains to be determined whether the 3′ region affects transcription, mRNA stability, or translation. Independent transgenic *F. bidentis* plants have also been generated with the 5′ sequences of the *F. trinervia* *C$_4$Me1* gene and the C$_3$ *F. pringlei* ortholog. The 5′ region of the *C$_4$Me1* gene is sufficient for the basipetal and BSC-specific expression patterns, whereas the 5′ sequence of the C$_3$ *Me1* gene directs expression in all cell types and the expression is turned off basipetally (T Nelson, personal communication). In maize, the *C$_4$Me1* promoter is active in MC (112). A nuclear run-on experiment with MC and BSC shows *C4Me* transcription in both cell types, implying posttranscriptional control for BSC specificity (J Sheen, unpublished data). These different results point to the possibility that the *C$_4$Me1* genes might have evolved by very different mechanisms even in two very closely related C$_4$ species, *F. trinervia* and *F. bidentis*.

The promoter of a C$_4$ *F. trinervia* *RbcS* gene gives a BSC-specific pattern very similar to that of the *F. trinervia* *C$_4$Me1* promoter in transgenic *F. bidentis*, whereas the promoter of a C$_3$ *F. pringlei* *RbcS* gene shows expression in both MC and BSC (T Nelson, personal communication). Thus there appear to be C$_4$-specific *cis*-acting elements for the expression of the *F. trinervia* *RbcS* gene in BSC. However, analyses of the *RbcS* promoters in maize and rice MC and the *RbcS* expression patterns in C$_4$/C$_3$ *Flaveria* hybrids imply the involvement of C$_4$-specific *trans*-acting factors (87, 109, 127). In the transgenic C$_4$-like *F. brownii*, a C$_3$ petunia *RbcS* gene can direct a high level of leaf-specific expression, although the cell-type pattern is not known (85). In addition, translation enhancer function of the 5′ (47-bp) and 3′ (130-bp) UTR sequences of the *RbcS* gene from amaranth can be detected when fused to GUS in transgenic *F. bidentis*. (AC Corey & JO Berry, personal communication). Thus the regulation of *RbcS* involves complex and diverse mechanisms in C$_4$ plants.

In addition to the analyses of gene regulation, the C$_4$ *F. bidentis* transformation system provides an unprecedented opportunity for the genetic manipulation of key photosynthetic enzymes in C$_4$ plants (42, 43). Antisense transformants for *C$_4$Pdk* and *RbcS* genes have been generated that can reduce C$_4$PPDK to 40% and RBC to 15% of the wild-type levels. Co-suppression has been used for the *C$_4$Mdh* gene to achieve lower than 2% of the wild-type C$_4$MDH activity. Under saturating illumination, RBC levels have the most significant effect on photosynthetic rates, but C_4MDH has a very minor impact. These valuable transgenic lines will aid our understanding and improvement of C_4 photosynthesis.

C_4 Monocot Maize

The first study of the C_4Ppc promoter in transgenic C_4 monocot maize has been accomplished (63). The 5′ flanking region (1.7 kb) of the maize C_4Ppc gene is sufficient to direct GUS expression in a MC-specific and light-inducible fashion identical to the endogenous gene. Similar to C_4 dicots, the adoption of DNA regulatory elements for C_4-specific gene expression is a crucial step in the C_4Ppc gene evolution in maize. The transgene is repressed by metabolic signals and is blocked completely when the biogenesis of chloroplasts is inhibited under light (63). The regulatory mechanisms of C_4 genes need to be integrated into the universal signaling network that modulates photosynthesis genes in both C_4 and C_3 plants (26, 40, 60, 140, 144). The results from transgenic maize studies fully support previous observations using isolated maize mesophyll protoplasts which do not dedifferentiate in culture and behave as bona fide MC (110, 112).

C3 Dicot Tobacco and C3 Monocot Rice

Several studies of C_4 gene expression in transgenic C3 dicot tobacco and C3 monocot rice provide valuable insights into the nature of the regulatory elements acquired by the C_4 genes. The 5′ flanking regions of the C_4Pdk (-1032 to $+71$), C_4Ppc (-1212 to $+78$), and $RbcS$ (-444 to $+66$) genes from maize have been introduced into transgenic rice (86, 87, 90, 91). The promoters of C_4Pdk and C_4Ppc display MC specificity, light inducibility, and high activity that are characteristic of the C_4 genes. Moreover, similar DNA-binding activities for cis-acting elements in the C_4Pdk and C_4Ppc promoters have been detected in the nuclear extracts of maize and rice (56, 87, 89, 91, 162, 163). Very high levels of C_4PPDK and C_4PEPC activities have also been ectopically expressed in transgenic rice (88), opening an avenue to genetically manipulate the C_4 cycle pathway in C_3 plants. However, the $RbcS$ promoter of the C_4 maize does not support the BSC-specific expression pattern in transgenic rice. It is suggested that differences in trans-acting factors exist between C_4 maize and C_3 rice for the regulation of $RbcS$ genes (87). The C_4 expression pattern of the F. trinervia C_4Ppc promoter in C_3 transgenic tobacco reinforces the concept that the evolution of this C_4 gene needs alternations only in the cis-acting elements. However, the evolution and regulatory mechanisms of the C_4Me gene are unique to C_4 plants and cannot be reproduced in C_3 transgenic tobacco (83).

TRANSIENT EXPRESSION ANALYSES OF GENES INVOLVED IN C$_4$ PHOTOSYNTHESIS

Many regulatory mechanisms can be more efficiently investigated by transient expression assays using cellular systems where gene expression patterns are faithfully retained. Maize mesophyll protoplasts currently represent the most sophisticated system for the study of gene expression and signal transduction relevant to photosynthesis in higher plants (22, 59, 109, 110, 112–114, 116, 163). Rice, tobacco and Arabidopsis protoplasts have also been applied (47, 87, 109). However, the protoplast system has a limitation as BSC are not easily isolated with high activity. An alternative transient expression assay is to transform whole tissues of maize and *Flaveria* with DNA-coated microprojectiles (4, 5; T Nelson, personal communication). Histochemical analysis of intact tissues or fluorometric assay of tissue extracts based on GUS activity is used in this transient expression assay.

Pdk Genes

The analysis of the three maize *Pdk* gene promoters in mesophyll protoplasts identified distinct *cis*-acting elements for leaf, stem, and root expression. Deletion analysis revealed that a 300-bp region of the *C$_4$Pdk* promoter (−347 to −44) shows strong leaf specificity, developmental regulation, and light inducibility even when it is fused to a heterologous basal promoter in etiolated and greening protoplasts. The same promoter sequences are important for sugar, acetate, and glycerol repression (112, 113). Deletion and site-directed mutational analyses have identified at least six distinct functional elements (K To & J Sheen, unpublished data). The most significant *cis*-acting element is a 29mer (−286 to −258) with a pair of 7-bp inverted repeats (Table 2). The promoter activity is reduced tenfold when this 29mer is deleted. Most significantly, only the nuclear extract isolated from greening leaves but not from etiolated and green leaves or roots can form a complex with the ^{32}P-labeled 29mer by a gel mobility shift assay (K To & J Sheen, unpublished data). However, this 29mer alone cannot activate transcription when fused to a heterologous basal promoter, which suggests that there is a synergistic cooperation between this element and other regulatory elements.

Another important finding is the dependence of the upstream elements on the 5′UTR elements (+1 to +139) for a high level of transcriptional activity that distinguish the *C$_4$Pdk* promoter from the *c$_y$Pdk1* promoter (113; P Leon & J Sheen, unpublished data). Extensive deletion and mutational analyses of this 5′UTR region have defined at least four functional CT and CA repeats whose deletion decreases the *C$_4$Pdk* promoter activity 80-fold (Table 2). As the *Pdk* gene in C$_3$ rice only differs from the *C$_4$Pdk* gene in maize at the transcriptional

Table 2   *Cis*-acting regulatory elements of genes involved in C$_4$ photosynthesis

Genes[a]	Regions	Sequences[b]	Species
C$_4$Pdk	5′ flanking	−33 TATAA −29	Maize
	5′ flanking	−286 CTGTAGC–GCTACAG −258 (29mer)	Maize
	5′ flanking	−308 AGTGGAGTCGTGCCGCGTGT −289	Maize
	5′ UTR	CCCCCTCTCC (CT repeats)	Maize
	5′ UTR	CACTCGCCACACACA (CA element)	Maize
	5′ flanking	−1212 to +279	F. t.[c]
C$_4$Ppc	5′ flanking[d]	−119 CCATCCCTATTT −107	Maize
	5′ flanking	CCCTCTCCACATCC (C14 repeats)	Maize
	5′ flanking	AAAAAGG (29A repeats)	Maize
	5′ flanking[d]	−2187 to −1	F. t.
C$_4$Me	5′ flanking	−2038 TO +323, −311 TO +83	F. b.[c]
	3′ flanking	900 bp, 5900 bp	F. b.
RbcSZm1	5′ flanking[d]	−24 CTATATATGCCGTCGGTG −7	Maize
	5′ flanking[d]	−105 GAACGGTGGCCACT-CCACA −83	Maize
	5′ flanking[d]	−123 CCGGGTGCGGCCAC −110	Maize
	5′ flanking[d]	−154 GCGCGCGT −147	Maize
	5′ flanking[d]	−179 GATAAG −174	Maize
	5′ flanking[d]	−588 to −183 (silencer)	Maize
	3′ UTR	281 b	Maize
RbcSZm3	5′ flanking[d]	−31 CTATATATGCCGTCGGTG −14	Maize
	5′ flanking[d]	−102 GTCCTGTCCTGTACT-GTCCT −81	Maize
	5′ flanking[d]	−133 GAACGGTGGCCACT-CCACA −111	Maize
	5′ flanking[d]	−151 CCGGGTGCGGCCAC −138	Maize
	5′ flanking[d]	−125 GATAAG −210	Maize
RbcS-m3	5′ flanking[d]	−885 to −229 (silencer), −907 to −455	Maize
	3′ UTR[d]	289 b, +720 to +957	Maize
RbcS	5′ & 3′ UTR	47 b and 130 b	Amaranth
Cab-m1	5′ flanking[d]	−1026 to −850, 359 to +14	Maize
	5′ flanking[d]	−949 AATATTTTTCT −937	Maize
CabZm1	5′ flanking[d]	−158 CGCGCCAAGTGTTCCAG −142	Maize
	5′ flanking[d]	−184 CCTCA-TGAGG −166	Maize
CabZm5	5′ flanking	−34 TATTTA −29 (TATA box)	Maize
	5′ flanking	−59 GATAAG −54 (I box)	Maize
	5′ flanking	−90 CCAAT −86 (CAT box)	Maize
	5′ flanking	−103 CACCTCCGGCGA −92	Maize
	5′ flanking	−115 ATCCGCCCACCT −104	Maize

[a]References: 4, 83, 89, 102, 109, 110, 112, 113, 114, 123, 131, 151.
[b]The *cis*-acting regulatory elements were defined by deletion, site-directed mutagenesis, or gain-of-function analyses.
[c]*F. t.*: *Flaveria trinervia*, *F. b.*: *Flaveria bidentis*.
[d]Numbers are based on the translation initiation site (+1).

activity, this specific 5'UTR may underlie the major evolutionary changes in the maize C$_4$Pdk gene (Figure 3). It is not clear why the two promoters show similar activity in maize mesophyll protoplasts, as reported by another group (56, 88). The differences could be attributed to protoplast activity and plasmid constructs. Interestingly, this 5'UTR can function at only one position but can interact with multiple promoters in leaves, roots, and stems in dicot or monocot and in C$_3$ or C$_4$ plants (P Leon & J Sheen, unpublished data). This is another example that a C$_4$ gene utilizes conserved machinery to enhance gene expression (12, 143). Distinct from the C$_4$Pdk promoter, the c$_y$Pdk1 and c$_y$Pdk2 promoters are expressed ubiquitously. The sequences upstream of the TATA box of the two promoters are divergent and responsible for a 20-fold difference in promoter strength (113). Deletion analysis of the c$_y$Pdk1 promoter has identified distinct cis-acting elements for leaf or root/stem expression, suggesting the use of combinatorial cis-acting elements for tissue-specific controls (113). Functional analysis of the maize C$_4$Pdk promoter (-327 to $+211$) has also been performed in maize green leaves by microprojectile bombardment (89). The sequence between -308 and -289 is important for the promoter activity in green leaves. Protein binding activity has been found with a shorter element (-301 to -296) using nuclear extracts isolated from green leaves but not from etiolated leaves or roots (89). This cis-acting element may be more important for the maize C$_4$Pdk promoter activity in green than in greening leaves (K To & J Sheen, unpublished data). In maize mesophyll protoplasts, the rice c$_y$Pdk promoter is more active than the rice promoter for the chloroplast PPDK (56, 86–88).

Ppc Genes

The regulatory elements of the Ppc promoters have been well characterized in maize by using transfected mesophyll protoplasts. Among the three maize Ppc promoters examined, only the C$_4$Ppc promoter shows leaf-specific, light, and developmental regulation (110). The 5' flanking sequences are sufficient to give C$_4$ expression patterns in isolated mesophyll protoplasts and in transgenic maize plants (63). The importance of two redundant cis-acting elements (A29 and C14) has been demonstrated by deletion and hybrid promoter analyses (Table 2). The C14 and A29 direct repeats act as enhancers in maize or rice protoplasts (87, 109). Two groups have shown that proteins in the maize and rice leaf nuclear extracts can bind to these elements (162, 87). A short promoter (-119 to $+1$) retains 20% of the full promoter activity. The significance of the TATTT sequence is confirmed by deletion and site-directed mutagenesis. The C$_4$Ppc and C$_4$Pdk promoters in maize do not share common regulatory elements, which indicates independent evolution.

Despite the identification of numerous cis-acting elements important for C$_4$ gene expression in maize, very little is known about the properties of the corresponding *trans*-acting factors. A recent study has focused on the function of a conserved zinc-finger binding protein Dof1 (DNA-binding with one finger) in the regulation of the maize C_4Ppc promoter (160, 161, 163). Dof proteins represent a new family of transcription factors carrying a conserved motif for DNA-binding, which seem to be ubiquitous in plants. The consensus AAAAGG core sequence motif bound by Dof1 is part of the A29 enhancer identified previously in the maize C_4Ppc promoter. Although Dof1 is constitutively expressed in the nuclei of etiolated and greening mesophyll protoplast, it only binds to DNA and activates transcription in greening protoplasts assayed with a synthetic promoter. In fact, Dof1 can only specifically activate the C_4Ppc promoter but not the maize *RbcS* and C_4Pdk promoters. Although Dof1 exhibits a ubiquitous expression pattern in maize, the function of Dof1 is leaf specific. The activity of Dof1 in other tissues is likely inhibited by a competitive transcriptional repressor Dof2 that is only expressed in roots and stems (163).

RbcS Genes

Nuclear run-on experiments and promoter analyses in maize mesophyll protoplasts and transfected leaves have indicated the involvement of both transcriptional and posttranscriptional regulation of *RbcS* genes (109, 151). Although *RbcS* transcripts do not accumulate in the MC of greening and green maize leaves, the promoters of *RbcSZm1* and *RbcSZm3* are very active and lightinducible in transfected mesophyll protoplasts. The C_4 maize and C_3 wheat *RbcS* promoters share three of five functional *cis*-acting elements (Table 2) that are distinct from the *cis*-acting elements found in the dicot *RbcS* promoters except the universal I box (GATAAG). The activity of the upstream regulatory elements absolutely depends on the specific TATA box, suggesting their synergism in function (109, 114). Further investigations show that the monocot *RbcS* promoters (maize, wheat, and rice) are only active in maize and wheat protoplasts regardless of their C_4 or C_3 origin, whereas the dicot *RbcS* promoters (pea, tobacco, and Arabidopsis) are only active in the dicot tobacco protoplasts (6, 109, 142, 143, 144). Deletion analyses have uncovered upstream silencers (−588 to −183 for *RbcSZm1* and −885 to −229 for *RbcSZm3*) and mRNA destablization elements (281 b in *RbcSZm1* and 289 b in *RbcSZm13*) in the 3′UTR that may mediate the repression of maize *RbcS* in MC (109; P Leon & J Sheen, unpublished data). Based on an in situ transient expression assay in which each blue spot represents a group of MC or BSC, the *RbcS-m3* promoter activity is also detected in both MC and BSC. However, further studies have identified the 5′ upstream region (−907 to −455) and the 3′ region (+720 to +957) that act together to partially suppress *RbcS-m3* expression in MC of illuminated dark-grown seedlings. As the 3′ region is equally active when located upstream of the *RbcS-m3* promoter, its photoregulated suppression is mediated by transcriptional control (100, 151).

Cab Genes

At least six members of the *Cab* gene family are differentially expressed in MC and BSC of the maize leaves (4, 118). Using the microprojectile-based in situ transient expression assay, the regulatory *cis*-acting elements of the MS-specific *Cab-m1* have been determined based on GUS reporter activity. Upstream sequences (-1026 to -850) containing four *cis*-acting elements (Table 2) and the core promoter (-359 to $+14$) are identified by deletion and gain-of-function experiments to be important for *Cab-m1* regulation in maize leaves. The core promoter directs high expression in MC and the upstream elements inhibit expression in BSC (4, 5, 123).

Deletion and mutational analyses have been carried out with the distinct *CabZm1* and *CabZm5* promoters using the maize mesophyll protoplast transient assay (59, 112; H Huang & J Sheen, unpublished data). The two promoters share extensive sequence similarity in the proximal elements from the TATA box, I box to the CCAAT box, but use different upstream regulatory elements (Table 2) that determine high level expression and light responses. The *CabZm5* promoter is active both in the dark and light, but the *CabZm1* promoter is only active under light. At least two upstream *cis*-acting elements have been defined in the *CabZm1* promoter that act synergistically (Table 2). In addition, mutations that alter the sequences of TATA, I, or CCAAT boxes in both promoters eliminate their activity. Unlike the strict monocot and dicot specificity of the *RbcS* promoters, the monocot (maize, wheat, and rice) *Cab* promoters can function in dicot tobacco or Arabidopsis mesophyll protoplasts but they exhibit more than tenfold higher activity in maize mesophyll protoplasts. Similarly, the *Cab* promoters from pea, tobacco, and Arabidopsis show significant activity and light regulation in maize protoplasts (22; J Sheen, unpublished data). Only minor changes are expected in the C$_4$ plants to adjust the expression patterns of various *Cab* gene members in MC and BSC.

CHLOROPLAST GENE REGULATION IN C$_4$ PLANTS

The functional coordination between the nucleus and the chloroplast is evident in C$_4$ plants as the differential expression of the nuclear-encoded RBCS in BSC is accompanied by a matching specificity of the chloroplast-encoded RBCL. Moreover, dimorphic chloroplasts are pronounced in maize with agranal chloroplasts in BSC that have greatly reduced photosystem II (*psb*) activity encoded by both nuclear and chloroplast genes. RNA and protein blot analyses with purified MC and BSC from maize show that the transcripts and proteins encoded by *psbA, B, C, D,* and *E/F* are predominant in MC but those encoded by *RbcL* are BSC specific. The differences are more than tenfold in mature greening (after 24 to 96 h of illumination) and green leaves (119, 122). However, in vitro transcription and slot blot assays using chloroplasts isolated from the same BSC and MC preparations reveal little difference in the transcription activity of *RbcL* and *psb* genes (J Sheen, unpublished data). Thus, posttranscriptional regulation plays a crucial role in the differential expression of chloroplast genes in BSC and MC of maize leaves. In dark-grown etiolated leaves, *RbcL* transcripts but not the protein can be detected in both MC and BSC. Illumination enhances its expression in BSC but causes repression in MC (75, 119). This result has been confirmed by in situ hybridization (75). The differential expression of *RbcL* and *psbA* genes studied by Kubicki et al was less significant in maize (threefold) than in sorghum (tenfold), perhaps due to differences in experimental procedures (68). In amaranth, the differential expression of *RbcL* in BSC and MC is also controlled at the posttranscriptional level (11). The signals and molecular mechanisms for the coordinated mRNA degradation in the nucleus and chloroplast in BSC and MC remain unknown (140).

REGULATION OF C$_4$ GENE EXPRESSION BY DIVERSE SIGNALS

Nitrogen and Cytokinin

Nitrogen is a limiting factor in photosynthetic capacity. C$_4$ plants not only have higher efficiency than C$_3$ plants for the utilization of light and water, but also use nitrogen more efficiently, perhaps owing to low photorespiration that affects carbon and nitrogen metabolism (99, 134). Although the effect of nitrogen on photosynthesis is well known, the underlying mechanisms remained illusive. The first evidence that nitrogen signals control C$_4$ gene expression was discovered in maize. In nitrogen-starved maize seedlings, nitrate, ammonium, or glutamine can induce the accumulation of transcripts encoding *C$_4$Ppc* and *Cah* in leaves (132–135). Both transcriptional and posttranscriptional mechanisms are involved. The signaling process does not require de novo protein synthesis but involves calcium and protein phosphorylation (106). Recent studies have demonstrated that nitrogen signals are sensed by roots and stimulate the accumulation of cytokinin, which activates C$_4$ gene expression in leaves (107, 134). This study illustrates the first link between nitrogen and cytokinin signaling in controlling photosynthesis in higher plants. Recent advances include the identification of early cytokinin-inducible genes in maize (*ZmCip1*) and Arabidopsis (*ARR3-7* and *IBC6&7*) (13, 57, 107, 137). Sequence analyses show that these proteins possess similarity to the response-regulators in the bacterial two-component signaling systems. It is proposed that *ZmCip1* could be a signaling component between cytokinin and transcription of *C₄Ppc*, *C₄Pdk*, and *Cah* in maize leaves (107, 134).

Metabolites

Metabolic repression of C_4 gene transcription has been extensively demonstrated in maize mesophyll protoplasts using C_4 gene promoters (59, 112). The studies revealed that the *C₄Pdk* promoter is specifically repressed by sucrose, glucose, fructose, acetate, and glycerol. Inhibition of positive elements but not activation of negative elements causes transcriptional repression. The examination of six other maize promoters has uncovered the global effects of sugars and acetate in gene repression. As these seven maize promoters are controlled by distinct *cis*-acting elements, no consensus sequences responsible for the sugar repression can be identified. Studies performed in C_3 plants support the concept that sugar repression of genes involved in photosynthesis is a universal phenomenon in higher plants. Biochemical, pharmacological, and molecular approaches have been taken to elucidate the sugar sensing and signaling mechanisms using the maize mesophyll protoplast system. The results have demonstrated the uncoupling of sugar signaling from sugar metabolism via glycolysis, and the involvement of protein kinases and phosphatases. Protein synthesis is not required. There is no evidence to suggest a role of calcium, cAMP, cGMP, or other second messengers in the signaling process (59, 60; J Jang & J Sheen, unpublished data). The implication that hexokinase is a sugar sensor has been supported by both biochemical experiments in transient expression assays and by genetic manipulation in transgenic plants (58, 59). Metabolic repression overrides light and developmental control and may serve as a mechanism for feedback regulation of photosynthesis and source-sink interactions in plants. The coupling of differential *Rbc* gene expression in BSC and source-sink transition discovered in amaranth may be an especially intriguing case of metabolic control (155).

Stress and Abscisic Acid

The amphibious leafless sedge *E. vivipara* develops Kranz anatomy and conducts C_4 photosynthesis under terrestrial conditions, but develops C_3-like traits and operates C_3 photosynthesis when submerged in water (147, 148). As the transition from water to land could signal a water-deficient condition for this unique plant, the anatomical development and gene expression could represent an adaptational response to water stress. Abscisic acid (ABA) is a stress hormone that is involved in the determination of leaf identity in some heterophyllic aquatic plants and the induction of crassulacean acid metabolism (CAM) in some succulent plants (148). The effect of ABA was tested in submerged

E. vivipara and shown to fully induce Kranz anatomy, C_4 gene expression, and C_4 photosynthesis. These experiments provide a new example for hormonal control of the development of the C_4 cycle pathway. Although the development of Kranz anatomy and C_4-type of gene expression are always tightly associated, recent studies suggest that the two processes can be uncoupled. In the transition region of newly sprouting culms after the terrestrial culm is submerged, the C_4 pattern of *RbcS* and *Ppc* expression can still be maintained when the development of BSC is repressed (146).

In contrast, ABA and environmental stress signals repress the expression of genes involved in C_4 photosynthesis. For example, the *RbcS* and *C_4Pdk* promoters are repressed significantly by ABA in maize (116, 117). This ABA signaling pathway is mediated by calcium-dependent protein kinases (CDPKs) and can be blocked by multiple protein phosphatases 2C (115, 116). It has been well documented that ABA and other stresses repress photosynthetic gene expression in C_3 plants (30, 65, 156, 158). Despite their recent evolution, the C_4 genes are subject to regulation of the conserved signaling pathways in plants. In the common ice plant, NaCl stress triggers complex responses in gene expression, including the induction of genes involved in CAM but repression of *RbcS* expression (25, 30).

Because of their ubiquitous expression patterns, it has been speculated that most ancestral genes for the C_4 genes have housekeeping functions, but their low expression may suggest that these functions may not be essential. A different explanation is suggested by the finding that the rice *cyPdk* gene is dramatically induced by low oxygene, ABA, and PEG, implicating a physiological function in stress responses in roots. The activities of PPDK, PEPC, and MDH are similarly induced in stressed rice plants (96). As plants cope with changing environments, their ability to alter carbon metabolism through gene expression offers great flexibility and adaptability for survival. Future experiments will confirm whether this stress response is conserved in plants.

Light

The most fascinating discoveries in the regulatory mechanisms for C_4 photosynthesis are the distinct posttranslational controls of the C_4 enzyme activities by light. The light activation of C_4PEPC requires protein phosphorylation. The use of protoplasts and biochemical tools has revealed a complex signaling cascade that includes cytosolic alkalination, calcium release, CDPK activation, protein synthesis, and the activation of a specific PEPCPK that is calcium-independent (34, 44, 62, 129, 149, 150). Although first discovered in C_4 plants, similar regulatory mechanisms are functional in diverse physiological contexts involving PEPC (44, 77, 150). Another light-regulated C_4 enzyme is C_4PPDK. C_4PPDK undergoes a diurnal light-dark regulation of activity mediated by an unusual bifunctional regulatory protein (RP). RP inactivates PPDK in the dark by an ADP-dependent phosphorylation and activates PPDK in the light by dephosphorylation of the same Thr (456 in maize) (19, 48, 49). The light activation of the C$_4$ NADP-MDH is mediated by another distinct mechanism, reductive process during which disulfides are reduced into dithiols by reduced thioredoxin (48, 95). The isolation of genes encoding PEPCPK and RP, and the identification of regulatory residues by site-directed mutagenesis coupled to enzymatic activity assays are the focus of current research.

The expression of most C$_4$ photosynthesis genes is controlled by light perceived by red, blue, and UV light photoreceptors (17, 100, 117). The study of the maize *C$_4$Pdk* promoter in mesophyll protoplasts has defined a distinct blue light signaling pathway that is insensitive to green and red light, and uncoupled from the phytochrome or cryptochrome action mediated by calcium (6, 18, 26, 55, 61, 100, 117, 123). This is a direct and fast light response that does not depend on the presence of chloroplasts. The requirement of low fluence light (activated by 0.1 μE/m^{2}s of white light and saturated by 2.8 μE/m^{2}s of blue light) separates it from the blue/UVA pathway activated by high-intensity white light (100 μE/m^{2}s) (18, 23, 41, 61, 117, 124). A similar response is observed with the *RbcS* and *CabZm1* promoters but not the *C$_4$Ppc* promoter. Inhibition of PP1 activity by okadeic acid blocks this signaling process, which is also found in barley (23, 114). However, PKs seem to be required as well because two PK inhibitors, K252a and staurosporine, can also abolish this blue light response (K To & J Sheen, unpublished data). It will be interesting to determine whether any of the isolated cryptochrome genes (18) and the light-activated PKs (53, 108) are involved in this unique blue light signaling process.

Unlike the *C$_4$Pdk, RbcS* and *CabZm1* promoters, the activity of the *C$_4$Ppc* promoter is not detected in etiolated maize mesophyll protoplasts even under illumination (110). It is proposed that activation of this promoter by light is indirect and tightly coupled to the light induction of chloroplast biogenesis. In transgenic maize, it takes 6 h of illumination to activate this promoter and the inhibition of chloroplast protein synthesis eliminates its activity completely (63). This indirect light effect also contributes to the activity of the *RbcS* and *CabZm1* promoters. As Dof1 only activates the *C$_4$Ppc* promoter, distinct transcription factors are likely involved in transmitting the chloroplast signals to the *RbcS* and *CabZm1* promoters (40, 140).

Using the transient in situ expression assay, it has been shown that the *RbcS-m3* reporter gene is stimulated in BSC via a red/far-red reversible phytochrome signaling pathway. However, blue light is required for its suppression in MC in illuminated maize leaves (100). Although the quality of light for the activation of the maize *Cab-m1* promoter in MC is not determined, the 54-bp fragment (−953 to −899) has been reported to mediate its photoregulation through calcium and calmodulin (6, 123, 136).

In amaranth, extensive studies indicate that translation initiation and elongation control the accumulation of RBC proteins during dark-to-light and light-to-dark transitions, respectively (7–10). Recent experiments have identified two 47-kDa proteins that specifically interact with the *RbcL* 5′UTR in light-grown but not in etiolated amaranth plants based on polysome heel printing, gel retardation, and UV cross-linking (DJ McCormac & JO Berry, personal communication). Whether these RNA binding proteins are universal for the regulation of RBCL translation but are modified to respond to different regulatory signals in diverse plants has not been determined.

Recent studies have described an effect of UV-B radiation on C_4 gene expression (17, 33). The transcript, protein, and enzymatic activity of C_4ME can all be increased by a short exposure (2–60 min) to low dose (0.5-2 $\mu E/m^2 s$) of UV-B but not UV-A in maize. In addition, a low level of red light (10 $\mu E/m^2 s$) exhibits similar effects as UV-B radiation. A 5-min exposure to far-red light (100 $\mu E/m^2 s$) following UV-B or red light treatment largely reverses the induction. Furthermore, the maize $C_4 Pdk$ promoter can be activated by UV-A and UV-B radiation in maize mesophyll protoplasts (J Sheen, unpublished data). These surprising results support a novel physiological function of the C_4 cycle pathway in UV responses and deserve further investigation. It is proposed that the induction of the C_4 enzymes may contribute to repair of UV-induced damage by providing reductive power, increasing pyruvate for respiration, and providing substrates for lipid synthesis and membrane repair (17, 33).

GENETIC ANALYSIS OF C_4 GENE REGULATION

Maize Mutants

To dissect the pathway leading to dimorphic photosynthetic cell differentiation, a powerful genetic approach has been taken to isolate maize mutants that exhibit perturbed leaf development (28, 29, 46, 70, 104, 111). Many *bundle sheath defective* (*bsd*) mutants that display specific disruption in BSC have been identified and characterized (46, 70, 104). The primary effect of the *bsd1* (allelic to *goden 2, g2*) mutation is on plastid biogenesis and gene expression (*RbcS, RbcL, C_4Me*) specifically in BSC in a light-independent manner (70). The G2 gene has been cloned by using Suppressor-mutator as a molecular tag (46). The leaf-specific expression pattern matches its function in leaves. G2 is the first member of a plant protein family with a TEA DNA-binding domain conserved from yeast, fly, to mammals. It has a functional nuclear localization signal and a proline-rich region that can act as a transcription activation domain. Although G2 expression is detected in both MC and BSC, it is proposed that G2 might be regulated posttranscriptionally or act with BSC partners to serve its BSC-specific function. However, in the C3-type sheath leaves (where *Rbc* is expressed in both MC and BSC), G2 seems to be important for *Rbc* expression in both MC and BSC (46). The BSC-specificity of this mutant could be due to its regulation of photosynthetic functions more essential in BSC. It will be informative to determine whether the expression of other genes encoding Calvin cycle enzymes and plastid-encoded genes are also affected by the *bsd1-g2* mutation. The only known MC-specific maize mutant is *high chlorophyll fluorescence3* (*hcf3*) that displays MC-specific defect in the photosystem II thylakoid complex that is less important in BSC (28).

The *bsd2* mutant shows specific defect in RBC protein translation, accumulation, or assembly. The expression of BS-specific ME and plastid biogenesis in the dark are normal. The BS-specific defects of chloroplasts under light and the ectopic accumulation of *RbcL* transcripts in MC may be secondary effects due to the lack of RBC for Calvin cycle activity (104). Another interesting maize mutant that shows preferential disruption in BSC is the *leaf permease1* (*lpe1*) mutant (111). The cloned *Lpe1* sequence indicates that it is a membrane protein sharing homology to the bacterial and eukaryotic pyrimidine and purine transporters or permeases (111). The BS preferential nature of the phenotype may be a consequence of the greater dependence of BSC on Lpe1-related metabolites. How Lpe1 affects early chloroplast development with a root- and dark-specific expression pattern is an intriguing question. As proposed, its effect should have occurred at the very early stage of plastid development in leaves. It is not clear whether the *lpe1* mutant exhibits aberrant phenotypes in roots or other organs.

Amaranth Mutants and Flaveria C$_3$-C$_4$ Hybrids

By screening plants able to grow under high CO_2 but showing symptoms of stress or chlorosis following exposure to normal air, some *Amaranthus edulis* mutants have been isolated. Further characterization confirms that these mutants show severely reduced PEPC or lack NAD-ME that are essential for C$_4$ photosynthesis. These mutants are very valuable for further investigation of the role and regulation of PEPC, NADP-ME, and photorespiration in C$_4$ photosynthesis (31, 69).

Taking advantage of the sexual compatibility between *Flaveria* C$_4$ and C$_3$ species and gene-specific molecular markers, it has been possible to genetically identify *cis*-acting elements and *trans*-acting factors that are important for the C$_4$ gene expression patterns (127). A cross between a C$_4$ *F. palmeri* and a non-C$_4$ *F. ramossisima* has been carried out. The F$_1$ hybrids and F$_1$ backcross segregates have been analyzed for *RbcS* and *Pdk* gene expression by using PCR markers, in situ hybridization, and immunolocalization. The results suggest that the BSC-specific trait of *RbcS* expression in the C_4 *F. palmeri* is recessive and may be controlled by one or few *trans*-acting factors. On the other hand, the C_4 *Pdk* gene expression trait is dominant and mainly mediated by *cis*-acting elements (127). The result of the C_4 *Pdk* gene expression in the C_4/C_3 hybrids is consistent with the similar analyses performed in transgenic C_4 and C_3 plants. However, the regulation of *RbcS* genes in BSC seems to be complex in different C_4 plants.

SUMMARY

The studies of C_4 gene expression have illustrated the complexity and flexibility of molecular mechanisms and the diversity of signal transduction pathways that are engaged in gene regulation in plants. It is clear that all C_4 genes originated from closely related ancestral genes by acquiring new *cis*-acting regulatory elements that are surprisingly divergent in sequences. The need of C_4-specific *trans*-acting factors is also evident for *C_4Me* and *RbcS* expression. C_4 photosynthesis research was initiated 30 years ago by the discovery of the unconventional CO_2 fixation pathway. New findings including the dimorphic BSC and MC in C_4 leaves, the three decarboxylation pathways, and the light regulation of C_4 enzymes by phosphorylation followed. The recent application of molecular, cellular, transgenic, and genetic approaches has revealed novel C_4 gene structure, specific cell-cell communications, complex transcriptional and posttranscriptional regulations, new DNA-binding transcription factors, and innovative signal transduction pathways.

The powerful combination of transgenic and transient expression tools will bring new insights into the molecular mechanisms of C_4 gene regulation. The genetic crosses and molecular analyses of the *Flaveria* C_3/C_4 hybrids provide new means to sort out the regulatory components involved in the C_4 gene expression patterns. The protoplast transient expression system is most useful for the elucidation of signal transduction pathways. Novel mutant screens for uncommon MC phenotypes and variations in leaf morphology and anatomy may uncover genetic components controlling C_4 photosynthesis and leaf development. With facile transformation methods and the knowledge of C_4 gene structure and regulation, genetic manipulation of the C_4 cycle pathway in C_4 and C_3 plants can now be fully enjoyed.

ACKNOWLEDGMENTS

This review is dedicated to MD Hatch for scientific inspiration and to my mother for spiritual inspiration. I thank my colleagues A Schäffner, H Huang, P Leon, K To, J-C Jang, A Kausch, and S Yanagisawa for numerous contributions. Special thanks to WC Taylor, M Matsuoka, JO Berry, L Bogorad, G Edwards, M Ku,

R Chollet, C Chastain, H Sakakibara, T Shugiyama, O Ueno, T Izui, T Nelson, N Dengler, E Kellogg, P Westhoff, L Mets, and G-P Shu for providing information. Work in the Sheen lab has been supported by the USDA photosynthesis program, NSF Integrated Plant Biology, and Hoechst AG.

Visit the *Annual Reviews home page* at
http://www.AnnualReviews.org

Literature Cited

1. Agarie S, Kai M, Takatsuji H, Ueno O. 1997. Expression of C3 and C4 photosynthetic characteristics in the amphibious plant *Eleocharis vivipara*: structure and analysis of the expression of isogenes for pyruvate, orthophosphate dikinase. *Plant Mol. Biol.* 34:363–69

2. Aoyagi K, Bassham JA. 1986. Appearance and accumulation of C4 carbon pathway enzymes in developing maize leaves and differentiating maize A188 callus. *Plant Physiol.* 80:322–33

3. Badger RM, Price GD. 1994. The role of carbonic anhydrase in photosynthesis. *Annu. Rev. Plant Physiol. Plant Mol. Biol.* 45:369–92

4. Bansal KC, Bogorad L. 1993. Cell type-preferred expression of maize cab-m1: repression in bundle sheath cells and enhancement in mesophyll cells. *Proc. Natl. Acad. Sci. USA* 90:4057–61

5. Bansal KC, Viret JF, Haley J, Khan BM, Schantz R, Bogorad L. 1992. Transient expression from cab-m1 and rbcS-m3 promoter sequences is different in mesophyll and bundle sheath cells in maize leaves. *Proc. Natl. Acad. Sci. USA* 89:3654–58

6. Barnes SA, McGrath RB, Chua N-H. 1997. Light signal transduction in plants. *Trends Cell Biol.* 7:21–26

7. Berry JO, Breiding DE, Klessig DF. 1990. Light-mediated control of translational initiation of ribulose-1,5-bisphosphate carboxylase in amaranth cotyledons. *Plant Cell* 2:795–803

8. Berry JO, Carr JP, Klessig DF. 1988. mRNA, encoding ribulose-1,5-bisphosphate carboxylase remain bound to polysomes but are not translated in amaranth seedlings transferred to darkness. *Proc. Natl. Acad. Sci. USA* 85:4190–94

9. Berry JO, McCormac DJ, Long JJ, Boinski J, Corey AC. 1997. Photosynthetic gene expression in amaranth, and NAD-ME type C4 dicot. *Aust. J. Plant Physiol.* 24:423–28

10. Berry JO, Niklou BJ, Carr JP, Klessing DF. 1986. Translational regulation of light-induced ribulose-1,5-bisphosphate carboxylase gene expression in amaranth. *Mol. Cell Biol.* 6:2347–53

11. Boinski JJ, Wang J-L, Xu P, Hotchkiss T, Berry JO. 1993. Post-transcriptional control of cell type-specific gene expression in bundle sheath and mesophyll chloroplasts of *Amaranthus hypochondriacus*. *Plant Mol. Biol.* 22:397–410

12. Bolle C, Sopory S, Lubberstedt T, Herrmann RG, Oelmuller R. 1994. Segments encoding 5'-untranslated leaders of genes for thylakoid proteins contain cis-elements essential for transcription. *Plant J.* 6:513–23

13. Brandstatter I, Kieber JJ. 1998. Two genes with similarity to bacterial response regulators are rapidly and specifically induced by cytokinin in Arabidopsis. *Plant Cell* 10:1009–19

14. Broglie R, Coruzzi G, Keith B, Chua N-H. 1984. Molecular biology of C4 photosynthesis in *Zea mays*: differential localization of proteins and mRNAs in the two leaf cell types. *Plant Mol. Biol.* 3:431–44

15. Burgener M, Suter M, Jones S, Brunold C. 1998. Cyst (e) ine is the transport metabolite of assimilated sulfur from bundle-sheath to mesophyll cells in maize leaves. *Plant Physiol.* 10:369–73

16. Burnell JN, Ludwig M. 1997. Characterisation of two cDNAs encoding carbonic anhydrase in maize leaves. *Aust. J. Plant Physiol.* 24:451–58

17. Casati P, Drincovich MF, Andreo CS, Donahue R, Edwards GE. 1998. UV-B, red and far red light regulate induction of the C4 isoform of NADP-malic enzyme in etiolated maize seedlings. *Aust. J. Plant Physiol.* 25:701–8

18. Cashmore AR. 1997. The cryptochrome

family of photoreceptors. *Plant Cell Environ.* 20:764–67

19. Chastain CJ, Lee ME, Moorman MA, Shameekumar P, Chollet R. 1997. Site-directed mutagenesis of maize recombinant C4-pyruvate, orthophosphate dikinase at the phosphorylatable target threonine residue. *FEBS Lett.* 413:169–73

20. Cheng W-H, Im KH, Chourey PS. 1996. Sucrose phosphate synthase expression at the cell and tissue level is coordinated with sucrose sink-to-source transitions in maize leaf. *Plant Physiol.* 111:1021–29

21. Chitty J, Furbank R, Marshall J, Chen Z, Taylor W. 1994. Genetic transformation of the C4 plant *Flaveria bidentis*. *Plant J.* 6:949–56

22. Chiu W-L, Niwa Y, Zeng W, Hirano T, Kobayashi H, Sheen J. 1996. Engineered GFP as a vital reporter in plants. *Curr. Biol.* 6:325–30

23. Christopher DA, Xinli L, Kim M, Mullet JE. 1997. Involvement of protein kinase and extraplastidic serine/threonine protein phosphatases in signaling pathways regulating plastid transcription and the psbD blue light-responsive promoter in barley. *Plant Physiol.* 113:1273–82

24. Chu C-C, Qu N, Bassuner B, Bauwe H. 1997. Genetic transformation of the C3–C4 intermediate species, *Flaveria pubescens* (Asteraceae). *Plant Cell Rep.* 16:715–18

25. Cushman JC, Bohnert HJ. 1997. Molecular genetics of crassulacean acid metabolism. *Plant Physiol.* 113:667–76

26. Deng XW. 1994. Fresh review of light signal transduction in plants. *Cell* 76:423–26

27. Dengler NG, Dengler RE, Hattersley PW. 1985. Differing ontogenetic origins of PCR ("Kranz") sheaths in leaf blades of C4 grasses (Poaceae). *Am. J. Bot.* 72:284–302

28. Dengler NG, Nelson T. 1998. Leaf structure and development in C4 plants. In *The Biology of C4 Photosynthesis*, ed. R Sage, R Monson. In press

29. Dengler NG, Taylor WC. 1998. Developmental aspects of C4 photosynthesis. In *Advances in Photosynthesis*, ed. R Leegood. In press

30. DeRocher EJ, Bohnert HJ. 1993. Development and environmental stress employ different mechanisms in the expression of a plant gene family. *Plant Cell* 5:1611–25

31. Dever LV, Bailey KJ, Leegood RC, Lea PJ. 1997. Control of photosynthesis in *Amaranthus edulis* mutants with reduced amounts of PEP carboxylase. *Aust. J. Plant Physiol.* 24:469–76

32. Dong L-Y, Masuda T, Kawamura T, Hata S, Izui K. 1998. Cloning, expression, and characterization of a root-form phosphoenolpyruvate carboxylase from *Zea mays*: comparision with the C4-form enzyme. *Plant Cell Physiol.* 39:865–73

33. Drincovich MF, Casati P, Andreo CS, Donahue R, Eduwards GE. 1998. UV-B induction of NADP-malic enzyme in etiolated and green maize seedlings. *Plant Cell Environ.* 21:63–70

34. Duff SMG, Giglioli-Guivarc'h N, Pierre J-N, Vidal J, Condon SA, Chollet R. 1996. In-situ evidence for the involvement of calcium and bundle-sheath-derived photosynthetic metabolites in the C4 phosphoenolpyruvate-carboxylase kinase signal-transduction chain. *Planta* 199:467–74

35. Edwards GE, Ku MSB, Monson RK. 1985. C4 photosynthesis and its regulation. In *Photosynthetic Mechanisms and the Environment*, ed. J Barber, NR Baker, pp. 289–327. New York: Elsevier

36. Edwards GE, Walker DA. 1983. C_3, C_4: *Mechanisms, and Cellular and Environmental Regulation of Photosynthesis*. Oxford: Blackwell Sci.

37. Ehleringer JR, Monson RK. 1993. Evolutionary and ecological aspects of photosynthetic pathway variation. *Annu. Rev. Ecol. Syst.* 24:411–39

38. Ernst K, Westhoff P. 1997. The phosphoenolpyruvate carboxylase (ppc) gene family of *Flaveria trinervia* (C4) and *F. pringlei* (C3): molecular characterisation and expression analysis of the ppcB and ppcC genes. *Plant Mol. Biol* 34:427–43

39. Ewing R, Jenkins G, Langdale J. 1998. Transcripts of maize RbcS genes accumulate differentially in C3- and C4-tissues. *Plant Mol. Biol* 36:593–99

40. Fankhauser C, Chory J. 1997. Light control of plant development. *Annu. Rev. Cell Dev. Biol.* 13:203–29

41. Fuglevand G, Jackson JA, Jenkins GI. 1996. UV-B, UV-A, and blue light signal transduction pathways interact synergistically to regulate chalcone synthase gene expression in Arabidopsis. *Plant Cell* 8:2347–57

42. Furbank RT, Chitty JA, Jenkins CLD, Taylor WC, Trevanion SJ, et al. 1997. Genetic manipulation of key photosynthetic enzymes in the C4 plant *Flaveria bidentis*. *Aust. J. Plant Physiol.* 24:477–85

43. Furbank RT, Taylor W. 1995. Regulation of photosynthesis in C3 and C4 plants: a molecular approach. *Plant Cell* 7:797–807

44. Giglioli-Guivarc'h J, Pierre J-N, Brown

S, Chollet R, Vidal J, Gadal P. 1996. The light-dependent transduction pathway controlling the regulatory phosphorylation of C$_4$ phosphoenolpyruvate carboxylase in protoplasts from *Digitaria sanguinalis. Plant Cell* 8:573–86

45. Glackin C, Grula J. 1990. Organ-specific transcripts of different size and abundance derive from the same pyruvate, orthophosphate dikinase gene in maize. *Proc. Natl. Acad. Sci. USA* 87:3004–8

46. Hall L, Rossini L, Cribb L, Langdale J. 1998. Golden 2, a novel transcriptional regulator of cellular differentiation in the maize leaf. *Plant Cell* 10:925–36

47. Hartmann U, Valentine WJ, Christie JM, Hays J, Jenkins GI, Weisshaar B. 1998. Identification of UV/blue light-response elements in the *Arabidopsis thaliana* chalcone synthase promoter using a homologous protoplast transient expression system. *Plant Mol. Biol.* 36:741–54

48. Hatch MD. 1987. C4 photosynthesis: a unique blend of modified biochemistry, anatomy and ultrastructure. *Biochim. Biphys. Acta* 895:81–106

49. Hatch MD. 1997. Resolving C4 photosynthesis: trials, tribulations and other unpublished stories. *Aust. J. Plant Physiol.* 24:413–22

50. Hatch MD, Burnell J. 1990. Carbonic anhydrase activity in leaves and its role in the first step in C4 photosynthesis. *Plant Physiol* 93:380–83

51. Hermans J, Westhoff P. 1990. Analysis of expression and evolutionary relationships of phosphoenolpyruvate carboxylase genes in *Flaveria trinervia* (C4) and *F. pringlei* (C3). *Mol. Gen. Genet.* 224: 459–68

52. Hermans J, Westhoff P. 1992. Homologous genes for the C4 isoform of phosphoenolpyruvate carboxylase in a C3 and a C4 *Flaveria* species. *Mol. Gen. Genet.* 234:275–84

53. Holappa LD, Walker-Simmons MK. 1997. The wheat protein kinase gene, TaPK3, of the PKABA1 subfamily is differentially regulated in greening wheat seedlings. *Plant Mol. Biol.* 33:935–41

54. Hudspeth RL, Grula JW. 1989. Structure and expression of the maize gene encoding the phosphoenolpyruvate carboxylase isozyme involved in C4 photosynthesis. *Plant Mol. Biol.* 12:579–89

55. Im C-s, Matters GL, Beale SI. 1996. Calcium and calmodulin are involved in blue light induction of the gsa gene for an early chlorophyll biosynthetic step in *Chlamydomonas. Plant Cell* 8:2245–53

56. Imaizumi N, Ku MS, Ishihara K, Same-jima M, Kaneko S, Matsuoka M. 1997. Characterization of the gene for pyruvate,orthophosphate dikinase from rice, a C3 plant, and a comparison of structure and expression between C3 and C4 genes for this protein. *Plant Mol. Biol.* 34:701–16

57. Imamura A, Hanaki N, Umeda H, Nakamura A, Suzuki T, et al. 1998. Response regulators implicated in His-to-Asp phosphotransfer signaling in Arabidopsis. *Proc. Natl. Acad. Sci. USA* 95:2691–96

58. Jang J-C, Leon P, Zhou L, Sheen J. 1997. Hexokinase as a sugar sensor in higher plants. *Plant Cell* 9:5–19

59. Jang J-C, Sheen J. 1994. Sugar sensing in higher plants. *Plant Cell* 6:1665–79

60. Jang J-C, Sheen J. 1998. Sugar sensing and signaling in higher plants. *Trends Plant Sci.* 2:208–14

61. Jenkins GI. 1997. UV and blue light signal transduction in Arabidopsis. *Plant Cell Environ.* 20:773–78

62. Jiao J-a, Echevarria C, Vidal J, Chollet R. 1991. Protein turnover as a component in the light/dark regulation of phosphoenolpyrutate carboxylase protein-serine kinase activity in C4 plants. *Proc. Natl. Acad. Sci. USA* 88:2712–15

63. Kausch A, Zachwieja S, Sheen J. 1998. Mesophyll-specific, light and metabolic regulation of the C4PPCZm1 promoter in transgeneic maize. Submitted

64. Kawamura T, Shigesada K, Toh H, Okumura S, Yanagisawa S, Izui K. 1992. Molecular evolution of phosphoenolpyruvate carboxylase for C4 photosynthesis in maize: comparison of its cDNA sequence with a newly isolated cDNA encoding an isozyme involved in the anaplerotic function. *J. Biochem.* 112:147–54

65. Kombrink E, Hahlbrock K. 1990. Rapid, systemic repression of the synthesis of ribulose 1,5-bisphosphate carboxylase small-subunit mRNA in fungus-infected or elicitor-treated potato leaves. *Planta* 181:216–19

66. Kopriva S, Chu CC, Bauwe H. 1996. H-protein of the glycine cleavage system in Flaveria: alternative splicing of the premRNA occurs exclusively in advanced C4 species of the genus. *Plant J.* 10:369–73

67. Ku MSB, Kano-Murakami Y, Matsuoka M. 1996. Evolution and expression of C4 photosynthesis genes. *Plant Physiol.* 111:949–57

68. Kubicki A, Steinmuller K, Westhoff P. 1994. Differential transcription of plastome-encoded genes in the mesophyll and bundle-sheath chloroplasts of the monocotyledonous NADP-malic enzyme-type

C4 plants maize and sorghum. *Plant Mol. Biol.* 25:27–52

69. Lacuesta M, Dever LV, Munoz-Rueda A, Lea PJ. 1997. A study of photorespiratory ammonia production in the C4 plant *Amaranthus edulis*, using mutants with altered photosynthetic capacities. *Physiol. Plant.* 99:447–55

70. Langdale JA, Kidner CA. 1994. Bundle sheath defective, a mutation that disrupts cellular differentiation in maize leaves. *Development* 120:673–81

71. Langdale JA, Lane B, Freeling M, Nelson T. 1989. Cell lineage analysis of maize bundle sheath and mesophyll cells. *Dev. Biol.* 133:128–39

72. Langdale JA, Metzler MC, Nelson T. 1987. The argentia mutation delays normal development of photosynthetic cell-types in *Zea mays*. *Dev. Biol.* 122:243–55

73. Langdale JA, Nelson T. 1991. Spatial regulation of photosynthetic development in C4 plants. *Trends Genet.* 7:191–96

74. Langdale JA, Rothermel BA, Nelson T. 1988. Cellular pattern of photosynthetic gene expression in developing maize leaves. *Genes Dev.* 2:106–15

75. Langdale JA, Zelitch I, Miller E, Nelson T. 1988. Cell position and light influence C4 versus C3 patterns of photosynthetic gene expression in maize. *EMBO J.* 7:3643–51

76. Lepiniec L, Keryer E, Philippe H, Gadal P, Cretin C. 1993. Sorghum phosphoenolpyruvate carboxylase gene family: structure, function and molecular evolution. *Plant Mol. Biol.* 21:487–502

77. Li B, Zhang X-Q, Chollet R. 1996. Phosphoenolpyruvate carboxylase kinase in tobacco leaves is activated by light in a similar but not identical way as in maize. *Plant Physiol.* 111:497–505

78. Link G, Coen DM, Bogorad L. 1978. Differential expression of the gene for the large subunit of ribulose bisphosphate carboxylase in maize leaf cell types. *Cell* 15:725–31

79. Long J, Berry J. 1996. Tissue-specific and light-mediated expression of the C4 photosynthetic NAD-dependent malic enzyme of amaranth mitochondria. *Plant Physiol.* 112:473–82

80. Luchetta P, Cretin C, Gadal P. 1990. Structure and characterization of the *Sorghum vulgare* gene encoding NADP-malate dehydrogenase. *Gene* 89:171–17

81. Ludwig M, Burnell JN. 1995. Molecular comparison of carbonic anhydrase from Flaveria species demonstrating different photosynthetic pathways. *Plant Mol. Biol.* 29:353–65

82. Lunn JE, Furbank RT. 1997. Localisation of sucrose-phosphate synthase and starch in leaves of C4 plants. *Planta* 202:106–11

83. Marshall J, Stubbs J, Chitty J, Surin B, Taylor W. 1997. Expression of the C4 Me1 gene from *Flaveria bidentis* requires an interaction between 5′ and 3′ sequences. *Plant Cell* 9:1515–25

84. Marshall J, Stubbs J, Taylor W. 1996. Two genes encode highly similar chloroplastic NADP-malic enzymes in *Flaveria*: implications for the evolution of C4 photosynthesis. *Plant Physiol.* 111:1251–61

85. Martineau B, Smith HJ, Dean C, Dunsmuir P, Bedbrook J, Mets LJ. 1989. Expression of a C3 plant Rubisco SSU gene in regenerated C4 *Flaveria* plants. *Plant Mol. Biol.* 13:419–26

86. Matsuoka M. 1995. The gene for pyruvate, orthophosphate dikinase in C4 plants: structure, regulation and evolution. *Plant Cell Physiol.* 36:937–43

87. Matsuoka M, Kyozuka J, Shimamoto K, Kano-Murakami Y. 1994. The promoters of two carboxylases in a C4 plant (maize) direct cell-specific, light-regulated expression in a C3 plant (rice). *Plant J.* 6: 311–19

88. Matsuoka M, Nomura M, Agarie S, Miyao-Tokutomi M, Ku MSB. 1998. Evolution of C4 photosynthetic genes and overexpression of maize C4 genes in rice. *J. Plant Res.* In press

89. Matsuoka M, Numazawa T. 1991. cis-acting elements in the pyruvate, orthophosphate dikinase gene in maize. *Mol. Gen. Genet.* 228:143–52

90. Matsuoka M, Sanada Y. 1991. Expression of photosynthetic genes from the C4 plant, maize, in tobacco. *Mol. Gen. Genet.* 225:411–19

91. Matsuoka M, Tada Y, Fujimura T, Kano-Murakami Y. 1993. Tissue-specific light-regulated expression directed by the promoter of a C4 gene, maize pyruvate, orthophosphate dikinase, in a C3 plant, rice. *Proc. Natl. Acad. Sci. USA* 90:9586–90

92. McCormac D, Boinski JJ, Ramsperger VC, Berry JO. 1997. C4 gene expression in photosynthetic and non-photosynthetic leaf regions of *Amaranthus tricolor*. *Plant Physiol.* 114:801–15

93. McGonigle B, Nelson T. 1995. C4 isoform of NADP-malate dehydrogenase. cDNA cloning and expression in leaves of C4, C3, and C3–C4 intermediate species of Flaveria. *Plant Physiol.* 108:1119–26

94. Metzler MC, Rothermel BA, Nelson T. 1989. Maize NADP-malate dehydrogenase: cDNA cloning, sequence, and

mRNA characterization. *Plant Mol. Biol.* 12:713–22

95. Miginiac-Maslow M, Issakidis E, Lemaire M, Ruelland E, Jacquot J-P, Decottignies P. 1997. Light-dependent activation of NADP-malate dehydrogenase: a complex process. *Aust. J. Plant Physiol.* 24:529–42

96. Moons A, Valcke R, Montagu MV. 1998. Low-oxygen stress and water deficit induce cytosolic pyrubate orthophosphate dikinase (PPDK) expression in roots of rice, a C3 plant. *Plant J.* 15:89–98

97. Nelson T, Harpster M, Mayfield SP, Taylor WC. 1984. Light-regulated gene expression during maize leaf development. *J. Cell Biol.* 98:558–64

98. Nelson T, Langdale J. 1992. Developmental genetics of C4 photosynthesis. *Annu. Rev. Plant Physiol. Plant Mol. Biol.* 43:25–47

99. Oaks A. 1994. Efficiency of nitrogen utilization in C3 and C4 cereals. *Plant Physiol.* 106:407–14

100. Purcell M, Mabrouk YM, Bogorad L. 1995. Red/far-red and blue light-responsive regions of maize rbcS-m3 are active in bundle sheath and mesophyll cells, respectively. *Proc. Natl. Acad. Sci. USA* 92:11504–8

101. Ramsperger VC, Summers RG, Berry JO. 1996. Photosynthetic gene expression in meristems and during initial leaf development in a C4 dicotyledonous plant. *Plant Physiol.* 111:999–1010

102. Rosche E, Chitty J, Westhoff P, Taylor WC. 1998. Analysis of promoter activity for the gene encoding pyruvate orthophosphate dikinase in stably transformed C4 Flaveria species. *Plant Physiol.* 117:821–29

103. Rosche E, Westhoff P. 1995. Genomic structure and expression of the pyruvate, orthophosphate dikinase gene of the dicotyledonous C4 plant *Flaveria trinervia* (Asteraceae). *Plant Mol. Biol.* 29:663–78

104. Roth R, Hall LN, Brutnell TP, Langdale JA. 1996. *bundle sheath defective2*, a mutation that disrupts the coordinated development of bundle sheath and mesophyll cells in the maize leaf. *Plant Cell* 8:915–27

105. Rothermel BA, Nelson T. 1989. Primary structure of the maize NADP-dependent malic enzyme. *J. Biol. Chem.* 264:19587–92

106. Sakakibara H, Kobayashi K, Deji A, Sugiyama T. 1997. Partial characterization of the signaling pathway for the nitrate-dependent expression of genes for nitrogen-assimilatory enzymes using

detached maize leaves. *Plant Cell Physiol.* 38:837–43

107. Sakakibara H, Suzuki M, Takei K, Deji A, Taniguchi M, Sugiyama T. 1998. A response-regulator homologue possibly involved in nitrogen signal transduction mediated by cytokinin in maize. *Plant J.* 14:337–44

108. Sano H, Youssefian S. 1994. Light and nutritional regulation of transcripts encoding a wheat protein kinase homolog is mediated by cytokinins. *Proc. Natl. Acad. Sci. USA* 91:2582–86

109. Schäffner AR, Sheen J. 1991. Maize rbcS promoter activity depends on sequence elements not found in dicot rbcS promoters. *Plant Cell* 3:997–1012

110. Schäffner AR, Sheen J. 1992. Maize C4 photosynthesis involves differential regulation of maize PEPC genes. *Plant J.* 2:221–32

111. Schultes NP, Brutnell TP, Allen A, Dellaporta SL, Nelson T, Chen J. 1996. Leaf permease1 gene of maize is required for chloroplast development. *Plant Cell* 8:463–75

112. Sheen J. 1990. Metabolic repression of transcription in higher plants. *Plant Cell* 2:1027–38

113. Sheen J. 1991. Molecular mechanisms underlying the differential expression of maize pyruvate, orthophosphate dikinase genes. *Plant Cell* 3:225–45

114. Sheen J. 1993. Protein phosphatase activity is required for light-inducible gene expression in maize. *EMBO J.* 12:3497–505

115. Sheen J. 1996. Specific Ca²⁺ dependent protein kinase in stress signal transduction. *Science* 274:1900–2

116. Sheen J. 1998. Mutational analysis of two protein phosphatases involved in ABA signal transduction in higher plants. *Proc. Natl. Acad. Sci. USA* 98:975–80

117. Sheen J, Jang J-C. 1998. Functional conservation between mammalian CaMKII and plant CDPK in blocking blue light signal transduction. Submitted

118. Sheen J-Y, Bogorad L. 1986. Differential expression of six light-harvesting chlorophyll a/b binding protein genes in maize leaf cell types. *Proc. Natl. Acad. Sci. USA* 83:7811–15

119. Sheen J-Y, Bogorad L. 1986. Expression of ribulose bisphosphate carboxylase large subunit and three small subunit genes in two cell types of maize leaves. *EMBO J.* 13:3417–22

120. Sheen J-Y, Bogorad L. 1987. Differential expression of C4 pathway genes in mesophyll and bundle sheath cells of greening

maize leaves. *J. Biol. Chem.* 262:11726–30

121. Sheen J-Y, Bogorad L. 1987. Differential expression of oxygen-evolving polypeptide genes in maize leaf cell types. *Plant Mol. Biol.* 9:217–26

122. Sheen J-Y, Bogorad L. 1988. Differential expression of genes for photosystem II components encoded by the plastid genome in bundle sheath and mesophyll cells of maize. *Plant Physiol.* 86:1020–26

123. Shiina T, Nishii A, Toyoshima Y, Bogorad L. 1997. Identification of promoter elements involved in the cytosolic Ca^{2+}-mediated photoregulation of maize cab-m1 expression. *Plant Physiol.* 115:477–83

124. Short TW, Briggs WR. 1994. The transduction of blue light signals in higher plants. *Annu. Rev. Plant Physiol. Plant Mol. Biol.* 45:143–71

125. Shu G, Dengler NG, Pontieri V, Mets L. 1998. Light induction of cell type differentiation and cell type-specific expression of rbcS, PEPCase, and PPDK genes in the post-germination cotyledons of a C4 plant, *Flaveria trinervia* (Asteraceae). Submitted

126. Shu G, Mets L. 1998. Expressed members of the rbcS gene families of a C4 plant, *Flaveria palmeri*, and a non-C4 plant, *F. ramosissima*: nucleotide sequence divergence and evolutionary implications. *Mol. Gen. Genet.* In press

127. Shu G, Reichardt M, Mets L. 1998. Genetic control of cell type-specific expression of rbcS genes in C4 plants: evidence from the F1 hybrids and backcross segregates of *Flaveria palmeria* (C_4) and *Flaveria ramossisima* (non-C_4). Submitted

128. Sinha N, Kellogg E. 1996. Parallelism and diversity in multiple origins of C4 photosynthesis in the grass family. *Am. J. Bot.* 83:1458–70

129. Smith LH, Langdale JA, Chollet R. 1998. A functional Calvin cycle is not indispensable for the light activation of C4 phosphoenolpyruvate carboxylase kinase and its target enzyme in the maize mutant bundle sheath defective2–mutable1. *Plant Physiol.* 118:191–97

130. Stockhaus J, Poetsch W, Steinmüller K, Westhoff P. 1994. Evolution of the C4 phosphoenolpyruvate carboxylase promoter of the C4 dicot *Flaveria trinervia*: an expression analysis in the C3 plant tobacco. *Mol. Gen. Genet.* 245:286–93

131. Stockhaus J, Schlue U, Koczor M, Chitty J, Taylor W, Westhoff P. 1997. The promoter of the gene encoding the C4 form of phosphoenolpyruvate carboxylase directs mesophyll specific expression in transgeneic C4 Flaveria. *Plant Cell* 9:479–89

132. Sugiharto B, Sugiyama T. 1992. Effects of nitrate and ammonium on gene expression of phosphoenolpyruvate carboxylase and nitrogen metabolism in maize leaf tissue during recovery from nitrogen stress. *Plant Physiol.* 100:153–56

133. Sugiharto B, Suzuki I, Burnell JN, Sugiyama T. 1992. Glutamine induces the N-dependent accumulation of mRNAs encoding phosphoenolpyruvate carboxylase and carbonic anhydrase in detached maize leaf tissue. *Plant Physiol.* 100:2066–70

134. Sugiyama T. 1998. Nitrogen-responsive expression of C4 photosynthesis genes in maize. In *Stress Responses of Photosynthetic Organisms*, ed. K Satoh, N Murata, pp. 167–80. Tokyo: Elsevier

135. Suzuki I, Cretin C, Omata T, Sugiyama T. 1994. Transcriptional and posttranscriptional regulation of nitrogen-responding expression of phosphoenolpyruvate carboxylase gene in maize. *Plant Physiol.* 105:1223–29

136. Szymanski DB, Liao B, Zielinski RE. 1996. Calmodulin isoforms differentially enhance the binding of cauliflower nuclear proteins and recombinant TGA3 to a region derived from the Arabidopsis Cam-3 promoter. *Plant Cell* 8:1069–77

137. Taniguchi M, Kiba T, Sakakibara H, Ueguchi C, Mizuno T, Sugiyama T. 1998. Expression of Arabidopsis response regulator homologs is induced by cytokinins and nitrate. *FEBS Lett.* 429:259–62

138. Taniguchi M, Mori J, Sugiyama T. 1994. Structure of genes that encode isozymes of aspartate aminotransferase in *Panicum miliaceum* L., a C4 plant. *Plant Mol. Biol.* 26:723–34

139. Taniguchi M, Sugiyama T. 1997. Expression of 2–oxoglutarate/malate translocator in the bundle-sheath mitochondria of *Panicum miliaceum*, a NAD-malic enzyme-type C4 plant, is regulated by light development. *Plant Physiol.* 114:285–93

140. Taylor WC. 1989. Regulatory interactions between nuclear and plastid genomes. *Annu. Rev. Plant Physiol. Plant Mol. Biol.* 40:211–33

141. Taylor WC, Rosche E, Marshall JS, Ali S, Chastain CJ, Chitty JA. 1997. Diverse mechanisms regulate the expression of genes coding for C4 enzymes. *Aust. J. Plant Physiol.* 24:437–42

142. Terzaghi WB, Cashmore AR. 1995.

Light-regulated transcription. *Annu. Rev. Plant Physiol. Plant Mol. Biol.* 46:445–74

143. Thompson W, White M. 1991. Physiological and molecular studies of light-regulated nuclear genes in higher plants. *Annu. Rev. Plant Physiol. Plant Mol. Biol.* 42:423–66

144. Tobin EM, Kehoe DM. 1994. Phytochrome regulated gene expression. *Semin. Cell Biol.* 5:335–46

145. Toh H, Kawamura T, Izui K. 1994. Molecular evolution of phosphoenolpyruvate carboxylase. *Plant Cell Environ.* 17:31–43

146. Uchino A, Sentoku N, Nemoto N, Ishii R, Samejima M, Matsuoka M. 1998. C4–type gene expression is not directly dependent on Kranz anatomy in an amphibious sedge *Eleocharis vivipara*. *Plant J.* 14:565–72

147. Ueno O. 1996. Structural characterization of photosynthetic cells in an amphibious sedge, *Eleocharis vivipara*, in relation to C3 and C4 metabolism. *Planta* 199:382–93

148. Ueno O. 1998. Induction of Kranz anatomy and C4–like biochemical characteristics in a submerged amphibious plant by abscisic acid. *Plant Cell* 10:517–83

149. Ueno Y, Hata S, Izui K. 1997. Regulatory phosphorylation of plant phosphoenolpyruvate carboxylase: role of a conserved basic residue upstream of the phosphorylation site. *FEBS Lett.* 417:57–60

150. Vidal J, Chollet R. 1997. Regulatory phosphorylation of C4 PEP carboxylase. *Trends Plant Sci.* 2:230–37

151. Viret J-F, Mabrouk Y, Bogorad L. 1994. Transcriptional photo-regulation of cell-type-preferred expression of maize rbcS-m3: 3′ and 5′ sequences are involved. *Proc. Natl. Acad. Sci. USA* 91:8577–81

152. Walbot V. 1977. The dimorphic chloroplasts of the C4 plant panicum maximum contain identical genomes. *Cell* 11:729–37

153. Walker RP, Acheson RM, Tecsi LI, Leegood RC. 1997. Phosphoenolpyruvate carboxykinase in C4 plants; its role and regulation. *Aust. J. Plant Physiol.* 24:459–68

154. Wang J-L, Long J, Hotchkiss T, Berry J. 1993. Regulation of C4 gene expression in light- and dark-grown amaranth cotyledons. *Plant Physiol* 102:1085–93

155. Wang J-L, Turgeon R, Carr J, Berry J. 1993. Carbon sink-to-source transition is coordinated with establishment of cell-specific gene expression in a C4 plant. *Plant Cell* 5:289–96

156. Weatherwax SC, Ong MS, Degenhardt J, Bray EA, Tobin EM. 1996. The interaction of light and abscisic acid in the regulation of plant gene expression. *Plant Physiol.* 111:363–70

157. Westhoff P, Svensson P, Ernst K, Bläsing O, Burscheidt J, Stockhaus J. 1997. Molecular evolution of C4 phosphoenolpyruvate carboxylase in the genus *Flaveria*. *Aust. J. Plant Physiol.* 24:429–36

158. Williams SA, Weatherwax SC, Bray EA, Tobin EM. 1994. NPR genes, which are negatively regulated by phytochrome action in *Lemna gibba* L. G-3, can also be positively regulated by abscisic acid. *Plant Physiol.* 105:949–54

159. Wyrich R, Dressen U, Brockmann S, Streubel M, Chang C, et al. 1998. The molecular basis of C4 photosynthesis in sorghum: isolation, characterization and RFLP mapping of mesophyll- and bundle-sheath-specific cDNAs obtained by differential screening. *Plant Mol. Biol.* 37:319–35

160. Yanagisawa S. 1996. Dof DNA binding proteins contain a novel zinc finger motif. *Trends Plant Sci.* 1:213–14

161. Yanagisawa S. 1997. Dof DNA-binding domains of plant transcription factors contribute to multiple protein-protein interactions. *Eur. J. Biochem.* 250:403–10

162. Yanagisawa S, Izui K. 1992. MNF1, a leaf tissue-specific DNA-binding protein of maize, interacts with the cauliflower mosaic virus 35S promoter as well as the C4 photosynthetic phosphoenolpyruvate carboxylase gene promoter. *Plant Mol. Biol.* 19:545–53

163. Yanagisawa S, Sheen J. 1998. Involvement of maize Dof zinc finger proteins in tissue-specific and light-regulated gene expression. *Plant Cell* 10:75–89

Annu. Rev. Plant Physiol. Plant Mol. Biol. 1999. 50:219–43

GENETIC ANALYSIS OF HORMONE SIGNALING

Peter McCourt

Department of Botany, University of Toronto, Toronto, Ontario, M5S 3B2, Canada;
e-mail: mccourt@botany.utoronto.ca

KEY WORDS: signal transduction, Arabidopsis mutants, genetic interactions

ABSTRACT

Phytohormones influence many diverse developmental processes ranging from seed germination to root, shoot, and flower formation. Recently, mutational analysis using the model plant *Arabidopsis thaliana* has been instrumental in determining the individual components of specific hormone signal transduction pathways. Moreover, epistasis and suppressor studies are beginning to explain how these genes and their products relate to one another. While no hormone transduction pathway is completely understood, the genes identified to date suggest that simple molecular rules can be established to explain how plant hormone signals are transduced. This review describes some of the shared characteristics of plant hormone signal transduction pathways and the properties for informational transfer common to many of the genes that specify the transduction of the signal.

CONTENTS

219

1040-2519/99/0601-0219$08.00

INTRODUCTION

Approximately 100 years ago, the German botanist Julian von Sachs suggested that plant growth and development may be controlled by specific endogenous plant substances. We now know that a small group of compounds, termed growth regulators or phytohormones, mediate many diverse plant processes. However, even with the protagonists in hand, the roles of these substances in plant development are still unclear. For example, within a plant, a single hormone can regulate many different processes and at the same time different hormones can influence a single process. Does the flexibility of hormone action contribute to the plasticity of plant development or is it a consequence? The complexities of hormonal responses have given rise to protracted speculation as to the logic of hormones' action, but this debate has contributed little to our understanding of plant hormone biology. Understanding the intricacies of plant hormones requires first the identification of the molecules that participate in transducing the hormone signal into a cellular response. With this information in hand, rather than questioning the logic, logical questions can be asked. For example, do different hormone pathways use similar signaling molecules and do different cells, tissues, or even species of plants use all or only some of the same steps? The application of genetic analysis to hormone signaling has begun to provide some answers. Genetics is a powerful tool for establishing in vivo links between the signal and the response because inferences can be made based solely on plant mutant phenotypes. Furthermore, when combined with molecular analysis, these functional relationships can be witnessed biochemically.

Genetic analysis requires an experimental system in which a large number of mutations affecting the response of the plant to a particular hormone can be easily identified. Coupled with the ability to cross, complement, and map mutations, these saturation mutagenesis experiments can determine the number, the types, and the nature of gene products involved in the signaling pathway. Although a number of genetically tractable plant species exist, the small crucifer *Arabidopsis thaliana* (Arabidopsis) has dominated hormone signal transduction studies. This review focuses on this body of work. The popularity of Arabidopsis for mutant screens involving hormones stems not only from its favorable genetics but also from the plant's ability to grow on petri plates under sterile conditions. This growth attribute allows large-scale screening of individuals under completely defined growth conditions, thereby allowing plant mutational screens to be achieved with a microbiologist's technique (10^3 seeds on a single petri plate). Mutants with increased or reduced sensitivity to a particular growth regulator are often easily identified in the uniform background of normally responding wild-type plants.

Excellent reviews on the genetic dissection of hormone signaling in Arabidopsis have recently been published (13, 16, 22, 25, 36, 52, 71) and, therefore, the use of genetics is not discussed in detail here. We have now reached a stage where the characteristics of a typical signaling pathway give some predictive value as to the types of genes that encode signaling components. This review aims to show that it is now possible to formulate simple rules to explain how genetic analysis can dissect a signal transduction pathway. First, I define the properties necessary for a hormone signal to be transduced and describe how allelic forms of genes encoding components involved in the pathway reflect these properties. Second, since an informational pathway by definition involves interactions, I present a range of examples of how genetic analysis has been used to understand these underlying interactions. Finally, I refer briefly to new methods to further our understanding of the genetic basis of hormone signaling transduction.

GENETIC IDENTIFICATION OF HORMONE SIGNALING GENES

Signal Transduction Pathways

Signal transduction designates a specific information pathway within a cell that translates an extrinsic signal into a specific cellular response. The initial phase before the signal is transduced requires high-affinity binding of the hormone to a specific receptor protein. Although the chemical nature of plant hormones does not predict the cellular location of the receptor, the binding event must cause the receptor to undergo a conformational change to initiate the transduction of the signal. Once activated, the receptor could alter gene expression directly by acting as a transcription factor, as is the case with the mammalian glucocorticoid receptor (6). Formally, in this scenario, there is no signal transduction pathway. Alternatively, the receptor may pass the signal to the nucleus through a series of intermediary steps that define the length of the transduction pathway. If a signal transduction pathway exists downstream of the receptor, the stimulation of the receptor must activate or inactivate relay components of the pathway through some type of cascading mechanism. Often the signaling components are modified by a phosphorylation event or by the binding and hydrolysis of a guanine nucleotide (54, 69). In these cases, the protein acts as a molecular switch, depending on its conformational state. These epigenetic changes in signaling proteins permit rapid response to the hormone signal and, more important, because they are epigenetic these changes are readily reversible. This therefore allows the signal to be shut down rapidly. This reversibility also allows the recycling of the components of the signaling system so that they can receive further signals.

The epigenetic nature of the signaling components of a transduction pathway allows predictions to be made as to the types of mutations that will disrupt normal signaling. In principle, it is possible to identify allelic states of a signaling component that mimic one of the two epigenetic conformations. Dominant alleles define high-level or constitutive activity, whereas null loss-of-function alleles would have reduced or no activity. For example, if a protein must be phosphorylated to activate the next step in the signaling pathway, loss-of-function mutations in this gene will cause loss of the signal. Conversely, mutations that lock the protein into an activated conformation will constitutively activate the pathway. If loss- and gain-of-function alleles have opposite effects on hormone sensitivity, during normal signaling the differential states of these gene products should specify the signal to turn on or turn off. Moreover, if components act as binary switches, it should be possible to determine how a switch is set and how that state regulates the signal. Thus, as mutations are identified for each protein in the pathway, they can be classified into either positive or negative regulators of that pathway.

Identification of Response Mutants Using Hormone Application

The simplest method to identify mutations in genes involved in hormone signaling is by assaying a mutagenized population for an altered response to supplied hormone. To be useful, a clear and reproducible response must occur in wild-type plants in the presence of the hormone. For example, ethylene gas mediates many plant processes ranging from seed germination to senescence of flowers, fruits, and leaves (36). Most of these responses, however, are variable in their penetrance and therefore are difficult to assay in genetic screens. By contrast, exposure of dark-grown seedlings to exogenous ethylene reproducibly inhibits root and hypocotyl elongation, and causes radial swelling of the hypocotyl and exaggerated apical hook growth (51). Collectively known as the ethylene triple response, this simple growth assay for ethylene action made it possible for a number of laboratories to design visual screens to obtain the desired mutants (5, 24). Screening mutagenized populations of dark-germinated Arabidopsis seedlings for plants displaying an absence of a triple response in the presence of ethylene yields mutants insensitive to the hormone. Conversely, dark-grown seedlings that show a triple response in the absence of ethylene identify constitutive response mutants. Analogous screens for mutations that alter the germination response to exogenous abscisic acid (ABA) (20, 40) or gibberellin (GA) biosynthetic inhibitors (33) and growth responses to auxin

(19, 47), brassinosteroid (BR) (15), cytokinin (18), and jasmonate (3) have allowed the identification of a large number of mutants in Arabidopsis for these hormonal responses. The use of hormone application as a selective agent, however, must be judicious since the screening procedure is artificial. In most of these screens, seeds and seedlings are exposed to higher concentrations of hormone than experienced under normal growth conditions. Mutations that confer insensitivity to such conditions may not always be specific to the hormone-dependent pathway of interest.

Identification of Response Mutants Using Auxotrophic Phenotypes

A second method to screen for mutations that affect hormone signaling involves identifying plants that display phenotypes similar to mutants that are deficient in the synthesis of a particular hormone. However, unlike auxotrophic mutants, these mutants are nonresponsive and, therefore, are not rescued by application of the hormone. For example, a number of mutants deficient in BR biosynthesis in Arabidopsis have a characteristic cabbage-like appearance and grow as dark green dwarves if not sprayed with BR (2, 46, 72). Based on this auxotrophic phenotype, Li & Chory (45) identified one complementation group, designated *bri1*, by screening mutagenized populations of Arabidopsis for nonresponsive BR dwarves. The *bri1* mutations are allelic to a previously characterized BR-insensitive mutant that had been identified by its ability to show normal root growth on exogenous BR (15). The BRI1 gene appears to encode a putative receptor kinase (45). Although binding of BR to this protein has not been shown, this particular screening method did indeed identify a candidate gene for a hormone transduction pathway. Similar approaches based on the dwarfed phenotype of GA auxotrophs have allowed the identification of potential GA-response mutants in a variety of plant species (71). The phenotypic similarities of nonresponsive mutants and biosynthetic auxotrophs can, however, be deceiving. Phenotypic analysis of the auxotroph must be precise. For example, a number of BR auxotrophic mutants in barley were originally classified as GA-response mutants because the plants did not respond to applied GA (59). Second, although mutations affecting biosynthesis may have dramatic effects on the development of many tissues in the plant, signaling components of a response pathway may be redundant. Thus, mutations in any one component gene may only give subtle phenotypes or perhaps influence only a subset of development, which in turn leads to unexpected phenotypes. This latter case may explain why the BR nonresponsive screen only identified *bri1*.

Altered response to applied hormone or nonresponsive mutants that phenocopy hormone auxotrophs are not a sufficient genetic criterion to identify a gene involved directly in signal transduction (i.e. a gene product whose primary function is involved in transducing the hormone signal). It is possible that mutations identified in such screens mark genes whose functions are necessary for a signaling event to occur but are not directly involved in the signal transduction pathway. Although a hormone may transduce its signal through a conserved pathway in different tissues, the developmental states of the tissues may attenuate or amplify the signal's current. For example, the amount of GA needed to rescue the germination defect of an Arabidopsis auxotroph is orders of magnitude lower than that needed to restore normal growth stature (41). Thus, genetically perturbing the developmental commitment of a tissue may change the ability of that tissue to respond to the hormone. However, as noted above, proteins involved in signaling pathways usually act as binary switches and therefore it should be possible to identify mutations that result in opposite phenotypes. Because a gain-of-function mutation does not eliminate gene function, this allelic form can exclude genes whose function is required for a developmental event to occur but is not necessary for the signal transduction pathway.

Hormone Response Mutants

Using the approaches mentioned above, a number of laboratories have obtained a variety of mutants affecting hormone responses in Arabidopsis. These mutations and the cloning of their wild-type alleles have been instrumental in identifying individual components involved in a signaling pathway. Below I describe a small number of these, concentrating on how such mutants allow the presence of switch genes in a hormone signaling pathway to be established.

ETHYLENE TRIPLE RESPONSE MUTANT, *etr1*   Dominant mutations in the *ETR1* gene of Arabidopsis were identified by the inability of dark-grown seedlings to perform the triple response in the presence of ethylene (5). Other phenotypes associated with the *etr1* mutation include poor seed germination, decreased senescence of detached leaves, and reduced ethylene-induced gene expression in vegetative tissue. The ETR1 gene encoded a protein with sequence similarity to bacterial two-component histidine kinases (11). Unlike their bacterial counterparts, however, the Arabidopsis kinase and response regulator domains are both localized to the carboxyl terminal of the ETR1 protein. All *etr1* mutations clustered to the amino termini of the protein, which contains a novel motif necessary for ethylene binding (21, 30, 65). These results strongly suggest that the ETR1 protein encodes an ethylene receptor. Four ETR1-like histidine kinases have been cloned in Arabidopsis, and mutations in a number of these genes confer ethylene insensitivity to the plant (27–29, 64). The dominant insensitivity phenotype of *etr1* in the presence of multiple wild-type ethylene receptor proteins may mean that the wild-type ETR1-like proteins function as negative regulators of ethylene responses in the absence of ethylene. Ethylene inhibits wild-type ETR1 function, whereas *etr1* mutants do not recognize ethylene and therefore show a dominant insensitivity. Alternatively, mutant ETR1 protein may function to poison the ETR1 family of wild-type complexes. The strong dominance of the *etr1* mutations over the complete ETR1 gene family, however, makes this latter possibility seem less likely (36).

CONSTITUTIVE TRIPLE RESPONSE MUTANT, *ctr1*   Recessive *ctr1* mutants of Arabidopsis display the ethylene triple response in dark-grown seedlings and constitutive expression of ethylene-regulated genes in the absence of ethylene (37). Mutant plants have compact dwarfed rosettes and smaller roots, both phenotypes that can be phenocopied by exposing wild-type plants to ethylene. The observation that *ctr1* mutants do not overproduce ethylene suggests that the CTR1 gene product acts as a negative regulator of ethylene signaling and as a consequence, loss-of-function mutations in this gene confer an ethylene constitutive phenotype. The amino acid sequence of the CTR1 gene most closely resembles the Raf family of mammalian protein kinases (37), which suggests a phosphorelay is involved in transduction of the ethylene signal. Although sequenced mutations in the CTR1 gene demonstrate that *ctr1* phenotypes are due to a loss-of-function, the lack of gain-of-function mutations in CTR1 has not allowed the assignment of CTR1 as switch gene. Furthermore, that the amino-terminal end of the CTR1 diverges from other Raf kinases suggests this protein may be regulated by novel mechanisms in plants. The recent report that CTR1 and ETR1 proteins can interact in a two-hybrid yeast assay (14) implies that, unlike other biological systems, in plants the bacterial-like ETR1 receptor may interact directly to the CTR1 mammalian-type kinase.

ETHYLENE-INSENSITIVE MUTANT, *ein3*   Recessive mutations in the *ein3* gene of Arabidopsis have similar phenotypes to *etr1* mutants in that plants show reduced response to ethylene (62). The *EIN3* gene encodes a novel nuclear-localized protein that shares similar protein domains to known eukaryotic transcriptional activators (12). All *ein3* alleles so far sequenced appear to be loss-of-function mutations, which suggests that this gene is a positive activator of ethylene response (12). Furthermore, transgenic plants overexpressing *EIN3* show constitutive ethylene responses similar to those observed in *ctr1* loss-of-function alleles (12). Thus, EIN3 fulfills the binary switch properties of a signaling molecule in that it is both necessary and sufficient for the ethylene response pathway in Arabidopsis. Ecker and co-workers have suggested a number of models to explain these results (12). The EIN3 protein may be positively activated by the ethylene signal and overexpression of the gene may result in increased EIN3 activity. Another possibility is that EIN3 is under negative regulation and exposure to ethylene alleviates this inhibition. If true, high EIN3 levels may titrate the negative regulators and lead to ethylene-independent activation of ethylene responses. These models predict that the EIN3 protein recognizes either a positive or negative signal. Whichever is the case, the outcome of specific mutations in the EIN3 gene could differentiate between these possibilities. Mutations that abolish the positive interaction should produce an EIN3 loss-of-function phenotype. Alternatively, if EIN3 is negatively regulated, disruption of this interaction may yield a constitutive ethylene response.

ABA-INSENSITIVE MUTANT, *abi1* AND *abi2*   ABA-insensitive mutants of Arabidopsis were identified by their ability to germinate on concentrations of ABA that normally inhibit wild-type (20, 40). One ABA-insensitive locus, designated *ABI1*, encodes a protein type 2C phosphatase, which indicates that protein phosphorylation and dephosphorylation are involved in ABA signaling (43, 48). Only one specific dominant mutation in ABI1 (*abi1-1*) confers reduced ABA responses to the plant at both embryonic and vegetative levels. Biochemical characterization of *abi1-1* protein suggests that this mutation acts in a dominant-negative fashion (4). Interestingly, an identical mutation in a second gene (*abi2-1*) that also encodes a type 2C phosphatase also confers an ABA-insensitive phenotype to the plant (44, 61). The functional and molecular redundancy of these genes may explain why mutations in either gene must be dominant to confer ABA insensitivity. In this scenario, *abi1-1* mutant protein would not only have reduced activity but would also interfere with an ABA signaling complex that might include the ABI2 wild-type protein. Unfortunately, loss-of-function mutations of both genes do not exist to verify this hypothesis. Such double mutants should also be insensitive to exogenous ABA. If *ABI1* and *ABI2* are molecular switch genes in the ABA signaling pathway, then overexpression of the *ABI1* or *ABI2* wild-type gene products should confer a supersensitive phenotype to the plant. Although this experiment has not been done, the identification of single mutations that enhance the sensitivity of the plant to exogenous ABA (17) implies that signal current in the ABA pathway can be increased by simple genetic variation.

ABA-INSENSITIVE MUTANT, *abi3*   Severe alleles of *abi3* are of interest because this mutant's phenotypes are restricted to late seed maturation, which suggests that this gene may mark a developmental branch in the ABA signaling pathway (50, 53). The *ABI3* gene appears to encode a seed-specific transcriptional activator whose expression patterns reflect the *abi3* mutant phenotypes (23). Misexpression of ABI3 protein causes seed-specific mRNA transcript accumulation in vegetative tissues, which indicates that this gene is sufficient to regulate gene expression in seed. Although this result supports a binary switch role for

ABI3, other criteria defining this gene as a component of ABA signaling are not met. First, many defects observed in the loss-of-function null alleles are not in processes that are perturbed in ABA auxotrophs (55). Second, seeds from transgenic plants overexpressing *ABI3* do not show the expected enhancement of ABA responses such as increased sensitivity to exogenous ABA or hyper-dormancy (7). These results have led to the suggestion that *ABI3* could encode a developmental regulator that is necessary for correct implementation of seed ABA signaling rather than encoding an integral component of the signaling pathway (Figure 1). In this instructive role, ectopic expression of ABI3 could alter the developmental state of cells, for example, by changing the stoichiome-try of ABA signaling components. Consistent with this hypothesis, loss of *ABI3* function would result in a cell state that would be unable to respond to ABA.

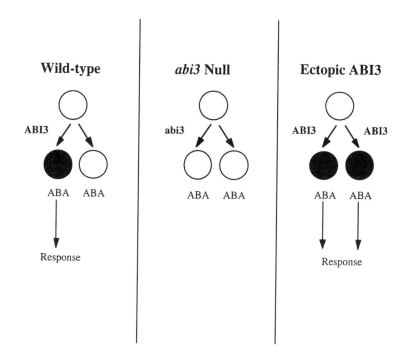

Figure 1   Model of an instructive role for ABI3 in ABA signal transduction. In this model expression of ABI3 is a prerequisite for the correct embryo response to the ABA signal. Loss of ABI3 function results in a developmental state that is unresponsive to the ABA signal. Ectopic expression of ABI3 expands the domain of ABI3 function into non-embryo cell types, thereby increasing their responsiveness to the ABA signal. Colors of circles represent different developmental states of a cell. *Open circles*; cells that are developmentally unresponsive to the ABA signal. *Black circles*; cells that are developmentally responsive to the ABA signal.

Consequently, *ABI3* may be a molecular switch gene in seed development but not necessarily a signaling component in ABA transduction.

GENETIC INTERACTION BETWEEN SIGNALING COMPONENTS

Mutational screens that perturb plant hormonal responses permit insights into the underlying mechanisms of how individual gene products contribute to transducing a hormone signal. However, how these genes interact to transduce the signal is often not obvious. In relaying the signal, the protein can either positively or negatively regulate the next component of the signaling pathway. Sometimes these interactions can be determined biochemically, but often these studies inform us only of the molecular mechanism of interaction and not the sign of the relay. The interplay between component states in a signaling pathway can be further determined by examining the phenotypes of plants containing two mutations that affect a signaling pathway. In some cases, this analysis is carried out by constructing double mutant strains between well-defined signal transduction mutants. Alternatively, if few mutations exist for a particular pathway, a mutant can be used as starting material to find new genetic interactions.

Epistasis

If one mutation can completely mask the phenotype of another and replace it with its own, it is termed epistatic. To be interpretable, epistatic analysis has rigorous genetic rules (1). First, the recessive mutations must be nulls because intermediate signaling states due to leaky alleles can give ambiguous results. Thus, the molecular basis of the mutations must be known. Second, the two mutations used must have clearly distinct phenotypes such as insensitivity in the presence of the hormone and constitutive response in the absence of the hormone. In cases where the two mutations confer opposite signaling states, the epistatic mutation will be genetically downstream (Figure 2). If the mutations have the same signaling state, for example, when one molecule activates another component that also activates the signaling response, then the epistatic mutation is genetically upstream. This example is analogous to epistasis in a simple linear metabolic pathway in which the earlier intermediate accumulates in the double mutant. However, mutations that confer similar signaling states in a hormone transduction pathway often have indistinguishable phenotypes and, therefore, cannot be interpreted. For example, epistatic analysis between two loss-of-function mutations that both lead to hormone insensitivity may not distinguish whether the genes act in one linear pathway or two converging branches of a parallel pathway.

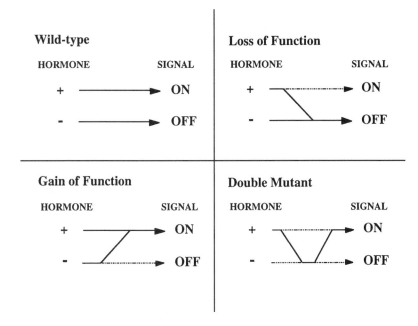

Figure 2   Epistasis between mutations that confer opposite signal states in a hormone transduction pathway. In wild-type when the hormone is present, the signal is ON; when it is absent, the signal is OFF. Loss-of-function insensitive mutants divert the signal state to OFF even in the presence of the hormone. Gain-of-function constitutive mutants divert the signal to ON even in the absence of hormone. In this double mutant example, the gain-of-function mutation affects a gene product that works downstream of the loss-of-function mutation. In the double mutant the signal is eventually diverted to ON and thus the gain-of-function mutation is epistatic to the loss-of-function mutation.

Even if constitutive gain-of-function and recessive loss-of-function mutations exist for each step of the pathway, epistatic analysis can be deceptive. Many mutations may interact through independent regulatory mechanisms rather than within a dependent pathway. For example, the failure of Arabidopsis embryo lethal mutants to germinate, which is a similar phenotype to a GA auxotroph, is not due to any direct involvement in GA signaling. Both mutations impinge on germination but not on the same dependent pathway. Moreover, even in a dependent pathway, the products of one signaling gene may modify the activity of a number of downstream components or may be modified by multiple regulators, which leads to branches or convergence. These nodal pathways add a further level of complexity to interpretations of double-mutant phenotypes. These examples demonstrate why epistatic analysis alone is not sufficient to determine molecular mechanisms of gene action. In general, a combination of molecular and classical genetics is required to confirm a genetic model.

The hormone signaling pathway that has been studied the most extensively at the genetic and molecular level is the ethylene transduction pathway, and the phenotypes of the mutants have been discussed above.

Ethylene Pathway Interactions

Interactions between various ethylene-response mutants have been studied by examining double mutant phenotypes between mutations that show opposite signaling states in the presence and absence of the hormone. For example, *ctr1* mutants that give a constitutive ethylene response are epistatic to ethylene-insensitive *etr1* mutants, which suggests that CTR1 is genetically downstream of ETR1 (37). Two possible models emerge based on this interaction. In one case, in the absence of ethylene, the ETR1 receptor constitutively activates CTR1, which inhibits ethylene responses downstream. The addition of ethylene inhibits the ETR1 activation of CTR1, thereby releasing ethylene-response genes from CTR1 inhibition (Figure 3A). In the second model, ETR1 is activated by ethylene to inactivate CTR1 function (Figure 3B). Again in this model CTR1 negatively regulates ethylene action downstream. An essential difference between these two models is that the former predicts that ethylene inhibits a positive regulator whereas in the latter, the gas activates a negative regulator. The identification of alternative allelic states of ETR1 can discriminate between these possibilities; however, the redundancy of the ETR1-like receptor kinases makes the phenotypic identification of loss-of-function ETR1 alleles difficult. Despite this problem, loss-of-function alleles for each ETR1 homologous gene were identified by screening directly for loss of ethylene insensitivity in each of the ETR1-like mutants (28). Selection against dominant insensitivity is advantageous since the phenotype is unambiguous and there is no preconceived bias on the phenotype of ETR1 loss-of-function alleles. As expected, each loss-of-function mutation by itself had no observable phenotype; nevertheless, when combined, quadruple mutant plants are phenotypically similar to a *ctr1* mutant. If ETR1 only regulates CTR1 in the presence of ethylene and has no function in the absence of the gas, as suggested in the second model (Figure 3B), genetic loss of the ethylene reception should have no effect on the plant's development in the absence of ethylene. Because the quadruple mutant mimics *ctr1* in the absence of ethylene, the ethylene receptor must activate CTR1 in the absence of the gas. It follows then that ethylene inhibits the ETR1 activation of CTR1 (Figure 3A).

Epistatic analysis of CTR1 with other ethylene-insensitive genes EIN2, EIN3, EIN5, EIN6, and EIN7 suggests that these genes act downstream of CTR1 (62). The recent cloning and demonstration that EIN3 overexpression confers an ethylene constitutive response in Arabidopsis have allowed the genetic placement of this gene in the ethylene signal transduction pathway. The constitutive ethylene

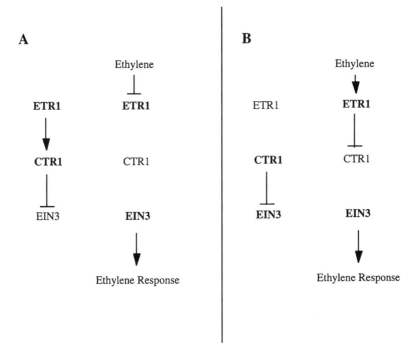

Figure 3 Two possible models of ethylene signaling in Arabidopsis. *Arrows* represent positive regulation and *bars* represent negative regulation. *A*. The ETR1 receptor activates CTR1, which inhibits downstream ethylene responses. In the presence of ethylene, the gas inactivates ETR1, which results in inactivation of CTR1, which in turn results in release of downstream ethylene responses from CTR1 inhibition. *B*. In the absence of ethylene, ETR1 is inactive, and therefore CTR1 can inactivate downstream ethylene responses. Ethylene gas activates ETR1 to inhibit CTR1, which results in release of downstream ethylene responses from CTR1 inhibition.

response observed in ectopically expressed EIN3 seedlings is independent of *ein2* loss-of-function alleles, which suggests that EIN3 acts genetically downstream of EIN2 (12). Since other ein mutants are recessive loss-of-function ethylene-insensitive mutations, the use of gain-of-function EIN3 overexpression plants should allow placement of the other EIN genes in the pathway.

These studies of the genetics of ethylene signal transduction indicate that epistatic analysis can be used to infer gene order in a dependent pathway in plant hormone signaling. Similar analysis should be possible for other hormone pathways as more mutations are identified. However, even for ethylene signaling, more allelic states are required for each of the genes involved. As genes are cloned, overexpression studies can function as dominant gain-of-function alleles, but caution must be used. Overexpression of regulatory genes is problematic since the gene product accumulates to artificial levels and is produced ubiquitously. In this sense, gain-of-function mutations that maintain wild-type levels and normal localization of protein may be less prone to artifact. For example, ectopic expression of a related gene to EIN3, designated EIL1, results in ethylene-independent activation of ethylene responses, and these responses are independent of EIN2 function (12). The EIL1 gene product, however, may act in a disparate pathway that when misexpressed interferes with ethylene signaling. Until loss-of-functional alleles of EIL1 are identified, a neomorphic function for EIL1 overexpression in ethylene signaling cannot be discounted.

Interactions Between Signaling Pathways

Physiological studies have demonstrated that plant growth and development require the coordinated action of multiple hormones and that these interactions are often reflected in the phenotypes of hormone-response mutants. The *axr2* mutation in Arabidopsis, for example, confers insensitivity to auxin, ethylene, and ABA (78). Altered sensitivities to more than one hormone can cause a dilemma as to which signaling pathway is perturbed and can confound genetic screens. In Arabidopsis, for example, the *ckr1* mutation confers a reduced response to exogenous cytokinin at the level of root elongation and was therefore catalogued as cytokinin insensitive (70). Subsequent genetic analysis has shown this mutation is allelic to the ethylene-insensitive *ein2* mutation (10). Cytokinin induces ethylene biosynthesis because this hormone positively regulates an isoform of 1-aminocyclopropane-1-carboxylate synthase, the first step in ethylene biosynthesis (74). Cytokinin insensitivity is due to the inability of *ckr1* to respond to the ethylene production induced by cytokinin, not to the cytokinin directly.

Although mutants with altered sensitivity to one hormone need not be defective in that particular signaling pathway, such mutants do establish relationships between signaling pathways and can lead to new genetic screens. The inability to induce ethylene in the presence of cytokinins results in a loss of the triple response in dark-grown Arabidopsis seedlings, and this assay has been the basis of a genetic screen for mutants with altered cytokinin responses (73). To date, four classes of mutants, designated *cin* for cytokinin insensitive, are unable to show a triple response in the presence of cytokinin. Although these mutants should offer insights into cytokinin signaling events, the ethylene-cytokinin interactions reinforce the importance of good phenotypic characterization of hormone response mutants.

Another, and potentially more interesting, mechanism of interacting hormone responses is the possibility that the same gene products are used in different signaling pathways. If signaling components are promiscuous, different hormones could fulfill similar functions due to cross talk. For example, auxin, GA, and cytokinin can all induce cell division in different tissues, which suggests the downstream signal output from activation by these hormones is funneled into control of the cell cycle. Based on what is currently known about mutations that confer changes in sensitivity to multiple hormones, it is difficult to test this possibility. However, genetic analysis of auxin response in Arabidopsis suggests that proteins involved in correct auxin response interact with cell cycle components to regulate the response of plant cells to this hormone (57).

Suppressor Mutations

Although epistasis is useful in determining the order of gene products in a dependent pathway, the architecture of a signaling pathway often does not allow for simple epistatic interactions. Divergence and convergence of pathways, for example, may not lead to clear double mutant phenotypes. A potential genetic method of determining relationships in complex interactions is by suppression analysis. Not to be confused with epistasis, a suppressor mutation either partially or completely restores the phenotype to wild-type but does not substitute its own. Although the criterion for genetic suppression is less restrained than that of epistasis, analysis of suppressor mutations can be laborious. If suppressor mutations have no obvious phenotypes on their own, segregation, complementation, and mapping experiments are limited to systems with good genetic maps. Second, although the suppressor net is cast widely, genes identified in these screens may have little to do with the transduction of the pathway of interest. For example, one suppressor mutation affecting cytokinin-dependent ethylene production is a new allele of *fus9/cop10* (73). Although this is a nice demonstration of the interaction between this hormone and light signaling, there is probably no direct role for *fus9/cop10* in cytokinin signaling per se.

Broadly speaking, in hormone signaling pathways the two most useful compensating changes that result in suppression are intragenic and extragenic mutations. Intragenic suppressors are often identified by selecting for loss of a dominant gain-of-function phenotype such as suppressing hormone insensitivity. As demonstrated in *etr1* suppressor screens (28), this strategy can be extremely useful when trying to generate alternative allelic states for a gene of interest.

Extragenic suppressors or second-site revertants restore normal function to a pathway by changing another function in the pathway. These mutations often uncover genes in a transduction pathway by causing a shift in the signal flux. Depending on the nature of the signaling genes being suppressed (i.e. positive or negative), mutations in genes elsewhere in the pathway may either decrease or increase the signal current. Moreover, if the pathway is exquisitely sensitive to perturbation in signaling flux, it may be possible to identify suppressor mutations in a heterozygous state. These mutations will be dominant suppressors of the original mutation and may display new phenotypes when homozygous in a wild-type genetic background. Suppressors demonstrating new phenotypes on their own are not only useful for identifying new gene functions but also for identifying new mutations in previously characterized genes. Genes encoding components of a particular signaling pathway may have other functions that may be missed by direct screening but can be identified genetically among suppressor mutations of signaling mutants.

Hormones affect many aspects of plant growth and development, with the result that mutations affecting early steps in the response pathway usually result in pleiotropic phenotypes. By contrast, if output branches exist further down the transduction pathway, mutations in these steps will only affect subsets of hormone-regulated functions. Comparably, extragenic suppressors that bypass all the phenotypes of an original mutation most probably identify genes that interact closely with the allele being suppressed. In the extreme case, allele-specific suppressor mutations indicate that the two gene products most likely interact biochemically. Conversely, extragenic mutations that only suppress a subset of original phenotypes probably are further away from the function being suppressed. For example, loss-of-function *hookless1* mutations in Arabidopsis only affect apical hook growth in response to ethylene, and both genetic and molecular studies confirm that this gene is downstream of early ethylene signaling events (42).

Suppressor Analysis of the GA Signaling Pathway

GA-deficient mutants have been extremely useful to understanding the roles of GA in plant growth and development (26). Mutants deficient in GA synthesis display a number of phenotypes, including poor germination, dwarfed growth habit due to reduced cell expansion, underdeveloped petal and stamen development, delayed flowering, and delayed senescence (26, 71). Using the phenotypes of GA auxotrophic mutants as a guide, GA-response mutants have been identified in a number of plant species including Arabidopsis (52, 71). Generally, these mutants fall into two categories, dominant or semidominant mutations that result in a reduced sensitivity to GA, and recessive mutations that confer a GA-independent phenotypes to the plants. Below I describe the phenotypes of a number of GA response mutants, focusing on how suppressor analysis has been used to further our understanding of how these genes function in GA signal transduction.

GIBBERELLIN-INSENSITIVE MUTANT, *gai*   A semidominant *gai1-1* mutation causes Arabidopsis plants to grow as dark green semidwarf, with reduced fertility and germination (38). Although these phenotypes are similar to GA-deficient mutants, the lack of rescue of *gai1-1* by GA application suggests this mutation results in reduced GA responsiveness. Intrallelic suppressor mutations of *gai1-1* were identified by screening for rescue of the *gai1-1* semidwarf phenotype. Early phenotypic characterization suggested these intrallelic mutations did not confer any profound developmental phenotypes to the plant (56, 79). Subsequent physiological studies, however, have demonstrated that these suppressor plants are weakly GA independent in that they are slightly insensitive to GA biosynthetic inhibitors (56). The GAI1 protein has sequence identity to a number of known transcriptional activators, and the semidominant *gai1-1* allele contains an inframe 51 base-pair insertion mutation (56). More important, the *gai1-1* intrallelic suppressor mutations all disrupted the GAI1 open reading frame, which indicates that these are loss-of-function alleles. Although the GA-independent phenotype of these alleles is weak, this loss-of-function phenotype suggests that this protein may have binary switch properties of a signal transduction component. With this in mind, Harbord and co-workers (56) have posited a model to explain GAI1 action in plant GA responses. Wild-type GAI1 may negatively regulate cell elongation and this function is inhibited by GA. In this model, the semidominant *gai1-1* mutant has lost the capacity to be antagonized by GA, thereby allowing the protein to continuously inhibit cell elongation. If GA interacts directly with GAI1, there may be no GA signal transduction pathway per se since GAI1 appears to encode a transcription factor. Alternatively, if GAI1 interacts with a GA signaling component, this interaction is at the end of the GA signaling pathway. Whichever is the case, the finding that *gai1-1* GA auxotrophic double mutants produce a more severe dwarf phenotype indicates that GA can still be perceived in the *gai1-1* mutant. Hence, GAI1 is not the sole component of GA signaling in Arabidopsis.

REPRESSOR OF *ga1* MUTANT, *rga1*   Recessive *rga1* mutations were identified in a suppressor screen of an Arabidopsis GA biosynthetic mutation *ga1-3* (67). Mutations in the *rga1* gene restore most of the vegetative phenotypes associated with the GA auxotrophy but do not suppress the germination defects (67). In a wild-type background, the *rga* mutations cause adult plants to be slightly etiolated; however, for the most part these mutations lack distinctive phenotypes. The *RGA1* gene shows high sequence similarity to *GAI1*, and this molecular redundancy may explain why single loss-of-function alleles of *GAI1* and *RGA1* are phenotypically subtle (66). Construction of *rga1 gai1* double mutants should resolve this issue. However, that loss-of-function alleles of GAI1 were never identified in the *ga1-3* suppressor screen argues that these genes may only overlap in a subset of functions (67).

Nevertheless, the functional and molecular redundancy of *RGA1* and *GAI1* demonstrates the difficulty of clearly identifying single switch genes in higher plants. Based on other genetic systems, it appears that a considerable number of loss-of-function mutations produce no obvious phenotype (9). Although many of these genes may simply have nonessential functions under laboratory conditions, it is equally likely that some of these loci are functionally redundant. Although dominant gain-of-function mutants have the advantage of identifying functionally redundant signaling genes, the loss-of-function alleles in these same genes may not give the predicted opposite phenotype because another gene may genetically cover this function. A genetic solution to this problem is to identify mutations that enhance the phenotypes conferred by the first mutation but when isolated, by themselves are phenotypically innocuous. As with phenotypically neutral suppressor mutations, analysis of these types of synthetic enhancers are constrained by the sophistication of the genetics of the system.

spindly1 MUTANT, *spy1*   The first *spy1* mutations were identified by the ability of Arabidopsis seed to germinate in the presence of inhibitory concentrations of a GA biosynthetic inhibitor (33). Although this type of screen formally does not qualify as a hunt for a genetic suppressor, subsequent screening for *ga1-3* auxotrophic suppressors did identify more *spy1* alleles (67), which demonstrated that these screens are functionally similar. Recessive mutations at the *spy1* locus result in partial suppression of all the defects associated with GA auxotrophy including defective germination and, therefore, this mutation confers a GA-independent phenotype to the plant (33). Moreover, unlike loss-of-function *rga1* and *gai1* mutations, *spy1* mutations in a wild-type genetic background have a reduced GA requirement for germination, increased internode length, and floral timing defects. The recessive nature of *spy1* alleles suggests this gene encodes a negative regulator of GA response. The SPY1 gene shows sequence similarity to Ser (Thr)-O–linked acetylglucosamine (O-GlcNAc) transferases (32). These enzymes glycosylate proteins and, in some cases, this modification can interfere with phosphorylation of a protein. Interestingly, RGA1 and GAI1 proteins both contain potential O-GlcNAc sites, hence, SPY1 may modify these two regulators.

One advantage of working on GA signal transduction pathways has been the elegant use of the barley aleurone cell system as a molecular assay for GA responses (26). In this system, α-amylase mRNA is readily induced by addition of GA whereas ABA interferes with this induction. Microbombardment of aleurone cells with the barley homolog of SPY1 (HvSPY) prevents GA activation of α-amylase, supporting the role of SPY1 as a negative regulator (60). Surprisingly, these transfection experiments also demonstrate a positive role for HvSPY in ABA responses. Possibly, the multiple effects of HvSPY occur because the O-GlcNAc transferase modifies separate target proteins in both GA and ABA response pathways. Alternatively, HvSPY may only enhance ABA signaling activity and this response reduces GA induced α-amylase induction. The demonstration that *spy1* mutant seeds are partially insensitive to ABA as assayed by germination supports this idea (68). However, it is also possible that increased GA signaling in the seed causes decreased ABA sensitivity. GA and ABA synthesis are temporally separated in Arabidopsis seeds, which indicates that the functions of these hormones on the establishment and breaking of dormancy is probably by different mechanisms (35, 39). Active GA signaling in *spy1* mutant embryos may be incompatible with ABA-induced seed functions and result in a less dormant seed. These issues cannot be resolved until the targets of the O-GlcNAc transferase are identified.

sleepy1 MUTANT, *sly1*   The duality of GA/ABA signaling responses in plants is often reflected in genetic screens. The first ABA-deficient mutants of Arabidopsis were originally identified by screening for suppressor mutations of the germination defect of a *ga1* auxotroph (39). By contrast, genetic screens for suppressors of the dominant ABA-insensitive mutation *ABI1-1* enriches for mutations that are defective in GA synthesis or response (68). The GA nonresponsive mutants are phenotypically indistinguishable from severe Arabidopsis GA auxotrophs and define one complementation group designated *sly1*. The alleles of *sly1* are the first recessive GA-insensitive mutations and they probably identify a key regulator in GA reception; however, the ability to find such mutants as suppressors of mutants defective in ABA action was unexpected. ABA acts first to establish seed dormancy in the embryo and later GA works to reverse ABA-induced dormancy (39). Mutants defective in GA signal transduction cannot germinate unless they have also acquired a mutation in a gene required for ABA synthesis or response. Physiological studies using an allelic series of ABA auxotrophic mutations have, however, demonstrated that the degree of ABA-induced seed dormancy is a reflection of the flux of the ABA signal (35). Thus, the amount of GA needed to germinate seed will depend on the level of dormancy established by ABA, which will, in turn, be determined by the severity of the mutation affecting ABA signal flux. By genetically decreasing the ABA sensitivity of the seed, it is possible to identify mutations that have decreased GA sensitivity and that would normally not germinate by themselves. The observation that *sly1* alleles do not germinate in an ABI1 wild-type background supports this model. The germination block imposed by normal ABA action in the seed may explain the dearth of recessive loss-of-function mutations in GA signaling genes. If true, suppressor screens against known ABA-response mutants under different selective conditions may identify new genes involved in GA signal transduction.

NEW TECHNOLOGIES

Although the use of classical genetic screens is a powerful tool in determining how genes specify the transduction of a hormone signal, this method is limited in the scope of genes it can identify. Traditional mutagenesis uses chemicals and ionizing radiation to create random mutations, which frequently cause a reduction or loss-of-gene function. These mutations are often difficult to recognize because they confer no obvious phenotype owing to genetic redundancy. In other model systems, an estimated two thirds of the genes have no phenotype under laboratory conditions (49). No technology will replace traditional genetic screens, although new molecular techniques are now making it possible to more directly and efficiently pinpoint genes in specific signaling pathways. The success of these new screens requires efficient transformation and heterologous expression systems, both of which are available in Arabidopsis.

Activation-Transferred DNA Tagging Mutants.

The ability to express a gene in a cell where it is not normally active has various potential outcomes to a signaling pathway. Ectopic expression may have no effect on the transduction of the signal, which would indicate that the concentrations of molecular components of the signaling pathway are not essential in determining the current. Usually for a signaling cascade, however, this is not the case. In *Saccharomyces cerevisiae*, for example, the components of the pheromone signaling pathway are maintained at specific levels (58, 75). Changing the stoichiometry of components in the pathway often causes the transduction current to increase or decrease depending on whether the specific component is positive or negative for the pathway. This observation has been the basis of screens in which increased expression of one gene can genetically suppress the phenotype of a mutation in another gene involved in the same process (58).

In Arabidopsis, large-scale transformations with a t-DNA plasmid containing multiple enhancers allowed Kakimoto (34) to identify genes that confer cytokinin-independent growth to calli. One of these overexpressed genes, designated *CKI1*, shares sequence similarity to the two-component regulator family of proteins of which *ETR1* is a member. *CKI1* may be a downstream gene in the cytokinin signaling pathway that requires cytokinin for expression, and therefore overexpression of the gene uncouples it from cytokinin induction. The identification of two new two-component regulator genes in Arabidopsis by virtue of their rapid induction in response to cytokinin application supports such a contention (8). It is also possible that CKI1 is a protein that requires cytokinin for activation and that increased expression of *CKI1* overrides this hormonal need. On this theme, Kakimoto (34) has hypothesized that CKI1 could be a cytokinin receptor molecule that, when overexpressed, increases the ability of the cells to respond to the normally low endogenous levels of cytokinin found in calli. Although CKI1 may be a cytokinin receptor, binding of cytokinin to this protein has not been shown, and that no loss-of-function alleles exist makes it difficult to distinguish between these or other unanticipated possibilities.

Although these genetic screens hold much promise, most misexpression systems are at present constructed to overexpress proteins ubiquitously. Aside from the inability to control where and how much target protein is made, if ectopic expression of the gene is toxic in some tissues, it may be impossible to establish genetically tractable lines. Many of these problems, however, can be circumvented by the development of a modular GAL4 inducible system similar to one recently reported in Drosophila (63).

Screens Involving Reporter Constructs

The molecular identification of genes induced by the application of a particular hormone was originally designed to identify genes that respond to the hormone after the signal has been transduced. However, these same downstream targets of the hormone signaling pathway can also be exploited to identify mutations that affect transduction of the hormone signal. The system first requires the fusion of an easily assayed reporter gene to the cis-acting control elements of a hormone-regulated promoter. Transgenic lines containing this construct are mutagenized and screened for second site mutations that alter expression of the reporter gene. Using this approach, the reporter gene luciferase was fused to the stress-induced responsive promoter RD29, and Arabidopsis seedlings with altered responses to low temperature, drought, salinity, and ABA were identified (31). The specificity of the reporter-gene screens have two advantages over traditional genetic screens. First, the specificity of the phenotype being sought (i.e. increased or decreased expression) allows easy identification of mutants with no a priori bias on what the plants should look like. Second, the effects of the second site mutations on expression of the reporter gene may be more easily quantified than by using a physiological assay such as hypocotyl curvature. This latter quality may be extremely important in cases where a number of signaling pathways contribute to overall response such that mutations in any one of these only give subtle whole-plant phenotypes.

CONCLUSION

The studies reviewed here describe some of the characteristics of plant hormone signal transduction pathways and how these can be exploited to identify mutations in individual components in a hormone signaling pathway. Our understanding of hormone signaling will continue to expand quickly as more genetic variation is generated. As more hormone pathways are resolved at the molecular level and the logic of the system becomes apparent, it will become possible to answer questions on the role of these substances in plant evolution. For example, the *never-ripe* (*Nr*) mutation of tomato identifies an *ETR1*-like gene that contains an identical mutation to the Arabidopsis *etr1-4* allele (77). However, unlike the Arabidopsis *ETR1*, the *Nr* gene lacks the response regulator domain and is induced by ethylene during fruit ripening. Thus, natural variation has produced an *etr1-4* allelic change in a gene fundamentally different in structure and regulation. Introduction of the *Nr* gene in its wild-type and mutant forms into various Arabidopsis ethylene signaling mutants may uncover interactions that would not normally be observed by traditional mutagenesis and may lead to new methods of regulating these pathways. The conservation of a number of different hormone signaling components between Arabidopsis and other plant species argues that such experiments will be possible for all the hormone response pathways (23, 60, 76). These and other experiments should provide new insights into how plant hormone signaling pathways have evolved and perhaps how these different strategies can be used to generate new functions for hormone signaling pathways.

Visit the *Annual Reviews home page* at
http://www.AnnualReviews.org

Literature Cited

1. Avery L, Wasserman S. 1992. Ordering gene function: the interpretation of epistasis in regulatory hierarchies. *Trends Genet.* 8:312–16
2. Azpiroz R, Wu Y, LoCascio JC, Feldmann KA. 1998. An Arabidopsis brassinosteroid-dependent mutant is blocked in cell elongation *Plant Cell* 10:219–30
3. Benedetti CE, Xie D, Turner JG. 1995. Coi1-dependent expression of an Arabidopsis vegetative storage protein in flowers and siliques and in response to coronatine or methyl jasmonate. *Plant Physiol.* 109:567–72
4. Bertauche N, Leung J, Giraudat J. 1996. Protein phosphatase activity of abscisic acid insensitive 1 (ABI1) protein from *Arabidopsis thaliana*. *Eur. J. Biochem.* 241: 193–200
5. Bleecker AB, Estelle MA, Somerville C, Kende H. 1988. Insensitivity to ethylene conferred by a dominant mutation in *Arabidopsis thaliana*. *Science* 241:1086–89

6. Bohen SP, Kralli A, Yamamoto KR. 1995. Hold 'em and fold 'em: chaperones and signal transduction. *Science* 268:1303–4
7. Bonetta D, McCourt P. 1998. Genetic analysis of ABA signal transduction pathways. *Trends Plant Sci.* 3:231–35
8. Brandstatter I, Kieber JJ. 1998. Two genes with similarity to bacterial response regulators are rapidly and specifically induced by cytokinin in Arabidopsis. *Plant Cell* 10:1009–20
9. Burns N, Grimwade B, Ross-Macdonald PB, Choi EY, Finberg K, et al. 1994. Large-scale analysis of gene expression, protein localization, and gene disruption in *Saccharomyces cerevisiae*. *Genes Dev.* 8:1087–105
10. Cary AJ, Liu W, Howell SH. 1995. Cytokinin action is coupled to ethylene in its effects on the inhibition of root and hypocotyl elongation in *Arabidopsis thaliana* seedlings. *Plant Physiol.* 107: 1075–82
11. Chang C, Kwok SF, Bleecker AB,

Meyerowitz EM. 1993. Arabidopsis ethylene-response gene ETR1: similarity of product to two-component regulators. *Science* 262:539–44

12. Chao Q, Rothenberg M, Solano R, Roman G, Terzaghi W, Ecker JR. 1998. Activation of the ethylene gas response pathway in Arabidopsis by the nuclear protein *ETHYLENE-INSENSITIVE3* and related proteins. *Cell* 89:1133–44

13. Chory J, Li J. 1997. Gibberellins, brassinosteriods and light-regulated development. *Plant Cell Environ.* 20:801–5

14. Clark KL, Larsen PB, Wang X, Chang C. 1998. Association of the Arabidopsis *CTR1* Raf-like kinase with the *ETR1* and ERS ethylene receptors. *Proc. Natl. Acad. Sci. USA* 95:5401–6

15. Clouse SD, Langford M, McMorris TC. 1996. A brassinosteroid-insensitive mutant in *Arabidopsis thaliana* exhibits multiple defects in growth and development. *Plant Physiol.* 111:671–78

16. Creelman RA, Mullet JE. 1997. Biosynthesis and action of jasmonates in plants. *Annu. Rev. Plant Physiol. Plant Mol. Biol.* 48:355–81

17. Cutler S, Ghassemian M, Bonetta D, Cooney S, McCourt. 1996. A protein farnesyl transferase involved in abscisic acid signal transduction in Arabidopsis. *Science* 273:1239–41

18. Deikman J, Ulrich M. 1995. A novel cytokinin-resistant mutant of Arabidopsis with abbreviated shoot development. *Planta* 195:440–49

19. Estelle MA, Somerville CR. 1987. Auxin resistant mutants of Arabidopsis with an altered morphology. *Mol. Gen. Genet.* 206: 200–6

20. Finkelstein RR. 1994. Mutations at two new Arabidopsis ABA response loci are similar to the *abi3* mutations. *Plant J.* 5: 765–71

21. Gamble RL, Coonfield ML, Schaller GE. 1998. Histidine kinase activity of the ETR1 ethylene receptor from Arabidopsis. *Proc. Natl. Acad. Sci. USA* 95:7825–29

22. Giraudat J. 1995. Abscisic acid signaling. *Curr. Opin. Cell Biol.* 7:232–38

23. Giraudat J, Hauge BM, Valon C, Smalle J, Parcy F, Goodman HM. 1992. Isolation of the Arabidopsis *ABI3* gene by positional cloning. *Plant Cell* 4:1251–61

24. Guzman P, Ecker JR. 1990. Exploiting the triple response of Arabidopsis to identify ethylene-related mutants. *Plant Cell* 2:513–23

25. Hobbie L, Timpte C, Estelle M. 1994. Molecular genetics of auxin and cytokinin. *Plant Mol. Biol.* 5:1499–519

26. Hooley R. 1994. Gibberellins: perception, transduction and responses. *Plant Mol. Biol.* 26:1529–55

27. Hua J, Chang C, Sun Q, Meyerowitz EM. 1995. Ethylene insensitivity conferred by Arabidopsis *ERS* gene. *Science* 269:1712–14

28. Hua J, Meyerowitz EM. 1998. Ethylene responses are negatively regulated by a receptor gene family in *Arabidopsis thaliana*. *Cell* 94:261–71

29. Hua J, Sakai H, Nourizadeh S, Chen QG, Bleecker AB, et al. 1998. *EIN4* and *ERS2* are members of the putative ethylene receptor gene family in Arabidopsis. *Plant Cell* 10:1321–32

30. Imamura A, Hanaki N, Umeda H, Nakamura A, Suzuki T, et al. 1998. Response regulators implicated in His-to-Asp phosphotransfer signaling in Arabidopsis. *Proc. Natl. Acad. Sci. USA* 95:2691–96

31. Ishitani M, Xiong L, Stevenson B, Zhu JK. 1997. Genetic analysis of osmotic and cold stress signal transduction in Arabidopsis: interactions and convergence of abscisic acid-dependent and abscisic acid-independent pathways. *Plant Cell* 9:1935–49

32. Jacobsen SE, Binkowski KA, Olszewski NE. 1996. SPINDLY, a tetratricopeptide repeat protein involved in gibberellin signal transduction in Arabidopsis. *Proc. Natl. Acad. Sci. USA* 93:9292–96

33. Jacobsen SE, Olszewski NE. 1995. Mutations at the *SPINDLY* locus of Arabidopsis alter gibberellin signal transduction. *Plant Cell* 5:887–96

34. Kakimoto T. 1996. *CKI1*, a histidine kinase homolog implicated in cytokinin signal transduction. *Science* 274:982–85

35. Karssen CM, Brinkhorst-van der Swan DLC, Breekland AE, Koornneef M. 1983 Induction of dormancy during seed development by endogenous abscisic acid: studies on abscisic acid-deficient genotypes of *Arabidopsis thaliana* (L.) Heynh. *Planta* 157:158–65

36. Kieber JJ. 1997. The ethylene response pathway in Arabidopsis. *Annu. Rev. Plant Physiol. Plant Mol. Biol.* 48:277–96

37. Kieber JJ, Rothenberg M, Roman G, Feldmann KA, Ecker JR. 1993. CTR1, a negative regulator of the ethylene response pathway in Arabidopsis, encodes a member of the raf family of protein kinases. *Cell* 72:427–41

38. Koornneef M, Elgersma A, Hanhart CJ, van Loenen-Martinet EP, van Rijn L, Zeevaart JAD. 1985. A gibberellin insensitive mutant of *Arabidopsis thaliana*. *Physiol. Plant.* 65:33–39

39. Koornneef M, Jorna ML, Brinkhorst-van der Swan DLC, Karssen CM. 1982. The isolation of abscisic acid (ABA)-deficient mutants by selection of induced revertants in non-germinating gibberellin-sensitive lines of *Arabidopsis* (L.) Heynh. *Theor. Appl. Genet.* 61:385–93

40. Koornneef M, Reuling G, Karssen CM. 1984. The isolation and characterization of abscisic acid-insensitive mutants of *Arabidopsis thaliana. Physiol. Plant.* 61:377–83

41. Koornneef M, van der Veen. 1980. Induction and analysis of gibberellin-sensitive mutants of *Arabidopsis thaliana* (l.) Heynh. *Theor. Appl. Genet.* 58:257–63

42. Lehman A, Black R, Ecker JR. 1996. *HOOKLESS1*, an ethylene response gene, is required for differential cell elongation in the Arabidopsis hypocotyl. *Cell* 85:183–94

43. Leung J, Bouvier-Durand M, Morris PC, Guerrier D, Chefdor F, Giraudat J. 1994. Arabidopsis ABA-response gene *ABI1*: features of a calcium-modulated protein phosphatase. *Science* 264:1448–52

44. Leung J, Merlot S, Giraudat J. 1997. The Arabidopsis *ABSCISIC ACID-INSENSITIVE2* (*ABI2*) and *ABI1* genes encode homologous protein phosphatases 2C involved in abscisic acid signal transduction. *Plant Cell* 9:759–71

45. Li J, Chory J. 1997. A putative leucine-rich repeat receptor kinase involved in brassinosteroid signal transduction. *Cell* 90:929–38

46. Li J, Nagpal P, Vitart V, McMorris TC, Chory J. 1996. A role for brassinosteroids in light-dependent development of Arabidopsis. *Science* 272:398–401

47. Maher EP, Martindale SJ. 1980. Mutants of *Arabidopsis thaliana* with altered responses to auxins and gravity. *Biochem. Genet.* 12:1041–53

48. Meyer K, Leube MP, Grill E. 1994. A protein phosphatase 2C involved in ABA signal transduction in *Arabidopsis thaliana. Science* 264:1452–55

49. Miklos GL, Rubin GM. 1996. The role of the genome project in determining gene function: insights from model organisms. *Cell* 86:521–29

50. Nambara E, Keith K, McCourt P, Naito S. 1995. A regulatory role for the *ABI3* gene in the establishment of embryo maturation in *Arabidopsis thaliana. Development* 121:629–36

51. Neljubov D. 1901. Uber die horizontale Nutation der Stengel von *Pisum sativum* und einiger anderen Pflanzen. *Beih. Bot. Zentralbl.* 10:128–239

52. Ogas J. 1998. Dissecting the gibberellin response pathway. *Curr. Biol.* 8:165–67

53. Ooms JJJ, Léon-Kloosterziel KM, Bartel D, Koornneef M, Karssen M. 1993. Acquisition of desiccation tolerance and longevity in seeds of *Arabidopsis thaliana*. A comparative study using abscisic acid-insensitive *abi3* mutants. *Plant Physiol.* 102:1185–91

54. Palme K, Bischoff F, Cvrckova F, Zarsky V. 1997. Small G-proteins in *Arabidopsis thaliana. Biochem. Soc. Trans.* 25:1001–5

55. Parcy F, Valon C, Raynal M, Gaubier-Comella P, Delseny M, Giraudat J. 1994. Regulation of gene expression programs during Arabidopsis seed development: roles of the *ABI3* locus and of endogenous abscisic acid. *Plant Cell* 6:1567–82

56. Peng J, Carol P, Richards DE, King KE, Cowling RJ, et al. 1997. The Arabidopsis *GAI* gene defines a signaling pathway that negatively regulates gibberellin responses. *Genes Dev.* 11:3194–205

57. Pozo JC, Timpte C, Tan S, Callis J, Estelle M. 1998. The ubiquitin-related protein RUB1 and auxin response in Arabidopsis. *Science* 280:1760–63

58. Ramer SW, Elledge SJ, Davis RW. 1992. Dominant genetics using a yeast genomic library under the control of a strong inducible promoter. *Proc. Natl. Acad. Sci. USA* 89:11589–93

59. Reid JB, Howell SH. 1995. The function of hormones in plant growth and development. In *Plant Hormones Physiology, Biochemistry and Molecular Biology*, ed. P Davies, pp. 448–85. Dordrecht: Kluwer

60. Robertson M, Swain SM, Chandler PM, Olszewski NE. 1998. Identification of a negative regulator of gibberellin action, *HvSPY*, in barley. *Plant Cell* 10:995–1008

61. Rodriguez PL, Benning G, Grill E. 1998. ABI2, a second protein phosphatase 2C involved in abscisic acid signal transduction in Arabidopsis. *FEBS Lett.* 421:185–90

62. Roman G, Lubarsky B, Kieber JJ, Rothenberg M, Ecker JR. 1995. Genetic analysis of ethylene signal transduction in *Arabidopsis thaliana*: five novel mutant loci integrated into a stress response pathway. *Genetics* 139:1393–409

63. Rorth P, Szabo K, Bailey A, Laverty T, Rehm, et al. 1998. Systematic gain-of-function genetics in Drosophila. *Development* 125:1049–57

64. Sakai H, Hua J, Chen QG, Chang C, Medrano LJ, et al. 1998. *ETR2* is an *ETR*-like gene involved in ethylene signaling in Arabidopsis. *Proc. Natl. Acad. Sci. USA* 95:5812–17

65. Schaller GE, Bleecker AB. 1995. Ethylene-binding sites generated in yeast expressing the Arabidopsis *ETR1* gene. *Science* 270:1809–11

66. Silverstone AL, Ciampaglio CN, Sun T-p. 1998. The Arabidopsis *RGA* gene encodes a transcriptional regulator repressing the gibberellin signal transduction pathway. *Plant Cell* 10:155–69

67. Silverstone AL, Mak PY, Martinez EC, Sun TP. 1997. The new *RGA* locus encodes a negative regulator of gibberellin response in *Arabidopsis thaliana*. *Genetics* 146:1087–99

68. Steber CM, Cooney SE, McCourt P. 1998. Isolation of the GA-response mutant *sly1* as a suppressor of *ABI1-1* in *Arabidopsis thaliana*. *Genetics* 149:509–21

69. Stone JM, Walker JC. 1995. Plant protein kinase families and signal transduction. *Plant Physiol* 108:451–57

70. Su W, Howell SH. 1992. A single genetic locus, *ckr1*, defines Arabidopsis mutants in which root growth is resistant to low concentrations of cytokinin. *Plant Physiol.* 99:1569–74

71. Swain SM, Olszewski NE. 1996. Genetic analysis of gibberellin signal transduction. *Plant Physiol.* 112:11–17

72. Szekeres M, Nemeth K, Koncz-Kalman Z, Mathur J, Kauschmann A, et al. 1996. Brassinosteroids rescue the deficiency of *CYP90*, a cytochrome P450, controlling cell elongation and de-etiolation in Arabidopsis. *Cell* 85:171–82

73. Vogel JP, Schuerman P, Woeste K, Brands-tatter I, Kieber JJ. 1998. Isolation and characterization of Arabidopsis mutants defective in the induction of ethylene biosynthesis by cytokinin. *Genetics* 149:417–427

74. Vogel JP, Woeste KE, Theologis A, Kieber JJ. 1998. Recessive and dominant mutations in the ethylene biosynthetic gene *ACS5* of Arabidopsis confer cytokinin insensitivity and ethylene overproduction, respectively. *Proc. Natl. Acad. Sci. USA* 95:4766–71

75. Whiteway M, Hougan L, Thomas DY. 1990. Overexpression of the *STE4* gene leads to mating response in haploid *Saccharomyces cerevisiae*. *Mol. Cell Biol.* 10:217–22

76. Wilkinson JQ, Lanahan MB, Clark DG, Bleecker AB, Chang C, et al. 1997. A dominant mutant receptor from Arabidopsis confers ethylene insensitivity in heterologous plants. *Nat. Biotechnol.* 15:444–47

77. Wilkinson JQ, Lanahan MB, Yen HC, Giovannoni JJ, Klee HJ. 1995. An ethylene-inducible component of signal transduction encoded by never-ripe. *Science* 270:1807–9

78. Wilson AK, Pickett FB, Turner JC, Estelle M. 1990. A dominant mutation in Arabidopsis confers resistance to auxin, ethylene and abscisic acid. *Mol. Gen. Genet.* 222:377–83

79. Wilson RN, Somerville CR. 1995. Phenotypic suppression of the gibberellin insensitive mutant (*gai*) of Arabidopsis. *Plant Physiol.* 108:495–502

Annu. Rev. Plant Physiol. Plant Mol. Biol. 1999. 50:245–76

CELLULOSE BIOSYNTHESIS:
Exciting Times for A Difficult
Field of Study

Deborah P. Delmer

Section of Plant Biology, University of California Davis, Davis, California 95616;
e-mail: dpdelmer@ucdavis.edu

KEY WORDS: glucan, callose, Acetobacter, microfibril, cytoskeleton

---

ABSTRACT

The past few decades have witnessed exciting progress in studies on the biosynthesis of cellulose. In the bacterium *Acetobacter xylinum*, discovery of the activator of the cellulose synthase, cyclic diguanylic acid, opened the way for obtaining high rates of in vitro synthesis of cellulose. This, in turn, led to purification of the cellulose synthase and for the cloning of genes that encode the catalytic subunit and other proteins that bind the activator and regulate its synthesis and degradation, or that control secretion and crystallization of the microfibrils. In higher plants, a family of genes has been discovered that show interesting similarities and differences from the gene in bacteria that encodes the catalytic subunit of the synthase. Genetic evidence now supports the concept that members of this family encode the catalytic subunit in these organisms, with various members showing tissue-specific expression. Although the cellulose synthase has not yet been purified to homogeneity from plants, recent progress in this area suggests that this will soon be accomplished.

CONTENTS

1040-2519/99/0601-0245$08.00

The Early Decades of Study (1950–1980)

These are exciting times to study the mechanism of cellulose biosynthesis. For three decades, workers in this field struggled with little success to identify and characterize a cellulose synthase—the enzyme responsible for the polymerization of β-1,4-glucan—and to understand better how these chains self-assemble to form the microfibrils that represent the characteristic form of cellulose found in nature. Since about 180 billion tons of cellulose are produced per year in nature (41), study of the pathway of synthesis of this polymer is relevant for basic and applied science. This review concentrates on advances in this field over the past few decades. Readers new to the field are referred to a recent review by French & Delmer (43) that describes the history of the discovery of cellulose, the gradual elucidation of its structure, and the early struggles in studies on biosynthesis. Other reviews of the topic have appeared recently (13, 16, 18, 36, 47, 68, 103, 132).

Hestrin's group in Israel pioneered study with the gram-negative bacterium *Acetobacter xylinum*. Only a few genera of bacteria synthesize cellulose; the best known are various species of Acetobacter, Agrobacteria, Rhizobia, and the gram-positive genus *Sarcina*. Of these, *A. xylinum* produces massive amounts of cellulose that is secreted, not as a cell wall polymer as found in eukaryotes, but as an extracellular pellicle. Hestrin's group (59) established conditions for growth of *A. xylinum* and optimization of cellulose production, and also described the basic pathways of carbon metabolism. In the early 1950s, Leloir et al made the seminal discovery of UDP-glucose, and showed that this "high-energy" molecule could serve as a donor for a number of glycosyltransferases (80). Later, Glaser (51) demonstrated that membrane preparations derived from *A. xylinum* could synthesize limited amounts of β-1,4-glucan. Thereafter, many workers attempted to further characterize the cellulose synthase of *A. xylinum* and higher plants with almost no success. In *A. xylinum*, it proved difficult to enhance the very low rates observed by Glaser. In higher plants, little or no cellulose synthase activity was detectable above a very high background of callose (β-1,3-glucan) synthase activity. This latter enzyme, located in the plasma membrane of plants, uses UDP-glc as substrate to synthesize callose in a reaction that is dependent upon the presence of Ca^{2+} and a β-glucoside. Furthermore, callose, like cellulose, can be highly insoluble in alkali, and many early reports claimed synthesis of cellulose ("alkali-insoluble glucan") when in fact the product was callose (see 34 for review).

However, some important discoveries were made prior to 1980. Advances in microscopy allowed visualization of the cellulose microfibrils in the walls of algae and plants, and, hence, the realization that the patterns of cellulose deposition can often be highly ordered. In expanding plant cells, microfibrils are usually deposited in arrays transverse to the direction of cell expansion, leading to the concept that whatever controls the pattern of microfibril deposition plays a key role in determining directions of cell expansion. In the thick cell walls of many cellulosic algae and secondary walls of higher plant cells, microfibril deposition often occurs in layers that alternate in direction, thus creating walls of great strength. Following the discovery of microtubules by Ledbetter & Porter (78) were numerous papers showing that the orientation of the cortical microtubule network often paralleled that of the most recently deposited cellulose microfibrils.

In addition, Preston (99) proposed the "ordered granule hypothesis" that envisioned cellulose synthases as multisubunit complexes in which an ordered array of catalytic subunits would function together to synthesize glucan chains that would then self-assemble to form a microfibril. His vision was realized as being most likely true with the discovery in the alga Oocystis by Brown & Montezinos (17) of so-called linear terminal complexes (TCs)—multisubunit arrays at the ends of microfibrils that were seen using the new technique of freeze-fracture of the plasma membrane. This was followed by the finding in plants of the first smaller, hexagonal structure called a rosette (89), now believed to represent the type of synthase found in higher plants and some algae. Until 1982, no breakthroughs were achieved in in vitro synthesis of cellulose; thus, no synthase was purified, nor genes identified that encode the synthase or any other important proteins involved in the process.

Recent Advances in the Understanding of Cellulose Structure

Knowledge of the structure of cellulose is critical for understanding the mechanism of biosynthesis. The confusion and controversy in this field closely parallels that in the work on biosynthesis. Perhaps the most important "argument" has concerned the directionality of the glucan chains within the microfibril (for review see 43). By 1970, techniques for determining the structure of polymers using X-ray crystallography had advanced considerably, and during the 1980s much effort was devoted to determining whether all the chains in one microfibril of native cellulose were parallel (all reducing ends pointing in the same direction) or antiparallel. Although some native polymers like DNA have antiparallel structures, this necessitates a complicated mechanism for synthesis; the simplest mode of microfibril synthesis possible would be one in which all chains are synthesized by the same mechanism, thus leading to a parallel structure. Indeed, Gardner & Blackwell (45) and Sarko & Muggli (107) both proposed such a model for the cellulose I structure characteristic of native cellulose, but this result continued to be questioned. Furthermore, it was recognized that mercerization of native celluloses, a process that involves swelling, but not total dissolution, in alkali, leads to a conversion to the more stable cellulose II structure for which an antiparallel structure had most often been proposed. How could the chains re-orient from parallel to antiparallel without going into solution? The problem in interpretation was that, from the diffraction patterns obtained at that time, either parallel or antiparallel structures could be plausibly proposed.

A major breakthrough came when Atalla & VanderHart (8, 130) used the technique of solid-state NMR to provide evidence that two allomorphs are present in native crystalline cellulose. The Iα allomorph predominates in the cellulose derived from bacteria and many algae, whereas the Iβ allomorph predominates in cellulose derived from plants. However, since all native cellulose is a mixture of the two forms, it is difficult to conclude that these are synthesized by different synthetic mechanisms. This finding was taken further by advances in the use of electron diffraction by Sugiyama et al (124), who carried out studies with the large microfibrils of the alga *Microdictyon tenuius*. Scanning along the length of such a huge microfibril, these workers found regions that were essentially pure Iα, other regions of pure Iβ, and regions in which a mixture occurred. These results demonstrated that all the early crystallographic studies had been complicated by the fact that diffraction patterns were derived from samples containing a mixture of these two allomorphs. When techniques allowed analyses of the individual forms, the model invoking parallel orientation of chains became increasingly accepted, with a triclinic and a monoclinic unit cell being proposed for Iα and Iβ, respectively (124). Since the triclinic unit cell has only one chain, this necessitates a parallel model, which has been further confirmed by other techniques. If all the reducing groups are in the same end of the microfibril, they should be identifiable by chemical techniques such as silver labeling, and this in fact was observed by Hieta et al (60). In addition, Chanzy & Henrissat (27) showed that only one end of the large microfibril of the alga Valonia becomes frayed and pointed following digestion with a cellobiohydrolase, an enzyme that degrades only from the nonreducing end.

It is still not entirely clear what occurs in the conversion of cellulose I to cellulose II. Cellulose II can, in rare cases, be made in nature; two examples are that of the marine alga *Halicystis* (117) and the gram-positive bacterium *Sarcina* (22). Other cases relate to conditions where the normal process is perturbed, either by mutation (77, 111) or by addition of the dye Tinopal, which alters crystal structure and may cause chain folding that could lead to an antiparallel structure (30). Thus, folding of chains, at least after some chemical treatments leading to conversion of native cellulose I to II, cannot be excluded (28, 30). Another potential mechanism might be that, if individual microfibrils are synthesized with parallel orientation, but adjacent synthases move in different directions [a case suggested by images in Oocystis (100)], when such a native cellulose mixture is swollen in alkali, these oppositely oriented microfibrils could interdigitate with each other to create a partially antiparallel structure.

In sum, the normal process of synthesis of native cellulose I seems to result in production of parallel chains. But from which direction do they grow? Recently, Koyama et al (75) first determined which of the proposed packing models for cellulose I was correct. In one model, "parallel up," the reducing groups are oriented "up" with respect to the crystallographic c axis, and "down" in the "parallel down" model. Using the large microfibrils of the alga *Clado-phera* that had been chemically treated to convert all the cellulose to the single Iβ allomorph and subjecting them to electron diffraction, Koyama et al easily distinguished directionality by changes in diffraction pattern when a microfibril was pointed in one direction or the other by tilting the plane in which they are viewed. They also used silver labeling and cellobiohydrolase digestions to distinguish which was the reducing end, and showed by tilting experiments that the "parallel up" model was correct. They then examined a ribbon of cellulose (a composite of microfibrils) that extended outward and was still attached to the site of synthesis on the surface of an *Acetobacter aceti* cell. By tilting the specimen, they deduced that the ribbon was synthesized with its c axis outward from the cell, leading to the conclusion that the reducing ends point away from the site of synthesis—and, therefore, glucan addition must occur from the nonreducing end. This conclusion, if correct, has important implications for the mechanism of cellulose synthesis since conventional synthesis of polysaccharides such as starch and glycogen that involve direct polymerization from a nucleotide sugar substrate involve growth from the nonreducing end. However, more complex mechanisms such as those involving lipid-linked intermediates, often occur by growth from the reducing end.

The interaction between the chains within a microfibril is usually attributed to both intra- and interchain hydrogen bonding. However, this may represent only a minor factor in the interaction. In current models, only one interchain H-bond per glc residue is predicted, and most of the energy comes from van der Walls forces (43, 44). Cousins & Brown (16, 30) proposed that the initial stage of cellulose I crystallization occurs through these latter forces to generate glucan chain sheets that further associate to form the crystalline fibril.

Other issues relating to structure are also very relevant to biosynthesis, including the questions of what controls chain length, crystallinity, and microfibril size, but discussion of these is deferred until later in the review since further information regarding synthase structure is helpful for understanding current thinking on these questions.

Acetobacter xylinum *as a Model Organism*

In 1982, Benziman et al obtained rates of in vitro synthesis of cellulose that approached those observed in vivo in *A. xylinum* (2). To obtain these high rates, membranes had to be prepared in the presence of polyethylene glycol (PEG) and supplied with GTP. In a series of landmark papers, Benziman's group showed that PEG was precipitating a soluble enzyme, a diguanylate cyclase, that converted GTP to a unique activator of the cellulose synthase—cyclic diguanylic acid (c-di-GMP) (104; reviewed in 103, 132).

The discovery of c-di-GMP paved the way for subsequent purification of the synthase by a technique called product entrapment. Kang et al (67) had found that when detergent-solubilized chitin synthase from yeast was incubated with substrate, the enzyme remained tightly associated with the chitin product and could be effectively purified by centrifugation of the enzyme with the product. This technique worked well for the *A. xylinum* synthase, and two groups (81, 85) obtained a synthase preparation of relatively high purity. A prominent polypeptide of 83-kD was shown to bind the substrate UDP-glc (82) and has been accepted to be the catalytic subunit. Using sequence information derived from this polypeptide, Brown's laboratory (112) isolated a gene from *A. xylinum*, *AcsA*, that encodes the catalytic subunit. In 1990, Wong et al (137) used the genetic approach of complementation of a cellulose-deficient mutant of *A. xylinum* to isolate an operon of four genes called *BcsA–D*. Analyses of mutant strains demonstrated that genes *A-C* are essential for cellulose synthesis. The *BcsA* gene isolated by Wong et al shows high homology to the *AcsA* gene isolated by Brown's group and therefore also encodes the catalytic subunit.

The groups of Brown and Benziman have used different names for genes involved in cellulose synthesis. Matthysse's group subsequently adopted the term *CelA* for the catalytic subunit for their work on Agrobacterium (83, 84), a name now carried over for the homolog of this gene in higher plants (see later discussion). *CelA* is also unfortunate, as cellulases are named using the *Cel* designation. I now propose that adoption of the term *CeS* for "Cellulose synthase," preceded by genus and species designations such as Ax for *Acetobacter xylinum*, or Gh for *Gossypium hirsutum* (See Table 1). If the gene is followed by an "A," this would designate that it is a homolog of the gene encoding the bacterial catalytic subunit. Thus, for *A. xylinum*, the previously named *BcsA* or *AscA* would become *AxCeSA*, and that from cotton would be *GhCeSA*. Since *Agrobacterium tumefaciens* and *Arabidopsis thaliana* would

Table 1   Proposed nomenclature for genes involved in cellulose synthesis[a]

Organism	Old name	New name	Ref.	Accession no.
Acetobacter xylinum	BcsA	AxCeSA-1	137	M37202[b]
	AcsA	AxCeSA-2	112	X54676[b]
	acsAII	AxCeSA-3	109	U15957
Agrobacterium tumefaciens	CelA	AgtCeSA-1	83, 84	L38609
Gossypium hirsutum	CelA-1	GhCeSA-1	96	U58283
	CelA-2, pcsA-2	GhCeSA-2	64, 96	U58284
Arabidopsis thaliana	Rsw1	AtCeSA-1	7	AF027172
	Ath-A	AtCeSA-2	7	AF027173
	Ath-B	AtCeSA-3	7	AF027174
	Irx3	AtCeSA-4	119, 128	None yet
	p1 clone MRH10.9	AtCeSA-5	62	AB006703
	AraxCelA	AtCeSA-6	137a	AF062485
	BAC F25I18	AtCeSA-7 [c]	—	AC002334
Oryza sativa	Est D39394	OsCeSA-1	7	AF030052
	Est D48636	OsCeSA-2	96	None yet
Populus alba X *P. tremula (hybrid)*	PxCell	Pa/tCeSA-1	—	AF081534

[a]For the present time, for the plant genes, this proposed nomenclature will cover only those genes with domain structures that closely resemble those outlined in Figure 2 and referred to as the *CeSA* genes. The following rules were applied for determining these names: First letters for Genus and species, followed by *CeS* (for cellulose synthase), followed by *A* (presumed catalytic subunit), followed by a number that is determined by the timing of the first report of the gene. A mutant form of the gene would be written *ces*, as opposed to *CeS*. If future research indicates some of these may not encode a catalytic subunit of cellulose synthase, then they should be renamed according to their function. The term "A" should be reserved only for genes encoding a catalytic subunit that catalyzes chain elongation. Since B, C, and D have been adopted for *A. xylinum* for other genes involved in cellulose synthesis, if homologs are found in plants for these genes, they could also use *CeS* followed by these letter designations. If genes with yet different functions are found relating to cellulose synthesis, they might use *CeS* followed by other letters. Until the role of genes and gene products for other related glycosyltransferases that lack these plant-specific domains are characterized, I suggest continuing to use the designation of Cutler & Somerville (31) (*Csl*) for these genes. Once a function is assigned, they could be named upon that basis, e.g. a callose synthase catalytic subunit gene (that is not also a cellulose synthase) would be named *CaSA*. The data contained in this table at present do not include ESTs unless they are near full-length cDNAs and have been completely sequenced. This Table is being continually corrected and updated as new sequences emerge, and it can be viewed on the Web Page for the author: (http://www-plb.ucdavis.edu/Faculty/Delmer/delmer.html).

[b]Since both these genes were identified as being the first gene in a cellulose synthase operon, they are apparently allelic genes but, because they come from different strains and show significant sequence differences, it seems preferable to give them separate names. The *AxCeSA-3* is definitely found on a different position of the genome (109).

[c]The assignment of this gene to the *CeSA* subfamily is still tentative since the domain structures for the encoded protein from this gene are mostly similar to all other CeSAs, but this protein would lack the zinc fingers in the N-terminal.

both have the designation At, I suggest adding the additional letter "g" for *A. tumefaciens*, thus giving such a gene a designation *AgtCeSA* in this organism. These designations are used throughout this review.

The "B" gene, or *AxCeSB*, the second gene in the operon, is now presumed to encode a regulatory subunit that binds c-di-GMP, although the evidence for this is not completely firm. In addition to the 83-kDa polypeptide, another polypeptide of 90 kDa co-purifies with the synthase in *A. xylinum* (85). Although sequencing of this polypeptide supports the notion that it is encoded by *AxCeSB*, the gene product does not bind c-di-GMP. However, a 67-kDa polypeptide, also found in the purified synthase preparation, did bind c-di-GMP, and this was presumed to be derived from the 90-kDa polypeptide since it cross-reacted to antibodies prepared against the larger polypeptide. However, it is unclear why the 90-kDa polypeptide does not bind c-di-GMP, and the relationship between the two is based only upon immunological cross-reactivity. One strain of *A. xylinum* used by Brown's group has the *AxCeSA* and *B* genes fused into a single open reading frame, but the protein is presumably cleaved to give the A and B proteins after translation (111).

Recently, Weinhouse et al (133) discovered a new c-di-GMP-binding protein distinct from the *AxCeSB* gene product in *A. xylinum*. They provide evidence that this protein also loosely associates with the synthase complex and binds c-di-GMP with high affinity (Kd = 20 nM). Binding is favored in the presence of K^+, and the protein has been suggested to play a key role in modulating the intracellular concentration of free c-di-GMP. Given the lingering uncertainties about whether the *AxCeSB* gene product binds c-di-GMP, it is unclear whether this new binding protein may uniquely play the role of c-di-GMP binding in the complex or whether it concentrates and channels it to the B gene product. The gene encoding this new protein has not yet been identified, nor has the molecular weight of the individual polypeptide been reported, although the molecular weight in the native state is about 200-kDa (133). The 67-kDa polypeptide that bound c-di-GMP in previous studies and was presumed to be a degradation product of the 90-kD *AxCeSB* gene product could instead be this new c-di-GMP binding protein, but this remains to be established.

The intracellular level of c-di-GMP in *A. xylinum* can also be controlled by the levels of the diguanylate cyclase (dgc), which catalyzes its formation, and by phosphodiesterase A (pdeA), which catalyzes its degradation (103, 132). Genes encoding these enzymes have recently been isolated and found to be located on three distinct yet highly homologous cdg operons (125). Within each *cdg* operon, a *pdeA* gene lies upstream of a *dgc* gene. The *cdg1* operon contains two additional flanking genes, *cdg1a*, which encodes a putative transcriptional activator of a type regulated by oxygen in other bacteria, and *cdg1d*, which encodes a protein of unknown function. It is interesting that the *N* termini of the deduced dgc and pdeA proteins contain putative oxygen-sensing domains. The fact that *A. xylinum* is a non-motile strict aerobe has often led to the suggestion that it synthesizes cellulose in order to float on the surface of liquid substrates to gain better access to oxygen. These new results also suggest that oxygen tension may indeed regulate the process. Genetic disruption of the *dgc* genes markedly reduced cellulose production in vivo, again demonstrating that c-di-GMP is important for the process.

The function of the protein encoded by the *AxCeSC* gene is not clear, but it bears some sequence homology to a pore-forming protein. Saxena et al (111) suggested it might encode a protein that forms a pore in the outer membrane of this gram-negative organism through which the cellulose must be secreted. There do not appear to be any homologs of the *AxCeSC* gene in the databanks of plant gene sequences, and one would not expect such a protein to be necessary in plants that lack an outer membrane.

Studies with the *AxCeSD* gene suggest an extremely interesting role in cellulose synthesis. When this gene is disrupted, *A. xylinum* still makes cellulose, but the mutant cells, when agitated, make cellulose II instead of cellulose I (111). This suggests that this gene product may play a role in crystallization of the microfibrils by assisting in aligning the glucan chains so that they can interact and crystallize to form the thermodynamically less stable cellulose I. To date, homologs of this gene have not been found in databanks of sequenced plant genes, and it is not clear if plants achieve crystallization of cellulose I using another mechanism.

Studies with *A. xylinum* have proven useful in other ways for studying how glucan chains become ordered and crystallized into microfibrils. Optical brighteners, called by various names such as Tinopal or Calcofluor, can bind to glucan chains, and, when added to the medium of cells synthesizing cellulose, produce interesting effects. First, they speed the rate of glucan polymerization while at the same time lowering the crystallinity of the cellulose produced (reviewed in 53). This implies that the rate of cellulose synthesis might be limited by the rate of crystallization of the glucan chains; and that there is a temporal gap, however brief, between the act of polymerization and that of crystallization. The type of effects on crystallization depend upon the concentration of the dye; at low concentrations, minicrystals are formed but these do not aggregate into true microfibrils; at higher concentrations, only sheets of glucan chains are produced (30, 53).

Two other genes found upstream of the *AxCeSA-D* operon also appear to be important for cellulose synthesis (121). The first of these is a gene encoding an enzyme with cellulase activity. This gene seems to be present only in strains of Acetobacter that synthesize cellulose, and may indeed play a role in the process (discussed below). The second gene encodes a proline-rich protein of unknown function, but disruption of this gene leads to loss of cellulose synthesis in vivo but not in vitro.

Cellulosic Algae—The Under-Exploited Superstars of Cellulose Synthesis

Cellulosic algae have evolved mechanisms to make the largest and most highly crystalline cellulosic microfibrils. *Valonia*, for example, synthesizes a highly crystalline microfibril of more than 1000 glucan chains. If the ordered-granule hypothesis holds for such a case, this synthase complex would have more than 1000 catalytic subunits! This is an amazing example of macromolecular assembly, and understanding the nature of the subunit interactions, how such a complex could be assembled (a chaperonin might be involved?), stabilized, and transported to the plasma membrane should be a fascinating question for future research. Little work has been done to date at the biochemical and molecular level with these organisms, because many are difficult to grow in large quantities and lack defined genetic systems. In addition to *Valonia*, other algae such as *Oocystis, Erythrocladia, Vaucheria, Boergesenia, Cladophera, Boodlea*, and *Halocynthia* all make similarly large microfibrils, ranging in size from about 100 up to more than 1000 chains per microfibril (65, 127). The highly ordered and very large linear TCs found in these algae vary in shapes and sizes that parallel the varying size and shape of the resultant microfibrils that they synthesize (reviewed in 16).

In addition to the large size of the microfibril, cellulose from these algae is characterized by its high crystallinity and high degree of polymerization (DP) of the glucan chains [e.g. in *Boergesinia*, a DP of up to 23,000 has been measured, one of the highest reported (16)]. The crystalline nature of a microfibril of Valonia is so nearly perfect that is has been characterized as one large single crystal (123).

Possible Roles for Cellulases in Cellulose Synthesis

In comparing cellulose from different sources, there seems to be good correlation between the degree of crystallinity and the DP of the chains. Chain termination, about which little is known, might be caused when one catalytic subunit gets out of step with its many other partners. Indeed, it is difficult to envision the mechanism by which a huge Valonia complex starts or stops making a microfibril, or how all the subunits function together. The rate of synthesis might be limited by the slowest subunit in the complex, since rates of synthesis in vivo go up in the presence of Tinopal in *A. xylinum* when chain interactions are minimized (11). In terms of what controls DP, the presence of microfibrils much longer than predicted by the DP of single chains implies that chain initiation and termination occurs many times during the synthesis of one microfibril. It can be hypothesized that a possible "editor/chain terminator" in the system recognizes a region that is disordered or under strain due to faulty catalytic subunit function, cleaves the chain, releases tension, and allows that subunit to get a fresh start. If the overall design of the complexes is highly ordered, as in the algae with linear TCs, then the editor/terminator would function less frequently, leading to a higher DP of the chains within the microfibril.

What could serve as such an editor/terminator? One possibility is a membrane-associated cellulase. In *A. xylinum* and *A. tumefaciens*, which make cellulose, there is a gene for cellulase either upstream (121) or within (83, 84) an operon of genes required for cellulose synthesis. In addition, a highly unusual cellulase with one membrane-spanning region that localizes the protein both to Golgi and the plasma membrane has been recently characterized in tomato (19). In Arabidopsis, a similar gene was isolated that, when mutated, leads to a highly distorted phenotype reminiscent of another mutant (*rsw1*, see later discussion) that is defective in cellulose synthesis (92). In the bacteria, these are cellulases that prefer carboxymethylcellulose as substrate, making them candidates for cleavage of β-1,4-linkages in noncrystalline regions. Coupled with their potential location in the periplasmic space, these could be candidates for editor/terminator. Chapple & Carpita (29) recently suggested a similar role for these enzymes. A more elaborate speculation is that an endo-1,3-β-glucanase might also be important to correct possible errors by the synthase in synthesizing the 1,3-linkage as opposed to the 1,4-linkage, a possibility to be discussed in more detail later in the review.

An alternative role for such a cellulase has been proposed by Matthysse et al (68, 83, 84). Using genetic and biochemical approaches, they formulated a model for this organism in which cellulose is synthesized by transfer of glc from UDP-glc, to form lipid-linked cello-oligosaccharides; final polymerization is then catalyzed by transglycosylation of cellobiose or larger units from the lipids through the action of the extracellular cellulase. An early step in this proposed pathway involves a homolog of *AxCeSA*, called *CelA* by these authors, and referred to herein by the new convention as *AgtCeSA*. The proposed mechanism in *A. tumefaciens* for this gene product involves the specific step of transfer of glc from UDP-glc to a lipid-linked glc to form lipid-linked cellobiose and/or higher-oligosaccharides. Given the relatively high homology between *AxCeSA* and *AgtCeSA*, it is unlikely that these two gene products could catalyze very different reactions, either to directly polymerize glucan as in *A. xylinum* or to synthesize a lipid intermediate. Furthermore, these authors (83, 84) stated that their cellulose synthase was not activated in vitro by c-di-GMP, whereas two other reports suggest that this compound does activate the *A. tumefaciens* enzyme (3, 126). Additional data are needed before this novel mechanism can be further proven.

Rosettes as Synthase Complexes in Plants

Hexagonal rosette structures are now believed to represent the sites of cellulose synthesis in mosses, ferns, and higher plants. In plants, most microfibrils are much smaller than in the algae, existing mostly in the classic form of "elementary fibrils" that contain about 36 glucan chains. The sixfold symmetry of the rosettes suggests that each large "subunit" seen by freeze-fracture would thus contain within it six *CeSA* catalytic subunits. In fact, Haigler (C Haigler, unpublished data) (Figure 1) recently detected more detailed substructure within the rosette that would support such a notion. An extreme lability to these structures has been noted—a finding that parallels the inability to preserve high rates of cellulose synthesis in vitro. Some algae such as *Micrasterias* also possess rosettes. In this case, the large microfibril size is created by an aggregation of rosettes into higher structures (48). Organisms with linear rows of synthases such as *A. xylinum*, or linear TCs as in most algae, tend to synthesize a much higher proportion of the Iα allomorph of cellulose, whereas organisms containing rosettes synthesize much more Iβ (see 43). However, since both forms are synthesized in each case, one cannot say that one type of complex is responsible for one type of allomorph. The Iα form might be more highly ordered and the CeSA subunits more highly ordered in linear TCs, which favor synthesis of this form.

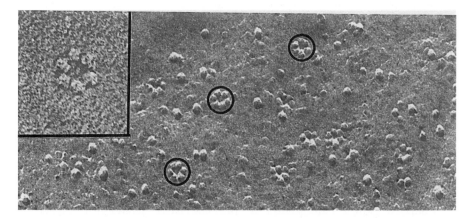

Figure 1 Freeze fracture replicas of rosettes associated with cellulose microfibril biogenesis. The rosettes after the fracture event exist in the leaflet of the plasma membrane bilayer that is nearest the cytoplasm (the PF face). In the main micrograph, several rosettes are shown (three surrounded by circles) in the plasma membrane of a differentiating tracheary element of *Zinnia elegans*; differentiating tracheary elements deposit abundant cellulose into patterned secondary wall thickenings. The inset shows one rosette at higher magnification and after high resolution rotary shadowing at ultracold temperature with a minimum amount of platinum/carbon. (Main micrograph, 222,000 x; inset, 504,545 x; both micrographs courtesy of Mark J Grimson and Candace H Haigler, Department of Biological Sciences, Texas Tech University, Lubbock, Texas.)

Genes at Last—A New Era for Studies on Cellulose Biosynthesis in Plants

Until recently, it was doubtful whether the significant progress made with *A. xylinum* could be applied to the more difficult questions involving cellulose synthesis in plants. Despite high rates of synthesis of cellulose in vitro with *A. xylinum*, similar activities in plasma membrane preparations derived from higher plants could not be demonstrated. The key in *A. xylinum* had been the discovery of c-di-GMP as an activator of the enzyme, and it was logical to test its effects in plants. However, to date, its effect as an activator of a plant glucan synthase has not been confirmed, although there are some indications that c-di-GMP may exist and function in plants. Mayer et al (85) presented data indicating that an antibody against the diguanyl cyclase of *A. xylinum* cross-reacted with a polypeptide of similar molecular weight in several plant species. Amor et al (5) showed that several polypeptides derived from cotton fiber membranes bound c-di-GMP with good specificity and affinity, although we have been unable to convincingly purify or characterize these proteins in subsequent studies. More recently, in a collaboration with Benziman, a novel triterpenoid saponin was purified from plants and shown to be a potent inhibitor of the diguanyl cyclase of *A. xylinum* (93, 94), but the exact function of the molecule in plants is unknown. However, the lack of effect of c-di-GMP on glucan synthase activity, coupled with the inability to detect convincing homologs of the diguanyl cyclase or *AxCeSB* genes in the databanks of plant gene sequences, cast doubt on the role of this compound in cellulose synthesis in plants. Here again, further work is warranted.

Once the *AxCeSA-D* operon was identified, at least four different groups tried unsuccessfully to screen cDNA libraries from plants to search for homologous genes, using the *AxCeSA* gene as probe. The genes were either vastly different in sequence or showed homology in only limited, conserved regions, as yet unidentified. The latter was determined to be true, but discovering this required further understanding of conserved motifs in these types of glycosyltransferases. In 1995, our group (36) and Saxena et al (111) identified a sequence in the *AxCeSA* genes that appeared to be roughly conserved in a number of glycosyltransferases that use UDP-glc or UDP-GlcNAc as substrate. Later, Saxena et al (110) used the technique of hydrophobic cluster analysis (HCA) to search for conserved motifs in glycosyltransferases that catalyze β-glycosyltransfer using either UDP-glc or similar nucleotide sugars as substrates. This technique, used very successfully in analyses of relationships among glycosylhydrolases, combines comparison of linear amino acid sequences with that of the protein secondary structures statistically centered on hydrophobic clusters (20). It is particularly useful for identifying relationships between proteins that share common functions and three-dimensional shapes, but that lack significant conservation at the primary amino acid sequence level. For this reason, it is also particularly useful in analysis of the β-glycosyltransferases (21, 46, 110). HCA allowed identification of two domains, A and B, found in a number of processive (synthesize polymers as opposed to single sugar additions) glycosyltransferases using UDP-sugars as substrates (110). Within domain A in particular, regions surrounding two conserved aspartate (D) residues were highly conserved, whereas in domain B, a region surrounding another D residue was also conserved, as was a QXXRW motif farther downstream. These regions were conserved in all the bacterial *CeSA* genes, as well as in other related enzymes such as chitin and hyaluronan synthases.

At about this same time, our group entered into collaboration with Stalker and colleagues at Calgene, Inc. to sequence random cDNA clones from our cotton fiber cDNA library. Two cDNA clones (*CelA-1* and *CelA-2*) were identified that, in preliminary analyses, showed some limited homology to the bacterial *CeSA* genes (96; see also 54, 64). Upon further analysis, it became clear that all the conserved regions surrounding the D,D,D and QXXRW motifs were highly conserved; however, surprisingly, plant-specific insertions of additional sequence separated these domains from each other compared to the *CeSA* genes of bacteria (see domain structures for the plant *CeSA*s in Figure 2, discussed in more detail later). The presence of these extra domains could explain why

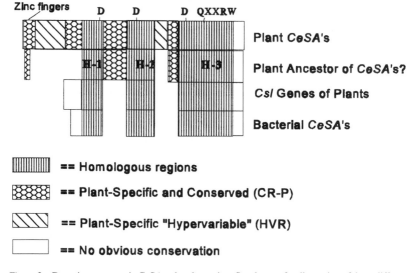

Figure 2   Domain structures in CeSA-related proteins. See the text for discussion of these different families of related proteins.

full-length *AxCeSA* probes could not be used successfully to isolate these genes previously. From hydropathy plots, the deduced proteins were predicted not to have a signal sequence but to have two transmembrane helices (TMHs) in the *N*-terminal region and six in the *C*-terminal region, characteristics similar to those predicted for the bacterial CeSA proteins. Such a prediction would place the entire central region comprised of all the domains surrounding the D,D,D, and QXXRW motifs plus two plant-specific domains within the cytoplasm, consistent with the notion that these synthases bind the substrate UDP-glc and carry out catalysis on the cytoplasmic face of the enzyme (see also Figure 3). We subcloned this central region, and the recombinant protein, when purified, subjected to SDS-PAGE, transferred to nitrocellulose, and renatured, bound the substrate UDP-glc (96). When the domain surrounding the first D was deleted, this recombinant protein no longer bound the substrate as might be predicted if these regions were involved in substrate binding and catalysis. Furthermore, levels of mRNA derived from these genes (by the new convention now referred to as *GhCeSA-1* and *GhCeSA-2*) began to be highly expressed in the fibers identified just at the time the rate of cellulose synthesis increased over 100-fold in vivo as fibers entered secondary wall synthesis (96). In cotton fibers at this stage of development, only two glucans are synthesized—some callose associated with the transition stage to secondary wall synthesis and massive amounts of cellulose. The expression patterns of these genes closely matched the patterns of cellulose, but not callose deposition. Furthermore, the binding studies showed that the recombinant protein bound UDP-glc in a Mg^{2+}-dependent manner, whereas the callose synthase of cotton fibers shows markedly increased affinity for UDP-glc in the presence of Ca^{2+} (56). Taken together, these results strongly support the notion that these two genes encode similar forms of the catalytic subunit of the cellulose synthase that function in secondary wall synthesis in cotton fibers. However, in that initial work, no proof of function in vivo was offered.

Strong genetic evidence that this class of genes encodes the catalytic subunit came recently from the work of Arioli et al (7; see also discussion of this work in 25, 35). This group had previously identified in Arabidopsis a set of mutants (designated *rsw1–5* for radial swelling) at five different alleles that all showed root tip swelling at high, but not low, temperature (10). Using a map-based cloning strategy, the gene mapping to the *Rsw1* (by the new convention used here, *AtCeSA-1* locus) was recently identified and shown to be a close homolog of the *GhCeSA* genes. The genomic sequence indicated that the gene, which maps to chromosome 4, contained 13 introns. Complementation of the mutant with this gene restored the wild-type phenotype. The gene appears to be expressed in a large number of tissues engaged in primary wall cellulose synthesis, and growth at high temperature results in a drastic phenotype in which many cells of the seedling are swollen, a situation so severe that the plants die within days. However, no phenotype is observed at the low temperature. The mutation is only expressed in the homozygous recessive state, a somewhat surprising result. If many CeSA subunits come together to form a rosette, a mixture of wild-type and mutant subunits might be predicted to disturb the process in the heterozygous state as well. Nevertheless, in the homozygous recessive mutant, the rosette structures observed by freeze fracture in the homozygous recessive mutant fall apart and eventually disappear. This is perhaps the best evidence to date that these structures are indeed involved in cellulose synthesis.

One additional finding by Arioli et al (7) was that the *AtcesA-1 (rsw-1)* mutant at high temperature accumulates a noncrystalline form of glucan concomitant with reduction in synthesis of crystalline cellulose. [Crystalline cellulose here is defined as that which resists digestion by the often-used "acetic-nitric reagent" as a treatment that destroys all wall polymers except the crystalline regions of cellulose (129)]. The glucan accumulated was claimed to be β-1,4-linked based upon the migration of the permethylated derivative with a 4-linked glc standard in GLC, although the solubility of this glucan in ammonium oxalate or KOH is not consistent with its being a β-1,4-glucan of substantial DP, since cellodextrins of DP > 7 are not soluble in aqueous solvents. This glucan may show unusual properties because it is linked to protein—perhaps either to the mutant cesA subunit itself or to a different primer protein. (No evidence currently exists for or against a role for a primer in cellulose synthesis.) Assuming that the glucan is 4-linked, Arioli et al (7) proposed that the nature of the *AtcesA-1 (rsw1)* mutation, which involves a single amino acid substitution of an alanine for a valine at position 549, causes the mutant cesA subunits to fall apart at high temperature, and thus prevents close association of the glucan chains, which cannot then crystallize properly.

A very interesting correlation to the above work is emerging in some recent studies being carried out at Novartis Crop Protection AG in Basel with an experimental thiatriazine herbicide called CGA 325'615 (K Kreuz, A Stoller, unpublished data). This herbicide is a very potent and specific inhibitor of cellulose synthesis in vivo with an IC50 in the nanomolar range. Recently, this group in collaboration with Herth (W Herth, unpublished data) showed that rosettes fall apart in Arabidopsis very quickly after application of this compound. CGA325'615 also causes an accumulation in cotton fibers of a noncrystalline glucan, at least some of which appears to be linked to protein (97). Although considerable confirming work on the glucan structure is still needed, taken together, these studies and those of Arioli et al (7) both suggest that CeSA subunits may be able to function in isolation to make a noncrystalline form of cellulose, and that the organization of the rosette dictates final assembly into crystalline microfibrils. A similar accumulation of such a glucan was not found in fibers treated with 2,6-dichlorobenzonitrile (DCB). DCB is another inhibitor of cellulose synthesis in plants (115, 131) or of other polysaccharides in some algae (6). These two herbicides may therefore inhibit cellulose synthesis by different mechanisms. Regarding DCB mode of action, an 18-kDa polypeptide was identified in cotton fibers that binds DCB (37) but its function is not clear. Very interesting are the recent results of Nakagawa & Sakurai (90) showing that an antibody against a conserved plant CeSA epitope was able to detect a CeSA protein in membrane fractions of tobacco BY-2 cells only after they were treated with DCB, suggesting that DCB may cause an elevation in the steady-state level of the catalytic subunit. Other inhibitors of cellulose synthesis have been reported, such as Isoxaben (58), phthoxazolin (95), and triazofenamide (57), although the mode of action of these compounds is not understood.

Three other mutants of Arabidopsis, called *irx1–3* by Turner & Somerville (128), have a phenotype that shows collapsed xylem vessels in the inflorescence stem. Chemical analyses indicated that these stems, particularly in *irx3*, show markedly reduced content of crystalline cellulose. The gene encoded by the *Irx3* locus has recently been identified and used to complement the mutant phenotype. When sequenced, the *Irx-3* gene (or by the new convention, *AtCeSA-5*; see Table 1) is a close homolog of the *GhCeSA-1 and -2* genes and the *AtCeSA-1* (*Rsw1*) gene, providing additional evidence for the function in vivo for this class of genes (119).

Strictly speaking, none of the studies to date on *CeSA* genes (e.g. 7, 96, 119) has definitely proven that the CeSA gene product is really the catalytic subunit of the synthase complex because no one has shown that the recombinant protein can catalyze the formation of β-1,4-glucan chains from UDP-glc. In the case of the studies with the *rsw1* and *irx3* mutant, one can argue that the data only show that the gene encodes a protein required for cellulose synthesis. In the case of the cotton *CeSA* genes, no genetic evidence was presented, the ability of the recombinant protein to bind the putative substrate at least was shown, and expression patterns for the mRNA are consistent with an important role in cellulose synthesis. The fact that all these genes possess motifs found in the bacterial catalytic subunit is also a strong argument in favor of a catalytic function involving glycosyl transfer. Nevertheless, other possibilities for function (e.g. a role in chain initiation as opposed to chain elongation or in synthesis of a lipid intermediate) still exist, and final proof of function in vivo still awaits further characterization.

Another mutant that does not deposit secondary wall cellulose in trichomes of Arabidopsis has been reported (98), but the nature of the mutated gene has not yet been determined. In addition, a mutant of barley with reduced cellulose content in the culm has been reported (73). Although the gene responsible for the mutation is not known, the fact that reduced cellulose content was associated with brittleness of the culm is interesting, and suggests that cellulose may play a key role in providing strength without conferring brittleness to plant tissues. In other studies with flax fibers, Girault et al (50) studied the rigidity of cellulose and matrix polymers by solid-state NMR and concluded that polymers in the adhesive matrix between microfibrils and/or cellulose layers play an important role in ensuring that cracks propagate along the matrix rather than across the fibers and in allowing flax fibers to approach the tensile strength of advanced synthetic fibers like carbon and Kevlar. The capability of engineering plants with altered cellulose levels will depend on studies such as these to provide clues to phenotypes that might be expected.

CeSA *Genes Comprise a Large Gene Family in Plants; Evidence for an Even Larger Superfamily of Glycosyltransferases*

The *CeSA* genes described above are all characterized by a domain structure, as outlined in Figure 2. Based upon comparisons of *CeSA* sequences available to date, these genes contain three domains (labeled H-1-3) showing good conservation of sequence with the bacterial *CeSA* genes, separated by other domains not found in the bacterial genes. Some of these extra domains are highly conserved in all the plant *CeSA* genes, whereas others surprisingly vary from gene to gene, and therefore are referred to as hypervariable regions. A survey of cDNA and genomic sequences from Arabidopsis and rice, coupled with the published sequences described above, indicates that the *CeSA* genes comprise a large family in these plants. In Arabidopsis there may be up to 10 *CeSA* genes, and a recent survey of ESTs indicates a similar number for corn (T Helentjaris, unpublished data). A phylogenetic comparison based upon amino acid sequence comparisons showed that these proteins can be further divided into 4–5 subclasses that are not based upon plant origin, but are largely distinguished by differences in the two hypervariable regions (N Holland, T Helentjaris, D Delmer, unpublished data).

A major issue being addressed now is why there are so many genes of this type. First, not all may function as catalytic subunits of cellulose synthase but might function in synthesis of other β-glycans. However, strong evidence exists for the *GhCeSA-1, AtCeSA-1* (*Rsw1*), and *AtCeSA-5* (*Irx3*) genes in the process of cellulose synthesis, which suggests that most, if not all, of these genes will share this common function, although this requires further study. If common function is indicated, the most obvious reason for so many genes is to regulate tissue-specific patterns of expression. Certainly, the *GhCeSA* genes are highly expressed in the fibers of cotton, whereas *AthCeSA-5* expression may be restricted to developing xylem. Our laboratory is examining the expression pattern for another Arabidopsis gene, *AthCeSA-4* (from genomic sequencing of P1 clone MRH10.9); its expression is also restricted to patches of developing vascular tissues, while our preliminary studies on one corn *CeSA* gene indicate expression in roots but not leaves (62). Thus, if this pattern holds true for most of these genes, they would resemble the situation for the plasma membrane proton ATPases, tubulin, and actin, other large gene families in Arabidopsis, the members of which show tissue-specific expression (55, 74, 87, 118).

In their surveys of Arabidopsis sequences, Cutler & Somerville (31) also noted another group of genes related to the *CeSA* genes, but lacking the plant-specific domains found in the *CeSA* genes. In this sense, they resemble more closely the bacterial *CeSA* genes, but, to date, there is no known function for these genes, and these authors have named them *Csl* genes for "cellulose-synthase-like" genes, and suggested they might function in synthesis of other related β-glycans (see Figure 2). Recently, from the Arabidopsis genomic sequencing of BAC T19F06, we have also found three other tandem sequences that are intermediate in structure between the plant *CeSA* and *Csl* genes in that they contain only small portions of some of the conserved plant-specific domains; we refer to these as the "CeSA Ancestor Genes" (see Figure 2), although their function and pattern of expression is still unknown.

Where Is the Gene for Callose Synthase?

One major unanswered question concerns the identity of the gene(s) in plants that encodes the catalytic subunit of the callose synthase. There might be two types of genes encoding this enzyme, one for the classic Ca^{2+}-activated form identified in a wide variety of tissues (36, 76) and a second found in pollen tubes that is not Ca^{2+}-dependent and activated by protease or detergent treatments in vitro (113). In other organisms, genes encoding two distinct β-1,3-glucan synthases have now been identified. In yeast and fungi, a gene encoding a large catalytic subunit of about 200–230 kDa has been identified that bears no relationship to the *CeSA* gene of either bacteria or plants (see 71). Shin & Brown (116) found a 170-kDa polypeptide among about five that were enriched in mung bean callose synthase preparations; sequencing of one tryptic peptide showed fairly good homology with the yeast catalytic subunit. Bacic's group (127a) also found a polypeptide of similar molecular weight in partially purified preparations of tobacco pollen tube callose synthase, although its sequence and function have not yet been determined. These preliminary findings hint at the existence in plants of a homolog of the yeast enzyme, but searches of the databanks of plant sequences have not yet convincingly uncovered such sequences. Of the *CeSA*-type genes compared to date, a recently isolated one that is highly expressed in tobacco pollen tube is the most different (38), and it will be important to determine the function of this gene product in vivo. Recently, a gene encoding a bacterial curdlan synthase (curdlan is also a β-1,3-glucan) has been identified (122). The gene was identified using a mutant unable to synthesize curdlan generated by insertion of a transposable element. The sequence of this gene, called *CrdS*, shows striking resemblance to the *CeSA* genes of bacteria and the *Csl* genes of plants, raising the possibility that one of the Csl genes of plants might encode a callose synthase. The concept that linkage specificity may be determined by slight modifications in amino acid composition around the active site (21, 79) may explain why genes of very similar sequence might catalyze different, but related, reactions.

Alternatively, *CeSA* genes might encode a catalytic subunit that is capable of synthesis of either cellulose or callose, depending upon its conformation. The idea that the cellulose and callose synthases might be the same enzyme was proposed many years ago (34, 66) but, lacking purified enzymes, no proof has been forthcoming for this hypothesis. However, the recent cloning of genes encoding hyaluronan synthases from bacteria and mammals showed that a single polypeptide chain can catalyze synthesis of both the β-1,3-glucuronyl and the β-1,4-GlcNAc residues that comprise the repeating unit of this polymer (39, 120). These enzymes are also clearly members of an ancient gene family that includes the *CeSA* genes, since they also contain the signature domains surrounding the D,D,D, and QXXRW residues. If a single polypeptide of this type can catalyze synthesis of two different linkages from two different substrates, the CeSA proteins might also be capable of a similar flexibility to make either cellulose or callose that might be modulated by changes in phosphorylation state, by a cation such as Ca^{2+} or, as in the case of lactose synthase, by the presence of a modifying protein (61). There can apparently be rapid modulation between the processes of cellulose and callose synthesis, two examples being upon wounding when callose synthesis is initiated and cellulose synthesis is presumably inhibited, and during the formation of the cell plate, which is characterized by deposition of callose in the early stages followed by synthesis of cellulose as the cell plates begin to fuse with the parental wall (105, 131). Perhaps also relevant is the observation that membrane preparations of *Dictyostelium discoideum*, an organism that makes cellulose but not callose in vivo, can catalyze the synthesis of both linkages in vitro, which suggests that the cellulose synthase might perform both functions (15). At this point, any or all of the possibilities for callose synthase genes exist, and the way is open now for studies to clarify these issues. The nature of the gene(s) encoding the Golgi-localized enzyme that encodes the β-1,3-β-1,4-glucan synthases of the graminaceous monocots (24) also awaits resolution.

Progress Also with Dictyostelium discoideum

This slime mold is an attractive model system for cellulose synthesis since so many developmental studies and genetic tools are available. Cellulose synthesis is highly regulated during development; it is found in the slime trail of slugs, in the sheath that surrounds the multicellular stages, in the stalk, and in the spore walls (12). Blanton showed that a cellulose synthase activity can be detected in vitro (15). Recently, a mutant incapable of synthesizing cellulose was isolated that had been generated by restriction enzyme mediated insertional (REMI) mutagenesis, allowing rapid cloning of the disrupted gene (14). This gene is highly similar to the bacterial *CeSA* sequences, containing the conserved 3 Ds and QXXRW signatures. (By analogy with the nomenclature proposed herein, this gene could be designated *DdCeSA-1*.) Given the fascinating patterns of cellulose expression and the powerful genetic tools available, further study with this system should provide more insights into the process in the future.

A Model for Topology of CeSA Proteins in the Plasma Membrane

Our current understanding of the mode of cellulose synthesis indicates that the substrate UDP-glc binds to an active site on the cytoplasmic face of the plasma membrane, whereas the glucan chain produced must traverse the membrane to be deposited within the cell wall in plants. This would fit with the predictions of TMH's in the deduced CeSA proteins that place the large central regions containing the putative substrate-binding and catalytic sites within the cytoplasm. Based upon these predictions, Figure 3 shows a tentative model for how a single CeSA subunit might be topologically oriented in the plasma membrane. The model predicts that the TMH's would interact to form a central channel through which the glucan could be secreted. Analysis of these TMHs reveals that many are predicted to be amphipathic, and therefore could be arranged such that the hydrophobic faces would interact with the lipids in the membrane, leaving the hydrophilic surfaces to face the central part of the pore structure. This model is strikingly similar to recent models of both proton and Ca^{2+} ATPases predicted by data generated from electron crystallography of two-dimensional crystal structures generated directly on electron microscope grids (9, 140; see also the review on membrane protein crystallization in this volume). This type of ATPase also shows a tendency to crystallize to form hexamers (26), a property expected for CeSA proteins that assemble into rosettes that have sixfold symmetry. Thus, genes encoding many of these channel-forming proteins may be derived from some very ancient and common ancestor.

Speculations on Mechanism of Catalysis

Recently, there has been much speculation concerning the nature of the active site in cellulose synthases. In the native, extended chains of cellulose, each residue is rotated approximately 180° with respect to its neighbor, raising the question of how a single active site could effect catalysis in two opposing orientations. Saxena et al (110) proposed a mechanism that involved a dual

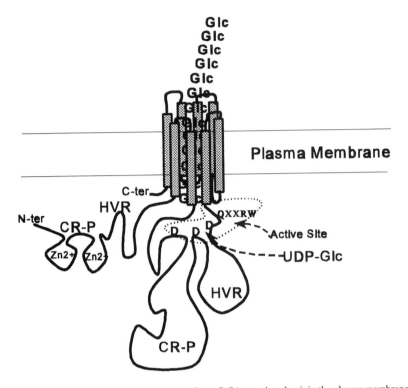

Figure 3 Hypothetical model for topology of one CeSA protein subunit in the plasma membrane of plants. The eight transmembrane helices are predicted to interact to form a pore through which the cellulose chain is secreted to the cell wall. The large central domain would fold in such a way as to bring together the conserved regions containing the 3-D residues and QXXRW motif that are believed to be important for substrate binding and catalysis. This would place the conserved (CR-P) and hypervariable (HVR) plant-specific regions also in the cytoplasm where they may serve to interact with other proteins. One CeSA subunit, such as the one shown here, must interact with other such subunits to form the synthase complex; it is not known how these interactions occur, although they might involve interactions between the transmembrane helices and/or some of the cytoplasmic domains.

addition of two UDP-glc molecules at two distinct active sites, with growth proceeding by addition at the reducing end. With the discovery of growth from the nonreducing end, Koyama et al (75) modified this model to accommodate growth from this direction. Similar models, including the possibility of transfer of cellobiose residues (from lipid- or protein-linked intermediates) have been put forward by Albersheim et al (1) and Carpita & Vergara (25). There is no evidence for duplicated active sites in the sequences characterized to date, so if these models were correct, synthesis would necessitate the cooperation of two adjacent CeSA subunits working together, an unlikely situation in view of the probable symmetry of subunits required for multiple subunit interactions in a rosette. Also, there can be as much as 120° rotation about the β-1,4-linkage (see 43), and thus all residues can probably be added by only one site in a particular orientation that can relax into the opposing orientations after they exit the catalytic region.

Interaction of CeSA Polypeptides with Other Proteins

Examination of the domain structure and predicted topology of CeSA proteins in Figures 2 and 3 raises questions about the possible function of the plant-specific domains. All of these would be predicted to face the cytoplasm and, as such, are excellent candidates to serve as domains that might interact with other proteins. The first conserved region in the N terminus surprisingly shows very high homology with several soybean transcription factors and contains two predicted zinc fingers; these most likely function as such since a recombinant protein of this region binds zinc [for more extensive discussion of this region, see (68, 69)]. This domain resembles the LIM or RING domains found in a number of animal proteins that seem to be involved in protein-protein interactions (e.g. see 106, 108). To date, we have not found significant homologies with other known proteins in the other plant-specific domains that are conserved nor with those that are hypervariable. The HVR domains suggest interactions that might be based upon the tissue-specific expression of these genes. The CeSA subunits presumably need to interact with each other to form synthase complexes, and one or more of these domains might serve this function, although interaction through the TMHs of the protein is another possibility. Several laboratories including our own are actively seeking to identify interacting partners for these domains but, to date, none has been conclusively identified.

Two obvious candidates for interaction with CeSA are the cytoskeleton or sucrose synthase (E.C. 2.4.1.13; reaction catalyzed is suc + UDP ↔ UDP-glc + fru). Neither sucrose synthase (SuSy) nor the cytoskeleton exists in prokaryotes, and so domains for interaction would not be expected on bacterial CeSAs. A membrane-associated form of SuSy has recently been identified in a number of plants (4, 23, 135). We proposed that this form of SuSy might interact with the cellulose synthase to directly channel carbon from sucrose via UDP-glc for the CeSA subunit (4, 36, 102). However, from our limited studies to date (P Hogan & D Delmer, unpublished data), we have found no evidence for interaction of SuSy with any of the plant-specific domains of CeSA. SuSy can be phosphorylated [maize, 63; soybean, 141; cotton, 33), and Winter et al (135) proposed that the non-phosphorylated form of SuSy may interact preferentially with the membrane. That study also found that the levels of membrane-associated SuSy increase about tenfold when corn pulvini initiate elongation following gravistimulation. However, in our studies of cotton fiber SuSy, we find in vivo that soluble and membrane-associated SuSy are about equally phosphorylated, so if this event plays a role, it may do so by preferential phosphorylation of distinct sites on the two forms of the enzyme (33). Recently, Winter et al (136) have shown an association of SuSy with actin, suggesting that the interaction may not be directly with a glucan synthase but rather with the cytoskeleton, another very interesting possible mechanism for localizing SuSy. Hayashi's group (91) showed that engineering of a plant SuSy gene into a strain of *A. xylinum* that can utilize sucrose as carbon source substantially increases the rate of cellulose production in vivo.

Space considerations limit extensive discussion of the role of the cytoskeleton in cellulose synthesis[for reviews see (32, 36, 49, 52, 139)]. As a broad generalization, there is a frequent correlation between the orientation of the cortical microtubule (MT) network and that of the most-newly synthesized microfibrils in the wall. Agents that disrupt MTs also frequently lead to disorganization of the pattern of microfibril deposition. Although some cases have been reported where there is no good correlation between these two networks (see 40), and others where MTs may only be required transiently (CH Haigler, personal communication), in most cases cellulose synthase complexes are believed to be somehow guided in their movement by the cortical MT network. It is unclear whether this occurs through a direct interaction, involves a motor protein such as a kinesin, or whether the MTs only provide constraining borders between which the complexes are free to move on their own. Other studies also indicate that the organization of the actin network may play a role in MT organization (72, 114). Thus, one cannot completely rule out interaction of cellulose synthases with either network or with other proteins that bind to those networks. The existence for strong wall-membrane associations in plant cells has received much attention (138), as has the possibility that the process of cellulose synthesis itself is required for proper MT organization (42, 86). These phenomena are not at all understood but raise the possibility of a continuum of association among the actin and MT networks, cellulose synthase complexes, and the cell wall. These types of associations will undoubtedly receive further attention in future.

Progress with In Vitro Synthesis of Cellulose in Plants

Recently, there has been significant progress in obtaining in vitro synthesis of cellulose using membranes isolated from higher plants. Kudlicka & Brown (76) separated activities for callose and cellulose synthesis by using native PAGE electrophoresis of detergent-solubilized proteins from membranes of mung beans. Using electron microscopy, both the shapes of the putative complexes and of the microfibrils synthesized differed considerably in preparations of these separated activities. Unfortunately, resolution of distinct polypeptides within these preparations using SDS-PAGE did not give definitive results, although more recent work (116) has identified enrichment of five polypeptides in the callose synthase preparation. This significant progress may eventually help to shed light on the relatedness of these two enzymes. Along different lines, the fact that a single amino acid change can result in apparent dissociation of rosettes (7) indicates just how potentially labile these complexes can be, a fact often discussed for freeze-fracture studies as well. Perhaps also relevant to problems with in vitro synthesis is the inability to detect an epitope-tagged version of the GhCeSA protein in Western blots derived from transgenic Arabidopsis and tobacco plants where the gene was introduced under control of the 35S promoter (70). This has led us to speculate that the steady-state level of the CeSA protein is low and/or subject to rapid degradation; alternatively, when functional, it may not co-extract with the plasma membrane because it remains associated with glucan chains or might aggregate in artefactual ways in vitro. These findings may also relate to the report (90) that CeSA protein was only detectable in BY-2 cells after treatment with DCB, and the authors suggested that DCB might stabilize the protein. Alternatively, DCB might inhibit synthesis of attached glucan chains and thus promote solubilization of the protein.

Mizuno (88) separated a callose and cellulose synthase activity by use of an affinity column prepared using an antibody directed against tubulins. When detergent-solubilized membrane proteins from Azuki bean are passed through this column, the callose synthase activity elutes unbound, while a cellulose synthase activity binds and can be eluted and still retain activity in vitro. This work may offer a clue to how the synthase might interact with the cytoskeleton, and also provide a way to isolate additional protein components of the complex.

Future Directions

An important immediate challenge is to determine the expression patterns and functions in vivo for the many newly discovered members of the *CeSA* gene superfamily. Such information can ultimately lead to development of strategies for altering the polysaccharide composition of plant cell walls in specific cell types of agronomically important plants (see also 29). Modification of the DP of glucan chains, of the crystallinity of microfibrils, and of the timing and extent of deposition of cellulose and other related polymers is now a feasible goal. Cellulose synthesis is also extremely sensitive to external perturbations such as changes in temperature (101) or wounding in ways that are not yet understood. Such modifications should also help to determine the role of these polymers in plant development. Studies on the evolution of these genes as well as the differences in organization between rosettes and terminal complexes should also yield interesting results. Further studies to determine what factors control substrate and linkage specificity in these enzymes are needed, as are more biochemical studies on isolation and optimization of activities of the synthases in vitro. It is also hoped that advances in ability to study the crystal structures of membrane proteins can be applied to these synthases to elucidate their topology and organization within the plasma membrane. One certainly suspects that the catalytic subunits will interact with other proteins, and determining what these proteins are, their function, and the domains on CeSA proteins with which they interact, are all important goals. Understanding how the cytoskeleton controls patterns of movement of cellulose synthase complexes should also be a fascinating field of study.

ACKNOWLEDGMENTS

The author acknowledges the support of her laboratory's research on cellulose synthesis through the years by grants from the United States Department of Energy, National Science Foundation, and The US-Israel Binational Agricultural Research and Development Fund (BARD).

> Visit the *Annual Reviews home page* at
> http://www.AnnualReviews.org

Literature Cited

1. Albersheim P, Darvill A, Roberts K, Staehelin LA, Varner JE. 1997. Do the structures of cell wall polysaccharides define their mode of synthesis? *Plant Physiol.* 113:1–3
2. Aloni Y, Delmer DP, Benziman M. 1982. Achievement of high rates of in vitro synthesis of 1,4-β-glucan: activation by cooperative interaction of the *Acetobacter xylinum* enzyme system with GTP, polyethylene glycol, and a protein factor. *Proc. Natl. Acad. Sci. USA* 79:6448–52
3. Amikan D, Benziman M. 1989. Cyclic diguanylic acid and cellulose synthesis in *Agrobacterium tumefaciens*. *J. Bacteriol.* 171:6649–55
4. Amor Y, Haigler CH, Johnson S, Wainscott M, Delmer DP. 1995. A membrane-associated form of sucrose synthase and its potential role in synthesis of cellulose and callose in plants. *Proc. Natl. Acad. Sci. USA* 92:9353–57
5. Amor Y, Mayer R, Benziman M, Delmer DP. 1991. Evidence for a cyclic diguanylic acid-dependent cellulose synthase in plants. *Plant Cell* 3:989–95

6. Arad SM, Kolani R, Simon-Berkovitch B, Sivan A. 1994. Inhibition by DCB of cell wall polysaccharide formation in the red microalga *Porphyridium* sp. (Rhodophyta). *Phycologia* 33:158–62
7. Arioli T, Peng LC, Betzner AS, Burn J, Wittke W, et al. 1998. Molecular analysis of cellulose biosynthesis in Arabidopsis. *Science* 279:717–20
8. Atalla RH, VanderHart DL. 1984. Native cellulose: a composite of two distinct crystalline forms. *Science* 223:283–85
9. Auer M, Scarborough GA, Kuhlbrandt W. 1998. Three-dimensional map of the plasma membrane H$^+$-ATPase in the open conformation. *Nature* 392:840–42
10. Baskin TI, Betzner AA, Hoggart R, Cork A, Williamson RE. 1992. Root morphology mutants in *Arabidopsis thaliana*. *Aust. J. Plant Physiol.* 19:427–37
11. Benziman M, Haigler CH, Brown RM Jr, White AR, Cooper KM. 1980. Cellulose biogenesis: polymerization and crystallization are coupled processes in *Acetobacter xylinum*. *Proc. Natl. Acad. Sci. USA* 77:6678–82
12. Blanton RL. 1993. Prestalk cells in

monolayer cultures exhibit two distinct modes of cellulose synthesis during stalk cell differentiation in Dictyostelium. *Development* 119:703–10

13. Blanton RL, Haigler CH. 1996. Cellulose biogenesis. In *Membranes: Specialized Functions in Plants*, ed. M Smallwood, JP Knox, DJ Bowles, pp. 57–76. Oxford, UK: BIOS Sci.

14. Blanton RL, Iranfar N, Fuller D, Grimson MJ, Loomis WF. 1998. Isolation and characterization of the gene for the catalytic subunit of cellulose synthase from *Dictyostelium discoideum. Proc. Int. Cell Wall Meet., 8th,* Norwich, UK. (Abstr.)

15. Blanton RL, Northcote DH. 1990. A 1,4-β-D-glucan synthase system from *Dictyostelium discoideum. Planta* 180:324–32

16. Brown RM Jr. 1996. The biosynthesis of cellulose. *J. Macromol. Sci. Pure Appl. Chem.* A33(10):1345–73

17. Brown RM Jr, Montezinos D. 1976. Cellulose microfibrils: visualization of biosynthetic and orienting complexes in association with the plasma membrane. *Proc. Natl. Acad. Sci. USA* 73:143–47

18. Brown RM Jr, Saxena IM, Kudlicka K. 1997. Cellulose biosynthesis in higher plants. *Trends Plant Sci.* 1:149–56

19. Brummell DA, Catala C, Lashbrook CC, Bennett AB. 1997. A membrane-anchored E-type endo-1,4-beta-glucanase is localized on Golgi and plasma membranes of higher plants. *Proc. Natl. Acad. Sci. USA* 94:4794–99

20. Callebaut I, Labesse G, Durand P, Poupon A, Canard L, et al. 1997. Deciphering protein sequence information through hydrophobic cluster analysis (HCA): current status and perspectives. *Cell. Mol. Life Sci.* 53:621–45

21. Campbell JA, Davies GJ, Bulone V, Henrissat B. 1997. A classification of nucleotide-diphospho-sugar glycosyltransferases based on amino acid sequence similarity. *Biochem. J.* 326:929–42

22. Canale-Parola E. 1970. Biology of the sugar-fermenting Sarcinae. *Bacteriol. Rev.* 34:82–97

23. Carlson SJ, Chourey PS. 1996. Evidence for plasma membrane-associated forms of sucrose synthase in maize. *Mol. Gen. Genet.* 252:303–10

24. Carpita NC. 1996. Structure and biogenesis of the cell walls of grasses. *Annu. Rev. Plant Physiol. Plant Mol. Biol.* 47:445–76

25. Carpita N, Vergara C. 1998. A recipe for cellulose. *Science* 279:672–73

26. Chadwick CC, Goormaghtigh E, Scarborough GA. 1987. A hexameric form of the *Neurospora crassa* plasma membrane H^+-ATPases. *Arch. Biochem. Biophys.* 252:348–56

27. Chanzy H, Henrissat B. 1985. Unidirectional degradation of Valonia cellulose microcrystals subjected to cellulase action. *FEBS Lett.* 184:285–88

28. Chanzy HD, Roche EJ. 1975. Fibrous mercerization of Valonia cellulose. *J. Polym. Sci. Polym. Phys. Ed.* 13:1859–62

29. Chapple C, Carpita N. 1998. Plant cell walls as targets for biotechnology. *Curr. Opin. Plant Biol.* 1:179–85

30. Cousins SK, Brown RM Jr. 1995. Cellulose I microfibril assembly: computational molecular mechanics energy analysis favors bonding by van der Waals forces as the initial step in crystallization. *Polymer* 36:3885–88

31. Cutler S, Somerville C. 1997. Cellulose synthesis: cloning *in silico. Curr. Biol.* 7:R108–11

32. Cyr RJ. 1994. Microtubules in plant morphogenesis: role of the cortical array. *Annu. Rev. Cell Biol.* 10:153–80

33. Datcheva M, Buster D, Vulliet R, Delmer D. 1998. Membrane-associated sucrose synthase: mechanism of membrane association and role in glucan synthesis. *Proc. Plant Polysaccharide Symp., Davis, CA* (Abstr.)

34. Delmer DP. 1987. Cellulose biosynthesis. *Annu. Rev. Plant Physiol.* 38:259–90

35. Delmer DP. 1998. A hot mutant for cellulose synthesis. *Trends Plant Sci.* 3:164–65

36. Delmer DP, Amor Y. 1995. Cellulose biosynthesis. *Plant Cell* 7:987–1000

37. Delmer DP, Read SM, Cooper G. 1987. Identification of a protein receptor in cotton fibers for the herbicide 2,6-dichlorobenzonitrile. *Plant Physiol.* 84:415–20

38. Doblin M, Newbigin E, Bacic A, Read SM. 1998. A glucan synthase gene expressed in pollen tubes of *Nicotiana alata. Proc. Int. Cell Wall Meet., 8th, Norwich, UK* (Abstr.)

39. Dougherty BA, van de Rijn I. 1994. Molecular characterizaton of hasA from an operon required for hyaluronic acid synthesis in Group A Streptococci. *J. Biol. Chem.* 269:169–75

40. Emons AMC. 1998. The making of the architecture of the plant cell wall: how

cells exploit geometry. *Proc. Natl. Acad. Sci. USA* 95:7215–19

41. Englehardt J. 1995. Sources, industrial derivatives, and commercial applications of cellulose. *Carbohydr. Eur.* 12:5–14

42. Fisher DP, Cyr RJ. 1998. Extending the microtubule/microfibril paradigm. Cellulose synthesis is required for normal cortical microtubule alignment in elongating cells. *Plant Physiol.* 116:1043–51

43. French AD, Delmer DP. 1999. The structure and mechanism of biosynthesis of cellulose. In *The Discoveries in Plant Biology Series*, Vol. 3, ed. SD Kung, SF Yang. Hong Kong: World Sci. In press

44. French AD, Miller DP, Aabloo A. 1993. Miniature crystal models of cellulose polymorphs and other carbohydrates. *Int. J. Biol Macromol.* 15:30–36

45. Gardner KH, Blackwell J. 1974. The structure of native cellulose. *Biopolymers* 13:1975–2001

46. Geremia RT, Petroni EA, Ielpi L, Henrissat B. 1996. Toward a classification of glycosyl transferases based on amino acid sequence similarities. Prokaryotic alpha-mannosyl-transferases. *Biochem. J.* 318:133–38

47. Gibeaut DM, Carpita NC. 1994. Biosynthesis of plant cell wall polysaccharides. *FASEB J.* 8:904–15

48. Giddings TH, Brower DL, Staehelin LA. 1980. Visualization of particle complexes in the plasma membrane of *Micrasterias denticulata* associated with the formation of cellulose fibrils in primary and secondary walls. *J. Cell Biol.* 84:327–39

49. Giddings TH, Staehelin LA. 1991. Microtubule-mediated control of microfibril deposition: a re-examination of the hypothesis. See Ref. 82a, pp. 85–99

50. Girault R, Bert F, Rihouey C, Jauneau A, Morvan C, Jarvis M. 1997. Galactans and cellulose in flax fibres: putative contributions to the tensile strength. *Int. J. Biol. Macromol.* 21:179–88

51. Glaser L. 1958. The synthesis of cellulose in cell-free extracts of *Acetobacter xylinum*. *J. Biol. Chem.* 232:627–36

52. Green PB, Selker JML. 1991. Mutual alignments of cell walls, cellulose and cytoskeletons: their role in meristems. See Ref. 82a, pp. 303–22

53. Haigler CH. 1991. Relationship between polymerization and crystallization in microfibril biogenesis. See Ref. 54a, pp. 99–124

54. Haigler CH, Blanton R. 1996. New hopes for old dreams. *Proc. Natl. Acad. Sci. USA* 93:12082–85

54a. Haigler CH, Weimer PJ, eds. 1991. *Biosynthesis and Biodegradation of Cellulose*. New York: Marcel Dekker

55. Harper JF, Manney L, Sussman MR. 1998. The plasma membrane H($^+$)-ATPase gene family in Arabidopsis: genomic sequence of AHA10 which is expressed primarily in developing seeds. *Mol. Gen. Genet.* 244:572–87

56. Hayashi T, Read SM, Bussell J, Thelen M, Lin FC, et al. 1987. UDP-glucose: (1,3)-β-glucan synthases from mung bean and cotton: differential effects of Ca^{2+} and Mg^{2+} on enzyme properties and on macromolecular structure of the glucan product. *Plant Physiol.* 83:1054–62

57. Heim DR, Larrinua IM, Murdoch MG, Roberts JL. 1998. Triazofenamide is a cellulose biosynthesis inhibitor. *Pestic. Biochem. Physiol.* 59:163–68

58. Heim DR, Skomp JR, Tschabold EE, Larrinua IM. 1990. Isoxaben inhibits the synthesis of acid insoluble cell wall materials in *Arabidopsis thaliana*. *Plant Physiol.* 93:695–700

59. Hestrin S. 1962. Synthesis of polymeric homopolysaccharides. In *The Bacteria*, ed. IC Gunsalus, RY Stanier, pp. 373–88. New York: Academic

60. Hieta K, Kuga S, Usuda M. 1984. Electron staining of reducing ends evidences a parallel-chain structure in Valonia cellulose. *Biopolymers* 23:1807–10

61. Hill RL, Brew K. 1975. Lactose synthetase. *Adv. Enzymol. Rel. Areas Mol. Biol.* 43:411–517

62. Holland N, Holland D, Delmer D. 1998. Tissue-specific expression of *CelA* and *Rac* genes. *Proc. Plant Polysaccharide Symp., Davis, CA* (Abstr. 25)

63. Huber SC, Huber JL, Liao PC, Gage DA, McMichael RW Jr, et al. 1996. Phosphorylation of serine-15 of maize leaf sucrose synthase. *Plant Physiol.* 112:793–802

64. Ihara Y, Fukumi S, Hayashi T. 1997. Cloning of homologs of *bcs* A from developing cotton fiber cells. *Wood Res.* 84:1–6

65. Itoh T. 1990. Cellulose synthesizing complexes in some giant marine algae. *J. Cell Sci.* 95:309–19

66. Jacob SR, Northcote D. 1985. *In vitro* glucan synthesis by membranes of celery petioles: the role of the membrane in determining the linkage formed. *J. Cell. Sci. Suppl.* 2:1–11

67. Kang MS, Elango N, Mattie E, Au-Young J, Robbins P, Cabib E. 1984. Isolation of chitin synthetase from *Saccharomyces cerevisiae*. Purification of an enzyme by entrapment in the reaction product. *J. Biol. Chem.* 259:14966–72

68. Kawagoe Y, Delmer DP. 1997. Pathways and genes involved in cellulose biosynthesis. *Genet. Eng.* 19:63–87

69. Kawagoe Y, Delmer DP. 1997. Cotton CelA1 has a LIM-like Zn binding domain in the N terminal cytoplasmic region. *Plant Physiol.* 114 (Suppl.):85 (Abstr.)

70. Kawagoe Y, Delmer DP. 1998. Where is the CelA protein in plants? *Proc. Plant Polysaccharide Symp., Davis, CA* (Abstr. 26)

71. Kelly R, Register E, Hsu M-J, Kurtz M, Nielsen J. 1996. Isolation of a gene involved in 1,3-β-glucan synthase in *Aspergillus nidulans* and purification of the corresponding protein. *J. Bacteriol.* 178:4381–91

72. Kobayashi H, Fukuda H, Shibaoka H. 1987. Reorganization of actin filaments associated with the differentiation of tracheary elements in *Zinnia* mesophyll cells. *Protoplasma* 138:69–71

73. Kokubo KA, Sakurai N, Kuraishi S, Takeda K. 1991. Culm brittleness of barley (*Hordeum vulgare* L.) mutants is caused by smaller number of cellulose molecules in the cell wall. *Plant Physiol.* 107:509–14

74. Kopczak SD, Haas NA, Hussey PJ, Silflow CD, Snustad DP. 1992. The small genome of Arabidopsis contains at least six expressed alpha-tubulin genes. *Plant Cell* 4:539–47

75. Koyama M, Helbert W, Imai R, Sugiyama J, Henrissat B. 1997. Parallel-up structure evidences the molecular directionality during biosynthesis of bacterial cellulose. *Proc. Natl. Acad. Sci. USA* 94:9091–95

76. Kudlicka K, Brown RM Jr. 1997. Cellulose and callose biosynthesis in higher plants. I. Solubilizaton and separation of (1–3)- and (1–4)-β-glucan synthase activities from mung bean. *Plant Physiol.* 115:643–56

77. Kuga S, Takagi S, Brown RM Jr. 1993. Native folded-chain cellulose II. *Polymer* 34:3293–97

78. Ledbetter MC, Porter KR. 1963. A 'microtubule' in plant cell fine structure. *J. Cell Biol.* 19:239–50

79. Legault DJ, Kelly RJ, Natsuka Y, Lowe JB. 1995. Human α(1,3/1,4–fucosyltransferases discriminate between different oligosaccharide acceptor substrates through a discrete peptide fragment. *J. Biol. Chem.* 270:20987–96

80. Leloir LF. 1971. Two decades of research on the biosynthesis of saccharides. *Science* 172:1299–303

81. Lin FC, Brown RM Jr. 1989. Purification of cellulose synthase from *Acetobacter xylinum*. In *Cellulose and Wood—Chemistry and Technology*, ed. C Scheurch, pp. 473–92. New York: Wiley

82. Lin FC, Brown RM Jr, Drake RR Jr, Haley BE. 1990. Identification of the uridine-5′-diphosphoglucose (UDP-glc) binding subunit of cellulose synthase in *Acetobacter xylinum* using the photoaffinity probe 5–azido-UDP-glc. *J. Biol. Chem.* 265:4782–84

82a. Lloyd CW, ed. 1991. *The Cytoskeletal Basis of Plant Growth and Development*. London: Academic

83. Matthysse AG, Thomas DOL, White AR. 1995. Mechanism of cellulose synthesis in *Agrobacterium tumefaciens*. *J. Bacteriol.* 177:1076–81

84. Matthysse AG, White S, Lightfoot R. 1995. Genes required for cellulose synthesis in *Agrobacterium tumefaciens*. *J. Bacteriol.* 177:1069–75

85. Mayer R, Ross P, Weinhouse H, Amikam D, Volman G, Ohana P, et al. 1991. Polypeptide composition of bacterial cyclic diguanylic acid-dependent cellulose synthase and the occurrence of immunologically cross reacting proteins in higher plants. *Proc. Natl. Acad. Sci. USA* 88:5472–76

86. McClinton RS. 1998. *Cytomorphogenesis in Arabidopsis*. PhD thesis. Univ. Calif. Berkeley

87. McKinney EC, Meagher RB. 1998. Members of the Arabidopsis actin gene family are widely dispersed in the genome. *Genetics* 149:663–75

88. Mizuno K. 1996. Tubulin-containing granules in the plasma membranes of Azuki bean epicotyls and their relationship to the synthesis of β-glucan. *Plant Cell Physiol.* 37 (Suppl.), (Abstr. S20)

89. Mueller SC, Brown RM Jr. 1980. Evidence for an intramembrane component associated with a cellulose microfibril synthesizing complex in higher plants. *J. Cell Biol.* 84:315–26

90. Nakagawa N, Sakurai N. 1998. Increase in the amount of celA-1 protein in tobacco BY-2 cells by a cellulose biosynthesis inhibitor, 2,6-dichlorobenzonitrile. *Plant Cell Physiol.* 39:779–85

91. Nakai T, Tonouchi N, Konishi T, Kojima Y, Tsuchida T, et al. 1999. Enhancement of cellulose production by expression of plant sucrose synthase in *Acetobacter xylinum*. In *Biosynthesis and Biodegradation of Polymers*, ed. A Steinbuchel. Weinheim: Wiley-VCH Verlag. In press

92. Nicol F, Hofte H. 1998. Plant cell expansion: scaling the wall. *Curr. Opin. Cell Biol.* 1:12–17

93. Ohana P, Delmer DP, Carlson RW, Glushka, Azadi P, et al. 1998. Identification of a novel triterpenoid saponin from *Pisum sativum* as a specific inhibitor of the diguanylate cyclase of *Acetobacter xylinum*. *Plant Cell Physiol.* 39:144–52

94. Ohana P, Delmer DP, Volman G, Benziman M. 1998. Glycosylated triterpenoid saponin: a specific inhibitor of diguanylate cyclase from *Acetobacter xylinum*. Biological activity and distribution. *Plant Cell Physiol.* 39:153–59

95. Omura S, Tanaka Y, Kanaya I, Shinose M, Takahashi Y. 1990. Phthoxazolin, a specific inhibitor of cellulose biosynthesis produced by a strain of *Streptomyces* sp. *J. Antibiot.* 43:1034–36

96. Pear J, Kawagoe Y, Schreckengost W, Delmer DP, Stalker D. 1996. Higher plants contain homologs of the bacterial CelA genes encoding the catalytic subunit of the cellulose synthase. *Proc. Natl. Acad. Sci. USA* 93:12637–42

97. Peng L-C, Delmer DP, Stoller A, Kreuz K. 1998. A comparison of the effects of two different herbicides on cellulose synthesis in cotton fibers. *Proc. Plant Polysaccharide Symp., Davis, CA* (Abstr. 27)

98. Potikha T, Delmer DP. 1995. Selection and characterization of a mutant of *Arabidopsis thaliana* which shows altered patterns of cellulose deposition. *Plant J.* 7:453–60

99. Preston RD. 1964. Structural and mechanical aspects of plant cell walls with particular reference to synthesis and growth. In *The Formation of Wood in Forest Trees*, ed. MH Zimmermann, pp. 169–201. New York: Academic

100. Quader H. 1991. Role of linear terminal complexes in cellulose synthesis. See Ref. 54a, pp. 51–69

101. Roberts EM, Nunna RR, Huang JY, Trolinder NL, Haigler CH. 1992. Effects of cycling temperatures on fiber metabolism in cultured cotton ovules. *Plant Physiol.* 100:979–86

102. Robinson D. 1996. SuSy *ergo* GluSy: new developments in the field of cellulose biosynthesis. *Bot. Acta* 109:261–63

103. Ross P, Mayer R, Benziman M. 1991. Cellulose biosynthesis and function in bacteria. *Microbiol. Rev.* 55:35–58

104. Ross P, Weinhouse H, Aloni Y, Michaeli D, Weinberger-Ohana P, et al. 1987. Regulation of cellulose synthesis in *Acetobacter xylinum* by cyclic diguanylic acid. *Nature* 325:279–81

105. Samuels AL, Giddings TH, Staehelin LA. 1995. Cytokinesis in tobacco BY-2 and root tip cells: a new model of cell plate formation in higher plants. *J. Cell Biol.* 130:1345–57

106. Sanchez-Garcia I, Rabbitts TH. 1994. The LIM domain: a new structural motif found in zinc-finger-like proteins. *Trends Genet.* 10:315–20

107. Sarko A, Muggli R. 1974. Packing analysis of carbohydrates and polysaccharides. III. *Valonia* cellulose and cellulose II. *Macromolecules* 7:486–94

108. Saurin AJ, Borden KLB, Boddy MN, Freemont PS. 1996. Does this have a familiar RING? *Trends Biochem. Sci.* 21:208–14

109. Saxena IM, Brown RM Jr. 1996. Identification of a second cellulose synthase gene [acsAII] in *Acetobacter xylinum*. *J. Bacteriol.* 177:5276–83

110. Saxena IM, Brown RM Jr, Fevre M, Geremia RA, Henrissat B. 1995. Multidomain architecture of β-glycosyl transferases: implications for mechanism of action. *J. Bacteriol.* 177:1419–24

111. Saxena IM, Kudlicka K, Okuda K, Brown RM Jr. 1994. Characterization of genes in the cellulose-synthesizing operon (acs Operon) of *Acetobacter xylinum*: implicaton for cellulose crystallization. *J. Bacteriol.* 176:5735–52

112. Saxena IM, Lin FC, Brown RM Jr. 1990. Cloning and sequencing of the cellulose synthase catalytic subunit gene of *Acetobacter xylinum*. *Plant Mol. Biol.* 15:673–83

113. Schlupmann H, Bacic A, Read SM. 1993. A novel callose synthase from pollen tubes of *Nicotiana*. *Planta* 191:470–81

114. Seagull RW. 1990. The effects of microtubule and microfilament disrupting agents on cytoskeletal arrays and wall deposition in developing cotton fibers. *Protoplasma* 159:44–59

115. Shedletzky E, Shmuel M, Trainin T, Kalman S, Delmer D. 1992. Cell wall

structure in cells adapted to growth on the cellulose-synthesis inhibitor 2,6-dichlorobenzonitrile. *Plant Physiol.* 100:120–30

116. Shin H, Brown RM Jr. 1998. The first biochemical identification of yeast β-1,3–glucan synthase homolog in cotton fibers (*Gossypium hirsutum*). *Proc. Keystone Symp. Plant Cell Biol., Taos, NM,* p. 45. (Abstr.)

117. Sisson WA. 1941. Some X-ray observations regarding the membrane structure of *Halicystis. Contrib. Boyce Thompson Inst.* 12:31–44

118. Snustad DP, Haas NA, Kopczak SD, Silflow CD. 1992. The small genome of Arabidopsis contains at least nine expressed beta-tubulin genes. *Plant Cell* 4:549–56

119. Somerville CR, Schieble W-R, Lukowitz W, Cutler S, Richmond T. 1998. Genetic dissection of cell wall composition in Arabidopsis. *Proc. Annu. Meet. Am. Soc. Plant Physiol, Madison, WI* (Abstr.)

120. Spicer A, McDonald JA. 1998. Characterization and molecular evolution of a vertebrate hyaluronan synthase gene family. *J. Biol. Chem.* 273:1923–32

121. Standal R, Iversen TG, Coucheron DH, Fjaervik E, Blatny JM, Valla S. 1994. A new gene required for cellulose production and a gene encoding cellulolytic activity in *Acetobacter xylinum* are colocalized with the *bcs* operon. *J. Bacteriol.* 176:665–72

122. Stasinopoulos SJ, Fisher PR, Stone BA, Stanisich VA. 1999. Detection of two loci involved in $(1 \rightarrow 3)$-β-glucan (curdlan) biosynthesis by *Agrobacterium* sp. ATCC31749, and comparative sequence analysis of the putative curdlan synthase gene. *Glycobiology.* In press

123. Sugiyama J, Harada H, Fujiyoshi Y, Uyeda N. 1985. Lattice images from ultrathin sections of cellulose microfibrils in the cell wall of *Valonia macrophysa* Kutz. *Planta* 166:161–68

124. Sugiyama J, Vuong R, Chanzy H. 1991. Electron diffraction study of two crystalline phases occurring in native cellulose from an algal cell wall. *Macromolecules* 24:4168–75

125. Tal R, Wong HC, Calhoon R, Gelfand D, Fear AL, et al. 1998. Three *cdg* operons control cellular turnover of cyclic-di-GMP in *Acetobacter xylinum.* Genetic organization and occurrence of conserved domains in isoenzymes. *J. Bacteriol.* 180:4416–25

126. Thelen MT, Delmer DP. 1986. Gel-electrophoretic separation, detection, and characterization of plant and bacterial UDP-glucose glucosyltransferases. *Plant Physiol.* 81:913–18

127. Tsekos I. 1996. The supramolecular organization of red algal cell membranes and their participation in the biosynthesis and secretion of extracellular polysaccharides: a review. *Protoplasma* 193:1–4

127a. Turner A, Bacic A, Harris PJ, Reed SM. 1998. Membrane fractionation and enrichment of callose synthase from pollen tubes of *Nicotiana alata* Link et Otto. *Planta* 205:380–88

128. Turner SR, Somerville CR. 1997. Collapsed xylem phenotype of Arabidopsis identifies mutants deficient in cellulose deposition in the secondary cell wall. *Plant Cell* 9:689–701

129. Updegraff DM. 1969. Semi-micro determination of cellulose in biological materials. *Anal. Biochem.* 32:420–24

130. VanderHart DL, Atalla RH. 1986. In *Cellulose: Structure, Modification and Hydrolysis,* ed. RA Young, RM Rowell, pp. 88–118. New York: Wiley-Intersci.

131. Vaughn KC, Hoffman JC, Hahn MG, Staehelin LA. 1996. The herbicide dichlobenil disrupts cell plate formation: immunogold characterization. *Protoplasma* 194:117–32

132. Volman G, Ohana P, Benziman M. 1995. Biochemistry and molecular biology of cellulose biosynthesis. *Carbohydr. Eur.* 12:20–27

133. Weinhouse H, Sapir S, Amikam D, Shilo Y, Volman G, et al. 1997. C-di-GMP-binding protein, a new factor regulating cellulose synthesis in *Acetobacter xylinum.* *FEBS Lett.* 416:207–11

134. Williamson RE. 1991. Orientation of cortical microtubules in interphase plant cells. *Int. Rev. Cytol.* 129:135–206

135. Winter H, Huber JL, Huber SC. 1997. Membrane-association of sucrose synthase: changes during gravitropic response and possible control by protein phosphorylation. *FEBS Lett.* 420:151–55

136. Winter H, Huber JL, Huber SC. 1998. Identification of sucrose synthase as an actin-binding protein. *FEBS Lett.* 430:205–8

137. Wong HC, Fear AL, Calhoon RD, Eichinger GH, Mayer R, et al. 1990. Genetic organization of the cellulose synthase operon in *Acetobacter xylinum.* *Proc. Natl. Acad. Sci. USA* 87:8130–34

137a. Wu L, Joshi CP, Chiang VL. 1998. AravCelA, a new member of cellulose synthase gene familiy from *Arabidopsis thaliana* (Accessioon No. AF 062485) (PGR98-114). *Plant Physiol.* 117:1125

138. Wyatt SE, Carpita NC. 1993. The plant cytoskeleton-cell wall continuum. *Trends Cell Biol.* 3:413–17

139. Wymer C, Lloyd C. 1996. Dynamic microtubules: implications for cell wall patterns. *Trends Plant Sci.* 1:222–28

140. Zhang P, Toyoshima C, Yonekura K, Green NM, Stokes DL. 1998. Structure of the calcium pump from sarcoplasmic reticulum at 8-Å resolution. *Nature* 392:835–39

141. Zhang XQ, Chollet R. 1997. Seryl-phosphorylation of sucrose synthase (nodulin-100) by a Ca^{2+}-dependent kinase. *FEBS Lett.* 410:126–30

Annu. Rev. Plant Physiol. Plant Mol. Biol. 1999. 50:277–303
Copyright © *1999 by Annual Reviews. All rights reserved*

NITRATE REDUCTASE STRUCTURE, FUNCTION AND REGULATION: Bridging the Gap between Biochemistry and Physiology

Wilbur H. Campbell
Phytotechnology Research Center and Department of Biological Sciences,
Michigan Technological University, Houghton, Michigan 49931-1295;
e-mail: wcampbel@mtu.edu

KEY WORDS: enzymology, 3-D structure, site-directed mutagenesis, molybdopterin cofactor, regulation, protein phosphorylation, 14-3-3 binding protein

---

ABSTRACT

Nitrate reductase (NR; EC 1.6.6.1-3) catalyzes NAD(P)H reduction of nitrate to nitrite. NR serves plants, algae, and fungi as a central point for integration of metabolism by governing flux of reduced nitrogen by several regulatory mechanisms. The NR monomer is composed of a ∼100-kD polypeptide and one each of FAD, heme-iron, and molybdenum-molybdopterin (Mo-MPT). NR has eight sequence segments: (*a*) *N*-terminal "acidic" region; (*b*) Mo-MPT domain with nitrate-reducing active site; (*c*) interface domain; (*d*) Hinge 1 containing serine phosphorylated in reversible activity regulation with inhibition by 14-3-3 binding protein; (*e*) cytochrome b domain; (*f*) Hinge 2; (*g*) FAD domain; and (*h*) NAD(P)H domain. The cytochrome b reductase fragment contains the active site where NAD(P)H transfers electrons to FAD. A complete three-dimensional dimeric NR structure model was built from structures of sulfite oxidase and cytochrome b reductase. Key active site residues have been investigated. NR structure, function, and regulation are now becoming understood.

CONTENTS

277

INTRODUCTION

Eukaryotic assimilatory nitrate reductase (NR) catalyzes the following reaction:

$$NO_3^- + NADH \rightarrow NO_2^- + NAD^+ + OH^-$$

$$\Delta G = -34.2 \, \text{kcal/mol} \, (-143 \, \text{kJ/mol}); \, \Delta E = 0.74 \, \text{V}$$

With such a large negative free energy—under standard conditions, reduction of nitrate to nitrite by pyridine nucleotides is, for all practical purposes, an irreversible reaction. NADH-specific NR forms (EC 1.6.6.1) exist in higher plants and algae; NAD(P)H-bispecific forms (EC 1.6.6.2) are found in higher plants, algae, and fungi; and NADPH-specific forms (EC 1.6.6.3) are found in fungi. NR catalyzes the first step of nitrate assimilation in all these organisms, which appears to be a rate-limiting process in acquisition of nitrogen in most cases. Since nitrate is the most significant source of nitrogen in crop plants, understanding the role of NR in higher plants has potential economic importance, especially in light of recent studies illuminating the enzyme as one focal point for integration of control of carbon and nitrogen metabolism. With nitrate triggering and NR responding to metabolic changes in plants, the literature on nitrate and NR is vast, and nitrate metabolism and NR in plants have been reviewed from many points of view over the past few years (9, 11, 12, 18, 19, 37, 39, 62, 78, 81, 87, 92, 93). The intent here is to focus on the biochemical aspects of NR and its regulation, with the emphasis on higher plant metabolism; this brings together many threads in plant physiology with our current understanding of NR structure and function.

Recent advancement of NR biochemistry has been dominated by molecular biology for over ten years. From the large number of NR cDNAs and genomic DNAs cloned and sequenced, a huge database of NR-deduced amino acid

a

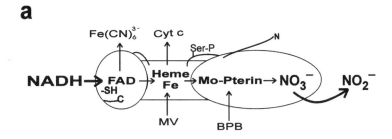

b

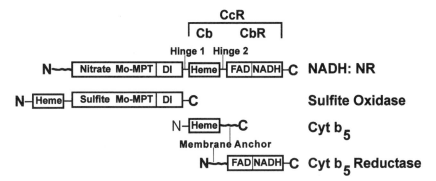

Figure 1 Nitrate reductase models. (*a*) Functional model of the enzyme; MV, methyl viologen; BPB, reduced bromphenol blue. (*b*) Sequence model of the enzyme; DI, dimer interface.

sequences has been produced (11, 18, 81, 93). This information on NR has confirmed many earlier discoveries in NR biochemistry and also led to new ones. To a great extent, the culmination of this combined biochemical knowledge is embodied in the two models depicted in Figure 1. The complex problem of studying the structure and function of NR has been simplified by expression of functional recombinant fragments of the enzyme. This resulted in the first 3-D structure information for NR (11, 56). With the problem of lack of expression of the holo-NR in recombinant form now overcome (28, 95, 96), NR biochemical discovery will return to its roots, and detailed analysis of the complete enzyme can begin in earnest. Molecular biology has been integrated as a tool to study NR biochemistry both as a mechanism to generate the large amounts of enzyme needed for the detailed studies of structure and function, and a means to produce the site-directed mutants that permit investigations not possible with studies limited to natural forms.

STRUCTURAL CHARACTERISTICS

Polypeptide Sequence

The early studies of NR identified it as a flavoprotein containing a heme-Fe and molybdenum complexed with a unique pterin or molybdopterin (Figure 1a). NR was shown to catalyze a number of partial reactions including a dehydrogenase functionality typified by NADH-dependent reduction of ferricyanide and mammalian cytochrome c, which could be inhibited at a sensitive thiol protected by NADH (11, 12, 18, 81, 94). It was also shown that nitrate reduction could be driven with reduced flavins, methyl viologen, and bromphenol blue, which could be inhibited independently from the dehydrogenase function. Subsequently, it was deduced that all these reactions catalyzed by NR, including the natural NADH/NADPH-driven nitrate reduction, were best understood by viewing the enzyme as a redox system with an internal electron transfer chain. The enzyme is a homodimer composed of two identical $\sim$100-kD subunits, each containing one equivalent of flavin adenine dinucleotide (FAD), heme-Fe, and Mo-molybdopterin (Mo-MPT). However, the amino acid sequence of the NR polypeptide was not revealed until the enzyme was cloned (11, 12, 18–20, 78, 81, 94). There are now more than 40 NR sequences in GenBank consisting of enzyme forms from higher plants, algae, and fungi and this number is growing every year. Comparison of NR sequences with those of known proteins and enzymes readily reveals conserved regions with similarity to a unique set of proteins (Figure 1b). The *Arabidopsis* NIA2 (hereafter called AtNR2), which contains 917 amino acid residues and has a molecular weight of 102,844 (20; GenBank Accession number J03240; Swiss Protein P11035), is a representative model for NR, and the numerical positions of the distinct sequence regions of AtNR2 are identified in Table 1. More recent data on the 3-D structure of NR, once thought to contain only 3 domains, indicate that it actually contains 5 structurally distinct domains: Mo-MPT, dimer interface, cytochrome b (Cb), FAD, and NADH (Figure 1b). When the FAD and NADH domains are combined, the cytochrome b reductase fragment (CbR) is formed, and when the Cb domain is joined to CbR, it is called the cytochrome c reductase fragment (CcR), as shown in Figure 1b. Three sequence regions with no similarity to another protein and varying in sequence among NR forms are: (a) the N-terminal region, which is rich in acidic residues but is also quite short in some NR forms; (b) Hinge 1, which contains the site of protein phosphorylation—Ser534 in AtNR2 and a trypsin proteolytic site; and (c) Hinge 2, which also contains a proteinase site. Recently, an NR was cloned from a marine diatom, *Heterosigma*

Table 1   Key invariant residues in Arabidopsis NIA2 (GenBank Accession No. J03240)[a]

Domain/region	Span (# residues)	Invariant (# residues/%)	Key residues	Function
N-terminal	1–90 (90)	0/0.0	None	Regulatory/Stability
Mo-MPT	91–334 (244)	33/13.5	8	Nitrate reducing active site
			Arg144	Nitrate binding
			Hist146	MPT binding
			Cys191	Mo ligand
			Arg196	Nitrate binding
			His294	MPT binding
			Arg229	MPT binding
			Gly308	Mo = O ligand
			Lys312	MPT binding
Dimer interface (DI)	335–490 (156)	10/6.4	2	Formation of stable dimer
			Glu360	Ionic bond at interface
			Lys399	Ionic bond at interface
Hinge 1	491–540 (50)	5/10.0	1	Regulatory
			Ser534	Phosphorylated
Cytochrome b (Cb)	541–620 (80)	10/12.5	2	Binds Heme-Fe
			His577	Heme-Fe ligand
			His600	Heme-Fe ligand
Hinge 2	621–660 (40)	0/0.0	None	Unknown
FAD	661–780 (120)	10/8.3	5	Binds FAD/active site
			Arg712	Binds FAD
			Tyr714	Binds FAD
			Gly745	Binds FAD
			Ser748	Binds FAD
			Lys 731	Binds NADH
NADH	781–917 (137)	9/6.6	3	Binds NADH/active site
			Gly794	Binds NADH
			Cys889	Active site
			Phe917	C-terminal

[a]See Figures 1*b* and 3.

*akashiwo.*Raphidophyceae, which contains a 116-residue hemoglobin domain (bacterial or protozoan type) inserted between the Cb and FAD domains of an otherwise typical NR sequence (Y Nakamura & T Ikawa, personal communication). Thus, the *C*-terminal region of this NR is similar to a bacterial enzyme called flavohemoglobin, which is a combination of a hemoglobin and an NADH flavo-reductase with structural features in common with NR's CbR (25). Clearly, NR is built from modular units that have evolved independently to some degree and are related to similar modular units existing in many modern enzymes and proteins. The CbR fragment of NR is most closely related to cytochrome b_5 reductase (66; EC 1.6.2.2), whereas NR's Cb is related to eukaryotic cytochrome b_5 (59), which is a redox partner of its reductase with both anchored in the endoplasmic reticulum, although soluble forms of these proteins also exist in mammalian systems (Figure 1b). The Mo-MPT and interface domains of NR are most closely related to sulfite dehydrogenase (EC 1.8.2.1), commonly called sulfite oxidase (SOX), which is a detoxification enzyme found in the intermembrane space of mitochondria where sulfite is reduced to sulfate by reduced cytochrome c (37, 51). SOX also contains a cytochrome b domain related to the Cb of NR.

Cofactors and Metal Ions

NR contains three internal cofactors (FAD, heme, and MPT) and two metal ions (Fe and Mo) in each subunit (11, 12, 18, 77, 81, 94). During catalytic turnover, the FAD, Fe, and Mo cyclically are reduced and oxidized. Thus, NR exists in oxidized and reduced forms, with the 12 to 18 possible oxidized and reduced forms (3 states for FAD, 2 states for Fe, and either 2 or 3 states for Mo) having only transient existence in vivo. Redox potentials for FAD, heme-Fe, and Mo-MPT in holo-NR and its proteolytic or recombinant fragments are -272 to -287 mV, -123 to -174 mV, and -25 to $+15$ mV, respectively (4, 11, 37, 75, 76, 93, 94, 98). This redox pattern is consistent with a "downhill" flow of electrons within the enzyme from NADH with a redox potential of -320 mV to the nitrate-reducing active site, where nitrate is reduced with a redox potential of $+420$ mV. The potential of the FAD in recombinant CbR is shifted more positive by 22 to 70 mV when NAD^+ is present and also by ADP and other NAD^+ analogs (4, 75, 94, 98). Since ADP produces a similar shift of potential in FAD as NAD^+ and the impact of the inhibitors are related to the strength of binding to CbR (4), these effects are probably transmitted through a conformational change in the NADH domain by its contacts with the FAD domain of CbR. However, NAD^+ forms a charge-transfer complex with reduced FAD with unique spectral properties (75, 94, 98). Site-directed mutagenesis of residues in either the FAD or NADH domains of CbR alters the redox potential of the FAD, but mutation of the active site Cys (Cys889 in AtNR2) does not affect the magnitude of the redox potential shift to a significant extent when NAD^+ is present (4, 75, 98). The redox potential of the heme-Fe in the enzyme's Cb depends greatly on the form of NR used for the measurement. Recombinant Cb or Cb fused to CbR in a CcR fragment has a redox potential of $+15$ or $+16$ mV, and a Cb fragment with an N-terminal extension containing the dimer interface domain and Hinge 1 yields -28 mV, whereas holo-NR's Cb is poised at about -100 to -200 mV (11, 13, 14, 76, 93, 94, 102). Thus, the Mo-MPT and interface domains of NR on the *N*-terminal side of the Cb interact with it to make the heme-Fe redox potential more negative by more than 100 mV. Mammalian cytochrome b_5 has a potential of $+5$ mV, whereas heme-Fe in flavocytochrome b_2 has a potential of -31 mV in the recombinantly expressed Cb fragment and $+5$ mV in the holoenzyme (7, 50, 79). The conclusion is that while the conformation of the heme-Fe in NR's Cb, as recently shown by NMR (102), is similar to that in cytochrome b_5 (59), the environment of the charged residues in the Cb domain in relation to the *N*-terminal domains of NR significantly influences the redox potential of heme-Fe in Cb. Perturbation of the interface between the Cb and *N*-terminal domains may therefore be a mechanism for "redox" regulation of NR activity.

Molybdopterin is a unique cofactor in NR found in only three other enzymes in plants, xanthine dehydrogenase, aldehyde oxidase, and SOX, with some question remaining about the existence of SOX in plants (62). These enzymes have a 31-amino acid residue sequence motif characteristic of eukaryotic molybdopterin oxidoreductases (40), containing the invariant Cys residue involved in binding to Mo-MPT, which is Cys191 in AtNR2 (20, 96). NR and SOX have a similar "oxy" form of Mo-MPT, whereas the other eukaryotic enzymes have a Mo-MPT with a terminal sulfur (37, 47, 51, 62). Prokaryotes contain a conjugated form of molybdopterin where MPT is linked to a nucleotide (37, 47). Although the chemical structure of the core of MPT was worked out some years ago by Johnson et al (47), 3-D structures of the cofactor in enzymes have revealed that the bicyclic pterin ring is fused to the side chain (37, 51, 62), presumably after the MPT is bound to the protein (Figure 2*a*). In the 3-D structure of chicken liver SOX (51), the "X" in MPT is the thiol side chain of Cys-185 (Figure 2*b*, *c*). This structure for Mo-MPT in SOX is consistent with X-ray absorption spectroscopy of human SOX where 3 S and 2 O atoms were liganded to the Mo (29, 30). Recent X-ray absorption spectroscopic analysis of NR reveals the same complement of ligands to Mo with bond distances similar to SOX (G George, J Mertens & WH Campbell, unpublished results). These results provide evidence that the thiol of Cys191 of AtNR2 is indeed a ligand to Mo in the Mo-MPT enzyme complex and suggest that Mo-MPT in NR may have a conformation very similar to the cofactor in SOX. Biosynthesis of MPT from guanosine in plants has been analyzed in detail recently, and the involvement of at least 7 enzymes and proteins has been identified (62).

Functional Fragments and Domains

Cloning of NR and discovery of the linear arrangement of the sequence regions for apparent binding of the cofactors in the enzyme's primary structure were major advances in our understanding of NR biochemistry (Figure 1*b*). A number of studies, including mild proteolytic degradation experiments, had shown that

a

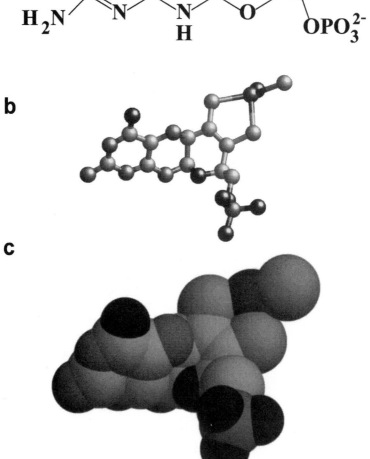

b

c

the cofactor binding fragments of NR, called "domains," represented functional subparts of the enzyme: a flavin-containing dehydrogenase fragment catalyzing either ferricyanide or cytochrome c reduction and the cytochrome b/Mo-MPT fragment catalyzing dye-dependent nitrate reduction (see Figure 1a). These experiments were the direct precursors of the recombinant expression of functional fragments of NR. The first recombinant fragments to be expressed were corn NR's CbR with NADH: ferricyanide reductase activity and *Chlorella* NR's Cb domain, which were expressed in *Escherichia coli* (11, 13, 14, 43). Subsequently, a fragment of NR containing Cb and CbR was expressed as a CcR fragment (11, 76, 90). A combined fragment of a "synthetic" rat cytochrome b5 and spinach NR's CbR was also expressed with cytochrome c reductase activity (73, 74, 98). These experiments unequivocally established that NR is built from modular units with stable structural integrity and catalytic functionality for partial reactions similar to the holo-NR. However, the Mo-containing fragment of NR has not been expressed as a recombinant independent fragment of the enzyme with functionality, but the holo-NR has now been expressed in active form in *Pichia pastoris*, a methylotrophic yeast (95, 96). NR of plants, algae, and fungi has also been expressed in recombinant systems where the NR gene is transformed into a NR-deficient mutant (21, 28, 33, 53, 68, 82–85). Finally and most recently, an extended form of the CcR fragment of corn NR with the putative dimer interface domain, called CcR-plus, has been expressed in *P. pastoris* and purified as an active cytochrome c reductase with a polypeptide size of 65 kD, which helps establish that the interface domain of NR is a structurally stable entity (JA Mertens & WH Campbell, unpublished results).

The CbR fragment of corn NR was crystallized and the 3-D structure determined by X-ray diffraction analysis (11, 56). The structure of CbR is composed of two domains: one for binding FAD and one for binding NADH. These results established NR's CbR as a member of the FNR structure family of flavoenzymes, which is named for ferredoxin NADP$^+$ reductase (6, 49). The interesting feature of this family of enzymes is that its members have little sequence similarity, sharing only a few key sequence motifs, while having a very similar conformation in their FNR-like fragment (17, 44, 49). Structures of many FNR family members have been determined, including spinach and *Anabaena* ferredoxin-NADP$^+$ reductases, corn NR's CbR, rat cytochrome P-450 reductase, pig

---

Figure 2   Structure of the molybdopterin cofactor of nitrate reductase. (*a*) Chemical structure of molybdopterin when bound to a Mo-containing enzyme where X is an undefined ligand atom. (*b*) Ball and stick model of the structure of the Mo-molybdopterin complex of sulfite oxidase with the thiol sulfur atom of Cys185 shown as "X" (51). (*c*) Space-filling model of sulfite oxidase Mo-molybdopterin (51).

cytochrome b_5 reductase, *Pseudomonas cepacia* phthalate dioxygenase reductase, *Alcaligenes eutrophus* flavohemoglobin, and *E. coli* flavodoxin reductase (6, 11, 17, 25, 46, 49, 56, 57, 66, 88, 101). The flavin-binding domains of these enzymes are 6-stranded anti-parallel β-barrels with a single α-helix. The pyridine nucleotide-binding domains are 5- or 6-stranded parallel β-sheets, which have similarity in conformation to the classic Rossman dinucleotide fold found in many dehydrogenases (49, 79). The relative position of the two nucleotide binding domains differs among these enzymes and this difference appears to be related to the electron-acceptor for the flavin, which is either another redox center within the same protein (heme-Fe, iron-sulfur, or another flavin) or in another protein (heme-Fe, ferredoxin, or flavodoxin). In corn NR's CbR, the active site sits between these two domains where two electrons are transferred from NADH to FAD. A structure for the complex of ADP with CbR has also been reported that identified part of the NADH binding site (57). In addition, an atom-replacement model, using mammalian cytochrome b_5 as a guide (59), was made for the cytochrome b domain of corn NR and docked to the CbR structure to generate a model for the CcR fragment of NR (57). Most recently, an apo-CbR (without bound FAD) complex with NAD^+ has been obtained, as well as a native CbR form with NAD^+ bound (54). Unfortunately, none of the NAD^+ complexes with CbR appear to show the cofactor bound into the active site with the nicotinamide portion positioned for electron transfer to FAD. An active site mutant was generated for corn CbR where the only invariant Cys in this sequence region of NR was replaced by Ser (called C242S), and it was shown by kinetic analysis that NADH bound normally to this mutant but that the transfer of electrons from NADH to FAD was greatly impaired (11, 24, 75). The 3-D model of the C242S mutant of CbR showed that the hydroxyl side chain of the replacement Ser hydrogen-bonded to the protein backbone, leaving a void in the active site where the SH group of the Cys in the wild-type enzyme normally sat and directed the positioning of the NADH for optimum electron transfer to the FAD (11, 57). A similar site-directed mutant was recently reported for the CbR fragment of spinach NR, and kinetic and redox potential analyses were carried out (4, 74, 98). These results identify this Cys residue (Cys889 in AtNR2) as the inhibitor-sensitive thiol of NR (see Figure 1a) and demonstrate that its role is in assisting electron transfer and not binding of NADH, as originally thought (9, 11, 12, 24, 75).

Working 3-D Structure Model of Holo-Nitrate Reductase

Crystallization of holo-NR has not yet been successful despite intensive efforts. Recently, Kisker et al (51) determined the 3-D structure of chicken liver SOX, which is the only known protein with a high degree of sequence similarity to the Mo-MPT-binding region of NR (11, 12, 18, 51, 81, 94). Since SOX and NR Mo-MPT fragments have almost 50% identity in sequence, a good quality atom replacement model for this region of NR has been generated using AtNR2 and the coordinates for SOX (WH Campbell & C Kisker, unpublished results). Furthermore, since SOX is a dimer like NR and has a cytochrome b domain with similarity to NR's Cb, it is possible to "dock" the CcR model of corn NR (57) in relation to the Mo-MPT and dimer interface domains to generate a complete working model for dimeric holo-NR (Figure 3: see color section at the end of the volume). The structure of the SOX/NR monomeric unit can be viewed thus: residues 2 to 84 = Cb domain with a 3-stranded anti-parallel β-sheet and 6 α-helixes; residues 96 to 323 = Mo-MPT/sulfite domain with 13 β-strands in 3 β-sheets and 9 α-helixes; and residues 324 to 466 = interface domain with 7-β strands in 2 β-sheets with similarity to the immunoglobulin structural family (51). The only difference between SOX and NR being the Cb domain in NR is *C-terminal* to the interface domain (Figure 1b). The contacts between the SOX monomers are composed almost entirely of residues from the interface domains bonding by hydrogen and ionic bonds (51). A single Mopterin cofactor is buried in the Mo-MPT domain with hydrogen bonds formed by residues that are conserved in both SOX and NR, indicating the cofactor may have the same conformation in the two enzymes (see Figure 2b, c). In addition, sulfite or sulfate was found near the Mo-MPT with the anion liganded to three positively charged residues (Arg-138, Arg-190, and Arg-450), which appear to form the substrate binding site with Trp-204, Tyr-322, and Lys-200 also contributing (51). Only some of these residues are conserved in NR as Arg-144, Arg-196, Trp-210, and Lys-206 in AtNR2, which may form the nitrate-binding pocket. An interchain disulfide bond between the monomers of NR was found in a higher plant form but not in algal and fungal NR forms (45), although no potential Cys residue(s) for this functionality could be identified at the subunit interface in the working model of NR. Three regions of NR cannot be modeled from SOX, including the *N*-terminal "acidic" region, Hinge 1, and Hinge 2. However, the relative position of these parts of NR can be suggested from the positions of the *N*- and *C*-terminal residues of the domains lying on either side of these regions, which are illustrated schematically in Figure 3b. The schematic model also clarifies several features of the holo-NR working model, such as the relative positions of the FAD- and NADH-binding domains of CbR to the Cb, Mo-MPT, and interface domains. The CbR fragment, especially in the short linker region between its FAD and NADH domains, may actually have structural contacts with either the interface or Mo-MPT domains, or both, as well as with the Cb domain. Confirmation of these and other aspects of the structure of holo-NR awaits determination of the structure of the complete enzyme.

FUNCTIONAL CHARACTERISTICS

Reactions Catalyzed by Nitrate Reductase and Its Functional Fragments

The physiological function of NR is to catalyze pyridine nucleotide-dependent nitrate reduction as a component of the nitrogen-acquisition mechanism in higher plants, fungi, and algae. We suggested that NR could also participate in iron reduction in vivo since it catalyzes NADH ferric citrate reduction, but many other enzymes also catalyze this reaction in plants (9, 78). The only other NR catalytic reactions in vivo are with the alternate substrates, chlorate, bromate, and iodate, which yield chlorite, bromite, and iodide (iodite is unstable), respectively. Chlorate reduction is, of course, deadly for plants (chlorate has been used commercially as a defoliant and herbicide) unless they have no NR or nitrate transport system. This property has been a useful tool for obtaining mutant plants (18, 19, 81, 94, 99). Reduction of iodate by algal NR is a likely mechanism for altering the iodate-iodide balance in the ocean. NR is unusual since it is a soluble protein that catalyzes a redox reaction involving an electron transport chain and has physically separated active sites: one for NADH to reduce FAD at the beginning of the electron transport chain and one for reduced NR by the Mo-MPT to reduce nitrate (Figure 1a). NR resembles the mitochondrial electron transport chain since electrons from NADH can leave the enzyme to other acceptors (ferricyanide, cytochrome c, etc) besides nitrate, other electron donors (reduced dyes and flavins) can provide electrons for nitrate reduction, and part of the enzyme can be inhibited while leaving the other part functional (11, 12, 16, 81, 94). However, the large free energy available in NADH-linked nitrate reduction is not conserved unless NR is membrane bound, as it is in bacterial respiratory forms (5). Many claims have been advanced over the years for membrane-bound NR forms in higher plants and algae, but the "membrane" fraction was always small compared to the soluble NR level. In addition, there is no solid evidence proving the existence of membrane NR nor is there any explanation of how membrane-bound NR would function any differently than the soluble form.

The "artificial" partial reactions catalyzed by NR have been very useful in studying the biochemistry of the enzyme and understanding its catalytic mechanism. The partial reactions allowed the clear identification of NR proteolytic fragments as functional units of the enzyme and have been instrumental in characterizing the recombinant fragments of the enzyme. Basically, the fragments of NR with functionality in separate portions of the polypeptide proved that NR has two physically separated active sites, which helps explain its unusual steady-state kinetic mechanism. The kinetic model of NR is a two-site ping-pong type where NADH reduced the FAD at the first active site, and the electrons are passed along the electron transport chain by the internal cytochrome b to the Mo-MPT where nitrate is reduced in the second active site (9, 11). A key concept in this mechanism is that NADH and NAD^+ bind to one active site in a mutually exclusive manner, while nitrate and nitrite act the same way at the other active site. Thus, nitrate can bind to the oxidized enzyme without impeding NR reduction by NADH, which may be important if the nitrate-reducing active site is buried deep in the Mo-MPT domain, as the sulfite reduction site is in the Mo-MPT domain of SOX (51). Although not completely accepted (94), the two-site ping-pong steady-state kinetic model for NR is the most logical description of the steady-state kinetics of NR. NR is a highly efficient catalyst with a turnover number of 200 s^{-1} and true K_m values of 1 to 5 μM for NADH and 20 to 40 μM for nitrate (4, 9, 24, 75, 76, 96, 97).

Pre-steady-state kinetic analysis of the complete NR reaction has not yet been carried out since the holo-enzyme was not available in sufficient amounts. Recombinant production of the CbR and CcR fragments has permitted analysis of the rates of reduction of CbR by NADH and transfer of electrons from reduced FAD to the heme-Fe in CcR (11, 75, 76). Rates of NADH reduction of FAD in CbR and CcR were 474 and 560 s^{-1}, respectively, at 10°C, which was used since the rates at 25°C were too fast to be evaluated by the available equipment. The Kd NADH was 3 μM for either CbR or CcR, which fits well with the K_m NADH for NR (75, 76). The steady-state kcat for NADH ferricyanide reduction catalyzed by CbR at 25°C is 1300 to 1400 s^{-1} (24, 74). Clearly, NADH reduction of NR is faster than overall turnover by a factor of 6 to 7, which indicates either internal electron transfer from $FADH_2$ to Mo-MPT by the cytochrome b or that the reduction of nitrate by Mo-MPT is the rate-limiting process in NR. Electron transfer from $FADH_2$ to heme-Fe in the CcR has a rate of 12 s^{-1} at 10°C (76), which is too slow to have any meaning in relation to electron transfer within holo-NR. This slow step in electron transfer within CcR was thought to be related to the release of NAD^+ from the active site after reduction of FAD and before electrons could move to the heme-Fe, which might also be a slow process in NR catalysis. In the structural model of CcR, the distance between FAD and heme-Fe is about 15 Å (57), whereas the distance between the heme-Fe and Mo-MPT in SOX is 32.3 Å (51). These long distances for electron transfer between the redox centers are surprising considering the rapid turnover rates of these enzymes.

Essential Amino Acid Residues for Functionality

When all the available NR sequences are compared with related proteins and enzymes in a multi-alignment, the number of invariant residues is reduced to ~77 from 917 in AtNR2, and this number can be further decreased to 21 specific residues using other available information (Table 1). Of these key residues, only four have been studied using site-directed mutagenesis. For example, there are only two invariant Cys residues, Cys191 and Cys889, in AtNR2. Cys191 was replaced by Ser and Ala in AtNR2 and expressed in *P. pastoris*; although both mutants produced a complete NR polypeptide, neither was active (96). Cys191 corresponds to Cys185 of SOX, a known ligand to Mo in the enzyme's active site (see Figure 2b), which suggests that Cys191 is essential for NR activity since it must be present to bind to Mo for functionality of this redox center. A Ser replacement mutant of the SOX Mo-liganded Cys has also been generated; it lacks activity and also lacked the sulfur ligand to Mo when examined by X-ray absorption spectroscopy (29, 30, 51). AtNR2 Cys889 is equivalent to the invariant Cys in the CysGly motif of FNR family enzymes (11, 17, 43, 49, 56). Cys242 of corn CbR, equivalent to C889 in AtNR2, was replaced by Ser (called C242S) and the purified, recombinant enzyme fragment lost most of its ferricyanide reductase activity (24). A study of mutants of this invariant Cys in *Neurospora crassa* NADPH:NR also found that all substitutions at this position lost NADPH nitrate-reducing activity as well as ferricyanide reductase activity in a recombinant CbR fragment of this enzyme expressed in *E. coli* (33). Clearly, this Cys is not absolutely required for activity of NR, but its presence makes electron transfer from NADH to FAD much more efficient (24, 75). The role of Cys889 in NR appears to be to position NADH for efficient reduction of FAD. Using the recombinant *A. nidulans* NADPH:NR expression system, several key conserved residues of NR have been investigated using site-directed mutagenesis (28), including replacement of the invariant Cys in the Mo-MPT domain with Ala (C150A in *A. nidulans*) and one of the His ligands of the heme-Fe of the cytochrome b domain with Ala (H547A in *A. nidulans*). The C150A mutant lost all nitrate-reducing activities, whereas H547A lost only the NADPH and methyl viologen NR activities while retaining the bromphenol blue NR, which does not depend on a functional heme-Fe (11, 28, 81, 94). A His-ligand to the heme-Fe of tobacco NR was mutated to an Asn in an NR-deficient plant (63). These results are entirely consistent with the model of NR functionality presented in Figure 1a, where the thiol group shown near the FAD represents Cys889 of AtNR2.

Several other residues in the FAD domain of *N. crassa* NADPH:NR have been mutated in both the CbR fragment and holoenzyme (33). The most notable mutation was the substitution of Gly809 and Thr812 with Val and Ala, respectively, which resulted in loss of both NADPH NR and ferricyanide reductase activities. These residues correspond to residues Gly745 and Ser748 in AtNR2 and Gly95 and Thr98 in recombinant corn CbR. In the 3-D structure of CbR, Gly95 and Thr98 are shown to be involved in binding FAD, with the Gly serving a structural role and the Ser/Thr hydrogen bonded to the pyrophosphate bridge in the middle of FAD (56, 57). Gly308 in AtNR2 was found to be mutated to Asp in a chlorate-resistant mutant plant and the NR was not only inactive but also not phosphorylated in vivo (52). This residue corresponds to Ala297 in SOX, which makes a polypeptide backbone hydrogen bond to the terminal oxygen of Mo in the active site (51). Substitution of Asp for Gly308 in AtNR2 could then abolish NR activity since the backbone might take a very different course in the mutant NR, but how this would preclude phosphorylation is less clear. Ser534 in AtNR2 has been identified as the site of phosphorylation of NR by replacement with Asp, which resulted in loss of inhibition of NR activity in an in vitro assay for regulation of NR (95). This residue in spinach NR (i.e. Ser543) has been studied extensively in relation to the protein kinase that catalyzes phosphorylation of NR and is discussed below. Obviously, Ser534 is not required for NR activity and is not even present in fungal and algal NR forms. Other nonessential, but key residues in NR, such as those at the interface between the two monomeric units of the dimer (Table 1), have not yet been studied. Several groups have investigated the N-terminal region of NR (53, 65, 68, 96), which is clearly not essential since it is virtually absent from some NR forms, and contains no invariant residues. These studies have focused on the role of this region in NR regulation and are discussed below.

Residues Determining Pyridine Nucleotide Specificity in NR

NR is unusual among oxidoreductases since it exists in NADH- and NADPH-specific forms as well as bispecific forms that accept electrons from either NADH or NADPH. Since bispecific NR forms were the first identified in soybeans and as secondary NR forms in monocots like corn, rice, and barley, they have fascinated investigators. Why do some plants have both NADH: and NAD(P)H:NR forms? How do bispecific NR forms differ from the more specific NADH: and NADPH:NR? The first question, which can also be asked for *Arabidopsis* with its two NADH:NR forms, is more physiological, and several answers have been found including tissue specificity and differential expression (16, 18, 19, 78, 81, 94). In soybean, the pH 6.5 NAD(P)H:NR is constitutively expressed without nitrate present, which suggests that this enzyme is involved in another plant process besides nitrogen acquisition. In fact, this NR also catalyzes reduction of nitrite to NOx gases, but this process has never been studied in detail nor is it obvious what benefit it would be to the plant unless NOx is a hormone like it is in animals (9). Corn seedlings appear to express three forms of NR:NADH-specific, NAD(P)H-bispecific with similarity to barley, and a unique NAD(P)H-bispecific with some similarity to green algal NR since it lacks the regulatory Ser found in Hinge 1 of all other known higher plant NR forms (3, 11, 78). The second question can be dealt with as a structural problem and investigated by sequence comparisons and structural determinations. The general conclusion is that NAD(P)H:NR forms appear to be similar to NADH:NR forms except that the barley enzyme has a shorter N-terminal region and minor differences occur in the pyridine nucleotide binding domain (11, 12, 18, 81, 94). Antibodies that readily distinguish the NR forms in barley or corn have not yet been obtained.

Since the 3-D structures of spinach FNR with 2′, 5′ADP bound and corn CbR with ADP bound are known (6, 11, 17, 57), these can serve as models for the design of site-directed mutants of NR forms to alter pyridine nucleotide specificity. The pyridine nucleotide binding domain of FNR family enzymes is a parallel β-sheet with the loops at the ends of the β-strands providing the ligands for binding the cofactor. In FNR and CbR, the third β-strand is followed by a loop containing the residues involved with determining if NADPH or NADH will bind. The distinguishing feature is that FNR binds the 2′ phosphate of NADPH with a positively charged Ser-Arg immediately after the β-strand, whereas CbR binds the 2′ hydroxyl of NADH with a negatively charged Asp (6, 17, 57). Since a Ser-Arg pair of residues is found in *N. crassa* NADPH:NR (Ser920-Arg921) in the same position as the Ser-Arg pair in FNR (43), mutants were prepared where Ala and Thr were substituted for Ser920, and Gly, Ala, and Thr for Arg921 (33). All these substitutions resulted in NR and CbR forms with NADPH nitrate and ferricyanide reducing activity that was decreased relative to wild type except for S920T, which had increased NR activity. To determine if the key residues in *N. crassa* NR are aligned like those in FNR or corn NR, mutant CbR forms were designed with Asp substituted for Ser920 and Ser and Thr for Arg921 and the mutant proteins purified for detailed kinetic analysis with both NADPH and NADH (91). Substituting Asp for Ser920 resulted in a virtual conversion of *N. crassa* CbR into an NADH-specific enzyme, whereas substitutions at Arg921 had no impact on specificity. Thus, it appeared that the 2′ phosphate of NADPH was not bound by Arg921, which indicates that the binding pocket in *N. crassa* NADPH:NR is more like that in NADH:NR than it is in FNR. An Arg more C-terminal (Arg932 in *N. crassa* NR), which is conserved in nearly all fungal NR forms and also found in monocot bispecific NR forms, is a candidate for supplying the positive charge in the binding site (91). Mutation of Arg932 did disrupt pyridine nucleotide binding but without showing a clear preference between NADPH and NADH (91). Birch NAD(P)H:NR is the only bispecific form to be investigated so far by site-directed mutagenesis, where residues in the binding site for the ribose and adenine of NADPH/NADH were altered (85). The Pro following the invariant Cys-Gly near the C terminus of NR (Pro891 in AtNR2), which is an Ala in birch NR and reverted to a Pro in the key mutant, favored NADH binding versus NADPH. The corresponding residue

Pro244 in corn CbR is indeed near the ribose of ADP and is well positioned to influence which cofactor binds to NR (57). Since other bispecific NR forms have not yet been studied in detail, the best explanation of their structural basis lies in the Ser or Lys residue found in the position immediately after the third β-strand of the pyridine nucleotide binding domain. If this is a negatively charged residue such as Glu854 in AtNR2 or Asp205 in corn CbR, then it appears that the negative charge on the 2′ phosphate of NADPH is repelled and the preferred cofactor is NADH.

Glutathione reductase (EC 1.6.4.2) and isocitrate dehydrogenase (EC 1.1.1.42) are enzymes where NADPH/NADP$^+$-specific forms have been converted into NADH/NAD$^+$-specific forms by site-directed mutagenesis and detailed kinetic analysis carried out (15, 86). The 3-D structures of the wild-type and mutant forms have been compared to analyze the fine structure of their pyridine nucleotide binding sites (40, 63). In both cases, there are many differences between the NADH and NADPH binding sites, and consequently the conformation of the cofactor differs significantly when bound to the enzyme. This is reflected in the fact that about seven mutations were needed in each of these enzymes to convert from an NADPH- to NADH-specific form (15, 41, 64, 86), unlike NR where *N. crassa* CbR was converted from NADPH-specific to NADH-specific by a single mutation (91). Although the structure needs to be determined for NR, it appears that NADPH and NADH are bound to the enzyme in a very similar conformation, with the major difference being focused on the residues involved in binding the 2′ phosphate of NADPH and 2′ hydroxyl group of NADH. A similar conclusion was drawn in comparing the structures of spinach NADP$^+$ FNR and bacterial NADH phthalate dioxygenase reductase (17). The general conclusion is that only small changes in the fine structure of the pyridine nucleotide domain of NR are required to change from an NADH-specific form to an NADPH-specific form or perhaps even fewer to make a bispecific NR from a monospecific one, which may explain the existence of so many bispecific NR forms in nature.

REGULATION

Molecular Mechanisms for Regulation

NR catalytic flux or the total nitrate-reducing capacity of a plant system depends on: (*a*) availability of the substrates in the cytoplasm (steady-state concentrations of NAD(P)H and nitrate); (*b*) the level of functional NR (amount of NR polypeptide and availability of cofactors and metal ions—FAD, heme, Fe, Mo-MPT, and Mo); and (*c*) the activity level of the functional NR. Each process is regulated either directly or indirectly, and the overall level of nitrate reduction capacity is controlled in relation to overall plant metabolic level by metabolic sensors and signal transduction pathways. Stitt and coworkers (82–84) have recently studied in detail these control systems in tobacco plants with the normal four copies of the NR gene, and they genetically manipulated plants with 0, 1, and 2 NR gene copies. The free Gln level and its ratio to free Glu, as well as the nitrate level, are probably the key metabolites governing the level of nitrate-reducing capacity in a plant (18, 81, 82, 94). When Gln levels are low and nitrate is available, NR level and nitrate-reducing capacity are boosted, whereas high Gln levels "throttle" nitrate reduction and decrease activity levels of NR. However, in transgenic plants where NR mRNA is expressed constitutively and posttranslational control is lost owing to deletion of the N-terminal region of NR, the control linked to the Gln/Glu balance is lost and nitrate-reducing capacity is controlled by NADH availability only (53, 68). In normal plants with optimum growth conditions and sufficient nitrate, the nitrate-reducing capacity is about two times greater than the plant needs, and NR activity levels cycle on a daily basis with low activity in the dark (82). Nitrate essentially acts as a hormone in plants by inducing functional NR and a host of other enzymes and proteins, perhaps including DNA regulatory proteins, involved in the metabolic response to the availability of a limiting nutrient, which includes changes in the root-to-shoot ratio and morphological changes such as root hair development (18, 78, 82–84). The expression of nitrate transporter genes are also induced by nitrate (18, 99). The degree of the plant response to nitrate depends on other environmental and genetic factors such as light and plant genotype, which influence NR and the other components of the nitrate metabolic mode. The response of NR to nitrate depends on a constitutively produced "nitrate-sensing" protein of unknown character, which presumably binds to regulatory regions in the NR gene and turns on expression of the NR mRNA (78). The nitrate box regulatory sequences in the promoter of the NR genes have been identified (42). Presumably, other regulatory boxes such as for light, tissue specificity, Gln/Glu balance, water and carbohydrate status, the photosynthesis rate of the plant, and other limiting conditions are present in the promoters of NR and related nitrate response genes (18, 78). These signals are integrated with the nitrate response by their specific DNA-binding proteins, which combine to influence the level of gene transcription by the strength of the initiation complex for binding RNA polymerase II to start sequences in the genes. While NR mRNA levels rise rapidly in response to nitrate treatment of plants and reach a steady-state level in a few hours (18, 19, 81, 94), in plants where nitrate levels cycle, NR mRNA levels also cycle (27, 78, 82). Efficiency of NR polypeptide translation may also be a site for regulation but this has not been carefully studied. NR polypeptide must be assisted in folding by various chaperones but none unique to NR has been identified. Although inhibition of heme biosynthesis blocks NR activity appearance, the small amounts of functional NR protein required to meet the nitrate-reducing needs of plants probably require little change in the normal cellular production of FAD and heme-Fe. Molybdate is probably transported into plants by the phosphate transport system and is probably not limiting. MPT biosynthesis requires seven gene products, but these are constitutively expressed and probably not influenced by the presence of nitrate (62). Analysis of NR protein levels in plants displaying a daily cycle of NR activity has revealed that NR is degraded daily (9, 11, 82). In summary, the level of functional NR is controlled at the transcriptional level, with fully complemented NR rapidly formed by combination of the NR polypeptide with the required cofactors. Functional NR, however, has a short half-life and the protein is degraded by proteolytic attack, perhaps involving a specific NR proteinase although no ubiquitous enzyme of this type has been identified. Superimposed on the de novo synthesis and irreversible degradation of NR are reversible controls of the level of enzyme activity.

NR Phosphorylation and Inhibition by the 14-3-3 Binding Protein

Kaiser and coworkers (48) discovered through the use of physiological studies of the level of NR activity in plants that NR is probably regulated by protein phosphorylation and dephosphorylation. Subsequently, NR was demonstrated to be a phospho-protein with the phosphorylated NR level linked to inhibition of NR activity in the dark in leaves, which depends on the presence of divalent cations such as Ca^{2+} or Mg^{2+} (38, 58). NR protein kinases that depend on calcium for activity have been characterized (3, 22, 39, 65). The specific site of NR phosphorylation is Ser534 of AtNR2 (Ser543 in spinach NR), which is found in Hinge 1 of virtually all higher plant NR forms (3, 23, 95). The identity of the target Ser has been confirmed in AtNR2 by directed mutation to Asp, which eliminated NR inhibition with an in vitro ATP-dependent system, and in spinach NR by sequencing of the phosphopeptides isolated after tryptic digestion and from model peptide studies (3, 23, 39, 65, 95). The sequence context of the Ser targeted for phosphorylation is important for protein kinase recognition and can be summarized as Leu-Lys-(Lys/Arg)-(Ser/Thr)-(Val/Ile/Ala)-target Ser-(Thr/Ser)-Pro-Phe-Met (3, 39). The reactivation of NR in the light depends on a type 2A protein phosphatase (inhibited by microcystin and okadaic acid), which catalyzes dephosphorylation of NR (38, 39, 58, 65). This apparently straightforward reversible regulatory mechanism for NR activity is complicated by the fact that highly purified NR is not inhibited by phosphorylation in vitro in the presence or absence of Mg^{2+} (39). This led to the discovery of an inhibitor protein in the extracts used to supply protein kinase activity, which has been identified as the widespread binding protein called 14-3-3 (2, 26, 39, 65, 87). The 3-D structure of mammalian 14-3-3 proteins has been determined; it contains nine α-helices and is a dimer with two binding grooves about 30 Å apart (55, 103). Complexes of 14-3-3 proteins bound with target peptides containing the sequence Arg-Xxx-Xxx-Ser(P)-Xxx-Pro have been analyzed; these show that the phosphate is bound to two Arg and a Lys residue in the binding groove on one side whereas the other binds hydrophobic residues of the target sequence (71). This sequence is similar to that required for protein kinase recognition of higher plant NR except that the initial Arg is often a Lys in NR. In addition, other unique sequences with a negative charge in place of the phospho-Ser can also bind to the groove in 14-3-3 (71). Since NR from which the N-terminal highly acidic region has been deleted does not bind 14-3-3 proteins as well as native NR (53, 65, 68), the existence of a secondary binding site for 14-3-3 in this region is possible and it might not require phosphorylation for binding. Since fusicoccin also binds to 14-3-3, this fungal toxin can reverse the inactivation of NR by the binding protein (65, 87). To summarize reversible regulation, the NR protein must be phosphorylated by Mg-ATP at the unique Ser in Hinge 1 as catalyzed by a "specific" NR protein kinase; then in the presence of Mg^{2+}, a binding protein called 14-3-3, which is already present in the cytoplasm, binds to phospho-NR and inhibits NR activity. The binding of 14-3-3 to phospho-NR appears to block electron flow from the Cb domain to the Mo-MPT by an unknown mechanism (38, 39). Since the 14-3-3 binding protein appears to bind in a region where the dimer interface and Mo-MPT domains meet the Cb domain (see Figure 3b), the inhibition may be due to modulation of the redox potential of the heme-Fe, which is altered by disturbance in this region.

PRACTICAL APPLICATIONS

The Nitrate Pollution Problem

By the 1970s, the accumulation of nitrate in some surface and groundwaters of the United States, Canada, and Europe had become a serious enough threat to human health for most countries to adopt a Maximum Contaminant Limit (MCL) for potable water (36, 67, 92). In the United States, the MCLs for nitrate as 10 ppm nitrate-N and nitrite as 1 ppm nitrite-N were set by the Clean Water and Safe Drinking Water Acts of 1974. The immediate threat posed by high concentrations of nitrate in drinking water is methemoglobinemia or blue-baby syndrome, caused by strong binding of nitrite to hemoglobin and oxidation of the iron center, which is more serious in infants, since fetal hemoglobin binds nitrite more strongly and can result in death. The linkage between other human health risks such as cancer and long-term exposure to high nitrate concentrations in drinking water are not well enough documented to justify further restriction of the nitrate MCL, which already takes about 10% of the potable water in the

United States out of the useable pool (36, 67). Nitrate pollution is probably caused by agricultural practices whereby excess N is applied to fields to maximize crop yield and animal wastes are released into the environment without having to undergo a tertiary process to remove nutrients. Industrial processes and air pollution, especially from automobile exhaust, also contribute significantly to nitrate pollution (36, 72, 92, 100). One initiative to control the use of excess fertilizer is the USDA's nitrate leaching and economic analysis package (NLEAP) computer program. How to balance the need for increased crop production with the control of pollution caused by underutilization of applied nutrients is an active area of research (60, 69, 89). Nevertheless, nitrate/nitrogen and other nutrient pollution has become a major ecological problem worldwide (92, 100). The increase in toxic algal blooms in many coastal waters, the "dead zone" in the Gulf of Mexico along the southern US coastline, and the massive fish kills in American estuaries often accompanied by toxic microalgae such as *Pfiesteria piscicida* (70) are illustrations in point. Although the causes of ecosystem changes in these complex systems are difficult to pinpoint, nutrient runoff is of growing concern worldwide.

Nitrate Reductase as an Environmental Biotechnology Tool

Regulations to monitor nitrate usage require a reliable method to detect and quantify nitrate in water and soil. Although methods for quantification of nitrate based on NR-driven reduction to nitrite and colorimetric nitrite analysis were described many years ago, chemical reductants such as cadmium and zinc are the most commonly used commercially (8). The availability of stable preparations of NR that can be shipped at room temperature has promoted the development of NR-based commercial nitrate detection tests (8). Nitrate testing based on NR is a less polluting alternative to tests based on toxic heavy metals and has become a standard in biomedical research where nitrate and nitrite are often monitored as indicators of nitric oxide production (8, 32, 34, 35). Nitrate must also be monitored in surface and ground waters on a real-time basis. Devices based on chemical reduction of nitrate have been described, but none has yet been widely adopted, and nitrate biosensors based on NR have attracted attention (1, 32). An optical method using immobilized NR to detect nitrate has been developed (1). Recently, a nitrate electrode was developed based on corn leaf NR, where electrons are directly supplied to the enzyme for nitrate reduction by an electrode coated with methyl viologen. It has the capability to quantify nitrate in fertilizer solutions (32). The rapid advances made in using enzymes in other biosensors such as those based on glucose oxidase for monitoring blood glucose promise the early advent of a commercial nitrate biosensor based on NR. Methods to remove nitrate from potable water would be a solution for many communities and individuals where nitrate pollution has contaminated their drinking water source. Nitrate is one of the most soluble anions and is difficult to remove by physical processes such as ion-exchange; although reverse osmosis is effective in removing nitrate, it is expensive. Furthermore, these removal methods concentrate the nitrate, which presents an additional significant waste-disposal problem. A better approach is to use denitrification, especially where the final product is environmentally benign dinitrogen gas. Denitrification can be accomplished by microorganisms (5), but the process can be slow and requires additional purification of the potable water. An alternative is to use an enzyme-based system with a combination of NR and bacterial denitrification enzymes (61). In the enzyme denitrification reactor, the reducing power was supplied to the immobilized enzymes by direct electric current through electron-carrying dyes, which resulted in complete removal of nitrate as dinitrogen with no additions to the water. This approach can be used for potable water at the point-of-use and generates no waste-disposal problem. The application of recombinant DNA technology to the production of NR and bacterial denitrification enzymes may make enzyme reactors for treating nitrate-polluted water commercially available. Clearly, basic research on NR has helped to advance the applications of this unique enzyme to address the nitrate pollution problem and genetically engineered plants may well promote more efficient use of N-fertilizers and thus reduce or prevent the pollution attributable to the application of excess nutrients.

EXPECTED DEVELOPMENTS

Within the next few years elucidation of a 3-D structure for plant holo-NR will guide its future study. Better understanding of the electron transfer process within NR is on the horizon, as is the identification of the rate-limiting process in nitrate reduction catalyzed by pyridine nucleotide-dependent NR. More detailed studies of various forms of NR including the unusual bispecific forms should provide a clearer understanding of pyridine nucleotide specificity. A better understanding of how the 14-3-3 binding protein inhibits phospho NR in higher plants and description of the details of a 3-D structural complex between phospho NR and 14-3-3 will be important in delineating the mechanism underlying regulation. Finally, more information is needed on how the nitrate signal is transmitted in plants and the character of DNA regulatory proteins that bind to the nitrate box in nitrate response genes. Considerable progress has been made recently in understanding the process of MPT biosynthesis, and the insertion of Mo-MPT into NR and related enzymes will surely be a focus of future research. Ultimately, validation of the public investment in basic research on NR may come in the form of commercial devices for monitoring nitrate levels in real-time and enzyme reactors for removing nitrate from water.

ACKNOWLEDGMENTS

I thank the US National Science Foundation and the US Department of Agriculture for the longstanding support of my research on nitrate reductase by means of various grants over the past 20 years. I also thank colleagues who shared their unpublished work with me.

Visit the *Annual Reviews home page* at
http://www.AnnualReviews.org

Literature Cited

1. Aylott JW, Richardson DJ, Russell DA. 1997. Optical biosensing of nitrate ions using a sol-gel immobilized nitrate reductase. *Analyst* 122:77–80
2. Bachmann M, Huber JL, Liao PC, Gage DA, Huber SC. 1996. The inhibitor protein of phosphorylated nitrate reductase from spinach. *Spinacia oleracea* leaves is a 14-3-3 protein. *FEBS Lett.* 387:127–31
3. Bachmann M, Shiraishi N, Campbell WH, Yoo BC, Harmon AC, et al. 1996. Identification of the major regulatory site as Ser-543 in spinach leaf nitrate reductase and its phosphorylation by a Ca^{2+}-dependent protein kinase in vitro. *Plant Cell* 8:505–17
4. Barber MJ, Trimboli AJ, Nomikos S, Smith ET. 1997. Direct electrochemistry of the flavin domain of assimilatory nitrate reductase: effects of NAD^+ and NAD^+ analogs. *Arch. Biochem. Biophys.* 345:88–96
5. Berks BC, Ferguson SJ, Moir JW, Richardson DJ. 1995. Enzymes and associated electron transport systems that catalyse the respiratory reduction of nitrogen oxides and oxyanions. *Biochim. Biophys. Acta* 1232:97–173
6. Bruns CM, Karplus PA. 1995. Refined crystal structure of spinach ferredoxin reductase at 1.7 Å resolution: oxidized, reduced and 2'-phospho-5'-AMP bound states. *J. Mol. Biol.* 247:125–45
7. Brunt CE, Cox MC, Thurgood AG, Moore GR, Reid GA, et al. 1992. Isolation and characterization of the cytochrome domain of flavocytochrome b_2 expressed independently in *Escherichia coli. Biochem. J.* 283:87–90
8. Campbell ER, Corrigan JS, Campbell WH. 1997. Field determination of nitrate using nitrate reductase. In *Proc. Symp. Field Anal. Methods Hazard. Wastes*

Toxic Chem., ed. E Koglin, pp. 851–60. Pittsburgh: Air & Waste Manage. Assoc.
9. Campbell WH. 1989. Structure and synthesis of higher plant nitrate reductase. See Ref. 102a, pp. 123–54
10. Campbell WH. 1992. Expression in *Escherichia coli* of cytochrome c reductase activity from a maize NADH:nitrate reductase cDNA. *Plant Physiol.* 99:693–99
11. Campbell WH. 1996. Nitrate reductase biochemistry comes of age. *Plant Physiol.* 111:355–61
12. Campbell WH, Kinghorn JR. 1990. Functional domains of assimilatory nitrate reductases and nitrite reductases. *Trends Biochem. Sci.* 15:315–19
13. Cannons AC, Barber MJ, Solomonson LP. 1993. Expression and characterization of the heme-binding domain of *Chlorella* nitrate reductase. *J. Biol. Chem.* 268:3268–71
14. Cannons AC, Iida N, Solomonson LP. 1991. Expression of a cDNA clone encoding the haem-binding domain of *Chlorella* nitrate reductase. *Biochem. J.* 278:203–9
15. Chen R, Greer A, Dean AM. 1995. A highly active decarboxylating dehydrogenase with rationally inverted coenzyme specificity. *Proc. Natl. Acad. Sci. USA* 92:11666–70
16. Cheng C-L, Acedo GN, Dewdney J, Goodman HM, Conkling MA. 1991. Differential expression of the two Arabidopsis nitrate reductase genes. *Plant Physiol.* 96:275–79
17. Correll CC, Ludwig ML, Bruns C, Karplus PA. 1993. Structural prototypes for an extended family of flavoprotein reductases: comparison of phthalate dioxygenase reductase with ferredoxin reductase and ferredoxin. *Protein Sci.* 2:2112–33

18. Crawford NM. 1995. Nitrate: nutrient and signal for plant growth. *Plant Cell* 7:859–68

19. Crawford NM, Arst HN. 1993. The molecular genetics of nitrate assimilation in fungi and plants. *Annu. Rev. Genet.* 27:115–46

20. Crawford NM, Smith M, Bellissimo D, Davis RW. 1988. Sequence and nitrate regulation of the *Arabidopsis thaliana* mRNA encoding nitrate reductase, a metalloflavoprotein with three functional domains. *Proc. Natl. Acad. Sci. USA* 85:5006–10

21. Dawson HN, Burlingame R, Cannons AC. 1997. Stable transformation of Chlorella: rescue of nitrate reductase-deficient mutants with the nitrate reductase gene. *Curr. Microbiol.* 35:356–62

22. Douglas P, Moorhead G, Hong Y, Morrice N, MacKintosh C. 1998. Purification of a nitrate reductase kinase from *Spinach oleracea* leaves, and its identification as a calmodulin-domain protein kinase. *Planta* 206:435–42

23. Douglas P, Morrice N, MacKintosh C. 1995. Identification of a regulatory phosphorylation site in the hinge 1 region of nitrate reductase from spinach. *Spinacea oleracea* leaves. *FEBS Lett.* 377:113–17

24. Dwivedi UN, Shiraishi N, Campbell WH. 1994. Identification of an "essential" cysteine of nitrate reductase via mutagenesis of its recombinant cytochrome b reductase domain. *J. Biol. Chem.* 269:13785–91

25. Ermler U, Siddiqui RA, Cramm R, Friedrich B. 1995. Crystal structure of the flavohemoglobin from Alcaligenes eutrophus at 1.75 Å resolution. *EMBO J.* 14:6067–77

26. Ferl RJ. 1996. 14-3-3 Proteins and signal transduction. *Annu. Rev. Plant Physiol. Plant Mol. Biol.* 47:49–73

27. Foyer CH, Valadier MH, Migge A, Becker TW. 1998. Drought-induced effects on nitrate reductase activity and mRNA and on the coordination of nitrogen and carbon metabolism in maize leaves. *Plant Physiol.* 117:283–92

28. Garde J, Kinghorn JR, Tomsett AB. 1995. Site-directed mutagenesis of nitrate reductase from *Aspergillus nidulans*. *J. Biol. Chem.* 270:6644–50

29. Garrett RM, Rajagopalan KV. 1996. Site-directed mutagenesis of recombinant sulfite oxidase. *J. Biol. Chem.* 271:7387–91

30. George GN, Garrett RM, Prince RC, Rajagopalan KV. 1996. XAS and EPR spectroscopic studies of site-directed mutants of sulfite oxidase. *J. Am. Chem. Soc.* 118:8588–92

31. Giovannoni G, Land JM, Keir G, Thompson EJ, Heales SJ. 1997. Adaptation of the nitrate reductase and Griess reaction methods for the measurement of serum nitrate plus nitrite levels. *Ann. Clin. Biochem.* 34:193–98

32. Glazier SA, Campbell ER, Campbell WH. 1998. Construction and characterization of nitrate reductase-based amperometric electrode and nitrate assay of fertilizers and drinking water. *Anal. Chem.* 70:1511–15

33. Gonzalez C, Brito N, Marzluf GA. 1995. Functional analysis by site-directed mutagenesis of individual amino acid residues in the flavin domain of *Neurospora crassa* nitrate reductase. *Mol. Gen. Genet.* 249:456–64

34. Granger DL, Taiutor RR, Broockvar KS, Hibbs JB. 1996. Measurement of nitrate and nitrite in biological samples using nitrate reductase and Griess reaction. *Methods Enzymol.* 268:142–51

35. Grisham MB, Johnson GG, Lancaster JR. 1996. Quantitation of nitrate and nitrite in extracellular fluids. *Methods Enzymol.* 268:237–45

36. Hallberg GR. 1989. Nitrate in ground water in the United States. In *Nitrogen Management and Ground Water Protection*, ed. RF Follet, pp. 35–74. Amsterdam: Elsevier

37. Hille R. 1996. The mononuclear molybdenum enzymes. *Chem. Rev.* 96:2757–816

38. Huber JL, Huber SC, Campbell WH, Redinbaugh MG. 1992. Reversible light/dark modulation of spinach leaf nitrate reductase activity involves protein phosphorylation. *Arch. Biochem. Biophys.* 296:58–65

39. Huber SC, Bachmann M, Huber JL. 1996. Post-translation regulation of nitrate reductase activity: a role for Ca^{2+} and 14-3-3 proteins. *Trends Plant Sci.* 1:432–38

40. Hughes RK, Doyle WA, Chovnick A, Whittle JR, Burke JF, et al. 1992. Use of rosy mutant strains of *Drosophila melanogaster* to probe the structure and function of xanthine dehydrogenase. *Biochem. J.* 285:507–13

41. Hurley JH, Chen R, Dean AM. 1996. Determinants of cofactor specificity in isocitrate dehydrogenase: structure of an engineered $ADP^+ \rightarrow NAD^+$ specificity-reversal mutant. *Biochemistry* 35:5670–78

42. Hwang CF, Lin Y, D'Souza T, Cheng CL. 1997. Sequences necessary for nitrate-dependent transcription of Arabidopsis nitrate reductase genes. *Plant Physiol.* 113:853–62

43. Hyde GE, Campbell WH. 1990. High-level expression in *Escherichia coli* of the catalytically active flavin domain of corn leaf NADH:nitrate reductase and its comparison to human NADH:cytochrome b_5 reductase. *Biochem. Biophys. Res. Commun.* 168:1285–91

44. Hyde GE, Crawford N, Campbell WH. 1991. The sequence of squash NADH:nitrate reductase and its relationship to the sequences of other flavoprotein oxidoreductases: a family of flavoprotein pyridine nucleotide cytochrome reductases. *J. Biol. Chem.* 266:23542–47

45. Hyde GE, Wilberding JA, Meyer AL, Campbell ER, Campbell WH. 1989. Monoclonal antibody-based immunoaffinity chromatography for purifying corn and squash NADH:nitrate reductases. Evidence for an interchain disulfide bond in nitrate reductase. *Plant Mol. Biol.* 13:233–46

46. Ingelman M, Bianchi V, Eklund H. 1997. The three-dimensional structure of flavodoxin reductase from *Escherichia coli* at 1.7 Å resolution. *J. Mol. Biol.* 268:147–57

47. Johnson JL, Bastian NR, Rajagopalan KV. 1990. Molybdopterin guanine dinucleotide: a modified form of molybdopterin identified in the molybdenum cofactor of dimethyl sulfoxide reductase from *Rhodobacter sphaeroides* forma *specialis denitrificans. Proc. Natl. Acad. Sci. USA* 87:3190–94

48. Kaiser WM, Brendle-Behnisch E. 1991. Rapid modulation of spinach leaf nitrate reductase activity by photosynthesis. *Plant Physiol.* 96:363–67

49. Karplus PA, Daniels MJ, Herriott JR. 1991. Atomic structure of ferredoxin-NADP$^+$ reductase: prototype for a structurally novel flavoenzyme family. *Science* 251:60–66

50. Kay CJ, Lippay EW. 1992. Mutation of the heme-binding crevice of flavocytochrome b_2 from *Saccharomyces cerevisiae*: altered heme potential and absence of redox cooperativity between heme and FMN centers. *Biochemistry* 31:11376–82

51. Kisker C, Schindelin H, Pacheco A, Wehbi WA, Garrett RM, et al. 1997. Molecular basis of sulfite oxidase deficiency from the structure of sulfite oxidase. *Cell* 91:973–83

52. LaBrie ST, Crawford NM. 1994. A glycine to aspartic acid change in the MoCo domain of nitrate reductase reduces both activity and phosphorylation levels in Arabidopsis. *J. Biol. Chem.* 269:14497–501

53. Lejay L, Quillere I, Roux Y, Tillard P, Cliquet J-B, et al. 1997. Abolition of posttranscriptional regulation of nitrate reductase partially prevents the decrease in leaf NO_3^- reduction when photosynthesis is inhibited by CO_2 deprivation, but not in darkness. *Plant Physiol.* 115:623–30

54. Lindqvist Y, Lu G, Schneider G, Campbell WH. 1997. Crystallographic studies of the FAD/NADH binding fragment of corn nitrate reductase. See Ref. 94a, pp. 899–907

55. Liu D, Bienkowska J, Petosa C, Collier RJ, Fu H, et al. 1995. Crystal structure of the zeta isoform of the 14-3-3 protein. *Nature* 376:191–94

56. Lu G, Campbell WH, Schneider G, Lindqvist Y. 1994. Crystal structure of the FAD-containing fragment of corn nitrate reductase at 2.5 Å resolution: relationship to other flavoprotein reductases. *Structure* 2:809–21

57. Lu G, Lindqvist Y, Schneider G, Dwivedi UN, Campbell WH. 1995. Structural studies on corn nitrate reductase. Refined structure of the cytochrome b reductase fragment at 2.5 Å, its ADP complex and an active site mutant and modeling of the cytochrome b domain. *J. Mol. Biol.* 248:931–48

58. MacKintosh C. 1992. Regulation of spinach-leaf nitrate reductase by reversible phosphorylation. *Biochim. Biophys. Acta* 1137:121–26

59. Mathews FS, Levine M, Argos P. 1971. The structure of calf liver cytochrome b_5 at 2.8 Å resolution. *Nat. New Biol.* 233:15–16

60. Matson PA, Naylor R, Ortiz-Monasterio I. 1998. Integration of environmental, agronomic and economic aspects of fertilizer management. *Science* 280:112–15

61. Mellor RB, Ronnenberg J, Campbell WH, Diekmann S. 1992. Reduction of nitrate and nitrite in water by immobilized enzymes. *Nature* 355:717–19

62. Mendel RR. 1997. Molybdenum cofactor of higher plants: biosynthesis and molecular biology. *Planta* 203:399–405

63. Meyer C, Levin JM, Roussel J-M, Rouze

P. 1991. Mutational and structural analysis of the nitrate reductase heme domain of *Nicotiana plumbaginifolia*. *J. Biol. Chem.* 266:20561–66

64. Mittl PR, Berry A, Scrutton NS, Perham RN, Schulz GE. 1993. Structural differences between wild-type NADP-dependent glutathione reductase from *Escherichia coli* and a redesigned NAD-dependent mutant. *J. Mol. Biol.* 231:191–95

65. Moorhead G, Douglas P, Morrice N, Scarable M, Aitken A, et al. 1996. Phosphorylated nitrate reductase from spinach leaves is inhibited by 14-3-3 proteins and activated by fusicoccin. *Curr. Biol.* 6:1104–13

66. Nishida H, Inaka K, Yamanaka M, Kaida S, Kobayashi K, et al. 1995. Crystal structure of NADH-cytochrome b_5 reductase from pig liver at 2.4 Å resolution. *Biochemistry* 34:2763–67

67. Nolan BT, Ruddy BC. 1996. *Nitrate in the ground waters of the United States—assessing the risk*. U.S. Geol. Surv. Fact Sheet FS-092–96 (http://wwwrvares.er.usgs.gov/nawqa/fs-092–96/fs-0092–96main.html)

68. Nussaume L, Vincentz M, Meyer C, Boutin J-P, Caboche M. 1995. Posttranscriptional regulation of nitrate reductase by light is abolished by an N-terminal deletion. *Plant Cell* 7:611–21

69. Pang XP, Gupta SC, Moncrief JF, Rosen CJ, Cheng HH. 1998. Evaluation of nitrate leaching potential in Minnesota glacial outwash soils using the CERES-maize model. *J. Environ. Qual.* 27:75–85

70. Pelley J. 1998. What is causing toxic algal blooms. *Environ. Sci. Technol.* 32:A26–30

71. Petosa C, Masters SC, Bankston LA, Pohl J, Wang B, et al. 1998. 14-3-3 zeta binds a phosphorylated Raf peptide and an unphosphorylated peptide via its conserved amphipathic groove. *J. Biol. Chem.* 273:16305–10

72. Puckett LJ. 1995. Identifying the major sources of nutrient water pollution. *Environ. Sci. Technol.* 29:408A–14A

73. Quinn GB, Trimboli AJ, Barber MJ. 1994. Construction and expression of a flavocytochrome b_5 chimera. *J. Biol. Chem.* 269:13375–81

74. Quinn GB, Trimboli AJ, Prosser IM, Barber MJ. 1996. Spectroscopic and kinetic properties of a recombinant form of the flavin domain of spinach NADH:nitrate reductase. *Arch. Biochem. Biophys.* 327:151–60

75. Ratnam K, Shiraishi N, Campbell WH, Hille R. 1995. Spectroscopic and kinetic characterization of the recombinant wild-type and C242S mutant of the cytochrome b reductase fragment of nitrate reductase. *J. Biol. Chem.* 270:24067–72

76. Ratnam K, Shiraishi N, Campbell WH, Hille R. 1997. Spectroscopic and kinetic characterization of the recombinant cytochrome c reductase fragment of nitrate reductase: identification of the rate limiting catalytic step. *J. Biol. Chem.* 272:2122–28

77. Redinbaugh MG, Campbell WH. 1985. Quaternary structure and composition of squash NADH:nitrate reductase. *J. Biol. Chem.* 260:3380–85

78. Redinbaugh MG, Campbell WH. 1991. Higher plant responses to environmental nitrate. *Physiol. Plant.* 82:640–50

79. Reid LS, Taniguchi VT, Gray HB, Mauk AG. 1982. Oxidation-reduction equilibrium of cytochrome b_5. *J. Am. Chem. Soc.* 104:7516–19

80. Rossmann MG, Mora D, Olsen KW. 1974. Chemical and biological evolution of a nucleotide-binding protein. *Nature* 250:194–99

81. Rouze P, Caboche M. 1992. Nitrate reduction in higher plants: molecular approaches to function and regulation. In *Inducible Plant Proteins*, ed. JL Wray, pp. 45–77. Cambridge, MA: Cambridge Univ. Press

82. Scheible WR, Gonzalez-Fontes A, Morcuende R, Lauerer M, Geiger M, et al. 1997. Tobacco mutants with a decreased number of functional nia genes compensate by modifying the diurnal regulation of transcription, post-translational modification and turnover of nitrate reductase. *Planta* 203:304–19

83. Scheible WR, Gonzalez-Fontes A, Morcuende R, Lauerer M, Muller-Rober B, et al. 1997. Nitrate acts as a signal to induce organic acid metabolism and repress starch metabolism in tobacco. *Plant Cell* 9:783–98

84. Scheible WR, Lauerer M, Schulze ED, Caboche M, Stitt M. 1997. Accumulation of nitrate in the shoot acts as signal to regulate shoot-root allocation in tobacco. *Plant J.* 11:671–91

85. Schondorf T, Hachtel W. 1995. The choice of reducing substrate is altered by replacement of an alanine by a proline in the FAD domain of a bispecific NAD(P)H-nitrate reductase from birch. *Plant Physiol.* 108:203–10

86. Scrutton NS, Berry A, Perham RN.

1990. Redesign of the coenzyme specificity of a dehydrogenase by protein engineering. *Nature* 343:38–43

87. Sehnke PC, Ferl RJ. 1996. Plant metabolism: enzyme regulation by 14-3-3 proteins. *Curr. Biol.* 6:1403–5

88. Serre L, Vellieux FM, Medina M, Gomez-Moreno C, Fontecilla-Camps JC, et al. 1996. X-ray structure of the ferredoxin:NADP$^+$ reductase from the cyanobacterium Anabaena PCC 7119 at 1.8 Å resolution, and crystallographic studies of NADP$^+$ binding at 2.25 Å resolution. *J. Mol. Biol.* 263:20–39

89. Shaffer MJ, Halvorson AD, Pierce FJ. 1991. Nitrate leaching and economic analysis package (NLEAP): model description and applications. In *Managing Nitrogen for Groundwater Quality and Farm Productivity*, ed. RF Follett, DR Keeney, RM Cruse, pp. 285–322. Madison, WI: Soil Sci. Soc. Am.

90. Shiraishi N, Campbell WH. 1997. Expression of nitrate reductase FAD-containing fragments in Pichia. See Ref. 94a, pp. 931–34

91. Shiraishi N, Croy C, Kaur J, Campbell WH. 1998. Engineering of pyridine nucleotide specificity of nitrate reductase: mutagenesis of recombinant cytochrome b reductase fragment of *Neurospora crassa* NADPH:nitrate reductase. *Arch. Biochem. Biophys.* 358:104–15

92. Smil V. 1997. Global population and the nitrogen cycle. *Sci. Am.* 277:76–81

93. Solomonson LP, Barber MJ. 1989. Algal nitrate reductases. See Ref. 102a, pp. 123–54

94. Solomonson LP, Barber MJ. 1990. Assimilatory nitrate reductase: functional properties and regulation. *Annu. Rev. Plant Physiol. Plant Mol. Biol.* 41:225–53

94a. Stevenson KJ, ed. 1996. *Flavins and Flavoproteins*. Calgary, Can: Univ. Calgary Press

95. Su W, Huber SC, Crawford NM. 1996. Identification in vitro of a posttranslational regulatory site in the hinge

1 region of Arabidopsis nitrate reductase. *Plant Cell* 8:519–27

96. Su W, Mertens JA, Kanamaru K, Campbell WH, Crawford NM. 1997. Analysis of wild-type and mutant plant nitrate reductase expressed in the methylotrophic yeast *Pichia pastoris*. *Plant Physiol.* 115:1135–43

97. Trimboli AJ, Barber MJ. 1994. Assimilatory nitrate reductase: reduction and inhibition by NADH/NAD$^+$ analogs. *Arch. Biochem. Biophys.* 315:48–53

98. Trimboli AJ, Quinn GB, Smith ET, Barber MJ. 1996. Thiol modification and site-directed mutagenesis of the flavin domain of spinach NADH:nitrate reductase. *Arch. Biochem. Biophys.* 331:117–26

99. Tsay YF, Schroeder JI, Feldmann KA, Crawford NM. 1993. The herbicide sensitivity gene CHL1 of Arabidopsis encodes a nitrate-inducible nitrate transporter. *Cell* 72:705–13

100. Vitousek PM, Mooney HA, Lubchenco J, Melillo JM. 1997. Human domination of Earth's ecosystems. *Science* 277:494–99

101. Wang M, Roberts DL, Paschke R, Shea TM, Masters BS, et al. 1997. Three-dimensional structure of NADPH-cytochrome P450 reductase: prototype for FMN- and FAD-containing enzymes. *Proc. Natl. Acad. Sci. USA* 94:8411–16

102. Wei X, Ming LJ, Cannons AC, Solomonson LP. 1998. 1H and 13C NMR studies of a truncated heme domain from *Chlorella vulgaris* nitrate reductase: signal assignment of the heme moiety. *Biochim. Biophys. Acta* 1382:129–36

102a. Wray J, Kinghorn J, eds. 1989. *Molecular and Genetic Aspects of Nitrate Assimilation*. New York: Oxford Univ. Press

103. Xiao B, Smerdon SJ, Jones DH, Dodson GG, Soneji Y, et al. 1995. Structure of a 14-3-3 protein and implications for coordination of multiple signaling pathways. *Nature* 376:188–91

Annu. Rev. Plant Physiol. Plant Mol. Biol. 1999. 50:305–32

CRASSULACEAN ACID METABOLISM: *Molecular Genetics*

John C. Cushman
Department of Biochemistry and Molecular Biology, Oklahoma State University, Stillwater, Oklahoma 74078-0454; e-mail: jcushman@biochem.okstate.edu

Hans J. Bohnert
Department of Biochemistry, The University of Arizona, Tucson, Arizona 85721-0088; e-mail: bohnerth@u.Arizona.edu

KEY WORDS: nocturnal CO_2 fixation, environmental stress, gene expression, evolution, signal transduction

ABSTRACT

Crassulacean acid metabolism (CAM) is an adaptation of photosynthesis to limited availability of water or CO_2. CAM is characterized by nocturnal CO_2 fixation via the cytosolic enzyme PEP carboxylase (PEPC), formation of PEP by glycolysis, malic acid accumulation in the vacuole, daytime decarboxylation of malate and CO_2 re-assimilation via ribulose-1,5-bisphosphate carboxylase (RUBISCO), and regeneration of storage carbohydrates from pyruvate and/or PEP by gluconeogenesis. Within this basic framework, the pathway exhibits an extraordinary range of metabolic plasticity governed by environmental, developmental, tissue-specific, hormonal, and circadian cues. Characterization of genes encoding key CAM enzymes has shown that a combination of transcriptional, posttranscriptional, translational, and posttranslational regulatory events govern the expression of the pathway. Recently, this information has improved our ability to dissect the regulatory and signaling events that mediate the expression and operation of the pathway. Molecular analysis and sequence information have also provided new ways of assessing the evolutionary origins of CAM. Genetic and physiological analysis of transgenic plants currently under development will improve our further understanding of the molecular genetics of CAM.

305

CONTENTS

INTRODUCTION

Crassulacean acid metabolism (CAM), named for the Crassulaceae family of succulent plants in which the pathway was first defined, represents one of three major modes of photosynthetic carbon fixation present in higher plants. The earliest observations of the peculiar features of succulent plant species, nocturnal CO_2 uptake and diurnal fluctuations of acidity, have been known for centuries (61, 85). It was not until the middle of this century, however, that the metabolic sequence of the CAM cycle, including diurnal, reciprocal flux of carbohydrate with organic acids, nocturnal stomatal opening and CO_2 uptake, and stomatal closure during the day, were understood and summarized in this series (132). Over the next decades, research focused on evolution, ecological physiology, and biochemistry of the CAM cycle; an emphasis on biochemical variations and regulation by environmental factors has been reviewed in this series (126, 156). Other aspects of CAM have also been summarized (40, 66, 78, 79, 92, 106, 107), including newly emerging areas of CAM research (174), with special emphasis on the molecular genetics of CAM (52). This chapter focuses on new information about the molecular genetic analysis of CAM. First, we review the extreme range of metabolic plasticity in CAM plants. Second, we consider the mechanisms controlling gene expression through transcriptional, posttranscriptional, translational, and posttranslational regulatory circuits. Third, we discuss CAM regulation by plant hormones, light, circadian rhythms, tissue specificity, and metabolite compartmentation and transport. Fourth, we analyze signal transduction resulting in changes in gene expression. Finally, we explore the utility of molecular sequence data in assessing the evolutionary origins of CAM and of genetic models that promise to improve our understanding of the molecular basis of CAM.

DEFINING CAM

The basic biochemical reactions of the CAM cycle are confined within single, chloroplast-containing cells wherein nighttime, primary fixation of CO_2 into C_4 acids (e.g. malate) is temporally separated from daytime, secondary incorporation of CO_2 into carbohydrates. Consequently, nocturnal malate formation and storage in the vacuole varies reciprocally with daytime accumulation of starch or extrachloroplastic carbohydrate (Figure 1). The patterns of CO_2 uptake, C_4 acid, and glucan formation can be pictured as the diurnal repetition of four phases (126). These may not be observed in perfect form because a complex interplay of environmental and developmental factors often alter or obscure certain features of each phase (20, 138).

During the nocturnal phase I, phospho*enol*pyruvate carboxylase (PEPC) assimilates atmospheric and respiratory CO_2 into oxaloacetate (OAA) by carboxylating phospho*enol*pyruvate (PEP) derived from glycolysis. OAA is then reduced to malate by cytosolic and/or mitochondrial NAD(P)-malate dehydrogenase and stored in the vacuole. In phase II, at dawn, both PEPC and ribulose 1,5-bisphosphate carboxylase/oxygense (RUBISCO) are active and malate is released from the vacuole. During most of the day, phase III, when stomata are closed, malate decarboxylation commences as the released CO_2 is refixed

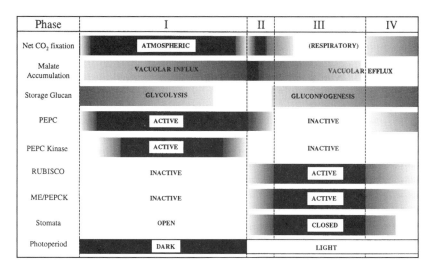

Phase	I	II	III	IV
Net CO_2 fixation	ATMOSPHERIC		(RESPIRATORY)	
Malate Accumulation	VACUOLAR INFLUX		VACUOLAR EFFLUX	
Storage Glucan	GLYCOLYSIS		GLUCONFOGENESIS	
PEPC	ACTIVE		INACTIVE	
PEPC Kinase	ACTIVE		INACTIVE	
RUBISCO	INACTIVE		ACTIVE	
ME/PEPCK	INACTIVE		ACTIVE	
Stomata	OPEN		CLOSED	
Photoperiod	DARK		LIGHT	

Figure 1 Major features of the temporal phases of CAM. The four major phases of CAM are indicated over the course of a 24-h photoperiod. Enzymatic and transport activities, and metabolite fluxes are indicated by *dark shading*. The absence or decline of these parameters is indicated by *white* or *light shading*.

by RUBISCO. The photosynthetic carbon reduction cycle incorporates the liberated 3-carbon compounds into storage carbohydrate via gluconeogenesis. In phase IV, malate reserves are eventually depleted, causing a decline in intracellular CO_2 availability. This results in stomatal opening and assimilation of atmospheric CO_2 until the onset of darkness. This phase resembles C_3 photosynthesis with RUBISCO serving as the primary carboxylase, although PEPC may become active before the end of the light period (20, 21, 80).

CAM plants separate into two distinct metabolic groups based on the type of C4-acid decarboxylases present (Figure 1). In malic enzyme (ME)-type CAM plants, malate is decarboxylated in either the cytosol or mitochondria utilizing NADP-ME or NAD-ME, respectively; the reaction generates pyruvate, which is subsequently converted to PEP in the chloroplast by pyruvate orthophosphate dikinase (PPDK), and to CO_2. In PEP carboxykinase (PEPCK)-type CAM plants, which lack PPDK activity, the OAA formed by NAD(P)-MDH is converted to PEP and CO_2 by PEPCK in the cytosol. In general, ME-type plants have higher ME activities than the PEPCK-type and completely lack PEPCK activity (15). These two decarboxylating systems are grouped within plant families, but not at higher taxonomic levels. CAM plants also display a great variety of storage carbohydrate reserves, which must be used to balance massive nocturnal acid accumulation. Some CAM plants accumulate predominantly starch or soluble sugars (e.g. sucrose, glucose, fructose) and polysaccharides (e.g. fructan, galactomannan) in extrachloroplastic compartments (15), while other species store both starch and glucose (44). Detailed studies have shown at least eight distinct combinations of malate decarboxylation and carbohydrate storage strategies in CAM plants (44). Diversity in carbohydrate accumulation patterns reflects evolutionary history rather than being governed by carbon flow constraints of the pathway (45).

PLASTICITY AND PERMUTATIONS OF CAM

Various criteria have been used to define different permutations of CAM such as net nocturnal CO_2 assimilation, diurnal fluctuations in organic acid accumulation (or titratable acidity), stomatal conductance, and abundant primary carboxylating (PEPC) or decarboxylating enzymes (e.g. ME/PPDK or PEPCK) (40, 78, 156). Each parameter can display considerable plasticity depending upon environmental conditions and plant development (Figure 2). In "obligate" or "constitutive" CAM species, net CO_2 uptake occurs almost exclusively at night, even under well-watered conditions accompanied by large diel malate fluctuations. For plants with weak CAM that appear "nearly-C_3", the term "CAM cycling" is used to describe C_3 gas exchange patterns concomitant with diel fluctuations in C_4 acids (19, 156), but with little or no net nocturnal carbon

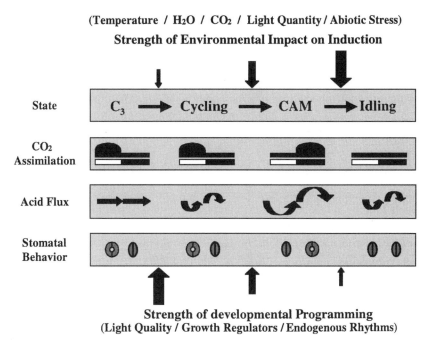

(Temperature / H_2O / CO_2 / Light Quantity / Abiotic Stress)
Strength of Environmental Impact on Induction

State

C_3 ⟶ Cycling ⟶ CAM ⟶ Idling

CO_2 Assimilation

Acid Flux

Stomatal Behavior

Strength of developmental Programming
(Light Quality / Growth Regulators / Endogenous Rhythms)

Figure 2   Impact of environmental and developmental factors on the major permutations of CAM. Sizes of *vertical arrows* indicate the relative strength of the impact that environmental factors or developmental programming have on different CAM states. Net CO_2 assimilation, indicated by *semicircular shape*, can occur predominantly in the light period (*white bar*) or dark period (*black bar*). The magnitude of diel acid fluctuations is indicated by arrow size.

assimilation. Plants that perform CAM-cycling typically grow in habitats with unpredictable daily water supply, and may derive benefits by being poised to enter CAM rapidly when called for by prevailing environmental conditions (112, 147). "Rapid-cycling CAM," a hypothetical variation of CAM in which both CO_2 acquisition and reduction phases of CAM are predicted to occur over time periods shorter than the normal diel cycle, has also been proposed (41).

Under severe drought conditions, plants may maintain closed stomata day and night yet continue to conduct diel fluctuations in organic acids. This behavior, designated "CAM-idling," may help preserve and maintain the photosynthetic apparatus by limiting photoinhibition, maintaining a positive carbon balance during severe droughts, and allowing rapid recovery upon rehydration (12, 19, 156). Between the extremes of cycling and idling lies a continuum of CAM modes tempered not only by the evolutionary history of the plant family, but also by ontogenetic and environmental factors (20). Other modes of CAM, such as "latent CAM" wherein basal organic acid concentrations are elevated above those normally present in C_3 plants, but do not fluctuate diurnally, may represent a C_3 to CAM progression in some species (101). The ecophysiological plasticity of CAM is further demonstrated by nocturnal CO_2 uptake and accumulation of organic acids by roots of epiphytic orchids (42, 173), terrestrial fern species of the Isoetid family (e.g. *Stylites andicola*) (93), and a large number of aquatic vascular species (reviewed in 92). These observations suggest that limited CO_2 may have provided the primary selective pressure for the evolution of CAM (40, 66, 126, 134).

CAM GENES, GENE FAMILIES, AND EXPRESSION

The characterization of CAM-specific genes, gene families, and their expression patterns in facultative CAM plants has begun to shed light on the molecular mechanisms underlying the recruitment and evolution of genes important in CAM. These genes and their regulatory properties, which are summarized in Table 1, form the basis of the metabolic plasticity typical of CAM. Recruitment of specific gene family members, which display elevated expression patterns, is a strategy that has evolved to satisfy the increased carbon flux demand of CAM. In *Mesembryanthemum crystallinum, Kalanchoe blossfeldiana*, and *Vanilla planifolia*, PEPC is encoded by a small multigene family of up to four members. CAM-specific isoforms exhibit enhanced expression in leaves of facultative or obligate CAM species (49, 54, 71, 74, 88, 127). The remaining isoforms, which presumably fulfill anapleurotic housekeeping or tissue-specific functional roles, undergo little change in expression and generally are expressed at low levels in CAM-performing tissues (50, 54, 74). Multigene families with at least one CAM-specific isogene have also been reported for NADP-malic enzyme (ME), NAD(P)-malate dehydrogenase (MDH), enolase, and phosphoglyceramuthase (PGM) in *M. crystallinum* (47, 48, 68, 69, 125) and for the V-ATPase c subunit genes in *M. crystallinum* (160) and *K. daigremontiana* (11). Cell-specific distribution patterns of PEPCK in leaves of the weak, CAM-inducible *Clusia apipoensis* suggest that one or more genes may encode this enzyme, although this has not been confirmed by direct characterization at the gene level (22). Gene duplication, recombination, or transposon-induced translocation events, followed by alteration of regulatory regions, are possible mechanisms to explain the recruitment of CAM-specific isogenes. Thus, one key to explaining how CAM evolved will be to understand how different gene expression patterns evolved by changes in control sequences.

Alternatively, a single gene may fulfill both C_3 photosynthesis- and CAM-specific functions, as appears to be the case for NAD-GAPDH (128) and PPDK (67). Other CAM enzymes are coordinately induced during CAM induction, but are localized to distinct subcellular compartments, including

Table 1 CAM associated genes and enzymes

Enzyme	Gene	Source organism	Subcellular location	Expression pattern (tissue/inducers)	Reference
Carboxylases/Decarboxylases					
Phospheno/pyruvate carboxylase	*Ppc1*	*M. crystallinum*	CYT	Leaf/NaCl, ABA, drought, BAP, light	39, 49, 54, 114, 127, 136, 153, 155
	Ppc2	*M. crystallinum*	CYT	Leaf, root/—	50
	Ppc1;1, 2	*K. blossfeldiana*	CYT	Leaf/ABA, drought, short day length	74, 151
	Ppc2;1, 2	*K. blossfeldiana*	CYT	?/?	74, 151
	Ppc1	*V. planifolia*	CYT	Leaf, stem/?	71
	Ppc2	*V. planifolia*	CYT	Root/? —	71
	Ppc1	*A. arborescens*	CYT	Green leaf/?	88
Decarboxylases					
NADP-Malic enzyme	*Mod1*	*M. crystallinum*	CYT	Leaf/NaCl	47
NAD-Malic enzyme	?	*M. crystallinum*	MT	Leaf/NaCl	87
PEP carboxykinase	?	*T. fasciculata, T. utriculata, N. fulgens, A. aristata, H. camosa*	CYT	?/?	165
Malate metabolism enzymes					
NADP-Malate dehydrogenase	*Mdh1*	*M. crystallinum*	CP	Leaf/NaCl	48
NAD-Malate dehydrogenase	*Mdh2*	*M. crystallinum*	CYT	Leaf/NaCl	125
Glycolytic/gluconeogenic enzymes					
Pyruvate orthophosphate dikinase	*Ppdk1*	*M. crystallinum*	CP	Leaf/NaCl	67
Enolase	*Pgh1;1, 2*	*M. crystallinum*	CYT	Leaf, root/NaCl, drought, cold, hypoxia, ABA, BAP	68
Phosphoglyceromutase	*Pgm1*	*M. crystallinum*	CYT	Leaf, root/NaCl, drought, ABA, BAP	69
Phosphoglyceratekinase	*Pgk1*	*M. crystallinum*	CYT	NaCl	JC Cushman, unpublished data
Phosphoglyceratekinase	?	*M. crystallinum*	CP	NaCl	172
NAD-Glyceraldehyde 3-phosphate dehydrogenase	*GapC1*	*M. crystallinum*	CYT	NaCl	128

(*Continued*)

Table 1 (*Continued*)

Enzyme	Gene	Source organism	Subcellular location	Expression pattern (tissue/inducers)	Reference
NADP-Glyceraldehyde 3-phosphate dehydrogenase	?	*M. crystallinum*	CP	NaCl	87
Phosphoglucomutase	Pglm1	*M. crystallinum*	CYT	?	HJ Bohnert, unpublished data
Fructose 1,6-bisphosphatase	Fbp1	*M. crystallinum*	CP	NaCl	87; JC Cushman, unpublished data
Phosphofructokinase	?	*M. crystallinum*	CP	—	87
Pyruvate kinase	?	*M. crystallinum*	?	—	87
Hexokinase	Hek1	*M. crystallinum*	CYT	NaCl	JC Cushman, unpublished data
Fructose-bisphosphatase aldolase	Fba1	*M. crystallinum*	CYT	NaCl	JC Cushman, unpublished data
Glucose-6-phosphate dehydrogenase	?	*M. crystallinum*	?	—	87
Glucose-6-phosphate isomerase	Gpi1	*M. crystallinum*	CYT/CP	NaCl	87; JC Cushman, unpublished data
Starch phosphorylase	Pho1	*M. crystallinum*	CP	NaCl	JC Cushman, unpublished data
Photorespiration enzymes					
Glycollate oxidase	Glx1	*M. crystallinum*	PER	?	HJ Bohnert, unpublished data
Serine hydroxylmethyl transferase	Sht1	*M. crystallinum*	MT	—	JC Cushman, unpublished data
Serine aminotransferase	Sat1	*M. crystallinum*	PER	—	JC Cushman, unpublished data
Glycine cleavage protein H	GcpH1	*M. crystallinum*	PER	?	HJ Bohnert, unpublished data
Glycine cleavage protein T	GcpT1	*M. crystallinum*	PER	?	HJ Bohnert, unpublished data

Protein kinases

Enzyme	Gene	Species	Localization	Expression/Induction	References
PEPC kinase	?	M. crystallinum, B. fedtschenkoi	CYT	Leaf/Circadian rhythm	83, 104
SNF1 kinase	MK1	M. crystallinum	CYT	?	JC Cushman, unpublished data
SNF1 kinase	MK9	M. crystallinum	CYT	Leaf/NaCl, Drought, ABA, 6-BAP	14; JC Cushman, unpublished data
'AGC' kinase	MK6	M. crystallinum	CYT	Leaf/NaCl, diurnal rhythm	B Baur, unpublished data
Ca^{2+}-dependent protein kinase	CPK1	M. crystallinum	CYT	Leaf, root/NaCl, drought?	T Taybi, JC Cushman, unpublished data
Tonoplast Enzymes					
H$^+$ATPase, A subunit	AtpvA	M. crystallinum	TP	Leaf, root/NaCl	105
H$^+$ATPase, B subunit	AtpvB	M. crystallinum	TP	Leaf, root/NaCl	105
H$^+$-ATPase, E subunit	AtpvE	M. crystallinum	TP	Leaf, root/NaCl	62
H$^+$-ATPase, F subunit	AtpvF	M. crystallinum	TP	Leaf, root/NaCl	JC Cushman, unpublished data
H$^+$-ATPase, G subunit	AtpvG	M. crystallinum	TP	Leaf, root/NaCl	JC Cushman, unpublished data
H$^+$-ATPase, c subunit	Atpvc	M. crystallinum, K. daigremontiana	TP	Leaf, root/NaCl, light, ABA	11, 105, 160
H$^+$-ATPase, 54-kDa subunit	Atpv	M. crystallinum	TP	Leaf, root/NaCl	JC Cushman, unpublished data
Malate transporter	?	K. daigremontiana	TP	?	133, 149
Pyrophosphatase	?	M. crystallinum, K. daigremontiana	TP	Leaf, root/—	110
Miscellaneous					
RNA-binding protein	Rbp1	M. crystallinum	CP	—	23
Ribosome inactivating protein	Rip1	M. crystallinum	CYT	NaCl, diurnal rhythm	137
Adenylate kinase	Adk1	M. crystallinum	CYT	NaCl	JC Cushman, unpublished data

Dashes (—) indicate that the genes are not induced. Question marks (?) indicate that the gene, its expression pattern, or the subcellular localization of the gene product have not been described in CAM species. Intracellular localization of gene products; CYT, cytoplasms; CP, chloroplast; MT, mitochondria; PER, peroxisome; TP, tonoplast; NC, nucleus; VAC, vacuole.

mitochondrial NAD-malic enzyme, and cytosolic NADP-ME (46, 47, 87, 142), cytosolic NAD-MDH, and chloroplastic NADP-malate dehydrogenase (48, 87, 125, 141). CAM induction in *M. crystallinum* also leads to large increases in the activities of starch-degrading enzymes and to increased rates of starch turnover (129). These increases also result in dramatic changes in the profile of starch degradation products exported from isolated chloroplasts. Chloroplasts isolated from plants performing C_3 photosynthesis export mainly maltose, whereas chloroplasts from plants performing CAM export mostly glucose-6-phosphate (119). To sort out the relative, functional contribution of each of these enzymes and transporters to the CAM pathway requires a comprehensive characterization of their genes or gene families.

Transcriptional Regulation

Transcriptional activation is the primary control point responsible for elevating the expression of many CAM-specific genes. This has been demonstrated by nuclear run-on transcription assays and transient assays that use promoter-reporter gene fusions (54, 143). mRNA accumulation can occur within hours following salinity or dehydration of plants or leaves of *M. crystallinum* (144) or *K. blossfeldiana* (24, 152). Transcription rates increase more than sixfold in salt-stressed *M. crystallinum* plants undergoing the C_3 to CAM switch (47, 48, 54, 56, 68, 69, 163). The degree of transcriptional activation depends on the type and severity of the stress and the developmental status of the plant (55, 84, 131, 146). For PEPC, increases in transcription are followed by subsequent increases in enzyme protein levels (87, 116, 172) that require de novo protein synthesis (70, 86).

The transcriptional activation of CAM-specific genes is mediated through the interaction of *cis*-acting DNA sequences and *trans*-acting factors. Regions sufficient for salt-inducible gene expression contain MYB consensus binding sites, which suggests that MYB-related transcription factors participate in controlling transcriptional activation events during CAM induction (143). Additional *cis*-acting elements for light- and ABA-responsive gene expression and putative binding sites for a variety of DNA-binding proteins are present in the 5′ flanking regions of the stress-inducible genes *Ppc1*, *Ppdk1*, and *GapC1* (57, 145). Serial 5′ deletions of the *Ppc1* and *GapC1* promoters disclosed the presence of *cis*-acting enhancer and silencer regulatory regions (143). Cooperation of two or more *cis*-acting elements is required for salt-inducible transcriptional activation (HJ Schaeffer, JC Cushman, unpublished data).

Multiple DNA-binding proteins interact with the *Ppc1* 5′ flanking region (51). One DNA-binding protein, PCAT-1, binds differentially to two AT-rich sites that have the consensus motif AARTAACWAKTTTY. PCAT-1 shows increased abundance or DNA-binding affinity following salt stress and shares characteristics of HMG-like proteins. Thus, PCAT-1 and associated binding proteins may play architectural roles in the assembly of active transcription complexes during CAM induction. Additional in vitro DNA-binding activities, designated PCAT-2 and 3, show either reduced or enhanced abundance or binding affinity, respectively, upon salt stress (51). DNA-binding proteins that interact with the "salt-enhancer" elements of the *Ppc1* and *GapC1* promoters also display altered abundance or DNA-binding affinity during the C_3 to CAM transition (143). Based on these limited analyses, the regulatory mechanisms that mediate transcriptional activation of CAM-specific genes remain unclear. The *Ppc1* promoter can direct high reporter gene expression in transgenic tobacco, but fails to exhibit salt inducibility,which suggests that tobacco lacks the requisite transcriptional machinery responsible for recognizing the specific salt-responsive promoter elements present in *M. crystallinum* (53). More detailed work characterizing *cis*- and *trans*-acting elements will be required to improve our understanding of transcriptional activation events leading to CAM induction and maintenance.

Posttranscriptional and Translational Regulation

Transcriptional induction and well-documented changes in mRNA populations following stress (127) may not be the only regulatory mechanisms governing changes in gene expression during the C_3 to CAM transition. *Ppc1* mRNA stability may be enhanced during CAM induction and in older plants (55), whereas *RbcS* transcripts encoding the small subunit of RUBISCO decline rapidly upon salt stress (18, 59). An increase in chloroplast RNA-binding proteins that may function to stabilize specific transcripts accompanies CAM induction (23). Also, transcription rates and mRNA accumulation for enolase increase in response to salinity stress without a corresponding increase in protein amounts (68). This suggests, but does not prove, that translation may control the expression of this and maybe other glycolytic enzymes.

Alterations in the translational efficiency of specific mRNA populations may also contribute significantly to the expression of key CAM enzymes such as PEPC and PPDK. In *M. crystallinum*, several lines of evidence support global changes in translational regulation during the C_3 to CAM transition. CAM induction is accompanied by a decline in total protein synthesis, whereas the synthesis of specific proteins, such as PEPC, increases (86). Such alterations in translational activity are correlated with rapid changes in mRNA populations associated with polyribosomes following stress (18). Furthermore, a ribosome-inactivating protein (RIP) is up-regulated following CAM induction with a dampened diurnal expression pattern in *M. crystallinum* (137). Activity of this RIP has been documented and is thought to alter translation profiles through the selective turnover of ribosomes in response to stress.

Posttranslational Regulation

To avoid futile cycles of carboxylation and decarboxylation, PEPC is tightly regulated by posttranslational control mechanisms on a diurnal or circadian basis (Figure 1). PEPC activity is influenced allosterically by both positive (glucose 6-phosphate, G6P), triose-phosphate (TP), and negative (L-malate) effectors (38, 164). In addition, PEPC undergoes reversible phosphorylation by PEPC kinase at a highly conserved serine residue present in the *N*-terminal domain (13, 29, 38, 95, 104, 121, 164, 167). In CAM plants, the dephosphorylated "day form" is more sensitive to malate inhibition, whereas the more active, phosphorylated "night form" has higher affinity for PEP and is more sensitive to G6P and TP (positive effectors), but less sensitive to L-malate (negative effector) (90, 121). PEPC kinase expression is induced coincidentally with its target substrate by salt stress in *M. crystallinum* (104). Circadian regulation of PEPC kinase activity is dependent upon mRNA and protein synthesis (33, 120). In contrast, protein synthesis-dependent dephosphorylation of PEPC by a protein phosphatase 2A (PP2A) does not appear to be regulated by an endogenous rhythm (32). PEPC kinase mRNA is approximately 20 times more abundant at night than during the day, which suggests that primary control occurs at the level of translatable mRNA (83). Activity disappears rapidly following the decline in translatable mRNA, indicating rapid turnover of the kinase (83). Inhibition of translation also blocks the appearance of translatable PEPC kinase mRNA, which suggests that the upstream signaling events controling circadian rhythmicity also require protein synthesis (83). Certain *Clusia* species show extended PEPC activity during daytime CAM phases II and IV, which suggests alterations in PEPC kinase activity that effect PEPC deactivation/reactivation properties (20, 21, 138). To date, the gene encoding PEPC kinase has not been cloned; however, a Ca^{2+}-independent kinase identified as two polypeptides (39 and 32 kDa) capable of phosphorylating purified PEPC from C_3, C_4, and CAM species has been reported (104). Whether the two proteins represent two isoforms or are the result of proteolytic processing of a single protein remains unclear (104).

PEPCK is phosphorylated at night and dephosphorylated during the day in *Tillandsia fasciculata* (165). This phosphorylation pattern is likely to modulate decarboxylase activity over the course of the diurnal cycle to avoid futile carboxylation cycles between PEPC and PEPCK as both enzymes are localized to the cytosol. However, it is not known how this occurs or whether PEPCK is regulated by light or in response to a circadian rhythm. Covalent modification of NAD-ME may also be responsible for diurnal changes in the kinetic properties of this enzyme (46). Posttranscriptional control of enolase activity by reversible phosphorylation has been suggested for enolase from *M. crystallinum* (68). Covalent modifications, such as the reduction of thiol groups of cytosolic

NAD-GAPDH and enolase from *M. crystallinum*, provide yet another type of posttranslational control for the reversible activation of CAM enzymes (7, 8).

Circadian Rhythms

One of the most extensively studied examples of circadian rhythms in plants is nocturnal CO_2 exchange in CAM species (16, 17, 108, 110, 169). CAM plants exhibit a persistent circadian rhythm of CO_2 exchange under constant environmental conditions (30, 31, 168). PEPC activity and properties are directly related to these circadian rhythms, which in turn parallel the in vitro and in vivo phosphorylation state of PEPC and PEPC kinase (33, 120–122). PEPC kinase activity increases for several hours following darkness and decreases well before the onset of the light period (33, 83). Under conditions of continuous illumination or darkness, PEPC phosphorylation status exhibits a persistent circadian rhythm (99, 122). Light and temperature extremes can perturb the regular rhythms of CO_2 fixation (4–6, 76, 77, 108) through modifications of PEPC kinase activity (34, 35, 83).

The origin and nature of the circadian oscillator controlling PEPC kinase activity remain elusive. Fluctuations in malate concentrations in the vacuole and cytoplasm and processes that determine malate transport may play a role in generating and regulating circadian rhythmicity (6, 96). Other important factors may include the regulation of PEPC kinase activity by reversible phosphorylation. In C_4 plants, PEPC kinase appears to be regulated by an upstream Ca^{2+}-dependent protein kinase (75). In CAM plants, however, the regulation of this enzyme is not well understood. Characterization of clock proteins and transcription factors in addition to the molecular cloning and biochemical characterization of CAM PEPC kinase should greatly facilitate the identification of the circadian oscillator and the signaling and control mechanisms by which it operates.

Tissue Specificity

Differences in CAM enzyme distribution between photosynthetically active and water-storing tissues suggested CAM biochemistry localized to all chloroplast-containing tissues, but not to hydrenchymous tissues without developed chloroplasts (64, 148, 171). In general, chloroplast-containing cells fit the "classical" CAM definition, whereas chlorophyll-deficient tissues express CAM weakly, suggesting that they function in a "CO_2 storage" capacity (148). In some species, histochemical analyses of CAM enzymes and gene expression patterns have shown cell-specific expression patterns within these tissues. In *Peperomia camptotricha*, leaves are organized into four cell layers: an upper epidermis composed of multiple cell layers, a middle one- to two-layered palisade mesophyll, a lower spongy mesophyll, and an abaxial epidermis (123). Analysis of these tissues indicated that CAM enzyme activities (PEPC and ME) and nocturnal CO_2 fixation predominate in the spongy mesophyll and to a lesser extent to the adaxial epidermis (123). Tissue printing of *Peperomia* leaves confirmed PEPC expression was localized preferentially to these two layers (157). In contrast, enzymes characteristic of light-dependent C_3 photosynthesis and RUBISCO activity are more abundant in palisade mesophyll (123, 124). RUBISCO is distributed throughout the leaf whereas *RbcS* mRNA is most abundant in the palisade layer (157). Although the leaves of *Clusia* have distinct palisade and spongy parenchymal mesophyll cell layers, there appears to be no clear division of photosynthetic labor as described in *Peperomia* leaves: PEPC, PEPCK, and RUBISCO were uniformly distributed throughout the palisade and spongy parenchyma (22). Comparisons of *Clusia* species differing in their CAM capacity indicate that the degree of CAM is highly correlated with leaf succulence, a morphological prerequisite for vacuolar storage capacity of organic acids. Given the morphological complexity of leaves in many CAM plants, additional examples of tissue-specific expression patterns of C_3 and CAM enzymes and morphological adaptations to distinct environmental niches can be expected.

Tissue differentiation also plays an important role in governing the extent to which CAM is present and/or induced. Photoperiodic induction of CAM in callus tissue of *K. blossfeldiana* is far less effective than in leaves (25), with the magnitude of induction showing a positive correlation with the degree of tissue organization (94). Similarly, salinity stress is far less effective at inducing malate accumulation and PEPC activity or expression in heterotrophic cell suspension cultures or green callus tissues from *M. crystallinum* than in intact plants (154, 161, 175). Photomixotrophic cell cultures of *M. crystallinum* grown on starch show only a weak (fourfold) induction of PEPC activity (175), although this was greater than the induction observed in heterotrophically grown cell suspensions. These results suggest that the presence of a functional photosynthetic apparatus is likely an important requisite for CAM (154). Photoautotrophic cell cultures from *M. crystallinum*, grown with CO_2 as the sole carbon source, showed significantly increased rates of CO_2 fixation in the light. When salt stressed, however, these cultures did not display CAM-related fluctuations or net nocturnal accumulations of malic acid (170). Nonetheless, they did exhibit slight increases (1.5- to 4-fold) in extractable PEPC, NADP ME, NAD- and NADP-MDH, and PPDK activities following salt stress (170). These increases could signal partial CAM induction in response to salinity at the cellular level without the corresponding fluctuations in C_4 acid or starch accumulation. Such fluctuations may depend on regulatory signals that are disrupted in the suspension cell cultures, which lack fully functional chloroplasts. Light-regulation of photosynthetic gene expression, an essential requirement for the metabolic flux of carbohydrates and other metabolites, may be lacking. Thus, it is likely that the signaling mechanisms associated with carbohydrate metabolism are disrupted in tissue cultures and that this prevents appropriate expression and regulation of CAM-specific enzymes.

Metabolite Compartmentation and Transport

Transporters governing the intracellular compartmentation and regulation of the diurnal organic acid/carbohydrate fluxes are central to the functioning of CAM (reviewed in 36). Vacuolar transport processes are energized by H^+-transporting, vacuolar type (V)-ATPase and/or H^+-PP_iase activities (reviewed in 109). Increases in V-ATPase activity and subunit accumulation correlate with increased CAM capacity (111). Genes encoding vacuolar H^+-ATPase subunits A, B, E, and c from *M. crystallinum* (62, 105, 160) as well as the c subunit from *K. daigremontiana* have been described (11). In *M. crystallinum*, ATPase-c transcripts accumulate following ABA and ionic (NaCl), but not osmotic (mannitol), stress (160). mRNAs for the ATPase-A, B, and E also increase in abundance following stress treatment (62, 105). Diurnal changes in transcript abundance for several of the subunit genes have been reported in leaves and photoautotrophic suspension cell cultures of *M. crystallinum* that may be related to CAM function (139, 140). Molecular characterization of vacuolar transport components awaits further investigation, although a putative malate transporter has been partially purified and reconstituted from *K. daigremontiana* (133, 149).

Pyruvate import and PEP export via the chloroplast membrane is required to support diurnal cycling of malate and starch pools in the malic enzyme-type CAM plants. During the C_3 to CAM transition, light-dependent pyruvate uptake by the chloroplast is enhanced in *M. crystallinum* (98). Increased enzyme activities associated with glycolysis/gluconeogenesis and starch degradation (129) and triose and hexose phosphate transport activities during CAM induction are accompanied by major changes in the profile of export products from the chloroplast (97, 119). Glucose 6-phosphate transport rates are reduced by illumination, suggesting that this transporter plays a role in the regulation of carbon flux. Molecular characterization of these transporters, as well as the malate, OAA, and pyruvate transport functions of the mitochondria, are needed to improve our understanding of the regulation of metabolite transport in CAM plants.

SIGNALING TRANSDUCTION AND CAM REGULATION

Relatively little is known about the signaling mechanisms that induce or maintain CAM. Split root experiments in *M. crystallinum* suggest that roots can perceive water stress and convey this information to leaves, triggering a switch from C_3 to CAM (65). Although the nature of the signal is unknown, experiments using detached leaves suggest that a reduction in cytokinin content may derepress CAM expression (130). However, other non-root-derived signals in detached leaves are sufficient to trigger CAM gene expression or enzyme activities in response to various stresses or plant growth regulator treatments (58, 130, 144, 146). Treatment of detached leaves with the Ca^{2+} ionophore, ionomycin, or with thapsigargin, a specific inhibitor of endomembrane Ca^{2+}-ATPases and stimulator of intracellular Ca^{2+} release, caused *Ppc1* transcript accumulation in the absence of stress or ABA treatments (151, 159). Similar results are observed for V-ATPase subunit c (159). These results suggest that changes in intracellular $[Ca^{2+}]$ derived from either extracellular or intracellular sources can stimulate *Ppc1* and *Atpc1*. Furthermore, EGTA completely abolished *Ppc1* mRNA increases in response to ABA and stress treatments (151). In contrast, thapsigargin and ionomycin treatments did not induce *Imt1*, which encodes inositol-O-methyltransferase, an enzyme involved in biosynthesis of ononitol (162). These results suggest that a signal transduction pathway distinct from that controlling CAM gene expression regulates the expression of *Imt1*, a key element of stress tolerance in *M. crystallinum* (159).

Specific inhibitors of kinases and phosphatases have provided a powerful approach to gauge the role of protein phosphorylations in controlling the cellular events leading to CAM induction. Okadaic acid, an inhibitor of protein phosphatase (PP) 2A and 1 activities, eliminated the salt-responsive induction of *Ppc1* and *Imt1* transcripts, but not the induction of *Atpc1* (159). Okadaic acid and the PP2A inhibitor, cantharidic acid, were also effective inhibitors of *Ppc1* transcript accumulation induced by dehydration and ABA. These results implicate PP2A and/or PP1 activity in CAM induction signaling; however, such results must be viewed with extreme caution given possible nonspecific effects of these inhibitors. Cyclosporin A (CsA), which specifically targets Ca^{2+}/calmodulin-activated PP2B (calcineurin), blocked the salt-induced expression of *Imt1*, but not that of *Ppc1* or *Atpc1* (159). CsA stimulated *Ppc1* transcript accumulation during stress or ABA treatments, suggesting that calcineurin may negatively regulate CAM induction (151). In contrast, treating detached *M. crystallinum* leaves with W7, a specific inhibitor of Ca^{2+}/calmodulin-dependent protein kinases, blocked *Ppc1* transcript accumulation in response to ionic, osmotic, and dehydration stress and ABA treatments (151), suggesting that Ca^{2+}/CaM protein kinases play a positive role in stress-induced increases in *Ppc1* expression.

A ser/thr protein kinase cDNA termed MK9 has been characterized from *M. crystallinum* and may play a role in carbohydrate signaling (14). A second ser/thr protein kinase cDNA cloned from *M. crystallinum*, designated MK6, is preferentially expressed at night in plants performing CAM, suggesting that it may be involved in the diurnal or circadian expression of CAM enzymes (B Baur, personal communication). Additional protein kinases related to cyclic-nucleotide-dependent (PKA and PKG) or Ca^{2+}-phospholipid-dependent kinases (PKC) have also been isolated from *M. crystallinum* ('AGC' kinases; Table 1) and may function downstream of cAMP or cGMP second messengers in signal transduction cascades. The isolation of ten members of the PP2C gene family is the first step in the comprehensive functional analysis of this class of phosphatase (117). Expression of at least one gene family member (MPC6) is consistent with a negative regulatory role in CAM induction.

Physiological and molecular studies have documented the role of phyto-chrome in light-sensing and signaling in CAM plants. Phytochrome is likely to control short-day CAM induction since interrupting the dark period with red light effectively inhibits CAM development (28, 118). Entrainment studies using *Bryophyllum* leaves pinpointed phytochrome as the sole photoreceptor regulating circadian rhythm of CO_2 exchange during CAM (81, 82). Phytochrome is also thought to enhance the appearance of a CAM-specific PEPC isoform in *Kalanchoe* (26, 27) and *M. crystallinum*, (43). These effects, amplified in salt-stressed plants, suggest that phytochrome and NaCl or ABA may act along the same or parallel signal transduction pathways to promote CAM induction (43, 114).

During the transition from C_3 to CAM, stomatal behavior is reversed relative to stomatal rhythms in C_3 and C_4 plants. The failure of red and/or blue light to induce stomatal opening in *Portulacaria afra* or *M. crystallinum* plants performing CAM suggests that the blue- and red-light photoreceptors become inactivated (100, 113). Inactivation of guard-cell photoreceptors may allow other factors, such as ABA or changes in $[CO_2]$, to dictate stomatal movements (113). Recent work has demonstrated that CAM induction abolishes both white- and blue-light–stimulated stomatal opening and light-dependent zeaxanthin formation, which suggests that light-dependent zeaxanthin formation in guard-cell chloroplasts may an important component that mediates nocturnal stomatal opening (150).

MOLECULAR EVOLUTION OF CAM

An ever-increasing body of DNA and protein sequence data for key CAM enzymes, such as PEPC, has allowed modeling of molecular phylogenies. These now confirm notions of the polyphyletic origins of CAM, which were based on the taxonomic distribution of CAM species (66, 134). This approach has been limited, owing to the lack of PEPC sequence information (88, 102, 103, 135, 158). The most recent analyses using up to 75 PEPC amino acid sequences have resulted in phylogenetic reconstructions that are largely consistent with evolutionary origins of higher plants based on taxonomic criteria (72; R Kämmerer, unpublished data). Comparisons of prokaryotic and eukaryotic PEPC point to a common ancestral gene arising from the γ-proteobacterial lineage, rather than from the cyanobacterial lineage (R Kämmerer, unpublished data). Among land plants, the precise relationships among various taxa are not well resolved owing to low sequence divergence and/or multiple speciation events. Nonetheless, phylogenetic reconstructions confirm the conclusion derived from comparative physiology and taxonomy that C_3 photosynthesis is ancestral to both CAM and C_4, with CAM appearing earlier than C_4 photosynthesis (66, 72, 134; R Kämmerer, unpublished data). Reconstructions with bona fide CAM-specific PEPC isoforms support at least three independent origins of the pathway in the Orchidaceae, Crassulaceae, and Caryophyllales. In the latter, a common lineage for *Pereskia* (*Cactaceae*) and *Mesembryanthemum* (*Aizoaceae*) is supported (R Kämmerer, unpublished data).

More precise analysis of C_4 and CAM evolution by phylogenetic inference requires both sequence information and knowledge about the specific metabolic role of a particular gene product. The assignment of specific physiological functions to distinct isoforms has been conducted only for a few species. In facultative CAM plants such as *M. crystallinum* and *K. blossfeldiana*, CAM-specific isogenes were assigned by their expression in response to environmental stress or changes in light quality or daylength (54, 74; T Taybi & H Gehrig, personal communication). In obligate CAM plants such as *V. planifolia* and *Aloe arborescens*, tissue-specific expression patterns have been used as indicators of CAM function (71, 88). Phylogenetic relationships among closely related CAM species could also be accurately assessed using DNA fingerprinting (73).

What are the specific mechanisms that might explain the evolution of the CAM-specific gene expression patterns? One possibility is that differences in controlling sequences arose by evolutionary changes in multipartite *cis*-regulatory regions present in 5' or 3' flanking regions following gene duplication events (10). The 5' flanking regions of C_3 and CAM isogenes from *M. crystallinum* are dissimilar, consistent with the ancient evolutionary origins of CAM (57). In addition to PEPC, various isoforms of NADP-malic enzyme have arisen during the evolution of C_4 photosynthesis, as demonstrated by immunocytological analysis of C_3, C_3-C_4 intermediate, and C_4 species from *Flaveria* (63). For the emergence of CAM-specific NADP-ME isoforms, for example (47, 142), modifications resulting in light, diurnal, circadian, and developmental regulation of expression are necessary. Additional modifications in coding regions directing intracellular localization and posttranslational changes for covalent modification to avoid futile and competing carboxylation reactions within a single cell are also expected (52). Molecular analysis of additional gene families encoding key CAM enzymes, including NADP-ME, NAD-ME, PPDK, and PEPCK, will be required to resolve the evolutionary relationships of the C_3, C_4, and CAM photosynthetic pathways.

FUTURE PERSPECTIVES

Despite the identification of many genes that play important structural roles in CAM, our understanding of the molecular mechanisms that control the complex assemblage of enzymes, transporters, and regulatory components required for CAM is still limited. Only 1% of all genes have been characterized from even the most intensively studied CAM model, *M. crystallinum*. One approach increasingly used to accumulate molecular genetic information rapidly is large-scale sequencing of anonymous cDNAs or expressed sequence tags (ESTs). Two EST projects have been initiated in *M. crystallinum*. One targets leaf tissue of unstressed and stressed plants (3) and the other transcripts specific for epidermal bladder cells (DE Nelson & HJ Bohnert, unpublished data), which seem to be essential for ion homeostasis under salt stress conditions. ESTs will provide an invaluable resource for rapid gene discovery, large-scale monitoring of gene expression changes using microarray or DNA chip technology, and future genome mapping and sequencing.

Our understanding of CAM has lagged behind that of C3 and C4 species mainly because there is no genetic model for studying CAM. Although mutant collections are essential tools for understanding plant structure, development, and metabolism, there are no reports of mutants defective in CAM or of mutants in any CAM species. A concerted effort by several laboratories to generate, characterize, and maintain CAM mutant populations should provide an important resource to address future questions in CAM molecular genetics. Facultative CAM species, such as *K. blossfeldiana* or *M. crystallinum*, will likely be the model species of choice. Mutant collections have been established in *M. crystallinum* following irradiation with fast neutrons or treatment with ethylmethane sulfonate (EMS). *M. crystallinum*, a rapidly growing, self-fertile species with a small genome (1, 60), has been well studied at the molecular level (see Table 1). The ability to miniaturize the plant and accelerate its normal life cycle from five months in its natural habitat to only seven weeks under extended photoperiods (37) will expedite genetic studies in *M. crystallinum*. Selection schemes have also been developed for screening putative CAM mutants that rely on detecting plants that fail to conduct nocturnal acidification and/or daytime starch accumulation (56). Availability and functional characterization of mutant lines will facilitate the analysis of key regulatory and structural components of the CAM pathway, including those involved in signal transduction, development, enzyme regulation, stomatal behavior, circadian oscillation of the cycle, and intracellular compartmentation and translocation of metabolites.

Central to a long-term strategy for developing a genetic model for CAM plants is the ability of a rapid and efficient transformation system. Several *Kalanchoe* species exhibit susceptibility to *Agrobacterium* infection and have been successfully transformed (2, 89). Transgenic hairy-root cultures or callus tissue from *M. crystallinum* have also been recovered following transformation by *Agrobacterium rhizogenes* and *A. tumefaciens*, respectively (9; JC Cushman, unpublished data). Regeneration protocols for *M. crystallinum* (115, 166) set the stage for future genetic engineering studies aimed at understanding the regulation and structure/function relationships of enzymes and other key regulatory components of CAM.

ACKNOWLEDGMENTS

Our work has been or is supported by the US Department of Agriculture (NRI) programs Plant Responses to the Environment and Developmental Mechanisms; by the National Science Foundation programs Integrative Plant Biology, Metabolic Biochemistry, and International Programs; by the US Department of Energy program Biological Energy Research; and by the Oklahoma and Arizona Agricultural Experiment Stations. Additional support has been provided to colleagues to work in our laboratories by the Deutsche Forschungsgemeinschaft (Germany), the Japanese Society for the Promotion of Science (JSPS), New Energy Development Organization (Japan), and the Rockefeller Foundation (USA). We thank many colleagues for making available unpublished data and for discussions, with special thanks to Bernhard Baur, Hans Gehrig, Howard Griffiths, Werner Herppich, Ralf Kämmerer, Manfred Kluge, Ulrich Lüttge, Craig Martin, Kate Maxwell, Barry Osmond, Andrew Smith, and Klaus Winter.

> **Visit the *Annual Reviews* home page at**
> **http://www.AnnualReviews.org**

Literature Cited

1. Adams P, Nelson DE, Yamada S, Chmara W, Jensen RG, et al. 1998. Growth and development of *Mesembryanthemum crystallinum* (Aizoaceae). *New Phytol.* 138:171–90
2. Aida R, Shibata M. 1996. Transformation of *Kalanchoe blossfeldiana* mediated by *Agrobacterium tumefaciens* and transgene silencing. *Plant Sci.* 121:175–85
3. Akselrod I, Landrith D, Stout L, Cushman JC. 1998. An ice plant (*Mesembryanthemum crystallinum*) expressed sequence tag (EST) database. *Plant Physiol. Suppl.* 117S:124 (Abstr.)
4. Anderson CM, Wilkins MB. 1989. Period and phase control by temperatures in the circadian rhythm of CO₂ exchange

in illuminated leaves of *Bryophyllum fedtschenkoi*. *Planta* 177:456–69
5. Anderson CM, Wilkins MB. 1989. Control of the circadian rhythm of carbon dioxide assimilation in *Bryophyllum* leaves by exposure to darkness and high carbon dioxide concentrations. *Planta* 177:401–8
6. Anderson CM, Wilkins MB. 1989. Phase resetting of the circadian rhythm of carbon dioxide assimilation *Bryophyllum fedtschenkoi* leaves in relation to their malate content following brief exposure to high and low temperatures, darkness, and 5% carbon dioxide. *Planta* 180:61–73
7. Anderson LE, Li D, Prakash N, Stevens FJ. 1995. Identification of potential

redox-sensitive cysteines in cytosolic forms of fructosebisphosphatase and glyceraldehyde-3-phosphate dehydrogenase. *Planta* 196:118–24

8. Anderson LE, Li D, Stevens FJ. 1998. The enolases of ice plant and Arabidopsis contain a potential disulphide and are redox sensitive. *Phytochemistry* 47:707–13

9. Andolfatto R, Bornhouser A, Bohnert HJ, Thomas JC. 1994. Transformed hairy roots of *Mesembryanthemum crystallinum*: gene expression patterns upon salt stress. *Physiol. Plant.* 90:708–14

10. Argüello-Astorga G, Herrera-Estrella L. 1998. Evolution of light-regulated plant promoters. *Annu. Rev. Plant Physiol. Plant Mol. Biol.* 49:525–55

11. Bartholomew DM, Rees DJG, Rambaut A, Smith JAC. 1996. Isolation and sequence analysis of a cDNA encoding the c subunit of a vacuolar-type H^+-ATPase from the CAM plant *Kalanchoë diagremontiana. Plant Mol. Biol.* 31:435–42

12. Bastide B, Sipes D, Hann J, Ting IP. 1993. Effect of severe water stress on aspects of crassulacean acid metabolism in *Xerosicyos. Plant Physiol.* 103:1089–96

13. Baur B, Dietz KJ, Winter K. 1992. Regulatory protein phosphorylation of phosphoenolpuruvate carboxylase in the facultative crassulacean-acid-metabolism plant *Mesembryanthemum crystallinum* L. *Eur. J. Biochem.* 209:95–101

14. Baur B, Fisher K, Winter K, Dietz KJ. 1994. cDNA sequences of a protein kinase from the halophyte *Mesembryanthemum crystallinum* L., encoding a SNF-1 homologue. *Plant Physiol.* 106:1225–26

15. Black CC, Chen J-Q, Doong RL, Angelov MN, Sung SJS. 1996. Alternative carbohydrate reserves used in the daily cycle of Crassulacean acid metabolism. See Ref. 173a, pp. 31–45

16. Blasius B, Beck F, Lüttge U. 1997. A model for photosynthetic oscillations in Crassulacean acid metabolism (CAM). *J. Theor. Biol.* 184:345–51

17. Blasius B, Beck F, Lüttge U. 1998. Oscillatory model crassulacean acid metabolism: structural analysis and stability boundaries with a discrete hysteresis switch. *Plant Cell Environ.* 21:775–84

18. Bohnert HJ, DeRocher EJ, Michalowski CB, Jensen RG. 1999. Environmental stress and chloroplast metabolism. In

Recent Advances in Plant Molecular Biology, ed. KC Bansal.New Delhi: Oxford/IBH Publ. In press

19. Borland AM. 1996. A model for the partioning of photosynthetically fixed carbon during the C-3-CAM transition in *Sedum telephium. New Phytol.* 134:433–44

20. Borland AM, Griffiths H. 1996. Variations in the phases of crassulacean acid metabolism and regulation of carboxylation patterns determined by carbon-isotope-discrimination techniques. See Ref. 173a, pp. 230–49

21. Borland AM, Griffiths H. 1997. A comparative study on the regulation of C_3 and C_4 carboxylation processes in the constitutive crassulacean acid metabolism (CAM) plant *Kalanchoë diagremontiana* and the C3–CAM intermediate *Clusia minor. Planta* 201:368–78

22. Borland AM, Tecsi LI, Leegood RC, Walker RP. 1998. Inducibility of crassulacean acid metabolism (CAM) in Clusia species; physiological/biochemical characterisation and intercellular localization of carboxylation and decarboxylation processes in three species which exhibit different degrees of CAM. *Planta* 205:342–51

23. Breiteneder H, Michalowski CB, Bohnert HJ. 1994. Environmental stress-mediated differential 3′ end formation of chloroplast RNA-binding protein transcripts. *Plant Mol. Biol.* 26:833–49

24. Brulfert J, Güclü S, Taybi T, Pierre JN. 1993. Enzymatic responses to water stress in detached leaves of the CAM plant *Kalanchoe blossfeldiana* Poelln. *Plant Physiol. Biochem.* 31:491–97

25. Brulfert J, Mricha A, Sossountzov L, Queiroz Q. 1987. CAM induction by photoperiodism in green callus cultures from a CAM plant. *Plant Cell Environ.* 10:443–49

26. Brulfert J, Müller D, Kluge M, Queiroz O. 1982. Photoperiodism and crassulacean acid metabolism I. Immunological and kinetic evidences for different patterns of phosphoenolpyruvate carboxylase isoforms in photoperiodically inducible and non-inducible crassulacean acid metabolism plants. *Planta* 154:326–31

27. Brulfert J, Queiroz O. 1982. Photoperiodism and crassulacean acid metabolism III. Different characteristics of the photoperiod-sensitive and non-sensitive isoforms of phosphoenolpyruvate carboxylase and in crassulacean acid

metabolism operation. *Planta* 154:339–43

28. Brulfert J, Vidal J, Keryer E, Thomas M, Gadal P, Queiroz O. 1985. Phytochrome control of phosphoenolpyruvate carboxylase synthesis and specific RNA level during photoperiodic induction in a CAM plant and during greening in a C$_4$ plant. *Physiol. Vég.* 23:921–28

29. Brulfert J, Vidal J, Le Marechal P, Gadal P, Queiroz Q, et al. 1986. Phosphorylation-dephosphorylation process as a probable mechanism for the diurnal regulatory changes of phosphoenolpyruvate carboxylase in CAM plants. *Biochem. Biophys. Res. Commun.* 136:151–59

30. Buchanan-Bollig IC, Smith JAC. 1984. Circadian rhythms in *Kalanchoë*: effects of the irradiance and temperature on gas exchange and carbon metabolism. *Planta* 160:264–71

31. Buchanan-Bollig IC, Smith JAC. 1984. Circadian rhythms in crassulacean acid metabolism: phase relationships between gas exchange, leaf water relations and malate metabolism in *Kalanchoë daigremontiana*. *Planta* 160:314–19

32. Carter PJ, Nimmo HG, Fewson CA, Wilkins MB. 1990. *Bryophyllum fedtschenkoi* protein phosphatase 2A can dephosphorylate phosphoenolpyruvate carboxylase. *FEBS Lett.* 263:233–36

33. Carter PJ, Nimmo HG, Fewson CA, Wilkins MB. 1991. Circadian rhythms in the activity of a plant protein kinase. *EMBO J.* 10:2063–68

34. Carter PJ, Wilkins MB, Nimmo HG, Fewson CA. 1995. The role of temperature in the regulation of the circadian rhythm of CO$_2$ fixation in *Bryophyllum fedtschenkoi*. *Planta* 196:381–86

35. Carter PJ, Wilkins MB, Nimmo HG, Fewson CA. 1995. Effects of temperature on the activity of phosphoenolpyruvate carboxylase and on the control of CO$_2$ fixation in *Bryophyllum fedtschenkoi*. *Planta* 196:375–80

36. Cheffings CM, Pantoja O, Ashcroft FM, Smith JAC. 1997. Malate transport and vacuolar ion channels in CAM plants. *J. Exp. Bot.* 48:623–31

37. Cheng S-H, Edwards GE. 1991. Influence of long photoperiods on plant development and expression of Crassulacean acid metabolism in *Mesembryanthemum crystallinum*. *Plant Cell Environ.* 14:271–78

38. Chollet R, Vidal J, O'Leary MH. 1996. Phospho*enol*pyruvate carboxylase: a

ubiquitous, highly regulated enzyme in plants. *Annu. Rev. Plant Physiol. Plant Mol. Biol.* 47:273–98

39. Chu C, Dai ZY, Ku MSB, Edwards GE. 1990. Induction of Crassulacean acid metabolism in the facultative halophyte *Mesembryanthemum crystallinum* by abscisic acid. *Plant Physiol* 93:1253–60

40. Cockburn W. 1985. Variations in photosynthetic acid metabolism in vascular plants: CAM and related phenomena. *New Phytol.* 101:3–24

41. Cockburn W. 1998. Rapid-cycling CAM; an hypothetical variant of photosynthetic metabolism. *Plant Cell Environ.* 21:845–51

42. Cockburn W, Goh CJ, Avadhani PN. 1985. Photosynthetic carbon assimilation in a shootless orchid, *Chiloschista usneoides* (DON) LDL. *Plant Physiol.* 77:83–86

43. Cockburn W, Whitelam GC, Broad A, Smith J. 1996. The participation of phytochrome in the signal transduction pathway of salt stress responses in *Mesembryanthemum crystallinum* L. *J. Exp. Bot.* 47:647–53

44. Christopher JT, Holtum JAM. 1996. Patterns of carbohydrate partitioning in the leaves of Crassulacean acid metabolism species during deacidification. *Plant Physiol.* 112:393–99

45. Christopher JT, Holtum JAM. 1998. Carbohydrate partitioning in the leaves of Bromeliaceae performing C3 photosynthesis or Crassulacean acid metabolism. *Aust. J. Plant Physiol.* 25:371–76

46. Cook RM, Lindsay JG, Wilkins MB, Nimmo HG. 1995. Decarboxylation of malate in the Crassulacean acid metabolism plant *Bryophyllum* (*Kalanchoë*) *fedtschenkoi*. *Plant Physiol.* 109:1301–7

47. Cushman JC. 1992. Characterization and expression of a NADP-malic enzyme cDNA induced by salt stress from the facultative CAM plant, *Mesembryanthemum crystallinum*. *Eur. J. Biochem.* 208:259–66

48. Cushman JC. 1993. Molecular cloning and expression of chloroplast NADP-malate dehydrogenase during crassulacean acid metabolism induction by salt stress. *Photosyn. Res.* 35:15–27

49. Cushman JC, Bohnert HJ. 1989. Nucleotide sequence of the gene encoding a CAM specific isoform of phospho*enol*pyruvate carboxylase from *Mesembryanthemum crystallinum*. *Nucleic Acids Res.* 17:6745–46

50. Cushman JC, Bohnert HJ. 1989. Nucleotide sequence of the *Ppc2* gene encoding a house-keeping isoform of phospho*enol*pyruvate carboxylase from *Mesembryanthemum crystallinum*. *Nucleic Acids Res.* 17:6743–44

51. Cushman JC, Bohnert HJ. 1992. Salt stress alters A/T-rich DNA-binding factor interactions within the phospho*enol*pyruvate carboxylase promoter from *Mesembryanthemum crystallinum*. *Plant Mol. Biol.* 20:411–24

52. Cushman JC, Bohnert HJ. 1997. Molecular genetics of Crassulacean acid metabolism. *Plant Physiol.* 113:667–76

53. Cushman JC, Meiners MS, Bohnert HJ. 1993. Expression of a phospho*enol*pyruvate carboxylase promoter from *Mesembryanthemum crystallinum* is not salt-inducible in mature transgenic tobacco. *Plant Mol. Biol.* 20:411–24

54. Cushman JC, Meyer G, Michalowski CB, Schmitt JM, Bohnert HJ. 1989. Salt stress leads to the differential expression of two isogenes of phospho*enol*pyruvate carboxylase during Crassulacean acid metabolism induction in the common ice plant. *Plant Cell* 1:715–25

55. Cushman JC, Michalowski CB, Bohnert HJ. 1990. Developmental control of Crassulacean acid metabolism inducibility by salt stress in the common ice plant. *Plant Physiol.* 94:1137–42

56. Cushman JC, Taybi T, Bohnert HJ. 1999. Induction of Crassulacean acid metabolism—molecular aspects. In *Photosynthesis: Physiology and Metabolism*, ed. RC Leegood, TD Sharkey, S von Caemmerer. Dordrecht: Kluwer. In press

57. Cushman JC, Vernon DM, Bohnert HJ. 1993. ABA and the transcriptional control of CAM induction during salt stress in the common ice plant. In *Control of Plant Gene Expression*, ed. DPS Verma, pp. 287–300. Boca Raton, FL: CRC Press

58. Dai Z, Ku MSB, Zhang DZ, Edwards GE. 1994. Effects of growth regulators on the induction of Crassulacean acid metabolism in the facultative halophyte *Mesembryanthemum crystallinum* L. *Planta* 192:287–94

59. DeRocher EJ, Bohnert HJ. 1993. Developmental and environmental stress employ different mechanisms in the expression of a plant gene family. *Plant Cell* 5:1611–25

60. DeRocher EJ, Harkins KR, Galbraith DW, Bohnert HJ. 1990. Developmentally regulated systemic endopolyploidy in succulents with small genomes. *Science* 250:99–101

61. De Saussure T. 1804. *Recherches Chimiques sur la Végétation*. Paris: Chez la V.ᶜ Nyon

62. Dietz KJ, Arbinger B. 1996. cDNA sequence and expression of subunit E of the vacuolar H$^{(+)}$-ATPase in the inducible Crassulacean acid metabolism plant *Mesembryanthemum crystallinum*. *Biochem. Biophys. Acta Gene Struct. Funct.* 1281:134–38

63. Drincovich MF, Casati P, Andreo CS, Chessin SJ, Franceschi VR, et al. 1998. Evolution of C4 photosynthesis in *Flaveria* species. Isoforms of NADP-malic enzyme. *Plant Physiol.* 117:733–44

64. Earnshaw MJ, Carver KA, Charlton WA. 1987. Leafy anatomy, water relations, and crassulacean acid metabolism in the chlorenchyma and colourless internal water-storing tissue of *Carpobrotus edulis* and *Senecio mandraliscae*. *Planta* 170:421–32

65. Eastmond PJ, Ross JD. 1997. Evidence that the induction of crassulacean acid metabolism by water stress in *Mesembryanthemum crystallinum* (L.) involves root signalling. *Plant Cell Environ.* 20:1559–65

66. Ehleringer JR, Monson RK. 1993. Evolutionary and ecological aspects of photosynthetic pathway variation. *Annu. Rev. Ecol. Syst.* 24:411–39

67. Fisslthaler B, Meyer G, Bohnert HJ, Schmitt JM. 1995. Age-dependent induction of pyruvate, orthophosphate dikinase in *Mesembryanthemum crystallinum* L. *Planta* 196:492–500

68. Forsthoefel NR, Cushman MAF, Cushman JC. 1995. Posttranscriptional and posttranslational control of enolase expression in the faculative crassulacean acid metabolism plant *Mesembryanthemum crystallinum* L. *Plant Physiol.* 108:1185–95

69. Forsthoefel NR, Vernon DM, Cushman JC. 1995. A salinity-induced gene from the halophyte *M. crystallinum* encodes a glycolytic enzymes, cofactor-independent phosphoglyceromutase. *Plant Mol. Biol.* 29:213–26

70. Foster JC, Edwards GE, Winter K. 1982. Changes in the levels of phospho*enol*pyruvate carboxylase with induction of CAM in *M. crystallinum* L. *Plant Cell Physiol.* 23:585–94

71. Gehrig H, Faist K, Kluge M. 1999. Identification of phospho*enol*pyruvate car-

boxylase isoforms in leaf, stem, and roots of the obligate CAM plant *Vanilla planifolia* SALIB. (Orchidaceae): a physiological and molecular approach. *Planta*. In press

72. Gehrig H, Heute V, Kluge M. 1998. Towards a better knowledge of the molecular evolution of phospho*enol*pyruvate carboxylase by comparison of partial cDNA sequences. *J. Mol. Evol.* 46:107–14

73. Gehrig H, Rosicke H, Kluge M. 1997. Detection of DNA polymorphisms in the genus Kalanchoe by RAPD-PCR fingerprint and its relationships to infrageneric taxonomic position and ecophysiological photosynthetic behaviour of the species. *Plant Sci.* 125:41–51

74. Gehrig H, Taybi T, Kluge M, Brulfert J. 1995. Identification of multiple PEPC isogenes in leaves of the facultative Crassulacean acid metabolism (CAM) plant *Kalanchoe blossfeldiana* Poelln. cv. Tom Thumb. *FEBS Lett.* 377:399–402

75. Giglioli-Guivarc'h N, Pierre JC, Brown S, Chollet R, Vidal J, Gadal P. 1996. The light-dependent transduction pathway controlling the regulatory phosphorylation of C_4 phospho*enol*pyruvate carboxylase in protoplasts from *Digitaria sanguinalis*. *Plant Cell* 8:573–86

76. Grams TEE, Beck F, Lüttge U. 1996. Generation of rhythmic and arrhythmic behaviour of Crassulacean acid metabolism in *Kalanchoë daigremontiana* under continuous light by varying the irradiance or temperature: measurements in vivo and model simulations. *Planta* 198:110–17

77. Grams TEE, Borland AM, Roberts A, Griffiths H, Beck F, Lüttge U. 1997. On the mechanism of reinitiation of endogenous Crassulacean acid metabolism rhythm by temperature changes. *Plant Physiol.* 113:1309–17

78. Griffiths H. 1988. Crassulacean acid metabolism: a re-appraisal of physiological plasticity in form and function. *Adv. Bot. Res.* 15:42–92

79. Griffiths H. 1989. Carbon dioxide concentrating mechanisms and the evolution of CAM in vascular epiphytes. In *Vascular Plants as Epiphytes: Evolution and Ecophysiology*, ed. U Lüttge, pp. 42–86. Berlin: Springer-Verlag

80. Griffiths H, Broadmeadow MSJ, Borland AM, Hetherington CS. 1990. Short-term changes in carbon isotope discrimination identify transitions between C_3 and C_4 carboxylation during Crassulacean acid metabolism. *Planta* 181:604–10

81. Harris PJC, Wilkins MB. 1978. Evidence for phytochrome involvement in the entrainment of the circadian rhythm of CO_2 metabolism in *Bryophyllum*. *Planta* 138:271–72

82. Harris PJC, Wilkins MB. 1978. The circadian rhythm in *Bryophyllum* leaves: phase control by radiant energy. *Planta* 143:323–28

83. Hartwell J, Smith LH, Wilkins MB, Jenkins GI, Nimmo HG. 1996. Higher plant phospho*enol*pyruvate carboxylase kinase is regulated at the level of translatable mRNA in response to light or a circadian rhythm. *Plant J.* 10:1071–78

84. Herppich W, Herppich M, von Willert DJ. 1992. The irreversible C_3 to CAM shift in well-watered and salt-stressed plants of *Mesembryanthemum crystallinum* is under strict ontogenetic control. *Bot. Acta* 105:34–40

85. Heyne B. 1815. On the deoxidation of the leaves of *Coltyledon calycina*. *Trans. Linn. Soc. London* 11:213–15

86. Höfner R, Vasquez-Moreno L, Winter K, Bohnert HJ, Schmitt JM. 1987. Induction of Crassulacean acid metabolism in *Mesembryanthemum crystallinum* by high salinity: mass increase and de novo synthesis of PEP carboxylase. *Plant Physiol.* 83:915–19

87. Holtum JAM, Winter K. 1982. Activities of enzymes of carbon metabolism during the induction of Crassulacean acid metabolism in *Mesembryanthemum crystallinum*. *Planta* 155:8–16

88. Honda H, Okamoto T, Shimada H. 1996. Isolation of a cDNA for a phospho*enol*pyruvate carboxylase from a monocot CAM-plant, *Aloe arborescens*: structure and its gene expression. *Plant Cell Physiol.* 37:881–88

89. Jia S-R, Yang M-Z, Ott R, Chua N-H. 1989. High frequency transformation of *Kalanchoe laciniata*. *Plant Cell Rep.* 8:336–40

90. Jiao J-A, Chollet R. 1991. Posttranslational regulation of phospho*enol*pyruvate carboxylase in C_4 and Crassulacean acid metabolism plants. *Plant Physiol.* 95:981–85

91. Deleted in proof

92. Keeley JE. 1998. CAM photosynthesis in submerged aquatic plants. *Bot. Rev.* 64:121–75

93. Keeley JE, Osmond CB, Raven JA. 1984. *Stylites*, a vascular plant without stomata absorbs CO_2 via its roots. *Nature* 310:694–95

94. Kluge M, Hell R, Pfeffer A, Kramer D. 1987. Structural and metabolic properties of green tissue cultures from a CAM plant, *Kalanchoë blossfeldiana* hybr. Montezuma. *Plant Cell Environ.* 10:451–62

95. Kluge M, Maier P, Brulfert J, Faist K, Wollny E. 1988. Regulation of phosph*oenol*pyruvate carboxylase in crassulacean acid metabolism (CAM). *J. Plant Physiol.* 133:252–56

96. Kluge M, Schomburg M. 1996. The tonoplast as a target of temperature effects in Crassulacean acid metabolism. See Ref. 173a, pp. 72–77

97. Kore-eda S, Kanai R. 1997. Induction of glucose 6–phosphate transport activity in chloroplasts of *Mesembryanthemum crystallinum* by the C_3-CAM transition. *Plant Cell Physiol.* 38:895–901

98. Kore-eda S, Yamashita T, Kanai R. 1996. Induction of light dependent pyruvate transport into chloroplasts of *Mesembryanthemum crystallinum* by salt stress. *Plant Cell Physiol.* 37:257–62

99. Kusumi K, Arata H, Iwasaki I, Nishimura M. 1994. Regulation of PEP-carboxylase by biological clock in a CAM plant. *Plant Cell Physiol.* 35:233–42

100. Lee DM, Assmann SM. 1992. Stomatal responses to light in the facultative Crassulacean acid metabolism species, *Portulacaria afra. Physiol. Plant.* 85:35–42

101. Lee HSJ, Griffiths H. 1987. Induction and repression of CAM in *Sedum telephium* L. in response to photoperiod and water stress. *J. Exp. Bot.* 38:834–41

102. Lepiniec L, Keryer E, Phillippe H, Gadal P, Cretin C. 1993. Sorghum phosph*oenol*pyruvate carboxylase gene family: structure, function, and molecular evolution. *Plant Mol. Biol.* 21:487–502

103. Lepiniec L, Vidal J, Chollet R, Gadal P, Cretin C. 1994. Phospho*enol*pyruvate carboxylase: structure, regulation, and evolution. *Plant Sci.* 99:111–24

104. Li B, Chollet R. 1994. Salt induction and the partial purification/characterization of phosphoenolpyruvate carboxylase protein-serine kinase from an inducible Crassulacean-acid-metabolism (CAM) plant, *Mesembryanthemum crystallinum* L. *Arch. Bioch. Biophys.* 314:247–54

105. Löw R, Rockel B, Kirsch M, Ratajczak R, Hörntensteiner S, et al. 1996. Early salt stress effects on the differential expression of vacuolar H^+-ATPase genes in roots and leaves of *Mesembryanthemum crystallinum. Plant Physiol.* 110:259–65

106. Lüttge U. 1987. Carbon dioxide and water demand; crassulacean acid metabolism (CAM), a versatile ecological adaptation exemplifying the need for integration in ecophysiological work. *New Phytol.* 106:593–629

107. Lüttge U. 1993. The role of crassulacean acid metabolism (CAM) in the adaptation of plants to salinity. *New Phytol.* 125:59–71

108. Lüttge U, Beck F. 1992. Endogenous rhythms and chaos in crassulacean acid metabolism. *Planta* 188:28–38

109. Lüttge U, Fischer-Schliebs E, Ratajczak R, Kramer D, Berndt E, Kluge M. 1995. Functioning of the tonoplast in vacuolar C-storage and remobilization in crassulacean acid metabolism. *J. Exp. Bot.* 46:1377–88

110. Lüttge U, Grams TEE, Hechler B, Blasius B, Beck F. 1996. Frequency resonances of the circadian rhythm of CAM under external temperature rhythms of varied period lengths in continuous light. *Bot. Acta* 109:422–26

111. Mariaux JB, Fischer-Schliebs E, Lüttge U, Ratajczak R. 1997. Dynamics of activity and structure of the tonoplast vacuolar-type H^+-ATPase in plants with differing CAM expression and in a C_3 plant under salt stress. *Protoplasma* 196:181–89

112. Martin CE. 1996. Putative causes and consequences of recycling CO_2 via Crassulacean acid metabolism. See Ref. 173a, pp. 192–203

113. Mawson BT, Zaugg MW. 1994. Modulation of light-dependent stomatal opening in isolated epidermis following induction of Crassulacean acid metabolism in *Mesembryanthemum crystallinum. J. Plant Physiol.* 144:740–46

114. McElwain EF, Bohnert HJ, Thomas JC. 1992. Light moderates the induction of Phospho*enol*pyruvate carboxylase by NaCl and abscisic acid in *Mesembryanthemum crystallinum. Plant Physiol.* 99:1261–64

115. Meiners MS, Thomas JC, Bohnert HJ, Cushman JC. 1991. Regeneration of multiple shoots and plants from *Mesembryanthemum crystallinum. Plant Cell Rep.* 9:563–66

116. Michalowski CB, Olson SW, Piepenbrock M, Schmitt JM, Bohnert HJ. 1989. Time course of mRNA induction elicited

by salt stress in the common ice plant (*M. crystallinum*). *Plant Physiol.* 89: 811–16

117. Miyazaki S, Koga R, Bohnert HJ, Fukuhara T. 1999. Tissue- and environmental response-specific expression of 10 PP2C transcripts in *Mesembryanthemum crystallinum*. *Mol. Gen. Genet.* In press

118. Mricha A, Brulfert J, Pierre JN, Queiroz O. 1990. Phytochrome-mediated responses of cells and protoplasts of green calli obtained from leaves of a CAM plant. *Plant Cell Rep.* 8:664–66

119. Neuhaus E, Schulte N. 1996. Starch degradation in chloroplasts isolated from C$_3$ or CAM (crassulacean acid metabolism)-induced *Mesembryanthemum crystallinum* L. *Biochem J.* 318: 945–53

120. Nimmo GA, Nimmo HG, Hamilton ID, Fewson CA, Wilkins MB. 1986. Purification of the phosphorylated night form and dephosphorylated day form of phospho*enol*pyruvate carboxylase from *Bryophyllum fedtschenkoi*. *Biochem. J.* 239:213–20

121. Nimmo GA, Nimmo HG, Hamilton ID, Fewson CA, Wilkins MB. 1987. Persistant circadian rhythms in the phosphorylation state of phospho*enol*pyruvate carboxylase from *Bryophyllum fedtschenkoi* leaves and in its sensitivity to inhibition by malate. *Planta* 170:408–15

122. Nimmo HG. 1998. Circadian regulation of a plant protein kinase. *Chronobiol. Int.* 15:109–18

123. Nishio JN, Ting IP. 1987. Carbon flow and metabolic specialization in the tissue layers of the crassulacean acid metabolism plant, *Peperomia camptotricha*. *Plant Physiol.* 84:600–4

124. Nishio JN, Ting IP. 1993. Photosynthetic characteristics of the palisade mesophyll and spongy mesophyll in the CAM/C$_4$ intermediate plant, *Peperomia camptotricha*. *Bot. Acta* 106:120–25

125. Ocheretina O, Scheibe R. 1997. Cloning and sequence analysis of cDNAs encoding cytosolic malate dehydrogenase. *Gene* 199:145–48

126. Osmond CB. 1978. Crassulacean acid metabolism: a curiosity in context. *Annu. Rev. Plant. Physiol.* 29:379–414

127. Ostrem JA, Olsen SW, Schmitt JM, Bohnert HJ. 1987. Salt stress increases the level of translatable mRNA for PEPC in *Mesembryanthemum crystallinum*. *Plant Physiol.* 84:1270–75

128. Ostrem JA, Vernon DM, Bohnert

HJ. 1990. Increased expression of a gene coding for NAD-glyceraldehyde-3-phosphate dehydrogenase during the transition from C$_3$ photosynthesis to crassulacean acid metabolism in *Mesembryanthemum crystallinum*. *J. Biol. Chem.* 265:3497–502

129. Paul MJ, Loos K, Stitt M, Ziegler P. 1993. Starch-degrading enzymes during the induction of CAM in *Mesembryanthemum crystallinum*. *Plant Cell Environ.* 16:531–38

130. Peters W, Beck E, Piepenbrock M, Lenz B, Schmitt JM. 1997. Cytokinin as a negative effector of phosphoenolpyruvate carboxylate induction in *Mesembryanthemum crystallinum*. *J. Plant Physiol.* 151:362–67

131. Piepenbrock M, Schmitt JM. 1991. Environmental control of phosphoenolpyruvate carboxylase induction in mature *Mesembryanthemum crystallinum* L. *Plant Physiol.* 97:998–1003

132. Ranson SI, Thomas M. 1960. Crassulacean acid metabolism. *Annu. Rev. Plant Physiol.* 11:81–110

133. Ratajczak R, Kemna I, Lüttge U. 1994. Characteristics and partial purification and reconstitution of the vacuolar malate transporter of the CAM plant *Kalanchoë daigremontiana* Hamet et Perrier de la Bâthie. *Planta* 195:226–36

134. Raven JA, Spicer RA. 1996. The evolution of Crassulacean acid metabolism. See Ref. 173a, pp. 360–85

135. Relle M, Wild A. 1996. Molecular characterization of a phosphoenolpyruvate carboxylase in the gymnosperm *Picea abies* (Norway spruce). *Plant Mol. Biol.* 32:923–36

136. Rickers J, Cushman JC, Michalowski CB, Schmitt JM, Bohnert HJ. 1989. Expression of the CAM-form of phospho(*enol*)pyruvate carboxylase and nucleotide sequence of a full length cDNA from *Mesembryanthemum crystallinum*. *Mol. Gen. Genet.* 215:447–54

137. Rippmann JF, Michalowski CB, Nelson DE, Bohnert HJ. 1997. Induction of a ribosome-inactivating protein upon environmental stress. *Plant Mol. Biol.* 35:701–9

138. Roberts A, Borland AM, Griffiths H. 1997. Discrimination processes and shifts in carboxylation during the phases of Crassulacean acid metabolism. *Plant Physiol.* 113:1283–92

139. Rockel B, Jia C, Ratajczak R , Lüttge U. 1998. Day-night changes of the amount of subunit-c transcript of the V-ATPase in suspension cells of *Mesembryan-*

themum crystallinum. J. Plant Physiol. 152:189–93

140. Rockel B, Lüttge U, Ratajczak R. 1998. Changes in message amount of V-ATPase subunits during salt-stress induced C3–CAM transition in *Mesembryanthemum crystallinum. Plant Physiol. Biochem.* 36:567–73

141. Saitou K, Agata W, Kawarabata T, Yamamoto Y, Kubota F. 1995. Purification and properties of NAD-malate dehydrogenase from *Mesembryanthemum crystallinum* L. exhibiting Crassulacean acid metabolism. *Jpn. J. Crop Sci.* 64:760–66

142. Saitou K, Agata W, Masui Y, Asakura M, Kubota F. 1994. Isoforms of NADP-malic enzyme from *Mesembryanthemum crystallinum* L. that are involved in C_3 photosynthesis and crassulacean acid metabolism. *Plant Cell Physiol.* 35:1165–71

143. Schaeffer HJ, Forsthoefel NR, Cushman JC. 1995. Identification of enhancer and silencer regions involved in salt-responsive expression of Crassulacean acid metabolism (CAM) genes in the facultative halophyte *Mesembryanthemum crytallinum. Plant Mol. Biol.* 28:205–18

144. Schmitt JM. 1990. Rapid concentration changes of phospho*enol*pyruvate carboxylase mRNA in detached leaves of *Mesembryanthemum crystallinum. Plant Cell Environ.* 13:845–50

145. Schmitt JM, Fisslthaler B, Sheriff A, Lenz B, Bässler M, Meyer G. 1996. Environmental control of CAM induction in *Mesembryanthemum crystallinum*—a role for cytokinin, abscisic acid and jasmonate? See Ref. 173a, pp. 159–75

146. Schmitt JM, Piepenbrock M. 1992. Regulation of phospho*enol*pyruvate carboxylase crassulacean acid metabolism induction in *Mesembryanthemum crystallinum* L. by cytokinin: modulation of leaf gene expression by roots? *Plant Physiol.* 99:1664–69

147. Sipes D, Ting IP. 1985. Crassulacean acid metabolism and crassulacean acid metabolism modifications in *Peperomia camptotricha. Plant Physiol.* 77:59–63

148. Springer SA, Outlaw WH. 1988. Histochemical compartmentation of photosynthesis in the crassulacean acid metabolism plant *Crassula falcata. Plant Physiol.* 88:633–38

149. Steiger S, Pfeifer T, Ratajczak R, Martinoia E, Lüttge U. 1997. The vacuolar malate transporter of *Kalanchoë daigremontiana*: a 32 kDa polypeptide? *J. Plant Physiol.* 151:137–41

150. Tallman G, Zhu JX, Mawson BT, Amodeo G, Nouhi Z, et al. 1997. Induction of CAM in *Mesembryanthemum crystallinum* abolishes the stomatal response to blue light and light-dependent zeaxanthin formation in guard cell chloroplasts. *Plant Cell Physiol.* 38:236–42

151. Taybi T, Cushman JC. 1998. Signal transduction events leading to Crassulacean acid metabolism (CAM) induction in the common ice plant, *Mesembryanthemum crystallinum. Plant Physiol. Suppl.* 1175:47 (Abstr.)

152. Taybi T, Sotta B, Gehrig H, Güclü S, Kluge M, Brulfert J. 1995. Differential effects of abscisic acid on phospho*enol*pyruvate carboxylase and CAM operation in *Kalanchoë blossfeldiana. Bot. Acta* 198:240–46

153. Thomas JC, Bohnert HJ. 1993. Salt stress perception and plant growth regulators in the halophyte, *Mesembryanthemum crystallinum. Plant Physiol.* 103:1299–304

154. Thomas JC, De Armond R, Bohnert HJ. 1992. Influence of NaCl on growth, proline, and phospho*enol*pyruvate carboxylase levels in *Mesembryanthemum crystallinum* suspension cultures. *Plant Physiol.* 98:626–31

155. Thomas JC, McElwain EF, Bohnert HJ. 1992. Convergent induction of osmotic stress-responses: abscisic acid, cytokinin, and the effect of NaCl. *Plant Physiol.* 100:416–23

156. Ting IP. 1985. Crassulacean acid metabolism. *Annu. Rev. Plant. Physiol.* 36: 595–622

157. Ting IP, Patel A, Sipes DL, Reid PD, Walling LL. 1994. Differential expression of photosynthesis genes in leaf tissue layers of *Peperomia* as revealed by tissue printing. *Am. J Bot.* 81:414–22

158. Toh H, Kawamura T, Izui K. 1994. Molecular evolution of phospho*enol*-pyruvate carboxylase. *Plant Cell Environ.* 17:31–43

159. Tsiantis MS. 1996. *Regulation of V-ATPase gene expression by ionic stress in higher plants.* DPhil thesis. Univ. Oxford, Oxford, UK

160. Tsiantis MS, Bartholomew DM, Smith JAC. 1996. Salt regulation of transcript levels for the c subunit of a leaf vacuolar H$^+$-ATPase in the halophyte *Mesembryanthemum crystallinum. Plant J.* 9:729–36

161. Vera-Estrella R, Barkla BJ, Bohnert HJ, Pantoja O. 1999. Salt-stress in *Mesembryanthemum crystallinum* L. cell sus-

pensions activates adaptive mechanisms similar to those observed in the whole plant. *Planta* 207:426–35

162. Vernon DM, Bohnert HJ. 1992. A novel methyl transferase induced by osmotic stress in the facultative halophyte *M. crystallinum. EMBO J.* 11:2077–85

163. Vernon DM, Ostrem JA, Bohnert HJ. 1993. Stress perception and response in a facultative halophyte: the regulation of salinity-induced genes in *Mesembryanthemum crystallinum. Plant Cell Environ.* 16:437–44

164. Vidal J, Chollet R. 1997. Regulatory phosphorylation of C_4 PEP carboxylase. *Trends Plant Sci.* 2:230–37

165. Walker RP, Leegood RC. 1996. Phosphorylation of phospho*enol*pyruvate carboxykinase in plants. Studies in plants with C_4 photosynthesis and Crassulacean metabolism and in germinating seeds. *Biochem. J.* 317:653–58

166. Wang B, Lüttge U. 1994. Induction and subculture of callus and regeneration of fertile plants of *Mesembryanthemum crystallinum* L. *Pol. J. Environ. Stud.* 3:55–57

167. Weigend M. 1994. In vivo phosphorylation of phospho*enol*pyruvate carboxylase from the facultative CAM plant *Mesembryanthemum crystallinum. J. Plant Physiol.* 144:654–60

168. Wilkins MB. 1984. A rapid circadian rhythm of carbon dioxide metabolism in *Bryophyllum fedtschenkoi. Planta* 161:381–84

169. Wilkins MB. 1992. Circadian rhythms: their origin and control. *New Phytol.* 121:347–75

170. Willenbrink ME, Huesemann W. 1995. Photoautotrophic cell suspension cultures from *Mesembryanthemum crystallinum* and their response to salt stress. *Bot. Acta* 108:497–504

171. Winter K. 1987. Gradient in the degree of crassulacean acid metabolism within leaves of *Kalanchoe daigremontiana. Planta* 172:88–90

172. Winter K, Foster JG, Edwards GE, Holtum JAM. 1982. Intracellular localization of enzymes of carbon metabolism in *Mesembryanthemum crystallinum* exhibiting C_3 photosynthetic characteristics or performing Crassulacean acid metabolism. *Plant Physiol.* 69:300–7

173. Winter K, Medina E, Garcia V, Mayoral ML, Muniz R. 1985. Crassulacean acid metabolism in roots of a leafless orchid, *Campylocentrum tyrridion* Caray & Dunsterv. *J. Plant Physiol.* 118:73–78

173a. Winter K, Smith JAC, eds. 1996. *Crassulacean Acid Metabolism. Biochemistry, Ecophysiology and Evolution*, Vol. 114. Berlin: Springer-Verlag

174. Winter K, Smith JAC. 1996. An introduction to Crassulacean acid metabolism. Biochemical principles and ecological diversity. See Ref. 173a, pp.1–13

175. Yen HE, Grimes HD, Edwards GE. 1995. The effects of high salinity, water deficit, and abscisic acid on phospho*enol*pyruvate carboxylase activity and proline accumulation in *Mesembryanthemum crystallinum* cell cultures. *J. Plant Physiol.* 145:557–64

Annu. Rev. Plant Physiol. Plant Mol. Biol. 1999. 50:333–59

PHOTOPROTECTION REVISITED:
Genetic and Molecular Approaches

Krishna K. Niyogi
Department of Plant and Microbial Biology, University of California, Berkeley, California 94720-3102; e-mail: niyogi@nature.berkeley.edu

KEY WORDS: antioxidant, oxidative stress, photosynthesis, photoinhibition, thermal dissipation

ABSTRACT
The involvement of excited and highly reactive intermediates in oxygenic photosynthesis poses unique problems for algae and plants in terms of potential oxidative damage to the photosynthetic apparatus. Photoprotective processes prevent or minimize generation of oxidizing molecules, scavenge reactive oxygen species efficiently, and repair damage that inevitably occurs. This review summarizes several photoprotective mechanisms operating within chloroplasts of plants and green algae. The recent use of genetic and molecular biological approaches is providing new insights into photoprotection, especially with respect to thermal dissipation of excess absorbed light energy, alternative electron transport pathways, chloroplast antioxidant systems, and repair of photosystem II.

CONTENTS

333

1040-2519/99/0601-0333$08.00

INTRODUCTION

Light is required for photosynthesis, yet plants need protection from light. Photosynthesis inevitably generates highly reactive intermediates and byproducts that can cause oxidative damage to the photosynthetic apparatus (16, 58). This photo-oxidative damage, if not repaired, decreases the efficiency and/or maximum rate of photosynthesis, termed "photoinhibition" (89; reviewed in 13, 98, 118, 129).

Oxygenic photosynthetic organisms have evolved multiple photoprotective mechanisms to cope with the potentially damaging effects of light, as diagrammed in Figure 1. Some algae and plants avoid absorption of excessive light by movement of leaves, cells (negative phototaxis), or chloroplasts. Within the chloroplast, regulation of photosynthetic light harvesting and electron transport

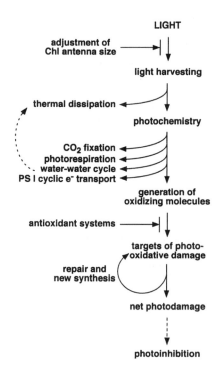

Figure 1   Schematic diagram of photoprotective processes occurring within chloroplasts.

balances the absorption and utilization of light energy. For example, adjustments in light-harvesting antenna size and photosynthetic capacity can decrease light absorption and increase light utilization, respectively, during relatively long-term acclimation to excessive light. Alternative electron transport pathways and thermal dissipation can also help to remove excess absorbed light energy from the photosynthetic apparatus. Numerous antioxidant molecules and scavenging enzymes are present to deal with the inevitable generation of reactive molecules, especially reactive oxygen species. However, despite these photoprotective defenses, damage to the photosynthetic machinery still occurs, necessitating turnover and replacement of damaged proteins. The overall goal of photoprotection is, therefore, to prevent net damage from occurring.

Because of its importance for maintaining photosynthesis and ultimately for survival of photosynthetic organisms in many natural environments, photoprotection has long been a topic of considerable interest in plant physiology and biochemistry. The general subject of photoprotection, as well as several specific photoprotective processes, have been extensively reviewed (9, 13, 18, 31, 37, 47–49, 51, 57, 65, 76, 77, 98, 112, 118, 128), so one might ask why the topic needs to be "revisited." Although genetic methods have long been used in analysis of photosynthesis (95, 142), only recently has there been widespread use of genetic and molecular techniques to dissect specific processes involved in photoprotection. The roles of specific cloned genes are being tested through reverse genetics, and classical (forward) genetics is uncovering new mutants and providing insights into the complexity of photoprotection. This article reviews several photoprotective processes that occur within chloroplasts of eukaryotic photosynthetic organisms, with a particular emphasis on the use of molecular and genetic approaches in intact green algae and plants. Space constraints direct my focus to results obtained with the most commonly used model organisms with good molecular genetics, the green alga *Chlamydomonas reinhardtii* (Table 1) and the C3 vascular plants *Arabidopsis thaliana* and tobacco (Table 2).

PHOTO-OXIDATIVE DAMAGE TO THE PHOTOSYNTHETIC APPARATUS

Before discussing specific photoprotective mechanisms, I briefly summarize the types of oxidizing molecules that are involved in damaging the photosynthetic machinery.

Generation of Oxidizing Molecules in Photosynthesis

Because of the large differences in redox potential between reactants and products and the involvement of excited intermediates, oxygenic photosynthesis

Table 1 Summary of *Chlamydomonas* mutants affecting photoprotection in the chloroplast

Photoprotective system	Mutant	Description	Photoprotection phenotype	Reference
Thermal dissipation and xanthophylls	*npq1*	Lacks zeaxanthin in HL[a]	Partly NPQ-deficient; grows in HL	109
	npq2	Accumulates zeaxanthin; lacks antheraxanthin, violaxanthin, and neoxanthin	Faster induction of NPQ; grows in HL	109
	lor1	Lacks lutein and loroxanthin	Partly NPQ-deficient; grows in HL	38, 110
	npq1 lor1	Lacks zeaxanthin in HL; lacks lutein and loroxanthin	NPQ-deficient; sensitive to HL	110
	npq4	Normal xanthophylls	NPQ-deficient; grows in HL	D Elrad, KK Niyogi & AR Grossman, unpublished results
	npq2 lor1	Accumulates zeaxanthin; lacks antheraxanthin, violaxanthin, neoxanthin, lutein, and loroxanthin	Grows in HL	KK Niyogi, unpublished results
Water-water cycle and scavenging enzymes	PAR19	Overexpresses FeSOD	n.d.[b]	85
PS II repair	*sr/spr*	Deficient in chloroplast protein synthesis	Sensitive to HL	73
	ag16.2	Accumulates D1 protein	Sensitive to HL	158
	*psbT*Δ	Deficient in PS II in HL	Sensitive to HL	105

[a]High light.
[b]Not determined.

poses unique problems for algae and plants with respect to the generation of reactive oxygen species and other oxidizing molecules. Potentially damaging molecules are generated at three major sites in the photosynthetic apparatus: the light-harvesting complex (LHC) associated with photosystem (PS) II, the PS II reaction center, and the PS I acceptor side.

Chlorophyll (Chl) molecules are critical participants in light-harvesting and electron transfer reactions in photosynthesis, but Chl can act as a potent endogenous photosensitizer in algae and plants. Absorption of light causes Chl to enter the singlet excited state (^{1}Chl), and the excitation energy is rapidly transferred (<ps time scale) between neighboring Chls in the LHC by resonance transfer. Before excitation energy is trapped in the reaction center, triplet Chl (^{3}Chl) can be formed from ^{1}Chl through intersystem crossing. This is an inherent physical property of ^{1}Chl (121), and the yield of ^{3}Chl formation

depends on the average lifetime of ^{1}Chl ($\sim$ns time scale) in the antenna (58). In contrast to ^{1}Chl, ^{3}Chl is relatively long-lived ($\sim$ms time scale) and can interact with O_2 to produce singlet oxygen ($^1O_2$) (56). Because the average lifetime of ^{1}Chl in the PS II LHC is several times longer than in the PS I LHC, the potential for generation of $^1O_2$ is greater in the PS II LHC.

In the PS II reaction center, trapping of excitation energy involves a primary charge separation between a Chl dimer (P680) and a pheophytin molecule (Pheo) that are bound to the D1 reaction center protein. The P680$^+$/Pheo$^-$ radical pair is reversible (44, 134), and the charge recombination (and other backreactions in PS II) can generate triplet P680 (^{3}P680) (11, 13, 115, 148). As in the LHC, energy exchange between ^{3}P680 and O_2 results in formation of $^1O_2$. The P680$^+$/Pheo$^-$ charge separation can be stabilized upon electron transfer from Pheo$^-$ to the quinone acceptor Q_A, and P680$^+$ is subsequently reduced by an electron derived from the oxidation of H_2O via the secondary donor Y_Z (tyrosine 161 of the D1 protein). Although the very high oxidizing potential of P680$^+$ and Y_Z^+ enables plants to use H_2O as an electron donor, P680$^+$ and Y_Z^+ themselves are also capable of oxidizing nearby pigments and proteins (11, 13).

In PS I, charge separation occurring between P700 and the primary acceptor Chl A_0 is stabilized by subsequent electron transfer to secondary acceptors, the phylloquinone A_1 and three iron-sulfur clusters (F_X, F_A, and F_B). In contrast to P680$^+$, P700$^+$ is less oxidizing and relatively stable; in fact, P700$^+$ is a very efficient quencher of excitation energy from the PS I LHC (44, 121). However, the acceptor side of PS I, which has a redox potential low enough to reduce NADP$^+$ via ferredoxin, is also capable of reducing O_2 to the superoxide anion radical (O_2^-) (102). O_2^- can be metabolized to H_2O_2, and a metal-catalyzed Haber-Weiss or Fenton reaction can lead to production of the hydroxyl radical (OH·), an extremely toxic type of reactive oxygen species (16).

Generation of these oxidizing molecules occurs at all light intensities, but when absorbed light energy exceeds the capacity for light energy utilization through photosynthesis, the potential for photo-oxidative damage is exacerbated. In excessive light, accumulation of excitation energy in the PS II LHC will increase the average lifetime of ^{1}Chl, thereby increasing the yield of ^{3}Chl and $^1O_2$. Higher excitation pressure in PS II can increase the frequency of direct damage by P680$^+$ and formation of $^1O_2$ as a result of P680$^+$/Pheo$^-$ recombination. Furthermore, the high ΔpH that builds up in excessive light can inhibit electron donation to P680$^+$ from the oxygen-evolving complex, resulting in longer-lived P680$^+$ and/or Y_Z^+. Overreduction of the PS I acceptor side in excessive light favors direct reduction of O_2 to form O_2^-. Analysis of transgenic tobacco plants with reduced levels of cytochrome b_6/f complex has pointed to the involvement of lumen acidification and/or overreduction of the PS I acceptor side in photoinhibition (80).

Table 2 Summary of Arabidopsis and tobacco mutants and transgenics affecting photoprotection in the chloroplast

Photoprotective system	Plant	Mutant/transgenic	Description	Photoprotection phenotype	Reference
Thermal dissipation and xanthophylls	Arabidopsis	npq1	Lacks zeaxanthin in HL[a]	NPQ-deficient; sensitive to short-term HL	111
	Arabidopsis	aba1 (npq2)	Accumulates zeaxanthin; lacks antheraxanthin, violaxanthin, and neoxanthin	Faster induction of NPQ	81, 111, 144
	Tobacco	aba2	Accumulates zeaxanthin; lacks antheraxanthin, violaxanthin, and neoxanthin	n.d.[b]	100
	Arabidopsis	lut2	Lacks lutein	Partly NPQ-deficient; grows in HL	127
	Arabidopsis	lut1	Accumulates zeinoxanthin; lacks lutein	Partly NPQ-deficient	127
	Arabidopsis	npq1 lut2	Lacks zeaxanthin in HL; lacks lutein	NPQ-deficient; older leaves sensitive to HL	O Björkman, C Shih, B Pogson, D DellaPenna & KK Niyogi, unpublished results
	Arabidopsis	ch1	Lacks Chl b	Partly NPQ-deficient	KK Niyogi, unpublished results
	Arabidopsis	npq4	Normal xanthophylls	NPQ-deficient	KK Niyogi, C Shih & O Björkman, unpublished results

Arabidopsis	*aba1 lut2*	Accumulates zeaxanthin; lacks antheraxanthin, violaxanthin, neoxanthin, and lutein	Partly NPQ-deficient; impaired growth	127
Photorespiration	Tobacco	35S-GS2	Overexpresses glutamine synthetase in chloroplast	Resistant to HL	91
Water-water cycle and scavenging enzymes	Arabidopsis	Cu/ZnSOD-deficient mutant	Lacks Cu/ZnSOD in chloroplast and cytosol	n.d.	DJ Kliebenstein & RL Last, unpublished results
	Arabidopsis	FeSOD-deficient mutant	Lacks FeSOD in chloroplast	n.d.	DJ Kliebenstein & RL Last, unpublished results
	Tobacco	35S-Cu/ZnSOD	Overexpresses Cu/ZnSOD in chloroplast	Resistant to HL + cold	137, 138
	Tobacco	35S-FeSOD	Overexpresses FeSOD in chloroplast	Same as wild type	147
PS I cyclic electron transport	Tobacco	*ndhC, ndhK, ndhJ*	Lacks thylakoid NDH complex	Slower induction of NPQ in water-stressed plants	36
	Tobacco	*ndhB*	Lacks thylakoid NDH complex	n.d.	139
Ascorbate	Arabidopsis	*vtc1*	Ascorbate-deficient	n.d.	40, 41
Glutathione	Arabidopsis	*cad2*	Glutathione-deficient	n.d.	39, 78
PS II repair	Tobacco	35S-glycerol-3-P acyltransferase	Decreased unsaturation of chloroplast lipids	Sensitive to HL + cold	106

[a]High light.
[b]Not determined.

Targets of Photo-Oxidative Damage

Although the exact mechanism(s) of damage has not been determined, PS II is a major target of photo-oxidative damage (10, 11, 13, 115). Interestingly, photoinhibition of PS II, measured as a decrease in either the flash yield of O_2 evolution or the Chl fluorescence parameter F_v/F_m, seems to depend on the number of absorbed photons rather than the rate of photon absorption. This implies that there is a constant probability of photodamage for each absorbed photon during conditions of steady-state photosynthesis (122, 123). However, the probability of photodamage can be modulated by changes in Chl antenna size and rate of electron transport (20, 124), which alter the excitation pressure on PS II (79) at a given light intensity.

Generation of $^1O_2$ within the LHCs can potentially lead to oxidation of lipids, proteins, and pigments in the immediate vicinity (87). Thylakoid membrane lipids are especially susceptible to damage by $^1O_2$ because of the abundance of unsaturated fatty acid side chains. Reaction between $^1O_2$ and these lipids produces hydroperoxides and initiates peroxyl radical chain reactions in the thylakoid membrane. Generation of $^1O_2$ and/or P680$^+$ in the PS II reaction center can result in damage to lipids, critical pigment cofactors, and protein subunits associated with PS II, especially the D1 protein, resulting in photo-oxidative inactivation of entire reaction centers (13, 18).

On the acceptor (stromal) side of PS I, the targets of oxidative damage by O_2^-, H_2O_2, and OH· include key enzymes of photosynthetic carbon metabolism such as phosphoribulokinase, fructose-1,6-bisphosphatase, and NADP-glyceraldehyde-3-phosphate dehydrogenase (16, 84). Photoinhibition due to damage to the PS I reaction center itself can be observed under some circumstances, especially during chilling stress (reviewed in 143).

ADJUSTMENT OF LIGHT-HARVESTING ANTENNA SIZE

Changes in the sizes of the Chl antennae associated with PS II and PS I are involved in balancing light absorption and utilization (reviewed in 7, 103). During long-term acclimation to growth in different light intensities, changes in antenna size are due to changes in LHC gene expression (54, 101, 150) and/or LHC protein degradation (96). No mutants affecting acclimatory changes in Chl antenna size have yet been reported.

Short-term alteration of the relative antenna sizes of PS II and PS I can occur because of state transitions. According to the state transition model (reviewed in 2), overexcitation of PS II relative to PS I reduces the plastoquinone pool and activates a kinase that phosphorylates the peripheral LHC associated with PS II. Subsequent detachment of phospho-LHC from PS II decreases the effective size of the PS II antenna, and the phospho-LHC may then transfer excitation energy to PS I. Although the state transition is often suggested to be photoprotective (for example, see 8, 13), there is no convincing evidence for its role in photoprotection, at least in excessive light. In fact, the LHC kinase system seems to be inactivated in high light (46, 130, 136). Mutants affecting the state transition would be useful to test the hypothesis.

THERMAL DISSIPATION OF EXCESS ABSORBED LIGHT ENERGY

In excessive light, an increase in the thylakoid ΔpH regulates PS II light harvesting by triggering the dissipation of excess absorbed light energy as heat (reviewed in 31, 47, 48, 49, 51, 65, 76, 77). Over 75% of absorbed photons can be eliminated by this process of thermal dissipation (50), which involves de-excitation of ^{1}Chl and which is measured (and often referred to) as nonphotochemical quenching of Chl fluorescence (NPQ). Thermal dissipation is thought to protect photosynthesis by (a) decreasing the lifetime of ^{1}Chl to minimize generation of $^1O_2$ in the PS II LHC and reaction center, (b) preventing overacidification of the lumen and generation of long-lived P680$^+$, and (c) decreasing the rate of O_2 reduction by PS I. Although thermal dissipation is usually reversible within seconds or minutes, sustained thermal dissipation (also called qI) can be manifested as photoinhibition, which may actually be due to a photoprotective mechanism (52, 77, 82, 92, 116). I focus on the rapidly reversible (pH-dependent) type of thermal dissipation (also called qE) that predominates under most circumstances.

pH-dependent thermal dissipation occurs in the PS II antenna pigment bed and involves specific de-epoxidized xanthophyll pigments (31, 47–49, 51; 65, 76, 77). The increase in the thylakoid ΔpH in excessive light is thought to result in protonation of specific LHC polypeptides associated with PS II, namely LHCB4 (CP29) and LHCB5 (CP26) (42, 75, 151, 152). The ΔpH also activates the enzyme violaxanthin de-epoxidase, which converts violaxanthin associated with the LHCs to zeaxanthin (and antheraxanthin) via the so-called xanthophyll cycle (see Figure 2) (55, 125). Binding of zeaxanthin and protons to the LHC may cause a conformational change, monitored by an absorbance change at 535 nm (23–25, 132), that is necessary for thermal dissipation. The actual mechanism of ^{1}Chl de-excitation may involve a direct transfer of energy from Chl to zeaxanthin (37, 62, 120). Alternatively, xanthophylls (and protons) may act as allosteric effectors of LHC structure, leading to "concentration quenching" by Chls (77) or quenching via Chl dimer formation (42).

Mutants of *Chlamydomonas* and Arabidopsis that affect thermal dissipation have been isolated by video imaging of Chl fluorescence quenching (109,

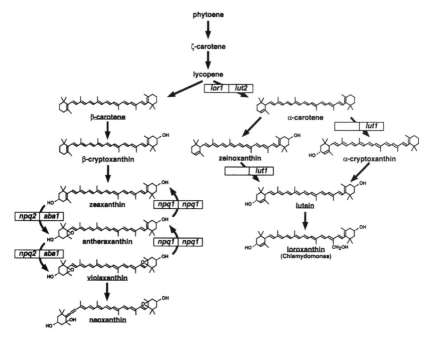

Figure 2 The pathway for carotenoid biosynthesis in green algae and plants. The steps blocked in mutants of *Chlamydomonas* and Arabidopsis are designated by the boxes, with the names of the mutants shown in the left half of the box for *Chlamydomonas* and the right half for Arabidopsis. Strains defective in some of these reactions have also been identified in *Scenedesmus obliquus* (26, 28). The xanthophyll cycle that operates in the β-carotene branch of the pathway involves zeaxanthin, antheraxanthin, and violaxanthin. The carotenoids that normally accumulate in the chloroplast are underlined; zeaxanthin accumulates in excessive light.

111, 140). Characterization of mutants affected in xanthophyll metabolism (Figure 2) has confirmed a role for zeaxanthin in thermal dissipation (81, 109, 111, 144). The *npq2* mutants of *Chlamydomonas* and Arabidopsis, together with the existing *aba1* mutant of Arabidopsis (90) and *aba2* mutant of tobacco (100), are defective in zeaxanthin epoxidase activity; they accumulate zeaxanthin and contain only trace amounts of antheraxanthin, violaxanthin, and neoxanthin (53, 100, 109, 111, 131). The constitutive presence of zeaxanthin in the PS II LHCs of *npq2* and *aba1* mutants is not sufficient for thermal dissipation, which also requires the ΔpH. However, induction of NPQ by illumination with high light is more rapid in the mutants compared to wild type (81, 109, 111, 144), presumably because it is driven solely by the build-up of the ΔpH. During short-term illumination with high light, some *aba1* mutants exhibited the same sensitivity to photoinhibition as the wild type

(81), whereas an *aba1* mutant with a different allele appeared more sensitive (144).

The *Chlamydomonas npq1* mutant, which is unable to convert violaxanthin to zeaxanthin, is partially defective in NPQ but retains substantial, pH-dependent NPQ, which suggests that some but not all thermal dissipation depends on operation of the xanthophyll cycle (109). Characterization of the *lor1* mutant, which lacks xanthophylls derived from α-carotene (38, 110), indicates a possible role for lutein in thermal dissipation (110). Neither *npq1* nor *lor1* is particularly sensitive to photoinhibition during growth in high light. However, an *npq1 lor1* double mutant lacks almost all pH-dependent NPQ and is very susceptible to photo-oxidative bleaching in high light. Although this result seems to provide evidence for the importance of thermal dissipation for photoprotection in vivo, the phenotype is complicated by the fact that xanthophylls are also involved in quenching of ^{3}Chl and ^{1}O$_2$ and inhibition of lipid peroxidation, as described below. Another mutant, *npq4*, lacks NPQ but has normal xanthophyll composition. It is able to survive in high light, suggesting that thermal dissipation itself is not required for photoprotection (D Elrad, KK Niyogi & AR Grossman, unpublished results).

Results with Arabidopsis xanthophyll mutants, although generally similar to results with *Chlamydomonas*, revealed that the relative contributions of different xanthophylls to thermal dissipation vary in different organisms. Like *npq1* of *Chlamydomonas*, Arabidopsis *npq1* mutants are unable to convert violaxanthin to zeaxanthin in high light. Genetic and molecular analyses (111) demonstrated that the phenotype of *npq1* is due to a recessive mutation in the Arabidopsis violaxanthin de-epoxidase gene (35). In contrast to the results with *Chlamydomonas*, induction of pH-dependent thermal dissipation in the Arabidopsis *npq1* mutant is almost completely inhibited, suggesting that most of the thermal dissipation in Arabidopsis depends on de-epoxidation of violaxanthin (111). Leaves of *npq1* plants sustain more photoinhibition than wild type following short-term illumination with high light, in agreement with experiments that used dithiothreitol as an inhibitor of violaxanthin de-epoxidase in detached leaves (23, 156). However, growth of *npq1* plants that are acclimated to high light is not noticeably different from that of the wild type, which suggests that, in the long term, other photoprotective processes can compensate for the defect in *npq1* (111).

As in *Chlamydomonas*, analysis of Arabidopsis mutants lacking lutein suggests that the residual pH-dependent NPQ in *npq1* may be attributable to a contribution of lutein-dependent NPQ (127). The *lut2* mutant, which lacks lutein owing to a mutation in the lycopene ε-cyclase gene (126), exhibits slower induction and a lower maximum extent of NPQ (127), and an *npq1 lut2* double mutant lacks almost all pH-dependent NPQ (O Björkman, C Shih, B Pogson,

D DellaPenna & KK Niyogi, unpublished results). Lutein may have a direct role in thermal dissipation, or the lack of lutein in *lor1* and *lut2* mutants may have an indirect effect. In mutants lacking lutein, zeaxanthin may be sequestered in binding sites that are normally occupied by lutein but inactive in thermal dissipation, thereby lowering the relative effectiveness of zeaxanthin (AM Gilmore, KK Niyogi & O Björkman, unpublished results).

The requirement for different proteins in the PS II LHC for thermal dissipation has been addressed using mutants lacking Chl *b*. These mutants, such as barley *chlorina f2* and Arabidopsis *ch1*, are impaired in the synthesis and/or assembly of LHC proteins, especially the peripheral LHC of PS II, due to the lack of Chl *b* (32, 93, 108). Measurements of NPQ and Chl fluorescence lifetime components in these mutants suggest that only the minor LHC proteins, such as LHCB4 and LHCB5, are necessary for thermal dissipation, although maximal NPQ requires the presence of the entire LHC (12, 34, 66, 68, 69, 97; KK Niyogi, unpublished results). To test the role of specific LHC proteins in thermal dissipation, mutants or antisense plants affecting individual genes will be very useful.

Several other *npq* mutants exhibit normal pigment composition and xanthophyll interconversions. These mutants presumably identify factors besides the ΔpH and xanthophylls (perhaps LHC components) that are required for thermal dissipation in *Chlamydomonas* (109; D Elrad, KK Niyogi & AR Grossman, unpublished results) and Arabidopsis (KK Niyogi, C Shih & O Björkman, unpublished results). For example, the Arabidopsis *npq4* mutant exhibits the same lack of NPQ as an *npq1 lut2* double mutant, which suggests that it is defective in all pH- and xanthophyll-dependent thermal dissipation, yet it is indistinguishable from wild type in terms of xanthophyll interconversion and growth in low light. In addition, *npq4* lacks the absorbance change that presumably reflects the conformational change that is necessary for thermal dissipation (KK Niyogi, C Shih & O Björkman, unpublished results). Determination of the molecular basis for mutations such as *npq4* will likely provide new insights into the molecular mechanism of thermal dissipation.

PHOTOPROTECTION THROUGH PHOTOCHEMISTRY

Assimilatory Linear Electron Transport

Much of the light energy absorbed by the LHCs is utilized through photochemistry that drives linear electron transport from H_2O to NADPH, resulting in O_2 evolution and reduction of CO_2, NO_3^-, and SO_4^{2-}. The maximum rate of photosynthesis is a dynamic parameter that can be altered during acclimation to growth in different light environments (reviewed in 30, 37) through changes in enzyme activities and gene expression. These acclimation responses generally occur during a period of several days, and there are few or no molecular genetic data that address the importance of long-term acclimation responses for photoprotection.

Oxygen-Dependent Electron Transport

There is abundant evidence that nonassimilatory electron transport to oxygen plays an important role in consuming excess excitation energy. Oxygen can function as an electron acceptor either through the oxygenase reaction catalyzed by Rubisco (photorespiration) or by direct reduction of oxygen by electrons on the acceptor side of PS I (102), and there is much debate about which process is more important for photoprotection (for example, see 22, 71, 119, 124, 155).

PHOTORESPIRATION   In C3 plants, especially under conditions of CO_2 limitation, photorespiratory oxygen metabolism is capable of maintaining considerable linear electron transport and utilization of light energy (reviewed in 117). The role of photorespiration in photoprotection can be conveniently assessed by varying the gas composition to inhibit the oxygenation reaction of Rubisco. Blocking photorespiration with mutations or inhibitors leads to inhibition of photosynthesis and photo-oxidative damage, and the cause of inhibition varies (14). Accumulation of photorespiratory metabolites or depletion of carbon intermediates can inhibit the Calvin cycle and shut down the photochemical sink for excitation energy. To assess the importance of photorespiration without the complications of accumulating toxic intermediates, mutants of Rubisco that lack oxygenase activity would be required.

Manipulation of glutamine synthetase activity, a rate-limiting step in photorespiratory metabolism, in transgenic tobacco plants has provided additional evidence for the role of photorespiration in photoprotection (91). Antisense plants with less glutamine synthetase are more sensitive to photo-oxidation under conditions of CO_2 limitation because of accumulation of photorespiratory NH_3, in agreement with previous results with mutant barley plants lacking glutamine synthetase (149). In contrast, plants that overexpress glutamine synthetase have a higher capacity for photorespiration and are more resistant to photoinhibition and photo-oxidative damage (91).

PHOTOREDUCTION OF OXYGEN BY PHOTOSYSTEM I   Direct reduction of O_2 by PS I is the first step in an alternative electron transport pathway that has been variously termed pseudocyclic electron transport, the Mehler-ascorbate peroxidase reaction, and the water-water cycle. Because a comprehensive review of this pathway appears in this volume (17), it is only briefly outlined here. The O_2^- produced on the acceptor side of PS I by reduction of O_2 is efficiently metabolized by thylakoid-bound isozymes of superoxide dismutase (SOD) and ascorbate peroxidase (APX) to generate H_2O and monodehydroascorbate, which can itself be reduced directly by PS I to regenerate ascorbate (15–17). Thus the four electrons generated by oxidation of H_2O by PS II are consumed by reduction of O_2 to H_2O by PS I. This pseudocyclic pathway generates a ΔpH for ATP synthesis, but neither NADPH nor net O_2 is produced.

Like photorespiration, the water-water cycle may help to dissipate excitation energy through electron transport. However, the capacity of this pathway to support electron transport is unclear; estimates range between 10% and 30% of normal linear electron transport in algae and C3 plants (22, 99, 119). The pathway may also be involved in maintaining a ΔpH necessary for thermal dissipation of excess absorbed light energy (135).

Genes encoding SOD and APX have been identified from several plants, including Arabidopsis (83, 86), and this should enable dissection of the water-water cycle by analysis of antisense plants or mutants. However, the mutant analysis will be complicated by the dual roles of these enzymes in scavenging reactive oxygen intermediates and in electron transport.

Cyclic Electron Transport

Within PS II, cyclic electron transport pathways, possibly involving cytochrome b_{559}, have been suggested to dissipate excitation energy (reviewed in 37, 154), but convincing evidence for their occurrence in vivo is lacking. Cytochrome b_{559} may function to oxidize Pheo$^-$ or reduce P680$^+$ to protect against photodamage to PS II (154). Site-directed mutagenesis of the chloroplast genes encoding the α and β polypeptides of cytochrome b_{559} in *Chlamydomonas* (1, 107) will be useful to dissect the possible role of this cytochrome in cyclic electron transport and photoprotection.

Cyclic electron transport around PS I is also suggested to have an important role in photoprotection. In addition to dissipating energy absorbed by PS I, cyclic electron transport may be involved in generating or maintaining the ΔpH that is necessary for downregulation of PS II by thermal dissipation of excess absorbed light energy (72). Biochemical approaches have led to the conclusion that there are at least two pathways of PS I cyclic electron transport, one involving a ferredoxin-plastoquinone oxidoreductase (FQR) and the other involving an NADPH/NADH dehydrogenase (NDH) complex (reviewed in 21). The FQR has not yet been identified, although the PsaE subunit of PS I is possibly involved. The NDH pathway involves a protein complex bound to the thylakoid membrane that is homologous to the NADH dehydrogenase complex I of mitochondria. Several subunits of this complex are encoded by genes on the chloroplast genome of many plants (reviewed in 63).

Mutants affecting the NDH complex have been generated by disrupting *ndh* genes in the chloroplast genome of tobacco by homologous recombination (36, 88, 139). These mutants have no obvious phenotype under normal growth conditions. However, measurements of Chl fluorescence and PS I reduction kinetics revealed that cyclic electron transport is partially impaired (36, 139). Induction of thermal dissipation upon sudden illumination was slightly delayed in mutants subjected to water stress, consistent with the idea that PS I cyclic electron transport is involved in maintaining a ΔpH for thermal dissipation (36). Thorough examination of photoprotection in these mutants under less favorable growth conditions should be very informative, although the existence of other cyclic pathways may complicate interpretations.

SCAVENGING OF REACTIVE OXYGEN SPECIES

Several antioxidant systems in the chloroplast can scavenge reactive oxygen species that are inevitably generated by photosynthesis (15, 57, 128). In many cases, increases in antioxidant molecules and enzymes have been observed during acclimation to excessive light (for example, see 67), and the specific roles of these molecules are starting to be tested using mutant and transgenic organisms (3–5, 57).

Antioxidant Molecules

CAROTENOIDS Carotenoids, including the xanthophylls, are membrane-bound antioxidants that can quench ^3Chl and $^1O_2$, inhibit lipid peroxidation, and stabilize membranes (reviewed in 51, 60, 61, 70). The genes and enzymes involved in their biosynthesis have been reviewed recently in this series (43). Xanthophylls bound to the LHC proteins are located in close proximity to Chl for efficient quenching of ^3Chl and $^1O_2$ (94). β-carotene in the PS II reaction center quenches $^1O_2$ produced from interaction of $^3P680$ and O_2 but is not thought to quench $^3P680$ itself (145). As discussed above, specific xanthophylls are also involved in quenching of ^1Chl during thermal dissipation.

Carotenoids in general have essential functions in photosynthesis and photoprotection, as demonstrated by the bleached phenotypes of algae and plants that are unable to synthesize any carotenoids owing to mutations affecting early steps of carotenoid biosynthesis (6, 133). However, characterization of xanthophyll mutants blocked in later steps in the carotenoid pathway (Figure 2) has demonstrated that no single xanthophyll is absolutely required for photoprotection (109–111, 126). For example, *Chlamydomonas* mutants lacking lutein and loroxanthin (*lor1*), antheraxanthin, violaxanthin, and neoxanthin (*npq2*), or zeaxanthin (*npq1*) are all able to grow as well as wild type in high light (109, 110).

Eliminating combinations of xanthophylls in double mutants has revealed a redundancy among xanthophylls in terms of photoprotection. Construction of an *npq1 lor1* double mutant has demonstrated that accumulation of either zeaxanthin or lutein is necessary for photoprotection and survival of

Chlamydomonas in high light (110). The light sensitivity of the *npq1 lor1* strain probably reflects the multiple roles of specific xanthophylls in photoprotection; quenching of ^{1}Chl, ^{1}O$_2$, and possibly also inhibition of lipid peroxidation are impaired in the absence of both zeaxanthin and lutein (110). A similar sensitivity to very high light is observed in older leaves of the Arabidopsis *npq1 lut2* double mutant, but the photo-oxidative damage is not lethal to the entire plant (O Björkman, C Shih, B Pogson, D DellaPenna & KK Niyogi, unpublished results).

Zeaxanthin is sufficient for photoprotection. Double mutants of *Chlamydomonas* (*npq2 lor1*) and Arabidopsis (*aba1 lut2*), as well as the C-2A'-67,3b strain of *Scenedesmus obliquus* (26), that contain zeaxanthin as the only chloroplast xanthophyll are viable, although the photosynthetic efficiency of the *Scenedesmus* strain is decreased (27), and growth of the Arabidopsis *aba1 lut2* is impaired (127). In the case of *Chlamydomonas npq2 lor1*, the mutant grows as well as wild type in high light (KK Niyogi, unpublished results). Selective elimination of all the xanthophylls derived from β-carotene to test whether lutein is sufficient for photoprotection may be difficult using mutants, because the same β-hydroxylase seems to be involved in synthesis of both zeaxanthin and lutein (43).

TOCOPHEROLS   Another important thylakoid membrane antioxidant is α-tocopherol (vitamin E), which can physically quench or chemically scavenge ^{1}O$_2$, O_2^- and OH· in the membrane to prevent lipid peroxidation (57). Whereas the xanthophylls are largely bound to proteins, α-tocopherol is free in the lipid matrix of the membrane and appears to have a role in controlling membrane fluidity and stability (64). In addition, α-tocopherol participates in the efficient termination of lipid peroxidation chain reactions with concomitant formation of its α-chromanoxyl radical (64).

Although α-tocopherol is the most abundant tocopherol in the chloroplast, other tocopherols such as β- and γ-tocopherols are also present at low levels. The minor tocopherols are intermediates in the synthesis of α-tocopherol that differ in the number of methyl groups on the chromanol head group. The relative abundance of the tocopherols ($\alpha > \beta > \gamma$) parallels their effectiveness as chemical scavengers of reactive oxygen species and as chain reaction terminators (64).

Unfortunately, there are few genetic data that address the importance of specific tocopherols, or tocopherols in general, in photoprotection. A *Scenedesmus* mutant lacking all tocopherols has been reported (29), but the lesion in the mutant appeared to affect the synthesis of phylloquinones and phytylbenzoquinones in addition to tocopherols (74). The Arabidopsis *pds1* and *pds2* mutants lack both tocopherols and quinones due to blocks in the biosynthetic pathway common to both types of molecules (113). Interestingly, the mutants also lacked carotenoids, uncovering a role for plastoquinone in the desaturation of phytoene (113). A gene encoding one of the later steps of tocopherol biosynthesis has recently been identified in cyanobacteria and Arabidopsis (140a), and application of reverse genetics promises insights into the photoprotective role of specific tocopherols in the near future.

ASCORBATE   The soluble antioxidant ascorbate (vitamin C) has a central role in preventing oxidative damage through direct quenching of $^1O_2$, O_2^- and OH·, in regeneration of α-tocopherol from the α-chromanoxyl radical, and as a substrate in both the violaxanthin de-epoxidase and APX reactions (112, 141). Although ascorbate is very abundant in chloroplasts ($\sim$25 mM), the biosynthetic pathway for ascorbate in plants was elucidated only very recently (153), and the importance of ascorbate in photoprotection has not been determined. Ascorbate-deficient mutants affecting the biosynthetic pathway have been identified recently in a screen for ozone-sensitive mutants in Arabidopsis (40, 41). One mutant, originally called *soz1* but renamed *vtc1*, accumulates approximately 30% as much ascorbate as wild type. In addition to its ozone sensitivity, the *vtc1* mutant is more sensitive to exogenous H_2O_2, SO_2, and illumination with UV-B, but sensitivity to high light has not yet been examined.

GLUTATHIONE   Another important soluble antioxidant in the chloroplast is glutathione, which is capable of detoxifying $^1O_2$ and OH·. Glutathione protects thiol groups in stromal enzymes, and it is also involved in α-tocopherol regeneration and ascorbate regeneration through the glutathione-ascorbate cycle (16, 59). Biosynthesis of glutathione proceeds by the reaction of glutamine and cysteine to form γ-glutamylcysteine, followed by the addition of glycine catalyzed by glutathione synthetase. The Arabidopsis *cad2* mutant is deficient in glutathione (78) owing to a defect in the gene encoding γ-glutamylcysteine synthetase (39). The level of glutathione in *cad2* leaves is approximately 30% of that in the wild type. As in the case of the ascorbate-deficient mutant, the sensitivity of *cad2* to photoinhibition remains to be determined.

Scavenging Enzymes

The enzymes SOD and APX are involved in scavenging reactive oxygen species in the chloroplast. As discussed above, O_2^- generated by reduction of O_2 by PS I is metabolized enzymatically by SOD to produce H_2O_2. The subsequent reduction of H_2O_2 by APX produces the monodehydroascorbate radical, which can be directly reduced by PS I (via ferredoxin) in the water-water cycle (17). Ascorbate can also be regenerated in the stroma by the set of enzymes comprising the glutathione-ascorbate cycle (16, 59).

Both SOD and APX exist in multiple isoforms within the chloroplast. The SOD isozymes are generally classified according to their active site metal (33): copper/zinc (Cu/ZnSOD), iron (FeSOD), or manganese (MnSOD). Most plants have FeSOD and Cu/ZnSOD in their chloroplasts, whereas most algae appear to lack the Cu/ZnSOD completely (45). Thylakoid-bound forms of SOD (114) and APX (104) may efficiently detoxify O_2^- and H_2O_2 at their site of production (16, 58) and prevent inactivation of Calvin cycle enzymes (84). Soluble forms of SOD and APX react with O_2^- and H_2O_2 that diffuse into the stroma from the thylakoid membrane.

Mutants or antisense plants have the potential to reveal roles of different SODs and APXs in photoprotection. Arabidopsis has at least seven SOD genes: three encoding Cu/ZnSOD, three encoding FeSOD, and one encoding MnSOD (86). One of the Cu/ZnSOD and all three of FeSOD gene products are located in the chloroplast. Mutants that are deficient in expression of plastidic and cytosolic Cu/ZnSODs or plastidic FeSODs have been isolated (DJ Kliebenstein & RL Last, unpublished results), but the phenotypes of these mutants in high light have not been tested. Genes encoding the stromal and thylakoid-bound forms of APX have been cloned from Arabidopsis (83), but mutants or transgenic plants with altered expression of chloroplast APX have not yet been reported.

Transgenic plants overexpressing SOD have been generated by several groups, and the results of these experiments were recently reviewed (3–5, 57). Photoprotection has been examined specifically in only a few cases. An enhancement of tolerance to a combination of high light and chilling stress has been observed in tobacco plants overexpressing chloroplast Cu/ZnSOD (137, 138) but not in plants overexpressing chloroplast FeSOD (147). This is consistent with the localization of Cu/ZnSOD at the thylakoid membrane where O_2^- is reduced by PS I (114). However, chloroplast APX activity is induced in the Cu/ZnSOD transgenic plants; the combination of increased Cu/ZnSOD and APX activities may result in enhanced photoprotection (138). The photoprotection phenotype of a *Chlamydomonas* mutant that overexpresses chloroplast-localized FeSOD has not been reported (85).

REPAIR OF PHOTODAMAGE

Despite multiple lines of defense, damage to the photosynthetic apparatus is an inevitable consequence of oxygenic photosynthesis, and the PS II reaction center is especially susceptible to photo-oxidative damage. Therefore, oxygenic photosynthetic organisms have evolved an elaborate but efficient system for repairing PS II that involves selective degradation of damaged proteins (primarily D1) and incorporation of newly synthesized proteins to reconstitute functional PS II (13). Under some circumstances, damaged PS II reaction centers may also be sites of thermal dissipation (qI) (92, 116).

The repair of PS II is an important photoprotective mechanism, because the rate of repair must match the rate of damage to avoid photoinhibition resulting from net loss of functional PS II centers. Therefore, continual new synthesis of chloroplast-encoded proteins, especially D1, is critical for photoprotection at all light intensities. Blocking chloroplast protein synthesis with inhibitors during exposure of algae or plants to light results in photoinhibition and net loss of D1 protein (for example, see 19, 146). Similarly, attenuation of chloroplast protein synthesis in rRNA mutants of *Chlamydomonas* leads to chronic photoinhibition and lower levels of D1 during growth in high light (73). To analyze the effect of a specific limitation of D1 synthesis, rather than a general inhibition of all chloroplast protein synthesis, it may be possible to isolate partial loss-of-function alleles of mutations such as *Chlamydomonas* F35, which specifically affects expression of D1 (157).

Other factors involved in repair of PS II are being identified by various genetic approaches. A screen for *Chlamydomonas* mutants that are more sensitive to PS II photodamage has uncovered mutants that may be defective in PS II repair, including a mutant that accumulates more D1 protein than wild type (158). A possible role for the *psbT* gene (originally known as *ycf8*) in the chloroplast genome of *Chlamydomonas* in protection of PS II in high light has been suggested by characterization of a *psbT* disruption mutant (105). In transgenic tobacco plants, decreasing the unsaturation of chloroplast glycerolipids results in greater sensitivity to low-temperature photoinhibition, apparently because of inhibition of PS II repair (106).

CONCLUSIONS AND PROSPECTS

Photoprotection of photosynthesis is a balancing act. Thermal dissipation and alternative electron transport pathways, together with changes in antenna size and overall photosynthetic capacity, help to balance light absorption and utilization in the constantly changing natural environment. The generation of reactive oxygen species is balanced by the capacity of antioxidant systems. The capacity for repair must match the damage that is not prevented by other photoprotective processes.

Multiple, partially redundant mechanisms acting in concert help to prevent net photo-oxidative damage that results from generation of reactive molecules through photosynthesis. The redundancy is not surprising given how critical photoprotection is for fitness and survival of algae and plants in most environments. Indeed, many of the molecules and enzymes involved in photoprotection have roles in more than one photoprotective mechanism. For example, xanthophylls are involved in both thermal dissipation and quenching of $^1O_2$, whereas SOD and APX enzymes scavenge reactive oxygen species while maintaining electron transport via the water-water cycle.

The use of molecular genetic approaches to study photoprotection is just beginning, but the potential for rapid progress is obvious. Mutants affecting many chloroplast processes are now available, and assessment of their relative importance for photoprotection is under way in several laboratories. Because photoprotective processes comprise several lines of defense against the damaging effects of light, construction of double and perhaps even triple mutants in some cases may be necessary in order to obtain clear phenotypes. In the future, other genetic approaches, such as screening for suppressors or enhancers of these phenotypes, may uncover previously unrecognized mechanisms of photoprotection.

ACKNOWLEDGMENTS

I thank Anastasios Melis and Catharina Casper-Lindley for comments on the manuscript and Dan Kliebenstein and Dean DellaPenna for communicating unpublished results. Unpublished work in the author's laboratory is supported by grants from the Searle Scholars Program/The Chicago Community Trust and the USDA NRICGP (#9801685).

> **Visit the** *Annual Reviews home page* **at**
> **http://www.AnnualReviews.org**

Literature Cited

1. Alizadeh S, Nechushtai R, Barber J, Nixon P. 1994. Nucleotide sequence of the *psbE, psbF* and *trnM* genes from the chloroplast genome of *Chlamydomonas reinhardtii. Biochim. Biophys. Acta* 1188:439–42
2. Allen JF. 1995. Thylakoid protein phosphorylation, state 1-state 2 transitions, and photosystem stoichiometry adjustment: redox control at multiple levels of gene expression. *Physiol. Plant.* 93:196–205
3. Allen RD. 1995. Dissection of oxidative stress tolerance using transgenic plants. *Plant Physiol.* 107:1049–54
4. Allen RD, Webb RP, Schake SA. 1997. Use of transgenic plants to study antioxidant defenses. *Free Rad. Biol. Med.* 23:473–79
5. Alscher RG, Donahue JL, Cramer CL. 1997. Reactive oxygen species and antioxidants: relationships in green cells. *Physiol. Plant.* 100:224–33

6. Anderson IC, Robertson DS. 1960. Role of carotenoids in protecting chlorophyll from photodestruction. *Plant Physiol.* 35:531–34
7. Anderson JM. 1986. Photoregulation of the composition, function and structure of thylakoid membranes. *Annu. Rev. Plant Physiol.* 37:93–136
8. Anderson JM, Andersson B. 1988. The dynamic photosynthetic membrane and regulation of solar energy conversion. *Trends Biochem. Sci.* 13:351–55
9. Anderson JM, Park Y-I, Chow WS. 1997. Photoinactivation and photoprotection of photosystem II in nature. *Physiol. Plant.* 100:214–23
10. Anderson JM, Park Y-I, Chow WS. 1998. Unifying model for the photoinactivation of Photosystem II *in vivo* under steady-state photosynthesis. *Photosynth. Res.* 56:1–13
11. Andersson B, Barber J. 1996. Mechanisms of photodamage and protein

degradation during photoinhibition of photosystem II. See Ref. 17a, pp. 101–21

12. Andrews JR, Fryer MJ, Baker NR. 1995. Consequences of LHC II deficiency for photosynthetic regulation in *chlorina* mutants of barley. *Photosynth. Res.* 44:81–91

13. Aro E-M, Virgin I, Andersson B. 1993. Photoinhibition of photosystem II. Inactivation, protein damage and turnover. *Biochim. Biophys. Acta* 1143:113–34

14. Artus NN, Somerville SC, Somerville CR. 1986. The biochemistry and cell biology of photorespiration. *CRC Crit. Rev. Plant Sci.* 4:121–47

15. Asada K. 1994. Mechanisms for scavenging reactive molecules generated in chloroplasts under light stress. See Ref. 17b, pp. 129–42

16. Asada K. 1994. Production and action of active oxygen species in photosynthetic tissues. In *Causes of Photooxidative Stress and Amelioration of Defense Systems in Plants*, ed. CH Foyer, PM Mullineaux, pp. 77–104. Boca Raton: CRC Press

17. Asada K. 1999. The water-water cycle. *Annu. Rev. Plant Physiol. Plant Mol. Biol.* 50:601–39

17a. Baker NR, ed. 1996. *Photosynthesis and the Environment*. Dordrecht: Kluwer

17b. Baker NR, Bowyer JR, eds. 1994. *Photoinhibition of Photosynthesis: From Molecular Mechanisms to the Field*. Oxford: BIOS Sci. Publ.

18. Barber J, Andersson B. 1992. Too much of a good thing: Light can be bad for photosynthesis. *Trends Biochem. Sci.* 17:61–66

19. Baroli I, Melis A. 1996. Photoinhibition and repair in *Dunaliella salina* acclimated to different growth irradiances. *Planta* 198:640–46

20. Baroli I, Melis A. 1998. Photoinhibitory damage is modulated by the rate of photosynthesis and by the photosystem II light-harvesting chlorophyll antenna size. *Planta* 205:288–96

21. Bendall DS, Manasse RS. 1995. Cyclic photophosphorylation and electron transport. *Biochim. Biophys. Acta* 1229:23–38

22. Biehler K, Fock H. 1996. Evidence for the contribution of the Mehler-peroxidase reaction in dissipating excess electrons in drought-stressed wheat. *Plant Physiol.* 112:265–72

23. Bilger W, Björkman O. 1990. Role of the xanthophyll cycle in photoprotection elucidated by measurements of light-induced absorbance changes, fluorescence and photosynthesis in leaves of *Hedera canariensis*. *Photosynth. Res.* 25:173–85

24. Bilger W, Björkman O. 1994. Relationships among violaxanthin deepoxidation, thylakoid membrane conformation, and nonphotochemical chlorophyll fluorescence quenching in leaves of cotton (*Gossypium hirsutum* L.). *Planta* 193:238–46

25. Bilger W, Björkman O, Thayer SS. 1989. Light-induced spectral absorbance changes in relation to photosynthesis and the epoxidation state of xanthophyll cycle components in cotton leaves. *Plant Physiol.* 91:542–51

26. Bishop NI. 1996. The β,ε-carotenoid, lutein, is specifically required for the formation of the oligomeric forms of the light harvesting complex in the green alga, *Scenedesmus obliquus*. *J. Photochem. Photobiol. B: Biol.* 36:279–83

27. Bishop NI, Bugla B, Senger H. 1998. Photosynthetic capacity and quantum requirement of three secondary mutants of *Scenedesmus obliquus* with deletions in carotenoid biosynthesis. *Bot. Acta* 111:231–35

28. Bishop NI, Urbig T, Senger H. 1995. Complete separation of the β,ε- and β,β-carotenoid biosynthetic pathways by a unique mutation of the lycopene cyclase in the green alga, *Scenedesmus obliquus*. *FEBS Lett.* 367:158–62

29. Bishop NI, Wong J. 1974. Photochemical characteristics of a vitamin E deficient mutant of *Scenedesmus obliquus*. *Ber. Dtsch. Bot. Ges.* 87:359–71

30. Björkman O. 1981. Responses to different quantum flux densities. In *Physiological Plant Ecology I. Responses to the Physical Environment*, ed. OL Lange, PS Nobel, CB Osmond, H Ziegler, pp. 57–107. Berlin: Springer-Verlag

31. Björkman O, Demmig-Adams B. 1994. Regulation of photosynthetic light energy capture, conversion, and dissipation in leaves of higher plants. In *Ecophysiology of Photosynthesis*, ed. E-D Schulze, MM Caldwell, pp. 17–47. Berlin: Springer

32. Bossmann B, Knoetzel J, Jansson S. 1997. Screening of *chlorina* mutants of barley (*Hordeum vulgare* L.) with antibodies against light-harvesting proteins of PS I and PS II: absence of specific antenna proteins. *Photosynth. Res.* 52:127–36

33. Bowler C, Van Camp W, Van Montagu M, Inzé D. 1994. Superoxide dismutase

in plants. *Crit. Rev. Plant Sci.* 13:199–218

34. Briantais J-M. 1994. Light-harvesting chlorophyll *a-b* complex requirement for regulation of Photosystem II photochemistry by non-photochemical quenching. *Photosynth. Res.* 40:287–94

35. Bugos RC, Hieber AD, Yamamoto HY. 1998. Xanthophyll cycle enzymes are members of the lipocalin family, the first identified from plants. *J. Biol. Chem.* 273:15321–24

36. Burrows PA, Sazanov LA, Svab Z, Maliga P, Nixon PJ. 1998. Identification of a functional respiratory complex in chloroplasts through analysis of tobacco mutants containing disrupted plastid *ndh* genes. *EMBO J.* 17:868–76

37. Chow WS. 1994. Photoprotection and photoinhibitory damage. In *Advances in Molecular and Cell Biology*, ed. EE Bittar, J Barber, 10:151–96. London: JAI Press

38. Chunaev AS, Mirnaya ON, Maslov VG, Boschetti A. 1991. Chlorophyll *b*- and loroxanthin-deficient mutants of *Chlamydomonas reinhardtii*. *Photosynthetica* 25:291–301

39. Cobbett CS, May MJ, Howden R, Rolls B. 1998. The glutathione-deficient, cadmium-sensitive mutant, *cad2–1*, of *Arabidopsis thaliana* is deficient in γ-glutamylcysteine synthetase. *Plant J.* 16:73–78

40. Conklin PL, Pallanca JE, Last RL, Smirnoff N. 1997. L-ascorbic acid metabolism in the ascorbate-deficient Arabidopsis mutant *vtc1*. *Plant Physiol.* 115:1277–85

41. Conklin PL, Williams EH, Last RL. 1996. Environmental stress sensitivity of an ascorbic acid-deficient Arabidopsis mutant. *Proc. Natl. Acad. Sci. USA* 93:9970–74

42. Crofts AR, Yerkes CT. 1994. A molecular mechanism for q_E-quenching. *FEBS Lett.* 352:265–70

43. Cunningham FX Jr, Gantt E. 1998. Genes and enzymes of carotenoid biosynthesis in plants. *Annu. Rev. Plant Physiol. Plant Mol. Biol.* 49:557–83

44. Dau H. 1994. Molecular mechanisms and quantitative models of variable photosystem II fluorescence. *Photochem. Photobiol.* 60:1–23

45. de Jesus MD, Tabatabai F, Chapman DJ. 1989. Taxonomic distribution of copper-zinc superoxide dismutase in green algae and its phylogenetic importance. *J. Phycol.* 25:767–72

46. Demmig B, Cleland RE, Björkman O.

1987. Photoinhibition, 77K chlorophyll fluorescence quenching and phosphorylation of the light-harvesting chlorophyll-protein complex of photosystem II in soybean leaves. *Planta* 172:378–85

47. Demmig-Adams B. 1990. Carotenoids and photoprotection in plants: a role for the xanthophyll zeaxanthin. *Biochim. Biophys. Acta* 1020:1–24

48. Demmig-Adams B, Adams WW III. 1992. Photoprotection and other responses of plants to high light stress. *Annu. Rev. Plant Physiol. Plant Mol. Biol.* 43:599–626

49. Demmig-Adams B, Adams WW III. 1996. The role of xanthophyll cycle carotenoids in the protection of photosynthesis. *Trends Plant Sci.* 1:21–26

50. Demmig-Adams B, Adams WW III, Barker DH, Logan BA, Bowling DR, Verhoeven AS. 1996. Using chlorophyll fluorescence to assess the fraction of absorbed light allocated to thermal dissipation of excess excitation. *Physiol. Plant.* 98:253–64

51. Demmig-Adams B, Gilmore AM, Adams WW III. 1996. In vivo functions of carotenoids in higher plants. *FASEB J.* 10:403–12

52. Demmig-Adams B, Moeller DL, Logan BA, Adams WW III. 1998. Positive correlation between levels of retained zeaxanthin + antheraxanthin and degree of photoinhibition in shade leaves of *Schefflera arboricola* (Hayata) Merrill. *Planta* 205:367–74

53. Duckham SC, Linforth RST, Taylor IB. 1991. Abscisic-acid-deficient mutants at the *aba* gene locus of *Arabidopsis thaliana* are impaired in the epoxidation of zeaxanthin. *Plant Cell Environ.* 14:601–6

54. Escoubas J-M, Lomas M, LaRoche J, Falkowski PG. 1995. Light intensity regulation of *cab* gene transcription is signaled by the redox state of the plastoquinone pool. *Proc. Natl. Acad. Sci. USA* 92:10237–41

55. Eskling M, Arvidsson P-O, Åkerlund H-E. 1997. The xanthophyll cycle, its regulation and components. *Physiol. Plant.* 100:806–16

56. Foote CS. 1976. Photosensitized oxidation and singlet oxygen: consequences in biological systems. In *Free Radicals in Biology*, ed. WA Pryor, 2:85–133. New York: Academic

57. Foyer CH, Descourvières P, Kunert KJ. 1994. Protection against oxygen radicals: an important defence mechanism

studied in transgenic plants. *Plant Cell Environ.* 17:507–23
58. Foyer CH, Harbinson J. 1994. Oxygen metabolism and the regulation of photosynthetic electron transport. In *Causes of Photooxidative Stress and Amelioration of Defense Systems in Plants*, ed. CH Foyer, PM Mullineaux, pp. 1–42. Boca Raton: CRC Press
59. Foyer CH, Lelandais M, Kunert KJ. 1994. Photooxidative stress in plants. *Physiol. Plant.* 92:696–717
60. Frank HA, Cogdell RJ. 1993. The photochemistry and function of carotenoids in photosynthesis. In *Carotenoids in Photosynthesis*, ed. A Young, G Britton, pp. 252–326. London: Chapman & Hall
61. Frank HA, Cogdell RJ. 1996. Carotenoids in photosynthesis. *Photochem. Photobiol.* 63:257–64
62. Frank HA, Cua A, Chynwat V, Young A, Gosztola D, Wasielewski MR. 1994. Photophysics of the carotenoids associated with the xanthophyll cycle in photosynthesis. *Photosynth. Res.* 41:389–95
63. Friedrich T, Steinmüller K, Weiss H. 1995. The proton-pumping respiratory complex I of bacteria and mitochondria and its homologue in chloroplasts. *FEBS Lett.* 367:107–11
64. Fryer MJ. 1992. The antioxidant effects of thylakoid Vitamin E (α-tocopherol). *Plant Cell Environ.* 15:381–92
65. Gilmore AM. 1997. Mechanistic aspects of xanthophyll cycle-dependent photoprotection in higher plant chloroplasts and leaves. *Physiol. Plant.* 99:197–209
66. Gilmore AM, Hazlett TL, Debrunner PG, Govindjee. 1996. Photosystem II chlorophyll *a* fluorescence lifetimes and intensity are independent of the antenna size differences between barley wild-type and *chlorina* mutants: photochemical quenching and xanthophyll cycle-dependent nonphotochemical quenching of fluorescence. *Photosynth. Res.* 48: 171–87
67. Grace SC, Logan BA. 1996. Acclimation of foliar antioxidant systems to growth irradiance in three broad-leaved evergreen species. *Plant Physiol.* 112:1631–40
68. Härtel H, Lokstein H. 1995. Relationship between quenching of maximum and dark-level chlorophyll fluorescence in vivo: dependence on Photosystem II antenna size. *Biochim. Biophys. Acta* 1228:91–94
69. Härtel H, Lokstein H, Grimm B, Rank B. 1996. Kinetic studies on the xanthophyll cycle in barley leaves. Influence

of antenna size and relations to nonphotochemical chlorophyll fluorescence quenching. *Plant Physiol.* 110:471–82
70. Havaux M. 1998. Carotenoids as membrane stabilizers in chloroplasts. *Trends Plant Sci.* 3:147–51
71. Heber U, Bligny R, Streb P, Douce R. 1996. Photorespiration is essential for the protection of the photosynthetic apparatus of C3 plants against photoinactivation under sunlight. *Bot. Acta* 109:307–15
72. Heber U, Walker D. 1992. Concerning a dual function of coupled cyclic electron transport in leaves. *Plant Physiol.* 100:1621–26
73. Heifetz PB, Lers A, Turpin DH, Gillham NW, Boynton JE, Osmond CB. 1997. *dr* and *spr/sr* mutations of *Chlamydomonas reinhardtii* affecting D1 protein function and synthesis define two independent steps leading to chronic photoinhibition and confer differential fitness. *Plant Cell Environ.* 20:1145–57
74. Henry A, Powls R, Pennock JF. 1986. *Scenedesmus obliquus* PS28: a tocopherol-free mutant which cannot form phytol. *Biochem. Soc. Trans.* 14:958–59
75. Horton P, Ruban AV. 1994. The role of light-harvesting complex II in energy quenching. See Ref. 17b, pp. 111–28
76. Horton P, Ruban AV, Walters RG. 1994. Regulation of light harvesting in green plants. Indication by nonphotochemical quenching of chlorophyll fluorescence. *Plant Physiol.* 106:415–20
77. Horton P, Ruban AV, Walters RG. 1996. Regulation of light harvesting in green plants. *Annu. Rev. Plant Physiol. Plant Mol. Biol.* 47:655–84
78. Howden R, Andersen CR, Goldsbrough PB, Cobbett CS. 1995. A cadmium-sensitive, glutathione-deficient mutant of Arabidopsis. *Plant Physiol.* 107:1067–73
79. Huner NPA, Öquist G, Sarhan F. 1998. Energy balance and acclimation to light and cold. *Trends Plant Sci.* 3:224–30
80. Hurry V, Anderson JM, Badger MR, Price GD. 1996. Reduced levels of cytochrome b_6/f in transgenic tobacco increases the excitation pressure on photosystem II without increasing the sensitivity to photoinhibition *in vivo*. *Photosynth. Res.* 50:159–69
81. Hurry V, Anderson JM, Chow WS, Osmond CB. 1997. Accumulation of zeaxanthin in abscisic acid-deficient mutants of Arabidopsis does not affect chlorophyll fluorescence quenching or

sensitivity to photoinhibition in vivo. *Plant Physiol.* 113:639–48

82. Jahns P, Miehe B. 1996. Kinetic correlation of recovery from photoinhibition and zeaxanthin epoxidation. *Planta* 198:202–10

83. Jespersen HM, Kjaersgard IVH, Ostergaard L, Welinder KG. 1997. From sequence analysis of three novel ascorbate peroxidases from *Arabidopsis thaliana* to structure, function and evolution of seven types of ascorbate peroxidase. *Biochem. J.* 326:305–10

84. Kaiser WM. 1979. Reversible inhibition of the Calvin cycle and activation of oxidative pentose phosphate cycle in isolated intact chloroplasts by hydrogen peroxide. *Planta* 145:377–82

85. Kitayama K, Kitayama M, Togasaki RK. 1995. Characterization of paraquat-resistant mutants of *Chlamydomonas reinhardtii*. In *Photosynthesis: From Light to Biosphere*, ed. P Mathis, 3:595–98. Dordrecht: Kluwer

86. Kliebenstein DJ, Monde R-A, Last RL. 1998. Superoxide dismutase in Arabidopsis: an eclectic enzyme family with disparate regulation and protein localization. *Plant Physiol.* 118:637–50

87. Knox JP, Dodge AD. 1985. Singlet oxygen and plants. *Phytochemistry* 24:889–96

88. Kofer W, Koop H-U, Wanner G, Steinmüller K. 1998. Mutagenesis of the genes encoding subunits A, C, H, I, J and K of the plastid NAD(P)H-plastoquinone-oxidoreductase in tobacco by polyethylene glycol-mediated plastome transformation. *Mol. Gen. Genet.* 258:166–73

89. Kok B. 1956. On the inhibition of photosynthesis by intense light. *Biochim. Biophys. Acta* 21:234–44

90. Koornneef M, Jorna ML, Brinkhorst-van der Swan DLC, Karssen CM. 1982. The isolation of abscisic acid (ABA) deficient mutants by selection of induced revertants in non-germinating gibberellin sensitive lines of *Arabidopsis thaliana* (L.) Heynh. *Theor. Appl. Genet.* 61:385–93

91. Kozaki A, Takeba G. 1996. Photorespiration protects C3 plants from photooxidation. *Nature* 384:557–60

92. Krause GH. 1988. Photoinhibition of photosynthesis. An evaluation of damaging and protective mechanisms. *Physiol. Plant.* 74:566–74

93. Król M, Spangfort MD, Huner NPA, Öquist G, Gustafsson P, Jansson S. 1995. Chlorophyll *a/b*-binding proteins, pigment conversions, and early light-induced proteins in a chlorophyll *b*-less barley mutant. *Plant Physiol.* 107:873–83

94. Kühlbrandt W, Wang DN, Fujiyoshi Y. 1994. Atomic model of plant light-harvesting complex by electron crystallography. *Nature* 367:614–21

95. Levine RP. 1969. The analysis of photosynthesis using mutant strains of algae and higher plants. *Annu. Rev. Plant Physiol.* 20:523–40

96. Lindahl M, Yang D-H, Andersson B. 1995. Regulatory proteolysis of the major light-harvesting chlorophyll *a-b* protein of photosystem II by a light-induced membrane-associated enzymic system. *Eur. J. Biochem.* 231:503–9

97. Lokstein H, Härtel H, Hoffmann P, Renger G. 1993. Comparison of chlorophyll fluorescence quenching in leaves of wild-type with a chlorophyll-b-less mutant of barley (*Hordeum vulgare* L.). *J. Photochem. Photobiol. B: Biol.* 19: 217–25

98. Long SP, Humphries S, Falkowski PG. 1994. Photoinhibition of photosynthesis in nature. *Annu. Rev. Plant Physiol. Plant Mol. Biol.* 45:633–62

99. Lovelock CE, Winter K. 1996. Oxygen-dependent electron transport and protection from photoinhibition in leaves of tropical tree species. *Planta* 198:580–87

100. Marin E, Nussaume L, Quesada A, Gonneau M, Sotta B, et al. 1996. Molecular identification of zeaxanthin epoxidase of *Nicotiana plumbaginifolia*, a gene involved in abscisic acid biosynthesis and corresponding to the *ABA* locus of *Arabidopsis thaliana*. *EMBO J.* 15:2331–42

101. Maxwell DP, Laudenbach DE, Huner NPA. 1995. Redox regulation of light-harvesting complex II and *cab* mRNA abundance in *Dunaliella salina*. *Plant Physiol.* 109:787–95

102. Mehler AH. 1951. Studies on reactions of illuminated chloroplasts. I. Mechanism of the reduction of oxygen and other Hill reagents. *Arch. Biochem. Biophys.* 33:65–77

103. Melis A. 1991. Dynamics of photosynthetic membrane composition and function. *Biochim. Biophys. Acta* 1058:87–106

104. Miyake C, Asada K. 1992. Thylakoid-bound ascorbate peroxidase in spinach chloroplasts and photoreduction of its primary oxidation product mondehydroascorbate radicals in thylakoids. *Plant Cell Physiol.* 33:541–53

105. Monod C, Takahashi Y, Goldschmidt-Clermont M, Rochaix J-D. 1994. The chloroplast *ycf8* open reading frame encodes a photosystem II polypeptide which maintains photosynthetic activity under adverse growth conditions. *EMBO J.* 13:2747–54

106. Moon BY, Higashi S-I, Gombos Z, Murata N. 1995. Unsaturation of the membrane lipids of chloroplasts stabilizes the photosynthetic machinery against low-temperature photoinhibition in transgenic tobacco plants. *Proc. Natl. Acad. Sci. USA* 92:6219–23

107. Mor TS, Ohad I, Hirschberg J, Pakrasi H. 1995. An unusual organization of the genes encoding cytochrome b-559 in *Chlamydomonas reinhardtii*: *psbE* and *psbF* are separately transcribed from different regions of the plastid chromosome. *Mol. Gen. Genet.* 246:600–4

108. Murray DL, Kohorn BD. 1991. Chloroplasts of *Arabidopsis thaliana* homozygous for the ch-1 locus lack chlorophyll *b*, lack stable LHCPII and have stacked thylakoids. *Plant Mol. Biol.* 16:71–79

109. Niyogi KK, Björkman O, Grossman AR. 1997. Chlamydomonas xanthophyll cycle mutants identified by video imaging of chlorophyll fluorescence quenching. *Plant Cell* 9:1369–80

110. Niyogi KK, Björkman O, Grossman AR. 1997. The roles of specific xanthophylls in photoprotection. *Proc. Natl. Acad. Sci. USA* 94:14162–67

111. Niyogi KK, Grossman AR, Björkman O. 1998. Arabidopsis mutants define a central role for the xanthophyll cycle in the regulation of photosynthetic energy conversion. *Plant Cell* 10:1121–34

112. Noctor G, Foyer CH. 1998. Ascorbate and glutathione: keeping active oxygen under control. *Annu. Rev. Plant Physiol. Plant Mol. Biol.* 49:249–79

113. Norris SR, Barrette TR, DellaPenna D. 1995. Genetic dissection of carotenoid synthesis in Arabidopsis defines plastoquinone as an essential component of phytoene desaturation. *Plant Cell* 7:2139–49

114. Ogawa K, Kanematsu S, Takabe K, Asada K. 1995. Attachment of CuZn-superoxide dismutase to thylakoid membranes at the site of superoxide generation (PSI) in spinach chloroplasts: detection by immuno-gold labeling after rapid freezing and substitution method. *Plant Cell Physiol.* 36:565–73

115. Ohad I, Keren N, Zer H, Gong H, Mor TS, et al. 1994. Light-induced degradation of the photosystem II reaction centre

116. Öquist G, Chow WS, Anderson JM. 1992. Photoinhibition of photosynthesis represents a mechanism for the long-term regulation of photosystem II. *Planta* 186:450–60

117. Osmond CB. 1981. Photorespiration and photoinhibition: some implications for the energetics of photosynthesis. *Biochim. Biophys. Acta* 639:77–98

118. Osmond CB. 1994. What is photoinhibition? Some insights from comparisons of shade and sun plants. See Ref. 17b, pp. 1–24

119. Osmond CB, Grace SC. 1995. Perspectives on photoinhibition and photorespiration in the field: quintessential inefficiencies of the light and dark reactions of photosynthesis? *J. Exp. Bot.* 46:1351–62

120. Owens TG. 1994. Excitation energy transfer between chlorophylls and carotenoids. A proposed molecular mechanism for non-photochemical quenching. See Ref. 17b, pp. 95–109

121. Owens TG. 1996. Processing of excitation energy by antenna pigments. See Ref. 17a, pp. 1–23

122. Park Y-I, Anderson JM, Chow WS. 1996. Photoinactivation of functional photosystem II and D1–protein synthesis in vivo are independent of the modulation of the photosynthetic apparatus by growth irradiance. *Planta* 198:300–9

123. Park Y-I, Chow WS, Anderson JM. 1995. Light inactivation of functional photosystem II in leaves of peas grown in moderate light depends on photon exposure. *Planta* 196:401–11

124. Park Y-I, Chow WS, Osmond CB, Anderson JM. 1996. Electron transport to oxygen mitigates against the photoinactivation of Photosystem II *in vivo*. *Photosynth. Res.* 50:23–32

125. Pfündel E, Bilger W. 1994. Regulation and possible function of the violaxanthin cycle. *Photosynth. Res.* 42:89–109

126. Pogson B, McDonald KA, Truong M, Britton G, DellaPenna D. 1996. Arabidopsis carotenoid mutants demonstrate that lutein is not essential for photosynthesis in higher plants. *Plant Cell* 8:1627–39

127. Pogson BJ, Niyogi KK, Björkman O, DellaPenna D. 1998. Altered xanthophyll compositions adversely affect chlorophyll accumulation and nonphotochemical quenching in *Arabidopsis* mutants. *Proc. Natl. Acad. Sci. USA* 95:13324–29

D1 protein *in vivo*: an integrative approach. See Ref. 17b, pp. 161–77

128. Polle A. 1997. Defense against photooxidative damage in plants. In *Oxidative Stress and the Molecular Biology of Antioxidant Defenses*, ed. JG Scandalios, pp. 623–66. Plainview, NY: Cold Spring Harbor Lab. Press

129. Powles SB. 1984. Photoinhibition of photosynthesis induced by visible light. *Annu. Rev. Plant Physiol.* 35:14–44

130. Rintamäki E, Salonen M, Suoranta U-M, Carlberg I, Andersson B, Aro E-M. 1997. Phosphorylation of light-harvesting complex II and photosystem II core proteins shows different irradiance-dependent regulation *in vivo*. *J. Biol. Chem.* 272:30476–82

131. Rock CD, Zeevaart JAD. 1991. The *aba* mutant of *Arabidopsis thaliana* is impaired in epoxy-carotenoid biosynthesis. *Proc. Natl. Acad. Sci. USA* 88:7496–99

132. Ruban AV, Young AJ, Horton P. 1993. Induction of nonphotochemical energy dissipation and absorbance changes in leaves. Evidence for changes in the state of the light-harvesting system of photosystem II in vivo. *Plant Physiol.* 102:741–50

133. Sager R, Zalokar M. 1958. Pigments and photosynthesis in a carotenoid-deficient mutant of *Chlamydomonas. Nature* 182:98–100

134. Schatz G, Brock H, Holzwarth AR. 1988. Kinetic and energetic model for the primary processes in photosystem II. *Biophys. J.* 54:397–405

135. Schreiber U, Neubauer C. 1990. O_2–dependent electron flow, membrane energization and the mechanism of nonphotochemical quenching of chlorophyll fluorescence. *Photosynth. Res.* 25:279–93

136. Schuster G, Dewit M, Staehelin LA, Ohad I. 1986. Transient inactivation of the thylakoid photosystem II light-harvesting protein kinase system and concomitant changes in intramembrane particle size during photoinhibition of *Chlamydomonas reinhardtii. J. Cell Biol.* 103:71–80

137. Sen Gupta A, Heinen JL, Holaday AS, Burke JJ, Allen RD. 1993. Increased resistance to oxidative stress in transgenic plants that overexpress chloroplastic Cu/Zn superoxide dismutase. *Proc. Natl. Acad. Sci. USA* 90:1629–33

138. Sen Gupta A, Webb RP, Holaday AS, Allen RD. 1993. Overexpression of superoxide dismutase protects plants from oxidative stress. Induction of ascorbate peroxidase in superoxide dismutase-overexpressing plants. *Plant Physiol.* 103:1067–73

139. Shikanai T, Endo T, Hashimoto T, Yamada Y, Asada K, Yokota A. 1998. Directed disruption of the tobacco *ndhB* gene impairs cyclic electron flow around photosystem I. *Proc. Natl. Acad. Sci. USA* 95:9705–9

140. Shikanai T, Shimizu K, Endo T, Hashimoto T. 1998. Screening of *Arabidopsis* mutants lacking down-regulation of photosystem II using an imaging system of chlorophyll fluorescence. In *Photosynthesis: Mechanisms and Effects*, ed. G Garah. Dordrecht: Kluwer. In press

140a. Shintani D, DellaPenna D. 1998. Elevating the vitamin E content of plants through metabolic engineering. *Science* 282:2098–100

141. Smirnoff N. 1996. The function and metabolism of ascorbic acid in plants. *Ann. Bot.* 78:661–69

142. Somerville CR. 1986. Analysis of photosynthesis with mutants of higher plants and algae. *Annu. Rev. Plant Physiol.* 37:467–507

143. Sonoike K. 1996. Photoinhibition of photosystem I: its physiological significance in the chilling sensitivity of plants. *Plant Cell Physiol.* 37:239–47

144. Tardy F, Havaux M. 1996. Photosynthesis, chlorophyll fluorescence, light-harvesting system and photoinhibition resistance of a zeaxanthin-accumulating mutant of *Arabidopsis thaliana. J. Photochem. Photobiol. B: Biol.* 34:87–94

145. Telfer A, Dhami S, Bishop SM, Phillips D, Barber J. 1994. β-carotene quenches singlet oxygen formed by isolated photosystem II reaction centers. *Biochemistry* 33:14469–74

146. Tyystjärvi E, Aro E-M. 1996. The rate constant of photoinhibition, measured in lincomycin-treated leaves, is directly proportional to light intensity. *Proc. Natl. Acad. Sci. USA* 93:2213–18

147. Van Camp W, Capiau K, Van Montagu M, Inzé D, Slooten L. 1996. Enhancement of oxidative stress tolerance in transgenic tobacco plants overproducing Fe-superoxide dismutase in chloroplasts. *Plant Physiol.* 112:1703–14

148. Vass I, Styring S, Hundal T, Koivuniemi A, Aro E-M, Andersson B. 1992. Reversible and irreversible intermediates during photoinhibition of photosystem II: Stable reduced Q_A species promote chlorophyll triplet formation. *Proc. Natl. Acad. Sci. USA* 89:1408–12

149. Wallsgrove RM, Turner JC, Hall NP,

Kendall AC, Bright SWJ. 1987. Barley mutants lacking chloroplast glutamine synthetase—biochemical and genetic analysis. *Plant Physiol.* 83:155–58

150. Walters RG, Horton P. 1994. Acclimation of *Arabidopsis thaliana* to the light environment: changes in composition of the photosynthetic apparatus. *Planta* 195:248–56

151. Walters RG, Ruban AV, Horton P. 1994. Higher plant light-harvesting complexes LHCIIa and LHCIIc are bound by dicyclohexylcarbodiimide during inhibition of energy dissipation. *Eur. J. Biochem.* 226:1063–69

152. Walters RG, Ruban AV, Horton P. 1996. Identification of proton-active residues in a higher plant light-harvesting complex. *Proc. Natl. Acad. Sci. USA* 93: 14204–9

153. Wheeler GL, Jones MA, Smirnoff N. 1998. The biosynthetic pathway of vitamin C in higher plants. *Nature* 393: 365–69

154. Whitmarsh J, Samson G, Poulson M. 1994. Photoprotection in photosystem II—the role of cytochrome b559. See Ref. 17b, pp. 75–93

155. Wiese C, Shi L-B, Heber U. 1998. Oxygen reduction in the Mehler reaction is insufficient to protect photosystems I and II of leaves against photoinactivation. *Physiol. Plant.* 102:437–46

156. Winter K, Königer M. 1989. Dithiothreitol, an inhibitor of violaxanthin de-epoxidation, increases the susceptibility of leaves of *Nerium oleander* L. to photoinhibition of photosynthesis. *Planta* 180:24–31

157. Yohn CB, Cohen A, Danon A, Mayfield SP. 1996. Altered mRNA binding activity and decreased translation initiation in a nuclear mutant lacking translation of the chloroplast *psbA* mRNA. *Mol. Cell. Biol.* 16:3560–66

158. Zhang L, Niyogi KK, Baroli I, Nemson JA, Grossman AR, Melis A. 1997. DNA insertional mutagenesis for the elucidation of a Photosystem II repair process in the green alga *Chlamydomonas reinhardtii. Photosynth. Res.* 53:173–84

Annu. Rev. Plant Physiol. Plant Mol. Biol. 1999. 50:361–89

MOLECULAR AND CELLULAR ASPECTS OF THE ARBUSCULAR MYCORRHIZAL SYMBIOSIS

Maria J. Harrison
The Samuel Roberts Noble Foundation, Ardmore, Oklahoma 73402;
e-mail: mjharrison@noble.org

KEY WORDS: root, fungus, plant-microbe interaction, phosphate transport, *Rhizobium*-legume symbiosis

ABSTRACT
Arbuscular mycorrhizae are symbiotic associations formed between a wide range of plant species including angiosperms, gymnosperms, pteridophytes, and some bryophytes, and a limited range of fungi belonging to a single order, the Glomales. The symbiosis develops in the plant roots where the fungus colonizes the apoplast and cells of the cortex to access carbon supplied by the plant. The fungal contribution to the symbiosis is complex, but a major aspect includes the transfer of mineral nutrients, particularly phosphate from the soil to the plant. Development of this highly compatible association requires the coordinate molecular and cellular differentiation of both symbionts to form specialized interfaces over which bi-directional nutrient transfer occurs. Recent insights into the molecular events underlying these aspects of the symbiosis are discussed.

CONTENTS

361

1040-2519/99/0601-0361$08.00

INTRODUCTION

The symbiotic associations of plant roots and fungi have intrigued many generations of biologists, and in the late 1880s these associations were given the name
mycorrhiza—derived from the Greek for fungus-root (68). Recent observations
of fossil plants from the Devonian era suggest that one type of mycorrhizal association, the arbuscular mycorrhiza, existed approximately 400 million years
ago (MYA), indicating that plants have formed associations with arbuscular
mycorrhizal (AM) fungi since they first colonized land (143, 151). Today, the
arbuscular mycorrhiza is the most widespread type of mycorrhizal association
and exists in ecosystems throughout the world where it creates an intimate link
between plants and the rhizosphere (6, 95, 180, 190). Despite the ubiquitous
occurrence of this association and its importance in nutrient movement between plants and the soil, our understanding of the mechanisms underlying the
development and functioning of the symbiosis is limited.

   This review builds on the information from two earlier *Annual Reviews* of
the arbuscular mycorrhizal symbiosis, in which the physiology and regulation
of the symbiosis were discussed (110, 178). In the conclusions to their 1988
review concerning the physiological interactions between the symbionts (178),
Smith & Gianinazzi-Pearson noted a need for approaches to provide fundamental information at a molecular level about the development and functioning of
the association. Over the past ten years there has been an explosion of molecular, genetic, and biochemical analyses of the AM fungi and the AM symbiosis.
This review emphasizes insights into aspects of the development of the symbiosis emerging from these studies. It focuses briefly on the AM fungi and the
advances in our understanding of these obligate symbionts and then on aspects
of development of the association, and finally briefly considers the mechanisms
underlying nutrient transport, which are still almost entirely unknown.

ARBUSCULAR MYCORRHIZAL FUNGI

The arbuscular mycorrhizal fungi are all members of the zygomycota and
the current classification contains one order, the Glomales, encompassing six
genera into which 149 species have been classified (27, 129). A major factor

hampering studies of the AM fungi, including the taxonomy, is their obligately biotrophic nature; so far, they have not been cultured in the absence of a plant host. Despite the lack of axenic culture, it is possible to grow them in sterile culture with plant roots, or with so-called hairy roots transformed with *Agrobacterium rhizogenes* (23, 56, 130, 132). Recent adaptations of these methods utilize petri plates in which the fungus and root are cultured together in one compartment while the external mycelium is permitted to ramify into a second compartment separate from the roots (185). Increasing numbers of fungal species are being established in this culture system (54, 57, 60), which is proving useful for studies of the fungal symbiont (12, 13, 25, 47, 106, 116). Such systems also provide access to pure, sterile fungal spores and mycelium that are essential for molecular analyses and useful for the taxonomy of these fungi.

Molecular Analyses of the Fungal Genome

There has been one report of sexual structures in AM fungi, but it remains unconfirmed. Recent genetic analyses also suggest that the fungi are asexual and reproduce clonally (153). The large resting spores formed by these fungi are unusual in that they are multinucleate and, depending on the species, may contain thousands of nuclei per spore (24). The nuclei in quiescent spores are arrested in the G0/G1 phase (29) and although initial studies suggested that DNA replication did not take place without a plant host (40), this was later demonstrated to be incorrect; both DNA replication and nuclear division occur during the initial growth of the hyphal germ tube, regardless of the presence or absence of a host plant (24). Using nuclear stains and flow cytometry, the DNA content of the nuclei has been estimated to range from 0.13 to 1.0 pg per nucleus for 12 species tested (30, 103). Analysis of the base composition of nine different species of glomalean fungi, containing representatives from four different genera, indicated that the genomes of these fungi have a relatively low GC content, averaging 33%, and in contrast to other fungal taxa, a high level of methylcytosine (2.23–4.26%), although not as high as that of plant genomes (102).

The first AM fungal genes to be sequenced were the small subunit rRNA genes (SSU) and the internal transcribed spacer (ITS) regions that were targets for phylogenetic analyses (119, 149, 169, 170). From the SSU sequence data it was possible to estimate a date for the origin of the AM fungi and the time at which further divergence within the group occurred. The origin of the AM fungi was placed between 462 and 353 MYA and the ancestral fungi were probably most like the extant Glomus species. The families Acaulosporaceae and Gigasporaceae emerged later and were estimated to have diverged from each other 250 MYA (170).

The SSU and ITS sequence data also allowed the design of specific primers that, when coupled with PCR amplification, enabled the identification of AM fungi from both spores and within plant roots in field situations (49, 55, 119, 149, 159, 160, 170, 171). Additional DNA-based methods of identification followed (120, 195, 209), including random amplification of polymorphic DNA (RAPD) analysis. This enabled the development of primer-pairs specific for a number of species including *Glomus mosseae, Gigaspora margarita*, and *Scutellospora castenea* (1, 114, 206). These primers have utility in taxonomic studies and in competitive PCR assays to quantify the amount of fungus within mycorrhizal roots (59, 66).

A particularly interesting finding that emerged during the molecular analyses of the ITS regions was the variability in the genetic composition within and between spores of a single fungal species. A single spore may contain more than one ITS sequence and individual spores have ITS sequences that differ from other spores of the same species (119, 159). This variability was further confirmed by analyses of other loci (153, 210). Using minisatellite-based markers, it was observed that the first generation of spores arising from single-spore cultures displayed a high level of variation, which suggests that the multinucleate spores are heterokaryotic. It seems likely that reassortment of genetically different nuclei provides a mechanism by which these fungi maintain genetic diversity (210).

Genomic libraries prepared from spores and cDNA libraries from spores and mycorrhizal roots have been constructed for a limited number of species (69, 192, 208), and the first cDNA clones representing genes other than the rRNAs have been identified. Glyceroldehyde-3-phosphate dehydrogenase (GAPDH), β-tubulin, ATPase, nitrate reductase, and DNA-binding protein sequences were among the first to be reported. Although most of these are considered housekeeping genes, they will be useful molecular markers for analysis of these fungi during the development of the symbiosis (42, 69, 70a, 107).

Future dissection of gene function in arbuscular mycorrhizal fungi will be difficult without the possibility to transform these fungi genetically. Progress toward this goal is being made and transient expression of a reporter gene construct in spores of *Gigaspora margarita* has been achieved. This in itself is an important technological achievement and will enhance molecular analyses of the AM fungi (67).

DEVELOPMENT OF THE ARBUSCULAR MYCORRHIZAL SYMBIOSIS

Early Signaling Events

Germination of AM fungal spores and the initial growth of hyphal germ tubes can occur in the absence of the plant root; however, both root exudates and volatiles such as CO_2 can stimulate both of these processes (15, 22, 25, 47, 84, 108, 135, 144, 152). In some cases, root exudates also elicit rapid and extensive branching of the hyphae as they enter the vicinity of the root (89, 91), a response that has been observed as hyphae approach the roots of host plants but not when they encounter non-host roots, which suggests that recognition of the host occurs. In this case, lack of recognition of the non-host could be due to lack of a signal; however, in other instances non-host status is probably due to inhibitory compounds (164, 165, 199).

The range of active components present in root exudates is unknown. However, some of the activities are probably due to flavonoid and phenolic compounds that stimulate the growth of some AM fungal species while inhibiting others (22, 62, 144, 172, 191). The particular molecule(s) responsible for eliciting hyphal branching is also unknown but, based on its estimated size, it could also be a phenolic or flavonoid derivative (90). Since the flavonoid compounds are active at very low concentrations, it is assumed that they do not have a nutritional effect but rather that they act as signals to stimulate or inhibit growth. Plant flavonoid/isoflavonoids bind to estrogen receptors, and recent experiments utilizing estrogens and anti-estrogens provide preliminary evidence for the presence of an AM fungal receptor capable of binding biochanin A and estrogens. Based on the structures of these molecules, it is suggested that the A and C rings of the isoflavonoid and the hydroxyl group at position A-7 are important features for recognition by the receptor (145). Although flavonoid derivatives can influence the initial stages of the fungal life cycle, experiments with flavonoid-deficient mutants of maize indicate that they are not essential for the development of the symbiosis (26). Maybe in natural environments, the flavonoid-mediated stimulation of growth and branching in the vicinity of the root helps to ensure contact with the root and the establishment of the symbiosis. The differential effects of flavonoids/isoflavonoids on different fungal species could be envisaged to influence the fungal populations associated with particular plants.

Appressorium Formation

Development of the symbiosis is initiated when a fungal hypha contacts the root of a host plant where it differentiates to form an appressorium. Although components of root exudates are capable of stimulating hyphal growth and branching, they are unable to elicit the formation of appressoria, which were initially only observed on intact plant roots (88). Recently, it was demonstrated that *Gigaspora margarita* could form appressoria in vitro on purified epidermal cell walls isolated from carrot roots, a host for *Gigaspora margarita*, but not on walls isolated from sugar beet, a non-host (134). The fungus also recognized specifically the epidermal cell walls and did not form appressoria on cortical or vascular cell walls. These experiments indicate that the signal for appressorium formation lies within the epidermal cell wall, a hypothesis suggested earlier by Tester et al (188). The experiments also confirm that the branching signal is either loosely bound to the wall or exuded from the roots, since the purified wall fragments did not elicit the extensive hyphal branching observed in intact roots. These purified walls probably consist of a mixture of polysaccharides, including cellulases and polygalacturonans and some proteins. Carbohydrate molecules act as signals in a number of other plant/fungal interactions and are likely candidates for the induction of appressoria in the AM symbiosis (121).

Penetration of the Root

Appressorium formation is followed by the development of a penetration hypha and penetration of the root. This can occur in different ways; in some species, the hypha enters by forcing its way between two epidermal cells, whereas in other cases, the hypha penetrates an epidermal or root hair cell wall and grows through the cell (34). The exact mechanisms involved in penetration are unknown; however, by analogy to a number of the biotrophic pathogens (127), it has been suggested that specific, localized production of cell wall–degrading enzymes, in combination with mechanical force, may facilitate entry of the hyphae without inducing defense responses (33). AM fungi produce exo- and endoglucanases, cellulases, xyloglucanases, and pectolytic enzymes including polygalacturonase (76–78, 80, 150), all of which expedite their passage through a cell wall.

Since the appressoria that developed on the purified epidermal cell wall fragments failed to form a viable penetration hypha and did not penetrate the wall, processes subsequent to appressorium formation likely require an intact cell (134). A wide range of plant mutants on which the AM fungi can form appressoria but cannot develop further are direct proof that the plant controls this developmental step in the association. Mutants blocked at this stage of this symbiosis have been described in *Pisum sativum* (63, 86), *Medicago sativa* (37, 38), *Vicia faba* (63), *Phaseolus vulgaris* (168), *Medicago truncatula* (155), *Lotus japonicus* (203), and *Lycopersicon esculentum* (21). The phenotypes of these mutant associations are fairly similar at the morphological level and fall into two broadly defined groups. In association with the *P. sativum, L. esculentum,* and *Medicago* mutants, the fungus forms appressoria that are frequently large and deformed and that become septate when the fungus fails to enter the root (21, 37, 86). In one of the *Medicago sativa* mutant lines, the number of appressoria formed on the mutant increases, a possible consequence of the failure to penetrate (37); however, increased numbers of appressoria were not reported for the *P. sativum* mutants (86). In *P. sativum*, the non-penetrating phenotype is referred to as myc$^{(-1)}$ and 21 such mutants have been described.

They belong to five complementation groups, which indicates that entry into the root is under complex genetic control. The traits segregate as single recessive loci and the condition is determined by the roots (86). Grafting wild-type scions onto mutant stocks did not rescue the mutants (86, 200). Hairy roots prepared from these genotypes also maintain their nonmycorrhizal status (14). Cytochemical analyses of one of the *P. sativum* and one of the *M. sativa* mutant interactions indicated that cell wall depositions, including callose and phenolics, were present in the walls of cells adjacent to the appressoria (93, 142). Such depositions were not seen in wild-type interactions, which suggests that a defense response has been elicited in these mutants. Based on these data, it is possible that a suppressor of defense responses has been mutated such that the plant now views the fungus as a pathogen. This situation is reminiscent of the barley/*Erysiphe graminis* interaction where mutation-induced recessive alleles of the *Mlo* locus confer resistance to a wide range of isolates of *Erysiphe* (71). Resistance is mediated by the formation of appositions on the cell walls below the appressoria and wild-type *Mlo* is a negative regulator of defense responses as well as leaf cell death (71). The *Mlo* gene has been cloned and the encoded protein is predicted to be an integral membrane protein; however, the function of the protein and mechanism of regulation of defense responses remain to be determined (43).

The *P. vulgaris* and *L. japonicus* mutants show a slightly different phenotype from the other species and appressorium formation is followed by penetration of the first cell layer (203). The association then aborts in the root epidermis where swollen and deformed hyphae are visible within these cells (203). In the *Lotus* mutants, the hyphae occasionally manage to overcome the block in the epidermis and growth from the deformed hyphae continues, producing normal internal mycorrhizal structures. Since these are indistinguishable from wild-type, it has been suggested that the mutated genes are not required for the later phases of growth (203). All of the legume mycorrhizal mutants are also affected in their ability to form a nitrogen-fixing symbiosis with *Rhizobium* species and thus define a set of genes, termed *sym* genes, essential for both symbioses. Similarities between these symbioses are just beginning to emerge, and some of the signaling pathways and downstream events occurring during the formation of the symbioses clearly are conserved (2, 194).

A nonpenetrating mutant identified in *L. esculentum* is the first mutant of this type to be identified from a nonleguminous species. This mutant shows a similar phenotype to the legume mutants, although slightly different responses were noted depending on the fungal symbiont involved. *Glomus mosseae* was unable to penetrate the *L. esculentum* mutant roots, whereas *Gigaspora margarita* was occasionally able to enter. In contrast to the *L. japonicus* mutants, this mutation appears to affect the internal stages of development of the mycorrhiza, and following entry, *G. margarita* did not develop extensively within the roots and was unable to form arbuscules (21). The future cloning of the mutated genes, which should be feasible for *L. esculentum* and also for the legumes, *L. japonicus* and *M. truncatula*, due to their small genome size, will provide insight into the controlling mechanisms.

Internal Growth and Development of Arbuscules in an Arum-Type Mycorrhiza

Following entry into the root, internal development of the fungus is influenced by the plant, and a single species of fungus may show significantly different morphological growth patterns depending on the plant partner in the association (82, 105). The two main patterns are referred to as the *Paris* and *Arum* types, named after the species in which they were first described (74, 174). Much of the laboratory research focuses on crop species that form the *Arum* type and molecular investigations of the *Paris* type have not been undertaken. Thus the *Arum* type is the focus of these discussions.

In the *Arum*-type mycorrhiza, penetration of the root is initially followed by intercellular hyphal growth, although in some instances the fungus will penetrate the exodermis and form hyphal coils in the exodermal cells as it passes through (34). On reaching the inner cortex, branches arising from the intercellular hyphae penetrate the cortical cell walls and differentiate terminally within the cell to form dichotomously branched structures known as arbuscules (Figure 1). Although an arbuscule develops within a cell, it remains essentially apoplastic as the plant plasma membrane extends to completely surround it. The fungal cell wall becomes progressively thinner as the arbuscule develops and consequently in these cells, there is an extensive intracellular interface in which the two symbionts are in extremely close contact, separated only by their membranes and a narrow plant-derived apoplast (35, 36, 178). This interface is thought to be the site at which phosphate and possibly carbon are transferred between symbionts, although some speculate that the intercellular hyphae might be responsible for carbon uptake (173, 176, 181). Despite the intensive effort expended by both symbionts to develop the arbuscule and the arbuscular interface, the life span of an arbuscule is only a few days, after which it collapses and decays leaving the cell undamaged and capable of hosting another arbuscule (4, 5). Both the variable growth patterns observed in the *Arum*-and *Paris*-type hosts and a *P. sativum* myc$^{(-2)}$ mutant in which the arbuscules barely develop, indicate that the plant also controls this stage of the association (83, 86).

Following formation of arbuscules, some species of AM fungi also form lipid-filled vesicles within the roots, which are presumed to act as a storage reserve for the fungus (178).

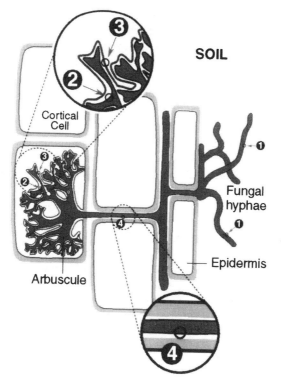

Figure 1   Diagram indicating possible sites of plant and fungal phosphate transporters and fungal glucose transporters in membranes of an arbuscular mycorrhiza (*Arum* type). Phosphate transport: 1, phosphate uptake across membranes in the external hyphae; 2, phosphate efflux across the arbuscule membrane; 3, phosphate uptake across the peri-arbuscular membrane. Carbon transport: 2 and 4, possible sites of glucose uptake by the fungus.

MOLECULAR AND CELLULAR ALTERATIONS IN THE CELLS DURING ARBUSCULE DEVELOPMENT   As the fungal hypha penetrates a cortical cell wall and begins to differentiate into an arbuscule, the invaded cell responds with fragmentation of the vacuole, migration of the nucleus to a central position within the cell, and an increase in the number of organelles (17, 33, 46). This response seems to be arbuscule specific and it does not occur in the exodermal cells during the development of coils. The plasma membrane extends approximately fourfold to form the peri-arbuscular membrane (5) that envelops the arbuscule, and therefore concomitant increases in membrane biosynthesis must be occurring. Although the peri-arbuscular membrane is connected to the peripheral membrane of the plant cell, cytochemical analyses indicate that it has high levels of ATPase activity, whereas the remainder of the plasma membrane shows little staining for ATPases (87). Since H^+-ATPase activity gives rise to the proton gradient required for many secondary active transporter processes, these data support the suggestion that active transport of nutrients may be occurring across this membrane. Currently, little else is known about activities associated with the peri-arbuscular membrane and the fact that it develops deep within the tissues of the root makes it recalcitrant to many of the modern techniques of membrane biology. It is assumed to retain the ability to synthesize and deposit cell wall components including β-1,4 glucans since these have been found in the new apoplastic compartment formed between the peri-arbuscular membrane and the arbuscule (33). Immunocytochemical analyses have also demonstrated the presence of a matrix composed of pectins, xyloglucans, nonesterified polygalacturonans, arabinogalactans, and hydroxyproline-rich glycoproteins (HRGP) in this compartment (20, 33, 92, 140a). Consistent with this composition, genes encoding both a putative arabinogalactan protein (AGP) and an HRGP have been shown to be induced in mycorrhizal roots of *M. truncatula* and maize, respectively, and the transcripts are localized specifically in the cells containing arbuscules (19, 193).

Although the interface compartment is continuous with the cell wall and the molecular content tends to reflect the composition of the cortical cell wall (18), the physical structure is considerably different. The components of the interface matrix do not become cross-linked and most closely resemble primary cell walls (33). The lack of cross-linking has been suggested to be the result of lytic enzymes released from the arbuscule, and in support of this hypothesis, endopolygalacturonase has been immunolocalized to the arbuscule and interface compartment (33, 139). In pea mycorrhizae, immunocytochemical analysis of the interface matrix indicates the presence of components in common with the peribacteroid compartment of pea nodule cells. These epitopes are not present in the peripheral cell walls, providing further evidence for the specialization of the interface matrices and reiterating similarities between the two symbioses (140a, 141). Whether the components of the arbuscular interface matrix have any symbiosis-specific functions is unclear; however, AGP-like proteins have been reported in two other symbiotic associations. They are induced both during nodule development in the *Rhizobium*-legume symbiosis (163) and also in the *Gunnera/Nostoc* association, where they are abundant in the *Gunnera* stem gland mucilage, which plays a central role in communication between the symbionts (147). In plant/biotrophic fungal pathogen associations, it was shown recently that the fungus produces a proline-rich glycoprotein that is very similar to the plant cell wall proline-rich and hydroxyproline-rich glycoproteins. The protein is deposited in the extrahaustorial matrix, a component analogous to the arbuscular interface matrix, where it is suggested to mimic plant cell wall proteins and prevent the plant from perceiving the pathogen (140). Components of the arbuscular interface matrix might conceivably function in a similar manner.

The specific changes associated with formation of the arbuscular interface clearly require significant reorganization of the cytoskeleton of the cell. Recent confocal microscopy studies confirm this hypothesis and demonstrate the presence of new cortical microtubule networks running along the hyphae and also between the hyphae and the nucleus of the cortical cell. This latter observation is consistent with earlier studies documenting the movement of the nucleus of the colonized cortical cell to a central position adjacent to the arbuscule (17, 81). A maize α-tubulin promoter is also activated specifically in cells in which arbuscules are developing, providing further evidence of an increase in cytoskeletal components in these cells (32).

It is predicted that development of the arbuscular interface will be accompanied by many other alterations in the cortical cells, and documentation of these changes is just beginning. In *M. truncatula*, transcripts encoding enzymes of the flavonoid biosynthetic pathway, phenylalanine ammonia lyase (PAL) and chalcone synthase (CHS), but not the defense-specific enzyme isoflavone reductase (IFR), are induced specifically in cells containing arbuscules (98). Since secondary metabolites from this pathway accumulate in mycorrhizal *M. truncatula* roots, the location of expression of these two genes has led to the speculation that they might be synthesized in these cells (97, 98). The potential function of the flavonoids/isoflavonoids within these cells is unclear; however, some of them stimulate growth of AM fungi both in the presymbiotic phase of their life cycle and during the symbiosis (47, 191, 207). In other cases, flavonoids act as auxin transport inhibitors (104) and therefore alter the hormone balance in the roots. Hormone levels are known to change in mycorrhizal roots and are probably responsible for the induction of expression of some of the mycorrhiza-induced genes (194). Transcription of an auxin-inducible gene encoding glutathione *S*-transferase (GST) is also induced in cells containing arbuscules (186). GSTs are stress-response enzymes that add glutathione (GSH) to a variety of compounds including phenylpropanoids/flavonoids to facilitate their transport via a specific carrier into the vacuole (53, 124). Sequestration in the vacuole prevents phytotoxic effects on the cell that can occur if high levels of these compounds accumulate. Although in this case the flavonoid and GST experiments were performed in different plants, it is not surprising that they are co-induced in the same cell type.

The External Mycelium

Following colonization of the root cortex, fungal hyphae develop extensively within the soil. This external mycelium plays a pivotal role in the AM symbiosis where its functions include the acquisition of mineral nutrients from the soil and their subsequent translocation to the plant, colonization of additional roots, and, in many cases, the production of spores. In addition to its role in the symbiosis, the extraradical mycelia contributes to soil stability by the aggregation of soil particles, probably mediated in part by glycoproteins produced by the hyphae (204, 205).

Studies of this phase of the symbiosis have lagged behind that of the internal phase, mainly owing to difficulties with observation, collection, and quantification of the mycelium. However, new methods that overcome many of these problems are now available (137, 185). Early studies of the external mycelium had indicated that it is comprised of different types of hyphae, including large runner hyphae and finer absorptive hyphae (72). Recently, these findings have been confirmed, and an ultrastructural examination of the fine absorptive hyphae has revealed features consistent with a role in nutrient absorption (10, 11). Measurements of both the internal pH of the hyphae and the external pH of the media surrounding the hyphae are also possible in this system; therefore, the physiological state of the hyphae can be monitored. (13, 106). Since appropriate tools are now available, studies of this phase of the symbiosis should be emphasized in the future.

IDENTIFICATION OF GENES INDUCED DURING THE ARBUSCULAR MYCORRHIZAL SYMBIOSIS

Over the past few years, a search for genes and proteins induced in mycorrhizal roots has been initiated in a number of laboratories. Analysis of protein profiles, differential screening (133, 187, 193), and differential display (125, 126) have all been used to provide insight into molecular changes occurring in both the plant and fungal symbionts during development of the symbiosis. The initial comparisons of the protein profiles of noncolonized and colonized roots indicated the presence of new proteins in mycorrhizal roots but did not permit differentiation between fungal proteins and newly induced plant proteins (8, 65, 79, 158, 162). These differences are more readily resolved via molecular analyses where comparison of cDNA and genomic sequences can be used to determine the genome of origin.

Consistent with the increased synthesis of plasma membrane in the mycorrhizal roots, many of the mycorrhiza-induced plant cDNAs identified so far encode membrane proteins. Differential screening of a cDNA library prepared from barley mycorrhizal roots resulted in the identification of a barley cDNA sharing sequence identity with a H^+-ATPase (133). This clone hybridizes to transcripts that are induced in mycorrhizal roots and although the cell and membrane location of the protein are still unknown, it is a potential candidate for the gene encoding the ATPase located on the peri-arbuscular membrane. A cDNA encoding a member of the membrane intrinsic protein (MIP) family is induced in mycorrhizal parsley roots. These integral membrane proteins facilitate the movement of small molecules across membranes and might be predicted to have a role in transport at the arbuscular interface (154). Differential display analyses of mycorrhizal pea roots resulted in the identification of a cDNA *psam1*, predicted to encode a novel protein sharing some similarities with phospholamban, a membrane-anchored protein that regulates the activity of Ca^{2+}-ATPase (125). The function of the *psam1* gene product in mycorrhizae is unknown.

Since the development of the matrix at the arbuscular interface requires the de novo synthesis of cell wall proteins, it might also be predicted that this class of proteins would be induced in mycorrhizal roots. Consistent with this hypothesis, genes encoding an HRGP, a putative AGP, and a member of the xyloglucan endo-transglycosylase (XET) family are all induced during development of the association (193). XETs are enzymes that cleave and reform xyloglucan bonds within the cell walls, and it is speculated that this activity might be employed to loosen cells walls to permit penetration of the fungal hyphae or, alternatively, to function in maintaining the structure of the arbuscular interface matrix (193).

In addition to the growing numbers of genes up-regulated during the symbiosis, there are clearly groups of genes that are specifically down-regulated in the mycorrhiza. In *Medicago truncatula*, it was observed that four phosphate starvation-inducible genes including two phosphate transporter genes, a gene of unknown function (Mt4), and an acid phosphatase homolog were all downregulated following colonization of the root by a mycorrhizal fungus (41, 118). Elevated levels of phosphate in the external medium down-regulate the expression of these genes, and since the symbiosis generally results in an increased level of phosphate in the roots, the fact that these genes are down-regulated in the symbiosis is not so surprising (41, 118). However, studies of the Mt4 gene indicate that down-regulation of this transcript also occurs in one of the *Medicago* mutants in which the fungus fails to penetrate the root and grows only on the external surface. In this association, it seems unlikely that the fungus delivers phosphate to the root. Therefore, this suggests that down-regulation of the Mt4 gene can occur also via a signal from the fungus and is consistent with the presence of two, initially independent pathways for the regulation of the Mt4 gene (41).

Although the identification of genes differentially regulated in mycorrhizal roots provides initial information about the changes that occur during development of the mycorrhiza, real insight into mechanisms underlying its development will occur in the future when the roles of these proteins in the symbiosis are elucidated. New technologies, such as confocal laser scanning microscopy coupled with the use of fluorescent reporter proteins such as green fluorescent protein (GFP) permit the in vivo monitoring of gene and protein expression within living cells and provide a powerful addition to the current arsenal of molecular tools with which to analyze the symbiosis.

Expression of Defense Responses

In many plant-fungal pathogen interactions, invasion of plant tissues by the fungus results in the induction and sustained expression of a varied battery of plant defenses that prevent further pathogen ingress (58). This is not the case, however, with some of the biotrophic pathogens that form ostensibly compatible but parasitic associations with their hosts and probably avoid eliciting defenses by the conservative and local production of hydrolytic wall-degrading enzymes and minimal damage to the plant cell (122, 127, 128). The arbuscular mycorrhiza seems to be the most highly attuned plant-fungal association. Data from many AM associations indicate that in the AM symbiosis plant defense responses generally show small transient increases in the early stages of the symbiosis, followed by suppression to levels well below those of noncolonized plants (85, 109).

In *Allium porrum*, chitinase and cell wall–bound peroxidase activities showed a transient increase in expression during the initial stages of development of the mycorrhiza; however, activity in a well-established mycorrhizal association was notably lower than in the controls (183, 184). In addition, immunocytochemical analysis suggested that the chitinase, when present, was located in the vacuole and was not in contact with the hyphae (183). Similar observations were later made in bean and tobacco roots where colonization by an AM fungus was accompanied by a transient increase and then a decrease in both the transcript levels and activity of chitinase, β-1,3 endoglucanase, and chalcone isomerase, an enzyme of isoflavonoid biosythesis (52, 112, 197, 201). These data indicated that regulation was occurring probably at the level of gene expression, a finding that is also supported by studies of the expression of genes encoding enzymes of the isoflavonoid biosynthetic pathway in *Medicago* species. *Medicago sativa* and *M. truncatula* respond to attack by fungal pathogens with the rapid induction of genes and enzymes of the isoflavonoid pathway, which results in the production of the defense compound, medicarpin. In contrast, colonization by AM fungi results in the down-regulation of the gene encoding isoflavone reductase (IFR), the penultimate enzyme of medicarpin biosynthesis, and medicarpin does not accumulate (97, 202). In situ hybridization revealed that down-regulation of the IFR transcript occurred exclusively in the areas of the root in which arbuscules had formed, indicating that this is a specific and local effect (98). Investigations of soybean roots colonized with different strains of *G. intraradices* suggested that the suppression of endochitinase expression was correlated with the infectivity of the different strains, with the most infective strains resulting in the most down-regulation (113). This finding might explain the results from parsley and bean roots where defense responses were not altered following colonization by mycorrhizal fungi (31, 70).

Although the suppression of plant defense responses seems to be a widespread occurrence in AM associations, the necessity for this suppression is challenged by data indicating that *Nicotiana tabacum* and *Nicotiana sylvestris* plants over-expressing various chitinases, glucanases or pathogenesis-related (PR) proteins were apparently entirely unaffected in their ability to form mycorrhizae (196, 198). In contrast, overexpression of these defense proteins inhibits the growth of fungal pathogens. Plants overexpressing chitinase were more resistant to the root pathogen *Rhizoctonia solani*, whereas those overexpressing PR-1a were more resistant to *Peronospora tabacina* and *Phytopthora parasitica* (3, 39, 198). The only gene whose overexpression was observed to inhibit colonization of plants by AM fungi was PR-2, a protein with β-1,3 glucanase activity (196). Although the data from these transgenic plants cast doubts on the requirement for suppression of these particular defense responses, there are other instances in which defense responses seem to be impeding the interaction. In *Salsola kali*, which is a non-mycotrophic species, AM fungi colonize the roots and initially form coils and arbuscules; however, the root responds very quickly with the production of autofluorescent compounds and the colonized areas of the root turn brown and die (7). It might be argued that this is a non-host response; however, similar responses are seen in alfalfa (a mycorrhizal species) in response to colonization by *G. margarita*. The fungus enters the root, but colonized cells show a hypersensitive response and become necrotic. Phenolic and isoflavonoid compounds characteristic of a defense response accumulate in these regions of the root (61). Thus, if defense responses are elicited in the appropriate cells, they can apparently prevent development of the association. These results also indicate that the AM fungi do not simply fail to elicit defense responses, as has been occasionally suggested, but rather that at some level, there is compatibility between the fungus and the plant, and recognition of compatibility prevents induction of defense responses. Incompatible combinations clearly exist, and as different genera and species of fungi are utilized for laboratory experiments, it might be predicted that more of the incompatible interactions will be identified. Additional evidence supporting the hypothesis that recognition and suppression of defense responses is required for successful colonization is provided by the pea locus *a* myc$^{(-1)}$ mutant in which defenses seem to be induced and prevent further development of the appressorium (93).

Although in general a sustained defense response is not induced during a successful AM symbiosis, there are some exceptions. For example, the defense gene, PR-1, is expressed in pea root cells containing arbuscules, and in a number of plants new symbiosis-related forms of chitinase are induced in the roots following colonization by mycorrhizal fungi (51, 64, 146). In these instances, the proteins could have roles other than classical defense. Chitinases released from spruce cells do not have deleterious effects on ectomycorrhizal fungi, but actually destroy fungal elicitors released from fungal cell walls by cleaving them into smaller, inactive units. In this way, the elicitation of defense responses can be prevented and development of the symbiosis can proceed (157). The new chitinase isoforms induced in arbuscular mycorrhizal roots may have a similar role.

Expression of Nodulation Genes in the Mycorrhizal Symbiosis

The emerging similarities between the *Rhizobium*-legume symbiosis and the arbuscular mycorrhizal symbiosis have stimulated investigations of the expression of nodulation genes during the AM symbiosis. Leghemoglobin transcripts were detected in mycorrhizal roots of *Vicia faba* (73), whereas in *Medicago sativa*, two nodulation genes, *Ms*ENOD40 and *Ms*ENOD2, were induced in mycorrhizal roots with similar tissue-specific patterns of expression as in roots inoculated with *Rhizobium* (194). Both genes can be induced in roots in the absence of a symbiosis via the application of cytokinin, and since cytokinin levels are elevated during nodulation and also in mycorrhizal roots (194), it is speculated that cytokinin is one component of the signal transduction pathway mediating induction of these genes during the symbioses. Further evidence of signal transduction pathways common to both symbioses is provided by studies of the *Ps*ENOD5 and *Ps*ENOD12A genes, which are induced in pea roots during interactions with either AM fungi or *Rhizobium*. In the pea *sym8* mutant, which is unable to form either of these symbioses, expression of both genes is blocked, suggesting that SYM8 functions in the signal transduction pathway for induction of these genes in both symbioses (2). Based on these studies and also on the legume symbiosis mutants, it is clear that some mechanisms are shared between the two symbioses and this has fuelled speculation that the *Rhizobium*-legume symbiosis arose by exploiting signaling pathways from arbuscular mycorrhizae (194). Since our understanding of the *Rhizobium*-legume symbiosis is more advanced than that of the arbuscular mycorrhizal symbiosis, it will be fruitful to exploit these overlaps to elucidate these aspects of the association.

NUTRIENT TRANSPORT ACROSS ARBUSCULAR MYCORRHIZAL INTERFACES

Nutrient transport between the symbionts is a central aspect of the symbiosis; however, the membrane transporters responsible for the movement of carbon or phosphate between the symbionts are unknown (Figure 1). In addition, there is speculation as to the interfaces involved in carbon transport between the plant and the fungus. [For in depth discussions of these issues, readers are referred to a number of comprehensive reviews (173, 179, 181).] A brief summary of the current opinions on transport in the symbiosis and insights obtained from recent data that will facilitate identification of the transport proteins are included here.

Carbon Transfer from the Plant to the Fungus

Although it has been known for over 20 years that carbon is transferred from the plant to the fungus, evidence as to the form of carbon has been lacking (28, 101). Recent in vivo ^{13}C nuclear magnetic resonance spectroscopy data strongly suggest that glucose is the form of carbon utilized by these fungi (167) and this is further supported by studies of isolated arbuscules that were observed recently to use glucose for respiration (182). Although carbon allocation to the roots increases during mycorrhizal associations, the amounts of carbon estimated to leak out of intact root cells into the apoplast are thought to be insufficient to account for the amount of fungal growth occurring in mycorrhizal roots. Therefore, enhanced efflux, or a decrease in the level of competing host uptake systems, has been proposed (138, 166, 173, 181). In *M. truncatula*, expression of a hexose transporter gene is induced in mycorrhizal roots, specifically in the cortical cells in the vicinity of the fungus, which suggests that in this case potentially competing host mechanisms are not suppressed (96). In other symbioses, enhanced efflux of nutrients, stimulated by the demand of the microsymbiont, has been observed and in some plant/fungal pathogen interactions, the fungi produce toxins that alter membrane transport processes to favor release of metabolites (75, 100, 123). Similar events may well occur during the AM symbiosis, and the fungal symbiont may possess the capability to stimulate efflux of carbon from the plant. So far, there is no molecular information about transport proteins responsible for the efflux of carbon out of plant cells; however, this is currently an active area of research since this type of transporter is also expected to exist in the mesophyll and vascular tissues where sugar export occurs (161).

Although the arbuscule presents a large area of close contact between the symbionts and was traditionally assumed to be the interface over which carbon would be transferred, the observation that the membrane of the arbuscule lacks ATPase activity has led to the suggestion that carbon uptake might occur via the intercellular hyphae, whose membranes have been observed to have a high ATPase activity and thus are energized for active transport processes (87, 176) (Figure 1). It is unclear whether uptake of carbon by the AM fungus requires an active transport mechanism similar to those of plant transporters, or whether concentrations of carbon at the interfaces could be sufficient to permit uptake by facilitated diffusion, as occurs in yeast (111). In the absence of information about the concentrations of carbon present in the various apoplastic interfaces, it is difficult to speculate on potential mechanisms of transport either out of plant cells or into the fungus. *Amanita muscaria*, an ectomycorrhizal fungus, utilizes both fructose and glucose but relies on plant invertases for release of these hexoses from sucrose (48, 156). A monosaccharide transporter was cloned recently from *Amanita muscaria* and is probably the transporter responsible for hexose uptake in both the free-living and symbiotic stages of its life (136). A similar transporter might be envisaged for AM fungi, although it seems unlikely that it will be present in the absence of the plant host. Sequence information from the *Amanita* transporter coupled with those from yeast and *Neurospora crassa* may facilitate the cloning of transporters from arbuscular mycorrhizal fungi.

Phosphate Transfer from Soil to the Plant via the Fungus

Phosphate movement in the symbiosis involves a number of membrane transport steps, beginning with uptake across membranes in the external hyphae. This is followed by translocation back to the internal fungal structures where it is thought to be released from the fungus across the arbuscule membrane and then taken up into the plant by transporters on the peri-arbuscular membrane (Figure 1) (173, 178, 181). There has been some progress toward understanding the mechanisms of uptake of phosphate by the external AM fungal hyphae and a high-affinity phosphate transporter has been cloned from *G. versiforme* (99). The transporter has a K_m of 18 μM as determined by expression in yeast cells, a value that is consistent with previous measurements of phosphate uptake by AM fungi (189). The transporter transcripts are present in the external mycelium and not in the structures of the fungus internal to the root and therefore the transporter may be responsible for the initial uptake of phosphate into the mycorrhiza (99). Unfortunately, the inability to perform gene disruption experiments in arbuscular mycorrhizal fungi currently prevents direct proof of its role in the symbiosis.

Phosphate flux across the symbiotic interfaces in the mycorrhiza has been estimated at 13 nmol $m^{-2}s^{-1}$, although this value increases if extra phosphate is supplied to the mycorrhiza (50, 177). In contrast, the general rate of efflux of phosphate from fungal hyphae growing in culture was measured at 12 pmol $m^{-2}s^{-1}$ (45). Based on these findings, the AM fungi likely have some type of specialized efflux mechanism operating in the arbuscule membrane to permit sufficient phosphate efflux to the arbuscular interface. Efflux of phosphate from hyphae of an ectomycorrhizal fungus was shown to be stimulated by divalent cations, and a similar mechanism, possibly triggered by a component of the interface matrix, might be envisaged for the AM fungi (45). Since the peri-arbuscular membrane has a high ATPase activity (87), the subsequent uptake of phosphate by the plant could then occur by proton-coupled transport mechanisms. However, as with the previous discussion of carbon transport, the lack of information about physiological conditions at the arbuscular apoplastic interface prevents informed speculation as to the type of transport mechanisms. Analysis of the conditions in the apoplastic interfaces in mycorrhizal roots is extremely challenging, although progress is being made (9).

In the past two years, phosphate transporters have been cloned from the roots of a number of plant species (115, 117, 118, 131, 175). These transporters are expressed during growth in low-phosphate environments and mediate high-affinity phosphate transport into the epidermal and cortical cells. In *Medicago truncatula*, the expression of these transporters is down-regulated in mycorrhizal roots (118). This suggests that the plant does not use these transporters during the symbiosis and therefore it seems unlikely that these transporters operate at the peri-arbuscular membrane. Studies of phosphate uptake in mycorrhizae have revealed that in some associations, phosphate uptake directly via the root cells was considerably reduced during the symbiosis and the bulk of the phosphate uptake occurred via the fungal symbiont (138a), which is consistent with the down-regulation of the root phosphate transporters during the symbiosis.

The cloning and functional analysis of the transport proteins operating in the symbiosis is just one of the challenges for the future. Part of this process might be accelerated by utilizing approaches that have been successful in other research fields, such as the plant/biotrophic fungal pathogen associations, where there have been recent advances in the identification of transporters operating in the haustorial membranes. The cloning of an amino acid transporter from a rust fungus was achieved from libraries prepared from isolated haustoria (94). With the refinements in methods for isolation of arbuscules, similar approaches might permit access to the fungal transport proteins operating at the arbuscule membrane.

CONCLUSIONS

Over the past few years, the first insights into molecular mechanisms underlying development of the symbiosis have been achieved. The initial similarities to, and differences from, other plant-microbe interactions have been explored and exploited. Despite this progress, there is a still a vast amount to learn, and the molecular approaches that have been initiated recently need to be integrated with other disciplines to address both development and functioning of the symbiosis.

Although the current range of mycorrhizal mutants is limited, they have played a pivotal role in proving that the plant controls various stages of development of the association. The complexity of this symbiosis renders genetic approaches essential, and the identification of additional mutants, particularly those that are specific for the mycorrhizal symbiosis, should be emphasized in the future.

ACKNOWLEDGMENTS

I thank Sally Smith (University of Adelaide, Australia) for critical reading of the manuscript and both Richard Dixon (Samuel Roberts Noble Foundation, USA) and Sally Smith for encouragement and advice. The work in the author's lab is funded by The Samuel Roberts Noble Foundation.

> Visit the *Annual Reviews home page* at
> http://www.AnnualReviews.org

Literature Cited

1. Abbas JD, Hetrick BAD, Jurgenson JE. 1996. Isolate specific detection of mycorrhizal fungi using genome specific primer pairs. *Mycologia* 88:939–46
2. Albrecht C, Geurts R, Lapeyrie F, Bisseling T. 1998. Endomycorrhizae and rhizobial nod factors activate signal transduction pathways inducing *PsENOD5* and *PsENOD12* expression in which Sym8 is a common step. *Plant J.* 15:605–15
3. Alexander D, Goodman RM, Gut-Rella M, Glascock C, Weymann K, et al. 1993. Increased tolerance to two oomycete pathogens in transgenic tobacco expressing pathogenesis-related protein-1a. *Proc. Natl. Acad. Sci. USA* 90:7327–31
4. Alexander T, Meier R, Toth R, Weber HC. 1988. Dynamics of arbuscule development and degeneration in mycorrhizas of *Triticum aestivum* L. and *Avena sativa* L. with reference to *Zea mays* L. *New Phytol.* 110:363–70
5. Alexander T, Toth R, Meier R, Weber HC. 1989. Dynamics of arbuscule development and degeneration in onion, bean and tomato with reference to vesicular-arbuscular mycorrhizae in grasses. *Can. J. Bot.* 67:2505–13
6. Allen MF. 1996. The ecology of arbuscular mycorrhizas: a look back into the 20th century and a peek into the 21st. *Mycol. Res.* 100:769–82
7. Allen MF, Allen EB, Friese CF. 1989. Responses of the non-mycotrophic plant *Salsola kali* to invasion by vesicular-arbuscular mycorrhizal fungi. *New Phytol.* 111:45–49
8. Arines J, Palma JM, Vilarino A. 1993.

Comparison of protein patterns in non-mycorrhizal and vesicular-arbuscular mycorrhizal roots of red clover. *New Phytol.* 123:763–68
9. Ayling SM, Smith SE, Smith FA, Kolesik P. 1997. Transport processes at the plant-fungus interface in mycorrhizal associations: physiological studies. *Plant Soil* 196:305–10
9a. Azcon-Aguilar C, Barea JM, eds. 1996. *Mycorrhizas in Integrated Systems from Genes to Plant Development.* Brussels: Eur. Comm.
10. Bago B, Azcón-Aguilar C, Goulet A, Piché Y. 1998. Branched absorbing structures (BAS): a feature of the extraradical mycelium of symbiotic arbuscular mycorrhizal fungi. *New Phytol.* 139:375–88
11. Bago B, Azcón-Aguilar C, Piché Y. 1998. Architecture and developmental dynamics of the external mycelium of the arbuscular mycorrhizal fungus *Glomus intraradices* grown under monoxenic conditions. *Mycologia* 90:52–62
12. Bago B, Chamberland H, Goulet A, Vierheilig H, Lafontaine J-G, Piché Y. 1996. Effect of Nikkomycin Z, a chitin-synthase inhibitor, on hyphal growth and cell wall structure of two arbuscular-mycorrhizal fungi. *Protoplasma* 192:80–92
13. Bago B, Vierheilig H, Piché Y, Azcón-Aguilar C. 1996. Nitrate depletion and pH changes induced by the extraradical mycelium of the arbuscular mycorrhizal fungus *Glomus intraradices* grown in monoxenic culture. *New Phytol.* 133:273–80
14. Balaji B, Ba AM, LaRue TA, Tepfer D,

Piché Y. 1994. *Pisum sativum* mutants insensitive to nodulation are also insensitive to invasion in vitro by the mycorrhizal fungus, *Gigaspora margarita*. *Plant Sci.* 102:195–203

15. Balaji B, Poulin MJ, Vierheilig H, Piché Y. 1995. Responses of an arbuscular mycorrhizal fungus, *Gigaspora margarita*, to exudates and volatiles from the Ri T-DNA-transformed roots of nonmycorrhizal and mycorrhizal mutants of *Pisum sativum* L. sparkle. *Exp. Mycol.* 19:275–83

16. Deleted in proof

17. Balestrini R, Berta G, Bonfante P. 1992. The plant nucleus in mycorrhizal roots: positional and structural modifications. *Biol. Cell* 75:235–43

18. Balestrini R, Hahn MG, Faccio A, Mendgen K, Bonfante P. 1996. Differential localization of carbohydrate epitopes in plant cell walls in the presence and absence of arbuscular mycorrhizal fungi. *Plant Physiol.* 111:203–13

19. Balestrini R, José-Estanyol M, Puigdoménech P, Bonfante P. 1997. Hydroxyproline-rich glycoprotein mRNA accumulation in maize root cells colonized by an arbuscular mycorrhizal fungus as revealed by *in situ* hybridization. *Protoplasma* 198:36–42

20. Balestrini R, Romera C, Puigdoménech P, Bonfante P. 1994. Location of a cell-wall hydroxyproline-rich glycoprotein, cellulose and β-1,3-glucans in apical and differentiated regions of maize mycorrhizal roots. *Planta* 195:201–9

21. Barker SJ, Stummer B, Gao L, Dispain I, O'Connor PJ, Smith SE. 1998. A mutant in *Lycopersicon esculentum* Mill with highly reduced VA mycorrhizal colonisation: isolation and preliminary characterisation. *Plant J.* 15:791–99

22. Bécard G, Douds DD, Pfeffer PE. 1992. Extensive *in vitro* hyphal growth of vesicular-arbuscular mycorrhizal fungi in the presence of CO_2 and flavonols. *Appl. Environ. Microbiol.* 58:821–25

23. Bécard G, Fortin JA. 1988. Early events of vesicular-arbuscular mycorrhiza formation on Ri T-DNA transformed roots. *New Phytol.* 108:211–18

24. Bécard G, Pfeffer PE. 1993. Status of nuclear division in arbuscular mycorrhizal fungi during in vitro development. *Protoplasma* 174:62–68

25. Bécard G, Piché Y. 1989. Fungal growth stimulation by CO_2 and root exudates in vesicular-arbuscular mycorrhizal symbiosis. *Appl. Environ. Microbiol.* 55:2320–25

26. Bécard G, Taylor LP, Douds DD Jr, Pfeffer PE, Doner LW. 1995. Flavonoids are not necessary plant signals in arbuscular mycorrhizal symbiosis. *Mol. Plant-Microbe Interact.* 8:252–58

27. Bentivenga SP, Morton JB. 1994. Systematics of Glomalean endomycorrhizal fungi: current views and future direction. In *Mycorrhizae and Plant Health*, ed. FL Pfleger, RG Linderman, pp. 283–308. St. Paul, MN: APS Press

28. Bevege DI, Bowen GD, Skinner MF. 1975. Comparative carbohydrate physiology of ecto and endomycorrhizas. In *Endomycorrhizas*, ed. FE Saunders, B Mossae, PB Tinker, pp. 149–74. London: Academic

29. Bianciotto V, Barbiero G, Bonfante P. 1995. Analysis of the cell cycle in arbuscular mycorrhizal fungus by flow cytometry and bromodeoxyuridine labelling. *Protoplasma* 188:161–69

30. Bianciotto V, Bonfante P. 1992. Quantification of the nuclear DNA content of two arbuscular mycorrhizal fungi. *Mycol. Res.* 96:1071–76

31. Blee KA, Anderson AJ. 1996. Defense-related transcript accumulation in *Phaseolus vulgaris* L. colonized by the arbuscular mycorrhizal fungus *Glomus intraradices*. *Plant Physiol.* 110:675–88

32. Bonfante P, Bergero R, Uribe X, Romera C, Rigau J, Puigdoménech P. 1996. Transcriptional activation of a maize α-tubulin gene in mycorrhizal maize and transgenic tobacco plants. *Plant J.* 9:737–43

33. Bonfante P, Perotto S. 1995. Strategies of arbuscular mycorrhizal fungi when infecting host plants. *New Phytol.* 130:3–21

34. Bonfante-Fasolo P. 1984. Anatomy and morphology of VA mycorrhizae. In *VA Mycorrhizae*, ed. CL Powell, DJ Bagyaraj, pp. 5–33. Boca Raton, FL: CRC Press

35. Bonfante-Fasolo P, Perotto S. 1990. Mycorrhizal and pathogenic fungi: Do they share any features? In *Electron Microscopy Applied in Plant Pathology*, ed. K Mendgen, DE Lesemann, pp. 265–75. Berlin: Springer-Verlag

36. Bonfante-Fasolo P, Perotto S. 1992. Plant and endomycorrhizal fungi: the cellular and molecular basis of their interaction. In *Molecular Signals in Plant-Microbe Communications*, ed. DPS Verma, pp. 445–70. Boca Raton, FL: CRC Press

37. Bradbury SM, Peterson RL, Bowley SR. 1991. Interaction between three alfalfa

nodulation genotypes and two *Glomus* species. *New Phytol.* 119:115–20

38. Bradbury SM, Peterson RL, Bowley SR. 1993. Further evidence for a correlation between nodulation genotypes in alfalfa (*Medicago sativa* L.) and mycorrhiza formation. *New Phytol.* 124:665–73

39. Broglie K, Chet I, Holliday M, Cressman R, Biddle P, et al. 1991. Transgenic plants with enhanced resistance to the fungal pathogen, *Rhizoctonia solani.* *Science* 254:1194–97

40. Burggraaf AJP, Beringer JE. 1988. Absence of nuclear DNA synthesis in vesicular-arbuscular mycorrhizal fungi during *in vitro* development. *New Phytol.* 111:25–37

41. Burleigh SH, Harrison MJ. 1997. A novel gene whose expression in *Medicago truncatula* roots is suppressed in response to colonization by vesicular-arbuscular mycorrhizal (VAM) fungi and to phosphate nutrition. *Plant Mol. Biol.* 34:199–208

42. Burleigh SH, Harrison MJ. 1998. A cDNA from the arbuscular mycorrhizal fungus *Glomus versiforme* with homology to a cruciform DNA-binding protein from *Ustilago maydis.* *Mycorrhiza* 7:301–6

43. Büschges R, Hollricher K, Panstruga R, Simons G, Wolter M, et al. 1997. The barley *Mlo* gene: a novel control element of plant pathogen resistance. *Cell* 88:695–705

44. Deleted in proof

45. Cairney JGW, Smith SE. 1993. Efflux of phosphate from the ectomycorrhizal basidiomycete *Pisolithus tinctorius*: general characteristics and the influence of intracellular phosphorus concentration. *Mycol. Res.* 97:1261–66

46. Carling DE, Brown MF. 1982. Anatomy and physiology of vesicular-arbuscular and nonmycorrhizal roots. *Phytopathology* 72:1108–14

47. Chabot S, Bel-Rhlid R, Chenevert R, Piché Y. 1992. Hyphal growth promotion *in vitro* of the VA mycorrhizal fungus, *Gigaspora margarita* Becker & Hall, by the activity of structurally specific flavonoid compounds under CO_2-enriched conditions. *New Phytol.* 122:461–67

48. Chen X-Y, Hampp R. 1993. Sugar uptake by protoplasts of the ectomycorrhizal fungus *Amanita muscaria.* *New Phytol.* 125:601–8

49. Clapp JP, Young JPW, Merryweather JW, Fitter AH. 1995. Diversity of fungal symbionts in arbuscular mycorrhizas from a natural community. *New Phytol.* 130:259–65

50. Cox G, Tinker PB. 1976. Translocation and transfer of nutrients in vesicular-arbuscular mycorrhizas. I. The arbuscule and phosphorus transfer: a quantitative ultrastructural study. *New Phytol.* 77:371–78

51. Dassi B, Dumas-Gaudot E, Asselin A, Richard C, Gianinazzi S. 1996. Chitinase and β-1,3-glucanase isoforms expressed in pea roots inoculated with arbuscular mycorrhizal or pathogenic fungi. *Eur. J. Plant Pathol.* 102:105–8

52. David R, Itzhaki H, Ginzberg I, Gafni Y, Galili G, Kapulnik Y. 1998. Suppression of tobacco basic chitinase gene expression in response to colonization by the arbuscular mycorrhizal fungus *Glomus intraradices.* *Mol. Plant-Microbe Interact.* 11:489–97

53. Dean JV, Devarenne TP, Lee I-S, Orlofsky LE. 1995. Properties of a maize glutathione S-transferase that conjugates coumaric acid and other phenylpropanoids. *Plant Physiol.* 108:985–94

54. Declerck S, Strullu DG, Plenchette C. 1996. *In vitro* mass-production of the arbuscular mycorrhizal fungus, *Glomus versiforme*, associated with Ri T-DNA transformed carrot roots. *Mycol. Res.* 100:1237–42

55. Di Bonito R, Elliott ML, Des Jardin EA. 1995. Detection of an arbscular mycorrhizal fungus in roots of different plant species with the PCR. *Appl. Environ. Microbiol.* 61:2809–10

56. Diop TA, Bécard G, Piché Y. 1992. Long-term *in vitro* culture of an endomycorrhizal fungus, *Gigaspora margarita*, on Ri T-DNA transformed roots of carrot. *Symbiosis* 12:249–59

57. Diop TA, Plenchette C, Strullu DG. 1994. Dual axenic culture of sheared-root inocula of vesicular-arbuscular mycorrhizal fungi associated with tomato roots. *Mycorrhiza* 5:17–22

58. Dixon RA, Harrison MJ, Lamb CJ. 1994. Early events in the activation of plant defense responses. *Annu. Rev. Phytopathol.* 32:479–501

59. Dodd JC, Rosendahl S, Giovannetti M, Broome A, Lanfranco L, Walker C. 1996. Inter- and intraspecific variation within the morphologically-similar arbuscular mycorrhizal fungi *Glomus mosseae* and *Glomus coronatum.* *New Phytol.* 133:113–22

60. Douds DD Jr. 1997. A procedure for the establishment of *Glomus mosseae*

in dual culture with Ri T-DNA-transformed carrot roots. *Mycorrhiza* 7:57–61
61. Douds DD Jr, Galvez L, Becard G, Kapulnik Y. 1998. Regulation of arbuscular mycorrhizal development by plant host and fungus species in alfalfa. *New Phytol.* 138:27–35
62. Douds DD Jr, Nagahashi G, Abney GD. 1996. The differential effects of cell wall-associated phenolics, cell walls, and cytosolic phenolics of host and non-host roots on the growth of two species of AM fungi. *New Phytol.* 133:289–94
63. Duc G, Trouvelot A, Gianinazzi-Pearson V, Gianinazzi S. 1989. First report of non-mycorrhizal mutants (Myc-) obtained in pea (*Pisum sativum* L.) and Fababean (*Vicia faba* L.). *Plant Sci.* 60:215–22
64. Dumas-Gaudot E, Furlan V, Grenier J, Asselin A. 1992. New acidic chitinase isoforms induced in tobacco roots by vesicular-arbuscular mycorrhizal fungi. *Mycorrhiza* 1:133–36
65. Dumas-Gaudot E, Guillaume P, Tahiri-Alaoui A, Gianinazzi-Pearson V, Gianinazzi S. 1994. Changes in polypeptide patterns in tobacco roots colonized by two *Glomus* species. *Mycorrhiza* 4:215–21
66. Edwards SG, Fitter AH, Young JPW. 1997. Quantification of an arbuscular mycorrhizal fungus, *Glomus mosseae*, within plant roots by competitive polymerase chain reaction. *Mycol. Res.* 101:1440–44
67. Forbes PJ, Millam S, Hooker JE, Harrier LA. 1998. Transformation of the arbuscular mycorrhizal fungus *Gigaspora rosea* by particle bombardment. *Mycol. Res.* 102:497–501
68. Frank B. 1885. Ueber die auf Wurzelsymbiose beruhende Ernährung gewisser Bäume durch unterirdische Pilze. *Ber. Dtsch. Bot. Ges.* 3:128–45
69. Franken P, Gianinazzi-Pearson V. 1996. Construction of genomic phage libraries of the arbuscular mycorrhizal fungi *Glomus mosseae* and *Scutellospora castanea* and isolation of ribosomal RNA genes. *Mycorrhiza* 6:167–73
70. Franken P, Gnädinger F. 1994. Analysis of parsley arbuscular endomycorrhiza: infection development and mRNA levels of defense-related genes. *Mol. Plant-Microbe Interact.* 7:612–20
70a. Franken P, Lapopin L, Meyer-Gauen G, Gianinazzi-Pearson V. 1997. RNA accumulation and genes expressed in spores of the arbuscular mycorrhizal fungus,

Gigaspora rosea. Mycologia 89:293–97
71. Freialdenhoven A, Peterhänsel C, Kurth J, Kreuzaler F, Schulze-Lefert P. 1996. Identification of genes required for the function of non-race-specific *mlo* resistance to powdery mildew in barley. *Plant Cell* 8:5–14
72. Friese CF, Allen MF. 1991. The spread of VA mycorrhizal fungal hyphae in the soil: inoculum types and external hyphal architecture. *Mycologia* 83:409–18
73. Frühling M, Roussel H, Gianinazzi-Pearson V, Pühler A, Perlick AM. 1997. The *Vicia faba* leghemoglobin gene VfLb29 is induced in root nodules and in roots colonized by the arbuscular mycorrhizal fungus *Glomus fasciculatum*. *Mol. Plant-Microbe Interact.* 10:124–31
74. Gallaud I. 1905. Etudes sur les mycorrhizes endotrophes. *Rév. Gen. Bot.* 17:5–48
75. Galun M, Bubrick P. 1984. Physiological interactions between partners of the lichen symbiosis. In *Cellular Interactions. Encyclopedia of Plant Physiology*, ed. HF Linskins, J Heslop-Harrison, 17:362–401. Berlin: Springer-Verlag
76. García-Garrido JM, Cabello MN, García-Romera I, Ocampo JA. 1992. Endoglucanase activity in lettuce plants colonized with the vesicular-arbuscular mycorrhizal fungus *Glomus fasciculatum*. *Soil Biol. Biochem.* 24:955–59
77. Garcia-Garrido JM, Garcia-Romera I, Ocampo JA. 1992. Cellulase production by the vesicular-arbuscular mycorrhizal fungus *Glomus mosseae* (Nicol. & Gerd.) Gerd. and Trappe. *New Phytol.* 121:221–26
78. García-Garrido JM, García-Romera I, Parra-García MD, Ocampo JA. 1996. Purification of an arbuscular mycorrhizal endoglucanase from onion roots colonized by *Glomus mosseae. Soil Biol. Biochem.* 28:1443–49
79. Garcia-Garrido JM, Toro N, Ocampo JA. 1993. Presence of specific polypeptide in onion roots colonized by *Glomus mosseae. Mycorrhiza* 2:175–77
80. Garcia-Romera I, Garcia-Garrido JM, Martinez-Molina E, Ocampo JA. 1991. Production of pectolytic enzymes in lettuce root colonized by *Glomus mosseae. Soil Biol. Biochem.* 23:597–601
81. Genre A, Bonfante P. 1997. A mycorrhizal fungus changes microtubule orientation in tobacco root cells. *Protoplasma* 199:30–38
82. Gerdemann JW. 1965. Vesicular-arbuscular mycorrhizas formed on maize and

tulip tree by *Endogone fasciculata*. *Mycologia* 57:562–75

83. Gianinazzi-Pearson V. 1996. Plant cell responses to arbuscular myorrhiza fungi: getting to the roots of the symbiosis. *Plant Cell* 8:1871–83

84. Gianinazzi-Pearson V, Branzanti B, Gianinazzi S. 1989. *In vitro* enhancement of spore germination and early hyphal growth of a vesicular-arbuscular mycorrhizal fungus by host root exudates and plant flavonoids. *Symbiosis* 7:243–55

85. Gianinazzi-Pearson V, Dumas-Gaudot E, Gollotte A, Tahiri-Alaoui A, Gianinazzi S. 1996. Cellular and molecular defence-related root responses to invasion by arbuscular mycorrhizal fungi. *New Phytol.* 133:45–57

86. Gianinazzi-Pearson V, Gianinazzi S, Guillemin JP, Trouvelot A, Duc G. 1991. Genetic and cellular analysis of resistance to vesicular arbuscular (VA) mycorrhizal fungi in pea mutants. In *Advances in Molecular Genetics of Plant-Microbe Interactions*, ed. H Hennecke, DPS Verma, 1:336–42. Dordrecht, Netherlands: Kluwer

87. Gianinazzi-Pearson V, Smith SE, Gianinazzi S, Smith FA. 1991. Enzymatic studies on the metabolism of vesicular-arbuscular mycorrhizas. *New Phytol.* 117:61–74

88. Giovannetti M, Avio L, Sbrana C, Citernesi AS. 1993. Factors affecting appressorium development in the vesicular-arbuscular mycorrhizal fungus *Glomus mosseae* (Nicol. & Gerd.) Gerd. & Trappe. *New Phytol.* 123:115–22

89. Giovannetti M, Sbrana C, Avio L, Citernesi AS, Logi C. 1993. Differential hyphal morphogenesis in arbuscular mycorrhizal fungi during pre-infection stages. *New Phytol.* 125:587–93

90. Giovannetti M, Sbrana C, Citernesi AS, Avio L. 1996. Analysis of factors involved in fungal recognition responses to host-derived signals by arbuscular mycorrhizal fungi. *New Phytol.* 133:65–71

91. Giovannetti M, Sbrana C, Logi C. 1994. Early processes involved in host recognition by arbuscular mycorrhizal fungi. *New Phytol.* 127:703–9

92. Gollotte A, Gianinazzi-Pearson V, Gianinazzi S. 1995. Immunodetection of infection thread glycoprotein and arabinogalactan protein in wild type *Pisum sativum* (L.) or an isogenic mycorrhiza-resistant mutant interacting with *Glomus mosseae*. *Symbiosis* 18:69–85

93. Gollotte A, Gianinazzi-Pearson V, Giovannetti M, Sbrana C, Avio L, Gianinazzi S. 1993. Cellular localization and cytochemical probing of resistance reactions to arbuscular mycorrhizal fungi in a 'locus *a*' myc-mutant of *Pisum sativum* L. *Planta* 191:112–22

94. Hahn M, Neef U, Struck C, Göttfert M, Mendgen K. 1997. A putative amino acid transporter is specifically expressed in haustoria of the rust fungus *Uromyces fabae*. *Mol. Plant-Microbe Interact.* 10: 438–45

95. Harley JL, Smith SE. 1983. *Mycorrhizal Symbiosis*. London/New York: Academic

96. Harrison MJ. 1996. A sugar transporter from *Medicago truncatula*: altered expression pattern in roots during vesicular-arbuscular (VA) mycorrhizal associations. *Plant J.* 9:491–503

97. Harrison MJ, Dixon RA. 1993. Isoflavonoid accumulation and expression of defense gene transcripts during the establishment of vesicular-arbuscular mycorrhizal associations in roots of *Medicago truncatula*. *Mol. Plant-Microbe Interact.* 6:643–54

98. Harrison MJ, Dixon RA. 1994. Spatial patterns of expression of flavonoid/isoflavonoid pathway genes during interactions between roots of *Medicago truncatula* and the mycorrhizal fungus *Glomus versiforme*. *Plant J.* 6:9–20

99. Harrison MJ, van Buuren ML. 1995. A phosphate transporter from the mycorrhizal fungus *Glomus versiforme*. *Nature* 378:626–29

100. Hinde R. 1983. Host release factors in symbioses between algae and invertebrates. In *Endocytobiology II. Intracellular Space as Oligogenetic System*, ed. HEA Schenk, W Schwemmler, pp. 709–26. Berlin/New York: Walter de Gruyter

101. Ho I, Trappe JM. 1973. Translocation of ^{14}C from *Festuca* plants to their endomycorrhizal fungi. *Nature* 244:30–31

102. Hosny M, de Barros J-PP, Gianinazzi-Pearson V, Dulieu H. 1997. Base composition of DNA from glomalean fungi: high amounts of methylated cytosine. *Fungal Genet. Biol.* 22:103–11

103. Hosny M, Gianinazzi-Pearson V, Dulieu H. 1998. Nuclear DNA contents of eleven fungal species in Glomales. *Genome* 41:422–28

104. Jacobs M, Rubery PH. 1988. Naturally occurring auxin transport regulators. *Science* 241:346–49

105. Jacquelinet-Jeanmougin S, Gianinazzi-Pearson V. 1983. Endomycorrhizas in

the Gentianaceae. I. The fungus associated with *Gentiana lutea* L. *New Phytol.* 95:663–66

106. Jolicoeur M, Germette S, Gaudette M, Perrier M, Becard G. 1998. Intracellular pH in arbuscular mycorrhizal fungi—a symbiotic physiological marker. *Plant Physiol.* 116:1279–88

107. Kaldorf M, Schmelzer E, Bothe H. 1998. Expression of maize and fungal nitrate reductase genes in arbuscular mycorrhiza. *Mol. Plant-Microbe Interact.* 11:439–48

108. Kape R, Wex K, Parniske M, Görge E, Wetzel A, Werner D. 1992. Legume root metabolites and VA-mycorrhiza development. *J. Plant Physiol.* 141:54–60

109. Kapulnik Y, Volpin H, Itzhaki H, Ganon D, Galili S, et al. 1996. Suppression of defence responses in mycorrhizal alfalfa and tobacco roots. *New Phytol.* 133:59–64

110. Koide RT, Schreiner RP. 1992. Regulation of the vesicular-arbuscular mycorrhizal symbiosis. *Annu. Rev. Plant Physiol. Plant Mol. Biol.* 43:557–81

111. Lagunas R. 1993. Sugar transport in *Saccharomyces cerevisiae*. *FEMS Microbiol. Rev.* 104:229–42

112. Lambais MR, Mehdy MC. 1993. Suppression of endochitinase, β-1,3–endoglucanase, and chalcone isomerase expression in bean vesicular-arbuscular mycorrhizal roots under different soil phosphate conditions. *Mol. Plant-Microbe Interact.* 6:75–83

113. Lambais MR, Mehdy MC. 1996. Soybean roots infected by *Glomus intraradices* strains differing in infectivity exhibit differential chitinase and β-1,3-glucanase expression. *New Phytol.* 134:531–38

114. Lanfranco L, Wyss P, Marzachi C, Bonfante P. 1995. Generation of RAPD-PCR primers for the identification of isolates of *Glomus mosseae*, an arbuscular mycorrhizal fungus. *Mol. Ecol.* 4:61–68

115. Leggewie G, Willmitzer L, Riesmeier JW. 1997. Two cDNAs from potato are able to complement a phosphate uptake-deficient yeast mutant: identification of phosphate transporters from higher plants. *Plant Cell* 9:381–92

116. Lei J, Bécard G, Catford JG, Piché Y. 1991. Root factors stimulate ^{32}P uptake and plasmalemma ATPase activity in vesicular-arbuscular mycorrhizal fungus, *Gigaspora margarita*. *New Phytol.* 118:289–94

117. Liu C, Muchhal US, Uthappa M, Kononowicz AK, Raghothama KG. 1998.

Tomato phosphate transporter genes are differentially regulated in plant tissues by phosphorus. *Plant Physiol.* 116:91–99

118. Liu H, Trieu AT, Blaylock LA, Harrison MJ. 1998. Cloning and characterization of two phosphate transporters from *Medicago truncatula* roots: regulation in response to phosphate and to colonization by arbuscular mycorrhizal (AM) fungi. *Mol. Plant-Microbe Interact.* 11:14–22

119. Lloyd-MacGilp SA, Chambers SM, Dodd JC, Fitter AH, Walker C, Yound JPW. 1996. Diversity of the ribosomal internal transcribed spacers within and among isolates of *Glomus mosseae* and related mycorrhizal fungi. *New Phytol.* 133:103–11

120. Longato S, Bonfante P. 1997. Molecular identification of mycorrhizal fungi by direct amplification of microsatellite regions. *Mycol. Res.* 101:425–32

121. Longman D, Callow JA. 1987. Specific saccharide residues are involved in the recognition of plant root surfaces by zoospores of *Pythium aphanidermatum*. *Physiol. Mol. Plant Pathol.* 30:139–50

122. Manners JM, Gay JL. 1983. The host-parasite interface and nutrient transfer in biotrophic parasitism. In *Biochemical Plant Pathology*, ed. JA Callow, pp. 163–68. Chichester, UK: Wiley

123. Marre E. 1979. Fusicoccin: a tool in plant physiology. *Annu. Rev. Plant Physiol.* 30:273–88

124. Marrs KA, Alfenito MR, Lloyd AM, Walbot V. 1995. A glutathione S-transferase involved in vacuolar transfer encoded by the maize gene *Bronze-2*. *Nature* 375:397–400

125. Martin-Laurent F, van Tuinen D, Dumas-Gaudot E, Gianinazzi-Pearson V, Gianinazzi S, Franken P. 1997. Differential display analysis of RNA accumulation in arbuscular mycorrhizy of pea and isolation of a novel symbiosis-regulated plant gene. *Mol. Gen. Genet.* 256:37–44

126. Martin-Laurent FA, Franken P, van Tuinen D, Dumas-Gaudot E, Schlichter U, et al. 1996. Differential display reverse transcription polymerase chain reaction: a new approach to detect symbiosis-related genes involved in arbuscular mycorrhiza. See Ref. 9a, pp. 195–98

127. Mendgen K, Deising H. 1993. Tansley Review No. 48. Infection structures of fungal plant pathogens—a cytological and physiological evaluation. *New Phytol.* 124:193–213

128. Mendgen K, Hahn M, Deising H. 1996. Morphogenesis and mechanisms of penetration by plant pathogenic fungi. *Annu. Rev. Phytopathol.* 34:367–86

129. Morton JB, Benny GL. 1990. Revised classification of arbusular mycorrhizal fungi (zygomycetes): a new order, glomales, two new suborders, glomineae and gigasporineae, and two new families, acaulosporaceae and gigasporaceae, with an amendation of glomaceae. *Mycotaxon* 37:471–91

130. Mosse B, Hepper CM. 1975. Vesicular-arbuscular mycorrhizal infections in root organ cultures. *Physiol. Plant Pathol.* 5:215–23

131. Muchhal US, Pardo JM, Raghathama KG. 1996. Phosphate transporters from the higher plant *Arabidopsis thaliana*. *Proc. Natl. Acad. Sci. USA* 93:101519–23

132. Mugnier J, Mosse B. 1987. Vesicular-arbuscular mycorrhizal infection in transformed root-inducing T-DNA roots grown axenically. *Phytopathology* 77:1045–50

133. Murphy PJ, Langridge P, Smith SE. 1997. Cloning plant genes differentially expressed during colonization of roots of *Hordeum vulgare* by the vesicular-arbuscular mycorrhizal fungus *Glomus intraradices*. *New Phytol.* 135:291–301

134. Nagahashi G, Douds DD Jr. 1997. Appressorium formation by AM fungi on isolated cell walls of carrot roots. *New Phytol.* 136:299–304

135. Nair MG, Safir GR, Siqueira JO. 1991. Isolation and identification of vesicular-arbuscular mycorrhiza stimulatory compounds from clover (*Trifolium repens*) roots. *Appl. Environ. Microbiol.* 57:434–39

136. Nehls U, Wiese J, Guttenberger M, Hampp R. 1998. Carbon allocation in ectomycorrhizas: identification and expression analysis of an *Amanita muscaria* monosaccharide transporter. *Mol. Plant-Microbe Interact.* 11:167–76

137. Olsson PA, Baath E, Jakobsen I. 1997. Phosphorus effects on the mycelium and storage structures of an arbuscular mycorrhizal fungus as studied in the soil and roots by analysis of fatty acid signatures. *Appl. Environ. Microbiol.* 63:3531–38

138. Patrick JW. 1989. Solute efflux from the host at plant-microorganism interfaces. *Aust. J. Plant Physiol.* 16:53–67

138a. Pearson JN, Jakobsen I. 1993. The relative contribution of hyphae and roots to phosphorus uptake by arbuscular mycorrhizal plants, measured by dual labeling with ^{32}P and [33]P. *New Phytol.* 124:489–94

139. Peretto R, Bettini V, Favaron F, Alghisi P, Bonfante P. 1995. Polygalacturonase activity and location in arbuscular mycorrhizal roots of *Allium porrum* L. *Mycorrhiza* 3:157–63

140. Perfect SE, O'Connell RJ, Green EF, Doering-Saad C, Green JR. 1998. Expression cloning of a fungal proline-rich glycoprotein specific to the biotrophic interface formed in the *Colletotrichum*-bean interaction. *Plant J.* 15:273–79

140a. Perotto S, Brewin N, Bonfante P. 1994. Colonization of pea roots by the mycorrhizal fungus *Glomus versiforme* and by *Rhizobium* bacteria: immunological comparison using monoclonal antibodies as probes for plant cell surface components. *Mol. Plant-Microbe Interact.* 7:91–98

141. Perotto S, Vandenbosch KA, Brewin NJ, Faccio A, Knox JP, Bonfante-Fasolo P. 1990. Modifications of the host cell wall during root colonization by Rhizobium and VAM fungi. In *Endocytobiology IV*, ed. P Nardon, V Gianinazzi-Pearson, AM Grenier, M Margulis, DC Smith, pp. 114–17. Paris: INRA Press

142. Peterson RL, Bradbury SM. 1995. Use of plant mutants, intraspecific variants, and non-hosts in studying mycorrhiza formation and function. In *Mycorrhiza: Structure, Function, Molecular Biology and Biotechnology*, ed. A Varma, B Hock, pp. 157–80. Berlin: Springer-Verlag

143. Pirozynski KA, Malloch DW. 1975. The origin of land plants: a matter of mycotrophism. *Biosystems* 6:153–64

144. Poulin M-J, Bel-Rhlid R, Piché Y, Chenevert R. 1993. Flavonoids released by carrot (*Daucus carota*) seedlings stimulate hyphal development of vesicular-arbuscular mycorrhizal fungi in the presence of optimal CO_2 enrichment. *J. Chem. Ecol.* 19:2317–27

145. Poulin M-J, Simard J, Catford J-G, Labrie F, Piché Y. 1997. Response of symbiotic endomycorrhizal fungi to estrogens and antiestrogens. *Mol. Plant-Microbe Interact.* 10:481–87

146. Pozo MJ, Dumas-Gaudot E, Slezack S, Cordier C, Asselin A, et al. 1996. Induction of new chitinase isoforms in tomato roots during interactions with *Glomus mosseae* and/or *Phytopthora nicotianae* va. parasitica. *Agronomie* 16:689–97

147. Rasmussen U, Johansson C, Renglin A, Peterson C, Bergman B. 1996. A molec-

ular characterization of the *Gunnera-Nostoc* symbiosis: comparison with *Rhizobium*- and *Agrobacterium*-plant interactions. *New Phytol.* 133:391–98

148. Deleted in proof

149. Redecker D, Thierfelder H, Walker C, Werner D. 1997. Restriction analysis of PCR-amplified internal transcribed spacers of ribosomal DNA as a tool for species identification in different genera of the order Glomales. *Appl. Environ. Microbiol.* 63:1756–61

150. Rejon-Palomares A, Garcia-Garrido JM, Ocampo JA, Garcia-Romera I. 1996. Presence of xyloglucan-hydrolyzing glucanases (xyloglucanases) in arbuscular mycorrhizal symbiosis. *Symbiosis* 21:249–61

151. Remy W, Taylor TN, Hass H, Kerp H. 1994. Four hundred-million-year-old vesicular arbuscular mycorrhizae. *Proc. Natl. Acad. Sci. USA* 91:11841–43

152. Rhlid RB, Chabot S, Piché Y, Chenevert R. 1993. Isolation and identification of flavonoids from Ri T-DNA-transformed roots (*Daucus carota*) and their significance in vesicular-arbuscular mycorrhiza. *Phytochemistry* 33:1369–71

153. Rosendahl S, Taylor JW. 1997. Development of multiple genetic markers for studies of genetic variation in arbuscular mycorrhizal fungi using AFLP. *Mol. Ecol.* 6:821–29

154. Roussel H, Bruns S, Gianinazzi-Pearson V, Hahlbrock K, Franken P. 1997. Induction of a membrane intrinsic protein-encoding mRNA in arbuscular mycorrhiza and elicitor-stimulated cell suspension cultures of parsley. *Plant Sci.* 126:203–10

155. Sagan M, Morandi D, Tarenghi E, Duc G. 1995. Selection of nodulation and mycorrhizal mutants in the model plant *Medicago truncatula* (Gaertn.) after γ-ray mutagenesis. *Plant Sci.* 111:63–71

156. Salzer P, Hager A. 1991. Sucrose utilisation of the ectomycorrhizal fungi *Amanita muscaria* and *Hebeloma crustuliniforme* depends on the cell wall-bound invertase activity of their host *Picea abies. Bot. Acta* 104:439–45

157. Salzer P, Hübner B, Sirrenberg A, Hager A. 1997. Differential effect of purified spruce chitinases and β-1,3-glucanases on the activity of elicitors from ectomycorrhizal fungi. *Plant Physiol.* 114:957–68

158. Samra A, Dumas-Gaudot E, Gianinazzi-Pearson V, Gianinazzi S. 1997. Changes in polypeptide profiles of two pea genotypes inoculated with the arbuscular my-

corrhizal fungus *Glomus mosseae.* See Ref. 9a, pp. 263–66

159. Sanders IR, Alt M, Groppe K, Boller T, Wiemken A. 1995. Identification of ribosomal DNA polymorphisms in spores of the Glomales: application to studies on the genetic diversity of arbuscular mycorrhizal fungal communities. *New Phytol.* 130:419–27

160. Sanders IR, Clapp JP, Wiemken A. 1996. The genetic diversity of arbuscular mycorrhizal fungi in natural ecosystems—a key to understanding the ecology and functioning of the mycorrhizal symbiosis. *New Phytol.* 133:123–34

161. Sauer N, Baier K, Gahrtz M, Stadler R, Stolz J, Truernit E. 1994. Sugar transport across the plasma membranes of higher plants. *Plant Mol. Biol.* 26:1671–79

162. Schellenbaum L, Gianniazzi S, Gianniazzi-Pearson V. 1992. Comparison of acid soluble protein synthesis in roots of endomycorrhizal wild-type *Pisum sativum* and corresponding isogenic mutants. *J. Plant Physiol.* 141:2–6

163. Scheres B, van Engelen F, van der Knaap E, van de Wiel C, van Kammen A, Bisseling T. 1990. Sequential induction of nodulin gene expression in the developing pea nodule. *Plant Cell* 2:687–700

164. Schreiner RP, Koide RT. 1992. Antifungal compounds from the roots of mycotrophic and non-mycotrophic plant species. *New Phytol.* 123:99–105

165. Schreiner RP, Koide RT. 1993. Mustards, mustard oils and mycorrhizas. *New Phytol.* 123:107–13

166. Schwab SM, Menge JA, Tinker PB. 1991. Regulation of nutrient transfer between host and fungus in vesicular-arbuscular mycorrhizas. *New Phytol.* 117:387–98

167. Shachar-Hill Y, Pfeffer PE, Douds D, Osman SF, Doner LW, Ratcliffe RG. 1995. Partitioning of intermediary carbon metabolism in vesicular-arbuscular mycorrhizal leek. *Plant Physiol.* 108:2979–95

168. Shirtliffe SJ, Vessey JK. 1996. A nodulation (Nod⁺/Fix⁻) mutant of *Phaseolus vulgaris* L. has nodule-like structures lacking peripheral vascular bundles (Pvb⁻) and is resistant to mycorrhizal infection (Myc⁻). *Plant Sci.* 118:209–20

169. Simon L. 1996. Phylogeny of the Glomales: deciphering the past to understand the present. *New Phytol.* 133:95–101

170. Simon L, Bousquet J, Lévesque RC, Lalonde M. 1993. Origin and diversification of endomycorrhizal fungi and co-

incidence with vascular land plants. *Nature* 363:67–69

171. Simon L, Lalonde M, Bruns TD. 1992. Specific amplification of 18S fungal ribosomal genes from vesicular-arbuscular endomycorrhizal fungi colonizing roots. *Appl. Environ. Microbiol.* 58:291–95

172. Siqueira JO, Safir GR, Nair MG. 1991. Stimulation of vesicular-arbuscular mycorrhiza formation and growth of white clover by flavonoid compounds. *New Phytol.* 118:87–93

173. Smith FA, Smith SE. 1989. Membrane transport at the biotrophic interface: an overview. *Aust. J. Plant Physiol.* 16:33–43

174. Smith FA, Smith SE. 1997. Structural diversity in (vesicular)-arbuscular mycorrhizal symbioses. *New Phytol.* 137:373–88

175. Smith FW, Ealing PM, Dong B, Delhaize E. 1997. The cloning of two *Arabidopsis* genes belonging to a phosphate transporter family. *Plant J.* 11:83–92

176. Smith SE. 1993. Transport at the mycorrhizal interface. *Mycorrhiza News* 5:1–3

177. Smith SE, Dickson S, Morris C, Smith FA. 1994. Transfer of phosphate from fungus to plant in VA mycorrhizas: calculation of the area of symbiotic interface and of fluxes of P from two different fungi to *Allium porrum* L. *New Phytol.* 127:93–99

178. Smith SE, Gianinazzi-Pearson V. 1988. Physiological interactions between symbionts in vesicular-arbuscular mycorrhizal plants. *Annu. Rev. Plant Physiol. Plant Mol. Biol.* 39:221–44

179. Smith SE, Gianinazzi-Pearson V, Koide R, Cairney JWG. 1994. Nutrient transport in mycorrhizas: structure, physiology and consequences for efficiency of the symbiosis. *Plant Soil* 159:103–13

180. Smith SE, Read DJ, eds. 1997. *Mycorrhizal Symbiosis.* San Diego, CA: Academic

181. Smith SE, Smith FA. 1990. Structure and function of the interfaces in biotrophic symbioses as they relate to nutrient transport. *New Phytol.* 114:1–38

182. Solaiman MDZ, Saito M. 1997. Use of sugars by intraradical hyphae of arbuscular mycorrhizal fungi revealed by radiorespirometry. *New Phytol.* 136:533–38

183. Spanu P, Boller T, Ludwig A, Wiemken A, Faccio A, Bonfante-Fasolo P. 1989. Chitinase in roots of mycorrhizal *Allium porrum*: regulation and localization. *Planta* 177:477–55

184. Spanu P, Bonfante-Fasolo P. 1988. Cell-wall-bound peroxidase activity in roots of mycorrhizal *Allium porrum*. *New Phytol.* 109:119–24

185. St-Arnaud M, Hamel C, Vimard B, Caron M, Fortin JA. 1996. Enhanced hyphal growth and spore production of the arbuscular mycorrhizal fungus *Glomus intraradices* in an *in vitro* system in the absence of host roots. *Mycol. Res.* 100:328–32

186. Strittmatter G, Gheysen G, Gianinazzi-Pearson V, Hahn K, Niebel A, et al. 1996. Infections with various types of organisms stimulate transcription from a short promoter fragment of the potato *gst1* gene. *Mol. Plant-Microbe Interact.* 9:68–73

187. Tahiri-Aloui A, Antoniw JF. 1996. Cloning of genes associated with the colonization of tomato roots by the arbuscular mycorrhizal fungus *Glomus mosseae*. *Agronomie* 16:699–707

188. Tester M, Smith SE, Smith FA. 1987. The phenomenon of "nonmycorrhizal" plants. *Can. J. Bot.* 65:419–31

189. Thomson BD, Clarkson DT, Brain P. 1990. Kinetics of phosphorus uptake by the germ-tubes of the vesicular-arbuscular mycorrhizal fungus, *Gigaspora margarita*. *New Phytol.* 116:647–53

190. Trappe JM. 1987. In *Ecophysiology of VA Mycorrhizal Plants*, ed. GR Safir, pp. 5–25. Boca Raton, FL: CRC Press

191. Tsai SM, Phillips DA. 1991. Flavonoids released naturally from alfalfa promote development of symbiotic *Glomus* spores *in vitro*. *Appl. Environ. Microbiol.* 57:1485–88

192. van Buuren ML, Lanfranco L, Longato S, Minerdi D, Harrison MJ, Bonfante P. 1998. Construction and characterisation of genomic libraries of two endomycorrhizal fungi: *Glomus versiforme* and *Gigaspora margarita*. *Fungal Genet. Biol.* In press

193. van Buuren ML, Maldonado-Mendoza IE, Trieu AT, Blaylock LA, Harrison MJ. 1998. Novel genes induced during an arbuscular mycorrhizal (AM) symbiosis between *M. truncatula* and *G. versiforme*. *Mol. Plant Microbe Interact.* In press

194. van Rhijn P, Fang Y, Galili S, Shaul O, Atzmon N, et al. 1997. Expression of early nodulin genes in alfalfa mycorrhizae indicates that signal transduction pathways used in forming arbuscular mycorrhizae and *Rhizobium*-induced nodules may be conserved. *Proc. Natl. Acad. Sci. USA* 94:5467–72

195. van Tuinen D, Jaquot E, Zhao B, Gollotte A, Gianinazzi-Pearson V. 1998. Characterization of root colonization profiles by a microcosm community of arbuscular mycorrhizal fungi using 25S rDNA-targeted nested PCR. *Mol. Ecol.* 7:879–87

196. Vierheilig H, Alt M, Lange J, Gut-Rella M, Wiemken A, Boller T. 1995. Colonization of transgenic tobacco constitutively expressing pathogenesis-related proteins by the vesicular-arbuscular mycorrhizal fungus *Glomus mosseae*. *Appl. Environ. Microbiol.* 61:3031–34

197. Vierheilig H, Alt M, Mohr U, Boller T, Wiemken A. 1994. Ethylene biosynthesis and activities of chitinase and β-1,3-glucanase in the roots of host and non-host plants of vesicular-arbuscular mycorrhizal fungi after inoculation with *Glomus mosseae*. *J. Plant Physiol.* 143:337–43

198. Vierheilig H, Alt M, Neuhaus JM, Boller T, Wiemken A. 1993. Colonization of transgenic *Nicotiana silvestris* plants, expressing different forms of *Nicotiana tabacum* chitinase, by the root pathogen *Rhizoctonia solani* and by the mycorrhizal symbiont *Glomus mosseae*. *Mol. Plant-Microbe Interact.* 6:261–64

199. Vierheilig H, Iseli B, Alt M, Raikhel N, Wiemken A, Boller T. 1996. Resistance of *Urtica dioica* to mycorrhizal colonization: a possible involvement of *Urtica dioica* agglutinin. *Plant Soil* 183:131–36

200. Vierheilig H, Piché Y. 1996. Grafts between peas forming the arbuscular mycorrhizal symbiosis (Myc$^+$) and pea mutants resistant to AM fungi (Myc$^-$) show the same colonization characteristics as ungrafted plants. *J. Plant Physiol.* 147:762–64

201. Volpin H, Elkind Y, Okon Y, Kapulnik Y. 1994. A vesicular arbuscular mycorrhizal fungus *Glomus intraradix* induces a defence response in alfalfa roots. *Plant Physiol.* 104:683–89

202. Volpin H, Phillips DA, Okon Y, Kapulnik Y. 1995. Suppression of an isofla-vonoid phytoalexin defense response in mycorrhizal alfalfa roots. *Plant Physiol.* 108:1449–54

203. Wegel E, Schauser L, Sandal N, Stougaard J, Parniske M. 1998. Mycorrhiza mutants of *Lotus japonicus* define genetically independent steps during symbiotic infection. *Mol. Plant-Microbe Interact.* 11:933-36

204. Wright SF, Franke-Snyder M, Morton JB, Upadhyaya A. 1996. Time-course study and partial characterization of a protein on hyphae of arbuscular mycorrhizal fungi during active colonization of roots. *Plant Soil* 181:193–203

205. Wright SF, Upadhyaya A. 1996. Extraction of an abundant and unusual protein from soil and comparison with hyphal protein of arbuscular mycorrhizal fungi. *Soil Sci.* 161:575–86

206. Wyss P, Bonfante P. 1993. Amplification of genomic DNA of arbuscular-mycorrhizal (AM) fungi by PCR using short arbitrary primers. *Mycol. Res.* 97:1351–57

207. Xie Z-P, Staehelin C, Vierheilig H, Wiemken A, Jabbouri S, et al. 1995. Rhizobial nodulation factors stimulate mycorrhizal colonization of nodulating and nonnodulating soybeans. *Plant Physiol.* 108:1519–25

208. Zézé A, Dulieu H, Gianinazzi-Pearson V. 1994. DNA cloning and screening of a partial genomic library from an arbuscular mycorrhizal fungus. *Mycorrhiza* 4:251–54

209. Zézé A, Hosny M, Gianinazzi-Pearson V, Dulieu H. 1996. Characterization of a highly repeated DNA sequence (SC1) from the arbuscular mycorrhizal fungus *Scutellospora castanea* and its detection in planta. *Appl. Environ. Microbiol.* 62:2443–48

210. Zézé A, Sulistyowati E, Ophel-Keller K, Barker S, Smith S. 1997. Intersporal genetic variation of *Gigaspora margarita*, a vesicular arbuscular mycorrhizal fungus, revealed by M13 minisatellite-primed PCR. *Appl. Environ. Microbiol.* 63:676–78

Annu. Rev. Plant Physiol. Plant Mol. Biol. 1999. 50:391–417

ENZYMES AND OTHER AGENTS THAT ENHANCE CELL WALL EXTENSIBILITY

Daniel J. Cosgrove
Department of Biology, 208 Mueller Laboratory, Pennsylvania State University, University Park, Pennsylvania 16802; e-mail: dcosgrove@PSU.EDU

KEY WORDS: endoglucanase, expansin, plant cell growth, xyloglucan endotransglycosylase, wall loosening

ABSTRACT

Polysaccharides and proteins are secreted to the inner surface of the growing cell wall, where they assemble into a network that is mechanically strong, yet remains extensible until the cells cease growth. This review focuses on the agents that directly or indirectly enhance the extensibility properties of growing walls. The properties of expansins, endoglucanases, and xyloglucan transglycosylases are reviewed and their postulated roles in modulating wall extensibility are evaluated. A summary model for wall extension is presented, in which expansin is a primary agent of wall extension, whereas endoglucanases, xyloglucan endotransglycosylase, and other enzymes that alter wall structure act secondarily to modulate expansin action.

CONTENTS

1040-2519/99/0601-0391$08.00

INTRODUCTION

From an assemblage of structurally diverse polysaccharides, proteins, phenolic compounds, and other materials, plant cells fashion a complex wall that serves many functions. These include structural support and cell shape; protection against pathogens, dehydration, and other environmental assaults; storage and release of signaling molecules; and storage of carbohydrates, metal ions, and other materials. In the formative phase of a cell's life, the wall plays a determinative role in establishing the size and shape of the cell. To grow, plant cells selectively extend their cell walls, resulting in larger cells and modified cell shapes. Each of the approximately two dozen cell types that comprise the plant body are recognizable, to a large extent, by their distinctive size, shape, and wall morphology—characteristics that emerge from the history of wall growth. Thus, plant cell growth and differentiation inevitably depend upon precise spatial and temporal patterns of wall growth, patterns that are uniquely regulated in each cell type.

The fundamental structure of the primary (growing) cell wall in all land plants appears to be very similar: cellulose microfibrils embedded in a hydrated matrix composed mostly of neutral and acidic polysaccharides and a small amount of structural proteins. Recent reviews describe the structure and function of wall polysaccharides (13) and proteins (14). In this review I discuss recent results dealing with the molecular control of wall extensibility. Other recent reviews deal with polysaccharide-modifying enzymes with other points of focus (42, 85, 86).

THE NOTION OF WALL "LOOSENING"

The concept that the growing cell wall must be "loosened" in order to expand its surface arises from various biophysical, biochemical, and physiological considerations (see 13, 24, 41). The growing wall behaves like a network of inextensible cellulose microfibrils laterally linked together via a complex matrix of flexible polysaccharides that may bind to cellulose and to each other. Current models of the wall envision hemicelluloses, such as xyloglucan, coating the surface of cellulose and directly linking microfibrils together (Figure 1). Such crosslinks have been observed microscopically (73), but whether there are enough of them to contribute substantially to wall strength is uncertain. An alternative model [e.g. (110)] that is equally consistent with experimental data shows microfibrils coated with hemicelluloses and embedded in additional layers of matrix polymers, that is, without direct

Model #1: Cellulose tethered by hemicelluloses

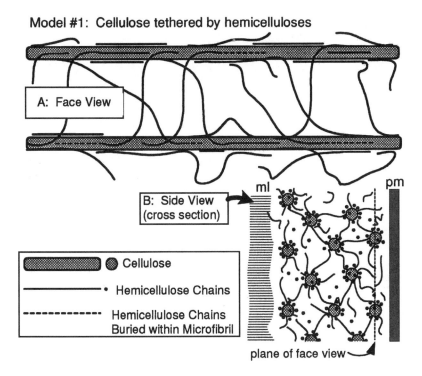

Figure 1 A model (#1) of the growing plant cell wall. *A*: View in the plane of the wall (face view, parallel to the plasma membrane). *B*: Side view, at right angle to *A*, giving a cross section of the wall. The plane of the face view in *A* is indicated by a dotted line. ml = middle lamella; pm = plasma membrane. This model attributes the mechanical strength of the wall to tethering of cellulose microfibrils by xyloglucans that are bonded noncovalently to the microfibril surface and entrapped within the microfibril. Pectins (not shown) are viewed as forming a co-extensive matrix in which the cellulose-xyloglucan network is embedded. This picture of the wall is found in many current discussions of cell walls, e.g. (13, 24, 41, 48, 72, 86).

microfibril-microfibril links (Figure 2). In this model, wall strength may depend largely on many noncovalent interactions between the laterally aligned matrix polymers.

Regardless of which picture of the wall is closer to reality, this microfibril/ matrix network is strong enough to resist the high tensile forces generated within the wall as it resists the outward force of cell turgor pressure. Wall enlargement requires controlled spreading of the cellulose/matrix network, evidently as a result of rearrangement of the matrix polymers (23). As a viscoelastic material, cell walls have inherent extension and yield properties that stem from the basic structure of the wall, and some attempts have been made to account for cell wall extension purely in terms of these viscoelastic properties (117, 123). However,

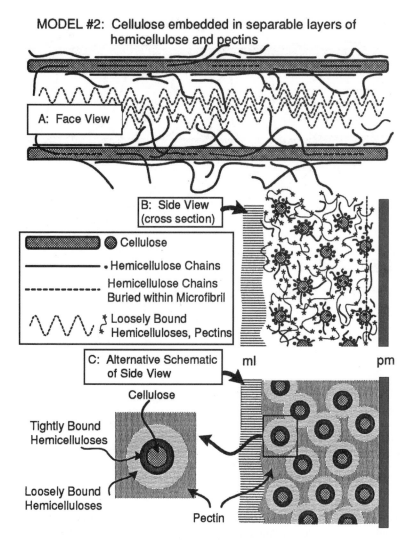

Figure 2   An alternative model (#2) of the growing plant cell wall. *A*: View in the plane of the wall (face view, parallel to the plasma membrane). *B*: Side view, at right angle to *A*, giving a cross section of the wall. The plane of the face view in *A* is indicated by a dotted line. *C*: An alternative depiction of the side view, showing more clearly the different "layers" in which the microfibrils are thought to be coated and embedded. This model differs from that in Figure 1 primarily by the lack of polymers that directly cross-link the microfibrils. Instead, the tightly bound hemicelluloses, such as xyloglucan, are pictured as sheathed in a layer of less tightly bound polysaccharides, which in turn are embedded in the pectin matrix that fills the spaces between the microfibrils. This model is adapted from Talbott & Ray (110), who cite older literature supporting similar models. ml = middle lamella; pm = plasma membrane.

simple wall viscoelasticity appears to be insufficient to account for the extension behaviors of growing walls (23). For instance, isolated walls treated so as to inactivate wall enzymes extend only transiently when put in tension. To a first approximation, such inactivated walls act like a viscoelastic solid. In contrast, the walls of growing cells exhibit a steady, long-term creep (a type of irreversible extension), and such creep may be mimicked in isolated walls, as long as their associated proteins are not inactivated (21, 22). Thus, protein-mediated loosening processes, perhaps in combination with integration of newly secreted polymers, seem to be required to catalyze and sustain anything more than a limited viscoelastic extension of the wall.

Several types of polymer rearrangements could plausibly lead to turgor-driven wall expansion. These include weakening of the noncovalent bonding between polysaccharides (as postulated for expansins), cleavage of the backbone of the major matrix polymers (e.g. by endoglucanases, pectinases, transglycosylases, and hydroxyl radicals), and breakage of crosslinks between matrix polymers (e.g. by esterases). It is a curious fact that the only agent (expansin) shown to catalyze wall extension in vitro acts by a still mysterious biochemical mechanism that appears to aid glycan-glycan slippage, whereas the enzymes with better defined biochemical actions, e.g. endoglucanases, do not appear capable of catalyzing wall extension in vitro by themselves (27). This result makes sense in the model shown in Figure 2, where backbone cleavage of matrix polymers would not directly lead to wall extension because it is the lateral, noncovalent bonding between adjacent polysaccharides that is primarily responsible for wall strength. In contrast, the result is difficult to rationalize in the model represented in Figure 1, where breakage of the tethers ought to lead to wall extension.

WALL SYNTHESIS   Synthesis, secretion, and integration of new wall polymers is, of course, necessary to maintain the structural integrity of the extending wall in the long term. Currently, there is no evidence that newly secreted polysaccharides directly contribute either to wall loosening or to wall extension. Thus, it is simpler at this stage to think of secretion and integration of wall polymers as separable from wall loosening and extension. This view is supported by the fact that sustained wall extension can indeed occur in vitro, without the direct need for wall synthesis (22, 94). However, it might be that the crucial experiments have not yet been tried to test whether wall extension in vitro can be induced by addition of polysaccharides, perhaps in the presence of enzymes that can assist their integration into the wall, e.g. xyloglucan endotransglycosylase.

Time Scales for Growth Control

It may be important to distinguish between short-term and long-term control of wall extension properties, as these likely involve different control mechanisms. We know that cell wall expansion can start and stop quickly (on the order of one minute or less), without substantial changes in the structure or composition of the wall. Notable examples include blue-light suppression of hypocotyl elongation (1), growth inhibition by metabolic inhibitors (93), and stimulation of growth by exogenous auxin or fusicoccin (20, 46). In this short time frame, these stimuli have little or no detectable effect on the viscoelastic properties or the composition of the wall, yet they nevertheless modulate wall extension behaviors (23). Modulation of wall-loosening agents seems the likeliest mechanism to account for these rapid growth changes. At the other end of the time scale, a growing cell typically attains a peak expansion rate followed by a gradual slowdown that may take hours or days to complete prior to maturation. During this time, the composition and structure of the wall may change substantially, as a result of secretion of new wall polymers and breakdown, turnover, crosslinking or other modifications of the polymers in muro. In this long time scale, walls become less extensible, as measured by various viscoelastic methods, e.g. (64). Thus there is reason to believe that different processes may contribute to rapid changes versus gradual changes in cell expansion, and that these two types of controls may function simultaneously.

In the sections that follow, I briefly summarize the major candidates for wall-loosening action.

EXPANSINS

An overview of expansin was recently published (25). These proteins were first isolated in 1992 as the mediators of "acid growth." Acid growth refers to the increase in growth rate that occurs when plant cells are placed in acidic solutions (95). Such growth stimulation comes about because the cell wall becomes more extensible at acidic pH. This effect on extensibility is not a direct pH effect on the wall polymers, e.g. on pH-sensitive bonding between wall polysaccharides (116). Instead, it is mediated by one or more protein factors that somehow catalyze wall extension. This conclusion is based on in vitro studies of walls that were treated so as to denature wall proteins (22). Denaturation caused wall extension properties to become insensitive to pH, at least as detected by wall creep and wall stress relaxation assays in the range of pH 4.5 to 7. However, pH sensitivity was largely restored by addition of purified expansins. The pH dependence of expansin-induced wall extension matches closely that of the acid-growth process, with an optimum at pH < 4, and a gradual decline in activity between pH 4 and 7. Expansin activity is also stimulated and inhibited by the same chemical agents that affect acid growth of walls (76). Hence, expansins appear to be the principal protein mediators of acid growth, at least in cucumber hypocotyl walls, where the most detailed work on expansins has been carried out.

Expansins were first defined in terms of their unique action on the rheological behavior of isolated cell walls: They induced long-term pH-dependent extension, and they enhanced stress relaxation of isolated walls over a broad time range, also in a pH-dependent manner (24). This functional definition of expansin has proved robust so far, inasmuch as the only proteins found to possess these properties are clearly homologous to the first expansins purified from cucumber hypocotyls. Since expansins were first cloned (102), many homologous sequences have been identified from a variety of dicots (8, 16, 78, 90, 96, 99, 103), from grasses (18), and even from pine (GenBank Accession No. U65981). Most of these expansins are known only from sequence, not from their biochemical or biological activities. It remains to be seen whether the functional definition of expansin and the sequence definition are fully congruent with each other, i.e. do all those expansin sequences encode proteins with expansin-like rheological effects on walls? The expression pattern of some expansins suggests that these proteins may serve additional functions besides growth, e.g. in fruit softening and cell separation (30, 99). Perhaps these expansins have variant effects on wall rheology. Direct characterization of protein properties are needed to test these possibilities.

Two families of expansins are now recognized, which we are calling α- and β-expansins (25, 26). The two types of expansin have only $\sim$25% amino acid identity to each other, but they appear to be homologous along the full length of their peptide backbones and they have very similar effects on cell wall rheology. However, they may be selective for different wall polymers and their binding properties to walls may differ. Our knowledge of how expansins make the wall more extensible is still limited, but most evidence points to a subtle mechanism, such as a destabilization of glycan-glycan interactions, rather than hydrolysis of matrix polymers.

α-Expansins

Most of what we know about the biochemical action of these proteins comes from studies of native α-expansins extracted from cucumber hypocotyls (75, 76), and the following summary is based primarily on this work. Limited characterization of expansin proteins in tomato leaves (62), maize roots (122), oat coleoptiles (67), soybean hypocotyls (66), and rice internodes (17) has also been published. It should be borne in mind that expansins may have additional roles besides their postulated function in cell growth and the properties of divergent α-expansins may differ from those described here.

Primary transcripts of α-expansins are predicted to encode a protein of $\sim$28 kD, which includes a secretory signal peptide (typically $\sim$23 amino acids) that is removed to make a mature protein of $\sim$25 kD. α-Expansins typically lack motifs for N-linked glycosylation, and biochemical tests indicate that glycosylation is indeed lacking both in the two α-expansins from cucumber hypocotyls (DJ Cosgrove & DM Durachko, unpublished data) and in two α-expansins from rice internodes (H-T Cho & H Kende, personal communication).

BINDING TO THE WALL   In terms of abundance, α-expansins are very minor components of the hypocotyl wall (we estimate $\sim$2 parts protein to 10,000 parts wall, on a dry-weight basis). The protein binds tightly to walls, apparently to noncrystalline surface regions of cellulose (75). Curiously, xyloglucan binding to cellulose neither augmented nor interfered with the capacity of cellulose to bind α-expansin. This observation, together with the finding that expansin binding and expansin activity showed parallel concentration dependencies, led us to suggest that α-expansin action in wall loosening might not involve xyloglucans (75). This inference is weakened, however, by the more recent finding that Zea m1, the maize pollen β-expansin, does not bind strongly to walls, yet it does have robust expansin activity (26). This result shows that tight binding is not essential for expansin activity. It is possible, although not demonstrated, that expansin binding and wall-loosening activity are separable functions, analogous to the separable binding and catalytic activities of many microbial cellulases (50, 115).

What, then, is the significance of expansin binding to the wall? Binding, or lack of it, may be crucial for determining whether expansin protein secreted by one cell can diffuse to adjacent cells and influence their wall properties. For a process such as cell growth, which needs to be regulated in a cell-specific manner, it may be vital to limit expansin diffusion to neighboring cells. In contrast, if the aim is to secrete a protein that will loosen the walls of neighboring cells (e.g. as a pollen tube penetrates stigma and style), then lack of tight wall binding may be important for proper protein function.

It is informative to compare the binding properties of α-expansin with that of another well-studied protein that binds to the surface of cellulose, namely the cellulose-binding domain (CBD) of microbial cellulases. In one study of a CBD from *Clostridium cellulovorans* (47), CBD adsorption to cellulose had an apparent Kd of about 1 μM and a maximum binding at saturation of about 1 μmol of CBD per g of cellulose. Comparison with expansin binding can only be approximate at this time because the two studies used different forms of cellulose; nevertheless, the general conclusions from the comparison are probably sound. With α-cellulose from cucumber hypocotyls (75), expansin binding was saturated at a protein:cellulose ratio of about 1:1100. This corresponds to a value of $\sim$40 nmol of expansin per g of cellulose, i.e. about 1/25th the density of CBD binding sites reported above. This low density of expansin binding does not begin to approach full coverage of the cellulose. Evidently, expansins bind to some specific and not very common feature of cellulose or to a minor contaminant of the α-cellulose fraction that was not extracted by strong base. What this binding site may be is unclear, but spectroscopic studies suggest that cellulose microfibril structure is more complex than commonly pictured (4, 118). The study on expansin binding did not report a Kd, but by the same kind of analysis used for the CBD study (47) the expansin Kd may be estimated from the data presented to be $\sim$6 nM, that is, about 166 times lower Kd than that of the CBD reported above. Thus, the α-expansin from cucumber hypocotyls appears to have much higher affinity and (probably) much lower binding at saturation than the *Clostridium* CBD.

As an aside, the *Clostridium* CBD was reported to have significant effects on growth of roots, root hairs, and pollen tubes (105); also biomass production by transgenic tobacco and poplar plants was enhanced when the CBD gene was expressed ectopically (104). The exact mechanism for these effects is unclear but might result from alteration of cellulose microfibril structure. Unlike the case for expansin, CBD and xyloglucan did compete for the same binding sites on the surface of cellulose. Presumably, CBD and expansin would occupy different sites in the wall, and synergistic effects on wall structure and rheology might result if both were applied simultaneously. I have tested this and other CBDs for expansin-like activity in wall extension and stress relaxation assays but failed to detect any activity (DJ Cosgrove, unpublished results).

There is no indication of cooperativity in binding of α-expansin to cellulose. Likewise, there is little or no cooperativity obvious in the extension-inducing action of these expansins. Hill coefficients were calculated from the dependence of wall extension rates on the concentration of applied expansin. Calculated values varied from $\sim$1.2 to 1.8 in different experiments, indicating little cooperativity (DJ Cosgrove, unpublished data).

MECHANISM OF ACTION   Studies to date indicate that α-expansin lacks significant hydrolytic activity against the major polysaccharides of the wall (75, 76). Confirmatory of this conclusion, α-expansin does not lead to a progressive, time-dependent weakening of the wall, as would be expected of a hydrolytic enzyme. Xyloglucan endotransglycosylase activity was not detected in purified expansin fractions (77). On the other hand, expansin can weaken pure cellulose papers (74) and enhance breakdown of cellulose by cellulases (28). The simplest interpretation of these results is that expansin weakens glucan-glucan binding. However, further experimental evidence in support of this hypothesis is needed.

Recent analysis of expansin genes suggests that they encode three functional domains, separated by introns (Figure 3). The N terminus of the primary transcript has a signal peptide of $\sim$22–25 amino acids. The C terminus has a series of conserved tryptophans with spacing similar to that of some bacterial CBDs

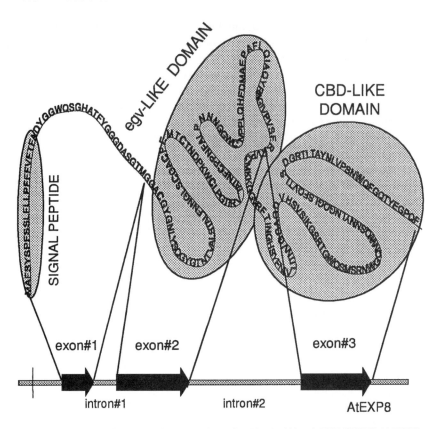

Figure 3 Structure of the expansin gene and protein. The Arabidopsis EXPANSIN8 (AtEXP8) gene contains three exons that encode three potential domains of the protein, including the signal peptide and the N terminus of the mature protein (exon#1), the endoglucanase-like core region (exon#2), and a domain (encoded by exon#3) with some structural resemblance to microbial cellulose-binding domains (CBD).

(35). Tryptophans and related aromatic amino acids are known to be important for protein-carbohydrate interactions; thus the conserved tryptophans and phenylalanines are prime suspects for expansin binding to walls. Although expansin's C terminus thus may function analogously to CBDs, there is no significant sequence homology between the two sequences.

The central part of the protein shows limited, but significant, sequence similarity to an unusual endo-1,4-β-glucanase cloned from *Trichoderma* (100). This endoglucanase, egl5, has been classified as a member of family-45 of glycosidases (52). The protein structure has been solved for the catalytic core of another endoglucanase ("cellulase") member of this family-45 (34). It is intriguing that what is conserved between the central part of expansin and the catalytic core of family-45 enzymes includes the cysteines and the key residues that make up the putative catalytic site of the endoglucanase. This suggests that the central region of expansin is homologous to the catalytic core of the family-45 endoglucanases. Following up on this molecular hint, we have found that some, but not all, expansins have a barely detectable level of endoglucanase activity when tested with barley 1,3:1,4-β-glucan (28). The endoglucanase activity, *per se*, cannot explain the rheological effect of expansins, because when family-45 endoglucanases were tested they had very high enzyme activity but did not induce wall extension. Thus, these endoglucanases are not expansins. Nevertheless, this sequence relatedness suggests homology and similarity of function, but perhaps using a different substrate or catalyzing a related reaction, such as a transglycosylation. Alternatively, the substrate distortion mechanism proposed for this class of endoglucanase (33, 34) may be conserved and modified in expansins, causing rheological effects, without hydrolytic activity. The search for expansin's mechanism of action goes on. . . .

EFFECTS ON LIVING CELLS   In addition to its effects on isolated cell walls, α-expansins have been shown to stimulate enlargement of living cells. All the work along these lines has been done using α-expansin extracted from cucumber hypocotyls. When α-expansin was applied to excised Arabidopsis hypocotyls, it significantly increased elongation ($\sim$50% elongation with expansin versus 14% elongation in the controls without expansin) (79). This growth effect was similar to that induced by 1 μM auxin. When α-expansin was applied to cucumber root hairs, it caused swelling and bursting (79). When applied via beads to incipient primordia on excised tomato shoot meristems, expansin caused distortions of the meristem and premature outgrowth of the primordium, at least in a small number of cases (40). When applied to tobacco suspension culture cells, α-expansins induced a threefold stimulation of the rate of cell enlargement (68). Thus it is clear that α-expansins from cucumber hypocotyls can stimulate cell expansion in several heterologous systems.

GENE EXPRESSION   Several studies of α-expansin gene expression have been published in the past two years. In rice, four α-expansins were studied by Cho & Kende (18, 19). Two α-expansin proteins were extracted from rice internodes and demonstrated to have classical expansin activity, e.g. they induced wall extension at acidic pH (17). The genes encoding these proteins are expressed in various organs of the rice plant and are induced differentially by treatments that stimulate growth, e.g. gibberellin and submergence (18, 19). In cotton, two studies have identified an α-expansin gene that is highly expressed during the period of maximal growth of the fiber (90, 103). In tomato there are different α-expansins expressed in different parts of the plant (8, 16, 96). A similar situation holds for expansin expression in Arabidopsis (29, 30). At the time of writing, nine different α-expansin genes have been identified in tomato, six in tobacco, and 22 in Arabidopsis (see the expansin web site at www.bio.psu.edu/expansins/ for an up-to-date census of published expansin sequences and related information). I currently estimate that Arabidopsis has $\sim$30 expansin genes; the exact count will be known when the genome sequencing is completed in the year 2000. Other plant species are likely to have at least this number.

An in situ hybridization study of the tomato shoot apical meristem indicates that an expansin gene is locally expressed in the cells underlying the site of future outgrowth of a leaf primordium (96). This result, in combination with the finding that local expansin application could cause premature primordium outgrowth and an alteration of subsequent phyllotaxy (40), was interpreted to support a variation of Green's model of the meristem (45). In this model, the pattern for leaf primordium initiation is postulated to depend on the pattern of physical forces in the epidermal cell layer, which in turn depends on the geometry of the meristem, the locations of existing local outgrowths, and the physical properties of the epidermal wall. Given this model, it is significant that expansin is not expressed in the epidermal layer of the incipient primordium, but rather in the underlying cells. This suggests that the outer epidermal wall is not loosened at the onset of primordium outgrowth, but rather that it is placed under increased local stress by expansin-induced wall loosening and growth of the underlying cells. Such loosening would tend to cause a displacement of the turgor force of the underlying cells to the restraining epidermal wall, thus increasing its tension. If Green's hypothesis is correct, cells respond to such physical forces by activating appropriate genes for subsequent differentiation processes. Reinhardt et al (96) show that an expansin gene is turned on in subepidermal cells (not in the epidermal cells) prior to primordium emergence; this appears to be the earliest marker of leaf initiation. What initiates expansin gene expression in these cells? Do they participate in the pattern of epidermal stress and strain envisioned in Green's model? If not, then either this earliest sign of primordium initiation is independent of epidermal tensions, or the epidermal cells somehow communicate their status to the underlying cells.

β-Expansins

Unlike α-expansins, which have a high degree of sequence relatedness ($\sim$75% amino acid identity, and up to 90% similarity), β-expansins—as we understand them today—are more divergent in sequence and perhaps in function. β-Expansins share only $\sim$25% amino acid identity with α-expansins, but they appear to be homologous to α-expansin in structure (26). Only one member of this family of expansins has been analyzed for its action on the cell wall, and the properties of this protein, Zea m1, may not be typical of the whole family.

GROUP-1 GRASS POLLEN ALLERGENS   Zea m1 is a member of a group of pollen proteins studied for years by immunologists because they (the proteins, not the immunologists) cause hay fever and seasonal asthma in humans. Known as the group-1 grass pollen allergens, the proteins are copiously secreted when grass pollen becomes hydrated. These proteins are highly soluble and are readily washed off of cell walls. Thus, in terms of abundance, solubility, and binding, Zea m1 greatly differs from α-expansins characterized in growing tissues (26). Zea m1 has similar rheological effects on cell walls as cucumber α-expansins, but it shows high specificity for grass cell walls over dicot walls. This suggests that it selectively acts on one of the matrix polymers specific to grasses, e.g. glucuronoarabinoxylan or mixed-linked 1,3:1:4-β-glucan. In contrast, α-expansins (whether of dicot or grass origin) are more active on dicot walls than on grass walls (17, 67, 76). Structural studies of other group-1 allergens have shown that they are glycosylated and two of the prolines near the N terminus are hydroxylated (55, 91). Such posttranslational modifications are not apparent in α-expansins. Zea m1 and its homologs in other grass species probably have a specialized function in pollination, namely to soften the walls of the stigma and style, thereby aiding pollen tube penetration of these tissues (26).

OTHER β-EXPANSINS   Although Zea m1 is the only β-expansin directly analyzed for its effects on walls, a search of GenBank reveals other related sequences that likely function as wall-loosening agents (26). First, there are a series of partial cDNAs in the rice EST database. We have sequenced some of these and confirmed their relatedness to the group-1 allergens. The proteins predicted from these cDNAs contain all of the structural hallmarks that are conserved between Zea m1 and α-expansins. Thus, they too may function as wall-loosening agents. What is particularly intriguing is that there are over 50 entries for this class of protein in the rice EST database, and these entries fall into 10 sequence classes (that is, they are derived from 10 different genes). Most of these sequences are from vegetative tissues, so we surmise they have wall-loosening functions that operate in parallel with α-expansins. It is also interesting to note that only a single Arabidopsis EST appears to be a β-expansin. The sequence data to date indicate that there are many more α-expansin genes than β-expansin genes in Arabidopsis, whereas in rice the numbers are more even. This hints that the evolution of the grass cell wall, with its peculiar composition of matrix polymers, went hand-in-hand with the duplication and divergence of the β-expansin family in the grasses. Much work remains to explore the functions of these numerous β-expansins in grasses.

β-Expansins from dicots are rare in GenBank, probably because dicots have few β-expansin genes. A cDNA known as CIM1 (for cytokinin-induced message-1) was originally identified in soybean cell cultures following depletion and return of cytokinin to the growth medium (32). The protein encoded by this message has all the hallmarks of a β-expansin, and its expression pattern is consistent with a role in wall loosening during cytokinin-induced cell proliferation and growth. One unusual feature of CIM1 is that the cytokinin effect on message abundance is regulated, at least in part, by posttranscriptional controls (37). A single Arabidopsis EST falls into the β-expansin family, and there are other sequences that share some remote relatedness to expansins [see discussion in (78)], though most of these remote sequences lack one or more of the key sequence features we currently recognize as diagnostic for expansins: (a) a signal peptide; (b) a set of conserved cysteines in the N-terminal half of the protein; (c) a conserved HFD (his/phe/asp) motif in the middle of the protein, and (d) a conserved set of tryptophans at the carboxy-terminal half of the protein (25).

POLYSACCHARIDE HYDROLASES

Cell walls contain numerous enzymes capable of hydrolyzing the major components of the wall matrix, and it is attractive to think that some of these enzymes might function to loosen the wall by breaking load-bearing links between cellulose microfibrils, thereby allowing the wall to extend (9, 42, 57, 85). Alternatively, wall hydrolytic enzymes might stimulate wall extension indirectly: By reducing the size and viscosity of matrix polymers, such enzymes could act synergistically to enhance the action of primary wall-loosening agents, such as expansin. A third alternative is that such hydrolytic enzymes have functions unrelated to wall loosening, e.g. in defense, in signaling, or in polysaccharide processing or breakdown to serve the cell's other metabolic or energy needs. Fry (42) identifies other possible functions for wall hydrolytic enzymes.

Given the model of the wall shown in Figure 1, it seems logical that hydrolysis of the matrix should lead to wall extension, but direct experiments have not shown this to occur. Several hydrolytic enzyme preparations active against cell walls were tested for their ability to directly induce extension of walls in vitro, in much the same way that expansins were demonstrated to act (27). These treatments either failed to cause any effect on wall extension, or they caused the wall to break without an intermediate period of prolonged extension. Thus, they weakened the wall but did not loosen it in the sense of permitting controlled wall yielding and expansion. Some of these enzyme preparations are very potent in hydrolyzing the major structural polysaccharides of the matrix. The *Trichoderma* cellulase preparation used in this study is probably most active on xyloglucan, which is commonly believed to tether microfibrils together (13, 41, 73, 86). It is therefore surprising, and I think a significant finding, that this and other cellulases did not induce wall extension. It can be argued, with some merit, that plant-derived enzymes have different properties from analogous enzymes obtained from *Trichoderma* or other microbial sources. However, there is no evidence that plant-derived endoglucanases are any more capable of inducing wall extension in vitro than their fungal or bacterial counterparts. Also, the fact that matrix hydrolysis by these enzymes does not cause significant wall extension must cast doubt on the hypothesis that matrix polysaccharide hydrolysis provides a reasonable basis for wall extension and similarly casts in doubt the structural significance of hypothetical glucan tethers between microfibrils, as illustrated in Figure 1.

This argument does not mean that wall hydrolytic activities are irrelevant for wall extension, however, because walls that were hydrolyzed for only a brief time were more sensitive to subsequent expansin-mediated extension (27). Notably, pectinase treatments were at least as effective in enhancing wall extension by expansin as were endoglucanase treatments. These results are more consistent with model B of the cell wall (Figure 2). Thus we might distinguish between two types of wall loosening: primary loosening agents, which are able directly to catalyze wall extension by themselves; and secondary loosening agents, which change the structure of the wall, making it more sensitive to the action of primary loosening agents.

Of the numerous hydrolytic activities in the wall, most growth-related studies have focused on enzymes capable of breaking down matrix glucans, specifically xyloglucan and the mixed-link $1,3:1,4-\beta$-glucan, which are abundant in the growing cell walls of dicots and grass seedlings, respectively.

Endoglucanases

Endoglucanases can hydrolyze glucosidic bonds at a site in the middle of the glucan. Microbial endoglucanases that can attack cellulose typically have long clefts in which single glucan chains can fit so they are precisely placed for catalytic attack by highly conserved residues in the catalytic site (36, 111). A long cleft and a conservation of catalytic amino acids has similarly been observed in a family of plant glucanases that degrade fungal $1,3-\beta$-glucans and grass wall $1,3:1,4-\beta$-glucans (56). No structure has yet been determined for plant-derived glucanases that can degrade β-glucans with a pure 1,4-linked backbone (e.g. cellulose and xyloglucan).

Numerous studies lend circumstantial support to the idea that endoglucanases may be involved in cell expansion. One of the earliest results on this theme was the 1966 report of stimulation of "cellulase" (endoglucanase) activity by high doses of auxin (38), and since then the topic of auxin-induced turnover of matrix glucans has attracted considerable attention. Several reviews deal with this topic in some detail (48, 60, 70, 98); here I focus mostly on recent contributions to this subject. To summarize briefly the older evidence: (*a*) auxin-induced growth of excised tissues is associated with an enhanced turnover of xyloglucans in dicots and mixed-link glucan in grasses; (*b*) mixed-linked glucans are synthesized and then degraded during the growth of grass coleoptiles and leaves; (*c*) antibodies directed against wall glucanases inhibit auxin-induced growth; and (*d*) lectins and antibodies that bind matrix glucans reduce wall autolysis and inhibit auxin-induced growth. I separate the discussion into work on grasses and dicots.

GRASSES   In young grass coleoptiles, the major matrix hemicelluloses include a mixed-link 1,3:1,4-β-glucan, which in the plant world is unique to grasses, and a branched xylan with arabinose and glucuronic acid branches (glucuronoarabinoxylan, or GAX) (12). Both polymers apparently bind to cellulose and could conceivably link cellulose together by binding to more than one microfibril. However, the exact structural role of these hemicelluloses is uncertain, and it has been proposed that the mixed-link glucan might deter wall crosslinking, e.g. as a filler, rather than function as a crosslinking polymer (69). This would account for the fact that this polymer is present only transiently in the growing coleoptile wall and is broken down and resorbed as the coleoptile ceases elongation. The branched xylan (GAX) appears to become debranched as the coleoptile ceases growth (11). Such debranching would likely make GAX adhere to the surface of cellulose and to other xylans more tightly, and this might be a mechanism for reducing wall extensibility. Very little is known about the endogenous enzymes that modify GAX, although enzymes from microbial sources have been characterized.

Recent plant growth work has focused on the enzymes mediating breakdown of mixed-link glucan in maize coleoptiles. The cloning of a cDNA for an unusual endoglucanase from maize coleoptiles was recently reported (112). This enzyme was previously characterized by Hatfield & Nevins (69), who found that it had high specificity for 1,3:1,4-β-glucans and that it appeared to cut the glucan at regular but infrequent spacings, chopping the glucan into 10–15-kD fragments. The cut sites may be places on the polymer consisting of more than five adjacent β-1,4 linkages. The enzyme has a pH optimum of 5 and activity decreases significantly at lower pH, thus the pH dependence does not match the acid growth phenomenon very closely. The reported cDNA sequence encodes a novel protein with a signal peptide of 2.1 kD and a mature polypeptide of 31 kD. This sequence does not fit into any of the 64 families of recognized glycosidases (51), so it is truly novel as an endoglucanase.

This endoglucanase is hypothesized to mediate wall loosening and cell expansion. However, the pattern of its activity raises some doubts on this point.

For example, endoglucanase activity reached maximum levels after coleoptile growth ceased, rather than at the time of maximal extension rate (61).

Although auxin stimulated wall autolysis and growth in excised coleoptiles, it did not increase the amount of the endoglucanase, as detected by immunoblots (61). To account for this discrepancy, Inouhe & Nevins (59) report that a non-enzymatic protein of ~40 kD, which they call an acid wall protein, enhances the activity of maize glucanases (both exo- and endo-type enzymatic activities). Treatment of excised coleoptiles with antibodies directed against the acid wall protein inhibited auxin-induced growth, inferring a role for this protein in growth control. However, the exact mechanism of action of this protein is uncertain; Inouhe & Nevins speculate that auxin might enhance its activity or stimulate its release from sequestration sites in the wall.

Perhaps the strongest argument in favor of wall-loosening activity of glucanases has been from experiments in which specific antibodies against these proteins were found to inhibit auxin-induced growth (e.g. 58). A recent study shows that antibodies against an endo-1,3-β-glucanase can similarly inhibit auxin-induced growth (80). Here's the rub: this study was carried out with mung bean hypocotyls, which are not known to contain 1,3-β-glucans as part of the load-bearing wall. Thus, the action of these antibodies in reducing auxin-induced growth is unlikely to be via direct interference with wall loosening. Instead, they may interfere with communication and translocation through cell walls (120). If this is the case for these mung bean experiments, might it not also hold for the similar experiments with glucanases in maize coleoptiles?

DICOTS   In dicots, endo-1,4-β-glucanases (EGs) from several species have been studied; for reviews, see (9, 63). Xyloglucan is the likeliest target for these enzymes, although it is possible that they could hydrolyze some surface 1,4-β-glucan chains that are in noncrystalline regions of the cellulose microfibril (48, 49). Cellulose is relatively stable in growing cells and exhibits negligible turnover, whereas xyloglucan turnover is significant (65, 88), although probably smaller than that of mixed-link glucan in the grasses.

In dicots EGs have been studied in the context of abscission, fruit softening, and auxin-induced growth (9, 97). In tomato, EGs comprise a multigene family of at least seven genes that are expressed in different organs and under the control of different stimuli (6). These genes are all classified in the same family of endoglucases (type E). One member of the tomato family, called *Cel4*, was found to be selectively expressed in the growing region of the tomato hypocotyl, and its message level was increased slightly (~twofold) when etiolated seedlings were treated with high doses of ethylene or auxin (2,4-D), causing lateral swelling of the hypocotyl. A second tomato EG, *Cel7*, was found to be expressed at low levels in both growing and nongrowing regions of the tomato hypocotyl, but was induced to high levels in both growing and nongrowing hypocotyl regions upon treatment with high doses of auxin (2,4-D) (15). In pea epicotyls and in poplar cell suspension cultures, related auxin-induced EGs were cloned and partially characterized (82, 121). These EG genes encode ~45–50-kD proteins with secretion signal peptides and without motifs for N-linked glycosylation.

Do these EGs function in wall loosening? The time course for induction of EGs by auxin shows that it requires many hours before EG mRNAs accumulate (15), whereas auxin stimulates growth after a lag of ~15 min. This timing argues against a direct role of EGs in the initial phases of auxin-induced growth. Catala et al (15) suggest that aside from a possible wall-loosening role, EGs may function to provide additional acceptor molecules for xyloglucan transglycosylation reactions (see below) or to generate oligosaccharides that may act as signals to modulate further growth (2, 31).

In a recent study, a potent microbial endo-1,4-β-glucanase was expressed in the cell walls of transgenic Arabidopsis (MT Ziegler, SR Thomas, KJ Danna, personal communication). Despite high levels of active enzyme in the cell walls, the plants grew normally and did not show obvious signs of growth disruption. This is a remarkable result because one would expect a significant weakening of the growing wall, if xyloglucans served as direct tethers to connect cellulose microfibrils, as depicted in some current models of the wall (see Figure 1). Either the xyloglucans are somehow protected from the action of the foreign enzyme, or the cell wall is able to compensate for the wholesale wall loosening this enzyme would be expected to inflict on the growing wall. Alternatively, this model may be missing important structural aspects of the wall that determine wall strength and wall growth.

In a similar study, transgenic tobacco plants expressing high levels of microbial xylanases in their walls grew normally, with little evidence of cell wall disturbance (53). Tobacco, as well as other dicots, have xylans in their walls, and it is curious that these potent enzymes did not have substantial phenotypic effects.

The dicot EGs are known principally from their nucleotide sequences, and only limited enzymatic characterization has been reported (119). To evaluate their possible function in wall loosening, it is important to know their substrate specificity, pH dependence, and wall-binding properties, as well as to make direct tests of their ability to induce cell wall extension in living tissues and isolated walls. Hayashi and colleagues have partially characterized EGs from pea epicotyls and poplar cell cultures (49, 81). These enzymes are able to bind cell walls, although binding properties have not been explored in detail. Because their pH optimum is ~6.5, they are unlikely to mediate acid growth. These enzymes have not been tested for their ability to alter cell wall extension properties by direct experiments.

New evidence in support of EG involvement in cell enlargement comes from cloning of the Arabidopsis *KORRIGAN* (*KOR*) gene (83). A plant with a mutation in this gene was found in a screen for short hypocotyl mutants, and it turned out that the growth of most organs was stunted in this mutant. Cytological examination showed that the *kor* mutants had irregular and disturbed patterns of cell expansion. Cell walls were thicker than wild-type walls and appeared disordered. Cloning of KOR showed it to encode an EG, related to the EGs described above, except that it contains a membrane anchor domain. An ortholog of this gene, called Cel3, was recently reported in tomato (7). Cell fractionations indicated that the KOR protein, as well as Cel3, was associated with the plasma membrane. Application of growth hormones had little effect on *KOR* or Cel3 expression. This report shows that *KOR* expression is required for normal cell enlargement, but the exact cellular and biochemical functions of KOR are unclear. If it is indeed anchored to the plasma membrane, then a traditional wall-loosening function is doubtful because it would not have access to most of the cell wall network. Because the newly secreted wall polymers that are closest to the plasma membrane are deposited in a relaxed state, they do not bear the mechanical stress generated by cell turgor and therefore can have little structural influence on cell expansion until the wall extends and they begin to take up some of the wall stress. At this point, they would likely be too far from the membrane-anchored KOR protein to interact with it. Nicol & Hofte (84) suggest several other possible functions of KOR. One speculative possibility is that KOR is part of the cellulose synthase complex and it functions in microfibril formation. This suggestion is consistent with the plasma membrane location of KOR and the fact that a similar membrane-bound EG is essential for cellulose synthesis in *Agrobacterium tumefaciens* (71).

XYLOGLUCAN ENDOTRANSGLYCOSYLASE (XET)

This enzyme, which has also been called endoxyloglucan transferase or EXGT (85), catalyzes a kind of molecular grafting reaction in which the backbone of a xyloglucan is cleaved and one of the resulting half-chains is added to the nonreducing end of a second xyloglucan chain. XET was proposed to have wall-loosening activity (44), but direct tests of its ability to catalyze wall extension in vitro have not supported this hypothesis (77). Several other possible functions for this enzyme, including incorporation of newly secreted xyloglucan into the wall, as well as rearrangement of the xyloglucan-cellulose network during wall assembly and growth, have been suggested by Nishitani (85, 86) and Fry (42, 77). XET may be responsible for the shifts in xyloglucan size recorded in several studies (87, 108, 109, 113). These shifts in size may have an effect on wall extension properties, but this has not been demonstrated directly.

Plant XET proteins are moderate sized (~33 kD) and, at least in some cases, are N-glycosylated. Glycosylation seems to be important for enzyme activity because when TCH4 (an Arabidopsis XET) was enzymatically deglycosylated it lost 98% of its activity (10). XETs contain a presumptive catalytic domain (containing the sequence: DEID-I/F-EFL) that is homologous with the catalytic domain of several bacterial endo 1,3:1,4-β-glucanases. Mutagenesis of the first glutamate (E) to glutamine in TCH4 resulted in an inactive protein (10). This result supports the functional role of the presumptive catalytic domain.

Plant XETs from different sources vary somewhat in their specificity for catalyzing a transglycosylation versus hydrolysis reaction. The difference is whether a saccharide or water can act as an acceptor for the glucan that is cleaved. For example, the enzyme from azuki bean displays only transglycosylation activity (89), whereas the related enzyme from nasturtium seeds mostly has hydrolytic activity (39), that is, it has endoglucanase activity. Four XETs from Arabidopsis lack hydrolytic activity (P Campbell & J Braam, in preparation). These differences in activity are probably due to slight structural changes in or around the catalytic site.

Recent studies have identified XETs as a subset of a larger family of related proteins, called XRPs, for xyloglucan-related proteins (85). Arabidopsis has more than 16 XRPs, which may have distinct patterns of expression in the plant (3, 85; P Campbell & J Braam, personal communication).

Sulová et al (107) showed that a stable xyloglucan-enzyme intermediate is formed during enzyme catalysis, and more recent work from the same group showed that XET can be readily purified by binding the stable XET-xyloglucan complex to paper, then releasing the enzyme by adding oligosaccharides that can serve as acceptors to complete the reaction, freeing the enzyme from the paper-bound xyloglucan (106).

Most biochemical work with XET activity has used solubilized substrates. To test whether newly synthesized xyloglucans in living cells actually formed hybrid molecules with older xyloglucans, Thompson et al (114) double-labeled xyloglucan in *Rosa* cell cultures with $^{13}C/^3H$, and then followed the fate of the newly synthesized xyloglucan by monitoring changes in xyloglucan density using isopycnic centrifugation. They found a gradual shift in the density of the newly synthesized xyloglucan over a period of seven days, consistent with the idea that it was integrated into the pre-existing xyloglucan-cellulose network.

Lastly for this section, transgenic tobacco plants were recently produced with reduced expression of an endogenous XET gene that is normally expressed in the leaf vasculature (54). Little change in phenotype was noted, but xyloglucan from the transgenic plants was increased in molecular size in comparison with wild-type plants.

HYDROXYL RADICALS

Fry (43) has proposed a novel mechanism of wall loosening, in which millimolar L-ascorbate nonenzymatically reduces O_2 to H_2O_2 and Cu^{2+} to Cu^+, which then react to form hydroxyl radicals ($\cdot$OH). Hydroxyl radicals are highly reactive and are able to cause oxidative cleavage of diverse polysaccharides nonspecifically. Thus, if $\cdot$OH were selectively released in the wall, it might cut xyloglucan or other potential polymers linking cellulose together. By way of evidence, Fry (43) showed that ascorbate-generated $\cdot$OH could indeed reduce the viscosity of xyloglucan solutions and that this scission activity has a pH optimum between 4.5 and 5.5; in addition, several free radical scavengers were effective in inhibiting the cleavage activity. Since ascorbate, H_2O_2, and Cu^{2+} are likely to be found in the wall at concentrations effective for $\cdot$OH generation, this powerful oxidant could potentially act to loosen walls, if it were produced very close to appropriate targets. Since H_2O_2 is also thought to function in oxidative cross-linking of wall phenolics substances, resulting in wall stiffening (5, 101), there might be a delicately controlled balance between cleavage and cross-linking activities resulting from H_2O_2 production in the wall. The physiological significance of these ideas remains to be determined.

A SUMMARY MODEL OF WALL EXTENSION

A satisfactory model of cell wall extension requires an adequate picture of the polymers and bonds that confer mechanical strength to the wall and that permit time-dependent extension, that is, creep of the wall network. These higher-order aspects of wall architecture are not well understood. Many current discussions of the wall depict its mechanical strength as due to a cellulose-xyloglucan network (as in Figure 1), whereas the physical properties of the wall, summarized by Preston (92), are more consistent with an assemblage of weak bonds gluing wall polysaccharides together, as in Figure 2. This alternative model is also supported by some wall extraction studies, summarized by Talbott & Ray (110), and by the effects of enzymatic treatments on wall extension behaviors (27).

For the reasons described previously, it may be important to distinguish between primary and secondary wall-loosening agents. I suggest an operational definition of primary loosening agents as those substances and processes that are competent and sufficient to induce extension of walls in vitro. Expansins fit this definition, whereas various wall enzymes with putative wall-loosening functions, such as endo-1,4-β-glucanase and XET, have not been shown to possess such activity. Secondary wall-loosening agents can be defined as those substances and processes that modify wall structure to enhance the action of primary agents. It is possible, although not actually demonstrated, that plant endoglucanases, XET, pectinases, as well as secretion of specific wall polymers and production of hydroxyl radicals, could function as secondary wall-loosening agents. The activity of primary and secondary wall-loosening agents could be modulated in various ways, e.g. by changes in wall pH, by secretion of ligands that activated or inactivated wall enzymes, by secretion of substrates, etc. Additionally, the wall could be modified by other enzymatic activities that cross-linked the wall or that changed the wall polymers such that they were no longer acted upon by wall-loosening agents. This latter notion suggests a method for assessing wall "stiffening" or "tightening" activities, that is, by testing whether walls pretreated with specific enzymes are reduced in their ability to respond in vitro to primary loosening agents.

Beyond in vitro experiments, it is important to extend the results to living cells and to whole plants and to test our knowledge of how the wall works by genetic manipulations. The limited transgenic experiments carried out so far with wall enzymes have shown that the plant cell wall is amazingly resilient to assault by these transgenic (mostly microbial) enzymes. Genetic engineering, or perhaps better stated, genetic tinkering, holds much potential for defining the cellular machinery needed for wall expansion.

ACKNOWLEDGMENTS

I am indebted to many past and present members of my laboratory and to many colleagues for their specific contributions and their challenging, but collegial, debates about how the plant cell wall extends. I gratefully acknowledge research support by grants from the National Science Foundation, the Department of Energy, and the National Aeronautics and Space Administration.

Visit the *Annual Reviews home page* at
http://www.AnnualReviews.org

Literature Cited

1. Addink C, Meijer G. 1972. Kinetic studies on the auxin effect and the influence of cycloheximide and blue light. In *Plant Growth Substances 1970*, ed. DJ Carr, pp. 68–75. Berlin/New York: Springer-Verlag
2. Aldington S, Fry SC. 1993. Oligosaccharins. *Adv. Bot. Res.* 19:1–101
3. Antosiewicz DM, Purugganan MM, Polisensky DH, Braam J. 1997. Cellular localization of Arabidopsis xyloglucan endotransglycosylase-related proteins during development and after wind stimulation. *Plant Physiol.* 115:1319–28
4. Atalla RH, Hackney JM, Uhlin I, Thompson NS. 1993. Hemicelluloses as structure regulators in the aggregation of native cellulose. *Int. J. Biol. Macromol.* 15:109–12
5. Brisson LF, Tenhaken R, Lamb C. 1994. Function of oxidative cross-linking of cell wall structural proteins in plant disease resistance. *Plant Cell* 6:1703–12
6. Brummell DA, Bird CR, Schuch W, Bennett AB. 1997. An endo-1,4-β-glucanase expressed at high levels in rapidly expanding tissues. *Plant Mol. Biol.* 33:87–95
7. Brummell DA, Catala C, Lashbrook CC, Bennett AB. 1997. A membrane-

anchored E-type endo-1,4-β-glucanase is localized on Golgi and plasma membranes of higher plants. *Proc. Natl. Acad. Sci. USA* 94:4794–99

8. Brummell DA, Harpster MH, Dunsmuir P. 1999. Differential expression of expansin gene family members during growth and ripening of tomato fruit. *Plant Mol. Biol.* In press

9. Brummell DA, Lashbrook CC, Bennett AB. 1994. Plant endo-1,4-β-D-glucanases. Structure, properties, and physiological functions. *ACS Symp. Ser. Enzym. Convers. Biomass Fuels Prod.* 566:100–29

10. Campbell P, Braam J. 1998. Co- and/or post-translational modifications are critical for TCH4 XET activity. *Plant J.* 5:553–61

11. Carpita NC. 1984. Cell wall development in maize coleoptiles. *Plant Physiol.* 76:205–12

12. Carpita NC. 1996. Structure and biogenesis of the cell walls of grasses. *Annu. Rev. Plant Physiol. Plant Mol. Biol.* 47:445–76

13. Carpita NC, Gibeaut DM. 1993. Structural models of primary cell walls in flowering plants: consistency of molecular structure with the physical properties of the walls during growth. *Plant J.* 3:1–30

14. Cassab GI. 1998. Plant cell wall proteins. *Annu. Rev. Plant Physiol. Plant Mol. Biol.* 49:281–309

15. Catala C, Rose JKC, Bennett AB. 1997. Auxin regulation and spatial localization of an endo-1,4-β-D-glucanase and a xyloglucan endotransglycosylase in expanding tomato hypocotyls. *Plant J.* 12:417–26

16. Chen F, Bradford KJ. 1998. Expansin genes are expressed in tomato seeds in association with germination. *Annu. Meet. Am. Soc. Plant Physiol. Abstr. 47*

17. Cho HT, Kende H. 1997. Expansins in deepwater rice internodes. *Plant Physiol.* 113:1137–43

18. Cho HT, Kende H. 1997. Expression of expansin genes is correlated with growth in deepwater rice. *Plant Cell* 9:1661–71

19. Cho HT, Kende H. 1998. Tissue localization of expansins in rice. *Plant J.* 15:805–12

20. Cleland RE. 1976. Fusicoccin-induced growth and hydrogen ion excretion of *Avena* coleoptiles: relation to auxin responses. *Planta* 128:201–6

21. Cleland RE, Cosgrove DJ, Tepfer M. 1987. Long-term acid-induced wall extension in an *in vitro* system. *Planta* 170:379–85

22. Cosgrove DJ. 1989. Characterization of long-term extension of isolated cell walls from growing cucumber hypocotyls. *Planta* 177:121–30

23. Cosgrove DJ. 1993. Wall extensibility: its nature, measurement, and relationship to plant cell growth. *New Phytol.* 124:1–23

24. Cosgrove DJ. 1997. Relaxation in a high-stress environment: the molecular bases of extensible cell walls and cell enlargement. *Plant Cell* 9:1031–41

25. Cosgrove DJ. 1998. Cell wall loosening by expansins. *Plant Physiol.* 118:333–39

26. Cosgrove DJ, Bedinger PA, Durachko DM. 1997. Group I allergens of grass pollen as cell wall loosening agents. *Proc. Natl. Acad. Sci. USA* 94:6559–64

27. Cosgrove DJ, Durachko DM. 1994. Autolysis and extension of isolated walls from growing cucumber hypocotyls. *J. Exp. Bot.* 45:1711–19

28. Cosgrove DJ, Durachko DM, Li L-C. 1998. Expansins may have cryptic endoglucanase activity and can synergize the breakdown of cellulose by fungal cellulases. *Annu. Meet. Am. Soc. Plant Physiol. Abstr. 171*

29. Cosgrove DJ, Durachko DM, Shcherban TY. 1998. The expansin super family in Arabidopsis. *9th Int. Conf. Arabidopsis Res.* 341 (Abstr.)

30. Cosgrove DJ, Shcherban TY, Durachko DM. 1998. Highly specific and distinct expression patterns for two alpha-expansin genes in Arabidopsis. *9th Int. Conf. Arabidopsis Res.* 166 (Abstr.)

31. Creelman RA, Mullet JE. 1997. Oligosaccharins, brassinolides, and jasmonates: nontraditional regulators of plant growth, development, and gene expression. *Plant Cell* 9:1211–23

32. Crowell DN. 1994. Cytokinin regulation of a soybean pollen allergen gene. *Plant Mol. Biol.* 25:829–35

33. Davies GJ. 1998. Structural studies on cellulases. *Biochem. Soc. Trans.* 26:167–73

34. Davies GJ, Tolley SP, Henrissat B, Hjort C, Schulein M. 1995. Structures of oligosaccharide-bound forms of the endoglucanase V from *Humicola insolens* at 1.9 Å resolution. *Biochemistry* 34:16210–20

35. Din N, Forsythe IJ, Burtnick LD, Gilkes NR, Miller RC, et al. 1994. The cellulose-binding domain of endoglucanase A (CenA) from *Cellulomonas fimi*: evidence for the involvement of tryptophan residues in binding. *Mol. Microbiol.* 11:747–55

36. Divne C, Ståhlberg J, Teeri TT, Jones TA.

1998. High-resolution crystal structures reveal how a cellulose chain is bound in the 50 Å long tunnel of cellobiohydrolase I from *Trichoderma reesei. J. Mol. Biol.* 275:309–25

37. Downes BP, Crowell DN. 1998. Cytokinin regulates the expression of a soybean β-expansin gene by a posttranscriptional mechanism. *Plant Mol. Biol.* 37:437–44

38. Fan D-F, Maclachlan GA. 1966. Control of cellulase activity by indoleacetic acid. *Can. J. Bot.* 44:1025–34

39. Fanutti C, Gidley MJ, Reid JSG. 1993. Action of a pure xyloglucan endo-transglycosylase (formerly called xyloglucan-specific endo-1,4-β-D-glucanase) from the cotyledons of germinated nasturtium seeds. *Plant J.* 3:691–700

40. Fleming AJ, McQueen-Mason S, Mandel T, Kuhlemeier C. 1997. Induction of leaf primordia by the cell wall protein expansin. *Science* 276:1415–18

41. Fry SC. 1989. Cellulases, hemicelluloses and auxin-stimulated growth: a possible relationship. *Physiol. Plant.* 75:532–36

42. Fry SC. 1995. Polysaccharide-modifying enzymes in the plant cell wall. *Annu. Rev. Plant Physiol. Plant Mol. Biol.* 46:497–520

43. Fry SC. 1998. Oxidative scission of plant cell wall polysaccharides by ascorbate-induced hydroxyl radicals. *Biochem. J.* 332:507–15

44. Fry SC, Smith RC, Renwick KF, Martin DJ, Hodge SK, Matthews KJ. 1992. Xyloglucan endotransglycosylase, a new wall-loosening enzyme activity from plants. *Biochem. J.* 282:821–28

45. Green PB. 1997. Expansin and morphology: a role for biophysics. *Trends Plant Sci.* 2:365–66

46. Green PB, Cummins WR. 1974. Growth rate and turgor pressure: auxin effect studied with an automated apparatus for single coleoptiles. *Plant Physiol.* 54:863–69

47. Hamamoto T, Foong F, Shoseyov O, Doi RH. 1992. Analysis of functional domains of endoglucanases from *Clostridium cellulovorans* by gene cloning, nucleotide sequencing and chimeric protein construction. *Mol. Gen. Genet.* 231:472–79

48. Hayashi T. 1989. Xyloglucans in the primary cell wall. *Annu. Rev. Plant Physiol. Plant Mol. Biol.* 40:139–68

49. Hayashi T, Wong YS, Maclachlan G. 1984. Pea xyloglucan and cellulose. II. Hydrolysis by pea endo-1,4-β-glucanases. *Plant Physiol.* 75:605–10

50. Hazlewood GP, Gilbert HJ. 1998. Structure and function analysis of *Pseudomonas* plant cell wall hydrolases. *Biochem. Soc. Trans.* 26:185–90

51. Henrissat B. 1998. Glycosidase families. *Biochem. Soc. Trans.* 26:153–56

52. Henrissat B, Teeri TT, Warren RA. 1998. A scheme for designating enzymes that hydrolyse the polysaccharides in the cell walls of plants. *FEBS Lett.* 425:352–54

53. Herbers K, Flint HJ, Sonnewald U. 1996. Apoplastic expression of the xylanase and β (1-3,1-4) glucanase domains of the xynD gene from *Ruminococcus flavefaciens* leads to functional polypeptides in transgenic tobacco plants. *Mol. Breeding* 2:81–87

54. Herbers K, Lorences EP, Barrachina C, Sonnewald U. 1998. Functional characterization of xyloglucan endotransglycosylase in transgenic tobacco: identification of a novel plant defense mechanism. *8th Int. Cell Wall Meet.* 7.11 (Abstr.)

55. Hiller KM, Esch RE, Klapper DG. 1997. Mapping of an allergenically important determinant of grass group I allergens. *J. Allergy Clin. Immunol.* 100:335–40

56. Hoj PB, Fincher GB. 1995. Molecular evolution of plant β-glucan endohydrolases. *Plant J.* 7:367–79

57. Hoson T. 1993. Regulation of polysaccharide breakdown during auxin-induced cell wall loosening. *J. Plant Res.* 103:369–81

58. Hoson T, Masuda Y, Nevins DJ. 1992. Comparison of the outer and inner epidermis. Inhibition of auxin-induced elongation of maize coleoptiles by glucan antibodies. *Plant Physiol.* 98:1298–303

59. Inouhe M, Nevins D. 1997. Regulation of cell wall glucanase activities by nonenzymic proteins in maize coleoptiles. *Int. J. Biol. Macromol.* 21:15–20

60. Inouhe M, Nevins DJ. 1997. Changes in the autolytic activities of maize coleoptile cell walls during coleoptile growth. *Plant Cell Physiol.* 38:161–67

61. Inouhe M, Nevins DJ. 1998. Changes in the activities and polypeptide levels of exo- and endoglucanases in cell walls during developmental growth of *Zea mays* coleoptiles. *Plant Cell Physiol.* 39:762–68

62. Keller E, Cosgrove DJ. 1995. Expansins in growing tomato leaves. *Plant J.* 8:795–802

63. Kemmerer EC, Tucker ML. 1994. Comparative study of cellulases associated with adventitious root initiation, apical buds, and leaf, flower, and pod abscission

zones in soybean. *Plant Physiol.* 104: 557–62

64. Kutschera U. 1996. Cessation of cell elongation in rye coleoptiles is accompanied by a loss of cell-wall plasticity. *J. Exp. Bot.* 47:1387–94

65. Labavitch JM, Ray PM. 1974. Relationship between promotion of xyloglucan metabolism and induction of elongation by IAA. *Plant Physiol.* 54:499–502

66. Li L-C, Wang X-C, Jing J-H. 1998. The existence of expansin and its properties in the hypocotyls of soybean seedlings. *Acta Bot. Sin.* 40:627–34

67. Li Z-C, Durachko DM, Cosgrove DJ. 1993. An oat coleoptile wall protein that induces wall extension in vitro and that is antigenically related to a similar protein from cucumber hypocotyls. *Planta* 191:349–56

68. Link BM, Cosgrove DJ. 1998. Acid growth response and α-expansins in suspension cultures of *Nicotiana tabacum* L. cv. BY2. *Plant Physiol.* 118:907–16

69. Luttenegger DG, Nevins DJ. 1985. Transient nature of a (1->3),(1->4)-β-D-glucan in *Zea mays* coleoptile cell walls. *Plant Physiol.* 77:175–78

70. Maclachlan G. 1988. β-glucanases from *Pisum sativum. Methods Enzymol.* 160: 382–91

71. Matthysse AG, White S, Lightfoot R. 1995. Genes required for celluose synthesis in *Agrobacterium tumefaciens. J. Bacteriol.* 177:1069–75

72. McCann MC, Roberts K. 1991. Architecture of the primary cell wall. In *Cytoskeletal Basis of Plant Growth and Form*, ed. C Lloyd, pp. 109–29. London/San Diego: Academic

73. McCann MC, Wells B, Roberts K. 1990. Direct visualization of cross-links in the primary plant cell wall. *J. Cell Sci.* 96:323–34

74. McQueen-Mason S, Cosgrove DJ. 1994. Disruption of hydrogen bonding between wall polymers by proteins that induce plant wall extension. *Proc. Natl. Acad. Sci. USA* 91:6574–78

75. McQueen-Mason S, Cosgrove DJ. 1995. Expansin mode of action on cell walls: analysis of wall hydrolysis, stress relaxation, and binding. *Plant Physiol.* 107:87–100

76. McQueen-Mason S, Durachko DM, Cosgrove DJ. 1992. Two endogenous proteins that induce cell wall expansion in plants. *Plant Cell* 4:1425–33

77. McQueen-Mason S, Fry SC, Durachko DM, Cosgrove DJ. 1993. The relationship between xyloglucan endotransglyco-sylase and in vitro cell wall extension in cucumber hypocotyls. *Planta* 190:327–31

78. Michael AJ. 1996. A cDNA from pea petals with sequence similarity to pollen allergen, cytokinin-induced and genetic tumour-specific genes: identification of a new family of related sequences. *Plant Mol. Biol.* 30:219–24

79. Moore RC, Flecker D, Cosgrove DJ. 1995. Expansin action on cells with tip growth and diffuse growth. *J. Cell. Biochem. Suppl.* 21A:457 (Abstr. J5–312)

80. Mutaftschiev S, Prat R, Pierron M, Devilliers G, Goldberg R. 1997. Relationship between cell-wall β-1,3-endglucanase activity and auxin-induced elongation in mung bean hypocotyl segments. *Protoplasma* 199:49–56

81. Nakamura S, Hayashi T. 1993. Purification and properties of an extracellular endo-1,4-β-glucanase from suspension-cultured poplar cells. *Plant Cell Physiol.* 34:1009–13

82. Nakamura S, Mori H, Sakai F, Hayashi T. 1995. Cloning and sequencing of a cDNA for poplar endo-1,4-β-glucanase. *Plant Cell Physiol.* 36:1229–35

83. Nicol F, His I, Jauneau A, Vernhettes S, Canut H, Höfte H. 1998. A plasma membrane-bound putative endo-1,4-β-D-glucanase is required for normal wall assembly and cell elongation in *Arabidopsis. EMBO J.* 17:5563–76

84. Nicol F, Hofte H. 1998. Plant cell expansion: scaling the wall. *Curr. Opin. Plant Biol.* 1:12–17

85. Nishitani K. 1997. The role of endoxyloglucan transferase in the organization of plant cell walls. *Int. Rev. Cytol.* 173:157–206

86. Nishitani K. 1998. Construction and restructuring of the cellulose-xyloglucan framework in the apoplast as mediated by the xyloglucan-related protein family—a hypothetical scheme. *J. Plant Res.* 111:1–8

87. Nishitani K, Masuda Y. 1982. Acid pH-induced structural changes in cell wall xyloglucans in *Vigna angularis* epicotyl segments. *Plant Sci. Lett.* 28:87–94

88. Nishitani K, Masuda Y. 1983. Auxin-induced changes in the cell wall xyloglucans: effects of auxin on the two different subfractions of xyloglucans in the epicotyl cell wall of *Vigna angularis. Plant Cell Physiol.* 24:345–55

89. Nishitani K, Tominaga T. 1992. Endo-xyloglucan transferase, a novel class of glycosyltransferase that catalyzes transfer of a segment of xyloglucan molecule

to another xyloglucan molecule. *J. Biol. Chem.* 267:21058–64

90. Orford SJ, Timmis JN. 1998. Specific expression of an expansin gene during elongation of cotton fibres. *BBA-Gene Struct. Expr.* 1398:342–46

91. Petersen A, Becker WM, Schlaak M. 1993. Characterization of grass group I allergens in timothy grass pollen. *J. Allergy Clin. Immunol.* 92:789–96

92. Preston RD. 1979. Polysaccharide conformation and cell wall function. *Annu. Rev. Plant Physiol.* 30:55–78

93. Ray PM, Ruesink AW. 1962. Kinetic experiments on the nature of the growth mechanism in oat coleoptile cells. *Dev. Biol.* 4:377–97

94. Rayle DL, Cleland RE. 1970. Enhancement of wall loosening and elongation by acid solutions. *Plant Physiol.* 46:250–53

95. Rayle DL, Cleland RE. 1992. The acid growth theory of auxin-induced cell elongation is alive and well. *Plant Physiol.* 99:1271–74

96. Reinhardt D, Wittwer F, Mandel T, Kuhlemeier C. 1998. Localized upregulation of a new expansin gene predicts the site of leaf formation in the tomato meristem. *Plant Cell* 10:1427–37

97. Rose JK, Hadfield KA, Labavitch JM, Bennett AB. 1998. Temporal sequence of cell wall disassembly in rapidly ripening melon fruit. *Plant Physiol.* 117:345–61

98. Rose JKC, Brummell DA, Bennett AB. 1996. Two divergent xyloglucan endotransglycosylases exhibit mutually exclusive patterns of expression in nasturtium. *Plant Physiol.* 110:493–99

99. Rose JKC, Lee HH, Bennett AB. 1997. Expression of a divergent expansin gene is fruit-specific and ripening-regulated. *Proc. Natl. Acad. Sci. USA* 94:5955–60

100. Saloheimo A, Henrissat B, Hoffren AM, Teleman O, Penttilä M. 1994. A novel, small endoglucanase gene, *egl5*, from *Trichoderma reesei* isolated by expression in yeast. *Mol. Microbiol.* 13:219–28

101. Schopfer P. 1996. Hydrogen peroxide-mediated cell-wall stiffening in vitro in maize coleoptiles. *Planta* 199:43–49

102. Shcherban TY, Shi J, Durachko DM, Guiltinan MJ, McQueen-Mason S, et al. 1995. Molecular cloning and sequence analysis of expansins—a highly conserved, multigene family of proteins that mediate cell wall extension in plants. *Proc. Natl. Acad. Sci. USA* 92:9245–49

103. Shimizu Y, Aotsuka S, Hasegawa O, Kawada T, Sakuno T, et al. 1997. Changes in levels of mRNAs for cell wall-related

enzymes in growing cotton fiber cells. *Plant Cell Physiol.* 38:375–78

104. Shoseyov O, Shpigel E, Shanil Z, Roiz L. 1998. Cellulose binding domain increases cellulose synthase activity and biomass of transgenic plants. *8th Int. Cell Wall Meet.* 1.55 (Abstr.)

105. Shpigel E, Roiz L, Goren R, Shoseyov O. 1998. Bacterial cellulose-binding domain modulates in vitro elongation of different plant cells. *Plant Physiol.* 117:1185–94

106. Sulová Z, Farkas V. 1998. A method for purification of XET based on affinity sorption of XET: xyloglucan complex on cellulose. *8th Int. Cell Wall Meet.* 7.41 (Abstr.)

107. Sulová Z, Takácová M, Steele NM, Fry SC, Farkas V. 1998. Xyloglucan endotransglycosylase: evidence for the existence of a relatively stable glycosylenzyme intermediate. *Biochem. J.* 330:1475–80

108. Talbott LD, Pickard BG. 1994. Differential changes in size distribution of xyloglucan in the cell walls of gravitropically responding *Pisum sativum* epicotyls. *Plant Physiol.* 106:755–61

109. Talbott LD, Ray PM. 1992. Changes in molecular size of previously deposited and newly synthesized pea cell wall matrix polysaccharides. *Plant Physiol.* 98:369–79

110. Talbott LD, Ray PM. 1992. Molecular size and separability features of pea cell wall polysaccharides. Implications for models of primary wall structure. *Plant Physiol.* 92:357–68

111. Teeri TT, Koivula A, Linder M, Wohlfahrt G, Divne C, Jones TA. 1998. *Trichoderma reesei* cellobiohydrolases: why so efficient on crystalline cellulose? *Biochem. Soc. Trans.* 26:173–78

112. Thomas BR, Simmons C, Inouhe M, Nevins DJ. 1998. Maize coleoptile endoglucanase is encoded by a novel gene family (Accession No. AF072326). PGR 98–143. *Plant Physiol.* 117:1525–25 (Abstr.)

113. Thompson JE, Fry SC. 1997. Trimming and solubilization of xyloglucan after deposition in the walls of cultured rose cells. *J. Exp. Bot.* 48:297–305

114. Thompson JE, Smith RC, Fry SC. 1997. Xyloglucan undergoes interpolymeric transglycosylation during binding to the plant cell wall *in vivo*: evidence from $^{13}C/^{3}H$ dual labelling and isopycnic centrifugation in caesium trifluoroacetate. *Biochem. J.* 327:699–708

115. Tomme P, Boraston A, McLean B, Kormos J, Creagh AL, et al. 1998. Characterization and affinity applications of

cellulose-binding domains. *J. Chromatogr. B* 715:283–96

116. Valent BS, Albersheim P. 1974. The structure of plant cell walls. V. On the binding of xyloglucan to cellulose fibers. *Plant Physiol.* 54:105–8

117. Veytsman BA, Cosgrove DJ. 1998. A model of cell wall expansion based on thermodynamics of polymer networks. *Biophys. J.* 75:2240–50

118. Vietor RJ, Ha MA, Aperley DC, Jarvis MC. 1998. Internal structure of cellulose microfibrils in primary cell walls. *8th Int. Cell Wall Meet.* 3.02 (Abstr.)

119. Wong YS, Fincher GB, Maclachlan GA. 1977. Kinetic properties and substrate specificities of two cellulases from auxin-treated pea epicotyls. *J. Biol. Chem.* 252:1402–7

120. Wong YS, Maclachlan GA. 1980. 1,3-β-D-Glucanases from *Pisum sativum* seedlings. III. Development and distribution of endogenous substrates. *Plant Physiol.* 65:222–28

121. Wu S-C, Blumer JM, Darvill AG, Albersheim P. 1996. Characterization of an endo-β-1,4-glucanase gene induced by auxin in elongating pea epicotyls. *Plant Physiol.* 110:163–70

122. Wu Y, Sharp RE, Durachko DM, Cosgrove DJ. 1996. Growth maintenance of the maize primary root at low water potentials involves increases in cell wall extensibility, expansin activity and wall susceptibility to expansins. *Plant Physiol.* 111:765–72

123. Yamamoto R, Sakurai N. 1990. A computer simulation of the creep process of the cell wall using stress relaxation parameters. *Biorheology* 27:759–68

Annu. Rev. Plant Physiol. Plant Mol. Biol. 1999. 50:419–46

LEAF DEVELOPMENT IN ANGIOSPERMS

Neelima Sinha

Section of Plant Biology, Division of Biological Sciences, University of California at Davis, Davis, California 95616; e-mail: nrsinha@ucdavis.edu

KEY WORDS: leaf morphogenesis, simple leaves, compound leaves, shoot meristems, homeobox genes

ABSTRACT

Leaves are produced in succession on the shoot apical meristem (SAM) of a plant. The three landmark stages in leaf morphogenesis include initiation, acquisition of suborgan identities, and tissue differentiation. The expression of various genes relative to these steps in leaf morphogenesis is described. KNOTTED-like homeobox (KNOX) genes, FLO/LFY, and floral homeotic genes may be involved in generation of leaf shape and complexity. The differences between compound leaves and simple leaves in gene expression characteristics and morphogenetic patterns are discussed.

CONTENTS

419

INTRODUCTION

The flowering plant body consists of an above-ground shoot system and a below-ground root system. The shoot system produces several kinds of organs with a diversity of functions. Cell divisions at the shoot apical meristem (SAM) followed by enlargement and differentiation of the derivative cells give rise to the shoot system. The central or axial part of the shoot consists of a generally radially symmetrical stem that can be branched or unbranched and bears lateral organs, the leaves. Leaves are typically bilaterally symmetrical and flattened and are borne on the flanks of the shoot apical meristem in a pattern characteristic for the species. SAMs have been discussed thoroughly elsewhere (32) as has acquisition of organ and tissue identity (145) and the specification of leaf identity during the life span of a shoot (68). Recent studies also seem to indicate that genes that play a role in meristem organization may also be involved (albeit sometimes by absence) in the formation or development of leaves.

Leaves are responsible for most of the fixed carbon in a plant and are critical to plant productivity and survival. They are also fascinating in their mode of initiation at the shoot apex, their arrangements on the shoot, and the diverse shapes and sizes they can attain. A great variability in leaf shapes and sizes is seen in nature. Leaves also exhibit varying degrees of complexity and range from simple to highly dissected. Leaves in most extant vascular plants have been termed megaphylls by evolutionary biologists. This distinguishes them from the earliest simple leaves in vascular plants, microphylls, that were proposed to have arisen as enations from the stem, had a simple vein, and were not associated with leaf gaps in the stem vasculature.

Within the seed plant lineage, both cycads and many angiosperm genera have compound leaves and the compound leaf of cycads possibly represents the ancestral condition in seed plants (30). It is generally assumed that the ancestral leaf form in angiosperms was simple (29, 148). The question naturally arises: Did complexity in leaf form evolve once or multiple times within angiosperms? Compound leaves occur in the palms, some aroids, and *Dioscorea* (among monocots) and in many unrelated lineages in the dicots. The scattered occurrences of compound leaves in families such as Solanaceae and Asteraceae, on the one hand, and Ranunculaceae on the other, point to numerous independent origins of this feature in the dicots (42).

MORPHOGENETIC EVENTS AT THE SHOOT APICAL MERISTEM

Developmental landmarks can be used to divide the process of leaf morphogenesis into three stages (145). At the organogenesis stage (stage 1), cells on the flank of the shoot apical meristem are set aside as the founder cells of the initiating leaf. Increased cell division rates characterize the region that will give rise to the leaf primordium. Stage 2 delimits the basic morphological domains for the growth and development of leaf parts. Cell and tissue differentiation occurs during the final stage (stage 3) of leaf development by coordinated processes of cell division, expansion, and differentiation (145). The latter two stages occur in the leaf primordium proper. Leaves can initiate at the SAM in one of several patterns. Numerous experimental studies indicate that preexisting primordia and the SAM itself can influence the placement of initiating primordia and that biophysical constraints may also play a role in primordium placement (36, 45, 137, 143, 162).

Leaf Initiation

The shoot apical meristem (SAM), once formed in the embryo or in an axillary position, initiates organ primordia throughout its life. The only exception may be SAMs under seasonal dormancy. After generating a series of leaf primordia, the SAM often terminates by producing carpel primordia in its function as a floral meristem. Appropriate environmental signals in *Impatiens* can cause the SAM to revert back to a vegetative phase after production of several floral organ whorls (7, 107).

Alteration in cell division activity at the SAM leads to leaf initiation. Periclinal cell divisions in the presumptive primordium region of the SAM rather than increase in cell division frequency have been suggested to be a major factor in leaf initiation (88). When cell divisions are completely suppressed in wheat SAMs by gamma irradiation, a bulge in the outermost or L1 layer of the SAM appears at the predicted leaf initiation site, suggesting that these sites can be set up without cell divisions (47). Further, mutations with compromised epidermal differentiation (like *crinkly4* in maize) or altered cell division patterns in the epidermis and other cell layers (like *pygmy-tangled* in maize) are capable of normal leaf initiation and morphogenesis. Biophysical studies have also suggested a role for wall loosening and cellulose microfibril reorganization in the L1 layer prior to leaf initiation (46, 126). This appears to be borne out by an experiment in which wall properties at the SAM were directly altered (36).

Gene Expression at the Time of Leaf Initiation

What are the genes that regulate these biophysical parameters at the SAM? Most genes involved in organ initiation events at the SAM appear to fall into two classes: They encode either for transcription factors or for receptor/signal molecules. Future research aimed at identifying downstream genes of these transcription factors should elucidate the signal transduction cascades that occur and connect gene expression to signaling events.

KNOTTED-LIKE GENES   The *KNOTTED*-like (*KNOX*) genes encode homeo-domain-containing proteins and have been subdivided into two classes (69, 158). *Kn1* and other Class I *KNOX* genes have also been associated with gain-of-function mutations that affect the maize leaf (37, 122). Mutations at the Class I *KNOX* gene *STM1* lead to absence of shoot meristems in Arabidopsis (6). The only known marker for leaf initiation appears to be the downregulation of Class I *KNOX* gene (like *KN1*, *KNAT1*, and *STM1*) expression in the P_0 primordium in maize and Arabidopsis (82, 85, 136). The maize SAM produces a series of repeated units called phytomers that comprise the internode, leaf, and axillary bud (38). Jackson and co-workers analyzed the expression of several *KNOX* genes in the SAM and the recently initiated phytomers in maize (53). The expression of *Rough Sheath1* (*Rs1*) and *KNOX3* in a ring below the initiating leaf corresponds to the incipient internode and axillary bud and is thought to also predict the site of leaf initiation and the basal limit of the vegetative phytomer (53). The function of Class II *KNOX* genes is as yet undetermined. 35S driven KN1 (or KNAT1) overexpression in Arabidopsis and tobacco leads to the production of lobed and puckered leaves and ectopic shoot apical meristems on leaves (18, 134). In contrast, not only are the Class II genes more ubiquitously expressed, but transgenic plants that overexpress or underexpress the *KNAT3* (*KNOX II*) gene display no obvious phenotypic abnormalities (127, 128). Other *KNOX* genes have been cloned from a number of flowering plant species (89, 91).

NO APICAL MERISTEM   The NAM family of transcription factors was identified by cloning genes causing mutations leading to fused organs and altered organ number. These genes are expressed in the SAM at organ boundaries and may delimit organs from each other and from the SAM. The *NAM* gene encodes a gene product of unknown function that is required for apical meristem formation in embryogenesis. Mutant seedlings attain viability by producing escape shoots (138). Arabidopsis has two *NAM* genes called *CUC* (*CUP SHAPED COTYLEDONS*) that have similar patterns of expression and function as the petunia *NAM* gene (1).

CLAVATA   *CLAVATA* genes control meristem size in Arabidopsis. Mutations at *CLV1* and *CLV3* lead to enlarged meristems and accumulation of excess undifferentiated cells in the meristems (20, 21). The *CLAVATA1* gene codes for a receptor kinase, which suggests a role for signal transduction events in shoot meristem function. It has been proposed that these genes serve to balance cell proliferation and organogenesis events. *CLV1* is expressed in the central zone and is absent from the flanks of the SAM (22). Genetic analyses indicate that the *CLV* genes and *STM* are each sensitive to dosage of the other and may play opposing roles in regulating the balance between cell division and cell differentiation at the SAM (19).

MGOUN   The *MGOUN (MGO)* genes also have a role in patterning at the SAM. Mutations at the *MGO* loci in Arabidopsis lead to reduced leaf number while larger SAMs are produced that tend to fragment into multiple meristems. The *MGO* genes act downstream of *STM1*, require meristematic tissue to act, and appear to be involved in primordium initiation (81).

TERMINAL EAR   The maize *terminal ear* mutation shows aberrant leaf initiation and development, irregular phyllotaxy, and altered internode length. The mutation is caused by absence of an RNA-binding protein similar to the yeast Mei2 protein. *TE1* RNA is expressed in semicircular bands just below the point of insertion of leaf primordia, and the gap in *TE1* expression marks the site of leaf initiation. It has been proposed that the *TE1* gene acts early in the process of leaf initiation by inhibiting cells that express it from becoming organizers of leaf development (156).

PHANTASTICA   Mutations at the *PHANTASTICA* locus (*PHAN*) in *Antirrhinum* lead to loss of dorsiventrality in leaves and floral organs (159). *PHAN* encodes a MYB domain protein. Although a role for the *PHAN* gene in leaf initiation events has not been determined, the gene is expressed in a pattern complementary to that seen for *STM1* (160). It has been proposed that *PHAN* serves to downregulate Class I *KNOX* gene expression in leaf primordia (123, 147).

LEAFY/FLORICAULA   The *LEAFY/FLORICAULA (FLO/LFY)* gene encodes a protein with a transcriptional activation domain (23, 163). Mutations in *FLO/LFY* result in replacement of flowers with leaf-bearing shoots and a reiteration of the inflorescence phase of development. The *FLO/LFY* gene product appears to be necessary for the production of determinate floral meristems. Although *FLO/LFY* expression is absent from vegetative meristems in Arabidopsis and *Antirrhinum*, the gene is expressed in newly initiated leaf primordia (11). In tobacco meristems, the *FLO/LFY* homologs *NFL1* and *NFL2* are expressed in vegetative SAMs in cells that may be precursors to procambium, as well as in the peripheral zone of the shoot apex. It has been proposed that the role of *NFL* may be to establish determinacy for recent derivatives of apical initial cells (66).

The expression patterns of these genes at the SAM and in the developing shoot are summarized in Figure 1. Other genes with roles in formative events in leaf initiation are expected to be identified. If biophysical properties of the SAM play a role in generating phyllotaxy one would expect to find some cytoskeletal and cell wall proteins that regulate phyllotaxy. This would also be

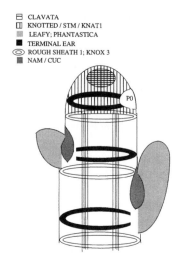

Figure 1 Gene expression related to simple leaf initiation at the shoot apex. The expression pattern of Class I *KNOX* genes *KN1* (136), *STM* (85), *KNAT1* (82), *CLAVATA1* (22), *LEAFY* (11), *PHANTASTICA* (159), *RS1* and *KNOX3* (53), *NAM/CUC* (1, 138) and *TERMINAL EAR* (156) as it relates to leaf initiation at the SAM and to the repeated phytomers on the shoot.

the prediction for leaf initiation events as early periclinal cell divisions are involved. It is also expected that plant hormones will be involved in this regulation. Hormone applications can cause shifts in phyllotaxy in ivy (114, 115). Application of auxin transport inhibitors such as triiodobenzoic acid (TIBA) not only leads to changes in leaf shape but also cause changes in phyllotaxy (124). Perhaps the most intriguing data that would connect gene expression levels to hormone signaling have come from measurement of plant hormone levels in transgenic plants overexpressing homeobox genes. These plants have reduced GA_1 content and suppression of GA 20-oxidase gene expression (72, 73). Thus, a complex network of events involving signaling at the SAM, hormone perception and transport, rearrangements of cytoskeletal and cell wall elements, and gene expression cascades occurs during the process of leaf initiation. The interconnecting threads between these various events remain to be identified.

LATER EVENTS IN LEAF MORPHOGENESIS

After leaf initiation, stage 2 in leaf morphogenesis proceeds, and suborgan identities or domains in the leaf are delimited. Along the three axes of the leaf these are the abaxial-adaxial (dorsiventral), apical-basal (proximodistal), and margin-blade-midrib (lateral) domains. In the maize leaf there are two major developmental gradients, one extending from the distal (leaf tip) to the proximal (leaf base) region and the other extending laterally out from the midrib. Developmental gradients are usually visible as a gradation of cell or tissue differentiation (especially trichomes on the epidermis) on the leaf (49). The homologies between leaf suborgans in dicot and monocot leaves are unclear. Analysis of leaf development in monocots by Kaplan (59) indicated that there is a distal unifacial upper leaf zone and a proximal bifacial sheath. Variation in leaf morphology results from a reciprocal elaboration and suppression of these two zones. In a number of monocot leaves the upper leaf zone is reduced to a fore-runner tip, whereas the lower leaf zone (the leaf base equivalent in a dicot leaf) gives rise to the rest of the leaf blade. Bharathan (10) examined leaf development in selected monocot genera and determined that some monocot species (for example, *Smilax, Aristolochia*) elaborate blade from the upper leaf zone and thus are more similar to dicots. Further detailed studies are needed to determine homologies of structures like the grass leaf sheath or dicot stipules.

Leaf Partitioning into Domains

Similar to the maize leaf, most other leaves also delimit domains along the three axes. The earliest reported acquisition of domain identity occurs in the maize leaf where cells that will give rise to the margin region of the leaf are morphogenetically distinct from other cells at the P_0 or founder cell stage (118, 119). In the *narrow sheath* mutation these cells fail to become incorporated into the founder cell population and a narrow leaf results. Differentiation of the leaf along its three axes also gives the leaf a shape characteristic of the species. Various physiological and genetic manipulations can lead to altered leaf shapes but their direct effects on differentiation along the three axes are unknown.

Dorsiventral patterning or acquisition of specific features along the adaxial/abaxial axis occurs very early in leaf primordia. While the terms adaxial and abaxial have very specific meanings, the terms dorsal and ventral have seen conflicting usage in plants (49, 59, 159). In this review, the more precisely defined terms adaxial and abaxial are used throughout to avoid any confusion. The abaxial surface of the leaf primordium grows faster (and shows earlier cellular differentiation) than the adaxial face and causes the primordium to arch over the SAM. The *phantastica* mutation in *Antirrhinum* produces leaves that are radially symmetrical and almost completely abaxialized. It has been proposed that PHAN plays a role in the acquisition of adaxial identity in lateral organs and that leaf blades arise at the junction between the adaxial and abaxial domains of the leaf. In the absence of PHAN function adaxial identity is not acquired, leading to loss of a boundary between the adaxial and abaxial domains and absence of blade (159). The *PHAN* transcript localizes to initiating leaf and floral organ primordia at the SAM and shows an expression pattern that is inverse of that for the KNOX gene *STM* (160). In the dominant *phabulosa-1d* (*phab-1d*) mutation in Arabidopsis, leaf polarity is also altered (92). However, in contrast to *phan*, adaxial features are present on the abaxial surface. These leaves fail to develop blades, supporting the hypothesis put forward by Waites & Hudson (159). The maize *rough sheath2* (*rs2*) mutant phenotype resembles that seen when KNOX genes (e.g. *Rs1, Kn1, Lg3*) are ectopically expressed in the maize leaf. Analysis of gene expression in *rs2* shoot apices indicates ectopic expression of the maize KNOX genes *Rs1* and *Kn1* in the leaf and a failure to downregulate KNOX 1 genes in the ring of founder cells. This leads to narrow and often bladeless leaves (123). The mutation is caused by alteration in a *PHAN*-like MYB gene (123, 147). Thus, *PHAN*-like genes may regulate the expression of *KNOX* genes. How this translates into the acquisition of dorsiventral identity remains to be elucidated. Mutations at the *ARGONAUTE* (*AGO*) locus in Arabidopsis also lead to radially symmetrical leaves (12). In maize, the *leafbladeless* (*lbl*) mutation also produces radially symmetrical leaves that are abaxialized. It has been proposed that *LBL* may have role in directly or indirectly downregulating *KN1* in leaf-like lateral organs (150). These results are summarized in Figure 2.

Partitioning in the proximo-distal dimension is more variable and depends on the gradient of differentiation in the leaf. In leaves with basipetal differentiation, the blade part of the leaf forms first and the petiole and base differentiate

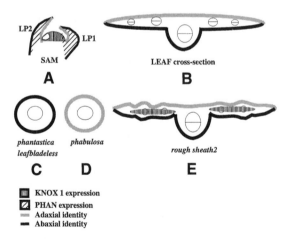

Figure 2   Adaxial-abaxial patterning in the leaf. *A*. The SAM with initiating leaf primordia showing KN1 expression (53, 136) and PHAN expression (159). The P_0 site has no KN1 expression. Adaxial-abaxial differentiation is apparent in initiating primordia and in the mature leaf in transverse section (*B*). *C*. Transverse section of completely abaxialized *phan* leaf with no blade (159). *D*. Transverse section of completely adaxialized *phab* leaf with no blade (92). *E*. Transverse section of a *rs2* leaf showing abnormal growth and ectopic KNOX expression with normal blade (123).

later. However, acropetal differentiation is also seen in a number of genera. Acquisition of proximo-distal identity in the maize leaf has been most thoroughly analyzed. The leaf primordium at plastochron 2 shows both transverse and longitudinal anticlinal divisions in the protoderm layer. During P_3 blade cells become differentiated from sheath cells and a preligular band forms between the two. During P_4 and P_5 ligule outgrowth can be seen as a ridge across the adaxial leaf surface and during P_5 the upper third of the blade completes its differentiation (144). The *LIGULELESS2* gene in maize encodes a basic leucine zipper (bZIP) transcription factor that is expressed in the SAM and developing ligule regions. Mutations in the gene lead to absence or incorrect positioning of both ligule and auricle tissue in the maize leaf (161). The *LIGULELESS1* gene in maize also encodes a nuclear localized protein similar to SQUAMOSA PROMOTER BINDING proteins 1 and 2 from snapdragon. The gene is expressed in leaf primordia at or prior to plastochron 6. Mutations at this locus also lead to loss of ligule and auricle and the formation of an imprecise blade sheath boundary (98). It has been proposed that *LG1* and *LG2* function in the same pathway. Early *LG2* function determines the precise positioning of the ligule and auricle and late *LG2* function interacts (either directly or indirectly) with *LG1* function to transmit and receive the make ligule/auricle signal (51). Analysis of chimeric leaves that ectopically express the *LG3 KNOX* gene in sectors indicates that there are competency states that all leaf cells go through: sheath, followed by auricle, immature blade, and mature blade. Expression of LG3 delays the acquisition of older cell fates in the leaf cells. Thus large sectors (early expression of LG3) remain sheath-like, whereas narrow small sectors (late expression of LG3) are almost normal (99). This confirms the maturation schedule hypothesis for the maize leaf put forward by Freeling (37).

In dicot leaves, partitioning of the leaf in the lateral dimension occurs concurrent with or a little delayed from proximodistal partitioning. Proximodistal leaf partitioning in tomato is initiated very early with the terminal leaflet tip differentiating at plastochron 1–2 (17). The first pair of lateral leaflets is produced basipetally when the primordium is about 300 microns in length at plastochron 3 (26). Procambium develops acropetally from the stem vascular cylinder and reaches the apex of the leaf primordium at approximately 300 micron length and is correlated with lamina development (24). This procambium will give rise to the leaf midvein. The marginal fimbriate vein, a characteristic feature of the tomato leaf margin, develops basipetally and continuously in the lamina of the terminal leaflet at about 500 micron leaf length and marks the end of marginal growth of the terminal leaflet (24). Later basipetally produced leaflets reiterate this pattern.

In contrast to the late development of marginal features in tomato leaves, recent studies indicate that maize leaves may be quite different. At plastochron 0 the presumptive leaf primordium is represented by a crescent of cells in the tunica and subepidermal regions that are undergoing cell divisions (129), referred to as the leaf founder cells. Development of the leaf midvein procambium marks the differentiation of the midrib region of the leaf and this event occurs early in leaf ontogeny, at plastochron 2 in maize. Procambial strands for the lateral veins that demarcate the blade reach the tip of the blade at plastochron 5 (129). This would suggest that differentiation in the lateral dimension to generate the blade occurs concurrently with proximodistal partitioning. The maize KN1 protein is absent from founder cells of the P_0 leaf primordium at the shoot apex. These cells form a complete ring that encircles the base of the SAM. Analysis of the narrow sheath mutation in maize indicates that there is a patch of cells that retain KN1 expression in the P_0 primordium, and are never initialized to make the margin domain of the maize leaf (120). Clonal analysis suggests that a region of the *ns* meristem is not utilized to generate founder cells in the margin domain (119). Therefore, partitioning in the lateral dimension (especially to generate the leaf margin) is determined in the SAM in maize and stage 1 and 2 are thus almost coincident. Whether this affects the peculiar linear and canalized way in which the maize leaves (and indeed all grass leaves) develop is not known. Perhaps very early lateral axis delimitation is the cause (or effect) of a very reduced upper leaf zone.

Acquisition of Leaf Shape, Size, and Complexity

Leaves can vary greatly in shape, size, and complexity along a single shoot. This variation can be of two types, heteroblastic or heterophyllic (2, 28, 41, 57, 93, 96 97, 121, 125, 139). Hormones like gibberellic acid (GA) or abscisic acid (ABA) appear to be involved in both heterophyllic and heteroblastic leaf development (3, 4, 27, 33, 40, 58, 113–115, 165, 166). The subject has been thoroughly reviewed by Kerstetter & Poethig (68).

Genetic Regulation of Leaf Shape, Size and Complexity

How leaf shape and size are regulated has also been investigated by using specific mutations that cause alterations in these parameters. Overexpression of Class I *KNOX* genes in tobacco and Arabidopsis leads to leaf lobing in an otherwise unlobed simple leaf (18, 82, 134). The *asymmetric* mutation in Arabidopsis causes the later leaves to become deeply lobed and asymmetric about the midrib. Based on a fewer number of hydathodes on the *as1* leaf, it was proposed that there was a loss of marginal segmentation in the mutation and that this was a late effect due to change in the direction in which new cells were supplied from the leaf base (155). The *curly leaf* mutation in Arabidopsis also has unusual leaf morphology including narrow, curled leaves of reduced length. The *CLF* gene is a member of the polycomb gene family. Genetic and molecular evidence suggests that *CLF* mRNA accumulates in leaf primordia and apical meristems where it serves to downregulate transcription of the *AGAMOUS* gene in leaves, inflorescence stems, and flowers (43, 130). Both the number and size of cells in the leaf are reduced in *clf* plants, which indicates a role for *CLF* in both the cell division and cell expansion phases of leaf development (71). Altered leaf shape and size are also seen in Arabidopsis plants transformed with an antisense construct of the Arabidopsis cDNA for methyltransferase (MET1). Ectopic expression of floral homeotic genes *AGAMOUS* and *APETALA3* is seen in leaf tissue in these plants (34). These results suggest that altered expression of homeobox genes or floral homeotic genes in leaves can lead to alterations in leaf shape, size, and growth patterns.

The role of cell division and cell expansion in organ shape generation has also been studied using a variety of mutations. Arabidopsis plants lacking microtubule preprophase bands show irregular cell division but still generate organs in their correct positions and differentiate appropriate tissue types (151). In the maize *warty* mutation abnormally large and improperly divided cells are produced and the cell cycle speeds up. Analysis of the mutation suggests that growth defects can be compensated for by alteration in cell cycle in neighboring cells and that cytokinesis may be linked to cell size ratios (110). The *diminuto* (*dim*) mutation in Arabidopsis produces plants with very short organs. The *DIM* gene encodes a protein with a putative nuclear localization signal, affects the expression of a β-tubulin gene, *TUB1*, and may play a critical role in plant cell elongation (146). The *ANGUSTIFOLIA* locus in Arabidopsis specifically regulates cell expansion in the transverse dimension, while the *ROTUNDIFOLIA3* locus affects cell expansion in the longitudinal dimension; these two loci act independently of each other (153). Mutations at these two loci produce leaves (and petals) that are narrower than normal or shorter than normal, respectively (153), indicating that defects in cell expansion may not always show compensatory mechanisms, as would be predicted by the organismal theory of plant development (64). In contrast, compensatory mechanisms are seen in the *pygmy/tangled* mutation in maize, where alterations in cell division along one axis may be compensated for by divisions in other dimensions so leaf shapes remain essentially unchanged (135). This suggests that morphogenesis can be uncoupled from the consequences of altered cell divisions but not altered cellular expansion.

Acquisition of Tissue Identities in the Leaf

During stage 3 of leaf development the leaf acquires tissue identities. Development of photosynthetic capability, a vascular system, and epidermal tissue all occur at this stage of leaf development. The various tissue identities are likely acquired in a coordinate manner. Furthermore, acquisition of the various tissue identities may not be hardwired as some tissues can develop even in the absence of others. An example is the acquisition of photosynthetic cell types even when vascular differentiation is aberrant. The earliest differentiation events are vascular, and the procambium of the future midrib often begins differentiating into the leaf primordium at its inception. The final events occur when the guard cells differentiate in the epidermis (24).

VASCULAR AND PHOTOSYNTHETIC DEVELOPMENT   Clonal analysis of maize leaf development indicates that the L1 layer gives rise to the abaxial and adaxial epidermises, while the L2 layer gives rise to all internal tissues in the leaf. This L2 layer divides once to form an abaxial and an adaxial layer. The adaxial layer most frequently divides once periclinally to give rise to the innermost tissue of the leaf that includes the vascular bundles, bundle sheaths, and the innermost mesophyll cells (76). As sector boundaries often lie in the middle of a vein, it appears that the veins are of mosaic origin with each half coming from a different progenitor cell (76). Ability to fix carbon via the C4 pathway in maize leaves seems closely associated with vascular development. RuBPCase downregulation occurs in cells closest to veins and bundle sheath cells. Organs that have widely spaced veins (e.g. husk leaves), or mutations that fail to develop proper bundle sheath cells, are C3/C4 intermediates with RuBPCASE expression in mesophyll cells (77, 132). Thus photosynthetic differentiation in the maize leaf appears to be intimately associated with vascular differentiation. This may not be true for dicot leaves as mutations that lead to a reduced vascular network differentiate a normal complement of photosynthetic cell types. However, chloroplast development affects differentiation of photosynthetic tissue in tomato. In the dcl (defective chloroplast and leaf) mutation in tomato, chloroplast development is aberrant and palisade cells fail to attain their characteristic columnar shape (65).

Several hypotheses have been put forward to explain vascular patterning in the leaf (reviewed in 100). Sachs (116) proposed a differentiation-dependent mechanism to explain patterning. Canalization of auxin flow through certain cells causes vascular patterns to develop as these cells become better and better transporters of auxin and eventually differentiate into vascular tissue. The diffusion-reaction prepattern hypothesis proposes that autocatalysis can lead to small random peaks in a homogeneous field, the peaks can increase by positive feedback, and long-range inhibition can prevent spread of the peaks (95). These two hypotheses have been put forward to explain vascular development in addition to stomatal and trichome patterning in the epidermis [for recent reviews see (78, 100, 145)].

EPIDERMAL DIFFERENTIATION   Epidermal identity is proposed to be established very early in embryonic development (13). Clonal analysis using revertant sectors generated in a mutable *gl1* mutation in maize indicates that the region of the epidermis between two adjacent lateral veins forms a compartment with an intermediate vein running in the middle. In this compartment, the founder cells are in the region closest to the midrib and undergo a very polarized mode of cell division in the protoderm of the maize leaf (14). Similar compartment boundaries have been proposed for the internal tissues in the maize leaf, but polarized cell divisions have not been described (76). The *cr4* mutation in maize shows aberrant epidermal development, defective aleurone layers, and epidermal fusions. The gene has been cloned and shown to be a maize homolog for the human tumor necrosis factor receptor (TNFR), suggesting that peptide signaling is probabaly involved in epidermal development (8). The *adherent* mutation in maize causes postgenital epidermal fusions without altering cellular identities within the leaf (131, 133). Recently, a large number of such mutations with abnormal epidermal fusions have been described in Arabidopsis and placed into nine complementation groups (84). In addition, a recently described mutation in maize, *Xcl 1*, causes extra cell layers to form in the leaf epidermis (70), and a multitude of both dominant and recessive loci appear to exist in maize that cause similar epidermal perturbations (S Kessler, personal communication). Taken together, these results indicate that multiple loci are involved not only in determining epidermal identity, but also in maintenance of this identity. Tissue patterning in the protoderm leads to the formation of specialized epidermal cells in a field of unspecialized or pavement cells. Cloning of the relevant genes from Arabidopsis indicates that MYB (GL1), basic helix-loop-helix (TTG), and homeobox (GL2) transcription factors are involved in regulation of patterning events, and that these factors may control trichome patterning and development events through cytoskeletal proteins such as kinesins (ZW1) (79, 80, 83, 101, 102, 109, 145). Stomatal patterning is less well understood but the future prospects in the field look bright (reviewed in 78, 145).

UNIQUE FEATURES OF COMPOUND LEAVES

Organs that bear separate foliar units called leaflets have been termed compound by some researchers (9, 140), whereas others treat leaves as a continuum between simple and highly dissected (60). The recent genetic and molecular treatments of leaves have used the compound terminology, which is also used here in the interest of homogeneity, with the caveat that the compound terminology is suggestive of such a leaf being a composite of many simple units. On the contrary, a highly dissected or compound leaf is thought by a majority of morphologists to be equivalent to a single simple leaf. However, compound leaves possess certain unique features that set them apart from simple leaves.

Morphogenetic Patterns

In dicots, compound leaves are very similar to simple leaves in initiation and growth patterns. Leaflets can be produced by one of three routes: acropetal, basipetal, or divergent (39). The most frequent pattern of leaflet initiation is basipetal. However, in bipinnately compound leaves the order in which secondary leaflets are produced is acropetal. Troll (152) proposed that in the divergent form basipetally initiated leaflets are homologs of lateral leaflets of the first order, while acropetally initiated leaflets are second-order leaflets developed from the terminal leaflet. According to Troll (152), acropetal leaflet initiation could be derived from the divergent form by suppression of basipetal leaflets.

The dicot simple leaf has been suggested to be derived phylogenetically from a compound leaf, and the basic leaf type is subdivided into segments (48). An opposing view suggested that that the simple leaf is the basic form that is maintained in ontogeny, with leaflets developing as do lobes in a simple leaf (31). It is likely that the ancestral state for seed plants as a whole was a compound leaf, whereas the ancestral state for the flowering plants was a simple leaf. It is unclear if later acquisitions of compound leaves represent a reversion to the ancestral (i.e. the cycad-like) condition.

There is likely to be similarity between simple and compound leaves in the patterning events at the shoot apex as no unique phyllotactic patterns are associated with compoundness. Similarly, morphological analyses of leaf initiation events do not reveal any unique features associated with compound structures in the early plastochrons. However, this may not indicate commonality of mechanisms.

Genetic and Morphological Analyses

The two species most thoroughly analyzed at both the genetic and molecular level are pea and tomato. Leaves in both these species are unipinnately compound. The pea leaf produces a pair of basal stipules, several pairs of unlobed lateral leaflets, two or more pairs of lateral tendrils, and a terminal tendril structure in acropetal succession. In contrast, the tomato leaf is unipinnately compound but leaflets are produced in basipetal succession on the rachis and are lobed.

GENETIC AND MORPHOLOGICAL ANALYSES OF PEA LEAVES   A number of leaf mutations and their interactions have been described by Marx (90). Three subdivisions in the pea leaf were proposed by Marx in the proximodistal axis—a proximal region consisting of basal stipules, a middle region consisting of paired lateral leaflets, and a distal region consisting of paired lateral leaflets and a single terminal tendril. Marx also proposed a further subdivision of the distal region into two compartments (characterized by *tac-tendrilled acacia* in which the upper leaf segment is divided into a terminal leaflet and with basal paired tendrils).

The stipules have to be accounted for when discussing leaf complexity in pea. The homology of stipules is unclear, and these structures have been variously treated as an integral part of the leaf, as extrafoliar appendages associated with the leaf, or as a unique organ (39). The presence of stipules per se does not make a leaf compound as many simple-leaved species also have stipules. However, in pea the stipules are markedly foliar. Combinations of mutations that affect the stipules and also alter waxes on either the abaxial or adaxial surface of the leaflets indicate that the stipules are an integral part of the pea leaf, and that stipule bases are similar to the abaxial leaflet surface, whereas stipule tips resemble the adaxial leaflet surface (90). Further, based on mutant combinations between *sinuate leaf* (*sil*) and *crispa*, and *sil, afila*, and *tendrilless*, Marx (90) suggested that stipules, while unambiguously a part of the pea leaf, are not simply homologous to leaflets but also have additional leaf-like features, as proposed by Jeune for the legumes (56).

In addition to making homeotic (90) or heterochronic (87, 157) conversions between tendrils and leaflets, certain mutations can either singly, or in combinations with other mutations, increase or decrease the degree of complexity seen in the pea leaf. The *uni* mutation leads to a leaf reduced to a terminal single, or two-leaflet structure with paired stipules still present at the base of the leaf. *uni* also produces abnormally proliferated floral meristems. When *afila* is placed in combination with *tendrilless*, a very complex parsley leaf–like phenotype called *pleiofila* is produced (87, 90). Lu and coworkers (87) suggested that the highly dissected parsley-like leaf of *af/af; tl/tl* represents the basal leaf form for pea. A heterochronic restriction of the developmental potential of this leaf by AF and TL leads to a unipinnate structure. Leaf development in pea was recently reviewed by Hofer & Ellis (52a).

Young (164) proposed a model that correlated meristem size to leaf development; primordia of decreasing sizes would give rise in order, to rachis, leaflet, or tendrils. Meicenheimer and coworkers (94) analyzed *af, st, tl*, and combinations of these mutations and concluded that what the primordium developed into depended on whether and for how long various defined meristems are active. Based on their results, they proposed independent genetic control over the formation of leaflet, stipule, and marginal meristems. These morphological studies show that differentiation occurs early and that development proceeds acropetally. While leaflet initiation in pea leaves is in acropetal succession, Villani & DeMason (157) found that the *pleiofila* form (*af/af; tl/tl*) showed bi-directional leaflet initiation in late postembryonic leaves.

GENETIC AND MORPHOLOGICAL ANALYSES OF TOMATO LEAVES   Compound leaves in the cultivated tomato have a terminal leaflet and three to four pairs of

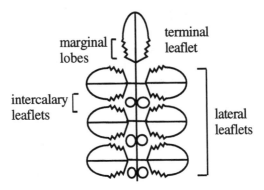

Figure 3   Wild-type tomato leaf.

lateral leaflets that are produced in a basipetal sequence. In addition, intercalary leaflets are produced between the major lateral leaflets (Figure 3). The leaflets are often deeply lobed, and the lobes themselves develop largely acropetally on each leaflet primordium (54). Thus, the tomato leaf shows two developmental gradients, an early basipetal gradient in the leaf primordium that generates leaflets, followed by a later acropetal gradient in each leaflet that leads to marginal lobe formation (54).

Based on the phenotypes produced, four general categories of mutations are seen in tomato (Figure 4). Type 1 mutations are defective in leaflet blade

Type 1 mutants	no laminar expansion	wiry wiry-3 wiry-6	
Type 2 mutants	simple or nearly simple leaf, or greatly reduced complexity	Lanceolate entire lyrata	
Type 3 mutants	leaflet margins reduced or unlobed	solanifolia potato leaf	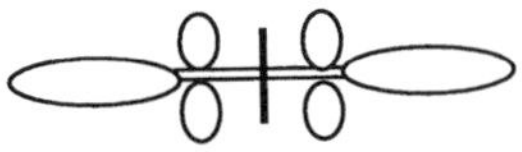
Type 4 mutants	leaves excessively subdivided	Mouse ears Curl Petroselenium bipinnata complicata clausa	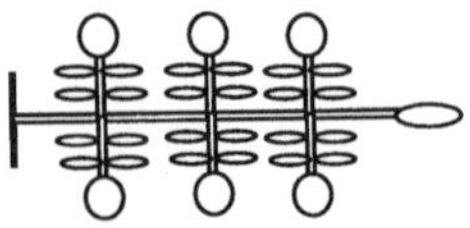

Figure 4   The four classes of leaf mutations seen in tomato. Mutant phenotypes are described against each class and typical leaves are diagrammed.

expansion. Type 2 mutations change the leaf into a simple or nearly simple leaf. Type 3 mutations alter the degree of leaflet lobing on a compound leaf. Type 4 mutations change the leaf into a more complex leaf with two or more orders of pinnation (N Sinha, unpublished observations).

The Type 1 mutations may be responsible for generating the leaf/leaflet blade, and any reduction in leaf complexity seen in these mutations will likely reflect the necessity of making a blade prior to leaflet initiation. This group of mutations is represented by the *wiry* series (142) and resembles the *phantastica* (*phan*) mutation in *Antirrhinum* (159). The several nonallelic *wiry* mutations in tomato could well represent PHAN and other upstream or downstream factors in the PHAN regulatory network.

The Type 2 mutations lead to reduced leaf dissection. Dengler (26) described the *Lanceolate* (*La*) and *entire* (*e*) mutants in tomato. In both *La* and *e*, lateral leaflet formation begins later than in wild type and is of shorter duration. Lamina expansion is faster in both these mutants compared to normal. Both *La* and *e* have reduced lobes in the leaf margins. The *Lanceolate* mutation was discovered in a primitive tomato cultivar in Peru and is an incomplete dominant. In the heterozygous stage small, simple, lanceolate leaves are produced. The homozygous mutant generally makes neither shoot apical meristem nor cotyledons and is a seedling lethal. Dosage analyses done by Stettler (141) show that extra wild-type doses of the gene cannot rescue the mutant phenotype, which suggests that the *La* mutation could be a dominant negative mutation.

Type 3 mutations have defective leaf margins and produce either unlobed leaflets or leaflets with very reduced marginal lobing. Chandra Shekhar & Sawhney (16) analyzed leaf development in the *solanifolia* (*sf*) mutant. In this recessive mutation, potato-like leaves are produced that lack lobes in the margins. Major differences between *sf* and wild-type leaves were seen at the time of leaflet initiation. In wild-type leaves, the first pair of leaflet primordia was produced at plastochron 3, whereas the first pair of leaflet primordia was produced at plastochron 5 in *sf*. Some of the mutant effects of the *SF* gene can be overcome by GA treatments or temperature-shift experiments (15). The *potato leaf* (*c*) mutation is very similar to the *sf* mutation in having unlobed leaflets. Rick & Harrison (112) tested for allelism between the two mutations and found that *c* was not allelic to *sf*. The double mutation *sf/sf*; *c/c* produced very reduced, unlobed and simple leaves.

Type 4 mutations increase the degree of dissection in the tomato leaf. The dominant *Mouse ears* and *Petroselenium* mutations, and the recessive *bipinnate*, *tripinnate*, and *clausa* mutations all show increased orders of leaflet proliferation leading to a bi- or tripinnate leaf. Since leaflets and marginal lobes are equivalent structures and leaflet initiation is in a basipetal sequence, proliferation of leaflets from the basal region or conversion of leaflet lobes into leaflets would result in excessively dissected leaves in the Type 4 mutations. Tomato plants overexpressing the Class I *KNOX* maize *KN1* gene, or the tomato *LeT6* gene, show leaves with increased dissection (50, 54). In mutations like *Lanceolate*, *entire*, and *trifoliate*, all with reduction in the number of leaflets, KN1 overexpression merely reiterates the basic architecture of the leaf. Thus, the *Lanceolate* leaf is lobed but not compound, while *trifoliate* reiterates the three-leaflet plan (50). This suggests that KN1-like function is insufficient to restore normal compound architecture in these leaves.

Molecular Analyses

Simple leaf development has been described in detail in maize and Arabidopsis (37, 106, 149, 154). It is unclear how common the morphogenetic mechanisms between simple and compound leaves will be. Indeed, whether the developmental principles derived from any of these model organisms will be generally applicable to all leaves remains to be seen. Determination of homologies between the various kinds of leaves and various suborgans of the leaves will be facilitated by comparison of the expression of molecular markers specific to developmental domains or key steps in morphogenesis.

CLASS I KNOX GENES   High levels of KNOX1 expression in the shoot apical meristems and downregulation or degradation of the gene product at presumptive initiation sites indicate a role for the *KNOX1* genes in the initiation and determination of lateral organs such as leaves or flowers (53, 82, 85, 136).

KNOX genes have also been cloned from compound-leaved species (89). The homolog of *STM1* has also recently been cloned from pea (C Gourlay & N Ellis, personal communication). In tomato, two Class I *KNOX* genes have been cloned and their orthology relationships determined. *TKN1*, and *LET6* (*TKN2*) are Class I genes, and a Class II gene, *LET12*, has also been identified (17, 50, 55, 103). Like the maize and Arabidopsis Class I *KNOX* genes, both *TKN1* and *LET6* (*TKN2*) are expressed at high levels in the shoot apical meristem. However, these genes are also expressed at presumed leaf initiation sites and in the leaf and leaflet primordia in tomato (17, 50, 55, 103). In contrast, in compound-leaved species in the Brassicaceae family, the leaf initiation sites show a downregulation of Class I *KNOX* gene expression. However, Class I *KNOX* expression is turned back on in leaves at later plastochrons prior to leaflet initiation (T Goliber & N Sinha, unpublished observations). A different situation occurs in pea leaves, where expression of the pea *SBH1* homolog is never seen in leaf primordia or older leaves (C Gourlay & N Ellis, personal communication). These results imply that the compound leaf in tomato, crucifers, and pea may not have arisen by similar mechanisms and are summarized in Figure 5.

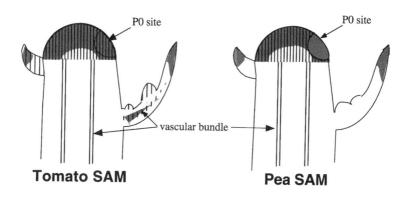

| | CLASS 1 KNOX expression

■ LFY/FLO expression

Figure 5   Gene expression during compound leaf development. Expression of Class I KNOX genes in tomato (17, 50, 54, 103) is compared to that described for pea (C Gourlay & N Ellis, personal communication). The expression of LFY/FLO is also compared between tomato (105) and pea (52).

Thus in tomato, the compound leaf program may diverge from that of simple leaves as early as stage 1 (leaf initiation), and the gene(s) controlling the difference between simple and compound leaf programs probably work upstream of *TKN1* and/or *LET6* (*TKN2*). This alteration in Class I *KNOX* expression within the shoot apical meristems and early determinate organ primordia is also seen in inflorescence meristems and in the Class II gene *LET12*, indicating that there may have been a global alteration in the regulation of *KNOX* genes in tomato (55). Both simple and compound leaves can occur in different species of the same genus (simple leaves in *Lepidium africanum* vs compound leaves in *Lepidium perfoliatum*) or even on the same shoot (*Neobeckia aquatica*), implying that determination of the degree of leaf dissection must involve only a small number of genes. While changes in the expression domain of genes like *TKN1*, *LET12*, and *LET6* (*TKN2*) in the SAM indicate that in tomato these changes may have involved regulatory genes, this cannot be the universal situation because of differences between the expression patterns of Class I *KNOX* genes in tomato on one hand and pea and *Lepidium* on the other.

FIMBRIATA/UNUSUAL FLORAL ORGANS   The pea homolog of *Antirrhinum* FIMBRIATA (Arabidopsis UFO) called STAMINA PISTILLOIDA (STP) has also recently been cloned. Severe mutant alleles of STP affect heteroblastic development of the pea compound leaf. In *stp* mutants, production of the first true leaf is delayed by a node compared to wild-type, and all leaves are more simple (possess fewer lateral structures) than leaves at a comparable node on wild-type plants (S Taylor & Ian Murfet, personal communication).

LEAFY/FLORICAULA   The pea leaf primordium generates second-order primordia in an acropetal series. In the *unifoliata* mutant, a pair of basal stipules subtend a terminal leaflet structure that may be single or subdivided into two units. The basic pinnate nature of the leaf is lost. Hofer and coworkers (52) have shown the *uni* mutation to be caused by deletions or alterations in the *PEAFLO* gene (the pea homolog of *LEAFY/FLORICAULA*). Alterations in flower development accompany leaf abnormalities in the *uni* mutation. *PEAFLO* is expressed in initiating leaf primordia and becomes restricted to the more distal (leaflet or tendril initiating) regions of the leaf in older primordia (summarized in Figure 5). While loss of FLO/LFY function leads to indeterminacy in inflorescence and floral meristems, loss of PEAFLO function prevents the acquisition of a transient phase of indeterminacy in pea leaves, preventing leaflet initiation and leading to production of a single lamina in the *uni* mutation (52). The role of *FLO/LFY* in tomato leaf morphogenesis remains to be investigated. However, expression studies indicate that the tomato *LFY/FLO* homolog *T-FLO* is expressed in vegetative apices and leaf primordia but expression is not seen in the central domain of the meristem (105). No known tomato mutations are associated with this gene. Thus, genes responsible for regulating reproductive development also function in the regulation of vegetative development. However, their roles in these different developmental phases of the plant may not be identical or similar and may even differ across families.

The Leaf Shoot Continuum Hypothesis

Based on patterns of leaflet initiation from the leaf primordium, some researchers considered compound leaves to be equivalent to shoot systems (5, 117). In *Murraya koenigii* and *Rhus typhina*, leaf primordia show an acropetal mode of leaflet initiation with leaflet primordia inserted transversely (with respect to the rachis) at the leaf primordium tip, thereby resembling leaf initiation at the shoot apex (74, 75, 117). Thus, the general issue of homology of compound leaves appears to be in dispute. Homeobox gene expression in the leaf and leaflet primordia in tomato resembles the expression pattern seen in the shoot apex itself (54, 55). Mutations that fail to maintain the shoot apical meristem and axillary meristems in tomato (111) initiate apical meristems at the junctions between the petiolule and rachis, the pseudoaxils. The mutation can be phenocopied by overexpressing homeobox genes (54), or pruning off all axillary and terminal meristems (CM Rick, personal communication). This suggests that the tomato compound leaf has some stem-like features and may be an intermediate structure between simple leaves and stems, although this finding may not be generalizable, and compound-leaved species will have to be evaluated case by case. Compound leaves in pea have been likened to flower-like determinate shoots (52). This parallel is based on the recent cloning of the pea leaf mutation *unifoliata*. Since loss-of-function mutations at FLO/LFY in Arabidopsis and snapdragon do not show any abnormalities in leaf development, it can be hypothesized that any leaf component to FLO/LFY function may be restricted to species with compound leaves. This function could be the presence of unique targets or activators present in compound leaves.

CONCLUSIONS

The utility of analyzing leaf development as a series of processes has been proven by recent genetic and molecular studies. While most of the processes discussed overlap to a certain degree, this reductionist approach has allowed simplified interpretations of what could otherwise be a complex series of events. However, in the final analysis, a synthesis of these various interpretations must be achieved so that a realistic picture emerges. Studies on simple leaves have led to the identification and cloning of numerous genes. However, parallels in morphogenetic processes between simple and compound leaves have been harder to derive.

In the two cases examined at the molecular level, compound leaves appear to show unique expression patterns of familiar genes. These are either the Class I *KNOX* genes (*KN1*, *STM1*, etc) that have a role in shoot meristem formation and maintenance, or *FLO/LFY* that has a role in suppressing indeterminacy in flowers. The "master regulator" of leaf dissection may be either one of these genes. However, alterations in some upstream regulator have more likely led to the expression of these shoot-specific genes in the leaf (or prevention of their turnover in leaf primordia), causing it to become compound. The coordinate upregulation of two Class I *KNOX* genes and one Class II *KNOX* gene in tomato leaf primordia is highly suggestive of the latter scenario. The fact that overexpression of Class I *KNOX* genes or *FLO/LFY* in species with simple leaves is insufficient to cause the generation of dissected leaves suggests that the master regulator is another factor or that the expression of more than one gene has to be altered to generate compound leaves. In any event, the analysis of compound leaves has led to the discovery of unknown functions for familiar genes.

Studies on leaf morphogenesis are rapidly progressing into the realm of gene cloning and functional analysis. Simultaneous investigation of model organisms such as maize, Arabidopsis, *Antirrhinum*, peas, and tomato should allow us not only to piece together the puzzle of how leaves are made, but also to discover equivalence relationships between different parts of these leaves. This area of investigation is poised to exploit the past genetic and morphological studies of leaves, and by combining such studies with molecular and functional analyses, generalizable principles for leaf morphogenesis can be derived.

ACKNOWLEDGMENTS

I thank Tom Goliber, J-J Chen, Sharon Kessler, Campbell Gourlay, Noel Ellis, Scott Taylor, and Ian Murfet for generously sharing unpublished data; Judy Jernstedt for sharing her knowledge of plant morphology and many helpful comments on the manuscript; members of my lab for their editorial suggestions; and Wynnelena Canio for help with processing the manuscript. Work in my lab on leaf development in tomato is supported by the National Science Foundation (IBN 9632013).

Visit the *Annual Reviews home page* at
http://www.AnnualReviews.org

Literature Cited

1. Aida M, Ishida T, Fukaki H, Fujisawa H, Tasaka M. 1997. Genes involved in organ separation in Arabidopsis: an analysis of the cup-shaped cotyledon mutant. *Plant Cell.* 9:841–57
2. Allsopp A. 1965. Land and water forms: physiological aspects. *Handb. Pflanzenphysiol.* 15:1236–55
3. Anderson LWJ. 1978. Abscisic acid induces formation of floating leaves in the heterophyllous aquatic angiosperm (*Potamogeton nodosus*). *Science* 201:1135–38
4. Anderson LWJ. 1982. Effects of abscisic acid on growth and leaf development in American pondweed (*Potamogeton nodosus* Poir.). *Aquat. Bot.* 13:29–44
5. Arber A. 1950. *The Natural Philosophy of Plant Form.* Cambridge: Cambridge Univ. Press
6. Barton MK, Poethig RS. 1993. Formation of the shoot apical meristem in *Arabidopsis thaliana*: an analysis of development in the wild type and in the *shoot meristemless* mutant. *Development* 119:823–31
7. Battey NH, Lyndon RF. 1990. Reversion of flowering. *Bot. Rev.* 56:162–89
8. Becraft PW, Stinard PS, McCarthy DR. 1996. Crinkly4: a TNFR-like receptor kinase involved in maize epidermal differentiation. *Science* 273:1406–9
9. Bell AD. 1991. *Plant Form: An Illus-*

trated Guide to Flowering Plant Morphology. Oxford/New York/Tokyo: Oxford Univ. Press
10. Bharathan G. 1996. Does the monocot mode of leaf development characterize all monocots? *Aliso* 14:271–79
11. Blazquez MA, Green R, Nilsson O, Sussman MR, Weigel D. 1998. Gibberellins promote flowering of Arabidopsis by activating the LEAFY promoter. *Plant Cell* 10:791–800
12. Bohmert K, Camus I, Bellini C, Bouchez D, Caboche M, Benning C. 1998. AGO1 defines a novel locus of Arabidopsis controlling leaf development. *EMBO J.* 17:170–80
13. Bruck DK, Walker DB. 1985. Cell determination during embryogenesis in *Citrus jambhiri*: II Epidermal differentiation as a one time event. *Am. J. Bot.* 72:1602–9
14. Cerioli S, Marocco A, Maddaloni M, Motto M, Salamini F. 1994. Early events in maize leaf epidermis formation as revealed by cell lineage studies. *Development* 120:2113–20
15. Chandra Shekhar KN, Sawhney VK. 1989. Regulation of the fusion of floral organs by temperature and gibberellic acid in the normal and *solanifolia* mutant of tomato (*Lycopersicon esculentum*). *Can. J. Bot.* 68:713–18

16. Chandra Shekhar KN, Sawhney VK. 1990. Leaf development in the normal and *solanifolia* mutant of tomato (*Lycopersicon esculentum*). *Am. J. Bot.* 77:46–53

17. Chen J-J, Janssen B-J, Williams A, Sinha N. 1997. A gene fusion at a homeobox locus: alternations in leaf shape and implications for morphological evolution. *Plant Cell* 9:1289–1304

18. Chuck G, Lincoln C, Hake S. 1996. *KNAT1* induces lobed leaves with ectopic meristems when overexpressed in Arabidopsis. *Plant Cell* 8:1277–89

19. Clark SE, Jacobsen SE, Levin JZ, Meyerowitz EM. 1996. The CLAVATA and SHOOT MERISTEMLESS loci competitively regulate meristem activity in Arabidopsis. *Development.* 122:1567–75

20. Clark SE, Running MP, Meyerowitz EM. 1993. CLAVATA1, a regulator of meristem and flower development in *Arabidopsis. Development* 121:2057–67

21. Clark SE, Running MP, Meyerowitz EM. 1995. CLAVATA3 is a specific regulator of shoot and floral meristem development affecting the same process as CLAVATA1. *Development* 121:2057–67

22. Clark SE, Williams RW, Meyerowitz EM. 1997. The CLAVATA1 gene encodes a putative receptor kinase that controls shoot and floral meristem size in Arabidopsis. *Cell* 89:575–585

23. Coen ES, Romero JM, Doyle S, Elliott R, Murphy G, Carpenter R. 1990. *floricaula*: a homeotic gene required for flower development in *Antirrhinum majus. Cell* 63:1311–22

24. Coleman WK, Greyson RI. 1976. The growth and development of the leaf in tomato (*Lycopersicon esculentum*). II. Leaf ontogeny. *Can. J. Bot.* 54:2704–17

25. Collier DE, Grodzinski B. 1996. Growth and maintenance respiration of leaflet, stipule, tendril, rachis, and petiole tissues that make up the compound leaf of pea (*Pisum sativum*). *Can. J. Bot.* 74:1331–37

26. Dengler NG. 1984. Comparison of leaf development in normal $(+/+)$, *entire* (*e/e*), and *Lanceolate* (*La*/+) plants of tomato, *Lycopersicon esculentum* 'Ailsa Craig'. *Bot. Gaz.* 145:66–77

27. Deschamp PA, Cooke TJ. 1983. Leaf dimorphism in aquatic angiosperms: significance of turgor pressure and cell expansion. *Science* 219:505–7

28. Deschamp PA, Cooke TJ. 1985. Leaf dimorphism in the aquatic angiosperm *Callitriche heterophylla. Am. J. Bot.* 72:1377–87

29. Donoghue MJ, Doyle JA. 1989. Phylogenetic analysis of angiosperms and the relationships of Hamamelidae. In *Evolution, Systematics, and Fossil History of the Hamamelidae*, ed. PR Crane, S Blackmore, 1:17–45. Oxford: Clarendon

30. Doyle JA, Donoghue MJ. 1986. Seed plant phylogeny and the origin of angiosperms: an experimental cladistic approach. *Bot. Rev.* 52:321–431

31. Eames AJ. 1961. *Morphology of Angiosperms*. New York: McGraw Hill

32. Evans MMS, Barton MK. 1997. Genetics of angiosperm shoot apical meristem development. *Annu. Rev. Plant Physiol. Plant Mol. Biol.* 48:673–701

33. Evans MMS, Poethig RS. 1995. Gibberellins promote vegetative phase change and reproductive maturity in maize. *Plant Physiol.* 108:475–87

34. Finnegan EJ, Peacock WJ, Dennis ES. 1996. Reduced DNA methylation in *Arabidopsis thaliana* results in abnormal plant development. *Proc. Natl. Acad. Sci. USA* 93:8449–54

35. Deleted in proof

36. Fleming AJ, McQueen-Mason S, Mandel T, Kuhlemeier C. 1997. Induction of leaf primordia by the cell wall protein expansin. *Science* 276:1415–18

37. Freeling M. 1992. A conceptual framework for maize leaf development. *Dev. Biol.* 153:44–58

38. Galinat WC. 1959. The phytomer in relation to the floral homologies in the American Maydea. *Bot. Mus. Lefl. Harvard Univ.* 19:1–32

39. Gifford EM, Foster AS. 1989. *Morphology and Evolution of Vascular Plants*. New York: Freeman

40. Goliber TE, Feldman LJ. 1989. Osmotic stress, endogenous abscisic acid, and the control of leaf morphology in *Hippuris vulgaris* L. *Plant Cell Environ.* 12:163–71

41. Goliber TE, Feldman LJ. 1990. Developmental analysis of leaf plasticity in the heterophyllous aquatic plant *Hippuris vulgaris. Am. J. Bot.* 77:399–412

42. Goliber TE, Kessler SA, Chen JJ, Bharathan G, Sinha N. 1998. Genetic, molecular and morphological analysis of compound leaf development. *Curr. Topics Dev. Biol.* 41:259–90

43. Goodrich J, Puangsomlee P, Martin M, Long D, Meyerowitz EM, Coupland G. 1997. A polycomb-group gene regulates homeotic gene expression in *Arabidopsis. Nature* 386:44–51

44. Deleted in proof

45. Green PB. 1987. Inheritance of pattern:

analysis from phenotype to gene. *Am. Zool.* 27:657–73

46. Green PB, Lang JM. 1981. Toward a biophysical theory of organogenesis: birefringence observations on regenerating leaves in the succulent, *Graptopetalum paraguayense* E. Walther. *Planta* 151:413–26

47. Haber AH. 1962. Nonessentiality of concurrent cell divisions for degree of polarization of leaf growth. I. Studies with radiation induced mitotic inhibition. *Am. J. Bot.* 49:583–89

48. Hagemann W. 1984. Morphological aspects of leaf development in ferns and angiosperms. In *Contemporary Problems in Plant Anatomy*, ed. RA White, WC Dickison, pp. 301–49. New York: Academic

49. Hagemann W, Gleissberg S. 1996. Organogenetic capacity of leaves: the significance of marginal blastozones in angiosperms. *Plant Syst. Evol.* 199:121–52

50. Hareven D, Gutfinger T, Parnis A, Eshed Y, Lifschitz E. 1996. The making of a compound leaf: genetic manipulation of leaf architecture in tomato. *Cell* 84:735–44

51. Harper L, Freeling M. 1996. Interactions of liguleless1 function during ligule induction in maize. *Genetics* 144:1871–82

52. Hofer J, Turner L, Hellens R, Ambrose M, Matthews P, et al. 1997. *UNIFOLIATA* regulates leaf and flower morphogenesis in pea. *Curr. Biol.* 7:581–87

52a. Hofer JMI, Ellis THN. 1998. The genetic control of patterning in pea leaves. *Trends Plant Sci.* 3:439–44

53. Jackson D, Veit B, Hake S. 1994. Expression of maize *KNOTTED-1* related homeobox genes in the shoot apical meristem predicts patterns of morphogenesis in the vegetative shoot. *Development* 120:405–13

54. Janssen B-J, Lund L, Sinha N. 1998. Overexpression of a homeobox gene, *LeT6*, reveals indeterminate features in the tomato compound leaf. *Plant Physiol.* 117:771–86

55. Janssen B-J, Williams A, Chen J-J, Mathern J, Hake S, Sinha N. 1998. Isolation and characterization of two knotted-like homeobox genes from tomato. *Plant Mol. Biol.* 36:417–25

56. Jeune B. 1981. Modèle empirique du développement des feuilles de Dicotyledons. *Bull. Mus. Natl. Hist. Nat. Sect. B Adansonia Bot. Phytochim.* 4:433–59

57. Jones CS. 1993. Heterochrony and heteroblastic leaf development in two subspecies of *Cucurbita argyrosperma* (Cucurbitaceae). *Am. J. Bot.* 80:778–95

58. Kane ME, Albert LS. 1987. Abscisic acid induces aerial leaf morphology and vasculature in submerged *Hippuris vulgaris* L. *Aquat. Bot.* 28:81–88

59. Kaplan DR. 1973. The monocotyledons: their evolution and comparative biology. VII. The problem of leaf morphology and evolution in the monocotyledons. *Q. Rev. Biol.* 48:437–57

60. Kaplan DR. 1975. Comparative developmental evaluation of the morphology of unifacial leaves in the monocotyledons. *Bot. Jahrb. Syst.* 95:1–105

61. Deleted in proof

62. Deleted in proof

63. Deleted in proof

64. Kaplan DR, Hagemann W. 1991 The relationship of cell and organism in vascular plants. *BioScience* 41:693–703

65. Keddie JS, Carroll B, Jones JDG, Gruissem W. 1996. The DCL gene of a tomato is required for chloroplast development and palisade morphogenesis in leaves. *EMBO J.* 15:4208–17

66. Kelly AJ, Bonnlander MB, Meeks-Wagner DR. 1995. NFL, the tobacco homolog of FLORICAULA and LEAFY, is transcriptionally expressed in both vegetative and floral meristems. *Plant Cell* 7:225–34

67. Deleted in proof

68. Kerstetter RA, Poethig RS. 1998. The specification of leaf identity during shoot development. *Annu. Rev. Cell Dev. Biol.* 14:373–98

69. Kerstetter R, Vollbrecht E, Lowe B, Yamaguchi J, Hake S. 1994. Sequence analysis and expression patterns divide the maize *knotted1*-like homeobox genes into two classes. *Plant Cell* 6:1877–87

70. Kessler S, Sinha N. 1997. Identity of extra cell-layers produced in the *glossy** mutation in maize. *Maize Gen. Coop. Newslett.* 71:30–31 (cited by permission)

71. Kim GT, Tsukaya H, Uchimiya H. 1998. The *CURLY LEAF* gene controls both division and elongation of cells during the expansion of the leaf blade in *Arabidopsis thaliana*. *Planta* 206:175–83

72. Kusaba S, Fukumoto M, Honda C, Yamaguchi I, Sakamoto T, et al. 1998. Decreased GA1 content caused by the overexpression of *OSH1* is accompanied by suppression of GA 20-oxidase gene expression. *Plant Physiol.* 117:1179–84

73. Kusaba S, Kano-Murakami Y, Matsuoka M, Tamaoki M, Sakamoto T, et al. 1998. Alteration of hormone levels in transgenic tobacco plants overexpressing a rice homeobox gene *OSH1*. *Plant Physiol.* 116:471–76

74. Lacroix CR. 1995. Changes in leaflet and leaf lobe form in developing compound and finely divided leaves. *Bot. Jahrb. Syst.* 117:317–31

75. Lacroix CR, Sattler R. 1994. Expression of shoot features in early leaf development of *Murraya paniculata* (Rutaceae). *Can. J. Bot.* 72:678–87

76. Langdale JA, Lane B, Freeling M, Nelson T. 1989. Cell lineage analysis of maize bundle sheath and mesophyll cells. *Dev. Biol.* 133:128–39

77. Langdale JA, Zelitch I, Miller E, Nelson T. 1988. Cell position and light influence C4 versus C3 patterns of photosynthetic gene expression in maize. *EMBO J.* 7:3643–51

78. Larkin JC, Marks MD, Nadeau J, Sack F. 1997. Epidermal cell fate and patterning in leaves. *Plant Cell* 9:1109–20

79. Larkin JC, Oppenheimer DG, Lloyd AM, Paparozzi ET, Marks MD. 1994. Roles of the *Glabrous1* and *Transparent Testa Glabra* genes in *Arabidopsis* trichome development. *Plant Cell* 5:1065–76

80. Larkin JC, Young N, Prigge M, Marks MD. 1996. The control of trichome spacing and number in Arabidopsis. *Development* 122:997–1005

81. Laufs P, Dockx J, Kronenberger J, Traas J. 1998. *Mgoun1* and *mgoun2*: two genes required for primordium initiation at the shoot apical and floral meristems in *Arabidopsis thaliana*. *Development* 125:1253–60

82. Lincoln C, Long J, Yamaguchi J, Serikawa K, Hake S. 1994. A *Knotted1*-like homeobox gene in Arabidopsis is expressed in the vegetative meristem and dramatically alters leaf morphology when overexpressed in transgenic plants. *Plant Cell* 6:1859–76

83. Lloyd AM, Walbot V, Davis RW. 1992. *Arabidopsis* and *Nicotiana* anthocyanin production activated by maize regulators *R* and *Cl. Science* 258:1173–75

84. Lolle SJ, Hsu W, Pruitt RE. 1998. Genetic analysis of organ fusion in *Arabidopsis thaliana*. *Genetics* 149:607–19

85. Long JA, Moan EI, Medford JI, Barton MK. 1996. A member of the KNOTTED class of homeodomain proteins encoded by the *SHOOTMERISTEMLESS* gene of *Arabidopsis*. *Nature* 379:66–69

86. Deleted in proof

87. Lu B, Villani PJ, Watson JC, Darllen AD, Cooke TJ. 1996. The control of pinna morphology in wild-type and mutant leaves of the garden pea (*Pisum sativum* L.). *Int. J. Plant Sci.* 157:659–73

88. Lyndon RF. 1982. Changes in polarity of

growth during leaf initiation in the pea, *Pisum sativum* L. *Ann. Bot.* 49:281–90

89. Ma H, McMullen MD, Finer JJ. 1994. Identification of a homeobox-containing genes with enhanced expression during soybean (*Glycine max* L.) somatic embryo development. *Plant Mol. Biol.* 24:465–73

90. Marx GA. 1987. A suite of mutants that modify pattern formation in pea leaves. *Plant Mol. Biol. Rep.* 5:311–35

91. Matsuoka M, Ichikawa H, Saito A, Tada Y, Fujimura T, Kano-Murakami Y. 1993. Expression of a rice homeobox gene causes altered morphology of transgenic plants. *Plant Cell* 18:1039–48

92. McConnell JR, Barton MK. 1998. Leaf polarity and meristem formation in *Arabidopsis*. *Development* 125:2935–42

93. McLellan T. 1993. The roles of heterochrony and heteroblasty in the diversification of leaf shapes in *Begonia dregei* (Begoniaceae). *Am. J. Bot.* 80:796–804

94. Meicenheimer RD, Muehlbauer FJ, Hindman JL, Gritton ET. 1983. Meristem characteristics of genetically modified pea (*Pisum sativun*) leaf primordia. *Can. J. Bot.* 61:3430–37

95. Meinhardt H. 1995. Development of higher organisms: how to avoid error propagation and chaos. *Physica* 86:96–103

96. Merrill EK. 1986. Heteroblastic seedlings of green ash: I. Predictability of leaf form and primordial length. *Can. J. Bot.* 64:2645–49

97. Merrill EK. 1986. Heteroblastic seedlings of green ash: II. Early development of simple and compound leaves. *Can. J. Bot.* 64:2650–61

98. Moreno MA, Harper LC, Krueger RW, Dellaporta SL, Freeling M. 1997. Liguleless1 encodes a nuclear-localized protein required for induction of ligules and auricles during maize leaf organogenesis. *Genes Dev.* 11:616–28

99. Muehlbauer G, Fowler JE, Freeling M. 1997. Sectors expressing the homeobox gene liguleless3 implicate a time-dependent mechanism for cell fate acquisition along the proximal-distal axis of maize leaf. *Development* 124:5097–106

100. Nelson T, Dengler N. 1997. Leaf vascular pattern formation. *Plant Cell* 9:1121–35

101. Oppenheimer DG, Herman PL, Sivakumaran S, Esch J, Marks MD. 1991. A *myb* gene required for leaf trichome differentiation in Arabidopsis is expressed in stipules. *Cell* 67:483–93

102. Oppenheimer DG, Pollock MA, Vacik J, Szymanski DB, Ericson B, et al. 1997.

Essential role of a kinesin-like protein in Arabidopsis trichome morphogenesis. *Proc. Natl. Acad. Sci. USA* 94:6261–66

103. Parnis A, Cohen O, Gutfinger T, Hareven D, Zamir D, Lifschitz E. 1997. The dominant developmental mutants of tomato, *Mouse-ear* and *Curl*, are associated with distinct modes of abnormal transcriptional regulation of a Knotted gene. *Plant Cell* 9:2143–58

104. Deleted in proof

105. Pneuli L, Carmel-Goren L, Hareven D, Gutfinger T, Alvarez J, et al. 1998. The *Self-Pruning* gene of tomato regulates vegetative to reproductive switching of sympodial meristems and is the ortholog of *CEN* and *TFL1*. *Development* 125:1979–89

106. Poethig S. 1984. Cellular parameters of leaf morphogenesis in maize and tobacco. In *Contemporary Problems in Plant Anatomy*, ed. RA White, WC Dickison, pp. 235–59. Orlando: Academic

107. Pouteau S, Nicholls D, Tooke F, Coen E, Battey N. 1997. The induction of flowering in *Impatiens*. *Development* 124:3343–51

108. Deleted in proof

109. Rerie WG, Feldmann KA, Marks MD. 1994. The *GLABRA2* gene encodes a homeodomain protein required for normal trichome development in Arabidopsis. *Genes Dev.* 8:1388–99

110. Reynolds JO, Eisses JF, Sylvester AW. 1998. Balancing division and expansion during maize leaf morphogenesis: analysis of the mutant, *warty-1*. *Development* 125:259–68

111. Rick CM, Butler L. 1956. Cytogenetics of tomato. *Adv. Genet.* 7:267–382

112. Rick CM, Harrison AL. 1959. Inheritance of five new tomato seedling characters. *J. Hered.* 50:91–98

113. Robbins WJ. 1957. Gibberellic acid and the reversal of adult Hedera to a juvenile state. *Am. J. Bot.* 44:743–46

114. Rogler CE, Hackett WP. 1975. Phase change in *Hedera helix*: induction of the mature to juvenile phase change by gibberellin A_3. *Physiol. Plant.* 34:141–47

115. Rogler CE, Hackett WP. 1975. Phase change in *Hedera helix*: stabilization of the mature form with abscisic acid and growth retardants. *Physiol. Plant.* 34:148–52

116. Sachs T. 1991. Cell polarity and tissue patterning in plants. *Dev. Suppl.* 1:83–93

117. Sattler R, Rutishauser R. 1992. Partial homology of pinnate leaves and shoots: orientation of leaflet inception. *Bot. Jahrb. Syst.* 114:61–79

118. Scanlon MJ, Freeling M. 1998. The *narrow sheath* leaf domain deletion: a genetic tool used to reveal developmental homologies among modified maize organs. *Plant J.* 13:547–61

119. Scanlon MJ, Freeling M. 1997. Clonal sectors reveal that a specific meristematic domain is not utilized in the maize mutant *narrow sheath*. *Dev. Biol.* 182:52–66

120. Scanlon MJ, Schneeberger RG, Freeling M. 1996. The maize mutant *narrow sheath* fails to establish leaf margin identity in a meristematic domain. *Development* 122:1683–91

121. Schmidt BL, Millington WF. 1968. Regulation of leaf shape in *Proserpinaca palustris*. *Bull. Torrey Bot. Club* 95:264–86

122. Schneeberger RG, Becraft PW, Hake S, Freeling M. 1995. Ectopic expression of the *knox* homeo box gene *rough sheath 1* alters cell fate in the maize leaf. *Genes Dev.* 9:2292–304

123. Schneeberger R, Tsiantis M, Freeling M, Langdale JA. 1998. The *rough sheath2* gene negatively regulates homeobox gene expression during maize leaf development. *Development* 125:2857–65

124. Schwabe WW. 1971. Chemical modification of phyllotaxis and its implications. *Symp. Soc. Exp. Biol.* 25:301–22

125. Sculthorpe CD. 1967. *The Biology of Aquatic Vascular Plants*. London: Arnold

126. Selker JML, Green PB. 1984. Organogenesis in *Graptopetalum paraguayense* E. Walther: Shifts in orientation of cortical microtubule arrays are associated with periclinal divisions. *Planta* 160:289–97

127. Serikawa KA, Martinez-Laborda A, Kim HS, Zambryski PC. 1997. Localization of expression of *KNAT3*, a class 2 *knotted1*-like gene. *Plant J.* 11:853–61

128. Serikawa KA, Zymbryski PC. 1997. Domain exchanges between KNAT3 and KNAT1 suggest specificity of the kn1-like homeodomains requires sequences outside the third helix and N-terminal arm of the homeodomain. *Plant J.* 11:863–69

129. Sharman BC. 1942. Developmental anatomy of the shoot of *Zea mays* L. *Ann. Bot. n.s.* 6:245–83

130. Sieburth LE, Meyerowitz EM. 1997. Molecular dissection of the *AGAMOUS* control region shows that *cis* elements for spatial regulation are located intragenically. *Plant Cell* 9:355–65

131. Sinha N. 1998. Organ and cell fusions in the *adherent1* mutant in maize. *Int. J. Plant Sci.* 159:702–15

132. Sinha N, Hake S. 1994. The Knotted leaf blade is a mosaic of blade, sheath, and auricle identities. *Dev. Genet.* 15:401–14

133. Sinha N, Lynch M. 1998. Fused organs in the *adherent*1 mutation in maize show altered epidermal walls with no perturbations in tissue identities. *Planta* 206:184–95

134. Sinha N, Williams RE, Hake S. 1993. Overexpression of the maize homeobox gene, *KNOTTED-1*, cause a switch from determinate to indeterminate cell fates. *Genes Dev.* 7:787–95

135. Smith LG, Hake S, Sylvester AW. 1996. The *tangled-1* mutation alters cell division orientations throughout maize leaf development without altering leaf shape. *Development.* 122:481–89

136. Smith LG, Greene B, Veit B, Hake S. 1992. A dominant mutation in the maize homeobox gene, *Knotted-1*, cause its ectopic expression in leaf cells with altered fates. *Development* 116:21–30

137. Snow M, Snow R. 1931. Experiments on phyllotaxis. I. The effect of isolating a primordium. *Phil. Trans. R. Soc. London Ser. B* 221:1–43

138. Souer E, Houwelingen AV, Kloos D, Mol J, Koes R. 1996. The *No Apical Meristem* gene of petunia is required for pattern formation in embryos and flowers and is expressed at meristem and primordia boundaries. *Cell* 85:159–70

139. Sparks PD, Postlethwait SN. 1967. Comparative morphogenesis of the dimorphic leaves of *Cyamopsis tetragonolobus. Am. J. Bot.* 54:281–85

140. Steeves TA, Sussex IM. 1989. *Patterns in Plant Development.* Cambridge/New York/New Rochelle/Melbourne/Sydney: Cambridge Univ. Press

141. Stettler RF. 1964. Dosage effects of the lanceolate gene in tomato. *Am. J. Bot.* 51:253–64

142. Stevens AM, Rick CM. 1986. Genetics and breeding. In *The Tomato Crop*, ed. JG Atherton, J Rudick, pp. 35–109. New York: Chapman & Hall

143. Sussex IM. 1955. Morphogenesis in *Solanum tuberosum* L. Apical structure and developmental pattern of the juvenile shoot. *Phytomorphology* 5:253–73

144. Sylvester AW, Cande WZ, Freeling M. 1990. Division and differentiation during normal and liguleless-1 maize leaf development. *Development* 110:985–1000

145. Sylvester AW, Smith L, Freeling M. 1996. Aquisition of identity in the developing leaf. *Annu. Rev. Cell Dev. Biol.* 12:257–304

146. Takahashi T, Gasch A, Nishizawa N, Chua NH. 1995. The *DIMINUTO* gene of Arabidopsis is involved in regulating cell elongation. *Genes Dev.* 9:97–107

147. Taylor CB. 1997. *Knox*-on effects on leaf development. *Plant Cell* 9:2101–5

148. Taylor DW, Hickey LJ. 1992. Phylogenetic evidence for the herbaceous origin of angiosperms. *Plant Syst. Evol.* 180:137–56

149. Telfer A, Poethig RS. 1994. Leaf development in Arabidopsis. In *Arabidopsis*, ed. EM Meyerowitz, CR Somerville, pp. 379–401. New York: Cold Spring Harbor Lab. Press

150. Timmermans MCP, Schultes NP, Jankovsky JP, Nelson T. 1998. *Leafbladeless1* is required for dorsoventrality of lateral organs in maize. *Development* 125:2813–23

151. Traas J, Bellini C, Nacry P, Kronenberger J, Bouchez D, Caboche M. 1995. Normal differentiation patterns in lacking microtubular preprophase bands. *Nature* 375:676–77

152. Troll W. 1935. Vergleichende Morphologie der Fiederblatter. *Nova Acta Leopoldina* 2:311–455

153. Tsuge T, Tsukaya H, Uchimiya H. 1996. Two independent and polarized processes of cell elongation regulate leaf blade expansion in *Arabidopsis thaliana (II)* Heynh. *Development* 122:1589–600

154. Tsukaya H. 1995. Developmental genetics of leaf morphogenesis in dicotyledonous plants. *J. Plant Res.* 108:407–16

155. Tsukaya H, Uchimiya H. 1997. Genetic analyses of the formation of the serrated margin of leaf blades in Arabidopsis: combination of a mutational analysis of leaf morphogenesis with the characterization of a specific marker gene expressed in hydathodes and stipules. *Mol. Gen. Genet.* 256:231–38

156. Veit B, Briggs SP, Schmidt RJ, Yanofsky MF, Hake S. 1998. Regulation of leaf initiation by the *terminal ear1* gene of maize. *Nature* 393:166–68

157. Villani PJ, DeMason DA. 1997. Roles of the *Af* and *Tl* genes in pea leaf morphogenesis: characterization of the double mutant (*afaftltl*). *Am. J. Bot.* 84:1323–36

158. Vollbrecht E, Veit B, Sinha N, Hake S. 1990. The developmental gene *Knotted-1* is a member of a maize homeobox gene family. *Nature* 350:241–43

159. Waites R, Hudson A. 1995. *phantastica*: a gene required for dorsoventrality of leaves in *Antirrhinum majus*. *Development* 121:2143–54

160. Waites R, Selvadurai HRN, Oliver IR, Hudson A. 1998. The Phantastica gene encodes a MYB transcription factor involved in growth and dorsoventrality

of lateral organs in *Antirrhinum*. *Cell* 93:779–89

161. Walsh J, Waters CA, Freeling M. 1998. The maize gene *liguleless 2* encodes a basic leucine zipper protein involved in the establishment of the leaf blade-sheath boundary. *Genes Dev.* 12:208–18

162. Wardlaw CW. 1949. Experiments on organogenesis in ferns. *Growth* 13:93–131 (Suppl.)

163. Weigel D, Alvarez J, Smyth D, Yanofsky MF, Meyerowitz EM. 1992. *LEAFY* con-trols floral meristem identity in *Arabidopsis*. *Cell* 69:843–59

164. Young JPW. 1983. Pea leaf morphogenesis: a simple model. *Ann. Bot.* 52:311–16

165. Young JP, Dengler NG, Horton RF. 1987. Heterophylly in *Ranunculus flabellaris*: the effect of abscisic acid on leaf anatomy. *Ann. Bot.* 60:117–25

166. Young JP, Horton RF. 1985. Heterophylly in *Ranunculus flabellaris* Raf: the effect of abscisic acid. *Ann. Bot.* 55:899–902

Annu. Rev. Plant Physiol. Plant Mol. Biol. 1999. 50:447–72

THE PRESSURE PROBE: A Versatile Tool in Plant Cell Physiology

A. Deri Tomos
Ysgol Gwyddorau Biolegol, Prifysgol Cymru Bangor, Bangor, Gwynedd LL57 2UW, United Kingdom; e-mail: a.d.tomos@bangor.ac.uk

Roger A. Leigh
Department of Plant Sciences, University of Cambridge, Downing Street, Cambridge CB2 3EA, United Kingdom; e-mail: RL225@cam.ac.uk

KEY WORDS: cell water relations, cell solutes, osmotic pressure, tissue mapping, turgor

ABSTRACT
This review discusses how the pressure probe has evolved from an instrument for measuring cell turgor and other water relations parameters into a device for sampling the contents of individual higher plant cells in situ in the living plant. Together with a suite of microanalytical techniques it has permitted the mapping of water and solute relations at the resolution of single cells and has the potential to link quantitatively the traditionally separate areas of water relations and metabolism. The development of the probe is outlined and its modification to measure root pressure and xylem tension described. The deployment of the pressure probe to determine and map turgor, hydraulic conductivity, reflection coefficient, cell rheological properties, solute concentrations and enzyme activities at the resolution of single cells is discussed. The controversy surrounding the interpretation of results obtained with the xylem-pressure probe is included. Possible further developments of the probe and applications of single cell sampling are suggested.

CONTENTS

447

INTRODUCTION

Molecular biology, biochemistry, and physiology meet at the resolution of the cell. At this level it is possible to relate chemical and molecular information gained in vitro to the behavior of the cell as the fundamental unit of physiology. To understand the functioning of the whole organism, it is necessary to have a description of the activities of its component cells in relation to time, space, and environment. This cannot be achieved by traditional whole-organism descriptions nor by biochemical approaches that use tissue homogenates. However, it is being addressed by the use of techniques such as reporter genes (8), in situ hybridization (70), ion-sensitive fluorescent probes (86), ion-selective microelectrodes (76), X-ray microanalysis (142), immunocytochemistry (39), and nuclear magnetic resonance imaging (143) that can be used to show that a particular component or compound is located in certain cell types. In some cases, however, quantitative precision of these techniques is poor and this can be a severe disadvantage. This is especially true where solutes are measured because these contribute to both metabolic and osmotic cell functions. Thus changes in solute concentrations can have effects both on the rate of flux through biochemical pathways and on turgor. Changes in the latter will have consequences for processes such as growth, movement, support, and long-distance transport, which in turn may affect metabolism. Where such interactions are likely to occur, it is important to develop and deploy techniques that allow integration of information from two areas of plant physiology that have traditionally progressed separately—metabolism and water relations. Here we review the contributions of one set of techniques, based on the cell-pressure probe of Zimmermann & Steudle (169), that now permit the appropriate parameters to be measured in individual plant cells. The pressure probe has evolved from an instrument for measuring cell–water relations parameters into a microsampling device that is providing quantitative information about the solute concentrations in individual cells and, potentially, could permit studies of their metabolism.

Development of the Pressure Probe and Its Variants

The first pressure probe was a water- and air-filled glass micro-capillary used to make direct measurements of turgor in the giant-celled alga, *Nitella* (34). The compression of the air bubble allowed turgor to be estimated using Boyle's Law. Zimmermann & Steudle (123, 158, 169, 171) improved on this by replacing the air bubble with an electronic pressure sensor attached to an oil-filled capillary and including a piston that allowed turgor to be varied. Thus parameters such as the half time for water exchange ($T_{1/2}$), hydraulic conductivity (L_p), and cell wall volumetric elastic modulus (ε) could also be measured. This device was used to measure these parameters for a range of giant-celled alga: *Nitella, Valonia, Chara, Halicystis, Lamprothamnium*, and *Acetabularia* (reviewed in 171).

Although measurements were performed on the large leaf bladder cells of *Mesembryanthemum crystallinum* using the original pressure probe (126), this device is generally unsuitable for application to most higher plant cells. This is because the material of the instrument, especially the rubber seals (*d* in Figure 1*A*), is elastic and allows the high pressure of the cell to press sap from the cell into the capillary tip. For large cells, such as those of giant-celled algae, the fraction of sap volume lost is small and this has negligible effect on its turgor. For most cells, however, the fraction of the sap lost is large. As a result, the cell will lose volume and pressure on insertion of the capillary tip. A partial recovery of pressure will follow as pure water enters the cell across the plasmamembrane driven osmotically by the drop of turgor creating a water potential gradient. A new osmotic equilibrium will be reached but at a lower turgor pressure since the osmotic solutes in the capillary have been lost from the cell. In most cases, this effect is so large as to make it impossible to obtain even a rough estimate of the pressure of the cell prior to the insertion of the probe.

This effect was partially overcome by minimizing the compressibility of the oil-filled chamber by reducing its volume and taking care over the design of the seals. The key development, however, was the use of the oil-cell sap interface (meniscus) (*c* in Figure 1*A*) as the datum point (44). By pushing the piston into the oil-reservoir, the meniscus can be brought as close as is practicable to the cell wall, thus resorting the original cell volume and turgor. This pressure compensation method enables the hydrostatic pressures of cells down to the volumes of stomatal guard cells (20–30 pl) to be determined. Guard cells themselves, however, have proven to be difficult subjects. The most convincing data for them have been provided with meniscus-free probes in which cell and probe were filled with the same fluid: either water (71) or silicone oil (23, 24). Neither of these techniques can, however, measure the in vivo pressure directly—each relates the measured pressure to the stomatal aperture. Their use is thus restricted to guard cells.

Usually, turgor drops transiently on insertion of the probe, and this can affect cytoplasmic streaming and plasmodesmatal function (84). A removable glass

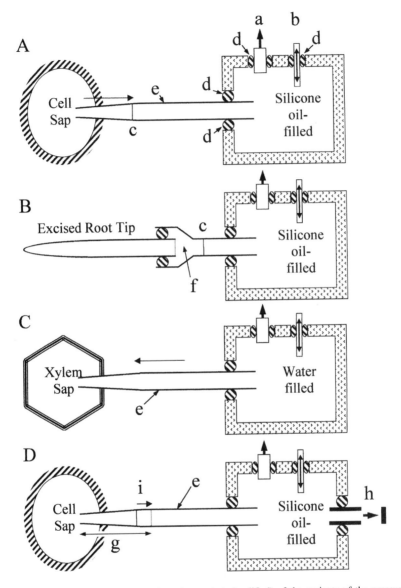

Figure 1 Diagrammatic representations (somewhat simplified) of the variants of the pressure probe currently being used. *A*, cell pressure probe; *B*, root pressure probe; *C*, xylem pressure probe; *D*, sampling pressure probe; *a*, pressure transducer and output; *b*, remote-controlled piston; *c*, water/oil interface (meniscus); *d*, compressible rubber seals; *e*, glass micro-capillary; *f*, solution placed in the probe prior to the attachment of the root; *g*, sample volume of cell sap that can be removed from cell in SiCSA technique; *h*, valve venting oil reservoir to atmospheric pressure; and *i*, movement of meniscus on opening of valve.

piston that blocks the capillary tip prior to its insertion into the cell has been devised to counter this (114). This arrangement also provides a good method of illuminating the meniscus and target cell with minimum application of heat from the illuminating source by using the glass piston as a light guide. Alternatively, Oparka et al (84) pressurized tips up to 0.2–0.3 MPa before insertion by relying simply on capillary forces in the tip, and they used this to minimize turgor changes upon injection of fluorescent tracers (49, 50, 85).

Measurements of cell volume are needed to calculate ε and L_p, and this is usually done by microscopy. Two methods for independent measurement of volume, one electronic (161), the other based on pressure relaxation (68), have been used. Also, several attempts have been made to automate monitoring of the sap/oil meniscus using either a resistance measurement (44) or image analysis (16, 17, 80). The latter is the more successful.

The study of water and solute relations have also been linked by combined use of the cell-pressure probe and micro-electrodes (170). This permitted observations of pressure-induced action potentials in *Chara corallina* (160), turgor-dependent movement of mobile charges within the membrane (162), turgor control of Cl^- channels (149), the interaction of K^+ and water transport (56, 112), and turgor regulation in the salt-tolerant giant alga, *Chara longifolia* (122). Zhu (155) described a single-barreled probe for measuring turgor and membrane potential, and its application to *Chara corallina*. Lew (61) used a similar instrument on root hair cells of *Arabidopsis thaliana*, but there are few such measurements on higher plants (however, see 50).

The higher plant pressure probe (Figure 1A) was initially called the micro-pressure probe. Since neither the pressures it measures nor the probe itself are any smaller than in the original version, the recent adoption of the term cell-pressure probe to distinguish the instrument from its derivatives the root (125) (Figure 1B) and the xylem-pressure probe (2) (Figure 1C) is to be encouraged.

In the root-pressure probe, the excised root replaces the glass capillary. Here, the meniscus is formed at the interface of the oil and a solution placed in the probe prior to the attachment of the root. Once the root is inserted into the apparatus (Figure 1B), the probe can be used to measure and vary root pressure and to determine the consequences for water flow across the root (123). Unlike the cell-pressure probe, this device measures the properties of a multicellular tissue. To relate the data to the contributions of individual cells or transport pathways requires measurements of additional parameters (e.g. 125). As the results obtained with the root-pressure probe do not themselves yield information at the single-cell level, we do not discuss work with this device in detail.

The most contentious derivative of the pressure probe has been its modification to measure the subatmospheric pressures of xylem vessels (3) (Figure 1C).

In this variant, de-gassed water replaces the low-viscosity oil of the cell-pressure probe in order to prevent cavitation at the oil-sap boundary (2, 3), although an oil-filled probe has been used successfully to measure negative pressures in the stelar apoplast (74). The xylem-pressure probe uses a feedback system to position automatically the capillary tip at a location within the tissue where hydrostatic pressure is below subatmospheric values (not shown in Figure 1C). This technique has attracted much interest in recent years, and some of the key conclusions drawn from it are discussed below.

THE MEASUREMENT OF WATER RELATIONS PARAMETERS

The pressure probe was initially applied to giant-celled algae. Measurements included not only turgor, but also L_p (170, 171), reflection coefficient (σ; 171), water and solute permeability coefficients (140), and ε (169). A range of empirical and theoretically predictable characteristics of these parameters have been measured including polarity of L_p (169), the temperature-dependence (activation energy) of L_p (112), and the dependence of L_p and ε on turgor and cell volume (171). The same fundamental water relations parameters have been measured for a variety of higher plant cells, which were treated as if isolated even though they were in tissues (e.g. 38, 44, 72, 128, 130, 136, 138, 165).

Turgor

Values of turgor in giant algal cells range from 0.1 to 0.6 MPa (see 133 for compilation). Values in well-watered higher plants range from below 0.1 MPa (*Suaeda maritima*, 13) to over 1 MPa (barley leaves, 28). In the case of the intertidal alga *Cladophora rupestris*, turgors in dilute seawater exceed 3.5 MPa and are the limits of the probe's current capabilities (64). Although this is not as high as the turgor proposed for the penetration organ of *Magnaporthe grisea* (8.0 MPa) into its host (43), it is already in the range of pressures found in the steam turbines of nuclear power stations!

These observations raise the question: What requirements determine the absolute values of turgor? In the case of *Cladophora*, it appears to be due to the absence of a turgor-regulation system (150). There is no obvious reason why such high turgors are needed. *Cladophora* cells do not use their turgor for physical work (as stomatal guard cells do) and do not have a supportive role. On the other hand, the proposed high turgors of the *Magnaporthe* appressorium are dictated by the requirement of penetrating the host epidermis (43). A similar situation occurs in the penetration of strong soil by roots (5, 12). It is usually assumed that high turgors are needed to support aerial organs but values are often higher than is necessary for this function (see 78). For instance, calculations (M Irving & AD Tomos, unpublished data) indicate that the mechanics of movement of the pulvinus of *Phaseolus* are such that a turgor of 0.1 MPa would be sufficient to support the leaf lamina, but the lower cortex that performs this function has a turgor of 0.5 MPa. Also, downward movement of the leaf is not simply a matter of collapse under gravity; the upper part of the pulvinus pushes the leaf down (45). Cell turgors in cortical cells on the upper and lower parts of the pulvinus change reciprocally by 0.4 to 0.5 MPa during each cycle of leaf movement. In general, few experiments have been done relating the absolute values of turgor to the physical demands of tissues. In view of the widely held view that turgor has an important structural role, there is a need for more quantitative investigations of this role.

Little exploitation has been made of the probe in field conditions (105), although there are several unpublished reports of such measurements (referred to in 163). A probe powered by a petrol-motor generator with a voltage stabilizer was used to measure turgor in the cladodes of *Retama sphaerocarpa* in a semiarid catchment in southern Spain (M Hansen, L Incoll & AD Tomos, unpublished data). Turgor declined from an initial value of between 0.1 and 0.2 MPa to zero within 20 min of local sunrise.

The pressure probe is most effective when it is used on surface cells. Generally, all of the experimental controls established using the giant-celled alga can be applied under these conditions. However, higher plant tissues are normally more than two cells thick. Measuring turgor (and taking sap samples—see below) from subsurface cells can be done, but it requires care. In leaves, cells can be accessed by puncturing through the epidermis (80, 165) or entering through a stomatal pore (54; see 29). Turgors in expanding zones of young graminaceous leaves were measured by cutting a small hole in the overlying older leaf bases (92, 131). In stems and roots it has been possible to construct turgor profiles across entire organs (15, 46, 73, 98, 109, 168). Some measurements on roots (46, 98) have shown a uniform turgor across the cortex. In sharp contrast, steep gradients (highest toward the stele) of turgor were found for the root cortex of two halophytes—*Mesembryanthemum crystallinum* (109) and *Aster tripolium* (168). The presence or absence of such gradients appears to be due to the state of transpiration of the plant, because in wheat and maize they were abolished by stopping transpiration or excising the roots (107). The ability of transpiration to directly affect cell turgor was recently shown (74). Radial pressure gradients in the root cortex, where the inner cells have a higher turgor than the outer cells, result in a driving force for outwardly directed water flow through plasmodesmata. This runs counter to the usually envisioned role of plasmodesmata as radial pathways for solute movement from the root surface to the stele. Perhaps under the conditions where these steps of turgor occur, the plasmodesmata do not support pressure-driven flow, i.e. they are shut (see 85).

In several cases, turgor values measured with the cell-pressure probe have been directly compared with those obtained with other techniques such as vapor point psychrometry (113, 136), isopiestic psychrometry (82), and the pressure bomb (80). In all cases there is good agreement between the values measured with the pressure probe and the other techniques.

Turgor Responses to Changing Conditions

The pressure probe is the only device available for measuring turgor changes in single cells in real time. Generally, this has been done by monitoring responses to osmotic or water stress (13, 77, 88, 90, 100, 116, 121, 132, 152), but responses to applied chemicals (21, 37, 46, 106), anaerobiosis (153), and heating (36) have also been measured. Osmotic changes can also be generated by the plants themselves. These have been analyzed during organic acid accumulation in Crassulacean acid metabolism (108) and during photosynthesis (55, 135).

In some cases there is little active adjustment of turgor following a water stress. In giant-celled algae, the responses range from full adjustment of turgor to no change (51). Higher plants have several different potential mechanisms of turgor adjustment—some of which are unavailable to giant algal cells. These mechanisms are (a) changes of internal osmotic pressure by polymerization and de-polymerization reactions; (b) transport of osmotic solutes across the plasma membrane, resulting in altered symplastic osmotic pressure; (c) exchange of osmotic solutes with distant parts of the plant through plasmodesmata; (d) transport of osmotic solutes across the plasma membrane in order to adjust the apoplast osmotic pressure (discussed in more detail below); (e) adjustment of the hydrostatic pressure of the apoplast by varying transpiration or root hydraulic conductivity; (f) passive osmotic adjustment due to elastic shrinkage or swelling to minimize turgor changes; and (g) dilution by growth. The following examples highlight the use of some of these mechanisms.

When cells of excised *Beta vulgaris* taproot are placed in different concentrations of mannitol, they undergo partial turgor recovery over a period of many hours, using hydrolysis and differential leakage of internal sugars to achieve either upward or downward changes in turgor (90). When 5 mM NaCl and KCl were added subsequently, turgor rose owing to net salt uptake. This second response resembles that of some giant-celled algal systems (51) and involves both plasma membrane and tonoplast transport. In some experiments with roots, mannitol-stressed cells show a turgor recovery with a $T_{1/2}$ of between 90 and 150 min (38, 100, 101), but in others, partial recovery is much faster (26). Although the faster times could be achieved by solute transport at the plasma membrane, Pritchard (94) proposed that expanding root cells obtain water and solutes from the phloem via symplastic connections. The majority of the solutes used in osmotic adjustment of the cortex of maize roots appear to arrive centrifugally rather than from the medium (95). Thorpe et al (132) suggested a similar explanation for the rapid osmotic adjustment of pea seed coats. In each case, the solutes must ultimately cross the tonoplast into the vacuole and water potential equilibrium must be maintained between the symplast and the vacuole without large changes in cytosolic volume (59).

In addition, the use of the pressure probe illustrated that higher plants can also employ their apoplast to adjust turgor, as suggested by Cram (18) and others (e.g. 17). For example, in the halophyte *Suaeda maritima*, the measured turgor of 0.1 MPa is not equal to the osmotic pressure of cell sap (approximately 2 MPa) and, as there is no significant apoplast hydrostatic tension and $\sigma = 1$, the water-free space of the apoplast must have a substantial osmotic pressure (13). Recently, these effects have been replicated and studied in vitro with excised *Suaeda* leaves, and this has shown that turgor control can be by both (rapid) apoplastic and (slower) symplastic/protoplastic osmotic adjustments (58). A similar situation is thought to occur in the taproot of *Beta vulgaris*, which maintains constant turgor throughout the growing season despite a large increase in cell osmotic pressure (135). Efflux-analysis experiments indicate that in red-beet the extracellular osmotic pressure adjustment involves K^+ (60), while turgor-sensitive H^+ efflux suggests that a plasma membrane H^+-pump may be responsible for the energizing apoplast osmotic adjustment (151).

Hydraulic Conductivity

Hydraulic conductivity is one of the determinants of how fast water moves into and out of cells and is usually measured by determining this flow following osmotic or hydrostatic perturbations. The usual way of doing this with the pressure probe is to change turgor and follow the relaxation of pressure as water enters or leaves the cell (169). Wendler & Zimmermann (146, 147) introduced a pressure-clamp technique in which volume flow is measured at constant pressure. Using this technique, they were able to distinguish between the tonoplast and plasma membrane L_p for *Chara corallina*—the former ranging from 2 to 4×10^{-6} m s^{-1} MPa^{-1} and the latter from 3 to 10 $\times$ 10^{-6} m s^{-1} MPa^{-1} (148). This technique has the advantage of not requiring independent knowledge of the cell volume, ε, or internal osmotic pressure, which are required if pressure relaxation is used. In higher plants, this technique has been used on sugarcane storage parenchyma (7) and wheat root cortical cells (153); Ortega et al (87) used it to measure transpiration rate in a fungal sporangiophore. Values of L_p in higher plants range from 0.1 $\times$ 10^{-6} to 6.1 $\times$ 10^{-6} m s^{-1} MPa^{-1} (see 133 for compilation), indicating that most higher plant cells will return to hydraulic equilibrium within seconds following a perturbation. Both the magnitude and the range of L_p have attracted much interest because of the possible involvement of water channel proteins (69). Henzler & Steudle (40) used $HgCl_2$ to show that water channels are responsible for 74% of the value of L_p of *Chara*. More recently, the sensitivity of water channels to $HgCl_2$ has been used to provide a complete set of water transport coefficients (41, 112).

Single-cell L_p values have been used in conjunction with whole-tissue L_p measurements to attempt to identify pathways of water through roots (reviewed by 123, 127). Three routes have been identified: the apoplastic, symplastic, and transcellular pathway. In the last of these, water flows across all membranes (plasmamembrane and tonoplast) in its path (127, 144). It has proved impossible to predict whole-root conductivity from single-cell values of L_p (47, 125), but Steudle (e.g. 127) proposed a composite model for water and solute flows across the root cortex in which all three pathways play a role in parallel and in series. Nothing equivalent has been done for leaves.

Reflection Coefficient

The reflection coefficient, σ, has a value between 0 and 1 and relates a solute gradient across a semipermeable membrane to its water-driving force. A value of zero indicates that an osmotic pressure gradient of the solute across a membrane will not drive water flow, whereas a solute with a σ of 1 will exert a maximum driving force equivalent to that predicated by the Van't Hoff equation (81). Potentially, σ would be a valuable parameter for a cell to control because it could be a mechanism for adjusting turgor and osmotically driven water flow by modifying membrane properties. For all physiological solutes that influence water relations of plant cells, measurements of σ yield a value close to 1 under all conditions tested (133, 171). For some nonphysiological solutes such as acetone and alcohol, the values are much lower (171) and can be modified by adding $HgCl_2$ (112), which is assumed to block water channels (see above).

In contrast to measurements at the single-cell level, measurements of σ for physiologically important solutes in whole roots and other tissues, using either the root-pressure probe or the xylem-pressure probe, often yield values substantially less than 1 (110, 111, 127, 156, 172). Explanations for this include separation of pathways of water and solute movement across the root (127) or concentration polarization of solutes (111).

Cell Rheological Properties

The pressure probe can be used to measure both the elastic and plastic properties of plant cells. Under conditions of changing turgor (ΔP), cell volume (V) will change by an amount (ΔV) determined by the volumetric elastic modulus,

$$\varepsilon = \frac{\Delta P}{\Delta V} \cdot V.$$

The importance of this for drought (water stress) tolerance has long been recognized (e.g. 9). Lower values of ε mean that changes in turgor in response to changes in plant water status are minimized. Elastic modulus is measured by determining the relative change in volume in response to an applied change in turgor. It is a visco-elastic property (133) and its time-dependence has been demonstrated using both transient and long-term changes in turgor. In the case of *Halicystis*, the two measurements diverged by an order of magnitude at higher turgors (164). The phenomenon of stress hardening of ε (a stiffening of the material in response to being stretched) observed for *Halicystis* has been noted by several workers for higher plants (e.g. 79, 99). It is not always so evident and was not observed in *Tradescantia* leaf epidermis (138) nor in wheat root cortex (47).

Measurements with the pressure bomb (see Reference 139 for a review of the apparatus) indicate that the plant can control the values of ε (9). It is not clear, however, if ε changes within each cell, or whether cells produced under different circumstances display different ε, although ε can certainly be under dynamic control. For instance, in expanding wheat roots it increases by an order of magnitude when the roots are excised (97). Recently, Findeklee & Goldbach (22) demonstrated the direct influence of boric acid on ε on the root cells of *Cucurbita pepo*. Not only does this throw light on the role of borate in cell wall structure, it also reveals a potentially physiological control over this key structural parameter.

The plastic (as opposed to elastic) behavior of cells is important for understanding expansion of plant cells and can be determined by simultaneous measurement of volume growth rate and turgor, although it has been known since the first experiments with the pressure probe that a strict physical description is complicated by the rapid metabolic responses of biological material (35). Attempts (e.g. 100) to assign even relative values to rheological parameters based on a linear stress-strain model (62) have proved indecisive (see 25). However, the pressure probe has contributed greatly to the study of expansion growth (for reviews see 14, 93, 137). The major finding is that the steady-state expansion rate of cells is modulated without a change in turgor that can be detected by the pressure probe, indicating that cell wall properties are regulating expansion (15, 93, 103, 137). This has led to attempts to relate changes in metabolism of cell wall polymers to changes in cell expansion rate (96).

SINGLE CELL SAMPLING AND ANALYSIS (SiCSA)

Mapping the Symplast

The cell-pressure probe measures turgor and other parameters in individual cells and thus has the potential to map these properties within tissues and organs. Jones et al (46) illustrated this in a simple way during an analysis of the influence of abscisic acid (ABA) on wheat roots. Although measurements of whole-tissue osmotic pressure indicated that ABA had no effect on this parameter, this hid differential changes in the inner and outer cortical cells. In other situations, however, turgor can be remarkably uniform within tissues (e.g. 15, 28, 98, 133).

Understanding the basis of similarities or differences in turgor between cells requires knowledge of the cell sap osmotic pressure and the solutes that contribute to it, hence these parameters must also be mapped at the resolution of single cells. This has been achieved using a suite of techniques collectively known a Single Cell Sampling and Analysis (SiCSA; 134), which involve the collection and analysis of the sample of cell sap that is driven into the pressure probe capillary by turgor when the cell is impaled (see below). Rapid collection of the sample is important, particularly in cells with short $T_{1/2}$, because water flow can dilute the sample (66). Rapid extraction of the sample can be achieved by fitting the probe with a solenoid-powered valve that instantly opens the oil reservoir to atmospheric pressure (66, 134). Using this method 10–50 pl of sap can be obtained and these can be stored under paraffin-oil saturated with water while awaiting further analysis. Water-saturated oil is used to prevent abstraction of water from the small sample. Measurements that can be made include osmotic pressure determined by freezing point depression (66, 83, 113), inorganic elements by X-ray microanalysis (67) or by capillary electrophoresis (4), and organic solutes and nitrate by enzyme-linked fluorescent microscope photometry (28, 29, 154). Details of most of these techniques are described by Tomos et al (134). The microscope-based fluorescent technique has also been used for measuring single-cell enzymatic activities such as malate dehydrogenase (29) and acid invertase (53). Other techniques that, potentially, can also be applied to the sap samples include ion-selective fluorescent probes (86), ion-selective electrodes (76), secondary isotope mass spectrometry (57), and immuno-detection (39). The limit is only set by having to accommodate such small samples that typically contain some 10^{-13} moles of analyte (based on 10 pl of 10 mM solution). The smallest sample that has been analyzed to date is the 0.5 pl Buller's drop of the ballistospores of *Itersonilia perplexans* (145).

An important aspect of SiCSA is that it includes intrinsic cross checks of data. Turgor cannot exceed protoplast osmotic pressure (66), the total charge of anions and cations must balance, and the sum of solutes (adjusted for osmotic coefficients) must be equal to the independently measured osmotic pressure (30, 95). Failure to meet any one of these criteria indicates either an artefact or, in the case of the last two, unidentified solutes.

To date, the most extensive application of SiCSA has been to barley leaves. It has shown differences in composition between epidermal, mesophyll, and bundle sheath cells (29); between adjacent cells in the upper epidermis (28, 32); and between the upper and lower epidermis (33). Subsequent measurements have investigated how NaCl (27, 30, 32), leaf age (28, 30, 32), high and low Ca^{2+} and phosphate nutrition (31), and ozone fumigation (19) affect these patterns.

Epidermal malate levels were found to depend on light intensity, with concentrations increasing at high light levels (28, 32). The absence of significant levels of organic solutes in the epidermis of low-light grown barley posed the question of sources of carbon for the growth of *Erysiphe graminis* (powdery mildew), the haustoria of which are only to be found in the epidermis. SiCSA measurements have shown that the presence of the haustorium in a cell increases the concentration of malate in that cell (104). The mechanism of this is unknown. Mapping of sucrose, glucose, fructose, and fructans in barley leaves has shown diurnal oscillations of these in the mesophyll and bundle sheath. By increasing the sugar load in leaves, it was found that the fructan synthesizing system becomes active at lower sucrose concentrations in the bundle sheath than in the mesophyll—thus maintaining a gradient of sucrose from mesophyll to bundle sheath even when phloem uptake is diminished (53, 54). Using SiCSA, Meshcheryakov et al (73) found turgor and osmotic gradients across the expanding *Ricinus* hypocotyl. Recently, the gradients of sucrose and glucose have also been measured in this tissue (J Verscht, E Komor & AD Tomos, unpublished data). Although mapping of cell composition is easiest in thin, easily accessible tissue, its use for more substantial organs is possible (36, 52, 72, 88, 115, 129).

Mapping the Apoplast

Attempts are currently under way to map individually both hydrostatic (P_{wall}) and osmotic pressure (π_{wall}) gradients within the apoplastic water-free space of leaves (Figure 2; see color section at the back of the volume). The basis of this is that at water potential equilibrium and if $\sigma = 1$, the water potential ($\psi_{wall} = P_{wall} - \pi_{wall}$) of the apoplast of a cell is the same as that of the cell's interior ($\psi_{cell} = P_{cell} - \pi_{cell}$). Both P_{cell} and π_{cell} can be measured by SiCSA, and thus maps of apoplast water potential ($P_{cell} - \pi_{cell} = \psi_{cell} = \psi_{wall} = P_{wall} - \pi_{wall}$) at the resolution of individual cells, can be made. The contribution of P_{wall} to the apoplastic water potential can be determined by repeating the measurements under nontranspiring conditions (with the leaf immersed in solution, or covered with paraffin grease) when P_{wall} rises to atmospheric pressure and so equals zero. Apoplastic water potential under this condition must be equal to π_{wall}. Infiltration of intact or excised tissues with solutions of known osmotic pressure can be used to displace this residual water potential—lending further support to its identification with freely diffusible solutes. Using this approach, Cuin (19) has shown that under transpiring conditions hydrostatic gradients of some 0.1–0.2 MPa occur in the epidermis of wheat radiating away from the stomata. These gradients are abolished by ozone fumigation. In contrast, no significant osmotic pressure gradients are observed in the apoplast of control plants, but they do appear following ozone fumigation—with the higher solute concentrations associated with the apoplast of the stomatal region. Koroleva et al (55) have looked at the equivalent gradients between the epidermis and the bundle sheath of barley leaves. They also showed transpiration-induced hydrostatic pressure gradients radiating from the epidermis. A similar approach applied by Nonami & Schulze (83) to the leaves of *Tradescantia* also reveals a transpiration-dependent water potential step from epidermis to mesophyll—more negative in the former. Standing osmotic gradients in the apoplast are the basis of metabolic pumping of water against water potential gradients proposed for water-stressed plants (48). Pritchard (94) used SiCSA and aphid stylets to map out osmotic pressure gradients in the symplast of growing barley roots.

MEASUREMENT OF XYLEM PRESSURE

The Cohesion Theory of the ascent of sap in plants predicts considerable negative pressures (tensions) in the xylem (157). Measurements with the pressure bomb (see 139 for review of method) have yielded tensions of the required values and have been taken as support for the Cohesion Theory (157). The adaptation of the pressure probe to detect and measure negative pressure (Figure 1*C*) (3) allowed direct determination of xylem pressures in situ. Pressures measured with the xylem-pressure probe under transpiring conditions in both small plants and tall trees were considerably less negative than those obtained with the pressure bomb and in many cases were positive, i.e. in the range between zero absolute and atmospheric pressure (2, 6, 7, 110, 111, 163, 166, 167, 172). Thus pressure probe measurements made on twigs and leaves of transpiring trees at a height of about 35 m yielded maximum xylem pressures of about -0.15 MPa, in contrast to -1 to -3 MPa measured with the pressure bomb (163, 166). The tensions measured with the probe were below those (approximately 0.7 MPa) needed to raise water to this height against gravity and frictional forces (157), whereas measurements made at different heights did not show the expected change with elevation (7, 167). In addition, several other observations made by Balling & Zimmermann (2) were difficult to reconcile with the accepted framework of xylem function. These included (*a*) an insensitivity of xylem tension to transpiration rate [but later work did show the expected relationship (6) or was interpreted on this basis (166)]; (*b*) no change in xylem tension measured outside a pressure bomb in response to pressurization of tissue inside the bomb; and (*c*) maintenance of an absolute xylem tension of about -0.1 MPa in response to pressurization of whole plants in a hyperbaric chamber, suggesting a mechanism for sensing and maintaining absolute pressures in the xylem irrespective of external pressure (2, 89). These latter points have been addressed (89, 159) and are not discussed further here. However, the small tensions measured by the xylem-pressure probe in relation to the requirements of the Cohesion Theory and the observation that the xylem sap cavitated at relatively low tensions (e.g. 2, 6, 7, 166, 172) have led to a major unresolved debate about the mechanism of ascent of xylem sap and of the veracity of the measurements made with the pressure probe compared with those made by other methods (11, 75, 89, 105, 120, 124, 159, 163, 166, 167, 172). Results with the xylem-pressure probe have not only called into question the established mechanism of sap ascent but have also raised doubts about the veracity of techniques such as the pressure bomb (139) and hygrometers (20) that have been central to many experiments on plant-water relations. Fortunately, this debate may be restricted to the xylem-pressure probe because turgor measurements made with the cell-pressure probe are in agreement with those made with the pressure bomb, provided account is taken of the osmotic pressure of the xylem sap (80, see also 113).

Important questions about the xylem-pressure probe are whether it is located in a transporting xylem vessel and whether it has the ability to measure large negative pressures if they exist. Injection of dye into the impaled cell has confirmed that the probe is in a xylem vessel and that the dye is able to move upward into interconnected vessels (2, 6, 163, 172). No dye could be detected below the point of probe insertion, suggesting that the dye movement was related to the transport activity of the vessel (2). This is important because approximately 1 μl of water (or dye-containing solution) is sucked into the xylem from the probe when it is inserted (2). Although Balling & Zimmermann (2) calculated that this is small in relation to the total volume of the xylem (approximately 1000 μl in the *Nicotiana* plants they were using) and the water-filled volume of the probe (500 μl), it is significant in terms of the length of xylem it could fill. For a vessel with an internal diameter of 50 μm, 1 μl will occupy 0.51 m (see also 65). However, in the absence of transport, this injected dye would move both up and down in the xylem and so the upward, but not downward, movement appears to confirm the transport competence of the injected vessel or at least its connection to such vessels (120). Nonetheless, the potential of the injected dye to fill large lengths of the xylem does question estimates of flow rates based on its movement (6). In addition, it has been suggested that dye movement could occur even in embolized vessels by being drawn through the cell walls into functional vessels, where it would then move upward in the transpiration stream (120).

The ability of the xylem-pressure probe to measure negative pressures has been demonstrated in both artificial and natural systems. These have included imposition of tensions in capillaries and xylem bundles of detached leaves using a Hepp-type osmometer (2, 3), in which the lumen of the xylem bundles are sealed to one side of a semipermeable membrane, the other side of which is bathed in a solution of high osmotic pressure. The osmotic gradient across the membrane results in water being drawn from the xylem vessels, with a resulting decrease in the pressure of their contents. As long as no leaks or cavitation occur, the pressure can drop into the range below zero absolute. Other controls include measurements in plants within a hyperbaric chamber (2); the alteration of xylem pressure through changes in the external salt concentration (110, 111, 156, 167); the imposition of drought (166); or changes in light intensity (111). In conditions of changing transpiration rate, the probe was able to measure changes in xylem pressure that were in the expected direction. Thus, when transpiration was decreased in tobacco, the xylem pressure rose from about +0.03 MPa to about +0.07 MPa and decreased when transpiration was increased (6). In trees and liana (*Tetrastigma voinierianum*), the xylem pressure decreased from a positive or slightly negative value early in the morning to more negative values later (7, 166). Also, when a second probe was inserted into an already probed vessel, the same pressure was measured, indicating that insertion of the probe does not apparently cause major perturbations to the tension in the xylem (110). However, no direct evidence (e.g. by dye injection) was presented to show that the two probes were in the same vessel. In all these experiments, cavitation prevented measurements at pressures more negative than −0.6 MPa. Whether this was due to perturbations caused by the probe remains unclear and is at the center of the present disagreements (75, 120, 141).

Several other lines of evidence have been invoked to support the presence of low tensions in the xylem (7, 11, 163, 166, 172). These include the apparent inability of xylem-feeding insects to withdraw sap at pressures more negative than −0.3 MPa (102) and the cavitation of water at pressures between −0.1 and −0.6 MPa when centrifuged in Z-shaped tubes (118). Cavitation at these pressures is not consistent with the large tensions required by the Cohesion Theory, at least in tall trees. As a result, several alternative mechanisms have been proposed to explain how sap ascends in the xylem; these include pressure from surrounding tissues (11); segmentation of the xylem into small, osmotically isolated segments (166); flow at the interface of the xylem sap and small air bubbles (Maragoni streaming) (163, 166, 172); and gel-supported transport involving the generation of a gradient of chemical activity by high-molecular-weight polymeric substances (7, 172). In most cases, no experimental tests have been suggested that might verify these ideas. In the case of the mechanism involving osmotically isolated segments, the requirement for solute reflecting membranes between these sections is not consistent with the observation that dyes injected by the xylem-pressure probe can move considerable distances (6).

Needless to say, these assaults on the Cohesion Theory have not gone unchallenged, and vigorous attempts have been made to show that the xylem-pressure probe measurements, and the other evidence that seems consistent with them, must be in error. Thus it has been suggested that the probe is incapable of measuring pressures more negative than about −0.6 MPa either because of an imperfect seal between the probe and the xylem wall (120) or the creation of micro-fissures in the xylem cell wall when the probe is inserted (75). In both cases, cavitation via "air-seeding" (157) is proposed to occur at pressures less negative than those normally sustained by the xylem. The experiments in which a second probe was inserted into a vessel without apparent disruption to the readings of the first (110) would seem to suggest that these perturbations are not as severe as suggested, but it is possible that the major damage is done by the first insertion and then both probes are measuring the same artefact. Another possible artefact is that the insertion of the probe causes cavitation in vessels with large tensions resulting in a positive pressure between 0 and atmospheric pressure (0.1 MPa), depending on the mixture of air and water vapor in the vessel (120). This would explain why so many measurements with the xylem-pressure probe are in this range. Only in plants with relatively less negative pressures in the xylem (e.g. well-watered plants with low transpiration rates) will the probe not cause cavitation, hence its ability sometimes to measure moderate negative pressures. However, this does not explain why negative pressures measured in tall trees were not of the expected size nor why vessels which initially have a slightly positive pressure become more negative as transpiration increases (7) or when the tissue is exposed to high external osmotic pressure (110, 111, 156). Milburn (75) suggested that the low tensions are because the probe is inserted into living, immature vessels that still have an intact plasma membrane and hence a significant turgor. Depending on their state of maturity, these would register pressures from high positive (immature cells undergoing vacuolation) to negative (at final stages of maturation). The range of values obtained with the probe would then reflect the state of development of the vessels but, in no case, would they be typical of mature xylem. However, measurements made on living late metaxylem in maize roots using a cell-pressure probe do not fully support this explanation. While these cells did have turgor, the values changed only by a small amount (about 0.1 MPa) as transpiration increased (74). Assuming that these immature cells are fully hydraulically linked to the mature, conducting xylem, the results indicate that only small tensions are generated in transpiring xylem, consistent with the observations made with the xylem-pressure probe. In the same work, insertion of the probe into the stele (location unclear but either in the xylem or the stelar apoplast) measured only small tensions, seemingly confirming the results with the xylem probe (74).

The apparent discrepancies between the xylem-pressure probe results and the Cohesion Theory have also led to new attempts to demonstrate the accuracy of the pressure chamber technique that has provided much of the evidence in favor of the Cohesion Theory. These have included measuring changes in the hydraulic conductivity of excised tree stems in response to applied air pressure

(119), centrifuging tissues to impose known tensions before measuring them in the pressure bomb (91, 120), using centrifugation to determine the tensions that induce cavitation in the xylem (1, 42), and injecting Hg to measure tensions (75). The results have confirmed the ability of the pressure bomb to accurately measure the imposed tensions and that cavitation occurs at pressures more negative than those measured with the xylem-pressure probe.

The apparent ability of water in the xylem to withstand large negative pressures before cavitating (1, 42, 119) is in disagreement with the results of experiments in which water was centrifuged in Z-shaped tubes and cavitated at low tensions (118). However, these centrifugation experiments were done in capillaries with a diameter of 3 mm, i.e. large in relation to that of xylem vessels [e.g. about 40 to 140 μm in *Vitis vinifera* (63)]. Recent results with a model system suggest that, in this range, the diameter of the vessel may be important in determining the pressure at which cavitation occurs. Brereton et al (10) measured cavitation of water supersaturated with O_2 as it was released from pre-compression at 100 MPa to atmospheric pressure along silica capillaries with diameters ranging from 10 to 100 μm. They found that cavitation occurred at higher pressures in the smaller-diameter tubes. This suggests that the wide-bore capillaries used by Smith (118) are not a good model for xylem vessels and that the results do not necessarily support those with the pressure probe. Instead, the results with the narrow-bore capillaries would seem to be consistent with the measurements indicating that cavitation in the xylem does not occur until relatively large tensions have developed.

At present, the questions raised by the measurements with the xylem-pressure probe remain unresolved. The measurements remain the only direct determinations of pressures in the xylem and although suggestions as to the possible source of errors in them have been made, none has been convincingly demonstrated. On the other hand, new tests of the accuracy of the pressure bomb and of the predictions of the Cohesion Theory lend support to the validity of these. There is an obvious need for groups other than Zimmermann's to make measurements with the xylem-pressure probe (see 141). This will not only lead to independent tests of the observations, but could also bring new insights into, or improvements of, the technique, perhaps highlighting artefacts or establishing the veracity of the method. Whichever way the present argument is resolved, the work with the xylem-pressure probe has resulted in new experimental tests of the Cohesion Theory that would probably not have been done otherwise. Ultimately, the resolution of these arguments may lie in the fact that the probe measures the properties of a single cell in an intact plant, whereas techniques such as the pressure bomb measure a bulk-averaged parameter derived from a multicellular excised tissue (120). Further exploration of the limitations of both approaches should lead to a stronger understanding of the physical basis of xylem transport.

FUTURE PROSPECTS

The pressure probe has contributed greatly to unraveling the water relations parameters of individual plant cells and, by its deployment as a tool of SiCSA, is linking quantitative measurements of cell water and cell solute relations. There are a number of seemingly logical developments of this technique that could extend the approach to the quantification of enzymes and other components at the single cell level. Thus, there would seem to be no obvious reason why SiCSA could not be extended to the quantitative extraction and measurement of enzymes from cells. By including an extraction buffer in the tip of the probe and deploying the device as a cell homogenizer, it ought to be possible to extract enzymes from cells in the same way as they are extracted by homogenization of whole tissue. Activities of these enzymes could then be measured by fluorescent microscope–based assays. This would include the use of NAD(P)H-linked assays and those using nonfluorescent substrates that yield a fluorescent product upon hydrolysis (e.g. resorufin or umbelifferyl-linked compounds). Similarly, it should be possible to quantify nucleic acids in cell extracts by RT-PCR (47a) or to measure proteins using immuno-based techniques. These methods would complement in situ hybridization or immunocytochemistry and, when linked to single-cell enzyme assays, have the potential not just to demonstrate gene expression and the presence of a protein but also to quantify its activity.

Another major development could come from combining the probe with other in situ measuring devices such as ion-selective microelectrodes (76) and optrodes (117). These have the added advantage of allowing quantification within cells without the need for extraction. Such devices could be particularly useful for measurements in more inaccessible cells such as the xylem and phloem. Xylem measurements would be particularly useful because there are presently no methods for measuring directly the concentrations of solutes in this pathway under transpiring conditions. For both extraction and in situ measurements, the use of tissues in which particular cells have been tagged with an intrinsic marker such as green fluorescent protein would also be an advantage so that the identity of the target cell could be confirmed.

Acknowledgments

We thank present and former colleagues in Bangor and at RAL's former workplace, IACR-Rothamsted, who contributed to the development of SiCSA and to the ideas mentioned in this review. The work in Bangor was supported by grants from the former Agricultural and Food Research Council and from the Biotechnology and Biological Sciences Research Council (BBSRC). IACR-Rothamsted is grant-aided by the BBSRC.

Visit the *Annual Reviews home page* at
http://www.AnnualReviews.org

Literature Cited

1. Alder NN, Pockman WT, Sperry JS, Nuismer S. 1997. Use of centrifugal force in the study of xylem cavitation. *J. Exp. Bot.* 48:665–74
2. Balling A, Zimmermann U. 1990. Comparative measurements of the xylem pressure of *Nicotiana* plants by means of the pressure bomb and pressure probe. *Planta* 182:325–38
3. Balling A, Zimmermann U, Büchner K-H. 1988. Direct measurement of negative pressure in artificial-biological systems. *Naturwissenschaften* 75:409–11
4. Bazzanella A, Lochmann H, Tomos AD, Bächmann K. 1998. Determination of inorganic cations and anions in single plant cells by capillary zone electrophoresis. *J. Chromatogra.* A. 809:231–39
5. Bengough AG, Croser C, Pritchard J. 1997. A biophysical analysis of root growth under mechanical stress. *Plant Soil* 189:155–64
6. Benkert R, Balling A, Zimmermann U. 1991. Direct measurements of the pressure and flow in the xylem vessels of *Nicotiana tabacum* and their dependence on flow resistance and transpiration rate. *Bot. Acta* 104:423–32
7. Benkert R, Zhu JJ, Zimmermann G, Türk R, Bentrup F-W, Zimmermann U. 1995. Long-term xylem pressure measurements in the liana *Tetrastigma voinierianum* by means of the xylem pressure probe. *Planta* 196:804–13
8. Berger F, Haseloff J, Schiefelbein J, Dolan L. 1998. Positional information in root epidermis is defined during embryogenesis and acts in domains with strict boundaries. *Curr. Biol.* 8:421–30
9. Bradford KJ, Hsiao TC. 1982. Physiological responses to moderate water stress. See Ref. 57a, 12B:263–324
10. Brereton GJ, Crilly RJ, Spears JR. 1998. Nucleation in small capillary tubes. *Chem. Phys.* 230:253–65
11. Canny MJ. 1995. A new theory for the ascent of sap—cohesion supported by tissue pressure. *Ann. Bot.* 75:343–57

12. Clark LJ, Whalley WR, Dexter AR, Barraclough PB, Leigh RA. 1996. Complete mechanical impedance increases the turgor of cells in the apex of pea roots. *Plant Cell Environ.* 19:1099–102
13. Clipson NJW, Tomos AD, Flowers TJ, Wyn Jones RG. 1985. Salt tolerance in the halophyte *Suaeda maritima* L. Dum. *Planta* 165:392–96
14. Cosgrove DJ. 1986. Biophysical control of plant cell growth. *Annu. Rev. Plant Physiol.* 37:377–405
15. Cosgrove DJ, Cleland RE. 1983. Solutes in the free space of growing stem tissues. *Plant Physiol.* 72:326–31
16. Cosgrove DJ, Durachko DM. 1986. Automated pressure probe for measurement of water transport properties of higher-plant cells. *Rev. Sci. Instrum.* 57:2614–19
17. Cosgrove DJ, Ortega JKE, Shropshire W. 1987. Pressure probe study of the water relations of *Phycomyces blakesleeanus* sporangiophores. *Biophys. J.* 51:413–23
18. Cram WJ. 1976. Negative feedback regulation of transporting cells. The maintenance of turgor, volume and nutrient supply. In *Encyclopedia of Plant Physiology*, ed. U. Lüttge, MG Pitman, 2B:284–316. Berlin: Springer
19. Cuin TA. 1996. *The effect of ozone-fumigation on single-cell water and solute relations of barley leaves*. PhD thesis. Univ. Wales
20. Dixon MA, Tyree MT. 1984. A new temperature corrected stem hygrometer and its calibration against the pressure bomb. *Plant Cell Environ.* 7:693–97
21. Eamus D, Tomos AD. 1983. The influence of abscisic acid on the water relations of leaf epidermal cells of *Rhoeo discolor*. *Plant Sci. Lett.* 31:253–59
22. Findeklee P, Goldbach HE. 1996. Rapid effects of boron deficiency on cell wall elasticity modulus in *Cucurbita pepo* roots. *Bot. Acta* 109:463–65
23. Franks PJ, Cowan IR, Farquhar GD. 1998. A study of stomatal mechanics using the

cell pressure probe. *Plant Cell Environ.* 21:94–100

24. Franks PJ, Cowan IR, Tyerman SD, Cleary AL, Lloyd J, Farquhar GD. 1995. Guard-cell pressure aperture characteristics measured with the pressure probe. *Plant Cell Environ.* 18:795–800

25. Frensch J. 1997. Primary responses of root and leaf elongation to water deficits in the atmosphere and soil solution. *J. Exp. Bot.* 48:985–99

26. Frensch J, Hsiao TC. 1994. Transient responses of cell turgor and growth of maize roots as affected by changes in water potential. *Plant Physiol.* 104:247–54

27. Fricke W. 1997. Cell turgor, osmotic pressure and water potential in the upper epidermis of barley leaves in relation to cell location and in response to NaCl and air humidity. *J. Exp. Bot.* 48:645–58

28. Fricke W, Hinde PS, Leigh RA, Tomos AD. 1995. Vacuolar solutes in the upper epidermis of barley leaves. Intercellular differences follow patterns. *Planta* 196:40–49

29. Fricke W, Leigh RA, Tomos AD. 1994. Concentrations of inorganic and organic solutes in extracts from individual epidermal, mesophyll and bundle-sheath cells of barley leaves. *Planta* 192:310–16

30. Fricke W, Leigh RA, Tomos AD. 1994. Epidermal solute concentrations and osmolality in barley leaves studied at the single-cell level. Changes along the leaf blade during leaf ageing and NaCl stress. *Planta* 192:317–23

31. Fricke W, Leigh RA, Tomos AD. 1995. Intercellular solute compartmentation in barley leaves. In *Sodium in Agriculture*, ed. CJC Phillips, PC Chiy, pp. 33–42. Canterbury: Chalcombe Publ.

32. Fricke W, Leigh RA, Tomos AD. 1996. The intercellular distribution of vacuolar solutes in the epidermis and mesophyll of barley leaves changes in response to NaCl. *J. Exp. Bot.* 47:1413–26

33. Fricke W, Pritchard J, Leigh RA, Tomos AD. 1994. Cells of the upper and lower epidermis of barley (*Hordeum vulgare* L) leaves exhibit distinct patterns of vacuolar solutes. *Plant Physiol.* 104:1201–8

34. Green PB. 1968. Growth physics of *Nitella*. A method for continuous in vivo analysis of extensibility based on a micromanometer techniques for turgor pressure. *Plant Physiol.* 43:1169–84

35. Green PB, Erickson RO, Bugg J. 1971. Metabolic and physical control of cell elongation rate. In vivo studies in *Nitella*. *Plant Physiol.* 47:423–30

36. Greve LC, Shackel KA, Ahmadi H, McArdle RN, Gohlke JR, Labavitch JM. 1994. Impact of heating on carrot firmness. Contribution of cellular turgor. *J. Agric. Food Chem.* 42:2896–99

37. Griffiths A, Jones HG, Tomos AD. 1997. Applied abscisic acid, root growth and turgor pressure responses of roots of wild-type and the ABA-deficient mutant, Notabilis, of tomato. *J. Plant Physiol.* 151:60–62

38. Griffiths A, Parry AD, Jones HG, Tomos AD. 1996. Abscisic acid and turgor pressure regulation in tomato roots. *J. Plant Physiol.* 149:372–76

39. Harris N. 1994. Immunocytochemistry for light and electron microscopy. See Ref. 39a, pp. 157–76

39a. Harris N, Oparka KJ, eds. 1994. *Plant Cell Biology—A Practical Approach*. Oxford: IRL Press

40. Henzler T, Steudle E. 1995. Reversible closing of water channels in *Chara* internodes provides evidence for a composite transport model of the plasma membrane. *J. Exp. Bot.* 46:199–209

41. Hertel A, Steudle E. 1997. The function of water channels in *Chara*. The temperature dependence of water and solute flows provides evidence for composite membrane transport and for a slippage of small organic solutes across water channels. *Planta* 202:324–35

42. Holbrook NM, Burns MJ, Field CB. 1995. Negative xylem pressures in plants: a test of the balancing pressure technique. *Science* 270:1193–94

43. Howard RJ, Valent B. 1996. Breaking and entering: host penetration by the fungal rice blast pathogen *Magnaporthe grisea*. *Annu. Rev. Microbiol.* 50:491–512

44. Hüsken D, Steudle E, Zimmermann U. 1978. Pressure probe technique for measuring water relations of cells in higher plants. *Plant Physiol.* 61:158–63

45. Irving MS, Ritter S, Tomos AD, Koller D. 1997. Phototropic response of the bean pulvinus: movement of water and ions. *Bot. Acta* 110:118–26

46. Jones H, Leigh RA, Tomos AD, Wyn Jones RG. 1987. The effect of abscisic acid on cell turgor pressures, solute content and growth of wheat roots. *Planta* 170:257–62

47. Jones H, Tomos AD, Leigh RA, Wyn Jones RG. 1983. Water relations parameters of epidermal and cortical cells of the primary root of *Triticum aestivum*. *Planta* 158:230–36

47a. Karrer EE, Lincoln JE, Hogenhout S, Bennett AB, Bostock RM, et al. 1995.

In situ isolation of mRNA from individual plant cells: creation of cell-specific cDNA libraries. *Proc. Natl. Acad. Sci. USA* 92:3814–18

48. Katou K, Furumoto M. 1986. A mechanism of respiration-dependent water uptake in higher plants. *Protoplasma* 130:80–82

49. Kempers R, Prior DAM, Van Bel AJE, Oparka KJ. 1993. Plasmodesmata between sieve element and companion cell of extrafascicular stem phloem of *Cucurbita maxima* permit passage of 3 kDa fluorescent-probes. *Plant J.* 4:567–75

50. Kempers R, Van Bel AJE. 1997. Symplasmic connections between sieve element and companion cell in the stem phloem of *Vicia faba* L. have a molecular exclusion limit of at least 10 kDa. *Planta* 201:195–201

51. Kirst GO. 1990. Salinity tolerance of eukaryotic marine algae. *Annu. Rev. Plant Physiol. Plant Mol. Biol.* 41:21–53

52. Korolev A, Tomos AD, Farrar JF, Kockenberger W, Hudson A, Bowtell R. 1998. Distribution of water and solutes in carrot taproot at single cell resolution. *J. Exp. Bot.* 49s:8

53. Koroleva OA, Farrar JF, Tomos AD, Pollock CJ. 1997. Solute patterns in individual mesophyll, bundle sheath and epidermal cells of barley leaves induced to accumulate carbohydrate. *New Phytol.* 136:97–104

54. Koroleva OA, Farrar JF, Tomos AD, Pollock CJ. 1998. Carbohydrates in individual cells of epidermis, mesophyll and bundle sheath in barley leaves with changed export or photosynthetic rate. *Plant Physiol.* 118:1525–32

55. Koroleva OA, Tomos AD, Farrar JF, Pollock CJ. 1998. Water relations of individual cells of barley source leaves. *J. Exp. Bot.* 49s:6

56. Kourie JI, Findlay GP. 1991. Ionic currents across the plasmalemma of *Chara inflata* cells. 3. Water-relations parameters and their correlation with membrane electrical properties. *J. Exp. Bot.* 42:151–58

57. Kuhn AJ, Bauch J, Schroder WH. 1995. Monitoring uptake and contents of Mg, Ca and K in Norway spruce as influenced by pH and Al, using microprobe analysis and stable-isotope labeling. *Plant Soil* 169:135–50

57a. Lange OL, Nobel PS, Osmond CB, Zeigler H, eds. 1982. *Encycopedia of Plant Physiology*. Berlin: Springer

58. Lawrence RA, Tomos AD. 1998. What is the role of the apoplast in the maintenance of cell turgor pressure in water stressed plants. *J. Exp. Bot.* 49s:6

59. Leigh RA. 1997. Solute composition of vacuoles. *Adv. Bot Res.* 25:171–94

60. Leigh RA, Tomos AD. 1983. An attempt to use isolated vacuoles to determine the distribution of sodium and potassium in cells of storage roots of red beet (*Beta vulgaris* L.). *Planta* 159:469–75

61. Lew RR. 1996. Pressure regulation of the electrical-properties of growing *Arabidopsis thaliana* L. root hairs. *Plant Physiol.* 112:1089–100

62. Lockhart JA. 1965. An analysis of irreversible plant cell elongation. *J. Theor. Biol.* 8:264–75

63. Lovisolo C, Schubert A. 1998. Effects of water stress on vessel size and xylem hydraulic conductivity in *Vitis vinifera* L. *J. Exp. Bot.* 49:693–700

64. Lützenkirchen G, Tomos AD. 1993. Water relations of the intertidal alga *Cladophora rupestris*. *J. Exp. Bot.* 44s:48

65. Malone M. 1993. Hydraulic signals. *Phil. Trans. R. Soc. London B* 341:33–39

66. Malone M, Leigh RA, Tomos AD. 1989. Extraction and analysis of sap from individual wheat leaf-cells: the effect of sampling speed on the osmotic pressure of extracted sap. *Plant Cell Environ.* 12:919–26

67. Malone M, Leigh RA, Tomos AD. 1991. Concentrations of vacuolar inorganic ions in individual cells of intact wheat leaf epidermis. *J. Exp. Bot.* 42:385–89

68. Malone M, Tomos AD. 1990. A simple pressure-probe method for the determination of volume in higher plant cells. *Planta* 182:199–203

69. Maurel C. 1997. Aquaporins and water permeability of plant membranes. *Annu. Rev. Plant Physiol. Plant. Mol. Biol.* 48:399–429

70. McFadden GI. 1994. In situ hybridization of RNA. See Ref. 39a, pp. 97–125

71. Meidner H, Edwards M. 1975. Direct measurements of turgor pressure potentials of guard cells. I. *J. Exp. Bot.* 26:319–30

72. Meshcheryakov AB, Lukyanova OA, Kholodova VP. 1992. Characteristics of water exchange in storage parenchyma of the sugar-beet (*Beta vulgaris* L.) root in relation to sugar accumulation. *Sov. Plant Physiol.* 39:740–46

73. Meshcheryakov A, Steudle E, Komor E. 1992. Gradients of turgor, osmotic pressure, and water potential in the cortex of the hypocotyl of growing *Ricinus* seedlings. Effects of the supply of water from the xylem and of solutes from the

phloem. *Plant Physiol.* 98:840–52
74. Meuser J, Frensch J. 1998. Hydraulic properties of living late metaxylem and interactions between transpiration and xylem pressure in maize. *J. Exp. Bot.* 49:69–77
75. Milburn JA. 1996. Sap ascent in vascular plants: challenges to the cohesion theory ignore the significance of immature xylem and the recycling of Münch water. *Ann. Bot.* 78:399–407
76. Miller AJ. 1994. Ion-selective microelectrodes. See Ref. 39a, pp. 283–96
77. Moore PH, Cosgrove DJ. 1991. Developmental changes in cell and tissue water relations parameters in storage parenchyma of sugarcane. *Plant Physiol.* 96:794–801
78. Moulia B, Fournier M, Guitard D. 1994. Mechanics and form of the maize leaf. In vivo qualification of flexural behavior. *J. Mater. Sci.* 29:2359–66
79. Murphy R, Ortega JKE. 1995. A new pressure probe method to determine the average volumetric elastic modulus of cells in plant tissue. *Plant Physiol.* 107:995–1005
80. Murphy R, Smith JAC. 1994. A critical comparison of the pressure-probe and pressure-chamber techniques for estimating leaf-cell turgor pressure in *Kalanchoë daigremontiana. Plant Cell Environ.* 17:15–29
81. Nobel PS. 1991. *Physicochemical and Environmental Plant Physiology.* San Diego: Academic
82. Nonami H, Boyer JB, Steudle E. 1987. Pressure probe and isopiestic psychrometer measure similar turgor. *Plant Physiol.* 83:592–95
83. Nonami H, Schulze E-D. 1989. Cell water potential, osmotic potential, and turgor in the epidermis and mesophyll of transpiring leaves: combined measurements with the cell pressure probe and nanoliter osmometer. *Planta* 177:35–46
84. Oparka KJ, Murphy R, Derrick PM, Prior DAM, Smith JAC. 1991. Modification of the pressure-probe technique permits controlled intracellular microinjection of fluorescent probes. *J. Cell Sci.* 98:539–44
85. Oparka KJ, Prior DAM. 1992. Direct evidence for pressure-generated closure of plasmodesmata. *Plant J.* 2:741–50
86. Oparka KJ, Read ND. 1994. The use of fluorescent probes for studies of living plant cells. See Ref. 39a, pp. 27–68
87. Ortega JKE, Bell SA, Erazo AJ. 1992. Pressure clamp method to measure transpiration in growing single plant cells. Demonstration with sporangiophores of *Phycomyces. Plant Physiol.* 100:1036–41
88. Palta JA, Wyn Jones RG, Tomos AD. 1987. Leaf diffusive conductance and tap root cell turgor pressure of sugarbeet. *Plant Cell Environ.* 10:735–40
89. Passioura JB. 1991. An impasse in plant water relations? *Bot. Acta* 104:405–11
90. Perry CA, Leigh RA, Tomos AD, Wyse RE, Hall JL. 1987. The regulation of turgor pressure during sucrose mobilisation and salt accumulation by excised storage root tissue of red beet. *Planta* 170:353–61
91. Pockman WT, Sperry JS, O'Leary JW. 1995. Sustained and significant negative water pressure in xylem. *Nature* 378:715–16
92. Pollock CJ, Tomos AD, Thomas A, Smith CJ, Lloyd EJ, Stoddart JL. 1990. Extension growth in a barley mutant with reduced sensitivity to low temperature. *New Phytol.* 115:617–23
93. Pritchard J. 1994. The control of cell expansion in roots. *New Phytol.* 127:3–26
94. Pritchard J. 1996. Aphid stylectomy reveals an osmotic step between sieve tube and cortical cells in barley roots. *J. Exp. Bot.* 47:1519–24
95. Pritchard J, Fricke W, Tomos D. 1996. Turgor-regulation during extension growth and osmotic stress of maize roots. An example of single-cell mapping. *Plant Soil* 187:11–21
96. Pritchard J, Hetherington PR, Fry SC, Tomos AD. 1993. Xyloglucan endotransglycosylase, microfibril orientation and the profiles of cell wall properties along growing regions of maize roots. *J. Exp. Bot.* 44:1281–89
97. Pritchard J, Tomos AD, Wyn Jones RG. 1987. Control of wheat root elongation growth. I. Effects of ions on growth rate, wall rheology and cell water relations. *J. Exp. Bot.* 38:948–59
98. Pritchard J, Williams G, Wyn Jones RG, Tomos AD. 1989. Radial turgor pressure profiles in growing and mature zones of wheat roots. A modification of the pressure probe technique. *J. Exp. Bot.* 40:567–71
99. Pritchard J, Wyn Jones RG, Tomos AD. 1988. Control of wheat root growth: effect of excision on water relations parameters and wall rheology. *Planta* 176:399–405
100. Pritchard J, Wyn Jones RG, Tomos AD. 1990. Measurements of yield threshold and cell wall extensibility of intact wheat roots under different ionic, osmotic and temperature treatments. *J. Exp. Bot.* 41:669–75

101. Pritchard J, Wyn Jones RG, Tomos AD. 1991. Turgor, growth and rheological gradients of wheat roots following osmotic stress. *J. Exp. Bot.* 42:1043–49

102. Raven JA. 1983. Phytophages of xylem and phloem: a comparison of animal and plant sap-feeders. *Adv. Ecol. Res.* 13:135–234

103. Rich TCG, Tomos AD. 1988. Turgor pressure and phototropism in *Sinapis alba* L. seedlings. *J. Exp. Bot.* 39:291–99

104. Richardson P, Tomos AD, Carver T. 1997. Effects of *Erysiphe graminis* on the water and solute relations of host leaf epidermal cells. *J. Exp. Bot.* 48s:40

105. Richter H. 1997. Water relations of plants in the field: some comments on the measurement of selected parameters. *J. Exp. Bot.* 48:1–7

106. Rygol J, Lüttge U. 1984. Effects of various benzene-derivatives, dodecylbenzenesulfonate and $HgCl_2$ on water-relation parameters at the cellular level. *Physiol. Végét.* 22:783–92

107. Rygol J, Pritchard J, Zhu JJ, Tomos AD, Zimmermann U. 1993. Transpiration induces radial turgor pressure gradients in wheat and maize roots. *Plant Physiol.* 103:493–500

108. Rygol J, Winter K, Zimmermann U. 1987. The relationship between turgor pressure and titratable acidity in mesophyll cells of intact leaves of a Crassulacean-acid-metabolism plant, *Kalanchoë daigremontiana* Hamet and Perr. *Planta* 172:487–493

109. Rygol J, Zimmermann U. 1990. Radial and axial turgor pressure measurements in individual root cells of *Mesembryanthemum crystallinum* grown under various saline conditions. *Plant Cell Environ.* 13:15–26

110. Schneider H, Witsuba N, Miller B, Gessner P, Thürmer F, et al. 1997. Diurnal variation in the radial reflection coefficient of intact maize roots determined with the xylem pressure probe. *J. Exp. Bot.* 48:2045–53

111. Schneider H, Zhu JJ, Zimmermann U. 1997. Xylem and cell turgor pressure probe measurements in intact roots of glycophytes: Transpiration induces a change in the radial and cellular reflection coefficients. *Plant Cell Environ.* 20:221–29

112. Schutz K, Tyerman SD. 1997. Water channels in *Chara corallina J. Exp. Bot.* 48:1511–18

113. Shackel KA. 1987. Direct measurement of turgor and osmotic potential in individual epidermal cells. Independent confirmation of leaf water potential as determined by in situ psychrometry. *Plant Physiol.* 83:719–22

114. Shackel KA, Brinckmann E. 1985. In situ measurement of epidermal cell turgor, leaf water potential, and gas-exchange in *Tradescantia virginiana* L. *Plant Physiol.* 78:66–70

115. Shackel KA, Greve C, Labavitch JM, Ahmadi H. 1991. Cell turgor changes associated with ripening in tomato pericarp tissue. *Plant Physiol.* 97:814–16

116. Shackel KA, Turner N. 1998. Seed coat cell turgor responds rapidly to air humidity in chickpea and faba bean. *J. Exp. Bot.* 49:1413–19

117. Shortreed MR, Dourado S, Kopelman R. 1997. Development of a fluorescent optical potassium-selective ion sensor with ratiometric response for intracellular applications. *Sens. Actu-B* 38:8–12

118. Smith AM. 1994. Xylem transport and the negative pressures sustainable by water. *Ann. Bot.* 74:647–51

119. Sperry JS, Saliendra NZ. 1994. Intra- and inter-plant variation in xylem cavitation in *Betula occidentalis*. *Plant Cell Environ.* 17:1233–41

120. Sperry JS, Saliendra NZ, Pockman WT, Cochard H, Cruiziat P, et al. 1996. New evidence for large negative xylem pressures and their measurement by the pressure chamber method. *Plant Cell Environ.* 19:427–36

121. Spollen WG, Sharp RE. 1991. Spatial distribution of turgor and root growth at low water potentials. *Plant Physiol.* 96:438–43

122. Stento N, Kiegle EA, Bisson MA. 1997. Regulation of turgor pressure in *Chara longifolia*: simultaneous measurements using pressure probe and electrophysiological measurements. *Plant Physiol.* 114s:197

123. Steudle E. 1993. Pressure probe techniques: basic principles and application to studies of water and solute relations at the cell, tissue and organ level. In *Water Deficits: Plant Responses from Cell to Community*, ed. JAC Smith, H Griffiths, pp. 5–36. Oxford: Bios

124. Steudle E. 1995. Trees under tension. *Nature* 378:663–64

125. Steudle E, Jeschke WD. 1983. Water transport in barley roots. *Planta* 158:237–48

126. Steudle E, Lüttge U, Zimmermann U. 1975. Water relations of the epidermal bladder cells of the halophytic species *Mesembryanthemum crystallinum*: direct

measurements of hydrostatic pressure and hydraulic conductivity. *Planta* 126:229–46

127. Steudle E, Peterson CA. 1998. How does water get through roots? *J. Exp. Bot.* 49:775–88

128. Steudle E, Smith JAC, Lüttge U. 1980. Water relations parameters of individual mesophyll cells of the CAM plant *Kalanchoë daigremontiana*. *Plant Physiol.* 66:1155–63

129. Steudle E, Wieneke J. 1985. Changes in water relations and elastic properties of apple fruit cells during growth and development. *J. Am. Soc. Hortic. Sci.* 110:824–29

130. Steudle E, Ziegler H, Zimmermann U. 1983. Water relations of the epidermal bladder cells of *Oxalis carnosa* molina. *Planta* 159:38–45

131. Thomas A, Tomos AD, Stoddart JL, Thomas H, Pollock CJ. 1989. Cell expansion rate, temperature and turgor pressure in growing leaves of *Lolium temulentum*. *New Phytol.* 112:1–5

132. Thorpe MR, Minchin PEH, Williams JHH, Farrar JF, Tomos AD. 1993. Carbon import into developing ovules of *Pisum sativum*: the role of the water relations of the seed coat. *J. Exp. Bot.* 44:937–45

133. Tomos AD. 1988. Cellular water relations of plants. In *Water Science Reviews*, ed. F Franks, 3:186–277. Cambridge: Cambridge Univ. Press

134. Tomos AD, Hinde P, Richardson P, Pritchard J, Fricke W. 1994. Microsampling and measurements of solutes in single cells. See Ref. 39a, pp. 297–314

135. Tomos AD, Leigh RA, Palta JA, Williams JHH. 1992. Sucrose and cell water relations. In *Carbon Partitioning Within and Between Organisms*, ed. CJ Pollock, JF Farrar, Gordon AJ, pp. 71–89. Oxford: Bios

136. Tomos AD, Leigh RA, Shaw CA, Wyn Jones RG. 1984. A comparison of methods for measuring turgor pressure and osmotic pressure of cells of red beet storage tissue. *J. Exp. Bot.* 35:1675–83

137. Tomos AD, Pritchard J. 1994. Biophysical and biochemical control of cell expansion in roots and leaves *J. Exp. Bot.* 45:1721–31

138. Tomos AD, Steudle E, Zimmermann U, Schulze E-D. 1981. Water relations of leaf epidermal cells of *Tradescantia virginiana*. *Plant Physiol.* 68:1135–45

139. Turner NC. 1988. Measurement of plant water status by the pressure chamber technique. *Irrig. Sci.* 9:289–308

140. Tyerman SD, Steudle E. 1984. Determination of solute permeability in *Chara* internodes by a turgor minimum method. Effects of external pH. *Plant Physiol.* 74:464–68

141. Tyree MT. 1997. The Cohesion-Tension theory of sap ascent: current controversies. *J. Exp. Bot.* 48:1753–65

142. van Steveninck RFM, van Steveninck ME. 1991. Microanalysis. In *Electron Microscopy of Plant Cells*, ed. JL Hall, CR Hawes, pp. 415–55. London: Academic

143. Verscht J, Kalusche B, Köhler J, Kockenberger W, Metzler A, et al. 1998. The kinetics of sucrose concentration in the phloem of individual vascular bundles of *Ricinus communis* seedlings measured by nuclear magnetic microimaging. *Planta* 205:132–39

144. Weatherley PE. 1982. Water uptake and flow into roots. See Ref. 57a, 12B:79–109

145. Webster J, Davey RA, Smirnoff N, Fricke W, Hinde R, et al. 1994. Mannitol and hexoses are components of Buller's drop. *Mycol. Res.* 99:833–38

146. Wendler S, Zimmermann U. 1982. A new method for the determination of hydraulic conductivity and cell volume of plant cells by pressure clamp. *Plant Physiol.* 69:998–1003

147. Wendler S, Zimmermann U. 1985. Determination of the hydraulic conductivity of *Lamprothamnium* by use of the pressure clamp. *Planta* 164:241–45

148. Wendler S, Zimmermann U. 1985. Compartment analysis of plant cells by means of turgor pressure relaxation. II. Experimental results on *Chara corallina*. *J. Membr. Biol.* 85:133–42

149. Wendler S, Zimmermann U, Bentrup FW. 1983. Relationship between cell turgor pressure, electrical membrane potential, and chloride efflux in *Acetabularia mediterranea*. *J. Membr. Biol.* 72:75–84

150. Wiencke C, Gorham J, Tomos AD, Davenport J. 1992. Incomplete turgor adjustment in *Cladophora rupestris* under fluctuating salinity regimes. *Est. Coast. Sci.* 34:413–27

151. Wyse RE, Zamski E, Tomos AD. 1986. Turgor regulation of sucrose transport in sugar beet taproot tissue. *Plant Physiol.* 81:478–81

152. Zhang WH, Atwell BJ, Patrick JW, Walker NA. 1996. Turgor-dependent efflux of assimilates from coats of developing seed of *Phaseolus vulgaris*. L: water relations of the cells involved in efflux. *Planta* 199:25–33

153. Zhang WH, Tyerman SD. 1991. Effect of low O_2 concentration and azide on hydraulic conductivity and osmotic volume

of the cortical cells of wheat roots. *Aust. J. Plant Physiol.* 18:603–13

154. Zhen R-G, Koyro HW, Leigh RA, Tomos AD, Miller AJ. 1991. Compartmental nitrate concentrations in barley root cells measured with nitrate-selective microelectrodes and by single-cell sap sampling. *Planta* 185:356–61

155. Zhu GL. 1996. A new turgor membrane-potential probe simultaneously measures turgor and electrical membrane-potential. *Bot. Acta* 109:51–56

156. Zhu JJ, Zimmermann U, Thürmer F, Haase A. 1995. Xylem pressure response in maize roots subjected to osmotic stress: determination of radial reflection coefficients by the use of the xylem pressure probe. *Plant Cell Environ.* 18:906–12

157. Zimmermann MH. 1983. *Xylem Structure and the Ascent of Sap.* Berlin: Springer

158. Zimmermann U. 1989. Water relations of plant cells: pressure probe technique. *Methods Enzymol.* 174:338–66

159. Zimmermann U, Balling A, Rygol J, Link A, Haase A. 1991. Comments on the article by J. B. Passioura "An impasse in plant water relations?" *Bot. Acta* 104:412–15

160. Zimmermann U, Beckers F. 1978. Generation of action potentials in *Chara corallina* by turgor pressure changes. *Planta* 138:173–79

161. Zimmermann U, Benz R, Koch H. 1981. A new electronic method for the determination of the true membrane area and the individual membrane resistances in plant cells. *Planta* 152:352–55

162. Zimmermann U, Büchner K-H, Benz R. 1982. Transport properties of mobile charges in algal membranes: influence of pH and turgor pressure. *J. Membr. Biol.* 67:183–97

163. Zimmermann U, Haase A, Langbein D, Meinzer F. 1993. Mechanisms of long-distance water transport in plants: a reexamination of some paradigms in the light of new evidence. *Phil. Trans. R. Soc. London* B 341:19–31

164. Zimmermann U, Hüsken D. 1980. Turgor pressure and cell volume relaxation in *Halicystis parvula. J. Membr. Biol.* 56:55–64

165. Zimmermann U, Hüsken D, Schulze E-D. 1980. Direct turgor pressure measurements in individual leaf cells of *Tradescantia virginiana. Planta* 149:445–53

166. Zimmermann U, Meinzer FC, Benkert R, Zhu JJ, Schneider H, et al. 1994. Xylem water transport: Is the available evidence consistent with the cohesion theory? *Plant Cell Environ.* 17:1169–81

167. Zimmermann U, Meinzer FC, Bentrup F-W. 1995. How does water ascend tall trees and other vascular plants? *Ann. Bot.* 76:545–51

168. Zimmermann U, Rygol J, Balling A, Klock G, Metzler A, Haase A. 1992. Radial turgor and osmotic pressure profiles in intact and excised roots of *Aster tripolium.* Pressure probe measurements and nuclear magnetic resonance-imaging analysis. *Plant Physiol.* 99:186–96

169. Zimmermann U, Steudle E. 1974. Hydraulic conductivity and volumetric elastic modulus in giant algal cells: pressure-and volume-dependence. In *Membrane Transport in Plants,* ed. U Zimmermann, J Dainty, pp. 64–71. Berlin: Springer

170. Zimmermann U, Steudle E. 1974. The pressure dependence of the hydraulic conductivity, the cell membrane resistance and the membrane potential during turgor pressure regulation in *Valonia utricularis. J. Membr. Biol.* 16:331–52

171. Zimmermann U, Steudle E. 1978. Physical aspects of water relations of plant cells. *Adv. Bot. Res.* 6:45–117

172. Zimmermann U, Zhu JJ, Meinzer FC, Goldstein G, Schneider H, et al. 1994. High molecular weight organic compounds in the xylem sap of mangroves: implications for long-distance water transport. *Bot. Acta* 107:218–29

Annu. Rev. Plant Physiol. Plant Mol. Biol. 1999. 50:473–503

THE SHIKIMATE PATHWAY

Klaus M. Herrmann

Department of Biochemistry, Purdue University, West Lafayette, Indiana 47907;
e-mail: Herrmann@biochem.purdue.edu

Lisa M. Weaver

Monsanto Company, St. Louis, Missouri 63198;
e-mail: Lisa.m.weaver@monsanto.com

KEY WORDS: aromatic amino acids, quinate, chloroplasts, plant secondary metabolism

ABSTRACT

The shikimate pathway links metabolism of carbohydrates to biosynthesis of aromatic compounds. In a sequence of seven metabolic steps, phosphoenolpyruvate and erythrose 4-phosphate are converted to chorismate, the precursor of the aromatic amino acids and many aromatic secondary metabolites. All pathway intermediates can also be considered branch point compounds that may serve as substrates for other metabolic pathways. The shikimate pathway is found only in microorganisms and plants, never in animals. All enzymes of this pathway have been obtained in pure form from prokaryotic and eukaryotic sources and their respective DNAs have been characterized from several organisms. The cDNAs of higher plants encode proteins with amino terminal signal sequences for plastid import, suggesting that plastids are the exclusive locale for chorismate biosynthesis. In microorganisms, the shikimate pathway is regulated by feedback inhibition and by repression of the first enzyme. In higher plants, no physiological feedback inhibitor has been identified, suggesting that pathway regulation may occur exclusively at the genetic level. This difference between microorganisms and plants is reflected in the unusually large variation in the primary structures of the respective first enzymes. Several of the pathway enzymes occur in isoenzymic forms whose expression varies with changing environmental conditions and, within the plant, from organ to organ. The penultimate enzyme of the pathway is the sole target for the herbicide glyphosate. Glyphosate-tolerant transgenic plants are at the core of novel weed control systems for several crop plants.

473

CONTENTS

INTRODUCTION

In this review the shikimate pathway is defined as the seven metabolic steps beginning with the condensation of phosphoenolpyruvate (PEP) and erythrose 4-phosphate (E4P) and ending with the synthesis of chorismate. This pathway was elucidated by BD Davis and DB Sprinson and their associates about 40 years ago. In the past decade a monograph (73) and a comprehensive review of the shikimate pathway (16) have appeared. During the same period, summaries of specific aspects emphasizing the role of the pathway in the biosynthesis of plant aromatic secondary products were published (84, 85, 168, 184). After a few brief historical remarks, we focus on the research results of the past ten years.

The shikimate pathway, the common route leading to production of the aromatic amino acids phenylalanine, tyrosine, and tryptophan, constitutes a part of metabolism that is found only in microorganisms and plants, never in animals. The pathway is therefore an important target for herbicides (100), antibiotics, and live vaccines (141). The penultimate step in this pathway is inhibited by N-[phosphonomethyl]glycine, the active ingredient of the broad-spectrum, nonselective herbicide glyphosate (190). Much effort has been put forward to understand not only the interaction of this herbicide with its target but also to seek other compounds with similar inhibitory capabilities. Chemical compounds that interfere with any enzyme activity in this pathway are considered "safe" for humans when handled in reasonable concentrations. In fact, glyphosate has been tested successfully in mice as a therapeutic agent against pathogenic protozoans that cause diseases like toxoplasmosis or malaria (161).

The seven enzymes of the shikimate pathway were originally discovered through studies on bacteria, mainly *Escherichia coli* and *Salmonella typhimurium*. Although the substrates and products of these enzymes, and thus the intermediates of the pathway, are identical for prokaryotic and eukaryotic organisms, sometimes great differences are found in the primary structure and properties of the prokaryotic and eukaryotic enzymes themselves. Also, how aromatic compounds serve as signals in the regulation of the pathway is well understood for some microorganisms, but is only now beginning to be investigated in higher plants. Therefore, we considered it useful to outline briefly the properties of bacterial enzymes, particularly when they differ a great deal from their plant homologues.

Chorismate, the endproduct of the shikimate pathway, is the precursor of the three aromatic amino acids and several other aromatic compounds of primary metabolism. In addition, the three aromatic amino acids are precursors to a large variety of plant secondary metabolites (34). Finally, the intermediates of the main trunk of the shikimate pathway also serve as starting points for biosynthesis of secondary products. Clearly, the shikimate pathway is of eminent importance to the biosynthesis of many compounds of commercial interest.

THE ENZYMES OF THE MAIN TRUNK

3-Deoxy-D-Arabino-Heptulosonate 7-Phosphate Synthase

The first step of the shikimate pathway is the condensation of PEP and E4P yielding 3-deoxy-D-*arabino*-heptulosonate 7-phosphate (DAHP) and inorganic phosphate. The basic structure of DAHP was confirmed by several different chemical syntheses. Fine structure analysis identified DAHP as 1-carboxy-2-deoxy-α-D-glucose-6-phosphate (49).

The enzymatic synthesis of DAHP is catalyzed by DAHP synthase (Figure 1), an enzyme discovered in *E. coli* and first purified to electrophoretic homogeneity from microorganisms. The most intensively investigated DAHP synthase is the enzyme from *E. coli* (83). Wild-type *E. coli* produces three feedback inhibitor-sensitive DAHP synthase isoenzymes: a Tyr-sensitive, a Phe-sensitive, and a Trp-sensitive enzyme. Their corresponding structural genes, *aroF*, *aroG*, and *aroH*, are scattered over the *E. coli* chromosome (18). The genes have been

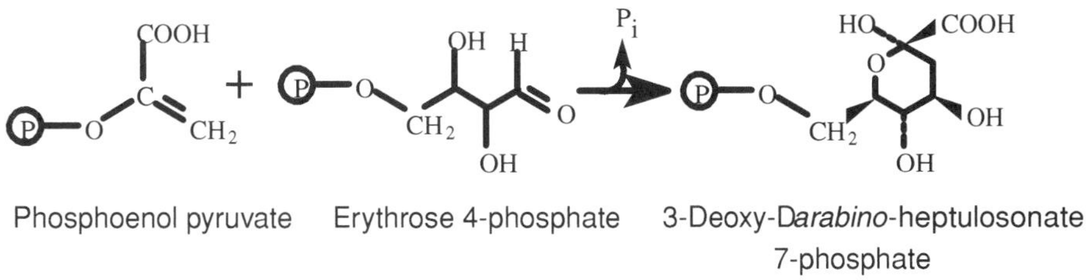

Phosphoenol pyruvate   Erythrose 4-phosphate   3-Deoxy-D*arabino*-heptulosonate
7-phosphate

Figure 1   The reaction catalyzed by 3-deoxy-D-*arabino*-heptulosonate 7-phosphate synthase.

cloned and the primary structures of the encoded isoenzymes have been obtained through a combination of protein and DNA sequencing efforts. The *E. coli* isoenzymes are oligomers with a subunit molecular weight of about 39,000. The Phe-sensitive isoenzyme has been crystallized as a binary complex with PEP or Phe and as a ternary complex with additional metal ions (179).

Plant DAHP synthases have been obtained in pure form from carrot (198) and potato (153) as oligomers with subunit molecular weights of about 54,000. Rabbit antibodies raised against the potato enzyme (152) were used to screen a cDNA library from potato cells grown in suspension culture (40). The resulting cDNA complements *E. coli* mutants devoid of DAHP synthase (215). The first cDNA served as a probe to isolate homologues from tobacco (212), *Arabidopsis* (95), tomato (58), and *Morinda citrifolia* (223). The *Arabidopsis* cDNA complements yeast mutants devoid of DAHP synthase (95). cDNA clones encoding a second DAHP synthase isoform have been reported from *Arabidopsis* (95), potato (227), and tomato (58). *M. citrifolia* (223) and *Arabidopsis* (B Keith, personal communication) actually contain three DAHP synthase isoenzymes. Although, like *E. coli*, some plants have three DAHP synthase-encoding genes, the structures of the encoded enzymes are quite different from the structures of the bacterial DAHP synthases.

Translation of plant cDNAs yields polypeptides with amino terminal signal sequences that direct plastid import. These sequences are clipped off during import into the plastid. The precise processing site cannot be determined by Edman degradation of the mature protein because the amino termini of plant DAHP synthases are blocked. Mature plant enzymes, purified from potato tuber or carrot roots, have subunits that are about 150 amino acid residues longer than their bacterial homologues. Typically, the comparison of the primary amino acid sequence of a plant DAHP synthase to a bacterial DAHP synthase shows a surprisingly low 20% pairwise identity. In fact, PRETTYBOX, a program used to make multiple sequence alignments, identifies only 24 invariant residues for all known DAHP synthases.

DAHP synthase in *E. coli* is a metallo protein that is inhibited by chelating agents (122). In vitro analyses of pure enzymes indicate that the metal requirement can be satisfied by several divalent cations (191). The metal content of the native enzyme may depend on general growth conditions. Under some conditions, DAHP synthase may be a Cu enzyme (7), but Fe and perhaps Zn are most likely the preferred metals in vivo (191). The metal plays a catalytic and, possibly, a structural role as well. A Cys residue in a Cys-X-X-His motif of the protein has been identified as part of a metal binding site (192).

A careful in vitro metal analysis comparable to the study on the Phe-sensitive isoenzyme from *E. coli* (191) has not yet been performed on a plant DAHP synthase. However, the activities of all plant DAHP synthases are enhanced by

Mn, and plant DAHP synthases also contain a Cys-X-X-His motif. In bacterial enzymes this motif is near the amino terminus, but in plant enzymes it is near the middle of the carboxy half of the enzyme. Thus, this motif is not positionally conserved in sequence alignments between bacterial and plant DAHP synthases, although it may well be part of a metal binding site for both (84).

Early studies on the plant enzymes suggested the existence of Mn- and Co-activated DAHP synthases. While both of these enzymes use PEP as one substrate, the Mn-activated enzyme has an absolute specificity for E4P as the second substrate. The Co-activated enzyme, on the other hand, is less specific, preferring glycolaldehyde as a second substrate (37). Thus it may be more properly called 4,5-dihydroxy-2-oxovalerate synthase. A correlation has been reported between this enzyme's activity and the biosynthesis of anthocyanin in suspension cultured cells of Vitis (201). In contrast to this result, it is the Mn-activated DAHP synthase that is induced in *Petroselinum crispum* suspension cultures by light (125) or by fungal elicitation, leading to enhanced phenylpropanoid biosynthesis; the Co-activated enzyme is not affected (124). A similar expression pattern is seen upon wounding of potato; the Mn-activated enzyme is wound-inducible, whereas the Co-activated enzyme is not (136). The Co-activated enzyme has recently been purified from cultured carrot cells (200) but its contribution to the overall synthesis of DAHP remains questionable because of its seemingly unphysiologically high K_m for E4P.

Structurally distinct true DAHP synthases with potentially different functions exist in the form of Mn-activated isoenzymes. These Mn-activated DAHP synthases, the Co-activated 4,5-dihydroxy-2-oxovalerate synthase, and the metal ion independent 3-deoxy-D-manno-octulosonate 8-phosphate synthase (224) all share PEP as one substrate and an aldehyde as the other. They catalyze similar condensation reactions that proceed with a C-O rather than a P-O bond cleavage on PEP (79) and appear to be evolutionarily related (196).

Because of the large differences in the primary structure, Walker and coworkers made a distinction between a small bacterial type I DAHP synthase of 39 kD and a large plant type II enzyme of 54 kD (210). However, a type II DAHP synthase was purified to homogeneity from the bacterium *Streptomyces rimosus* (193) and partial sequence data from this protein show substantial similarity to plant DAHP synthases (210). Also, both yeast DAHP synthases (108, 146, 148) are type I enzymes. Thus, the "plant type II" DAHP synthase is also found in prokaryotes and the "bacterial type I" in eukaryotic microorganisms.

DAHP synthase in *E. coli* is regulated at the transcriptional level by repression and at the protein level by feedback-inhibition. Both the Phe- and the Tyr-sensitive isoenzymes can be completely inhibited by about 0.1 mM of the corresponding amino acid. In contrast, the Trp-sensitive isoenzyme is only partially inhibited by Trp (1, 158). The inability of Trp to totally inhibit DAHP synthase apparently ensures a sufficient supply of chorismate for the biosynthesis of other aromatic compounds when Tyr, Phe, and Trp are present in excess in the growth medium. Amino acid residues within the enzymes' allosteric sites have been identified through structural analysis of feedback-insensitive mutant enzymes (54, 96, 159, 213, 214). It appears that these sites partially overlap the enzymes' active sites, in good agreement with kinetic data (1, 123, 173). Recently, an overlap of active and allosteric sites has also been shown for the Phe- and the Tyr-sensitive DAHP synthases from *Saccharomyces cerevisiae* (172).

Like their bacterial counterparts, plant DAHP synthases are oligomers and subject to metabolic regulation. Surprisingly, plant DAHP synthases are not feedback inhibited by any aromatic amino acid. In fact, tryptophan is a hysteretic activator of the enzymes (198). Arogenate, a post-chorismate intermediate of phenylalanine and tyrosine biosynthesis (93), inhibits the enzyme; 155 μM arogenate causes 50% inhibition of the bean DAHP synthase (163). However, since arogenate does not accumulate in plants to detectable levels, it seems unlikely that it can play a role as a physiological regulator (160). No classical allosteric inhibitor of a plant DAHP synthase has yet been identified.

The expression of the three *E. coli* genes encoding DAHP synthases is subject to repression by the *tyr*- and *trp*-repressors complexed to the aromatic amino acids (67, 228). *Cis*-acting regulatory mutants with lesions in the repressor target sites were identified in the regulatory regions of *aroH* (104, 228) and *aroF* (50). Furthermore, the *tyr*- and the *trp*-regulons are functionally connected by *aroH*, because the expression of this gene is controlled by both the *trp*- and the *tyr*-repressors (137). Transcriptional control for the expression of the three genes encoding DAHP synthases is complex; however, feedback inhibition of the three isoenzymes is quantitatively the major regulatory mechanism in vivo, as demonstrated by noninvasive nuclear magnetic resonance spectroscopy on whole living *E. coli* cells (142).

In plants, metabolic regulation of DAHP synthase appears to occur preferentially at the genetic level. Sublethal doses of glyphosate cause an increase in DAHP synthase activity in vivo (152). The target of the herbicide is the penultimate enzyme in the pathway. In vitro, the herbicide has no effect on the DAHP synthase enzyme activity. However, by inhibiting the penultimate enzyme, the herbicide reduces chorismate production in vivo, thereby initiating a signal that leads to increased carbon flow into the shikimate pathway through elevation of DAHP synthase activity. The nature of the signal is not known.

This elevation of the enzyme activity is due to an increase in the amount of the polypeptide chain. Modulation of the overall DAHP synthase activity is the result of differential isoenzyme expression (60, 62, 81, 95, 223). Developmental or environmental stimuli like light, mechanical wounding, or elicitation by microorganisms all influence DAHP synthase isoenzyme expression

(40, 60, 81, 124, 207). Gibberellic acid and jasmonate induce the enzyme in *Coptis japonica* and *Nicotiana tabacum*, respectively (72, 199). Although *cis*-acting elements have been identified in the regulatory region of one potato gene encoding DAHP synthase (213), to date none of the plant DAHP synthase promoters has been studied in detail.

In eukaryotic microorganisms, amino acid biosynthesis is subject to general control (87). For example, in yeast (109, 147) and the pathogen *Candida albicans* (150), the general control activator protein GCN4 regulates the expression of the two DAHP synthase genes. There is some evidence for such regulation in higher plants as well (69).

3-Dehydroquinate Synthase

In bacteria, reactions two to six of the shikimate pathway are catalyzed by five separate enzymes, but in fungi, a single polypeptide called the multifunctional AROM complex serves the same purpose. Space constraints prevent us from dealing in detail with this highly interesting system that was discovered three decades ago by NH Giles and his associates. AROM complexes have been studied from *Neurospora crassa, Aspergillus nidulans, Saccharomyces cerevisiae, Schizosaccharomyces pombe*, and *Pneumocystis carinii*. In these complexes, the enzymes do not appear in the order of the pathway reactions. Protein domains for DHQ synthase and EPSP synthase form the amino terminal part, domains for shikimate kinase, DHQ dehydratase, and shikimate dedydrogenase the carboxy terminal part (77). It appears that the *arom* locus evolved by gene fusion. DNA encoding the entire AROM complex has been cloned and sequenced from *A. nidulans* (29), yeast (39), and *P. carinii* (9), the pathogen that is the principle cause of death for patients with AIDS. A very interesting evolutionary aspect of fungal aromatic metabolism is a gene duplication of *arom* that resulted in genes encoding regulatory proteins for quinate degradation (see section on 3-dehydroquinate and 3-dehydroshikimate). In higher plants, reactions three and four are catalyzed by a bifunctional enzyme, the remaining three reactions of the shikimate pathway by separate enzymes that are structurally rather similar to their prokaryotic homologues.

The second reaction of the shikimate pathway is the elimination of phosphate from DAHP to generate 3-dehydroquinate (DHQ). The reaction is catalyzed by DHQ synthase (Figure 2), a monomeric enzyme with a molecular weight of 39,000. The enzyme from *E. coli* (47) requires divalent cations for activity; Co is the most active metal but Zn may be the ion used in vivo (13). There is evidence for two functionally distinct metal binding sites per polypeptide chain (185). DHQ synthase is activated by inorganic phosphate, one of the reaction products, and the enzyme also requires catalytic amounts of NAD for activity, even though the enzyme catalyzed reaction is redox neutral.

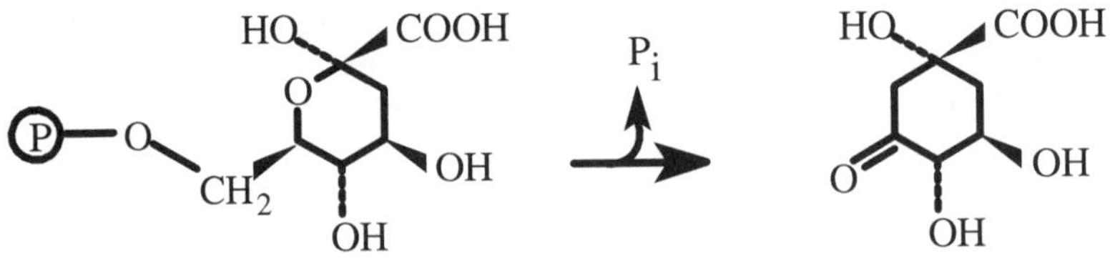

3-Deoxy-D-*arabino*-heptulosonate   3-Dehydroquinate
7-phosphate

Figure 2 The reaction catalyzed by 3-dehydroquinate synthase. The enzyme requires catalytic amounts of NAD for activity.

Conversion of DAHP to DHQ proceeds by way of an intramolecular exchange of the DAHP ring oxygen with carbon 7, driven by the cleavage of the phosphoester. The reaction involves an oxidation, a ß-elimination of inorganic phosphate, a reduction, a ring opening, and an intramolecular aldol condensation. The mechanism of this complicated reaction, already outlined by Sprinson and coworkers, was elucidated by PA Bartlett & JR Knowles and associates. The true substrate for the enzyme is apparently the pyran form of DAHP (49). The ß-elimination of phosphate proceeds with syn stereochemistry (11, 14, 221) and there is compelling evidence that the enzyme is a simple oxidoreductase (11). The phosphate monoester may either mediate its own elimination (132) or may be aided by the enzyme (131). The remaining partial reactions proceed spontaneously (221). The enzyme itself provides a potential conformational template to prevent formation of undesirable side products (10, 149). In these studies, DAHP analogs were used that can only undergo part of the overall reaction. Among these analogs is a carbacyclic phosphonate that inhibits the enzyme with a Ki of 0.8 nM (220). A similar enzyme mechanism was later found for the 2-deoxy-*scyllo*-inosose synthase of *Streptomyces fradiae*, an enzyme involved in antibiotic biosynthesis (226).

Plant DHQ synthases have been purified from *Phaseolus mungo* (225) and *Pisum sativum* (154). The primary structures of the bacterial (128) and plant (17) enzymes were deduced by translation of the corresponding DNA sequences. Interestingly, DHQ synthase sequences from *E. coli* and tomato are 52.5% identical, whereas bacterial and plant DAHP synthases are only 20% identical. Plant DHQ synthase is more closely related to the bacterial than to the fungal enzyme, unlike the case for DAHP synthase.

Like the plant cDNA encoding DAHP synthases, the tomato cDNA encodes a DHQ synthase with an amino terminal signal sequence for plastid import (17). The cDNA complements *E. coli* mutants lacking DHQ synthase. RNA

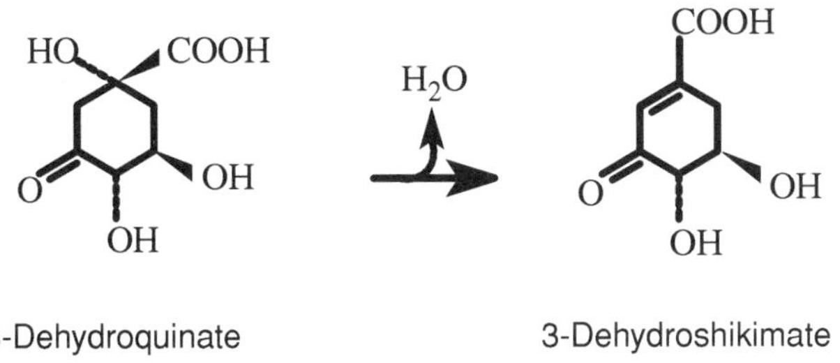

3-Dehydroquinate   3-Dehydroshikimate

Figure 3   The reaction catalyzed by 3-dehydroquinate dehydratase.

blots show that gene expression of DHQ synthase is organ specific and follows the pattern of gene expression for one of the DAHP synthase isoenzymes (17). Furthermore, DHQ synthase mRNA is induced when tomato cells grown in suspension culture are elicited with extracts from *Phytophthora megasperma* (17). In contrast, the DHQ synthase enzyme activity does not change when potato cells are exposed to glyphosate (152).

3-Dedydroquinate Dehydratase-Shikimate Dehydrogenase

The third step of the shikimate pathway, dehydration of DHQ to give 3-dehydroshikimate (DHS), is catalyzed by DHQ dehydratase (Figure 3) that exists in two forms: type I and II. Some bacteria, like *E. coli* (32) or *S. typhi* (133), have type I enzyme, whereas others, like *Streptomyces coelicor* (219), *Mycobacterium tuberculosis* (48, 134), or *Helicobacter pylori* (22), have type II. Type I DHQ dehydratase catalyzes syn elimination and type II anti-elimination of water (177). The mechanistic differences are reflected in the structures of these proteins; there is no sequence similarity between the type I and II enzymes, a rare example of convergent evolution (102). Both enzymes have been crystallized (23, 63). Analysis of diffraction data should reveal some fundamental differences in the reaction mechanisms of enzyme-catalyzed syn and anti-eliminations.

The best-studied type I DHQ dehydratase is from *E. coli* (38). The enzyme is a dimer with a subunit molecular weight of 27,000. The reaction proceeds by way of a Schiff base mechanism in which Lys-170 serves as the amino donor (30). The imine intermediate has been directly observed by electrospray mass spectrometry (178). In addition to Lys-170, other active site residues are His-143 (32, 115), Met-205 (101), and Arg-213 (106). Since the amino terminus shows sequence similarity to a region within DHQ synthase, it is thought to be part of the substrate binding site (101). There appears to be a VDL sequence motif among enzymes of the shikimate pathway (26).

The *S. coelicor* type II enzyme is a dodecamer with a subunit molecular weight of 16,000. The active site contains Arg-23 (107), Tyr-28 (106), a His (102), and a Trp (19) but no Lys residue. No imine intermediate has been identified, ruling out formation of a Schiff base. Arg-23 of *S. coelicor* corresponds to Arg-213 in type I enzymes, Tyr-28 of type II to Phe-219 in type I enzymes. Arg-23 and Tyr-28 of type II are in a nine residue motif that corresponds to a similar motif in type I containing residues Arg-213 and Phe-219. This motif is positionally not conserved but is considered part of the substrate binding sites for the two types of enzymes, a situation that is very similar to the positioning of the Cys-X-X-His motifs in pro- and eukaryotic DAHP synthases.

Some fungi have both type I and II DHQ dehydratases. In such organisms, the type I enzyme is considered the anabolic form in the main trunk of the shikimate pathway and the type II enzyme the catabolic form in the pathway that soil microorganisms use to degrade the abundant plant metabolite quinate. In contrast, the bacterium *Acinetobacter calcoaceticus* has a catabolic type I and an anabolic type II DHQ dehydratase (43). Finally, in the bacterium *Amycolatopsis* (formerly *Nocardia*) *methanolica*, DHQ dehydratase II serves both an anabolic and a catabolic function (44).

The fourth step in the shikimate pathway is the reduction of DHS to shikimate. In *E. coli*, the reaction is catalyzed by an NADP–dependent shikimate dehydrogenase (Figure 4) of molecular weight of 29,000 (5). Some microorganisms have a shikimate dehydrogenase that is pyrrolo-quinoline quinone dependent (42).

In plants, step three and four of the shikimate pathway are catalyzed by the bifunctional DHQ dehydratase-shikimate dehydrogenase. A 59-kD enzyme from *Pisum sativum* has been purified to electrophoretic homogeneity (135). The dehydratase activity resides in the amino terminal half of the polypeptide. A partial cDNA has been obtained that encodes a type I dehydratase with all the

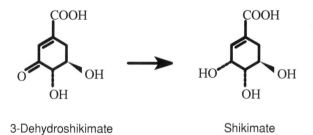

3-Dehydroshikimate   Shikimate

Figure 4 The reaction catalyzed by shikimate dedydrogenase. The plant enzyme uses NADP, whereas some microorganisms use pyrrolo quinoline quinone as cofactor.

identified active site residues and that is inhibited by borohydride as well (31). A longer, although also incomplete, cDNA has been obtained for the *Nicotiana tabacum* homologue (20). This sequence translates also into a protein with both activities, the dehydratase in the amino terminal and the dehydrogenase in the carboxy terminal half. The ratio of the turnover numbers for these two plant enzymes is about 1:10, meaning that DHS never accumulates (171).

The sequences for both plant enzymes are more similar to bacterial than to lower eukaryotic homologues. In addition, the 5'-end of the tobacco cDNA translates into a putative signal sequence for plastid import (20). This sequence is rich in hydroxylated amino acid residues but has a net negative charge, a feature that has not been seen previously for plastid transit peptides. There is evidence for isoenzymes but no isoenzyme-encoding DNA has been cloned yet.

Shikimate dehydrogenase enzyme activity increases about 20% within six days after mechanical wounding of *Capsicum annum* L. leaves (36). If this increase is significant, the dehydrogenase would join DAHP synthase (40) as the second enzyme of the shikimate pathway that is wound-inducible.

Although the plant bifunctional DHQ dehydratase-shikimate dehydrogenase has not been fully characterized at the molecular level, shikimate dehydrogenase has been widely used as a marker in genetics and in crop breeding for many years. Because this enzyme can be detected in starch or isoelectric focusing gels by activity staining, it is one of a handful of enzymes used for studies of heritable variation. The pIs of DHQ dehydratase-shikimate dehydrogenase isozymes differ sufficiently to create characteristic and unique gel patterns when the isoenzymes are separated by charge and stained for activity. The activity patterns, or zymograms, can serve as fingerprints for plant genotypes used to identify parents, clones, and seed lots (203).

While the analysis of random, amplified, polymorphic DNA (RAPD) has begun to supplant the use of zymograms, comparison of zymogram to RAPD technology has shown good correlation in some applications (88, 208). Since zymogram analysis uses readily available reagents and a simple apparatus, the use of shikimate dehydrogenase activity as a genetic marker is still prevalent. Recently, shikimate dehydrogenase zymograms have been used to determine the validity or extent of outcrossing (27), to evaluate genetic variation within a population (71, 90), and to derive evolutionary relationships between cultivars, ecotypes, and species (12, 164). In a few cases, a particular pattern of shikimate dehydrogenase isozymes has been linked to a phenotypic trait (110, 222).

Shikimate Kinase

In the fifth step of the shikimate pathway, shikimate kinase (Figure 5) catalyzes the phosphorylation of shikimate to yield shikimate 3-phosphate (S3P). *E. coli* has two shikimate kinases; isoenzymes I and II are monomers of molecular

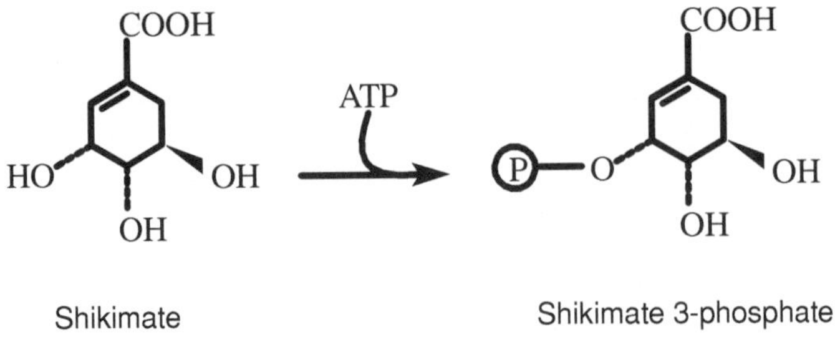

Figure 5   The reaction catalyzed by shikimate kinase.

weight 19,500 but share only 30% sequence identity (66, 217). Both enzymes can function in vitro in aromatic amino acid biosynthesis with isoenzyme II having a K_m for shikimate of 200 μM; the corresponding K_m for I is more than 100 times larger. Thus in vivo, isoenzyme I may not be an enzyme of the shikimate pathway at all; it may have a function in cell division (209). The expression of the gene encoding shikimate kinase II, but not I, is controlled by a synergism between the *trp*- and *tyr*-repressors (78) and possibly also by integration host factor (114). The three-dimensional structure of a type II shikimate kinase from *Erwinia chrysanthemi* has been solved to high resolution for two different enzyme-substrate complexes (105). The enzyme seems to undergo large conformational changes during catalysis.

Plant shikimate kinase has been purified to near electrophoretic homogeneity from spinach (170) and pepper (35). The enzyme is inhibited by 2,4-D; the herbicidal activity of 2,4-D may be due in part to its ability to inhibit shikimate kinase. The chloroplast-localized enzyme is regulated by energy charge but not by light. Plant cDNA encoding shikimate kinase has been cloned from tomato; this plant has only one gene encoding this protein (169). The cDNA-deduced amino acid sequence is quite similar to the sequences of microbial homologues but contains in addition an amino terminal extension that resembles a transit sequence for chloroplast import. In vitro synthesized tomato shikimate kinase precursor is processed and taken up by isolated chloroplasts (169). These results are consistent with the assumption of a single shikimate kinase in tomato that is localized exclusively in chloroplasts.

5-Enolpyruvylshikimate 3-Phosphate Synthase

In the sixth step of the shikimate pathway, a second PEP enters the pathway. It is condensed with S3P to yield 5-enolpyruvylshikimate 3-phosphate (EPSP) and inorganic phosphate. This reversible reaction is catalyzed by EPSP synthase (Figure 6), a monomeric enzyme of molecular weight 48,000. The enzyme has

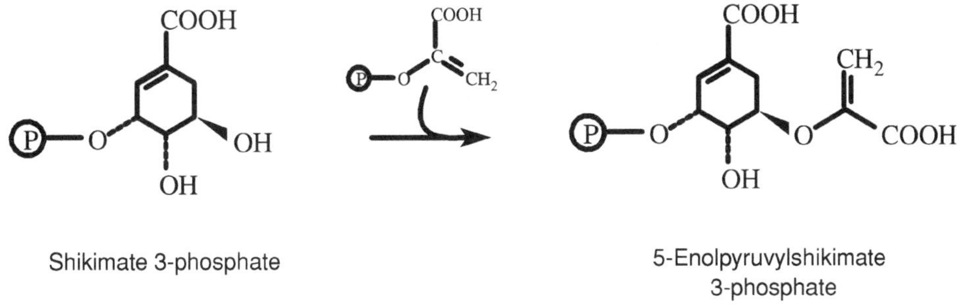

Shikimate 3-phosphate

5-Enolpyruvylshikimate
3-phosphate

Figure 6   The reaction catalyzed by 5-enolpyruvylshikimate 3-phosphate synthase.

been purified from prokaryotes and eukaryotes and the *E. coli* EPSP synthase has been crystallized. X-ray structure analysis indicates two domains with the active site near the interdomain crossover segment (189). Enzyme-substrate complexes were also characterized by NMR spectroscopy (194).

Separate studies proposed an ordered (4) and a random (68) kinetic mechanism for the enzyme-catalyzed reaction that proceeds through a tetrahedral intermediate. This tetrahedral intermediate had already been suggested by Sprinson and coworkers and was verified through extensive physicochemical investigations (174, 183). It has been detected directly by interfacing a rapid mixing apparatus with an electrospray mass spectrometer (145). Reaction intermediates were also demonstrated by solid-state NMR spectroscopy (91, 195) and the stereochemical course of the reaction was detailed in studies with fluoro derivatives of PEP (99).

EPSP synthase is the only cellular target for the herbicide glyphosate (190). Glyphosate does not bind to the naked enzyme but rather to the enzyme-S3P complex. Glyphosate binding is competitive with PEP. For a long time, the ternary complex enzyme-S3P-glyphosate has been considered a transition state analog in which glyphosate takes the place of PEP. Surprisingly, other enzymes that have PEP as substrates are not inhibited by glyphosate. Recently, enzyme-ligand distances within the glyphosate-containing ternary complex were measured by NMR (127), and now reports from several laboratories indicate that the complex may not be a transition state analog (165), because PEP and glyphosate binding are apparently not identical (126). In support of these findings, it has been noticed for some time that enzymes from different organisms show great variations in their sensitivities to the herbicide. Inhibitor constants may vary by as much as three orders of magnitude when some plant and bacterial enzymes are compared, while variations in the K_m's for PEP are much less. Thus, in spite of the competitive nature, glyphosate and PEP binding is not totally equivalent. This nonequivalency is corroborated through mutant enzyme studies (120, 144, 180, 181) that show no direct correlation between glyphosate inhibitor potency and the loss of catalytic efficiency. Since glyphosate is a competitive inhibitor with respect to PEP but does not bind in the same fashion as PEP, Sikorski & Gruys now call the inhibition an "adventitious allosteric interaction" (183). Under specific in vitro conditions, the EPSP synthase from *Bacillus subtilis* is an oligomeric protein with two non-equivalent PEP binding sites. Glyphosate binding is competitive with respect to one site only; thus this enzyme has been considered a classical allosteric protein (120).

Glyphosate is the only commercial herbicide that acts on EPSP synthase despite tremendous efforts to find equally effective inhibitory compounds (2, 211). However, recently discovered transition state analogs in which the shikimate ring is replaced by a benzene ring are very effective inhibitors of EPSP synthase (129, 130, 175). Since nM concentrations of some of these compounds inhibit the enzyme effectively, and since these compounds are readily accessible through organic synthesis, they may represent the next generation of commercial herbicides that function through inhibition of the shikimate pathway.

Glyphosate-tolerant cell lines from several different organisms have elevated levels of DAHP synthase (3, 41, 57, 139, 182, 186, 187). The elevation is due to gene amplification (197). Maintenance of these lines in the absence of the herbicide causes a time-dependent loss of tolerance (138). Plantlets regenerated from tolerant lines also show elevated EPSP synthase levels. The degree of herbicide tolerance is directly correlated to the enzyme levels (92).

Plant DNAs encoding EPSP synthases have been isolated from petunia (176), Arabidopsis (103), tomato (52), and *Brassica napus* (51). All these cDNAs encode precursor proteins with amino terminal transit sequences. In vitro uptake experiments with isolated chloroplasts show that these sequences direct plastid import of the enzyme (33). In petunia, EPSP synthase expression is tissue specific and developmentally regulated (15). A transcription activator involved in this regulation is a Cys/His type zinc finger protein (202).

Zea mays has two plastidic EPSP synthase isoforms; one appears to be constitutive, whereas the other may be subject to regulation (45, 46). The single tomato enzyme is induced in elicitor-treated cell cultures (60). Generally, plant EPSP synthases have lower K_i values for glyphosate than their bacterial homologues. Thus, glyphosate-tolerant transgenic plants have been obtained carrying mutant alleles that either encode bacterial enzymes (188) or that cause overproduction of the plant enzyme (57, 176, 186).

Commercially successful glyphosate-tolerant crop plants carry a naturally occurring *Agrobacterium tumefaciens* allele that encodes a glyphosate-insensitive EPSP synthase. Glyphosate-tolerant soybean (143) and cotton (140) plants have a cauliflower 35S promoter-driven transgene construct that encodes the

Petunia hybrida EPSP synthase signal sequence for plastid import, the above-mentioned *Agrobacterium tumefaciens* EPSP synthase coding region, and the nopaline synthase terminator. Glyphosate-tolerant sugar beet plants (121) carry in addition to this transgene construct a bacterial gene encoding glyphosate oxidase reductase that degrades glyphosate to nontoxic compounds. These three glyphosate-tolerant crop plants are the vanguard in weed control systems that should reduce not only production costs, but also the amount of chemicals in soil and water run-off. We can anticipate that more such crops will appear, particularly when transformation systems for grasses become easier to manipulate.

Chorismate Synthase

The seventh and final step in the main trunk of the shikimate pathway is the trans-1,4 elimination of phosphate from EPSP to yield chorismate (8, 74). In this reaction, the second of the three double bonds of the benzene ring is introduced. The reaction is catalyzed by chorismate synthase (Figure 7) and requires reduced flavin for activity even though the overall reaction is redox neutral. In this respect the enzyme is similar to DHQ synthase, the second enzyme in the shikimate pathway.

In chorismate synthase catalysis, the reduced flavin is apparently directly involved in the mechanism of the reaction (155, 156). The binary flavin-enzyme and the ternary flavin-enzyme-substrate complexes have been characterized in detail (118). A careful kinetic analysis (21) shows that the flavin reaction intermediate is formed after EPSP is bound but before it is consumed. The flavin intermediate decays after EPSP is converted to chorismate and after phosphate is released from the enzyme. The flavin may serve as an electron donor to EPSP, thereby initiating C-O bond cleavage. The intermediate may be a radical in what is most likely a non-concerted reaction.

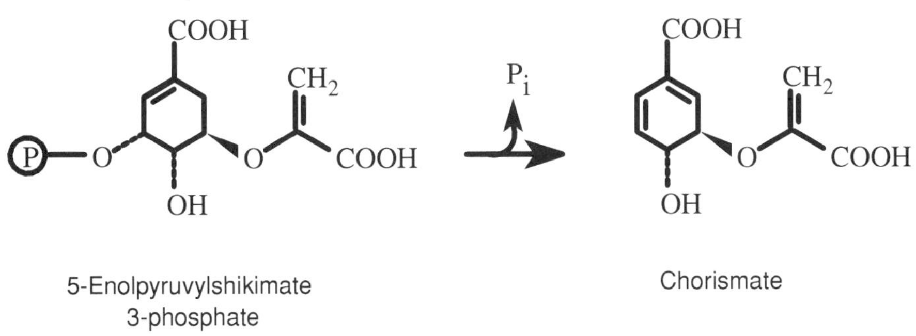

5-Enolpyruvylshikimate
3-phosphate

Chorismate

Figure 7   The reaction catalyzed by chorismate synthase. The enzyme requires catalytic amounts of reduced flavin for activity.

Depending upon the organism, chorismate synthase is either monofunctional, requiring the addition of reduced flavin to in vitro enzyme assays, or bifunctional, with an associated NADP-driven flavin reductase within the same polypeptide chain. The best-studied bifunctional chorismate synthase is from *Neurospora crassa*. The enzymes from *E. coli* and higher plants are monofunctional. The *E. coli* (218) and the plant enzyme from *Corydalis sempervirens* (167) have subunit molecular weights of around 40,000; either two or four polypeptides form the active enzyme complex. The bifunctional *N. crassa* subunit homolog is bigger, but shows extensive sequence identity to the other enzymes. Interestingly, no additional domain is found that would account for the flavin reductase activity. The active site residues for this second function of the polypeptide are apparently interspersed with residues that are conserved between mono- and bifunctional chorismate synthases. Thus the larger size of the bifunctional enzyme is not the consequence of an additional domain encoding a flavin reductase (80).

The first plant cDNA encoding chorismate synthase was obtained from *C. sempervirens* (166); it has a 5′ sequence that translates into a typical plastid import signal. Both the full-length cDNA and the truncated form encoding the presumed mature enzyme have been expressed in *E. coli*. Only the truncated, mature form of the enzyme is able to catalyze the synthesis of chorismate in vitro or complement the function in vivo in a heterologous system (82). Since *C. sempervirens* contains only one gene encoding chorismate synthase, these findings indicate that the enzyme activity of plant chorismate synthase is dependent upon plastid import and that there is no chorismate synthesis outside the plastids of at least this higher plant.

The *C. sempervirens* cDNA was used to identify two homologs in tomato (61) encoding chorismate synthase. The corresponding cDNA sequences encode polypeptides with amino terminal signal sequences for plastid import. A third tomato mRNA encoding chorismate synthase is generated through differential splicing of one of the two gene products (59). All three encoded proteins were heterologously expressed in *E. coli* as precursors and as mature, processed forms. Only the mature forms of two isoenzymes were enzymatically active, confirming the results from *C. sempervirens* that plant chorismate synthesis proceeds exclusively in plastids (24). The two active isoenzymes complement an *E. coli* mutant devoid of chorismate synthase. The two isoenzymes differ with respect to their K_m values for EPSP and their specific enzyme activities.

Differential expression of chorismate synthase in various parts of the plant shows highest amounts of mRNA in flowers and roots (61), with only the more active form of the enzyme occurring in higher amounts. Fungal elicitation further increases the amount of the more active isoenzyme (60). Thus in tomato, apparently just one of the three isoenzymes supplies the bulk of chorismate synthase activity under all physiological conditions.

BRANCH POINTS IN THE MAIN TRUNK

All intermediates of the main trunk of the shikimate pathway are potential branch points leading to other metabolic pathways (16). DAHP has long been suspected to be a precursor for an aromatic moiety of certain antibiotics that consists of seven carbons and a nitrogen (mC_7N). DHQ is readily converted to quinate, a ubiquitous plant building block for phytoalexins and UV protectants. Some organisms can use DHS as a sole carbon source by converting it to tricarboxylic acid cycle intermediates via protocatechuate. Shikimate is a direct degradation product of quinate and that reaction may be reversible. S3P and EPSP have been considered as precursors for cyclohexane carboxylate moieties of antibiotics.

3-Deoxy-D-Arabino-Heptulosonate 7-Phosphate

Amycolatopsis mediterranei produces rifamycin B. Part of the structure of this and other ansamycin antibiotics is a seven-carbon aromatic moiety derived from 3-amino-5-hydroxy benzoic acid (AHBA). This is the initiator for the formation of the polyketide chain that is eventually cyclized to form the mature antibiotic. AHBA is presumably derived from a shikimate pathway intermediate (28, 89). Mutant studies with rifamycin-producing bacteria seem to point to DAHP as a precursor, since DHQ and DHS could not serve as precursors for AHBA, but a DHQ synthase lacking mutant was still able to produce rifamycin (70).

Some indirect, supporting evidence for DAHP as a precursor of AHBA came from studies with *Streptomyces hygroscopicus*, another antibiotic-producing bacterium, that revealed a large gene cluster for the polyketide synthase. This cluster was physically in close proximity to an open reading frame encoding a plant-like DAHP synthase that could potentially catalyze the first step in the biosynthesis of the antibiotic (162). Such a gene was also found in the phenazine producer *Pseudomonas aureofaciens* (151). DAHP synthase has been purified to apparent electrophoretic homogeneity from *A. mediterranei*. A gene encoding a plant-like DAHP synthase was found within a cluster of *A. mediterranei* genes involved in rifamycin biosynthesis (6).

Studies by Floss and coworkers with *A. mediterranei* and *Streptomyces collinus* point to aminoDAHP as the precursor for AHBA (97). The ability to convert aminoDAHP, but not DAHP, in vitro, into AHBA led to the proposal of a novel shikimate pathway. In this pathway, PEP and E4P together with a still unknown nitrogen source are converted to aminoDAHP that serves as a precursor for aminoDHQ, aminoDHS, and eventually AHBA (97). Potential candidates for genes encoding enzymes for this pathway have been identified in the gene cluster mentioned above. This cluster contains a number of open reading frames with sequences that are very similar to genes encoding authentic shikimate pathway enzymes (6). The first step in the novel pathway, the biosynthesis of aminoDAHP, seems more complicated than originally assumed. While an enzyme catalyzing this step has not yet been described, the last step in AHBA biosynthesis is catalyzed by an enzyme that has been purified, its cDNA cloned, and expressed in *E. coli* (98).

It would be interesting to see if the plant DAHP synthase-like genes of *S. hygroscopicus* or of *A. mediterranei*, expressed in a heterologous system and supplied with the proper nitrogen donor or with additional polypeptides, could yield aminoDAHP. If *Streptomyces* and *Amycolatopses* indeed have only one DAHP synthase, namely a plant-like enzyme, it would then have to fulfill a dual function. At present, it is not known if DAHP synthase from the shikimate pathway produces a precursor for antibiotics or if a related but separate enzyme synthesizes aminoDAHP directly. The identification and characterization of enzymes involved in rifamycin biosynthesis will certainly be of great value for the generation of genetically engineered microorganisms that can produce clinically useful antibiotic variants.

3-Dehydroquinate and 3-Dehydroshikimate

DHQ can be converted to DHS or to quinate, the precursor to the ubiquitous plant secondary product chlorogenate, a condensation product of quinate and caffeate. Thus, chlorogenate is made by combining an early intermediate of the shikimate pathway with a late intermediate of phenylpropanoid metabolism that is many steps removed. Because the precursors are made in two different cell organelles, elucidating the mode of regulation of chlorogenate biosynthesis will be a challenging problem. Chlorogenate protects plants against fungal attack (119) and, through accumulation to substantial levels, serves as a UV protectant in several plants. Quinate itself also accumulates in some plants and has been considered an alicyclic carbon reservoir for aromatic compound biosynthesis.

The fully reversible conversion of DHQ to quinate is catalyzed by quinate dehydrogenase, an NAD-dependent enzyme of 42 kD that is subject to regulation by reversible phosphorylation of a Ser residue (94, 157) by a Ca-calmodulin-dependent protein kinase. The phospho protein is enzymatically active, whereas dephosphorylated enzyme is inactive. The quinate dehydrogenase is associated with a Ca-sensing 60-kD regulatory subunit (65).

Ca flux, dependent upon voltage-gated Ca channels (205), can initiate signal transduction in carrot cells. The cytoskeleton plays an important role in the establishment and stability of such channels (204). How these channels are activated in whole plants and how their activation regulates the flow of DHQ to DHS or quinate has yet to be addressed.

Plants also have a quinate hydrolyase that catalyzes the conversion of quinate to shikimate (116). Thus quinate can reenter the main trunk of the shikimate pathway by being converted either to DHQ or to shikimate. If quinate hydrolyase catalyzes a freely reversible reaction, plants would also have at least two ways to synthesize quinate: from DHQ, using the quinate dehydrogenase and from shikimate, using the quinate hydrolyase. Both enzymes are localized in the plastids. Nothing is known about the regulation of the latter enzyme or about the relative activities of these two pathways.

Fungi and some bacteria can use quinate as their sole carbon source (56) by degrading it via DHQ and DHS to protocatechuate and on to succinate and acetyl CoA. This catabolic sequence, described for *N. crassa* (56), *A. nidulans* (64), *Rhodococcus rhodochrous* (25), and *A. calcoaceticus* (55), is actually induced by quinate. Regulation of quinate catabolism is by transcriptional control mediated through an activator and a repressor (53). The activator is homologous to the amino terminal half and the repressor to the carboxy terminal half of the biosynthetic AROM complex (75, 76, 117). Quinate blocks the activity of the repressor (112). These regulatory molecules are apparently very similar in structure and function when related organisms are compared, because the activator from *N. crassa* recognizes promoters of *A. nidulans* (86).

The first three enzymes of quinate degradation are quinate dehydrogenase, DHQ dehydratase, and DHS dehydratase. The NAD-dependent quinate dehydrogenase that also oxidizes shikimate to DHS is different from the NADP-dependent shikimate dehydrogenase of the main trunk of the shikimate pathway (5). Most quinate-induced catabolic quinate dehydrogenases are NAD-dependent; however, the enzyme from *A. calcoaceticus* uses pyrrolo quinoline quinone as a cofactor (42).

The pathways of quinate degradation and chorismate biosynthesis have clearly different enzymes, even though they share at least two intermediates, DHQ and DHS. In *A. nidulans*, quinate degradation to protocatechuate is regulated at the level of quinate entry into the fungus that is facilitated by a permease (216). Simultaneous metabolic flux of quinate to chorismate and protocatechuate has also been studied in this organism (111, 112). Overexpression of the catabolic DHS dehydratase inhibits growth of the organism through interference with chorismate synthesis (113).

OUTLOOK

A combination of protein and DNA sequencing efforts have yielded primary structures for all the enzymes of the shikimate pathway from several organisms, including higher plants. Some of the enzymes have been crystallized. The elucidation of tertiary or quaternary structures by X-ray analysis and other physicochemical methods is ongoing. Together with molecular biological investigations, in particular site directed mutagenesis, these studies will refine our knowledge of the reaction mechanisms for enzymes of the shikimate pathway. Such studies are vital for the design of new inhibitors of shikimate pathway enzymes that are likely to attain importance as drugs or herbicides.

The coming years will see an increased emphasis on the study of the regulation of this prominent plant pathway. Since the main regulatory mechanism appears to be transcriptional, a detailed analysis of the regulatory regions for genes encoding shikimate pathway enzymes would be a good start. Isoenzymes have been described for some shikimate pathway enzymes. A DNA motif to bind a regulatory protein has been identified in an EPSP synthase promoter and the regulatory protein binding to this motif was obtained. It is safe to assume that other regulatory proteins will be identified, among them factors that are responsible for induction of the expression of specific isoenzymes. Such studies will contribute to the elucidation of signal transduction pathways involved in responses to mechanical wounding or pathogen attack. Investigations on the genetic control mechanisms should include mutant hunts, even though it may be very difficult to generate knock-out mutants in genes of the shikimate pathway, given the many products derived from this metabolic sequence.

A detailed analysis of the signal peptides required for import into plastids should also be undertaken. All shikimate pathway enzymes, including those of non-green tissues, are synthesized as precursors in the cytosol and imported into plastids. Such import requires signal sequences. During import these signals are processed off the protein precursors to give rise to the mature proteins. There are two general characteristics of signal peptides: They are rich in hydroxyl amino acid residues and positively charged. The large variation in the primary sequences of these signals seems to indicate a lack of any specific sequence requirement. That can hardly be the case, since primary structure dictates function. Also, one would expect specific motifs or structural differences for transport into different plastids.

As one of the most active plant pathways in terms of carbon flow, the shikimate metabolic sequence will continue to be of interest. New basic insights into plant metabolism and its regulation will be generated through further studies in this area.

ACKNOWLEDGMENTS

We thank Michael Poling for a critical reading of the manuscript. This is journal paper number 15856 of the Purdue University Agricultural Experiment Station.

Literature Cited

1. Akowski JP, Bauerle R. 1997. Steady-state kinetics and inhibitor binding of 3-deoxy-D-*arabino*-heptulosonate 7-phosphate synthase (tryptophan sensitive) from *Escherichia coli. Biochemistry* 36:15817–22

2. Alberg DG, Lauhon CT, Nyfelder R, Fässler A, Bartlett PA. 1992. Inhibition of EPSP synthase by analogues of the tetrahedral intermediate and of EPSP. *J. Am. Chem. Soc.* 114:3535–46

3. Amrhein N, Johänning D, Schab J, Schulz A. 1983. Biochemical basis for glyphosate tolerance in a bacterium and a plant tissue culture. *FEBS Lett.* 157:191–96

4. Anderson KS, Johnson KA. 1990. Kinetic and structural analysis of enzyme intermediates: lessons from EPSP synthase. *Chem. Rev.* 90:1131–49

5. Anton IA, Coggins JR. 1988. Sequencing and overexpression of the *Escherichia coli aroE* gene encoding shikimate dehydrogenase. *Biochem. J.* 249:319–26

6. August PR, Tang L, Yoon YJ, Ning S, Müller R, et al. 1998. Biosynthesis of the ansamycin antibiotic rifamycin: deductions from the molecular analysis of the *rif* biosynthetic gene cluster of *Amycolatopsis mediterranei* S699. *Chem. Biol.* 5:69–79

7. Baasov T, Knowles JR. 1989. Is the first enzyme of the shikimate pathway, 3-deoxy-D-*arabino*-heptulosonate 7-phosphate synthase (tyrosine sensitive), a copper metalloenzyme? *J. Bacteriol.* 171:6155–60

8. Balasubramanian S, Abell C, Coggins JR. 1990. Observation of an isotope effect in the chorismate synthase reaction. *J. Am. Chem. Soc.* 112:8581–83

9. Banerji S, Wakefield AE, Allen AG, Maskell DJ, Peters SE, Hopkin JM. 1993. The cloning and characterization of the *arom* gene of *Pneumocystis carinii. J. Gen. Microbiol.* 139:2901–14

10. Bartlett PA, McLaren KL, Marx MA. 1994. Divergence between the enzyme-catalyzed and noncatalyzed synthesis of 3-dehydroquinate. *J. Org. Chem.* 59:2082–85

11. Bartlett PA, Satake K. 1988. Does dehydroquinate synthase synthesize dehydro-quinate? *J. Am. Chem. Soc.* 110:1628–30

12. Belletti P, Lotito S. 1996. Identification of runner bean genotypes (*Phaseolus cossineus* L.) by isoenzyme analysis. *J. Genet. Breed.* 50:185–90

13. Bender SL, Mehdi S, Knowles JR. 1989. Dehydroquinate synthase: the role of divalent metal cations and of nicotinamide adenine dinucleotide in catalysis. *Biochemistry* 28:7555–60

14. Bender SL, Widlanski T, Knowles JR. 1989. Dehydroquinate synthase: the use of substrate analogues to probe the early steps of the catalyzed reaction. *Biochemistry* 28:7560–72

15. Benfey PN, Takatsuji H, Ren L, Shah DM, Chua NH. 1990. Sequence requirements of the 5-enolpyruvylshikimate 3-phosphate synthase 5'-upstream region for tissue-specific expression in flowers and seedlings. *Plant Cell* 2:849–56

16. Bentley R. 1990. The shikimate pathway—a metabolic tree with many branches. *CRC Crit. Rev. Biochem. Mol. Biol.* 25:307–84

17. Bischoff M, Rösler J, Raesecke HR, Görlach J, Amrhein N, Schmid J. 1996. Cloning of a cDNA encoding a 3-dehydroquinate synthase from a higher plant, and analysis of the organ-specific and elicitor-induced expression of the corresponding gene. *Plant Mol. Biol.* 31:69–76

18. Blattner FR, Plunkett G III, Bloch CA, Perna NT, Burland V, et al. 1997. The complete genomic sequence of *Escherichia coli* K-12. *Science* 277:1453–74

19. Boam DJ, Price NC, Kelly SM, Krell T, Coggins JR. 1997. Evidence that the active site in type II dehydroquinase from *Streptomyces coelicolor* is near the single tryptophan. *Biochem. Soc. Trans.* 25: S93

20. Bonner CA, Jensen RA. 1994. Cloning of cDNA encoding the bifunctional dehydroqinase shikimate dehydrogenase of aromatic amino acid biosynthesis in *Nicotiana tabacum. Biochem. J.* 302:11–14

21. Bornemann S, Lowe DJ, Thorneley RNF. 1996. The transient kinetics of *Escherichia coli* chorismate synthase:

substrate consumption, phosphate dissociation, and characterization of a flavin intermediate. *Biochemistry* 35:9907–16

22. Bottomley JR, Clayton CL, Chalk PA, Kleanthous C. 1996. Cloning, sequencing, expression, purification and preliminary characterization of a type II dehydroquinase from *Helicobacter pylori*. *Biochem. J.* 319:559–65

23. Boys CWG, Bury SM, Sawyer L, Moore JD, Charles IG, et al. 1992. Crystallization of a type I 3-dehydroquinase from *Salmonella typhi*. *J. Mol. Biol.* 227:352–55

24. Braun M, Henstrand JM, Görlach J, Amrhein N, Schmid J. 1996. Enzymatic properties of chorismate synthase isoenzymes of tomato (*Lycopersicon esculentum* Mill.). *Planta* 200:64–70

25. Bruce NC, Cain RB. 1990. Hydroaromatic metabolism in *Rhodococcus rhodochrous*: purification and characterization of its NAD-dependent quinate dehydrogenase. *Arch. Microbiol.* 154:179–96

26. Bugg TDH, Alefounder PR, Abell C. 1991. An amino acid sequence motif observed amongst enzymes of the shikimate pathway. *Biochem. J.* 276:841–43

27. Carre S, Tasei JN, Guen JL, Mesquida J, Morin G. 1993. The genetic control of seven isoenzymic loci in *Vicia faba* L. Identification of lines and estimates of outcrossing rates between plants pollinated by bumblebees. *Ann. Appl. Biol.* 122:555–68

28. Casati R, Beale JM, Floss HG. 1987. Biosynthesis of ansatrienin. Nonincorporation of shikimic acid into the mC7N unit and stereochemistry of its conversion to the cyclohexanecarboxylic acid moiety. *J. Am. Chem. Soc.* 109:8102–4

29. Charles IG, Keyte JW, Brammar WJ, Smith M, Hawkins AR. 1986. The isolation and nucleotide sequence of the complex *AROM* locus of *Aspergillus nidulans*. *Nucleic Acids Res.* 14:2201–13

30. Chaudhuri S, Duncan K, Graham LD, Coggins JR. 1991. Identification of the active-site lysine residue of two biosynthetic 3-dehydroquinases. *Biochem. J.* 275:1–6

31. Deka RK, Anton IA, Dunbar B, Coggins JR. 1994. The characterization of the shikimate pathway enzyme dehydroquinase from *Pisum sativum*. *FEBS Lett.* 349:397–402

32. Deka RK, Kleanthous C, Coggins JR. 1992. Identification of the essential histidine residue at the active site of *Escherichia coli* dehydroquinase. *J. Biol. Chem.* 267:22237–42

33. Della-Cioppa G, Bauer SC, Klein BK, Shah DM, Fraley RT, Kishore GM. 1986. Translocation of the precursor of 5-enolpyruvylshikimate 3-phosphate synthase into chloroplasts of higher plants in vitro. *Proc. Natl. Acad. Sci. USA* 83:6873–77

34. Dewick PM. 1998. The biosynthesis of shikimate metabolites. *Nat. Prod. Rep.* 15:17–58

35. Diaz J, Merino F. 1997. Shikimate dehydrogenase from pepper (*Capsicum annum*) seedlings. Purification and properties. *Physiol. Plant.* 100:147–52

36. Diaz J, Merino F. 1998. Wound-induced shikimate dehydrogenase and peroxidase related to lignification in pepper (*Capsicum annum* L.). *J. Plant Physiol.* 152:51–57

37. Doong RL, Gander JE, Ganson RJ, Jensen RA. 1992. The cytosolic isoenzyme of 3-deoxy-D-*arabino*-heptulosonate 7-phosphate synthase in *Spinacia oleracea* and other higher plants: extreme substrate ambiguity and other properties. *Physiol. Plant.* 84:351–60

38. Duncan K, Chaudhuri S, Campbell MS, Coggins JR. 1986. The overexpression and complete amino acid sequence of *Escherichia coli* 3-dehydroquinase. *Biochem. J.* 238:475–83

39. Duncan K, Edwards MR, Coggins JR. 1987. The pentafunctional *arom* enzyme of *Saccharomyces cerevisiae* is a mosaic of monofunctional domains. *Biochem. J.* 246:375–86

40. Dyer WE, Henstrand JM, Handa AK, Herrmann KM. 1989. Wounding induces the first enzyme of the shikimate pathway in Solanaceae. *Proc. Natl. Acad. Sci. USA* 86:7370–73

41. Dyer WE, Weller SC, Bressan RA, Herrmann KM. 1988. Glyphosate tolerance in tobacco (*Nicotiana tabacum* L.). *Plant Physiol.* 88:661–66

42. Elsemore DA, Ornston LN. 1994. The *pca-pob* supraoperonic cluster of *Acinetobacter calcoaceticus* contains *quiA*, the structural gene for quinate-shikimate dehydrogenase. *J. Bacteriol.* 176:7659–66

43. Elsemore DA, Ornston LN. 1995. Unusual ancestry of dehydratases associated with quinate catabolism in *Acinetobacter calcoaceticus*. *J. Bacteriol.* 177:5971–78

44. Euverink GJW, Hessels GI, Vrijbloed JW, Coggins JR, Dijkhuizen L. 1992. Purification and characterization of a dual function 3-dehydroquinate dehydratase from *Amycolatopsis methanolica*. *J. Gen. Microbiol.* 138:2449–57

45. Forlani G. 1997. Properties of the 5-enol-pyruvyl-shikimate 3-phosphate synthase isoforms isolated from maize cultured cells. *J. Plant Physiol.* 150:369–75

46. Forlani G, Parisi B, Nielsen E. 1994. 5-enol-pyruvyl-shikimate 3-phosphate synthase from *Zea mays* cultured cells. *Plant Physiol.* 105:1107–14

47. Frost JW, Bender JL, Kadonaga JT, Knowles JR. 1984. Dehydroquinate synthase from *Escherichia coli*: purification, cloning, and construction of overproducers of the enzyme. *Biochemistry* 23:4470–75

48. Garbe T, Selvos S, Hawkins A, Dimitriadis G, Young D, et al. 1991. The *Mycobacterium tuberculosis* shikimate pathway genes: evolutionary relationship between biosynthetic and catabolic 3-dehydroquinases. *Mol. Gen. Genet.* 228:385–92

49. Garner CC, Herrmann KM. 1984. Structural analysis of 3-deoxy-D-*arabino*-heptulosonate 7-phosphate by ¹H- and natural-abundance ¹³C-n.m.r. spectroscopy. *Carbohydr. Res.* 132:317–22

50. Garner CC, Herrmann KM. 1985. Operator mutations of the *Escherichia coli aroF* gene. *J. Biol. Chem.* 260:3820–25

51. Gasser CS, Klee HJ. 1990. A *Brassica napus* gene encoding 5-enolpyruvyl-shikimate 3-phosphate synthase. *Nucleic Acids Res.* 18:2821

52. Gasser CS, Winter JA, Hironaka CM, Shah DM. 1988. Structure, expression, and evolution of the 5-enolpyruvyl-shikimate 3-phosphate synthase genes of petunia and tomato. *J. Biol. Chem.* 263:4280–89

53. Geever RF, Huiet L, Baum JA, Tyler BM, Patel VB, et al. 1989. DNA sequence, organization and regulation of the *qa* gene cluster of *Neurospora crassa*. *J. Mol. Biol.* 207:15–34

54. Ger YM, Chen SL, Chiang HJ, Shiuan D. 1994. A single Ser-180 mutation desensitizes feedback inhibition of the phenylalanine-sensitive 3-deoxy-D-*arabino*-heptulosonate 7-phosphate (DAHP) synthetase in *Escherichia coli*. *J. Biochem.* 116:986–90

55. Gerischer U, Segura A, Ornston LN. 1998. PcaU, a transcriptional activator of genes for protocatechuate utilization in *Acinetobacter*. *J. Bacteriol.* 180:1512–24

56. Giles NH, Case ME, Baum J, Geever R, Hulet L, et al. 1985. Gene organization and regulation in the *qa* (quinic acid) gene cluster of *Neurospora crassa*. *Microbiol. Rev.* 49:338–58

57. Goldsbrough PB, Hatch EM, Huang B, Kosinsky WG, Dyer WE, et al. 1990. Gene amplification in glyphosate tolerant tobacco cells. *Plant Sci.* 72:53–62

58. Görlach J, Beck A, Henstrand JM, Handa AK, Herrmann KM, et al. 1993. Differential expression of tomato (*Lycopersicon esculentum* L.) genes encoding shikimate pathway isoenzymes. I. 3-deoxy-D-*arabino*-heptulosonate 7-phosphate synthase. *Plant Mol. Biol.* 23:697–706

59. Görlach J, Raesecke HR, Abel G, Wehli R, Amrhein N, Schmid J. 1995. Organ-specific differences in the ratio of alternatively spliced chorismate synthase (LeCS2) transcripts in tomato. *Plant J.* 8:451–56

60. Görlach J, Raesecke HR, Rentsch D, Regenass M, Roy P, et al. 1995. Temporally distinct accumulation of transcripts encoding enzymes of the prechorismate pathway in elicitor-treated, cultured tomato cells. *Proc. Natl. Acad. Sci. USA* 92:3166–70

61. Görlach J, Schmid J, Amrhein N. 1993. Differential expression of tomato (*Lycopersicon esculentum* L.) genes encoding shikimate pathway isoenzymes. II. Chorismate synthase. *Plant Mol. Biol.* 23:707–16

62. Görlach J, Schmid J, Amrhein N. 1994. Abundance of transcripts specific for genes encoding enzymes of the prechorismate pathway in different organs of tomato (*Lycopersicon esculentum* L.) plants. *Planta* 193:216–23

63. Gourley DG, Coggins JR, Isaacs NW, Moore JD, Charles IG, Hawkins AR. 1994. Crystallization of a type II dehydroquinase from *Mycobacterium tuberculosis*. *J. Mol. Biol.* 241:488–91

64. Grant S, Roberts CF, Lamb H, Stout M, Hawkins AR. 1988. Genetic regulation of the quinic acid utilization (*qut*) gene cluster in *Aspergillus nidulans*. *J. Gen. Microbiol.* 134:347–58

65. Graziana A, Dillenschneider M, Ranjeva R. 1984. A calcium binding protein is a regulatory subunit of quinate:NAD oxidoreductase from dark-grown carrot cells. *Biochem. Biophys. Res. Commun.* 125:774–83

66. Griffin HG, Gasson MJ. 1995. The gene (*aroK*) encoding shikimate kinase I from *Escherichia coli*. *DNA Sequence* 5:195–97

67. Grove CL, Gunsalus RP. 1987. Regulation of the *aroH* operon of *Escherichia coli* by the tryptophan repressor. *J. Bacteriol.* 173:3601–4

68. Gruys KJ, Marzabadi MR, Pansegrau PD, Sikorski JA. 1993. Steady-state kinetic evaluation of the reverse reaction for *Escherichia coli* 5-enolpyruvylshikimate 3-phosphate synthase. *Arch. Biochem. Biophys.* 304:345–51

69. Guyer D, Patton D, Ward E. 1995. Evidence for cross-pathway regulation of metabolic gene expression in plants. *Proc. Natl. Acad. Sci. USA* 92:4997–5000

70. Gygax D, Ghisalba O, Treichler H, Nuesch J. 1990. Study of the biosynthesis of rifamycin-chromophore in *Nocardia mediterranei*. *J. Antibiot.* 43:324–26

71. Hackenberg EM, Kohler W. 1996. Use of isoenzyme analysis in breeding of synthetic rapeseed cultivars. *Plant Breed.* 115:474–79

72. Hara Y, Laugel T, Morimoto T, Yamada Y. 1994. Effect of gibberellic acid on berberine and tyrosine accumulation in *Coptis japonica*. *Phytochemistry* 36:643–46

73. Haslam E. 1993. *Shikimic Acid Metabolism and Metabolites*. Chichester, UK: Wiley. 387 pp.

74. Hawkes TR, Lewis T, Coggins JR, Mousedale DM, Lowe DJ, Thorneley RNF. 1990. Chorismate synthase, presteady-state kinetics of phosphate release from 5-enolpyruvylshikimate 3-phosphate. *Biochem. J.* 265:899–902

75. Hawkins AR, Lamb HK, Moore JD, Roberts CF. 1993. Genesis of eukaryotic transcriptional activator and repressor proteins by splitting a multidomain anabolic enzyme. *Gene* 136:49–54

76. Hawkins AR, Lamb HK, Roberts CF. 1992. Structure of the *Aspergillus nidulans qut* repressor-encoding gene: implications for the regulation of transcription initiation. *Gene* 110:109–14

77. Hawkins AR, Smith M. 1991. Domain structure and interaction within the pentafunctional arom polypeptide. *Eur. J. Biochem.* 196:717–24

78. Heatwole VM, Somerville RL. 1992. Synergism between the *trp* repressor and *tyr* repressor in repression of the *aroL* promoter of *Escherichia coli* K-12. *J. Bacteriol.* 174:331–35

79. Hedstrom L, Abeles R. 1988. 3-deoxy-D-*manno*-octulosonate 8-phosphate synthase catalyzes the C-O bond cleavage of phosphoenolpyruvate. *Biochem. Biophys. Res. Commun.* 157:816–20

80. Henstrand JM, Amrhein N, Schmid J. 1995. Cloning and characterization of a heterologously expressed bifunctional chorismate synthase/flavin reductase from *Neurospora crassa*. *J. Biol. Chem.* 270:20447–52

81. Henstrand JM, McCue KF, Brink K, Handa AK, Herrmann KM, Conn EE. 1992. Light and fungal elicitor induce 3-deoxy-D-*arabino*-heptulosonate 7-phosphate synthase mRNA in suspension cultured cells of parsley (*Petroselinum crispum* L.). *Plant Physiol.* 98:761–63

82. Henstrand JM, Schmid J, Amrhein N. 1995. Only the mature form of the plastidic chorismate synthase is enzymatically active. *Plant Physiol.* 108:1127–32

83. Herrmann KM. 1983. The common aromatic biosynthetic pathway. In *Amino Acids: Biosynthesis and Genetic Regulation*, ed. KM Herrmann, RL Somerville, 17:301–22. London: Addison-Wesley. 453 pp.

84. Herrmann KM. 1995. The shikimate pathway as an entry to aromatic secondary metabolism. *Plant Physiol.* 107:7–12

85. Herrmann KM. 1995. The shikimate pathway: early steps in the biosynthesis of aromatic compounds. *Plant Cell* 7:907–19

86. Hiett KL, Case ME. 1990. Induced expression of the *Aspergillus nidulans QUTE* gene introduced by transformation into *Neurospora crassa*. *Mol. Gen. Genet.* 222:201–5

87. Hinnebusch AG. 1988. Mechanisms of gene regulation in the general control of amino acid biosynthesis in *Saccharomyces serevisiae*. *Microbiol. Rev.* 52:248–73

88. Hoey BK, Crowe KR, Jones VM, Polans NO. 1996. A phylogenetic analysis of *Pisum* based on morphological characters, allozyme and RAPD markers. *Theor. Appl. Genet.* 92:92–100

89. Hornemann R, Kehrer JP, Eggen JH. 1974. Pyruvic acid and D-glucose precursors in mytomycin biosynthesis by *Streptomyces verticillatus*. *J. Chem. Soc. Chem. Commun.* 1974:1045–46

90. Huang H, Layne DR, Peterson RN. 1997. Using isoenzyme polymorphisms for identifying and assessing genetic variation in cultivated pawpaw. *J. Am. Soc. Hortic. Sci.* 122:504–11

91. Jakeman DL, Mitchell DJ, Shuttleworth WA, Evans JN. 1998. On the mechanism of 5-enolpyruvylshikimate 3-phosphate synthase. *Biochemistry* 37:12012–19

92. Jones JD, Goldsbrough PB, Weller SC. 1996. Stability and expression of amplified EPSPS genes in glyphosate resistant tobacco cells and plantlets. *Plant Cell Rep.* 15:431–36

93. Jung E, Zamir LO, Jensen RA. 1986. Chloroplasts of higher plants synthesize L-phenylalanine via L-arogenate. *Proc. Natl. Acad. Sci. USA* 83:7231–35

94. Kang X, Neuhaus HE, Scheibe R. 1994. Subcellular localization of quinate:oxidoreductase from *Phaseolus mungo* L. sprouts. *Z. Naturforsch. Teil C* 49:415–20

95. Keith B, Dong X, Ausubel FM, Fink GR. 1991. Differential induction of 3-deoxy-D-*arabino*-heptulosonate 7-phosphate synthase genes in *Arabidopsis thaliana* by wounding and pathogenic attack. *Proc. Natl. Acad. Sci. USA* 88:8821–25

96. Kikuchi Y, Tsujimoto K, Kurahashi O. 1997. Mutational analysis of the feedback sites of phenylalanine-sensitive 3-deoxy-D-*arabino*-heptulosonate 7-phosphate synthase of *Escherichia coli*. *Appl. Environ. Microbiol.* 63:761–62

97. Kim CG, Kirschning A, Bergon P, Zhou P, Su E, et al. 1996. Biosynthesis of 3-amino-5-hydroxybenzoic acid, the precursor of mC7N units in ansamycin antibiotics. *J. Am. Chem. Soc.* 118:7486–91

98. Kim CG, Yu TW, Fryhle CB, Handa S, Floss HG. 1998. 3-Amino-5-hydroxybenzoic acid synthase, the terminal enzyme in the formation of the precursor of mC7N units in rifamycin and related antibiotics. *J. Biol. Chem.* 273:6030–40

99. Kim DH, Tucker-Kellog GW, Lees WJ, Walsh CT. 1996. Analysis of fluoromethyl group chirality establishes a common stereochemical course of the enolpyruvyl transfers catalyzed by EPSP synthase and UDP-GlcNAc enolpyruvyl transferase. *Biochemistry* 35:5435–40

100. Kishore GM, Shah DM. 1988. Amino acid biosynthesis inhibitors as herbicides. *Annu. Rev. Biochem.* 57:627–63

101. Kleanthous C, Campbell DG, Coggins JR. 1990. Active site labeling of the shikimate pathway enzyme dehydroquinase. *J. Biol. Chem.* 265:10929–34

102. Kleanthous C, Deka R, Davis K, Kelly SM, Cooper A, et al. 1992. A comparison of the enzymological and biophysical properties of two distinct classes of dehydroquinase enzymes. *Biochem. J.* 282:687–95

103. Klee HJ, Muskopf YM, Gasser CS. 1987. Cloning of an *Arabidopsis thaliana* gene encoding 5-enolpyruvylshikimate 3-phosphate synthase: sequence analysis and manipulation to obtain glyphosate tolerant plants. *Mol. Gen. Genet.* 210:437–42

104. Klig LS, Carey J, Yanofsky C. 1988. *trp* repressor interactions with the *trp, aroH*, and *trpR* operators. *J. Mol. Biol.* 202:769–77

105. Krell T, Coggins JR, Lapthorn AJ. 1998. The three-dimensional structure of shikimate kinase. *J. Mol. Biol.* 278:983–97

106. Krell T, Horsburgh MJ, Cooper A, Kelly SM, Coggins JR. 1996. Localization of the active site of type II dehydroquinases. *J. Biol. Chem.* 271:24492–97

107. Krell T, Pitt AR, Coggins JR. 1995. The use of electrospray mass spectrometry to identify an essential arginine residue in type II dehydroquinases. *FEBS Lett.* 360:93–96

108. Künzler M, Paravicini G, Egli CM, Irniger S, Braus GH. 1992. Cloning, primary structure and regulation of the *ARO4* gene, encoding the tyrosine-inhibited 3-deoxy-D-*arabino*-heptulosonate 7-phosphate synthase from *Saccharomyces cerevisiae*. *Gene* 113:67–74

109. Künzler M, Springer C, Braus GH. 1995. Activation and repression of the yeast *ARO3* gene by global transcription factors. *Mol. Microbiol.* 15:167–78

110. Kuramoto N, Tomaru N, Murai M, Ohba K. 1997. Linkage analysis of isoenzyme and dwarf loci, and detection of lethal genes in sugi. *Breed. Sci.* 47:259–66

111. Lamb HK, Bagshaw CR, Hawkins AR. 1991. In vivo overproduction of the pentafunctional *arom* polypeptide in *Aspergillus nidulans* affects metabolic flux in the quinate pathway. *Mol. Gen. Genet.* 227:187–96

112. Lamb HK, Newton GH, Levett LJ, Cairns E, Roberts CF, Hawkins AR. 1996. The QUTA activator and QUTR repressor proteins of *Aspergillus nidulans* interact to regulate transcription of the quinate utilization pathway genes. *Microbiology* 142:1477–90

113. Lamb HK, van den Hombergh JPTW, Newton GH, Moore JD, Roberts CF, Hawkins AR. 1992. Differential flux through the quinate and shikimate pathways. Implications for the channelling hypothesis. *Biochem. J.* 284:181–87

114. Lawley B, Pittard AJ. 1994. Regulation of *aroL* expression by *tyrR* protein and *trp* repressor in *Escherichia coli* K-12. *J. Bacteriol.* 176:6921–30

115. Leech AP, James R, Coggins JR, Kleanthous C. 1995. Mutagenesis of active site residues in type I dehydroquinase

from *Escherichia coli*. *J. Biol. Chem.* 270: 25827–36

116. Leuschner C, Herrmann KM, Schultz G. 1995. The metabolism of quinate in pea roots: purification and partial characterization of a quinate hydrolyase. *Plant Physiol.* 108:319–25

117. Levesley I, Newton GH, Lamb HK, van Schothorst E, Dalgleish RWM, et al. 1996. Domain structure and function within the QUTA protein of *Aspergillus nidulans*: implications for the control of transcription. *Microbiology* 142:87–98

118. Macheroux P, Petersen J, Bornemann S, Lowe DJ, Thorneley RNF. 1996. Binding of the oxidized, reduced, and radical flavin species to chorismate synthase. An investigation by spectroscopy, fluorimetry, and electron paramagnetic resonance and electron nuclear double resonance spectroscopy. *Biochemistry* 35:1643–52

119. Maher EA, Bate NJ, Ni W, Elkind Y, Dixon RA, Lamb CJ. 1994. Increased disease susceptibility of transgenic tobacco plants with suppressed levels of preformed phenylpropanoid products. *Proc. Natl. Acad. Sci. USA* 91:7802–6

120. Majumder K, Selvapandiyan A, Fattah FA, Arora N, Ahmad S, Bhatnagar RK. 1995. 5-enolpyruvyl-shikimate 3-phosphate synthase of *Bacillus subtilis* is an allosteric enzyme. *Eur. J. Biochem.* 229:99–106

121. Mannerlöf M, Tuvesson S, Steen P, Tenning P. 1997. Transgenic sugar beet tolerant to glyphosate. *Euphytica* 94:83–91

122. McCandliss RJ, Herrmann KM. 1978. Iron, an essential element for biosynthesis of aromatic compounds. *Proc. Natl. Acad. Sci. USA* 75:4810–13

123. McCandliss RJ, Poling MD, Herrmann KM. 1978. 3-deoxy-D-*arabino*-heptulosonate 7-phosphate synthase. Purification and molecular characterization of the phenylalanine-sensitive isoenzyme from *Escherichia coli*. *J. Biol. Chem.* 253:4259–65

124. McCue KF, Conn EE. 1989. Induction of 3-deoxy-D-*arabino*-heptulosonate 7-phosphate synthase by fungal elicitor in cultures of *Petroselinum crispum*. *Proc. Natl. Acad. Sci. USA* 86:7374–77

125. McCue KF, Conn EE. 1990. Induction of shikimic acid pathway enzymes by light in suspension cultured cells of parsley (*Petroselinum crispum*). *Plant Physiol.* 94:507–10

126. McDowell LM, Klug CA, Beusen DD, Schaefer J. 1996. Ligand geometry of the ternary complex of 5-enolpyruvyl

shikimate 3-phosphate synthase from rotational-echo double-resonance NMR. *Biochemistry* 35:5395–403

127. McDowell LM, Schmidt A, Cohen ER, Studelska DR, Schaefer J. 1996. Structural constraints on the ternary complex of 5-enolpyruvyl shikimate 3-phosphate synthase from rotational-echo double-resonance NMR. *J. Mol. Biol.* 256:160–71

128. Millar G, Coggins JR. 1986. The complete amino acid sequence of 3-dehydroquinate synthase of *Escherichia coli* K12. *FEBS Lett.* 200:11–17

129. Miller MJ, Braccolino DS, Cleary DG, Ream JE, Walker MC, Sikorski JA. 1994. EPSP synthase inhibitor design IV. New aromatic substrate analogs and symmetrical inhibitors containing novel 3-phosphate mimics. *Bioorg. Med. Chem. Lett.* 21:2605–8

130. Miller MJ, Cleary DG, Ream JE, Snyder KR, Sikorski JA. 1995. New EPSP synthase inhibitors: synthesis and evaluation of an aromatic tetrahedral intermediate mimic containing a 3-malonate ether as a 3-phosphate surrogate. *Bioorg. Med. Chem.* 3:1685–92

131. Montchamp JL, Frost JW. 1997. Cyclohexenyl and cyclohexylidene inhibitors of 3-dehydroquinate synthase: active site interactions relevant to enzyme mechanism and inhibitor design. *J. Am. Chem. Soc.* 119:7645–53

132. Montchamp JL, Peng JR, Frost JW. 1994. Inversion of an asymmetric center in carbocyclic inhibitors of 3-dehydroquinate synthase: examining and exploiting the mechanism for *syn*-elimination during substrate turnover. *J. Org. Chem.* 59:6999–7007

133. Moore JD, Hawkins AR, Charles IG, Deka R, Coggins JR, et al. 1993. Characterization of the type I dehydroquinase from *Salmonella typhi*. *Biochem. J.* 295: 277–85

134. Moore JD, Lamb HK, Garbe T, Servos S, Dougan G, et al. 1992. Inducible overproduction of the *Aspergillus nidulans* pentafunctional AROM protein and the type-I and -II 3-dedydroquinases from *Salmonella typhi* and *Mycobacterium tuberculosis*. *Biochem. J.* 287:173–81

135. Mousdale DM, Campbell MS, Coggins JR. 1987. Purification and characterization of bifunctional dehydroquinase-shikimate:NADP oxidoreductase from pea seedlings. *Phytochemistry* 26:2665–70

136. Muday GK, Herrmann KM. 1992. Wounding induces one of two isoenzymes

of 3-deoxy-D-*arabino*-heptulosonate 7-phosphate synthase in *Solanum tuberosum* L. *Plant Physiol.* 98:496–500

137. Muday GK, Johnson DI, Somerville RL, Herrmann KM. 1991. The tyrosine repressor negatively regulates *aroH* expression in *Escherichia coli*. *J. Bacteriol.* 173:3930–32

138. Murata M, Ryu JH, Caretto S, Rao D, Song HS, Widholm JM. 1998. Stability and culture medium limitations of gene amplification in glyphosate resistant carrot cell lines. *J. Plant Physiol.* 152:112–17

139. Nafziger ED, Widholm JM, Steinrücken HC, Killmer JL. 1984. Selection and characterization of a carrot cell line tolerant to glyphosate. *Plant Physiol.* 76:571–74

140. Nida DL, Kolacz KH, Buehler RE, Deaton WR, Schuler WR, et al. 1996. Glyphosate-tolerant cotton: genetic characterization and protein expression. *J. Agric. Food Chem.* 44:1960–66

141. O'Callaghan D, Maskell D, Liew FY, Easmon CS, Dougan G. 1988. Characterization of aromatic- and purine-dependent *Salmonella typhimurium*: attenuation, persistence, and ability to induce protective immunity in BALB/c mice. *Infect. Immun.* 56:419–23

142. Ogino T, Garner C, Markley JL, Herrmann KM. 1982. Biosynthesis of aromatic compounds: ^{13}C NMR spectroscopy of whole *Escherichia coli* cells. *Proc. Natl. Acad. Sci. USA* 70:5828–32

143. Padgette SR, Kolacz KH, Delannay X, Re DB, LaVallee BJ. 1995. Development, identification, and characterization of a glyphosate-tolerant soybean line. *Crop Sci.* 35:1451–61

144. Padgette SR, Re DB, Gasser CS, Eichholtz DA, Frazier RB, et al. 1991. Site-directed mutagenesis of a conserved region of the 5-enolpyruvylshikimate 3-phosphate synthase active site. *J. Biol. Chem.* 266:22364–69

145. Paiva AA, Tilton RF, Crooks GP, Huang LQ, Anderson KS. 1997. Detection and identification of transient enzyme intermediates using rapid mixing, pulsed-flow electrospray mass spectrometry. *Biochemistry* 36:15472–76

146. Paravicini G, Braus G, Hütter R. 1988. Structure of the *ARO3* gene of *Saccharomyces cerevisiae*. *Mol. Gen. Genet.* 214:165–69

147. Paravicini G, Mösch HU, Schmidheini T, Braus G. 1989. The general control activator protein GCN4 is essential for a basal level of *ARO3* gene expression in *Saccharomyces cerevisiae*. *Mol. Cell. Biol.* 9:144–51

148. Paravicini G, Schmidheini T, Braus G. 1989. Purification and properties of the 3-deoxy-D-*arabino*-heptulosonate 7-phosphate synthase (phenylalanine-inhibitable) of *Saccharomyces cerevisiae*. *Eur. J. Biochem.* 186:361–66

149. Parker EJ, Coggins JR, Abell C. 1997. Derailing dehydroquinate synthase by introducing a stabilizing stereoelectronic effect in a reaction intermediate. *J. Org. Chem.* 62:8582–85

150. Pereira SA, Livi GP. 1996. Aromatic amino-acid biosynthesis in *Candida albicans*: identification of the *ARO4* gene encoding a second DAHP synthase. *Curr. Genet.* 29:441–45

151. Pierson LS III, Gaffney T, Lam S, Gong FC. 1995. Molecular analysis of genes encoding phenazine biosynthesis in the biological control bacterium *Pseudomonas aureofaciens* 30–84. *FEMS Microbiol. Lett.* 134:299–307

152. Pinto JEBP, Dyer WE, Weller SC, Herrmann KM. 1988. Glyphosate induces 3-deoxy-D-*arabino*-heptulosonate 7-phosphate synthase in potato (*Solanum tuberosum* L.) cells grown in suspension culture. *Plant Physiol.* 87:891–93

153. Pinto JEBP, Suzich JA, Herrmann KM. 1986. 3-deoxy-D-*arabino*-heptulosonate 7-phosphate synthase from potato tuber (*Solanum tuberosum* L.). *Plant Physiol.* 82:1040–44

154. Pompliano DL, Reimer LM, Myrvold S, Frost JW. 1989. Probing lethal metabolic perturbations in plants with chemical inhibition of dehydroquinate synthase. *J. Am. Chem. Soc.* 111:1866–71

155. Ramjee MN, Balasubramanian S, Abell C, Coggins JR, Davies GM, et al. 1992. Reaction of (6R)-6–F-EPSP with recombinant *Escherichia coli* chorismate synthase generates a stable flavin mononucleotide semiquinone radical. *J. Am. Chem. Soc.* 114:3151–53

156. Ramjee MN, Coggins JR, Hawkes TR, Lowe DJ, Thorneley RNF. 1991. Spectrophotometric detection of a modified flavin mononucleotide (FMN) intermediate formed during the catalytic cycle of chorismate synthase. *J. Am. Chem. Soc.* 113:8566–67

157. Ranjeva R, Refeno G, Boudet AM, Marme D. 1983. Activation of plant quinate:NAD 3-oxidoreductase by Ca and calmodulin. *Proc. Natl. Acad. Sci. USA* 80:5222–24

158. Ray JM, Bauerle R. 1991. Purification

and properties of tryptophan-sensitive 3-deoxy-D-*arabino*-heptulosonate 7-phosphate synthase from *Escherichia coli. J. Bacteriol.* 173:1894–901

159. Ray JM, Yanofsky C, Bauerle R. 1988. Mutational analysis of the catalytic and feedback sites of the tryptophan-sensitive 3-deoxy-D-*arabino*-heptulosonate 7-phosphate synthase of *Escherichia coli. J. Bacteriol.* 170:5500–6

160. Razal RA, Lewis NG, Towers GHN. 1994. Pico-tag analysis of arogenic acid and related free amino acids from plant and fungal extracts. *Phytochem. Anal.* 5:98–104

161. Roberts F, Roberts CW, Johnson JJ, Kyle DE, Krell T, et al. 1998. Evidence for the shikimate pathway in apicomplexan parasites. *Nature* 393:801–5

162. Ruan XA, Stassi D, Lax SA, Katz L. 1997. A second type-I PKS gene cluster isolated from *Streptomyces hygroscopicus* ATCC 29253, a rifamycin-producing strain. *Gene* 203:1–9

163. Rubin JL, Jensen RA. 1985. Differentially regulated isozymes of 3-deoxy-D-*arabino*-heptulosonate 7-phosphate synthase from seedlings of *Vigna radiata* [L.] Wilczek. *Plant Physiol.* 79:711–18

164. Saha S, Zipf A. 1998. Genetic diversity and phylogenetic relationships in cotton based on isoenzyme markers. *J. Crop. Prod.* 1:79–93

165. Sammons RD, Gruys KJ, Anderson KS, Johnson KA, Sikorski JA. 1995. Reevaluating glyphosate as a transition state inhibitor of EPSP synthase: identification of an EPSP synthase-EPSP-glyphosate ternary complex. *Biochemistry* 34:6433–40

166. Schaller A, Schmid J, Leibinger U, Amrhein N. 1991. Molecular cloning and analysis of a cDNA coding for chorismate synthase from the higher plant *Corydalis sempervirens* Pers. *J. Biol. Chem.* 266:21434–38

167. Schaller A, Windhofer V, Amrhein N. 1990. Purification of chorismate synthase from a cell culture of the higher plant *Corydalis sempervirens* Pers. *Arch. Biochem. Biophys.* 282:437–42

168. Schmid J, Amrhein N. 1995. Molecular organization of the shikimate pathway in higher plants. *Phytochemistry* 39:737–49

169. Schmid J, Schaller A, Leibinger U, Boll W, Amrhein N. 1992. The in vitro synthesized tomato shikimate kinase precursor is enzymatically active and is imported and processed to the mature

enzyme by chloroplasts. *Plant J.* 2:375–83

170. Schmidt CL, Danneel HJ, Schultz G, Buchanan BB. 1990. Shikimate kinase from spinach chloroplasts. Purification, characterization, and regulatory function in aromatic amino acid biosynthesis. *Plant Physiol.* 93:758–66

171. Schmidt CL, Gründemann D, Groth G, Müller B, Hennig H, Schultz G. 1991. Shikimate pathway in non-photosynthetic tissues. Identification of common enzymes and partial purification of dehydroquinate hydrolyase-shikimate oxidoreductase and chorismate mutase from roots. *J. Plant Physiol.* 138:51–56

172. Schnappauf G, Hartmann M, Künzler M, Braus GH. 1998. The two 3-deoxy-D-*arabino*-heptulosonate 7-phosphate synthase isoenzymes from *Saccharomyces cerevisiae* show different kinetic modes of inhibition. *Arch. Microbiol.* 169:517–24

173. Schoner R, Herrmann KM. 1976. 3-deoxy-D-*arabino*-heptulosonate 7-phosphate synthase. Purification, properties, and kinetics of the tyrosine-sensitive isoenzyme from *Escherichia coli. J. Biol. Chem.* 251:5440–47

174. Seto CT, Bartlett PA. 1994. (Z)-9-Fluoro-EPSP is not a substrate for EPSP synthase: implications for the enzyme mechanism. *J. Org. Chem.* 59:7130–32

175. Shah A, Font JL, Miller MJ, Ream JE, Walker MC, Sikorski JA. 1997. New aromatic inhibitors of EPSP synthase incorporating hydroxymalonates as novel 3-phosphate replacements. *Bioorg. Med. Chem.* 5:323–34

176. Shah DM, Horsch RB, Klee HJ, Kishore GM, Winter JA, et al. 1986. Engineering herbicide tolerance in transgenetic plants. *Science* 233:478–81

177. Shneier A, Harris J, Kleanthous C, Coggins JR, Hawkins AR, Abell C. 1993. Evidence for opposite stereochemical courses for the reactions catalyzed by type I and type II dehydroquinases. *Bioorg. Med. Chem. Lett.* 3:1399–402

178. Shneier A, Kleanthous C, Deka R, Coggins JR, Abell C. 1991. Observation of an imine intermediate on dehydroquinase by electrospray mass spectrometry. *J. Am. Chem. Soc.* 113:9416–18

179. Shumilin IA, Kretsinger RH, Bauerle R. 1996. Purification, crystallization, and preliminary crystallographic analysis of 3-deoxy-D-*arabino*-heptulosonate 7-phosphate synthase from *Escherichia coli. Proteins* 24:404–6

180. Shuttleworth WA, Evans JN. 1994. Site

directed mutagenesis and NMR studies of histidine385 mutants of 5-enolpyruvylshikimate 3-phosphate synthase. *Biochemistry* 33:7062–68

181. Shuttleworth WA, Evans JN. 1996. The H385N mutant of 5-enolpyruvylshikimate 3-phosphate synthase: kinetics, fluorescence, and nuclear magnetic resonance studies. *Arch. Biochem. Biophys.* 334:37–42

182. Shyr YJ, Caretto S, Widholm JM. 1993. Characterization of the glyphosate selection of carrot suspension cultures resulting in gene amplification. *Plant Sci.* 88:219–28

183. Sikorski JA, Gruys KJ. 1997. Understanding glyphosate's molecular mode of action with EPSP synthase: evidence favoring an allosteric inhibitor model. *Acc. Chem. Res.* 30:2–8

184. Singh BK, Siehl DL, Connelly JA. 1991. Shikimate pathway: Why does it mean so much to so many? *Oxford Surv. Plant Mol. Cell. Biol.* 7:143–85

185. Skinner MA, Günel-Ozcan A, Moore J, Hawkins AR, Brown KA. 1997. Dehydroquinate synthase binds divalent and trivalent cations: role of metal binding in catalysis. *Biochem. Soc. Trans.* 25:S609

186. Smart CC, Johänning D, Müller G, Amrhein N. 1985. Selective overproduction of 5-enolpyruvylshikimate 3-phosphate synthase in a plant cell culture which tolerates high doses of the herbicide glyphosate. *J. Biol. Chem.* 260:16338–46

187. Smith CM, Pratt D, Thompson GA. 1986. Increased 5-enolpyruvylshikimic acid 3-phosphate synthase activity in a glyphosate-tolerant variant strain of tomato cells. *Plant Cell Rep.* 5:298–301

188. Stalker DM, Hiatt WR, Comai L. 1985. A single amino acid substitution in the enzyme 5-enolpyruvylshikimate 3-phosphate synthase confers resistance to the herbicide glyphosate. *J. Biol. Chem.* 260:4724–28

189. Stallings WC, Abdel-Meguid SS, Lim LW, Shieh HS, Dayringer HE. et al. 1991. Structure and topological symmetry of the glyphosate target 5-enolpyruvylshikimate 3-phosphate synthase: a distinctive protein fold. *Proc. Natl. Acad. Sci. USA* 88:5046–50

190. Steinrücken HC, Amrhein N. 1980. The herbicide glyphosate is a potent inhibitor of 5-enolpyruvylshikimic acid 3-phosphate synthase. *Biochem. Biophys. Res. Commun.* 94:1207–12

191. Stephens CM, Bauerle R. 1991. Analysis of the metal requirement of 3-deoxy-D-*arabino*-heptulosonate 7-phosphate synthase from *Escherichia coli. J. Biol. Chem.* 266:20810–17

192. Stephens CM, Bauerle R. 1992. Essential cysteines in 3-deoxy-D-*arabino*-heptulosonate 7-phosphate synthase from *Escherichia coli. J. Biol. Chem.* 267:5762–67

193. Stuart F, Hunter IS. 1993. Purification and characterization of 3-deoxy-D-*arabino*-heptulosonate 7-phosphate synthase from *Streptomyces rimosus. Biochim. Biophys. Acta* 1161:209–15

194. Studelska DR, Klug CA, Beusen DD, McDowell LM, Schaefer J. 1996. Long-range distance measurements of protein binding sites by rotational-echo double-resonance NMR. *J. Am. Chem. Soc.* 118:5476–77

195. Studelska DR, McDowell LM, Espe MP, Klug CA, Schaefer J. 1997. Slowed enzymatic turnover allows characterization of intermediates by solid-state NMR. *Biochemistry* 36:15555–60

196. Subramaniam PS, Xie G, Xia T, Jensen RA. 1998. Substrate ambiguity of 3-deoxy-D-*manno*-octulosonate 8-phosphate synthase from *Neisseria gonorrhoeae* in the context of its membership in a protein family containing a subset of 3-deoxy-D-*arabino*-heptulosonate 7-phosphate synthases. *J. Bacteriol.* 180:119–27

197. Suh H, Hepburn AG, Kriz AL, Widholm JM. 1993. Structure of amplified 5-enolpyruvylshikimate 3-phosphate synthase gene in glyphosate-resistant carrot cells. *Plant Mol. Biol.* 22:195–205

198. Suzich JA, Dean JFD, Herrmann KM. 1985. 3-deoxy-D-*arabino*-heptulosonate 7-phosphate synthase from carrot root (*Daucus carota*) is a hysteretic enzyme. *Plant Physiol.* 79:765–70

199. Suzuki K, Fukuda Y, Shinshi H. 1995. Studies on elicitor-signal transduction leading to differential expression of defense genes in cultured tobacco cells. *Plant Cell Physiol.* 36:281–89

200. Suzuki N, Sakuta M, Shimizu S. 1996. Purification and characterization of a cytosolic isoenzyme of 3-deoxy-D-*arabino*-heptulosonate 7-phosphate synthase from cultured carrot cells. *J. Plant Physiol.* 149:19–22

201. Suzuki N, Sakuta M, Shimizu S, Komamine A. 1995. Changes in the activity of 3-deoxy-D-*arabino*-heptulosonate 7-phosphate (DAHP) synthase in suspension-cultured cells of *Vitis. Physiol. Plant.* 94:591–96

202. Takatsuji H, Mori M, Benfey PN, Ren

L, Chua NH. 1992. Characterization of a zinc finger DNA-binding protein expressed specifically in *Petunia* petals and seedlings. *EMBO J.* 11:241–49

203. Tanksley SD, Orton TJ, eds. 1983. *Developments in Plant Genetics and Breeding.* Vol. I. *Isozymes in Plant Genetics and Breeding* (part A). New York: Elsevier

204. Thion L, Mazars C, Thuleau P, Graziana A, Rossignol M, et al. 1996. Activation of plasma membrane voltage-dependent calcium-permeable channels by disruption of microtubules in carrot cells. *FEBS Lett.* 393:13–18

205. Thuleau P, Moreau M, Schroeder JI, Ranjeva R. 1994. Recruitment of plasma membrane voltage-dependent calcium-permeable channels in carrot cells. *EMBO J.* 13:5843–47

206. Deleted in proof

207. Tieman DM, Handa AK. 1996. Molecular cloning and characterization of genes expressed during early tomato (*Lycopersicon esculentum* MILL.) fruit development by mRNA differential display. *J. Am. Soc. Hortic. Sci.* 121:52–56

208. Vicario F, Vendramin GG, Rossi P, Lio P, Giannini R. 1995. Allozyme, chloroplast DNA, and RAPD markers for determining genetic relationships between *Abies alba* and the relic population of *Abies nebrodensis. Theor. Appl. Genet.* 90:1012–18

209. Vinella D, Gagny B, Joseleau-Petit D, D'Ari R, Cashel M. 1996. Mecillinam resistance in *Escherichia coli* is conferred by loss of a second activity of the *aroK* protein. *J. Bacteriol.* 178:3818–28

210. Walker GE, Dunbar B, Hunter IS, Nimmo HG, Coggins JR. 1996. Evidence for a novel class of microbial 3-deoxy-D-*arabino*-heptulosonate 7-phosphate synthase in *Streptomyces coelicolor* A3(2), *Streptomyces rimosus* and *Neurospora crassa. Microbiology* 142:1973–82

211. Walker MC, Jones CR, Somerville RL, Sikorski JA. 1992. (Z)-3-fluorophosphoenol-pyruvate as a pseudosubstrate of EPSP synthase: enzymatic synthesis of a stable fluoro analog of the catalytic intermediate. *J. Am. Chem. Soc.* 114:7601–3

212. Wang Y, Herrmann KM, Weller SC, Goldsbrough PB. 1991. Cloning and nucleotide sequence of a complementary DNA encoding 3-deoxy-D-*arabino*-heptulosonate 7-phosphate synthase from tobacco. *Plant Physiol.* 97:847–48

213. Weaver LM. 1990. *Aromatic amino acid insensitive DAHP synthases.* PhD thesis. Purdue Univ., Ind.

214. Weaver LM, Herrmann KM. 1990. Cloning of an *aroF* allele encoding tyrosine-insensitive 3-deoxy-D-*arabino*-heptulosonate 7-phosphate synthase. *J. Bacteriol.* 172:6581–84

215. Weaver LM, Pinto JEBP, Herrmann KM. 1993. Expression of potato DAHP synthase in *Escherichia coli. Bioorg. Med. Chem. Lett.* 3:1421–28

216. Wheeler KA, Lamb HK, Hawkins AR. 1996. Control of metabolic flux through the quinate pathway in *Aspergillus nidulans. Biochem. J.* 315:195–205

217. Whipp MJ, Pittard AJ. 1995. A reassessment of the relationship between *aroK*- and *aroL*-encoded shikimate kinase enzymes of *Escherichia coli. J. Bacteriol.* 177:1627–29

218. White PJ, Millar G, Coggins JR. 1988. The overexpression, purification and complete amino acid sequence of chorismate synthase from *Escherichia coli* K12 and its comparison with the enzyme from *Neurospora crassa. Biochem. J.* 251:313–22

219. White PJ, Young J, Hunter IS, Nimmo HG, Coggins JR. 1990. The purification and characterization of 3-dehydroquinase from *Streptomyces coelicolor. Biochem. J.* 265:735–38

220. Widlanski T, Bender SL, Knowles JR. 1989. Dehydroquinate synthase: a sheep in wolf's clothing? *J. Am. Chem. Soc.* 111:2299–300

221. Widlanski T, Bender SL, Knowles JR. 1989. Dehydroquinate synthase: the use of substrate analogues to probe the late steps of the catalyzed reaction. *Biochemistry* 28:7572–82

222. Williamson SC, Yu H, Davis TM. 1995. Shikimate dehydrogenase allozymes: inheritance and close linkage to fruit color in the diploid strawberry. *J. Hered.* 86:74–76

223. Wind JC, van Eldik GL, van der Plas LHW, Herrmann KM, Croes AF, Wullems GJ. 1999. Three differentially expressed 3-deoxy-D-*arabino*-heptulosonate 7-phosphate synthase genes in *Morinda citrifolia. Plant Physiol.* In press

224. Woisetschläger M, Högenauer G. 1986. Cloning and characterization of the gene encoding 3-deoxy-D-*manno*-octulosonate 8-phosphate synthase from *Escherichia coli. J. Bacteriol.* 168:437–39

225. Yamamoto E. 1980. Purification and metal requirements of 3-dehydroquinate synthase from *Phaseolus mungo* seedlings. *Phytochemistry* 19:779–81

226. Yamauchi N, Kakinuma K. 1995. Enzymatic carbocycle formation in microbial

secondary metabolism. The mechanism of the 2-deoxy-*scyllo*-inosose synthase reaction as a crucial step in the 2-deoxy-streptamine biosynthesis in *Streptomyces fradiae*. *J. Org. Chem.* 60:5614–19

227. Zhao J, Herrmann KM. 1992. Cloning and sequencing of a second cDNA encoding 3-deoxy-D-*arabino*-heptuloso-nate 7-phosphate synthase from *Solanum*

tuberosum L. *Plant Physiol.* 100:1075–76

228. Zurawski G, Gunsalus RP, Brown KD, Yanofsky C. 1981. Structure and regulation of *aroH*, the structural gene for the tryptophan-repressible 3-deoxy-D-*arabino*-heptulosonate 7-phosphate synthase of *Escherichia coli*. *J. Mol. Biol.* 145:47–73

Annu. Rev. Plant Physiol. Plant Mol. Biol. 1999. 50:505–37

ASYMMETRIC CELL DIVISION IN PLANTS

Ben Scheres[1] and Philip N. Benfey[2]
[1]Department of Molecular Cell Biology, Utrecht University, 3584 CH Utrecht, The Netherlands; and [2]Department of Biology, New York University, New York, NY 10003; e-mail: b.scheres@bio.uu.nl

KEY WORDS: *Arabidopsis*, cell fate determinant, polarity, intrinsic cue, extrinsic cue

ABSTRACT

Asymmetric cell divisions generate cells with different fates. In plants, where cells do not move relative to another cell, the specification and orientation of these divisions is an important mechanism to generate the overall cellular pattern during development. This review summarizes our knowledge of selected cases of asymmetric cell division in plants, in the context of recent insights into mechanisms underlying this process in bacteria, algae, yeast, and animals.

CONTENTS

INTRODUCTION

During the life cycle of plants and animals, a single cell produces a multi-cellular organism with many different specialized cell types. To generate the multitude of different cell types requires cell divisions in which daughter cells have different fates. These are called asymmetric cell divisions, whether or not asymmetry is morphologically evident at the time of division (48). Asymmetric cell divisions are traditionally divided into two flavors. The difference in daughter cells may be due to unequal partitioning of factors in the mother cell such that all or most are inherited by only one daughter. Alternatively, the division may result in daughter cells that have equal developmental potential at first but become different subsequently through interactions with each other or with different neighboring cells. In this case, extrinsic cues determine cell fate. The first can be considered a monarchy in which fate is passed down based on lineage, the second a modern democracy in which fate is determined by interest groups. Note that both mechanisms can involve external spatial information. In the intrinsic case, spatial information may direct the orientation of an asymmetric cell division prior to its occurrence, whereas in the extrinsic case this information acts to determine fate after the cell division. As highlighted in this review, actual development often involves a combination of these two strategies.

In plants, the cell wall prevents extensive rearrangement of cells, and common sense predicts that asymmetric division must be important to generate the abundance of diversified cells in the plant body. However, approximately correct tissue patterns can emerge from plants mutated in their ability to divide in the appropriate plane (102, 103). Moreover, the flexible fates of many plant cells (77a, 107) indicate that the determination of cell fates is not restricted to narrow time windows, as is often the case in animal development. Are these differences reflecting new mechanisms, or do they arise from subtle changes in the same underlying basic mechanisms? This review probes this question, making use of the wealth of information that has emerged recently on asymmetric cell division in organisms as diverse as bacteria, yeast, flies, and nematodes. We briefly highlight these findings and subsequently turn to selected cases in plants (for other recent reviews see 36, 51b, 84).

YEAST AND DROSOPHILA: INTRINSIC DETERMINANTS

Two criteria should be met by an intrinsic factor that mediates asymmetric cell division: (a) The factor is preferentially distributed into one daughter cell during the cell division; (b) its presence or absence determines cell fate. The second criterion predicts that loss-of-function mutants in the factor-encoding genes lead to two cells of one type, and ectopic expression of the factor in both cells leads to two cells of the other type. In both yeast and *Drosophila*, gene products that meet these criteria have been recently identified. The common elements found in all cases are: (a) localized cues initiate polarization of the cytoskeleton; (b) factors are partitioned to one end of the cell; and (c) the spindle is oriented so that when the cell divides the factors end up predominantly in only one of the daughters (Figure 1). In each case we first discuss the intrinsic

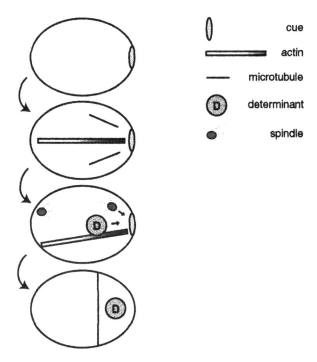

Figure 1   A schematic representation of an asymmetric cell division that involves partitioning of intrinsic determinants. An asymmetrically localized cue provides initial polarity; the cytoskeleton is oriented with respect to this polarity, leading to the preferential partitioning of cell fate determinants to one daughter cell, and concomitant orientation of the mitotic spindle.

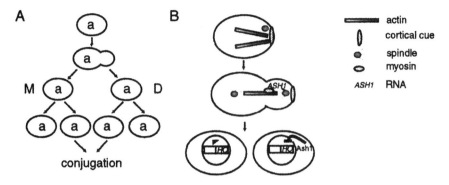

Figure 2 Asymmetric division in budding yeast results in cells with different mating-type switching abilities. *A.* Division pattern displaying the switches in mating types a and *α*. *B.* Concomitant distribution of *ASH1* RNA and spindle orientation.

factors or determinants, then associated proteins involved in their asymmetric distribution, and finally initial cues and their relation to spindle orientation.

Yeast: Ash1p as an Intrinsic Determinant

In the budding yeast *Saccharomyces cerevisiae*, cells of two mating types exist. Upon spore germination, cell division is asymmetric: It gives rise to a "mother" cell capable of mating-type switching during the next cell cycle, and a "daughter" cell—which originates as a growing bud on the mother cell—that is not able to switch (Figure 2*A*). In this way, yeast spores generate cells of two mating types that can subsequently conjugate to produce a diploid cell.

This asymmetric cell division appears to be intrinsically determined (97). Mating-type switching is initiated by a site-specific endonuclease encoded by the *HO* gene. This gene is transcribed in mother cells only in a narrow time window at the G1 to S phase transition (72*, 98). The difference in cell fates, with one daughter able to switch and the other not, is therefore determined by differential regulation of *HO* gene transcription. Recently, the *ASH1* gene was identified in two different genetic screens as a negative regulator of *HO* transcription (8, 93). The Ash1 protein, which is related to GATA transcriptional regulators, is partitioned into daughter cells. Its presence down-regulates *HO* expression and thereby determines fate, and loss-of-function and gain-of-function experiments demonstrate that this factor meets the criteria for being a cell-intrinsic determinant of cell fate.

The partitioning of Ash1 requires localization of *ASH1* mRNA, which is performed by an actin-based mechanism involving a myosin motor (62, 51). Although the precise cue responsible for Ash1 localization is unknown, it has been shown that *ASH1* RNA becomes anchored to a region of the cell cortex that will end up in the daughter cell (62). This may be a direct consequence of the polarization of the actin cytoskeleton in response to small GTPases that are asymmetrically activated in the cell cortex. The Bud3 protein, which marks a ring around the site of the previous cell division, initiates the activation of these GTPases (15). Bud3-dependent processes also orient the mitotic spindle by positioning one spindle pole body. Use of the same Bud3-dependent cue for localization of Ash1 and orientation of the spindle would ensure that Ash1 becomes partitioned into only the daughter cell (reviewed in 14, 86a).

In summary, in yeast a single (nonswitching) cell gives rise to two different cell types (switching and nonswitching) by actin-dependent preferential distribution of the Ash1 mRNA/protein, which acts as a cell fate determinant by repressing transcription of the *HO* gene. Initial polarity cues ultimately mediate partitioning of Ash1p and coordinate the orientation of the mitotic spindle to ensure proper segregation of fates (Figure 2*B*).

Drosophila: Numb and Prospero as Intrinsic Determinants

In the fruit fly, precursor cells for both the central (CNS) and the peripheral nervous system (PNS) undergo asymmetric cell divisions. Two genes, *numb* and *prospero*, encode intrinsic determinants essential for the acquisition of cell fate.

During the development of the peripheral nervous system, sensory organ precursor (SOP) cells divide asymmetrically to produce a neuron/sheath precursor and an outer support cell precursor (Figure 3*A*). The membrane-associated Numb protein was shown to be both partitioned into one daughter cell and responsible for the eventual differences in cell fate (82, 105).

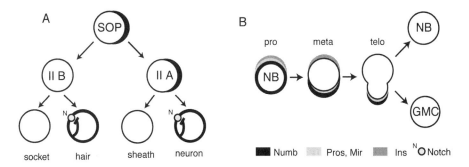

Figure 3   Asymmetric cell divisions during *Drosophila* neurogenesis. A. External sensory organ formation. Distribution of Numb is shown (thick outline). SOP: sensory organ precursor cell. B. Neuroblast (NB) division to generate ganglion mother cells (GMC). Distribution of Notch, Prospero, Miranda, and Inscuteable are shown at three stages of mitosis.

While Numb itself is an intrinsic determinant, the mechanism by which it determines cell fate is through an interaction with neighboring cells. The fate of the SOP daughters requires signaling mediated by the transmembrane receptor Notch: Without Notch activity, only neurons arise (44a). Notch activity is inhibited in the daughter cell that inherits the Numb protein, so Numb biases cell fate decisions by interfering with Notch signaling (33, 42). This, then, is an example of an intrinsic determinant that acts by antagonizing a signal received from neighboring cells.

Recently, it has been shown that the orientation of Numb partitioning as well as the orientation of the SOP cell divisions along the fly's antero-posterior axis depends on signaling from the Frizzled transmembrane receptor, which binds ligands of the Wnt family. This indicates that extracellular signals provide the polarizing cue for both Numb localization and antero-posterior spindle orientation (37).

The second intrinsic determinant in neural precursor cells, the homeodomain protein Prospero, is expressed in the precursors of the *Drosophila* central nervous system. Neuroblasts divide asymmetrically to produce a ganglion mother cell and another neuroblast (Figure 3*B*). During neuroblast division, Prospero protein is retained in the cytoplasm and partitioned to the ganglion mother cell only, whereupon it is translocated to the nucleus (46). Prospero most likely controls ganglion mother cell fate by transcriptional regulation (25, 106).

What is the mechanism for partitioning Prospero and Numb to the appropriate daughter cell? During asymmetric divisions in CNS and PNS, which occur in apical-basal orientation, Numb and Prospero colocalize and form a crescent at the basal cell cortex, suggesting a common mechanism for their partitioning (53a) (Figure 2*B*). The actin cytoskeleton is essential for localization of Numb and Prospero (12, 54) whereas disruption of microtubules affects the placement of the crescent with respect to the division plane (53a, 55). Therefore, actin is involved in determining partitioning itself, whereas the microtubular network serves an as yet undefined role in aligning this segregation mechanism with the orientation of the spindle. Tethering of the nuclear protein Prospero to the membrane is mediated by its association with an asymmetrically distributed adapter protein, Miranda, which also interacts with Numb (90, 50, 89). Localization of Numb, Prospero, and Miranda in the basal crescent requires the *inscuteable* gene, encoding a protein with a putative SH3 binding domain, frequently found in proteins that interact with the cytoskeleton (55). Surprisingly, Inscuteable protein is localized apically, at the opposite side of the future Numb/Prospero/Miranda crescent (Figure 3*B*), and its localization is actin dependent. Furthermore, Miranda can also bind Inscuteable through a domain that is required for its asymmetric localization. Inscuteable is also required to orient the mitotic spindle, suggesting that it coordinates asymmetric protein localization with spindle orientation (Figure 3B). Thus, Inscuteable localizes cell fate determinants in the apical-basal plane through its capacity to bind Miranda, and it orients the plane of cell division by anchoring centrosomes via an unknown mechanism.

In summary, in fly neuroblasts two intrinsic determinants, Numb and Prospero, are partitioned basally through an actin-dependent mechanism that involves several other proteins that are themselves partitioned. The cue responsible for the initial polarization of this machinery is still unknown.

CAENORHABDITIS ELEGANS: EXTERNAL SIGNALS REGULATE ASYMMETRIC CELL DIVISION

The key criteria for the involvement of external cell fate determination in an asymmetric division process are (a) inductive signals from neighboring cells determine cell fate and (b) both cells resulting from the division are equally responsive to these signals. An important prediction from the first criterion is that alterations in the cellular environment after the cell under investigation has divided can change cell fate. Manipulation of the cellular environment is feasible in the soil nematode C. elegans, and it has been demonstrated that inductive interactions can determine the fate of particular cells in the context of a nearly invariant cell division pattern. It will become evident that inductive processes can regulate asymmetric divisions in different ways. In the first case that we discuss, the division is initially symmetric and cellular interactions then bias one of the daughters to take a fate different from its sibling. The second case is highly analogous to the Drosophila neuroblast divisions, in that an external signal regulates the partitioning of intrinsic factors and spindle orientation.

GLP-1 Signaling Determines Daughter Cell Fate After Division

The first division of the zygotic embryo produces two cells AB and P_1 (Figure 4). This asymmetric division partitions cell-intrinsic factors that are not discussed here, but it is interesting to note that the gene products required for the asymmetry again are localized in the cell cortex (reviewed in 10, 43, 86). The AB cell then divides to give two cells ABa and Abp with different fates. Only the ABp cell directly contacts P_2, one of the daughters of P_1. If ABa and ABp are switched such that ABa now contacts P_2, a normal worm will result (79). Laser ablation of P_2 results in both AB daughter cells having the ABa fate (11). Furthermore, if ABp is not allowed to contact P_2 it will produce tissues normally produced by its sister, and if ABa is placed in contact with P_2 it will produce tissues normally made by ABp (68, 64). The conclusion from these

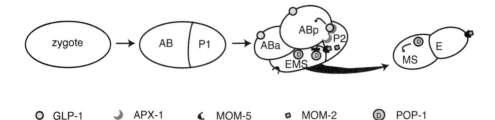

Figure 4 Asymmetric cell divisions of blastomeres in early *C. elegans* embryos. Receptors and ligands of two extrinsic signaling pathways are shown, which are involved in mediating cell fate decisions. From left to right: zygote, two-cell embryo, four-cell embryo, and the two EMS daughters at the eight-cell stage.

experiments is that ABa and ABp initially have equivalent potential to adopt either fate and that interaction with P_2 is the determining factor in the fate adopted.

Nuclear factors that determine the difference between the Aba and Abp fates are not known, but some of the molecules involved in the interaction that establishes the difference between these cells have been identified. Mutations in the *glp-1* gene result in ABp following an ABa fate (4, 78). *glp-1* is a homolog of the Drosophila *Notch* gene and encodes a receptor that accumulates on the membrane of AB descendants but not P_1 descendants. The ligand for GLP-1 appears to be APX-1, which is a homolog of the Notch ligand, Delta (68). APX-1 is membrane-tethered to the P_2 cell and is localized to the junction between ABp and P_2 (69). The localization of these two interacting proteins, which is in turn determined by previous asymmetric divisions as well as membrane polarization, therefore determines which cell will take on the ABp fate.

MOM-2 Signaling Directs Asymmetric Divisions

Inductive interactions can also occur prior to cell division. EMS, the sister cell to P_2, normally divides to give an MS cell whose descendants will form muscle and pharyngeal tissue, and an E cell whose descendants generate intestine (Figure 4). If EMS is removed from the embryo and allowed to divide in culture, both daughters have the MS fate (38). It has been demonstrated that P_2 is the source of the information necessary for the E fate, and that its position relative to EMS determines which of the daughters will follow that fate (reviewed in 44). In this case, P_2 is responsible for setting up an asymmetry within the EMS cell.

One intrinsic factor that is required to establish the E fate, in conjunction with others, is the product of the *POP-1* gene, a nuclear protein with an HMG domain (59). Interestingly, POP-1 is required for the fate of the anterior cell in many cell divisions in the antero-posterior orientation, and it presumably interacts with different partners in each of these cases.

Analogous to their involvement in asymmetric cell division of the SOP precursor cell in the fly, components of the Wnt signal transduction pathway have been shown to be involved in the determination of E fate. *mom-2* encodes a protein with homology to Wnt, while *mom-5* encodes its probable receptor, a member of the Frizzled gene family (83, 100). *mom-1* has homology to a Drosophila protein involved in the processing and secretion of Wnt (83). It has been shown that the putative ligand *mom-2* and the processing factor *mom-1* act in P_2, and it is assumed that the putative receptor *mom-5* acts in EMS (100). Downstream of the receptor/ligand interaction are the β-catenin homologue, *wrm-1*, and an APC (human adenomatous polyposis) related gene, *apr-1*, both implicated in Wnt signal transduction. It is hypothesized that the *wrm-1* and *apr-1* products form a complex with POP-1 that results in the latter's inactivation (44). One model is that the Wnt signal acts directly on the cytoskeleton to polarize the localization of POP-1 in the EMS mother cell, which results in the MS daughter having a higher concentration of POP-1 in its nucleus (44, 99). Interestingly, *mom-1,2,3,5* all affect both POP-1 localization and spindle orientation of the EMS cell division, whereas *wrm-1* only affects POP-1 localization (83, 100). Thus, the Wnt signal transduction pathway branches to coordinate spindle orientation and segregation of cell fate determinants.

In summary, the Notch signaling pathway can determine cell fate after cell division in *C. elegans*, and the Wnt signaling pathway can orient asymmetric cell divisions. Both pathways are also associated with asymmetric cell divisions in flies. Note the different utilization of the Notch pathway: In worms the position of a cell in relation to neighboring cells determines whether it receives a signal, whereas in flies an intrinsic determinant (Numb) is partitioned that inhibits Notch signaling.

BACILLUS SUBTILIS: A BACTERIAL SOLUTION TO ASYMMETRIC CELL DIVISION

During vegetative growth, the bacterium *B. subtilis* divides symmetrically to form two equal-sized daughters with identical fates. Under conditions of nutritional stress, division in *B. subtilis* becomes asymmetric. Instead of a septum forming in the middle of the cell, a choice is made between one of two sites at either end of the cell (Figure 5). Division results in a larger mother cell and a smaller forespore that is eventually engulfed by the mother cell.

The specialized sigma factors act as determinants for mother cell and forespore fate. The association of a sigma factor with RNA polymerase enables the regulation of transcription of cell-specific genes, among which are those that lead to new sigma factors (87, 110).

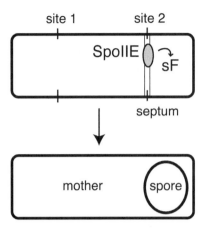

Figure 5 Asymmetric cell division in *B. subtilis*. Membrane-bound SpoIIE differentially activates the release of a sigma factor that mediates forespore-specific gene expression. Site 1, 2, the locations of two potential division sites, of which one is chosen.

One of the first steps in asymmetric cell division is the localization of a membrane-bound phosphatase, SpoIIE, to one of two potential division sites (Figure 5) (110). The choice of division site may be influenced by the previous site of vegetative division, or a case of "spontaneous symmetry breaking" (28, 47; reviewed in 110). In addition to marking the site of septum formation, SpoIIE initiates cell-specific gene expression after cell division. Its phosphatase activity dephosphorylates SpoIIAA, which then binds a kinase, SpoIIAB. This prevents the binding of SpoIIAB to sigma-F, releasing the specialized sigma factor subunit of RNA polymerase to activate gene expression in the forespore (1, 27, 31).

The pivotal question to understand asymmetry of the *B. subtilis* division is by what cue SpoIIE is differentially activated in the pre-spore cell. Preferential gene activation in one compartment is dependent upon accurate targeting of the SpoIIE protein to the septum membrane, which suggests that there are recognizable membrane domains in the bacterium (87). It is unknown whether SpoIIE is asymmetrically distributed in the septum membrane with its C-terminal phosphatase moiety facing the forespore compartment, or whether it is perhaps symmetrically distributed and its concentration (per cytoplasmic volume) is higher in the forespore due its smaller size (110). The latter possibility bears some relevance to asymmetric cell division in a distant plant relative, the multicellular alga *Volvox carteri*. Here, cell size appears to be the critical factor to distinguish between reproductive and somatic cell fate (53).

LESSONS: A COMMON THEME AND VARIATION

What are we able to discern from this overview of asymmetric divisions? One obvious common theme can be extracted from the eukaryotic examples of intrinsic fate determination: Localized cues initiate the partitioning of intrinsic cell fate–determining factors and coordinate these with spindle orientation. It seems relevant that these cues depend on or organize the actin cytoskeleton in all cases discussed. Given the involvement of actin and its associated motor proteins in active directional transport, their role in asymmetric cell division can be easily pictured. However, the role of actin in the various examples of asymmetric cell division appears to vary, as microfilaments are sometimes polarized (as in yeast) and sometimes not. This may reflect an underlying diversity of mechanisms.

Important variations on the simple subdivision of intrinsic versus extrinsic asymmetric divisions have emerged. Intrinsic mechanisms for asymmetric cell division sometimes adhere closely to the simple definition, but in other instances represent a mixed case. Yeast budding and the role of Prospero in CNS division represent the simple case: A transcription factor is localized to one part of the cell prior to division; the factor then activates genes in the daughter cell to which it is partitioned, determining the fate of that cell. However, the activity of the Numb protein, itself partitioned by an intrinsic mechanism, depends on cell-cell signaling. Furthermore, before intrinsic factors establish differences prior to cell division, surrounding cells often play a role in directing the orientation of the asymmetry, and we have seen examples in *Drosophila* as well as in *C. elegans*, where cell-cell communication organizes partitioning of an intrinsic factor in the mother cell. It is noteworthy that these cell signaling events are mediated by receptors like Frizzled and Notch, whose use is not restricted to orientation of asymmetric cell division because they are involved in a large variety of signaling events.

The molecular identity of localized cues involved in intrinsic asymmetric cell divisions appears to be diverse. It remains to be seen whether components of the yeast machinery for budding site selection, the nematode PAR proteins, or *Drosophila* cues have plant counterparts. It is even conceivable that the asymmetric activation mechanism of fate-determining factors in *Bacillus* turns out to be more relevant to plants, given the resemblance of their division process.

ASYMMETRIC CELL DIVISIONS IN PLANTS

Despite variations in cell division sequences, all plants generate a spectrum of different, regularly spaced cell types. Any cell division that generates daughters with different fates is asymmetric by definition. Cell movement is limited in plants, and thus control of the cell division plane has traditionally been considered important for the formation of regular patterns. Oriented cell divisions certainly have a role in generating ordered files of cells. Is control of the orientation of cell division also related to mechanisms of cell fate determination? Below, we first compare and contrast mechanisms for orienting cell divisions in plants and animals. We then survey examples of asymmetric cell divisions in plants to probe whether they are regulated by intrinsic or extrinsic fate determining processes.

Orientation of Cell Division in Plants: Role of the Cytoskeleton

In animals, the location of the centrosomes—microtubule organizing centers (MTOCs) for the nucleation of the spindle microtubules (80a)—determine the direction of chromosome segregation and the orientation of cell division. Spindle alignment is an important aspect of asymmetric cell division in yeast and animals. It is thought to be regulated by capture of the microtubules emanating from one centrosome, or its yeast analog, by regions of the cell cortex (14, 112). In nematodes, stable actin patches, possibly remnants of the previous cell division site, are present in specific regions of the cortex, and the Bud3p protein in yeast similarly marks the previous division site.

Plants contain no centrosomes. Instead, antibody staining suggests that the plant nuclear surface has MTOC properties (17, 30, 108). Thus, a centrosome-based mechanism to orient cell division cannot operate in plants. Nevertheless, several lines of evidence point to a role of the microtubular network in plant cell division. Microtubules of nuclear origin can connect to, and may even contribute to, the formation of the preprophase band (PPB), a dense cortical array of microtubules that transiently marks the site of division in plant cells (41, 61a, 77).

The finding that the PPB predicts the orientation of cell division has been followed up by several studies on its formation, in an attempt to uncover mechanisms that operate in cell division site selection. A less condensed cortical microtubule network is present in the interphase cell well before PPB formation (23). These cortical microtubules are already oriented in the same plane as the future PPB, indicating that cytoskeletal polarity coinciding with the cell division plane is present before the division site is selected. Apart from the possible involvement of the nucleus in the deposition of the cortical microtubular network and the PPB, the principal spatial cues for division site selection remain unknown. Centrifugation experiments have shown that altering the location of the nucleus can direct the formation of a second PPB (69a, 71), which suggests that the mother cell nucleus may direct the position of PPB formation.

It is unclear whether this involves the nuclear MTOC activity or other polarized domains on the nuclear surface.

There is evidence that the PPB, in turn, may be important in regulating spindle formation (69a, 71). However, *Arabidopsis* mutants in preprophase band formation demonstrate that the PPB is not required for spindle formation per se. In *tonneau/fass* mutants there is a total lack of the preprophase band (103), but spindles are formed nevertheless. Division planes in the mutants appear random from the earliest stages of embryogenesis onward, and hence the orientation of cell division is severely affected (102, 103). One might expect that cell fate would be severely affected in these mutants. However, all tissues are present in the right places and cellular differentiation appears normal, although organ morphology is affected. The finding that random cell divisions do not fundamentally alter embryonic patterning can be explained in at least two different ways. First, asymmetric divisions in plants may be guided by extrinsic rather then intrinsic cues. Second, asymmetric divisions may not be dependent on PPB formation.

The PPB is not the single decisive cue for orientation of the spindle. In particular cells, the spindle axis may rotate during division due to space constraints. However, in general, the cell plate corrects for this so that a curved cell wall is again attached to the PPB-marked site (21, 75). In extreme cases, a new division site may be chosen, resulting in a change of the orientation of cell division (74a). The emerging, sketchy, picture of orientation of cell division in plants is that the nucleus may regulate cortical array and PPB formation. The PPB provides the cue for the direction of division, and geometric constraints, which may be dictated by the cell's autonomous elongation program but also by its neighbors, may in some cases influence the division plane at a later stage. The identification of the principal cues that guide the cortical microtubule array in plant cells is a major challenge for the future.

The PPB disappears prior to the onset of mitosis. Yet there is evidence for a "memory" function that may involve the phosphorylation of cortex proteins by a P34[cdc2] kinase that localizes to the PPB (21a, 69a), an "actin depleted zone" (5, 20, 60), and the formation of an actin network between nucleus and division site.

While most animal cells form a contractile ring that, as its name suggests, contracts to pinch off the two daughter cells, most higher plant cells form a cell plate in between daughter nuclei. The cell plate serves as the assembly site of a new cell wall (reviewed in 2, 32, 96). The first step in cell plate formation is the assembly of a phragmoplast, another plant-specific structure, which consists of endoplasmic reticulum, Golgi-derived vesicles, microtubules, and microfilaments. Maturation of the cell plate requires continuous delivery of material via vesicle transport. The plate grows from the center outward until it reaches the sites previously marked by the PPB. The maturation process involves several stages: First there is a tubular network, then a fenestrated sheet, and eventually new cell walls are formed with plasma membrane on either side. A notable exception to this process is found in microspore mother cells that use a contractile ring for division (74b, 75a).

In the Arabidopsis mutants *knolle* (63) and *keule* (3) as well as in the pea mutant *cyd* (61), there appears to be a defect in completion of the cell plate resulting in cells with multiple nuclei. The *KNOLLE* gene encodes a member of the syntaxin family of integral membrane proteins involved in vesicle docking and fusion (63). Immunolocalization of KNOLLE protein and ultrastructural analysis indicates that it serves a cell plate–localized role in vesicle fusion (58). The availability of single components of the cell division machinery such as the KNOLLE protein will enhance our abilities to investigate molecular aspects of cytokinesis in plants.

A Survey of Asymmetric Cell Divisions During the Plant Life Cycle

Starting from the plant zygote, a large number of asymmetric divisions occurs to generate the mature plant with female and male gametes. Some of these not only give rise to different progeny but they also generate cells of different sizes. Examples of such asymmetric cell divisions in *Arabidopsis* are: (*a*) the first division of the zygote (65); (*b*) the embryonic division that gives rise to the lens-shaped progenitor cell of the quiescent centre (25a); (*c*) the male microspore division; and (*d*) divisions during stomatal complex formation (57). The physical asymmetry of these cell divisions indicates that the differences in fate of the daughters are defined during the process of cell division, which makes these divisions candidates for being determined by intrinsic factors. In later paragraphs, we discuss the evidence for intrinsic mechanisms.

Other asymmetric cell divisions lead to different cell types, but the differences are not yet evident during cell division. Important divisions of this class are the oriented periclinal divisions in the early embryo that separate the progenitor cells for the three main tissues, epidermis, ground tissue, and vascular tissue (51c). The stem cell divisions that separate differentiation-competent daughter cells and new stem cells in the root meristem are another example in this category (25a, 107). For this class of asymmetric cell divisions, the importance of intrinsic versus extrinsic mechanisms cannot be assessed without further analysis, and we discuss relevant experiments later.

The separation of the main organs (cotyledons, shoot apical meristem, hypocotyl, root, and root apical meristem) during embryogenesis is not correlated with early cell divisions, even in species like *Arabidopsis* with regular cell lineages (85). Thus, organ boundaries are not caused by asymmetric cell divisions but rather arise from the patterning of groups of cells, as in the well-investigated case of floral organ patterning. This does not preclude the involvement of asymmetric cell divisions in the generation of a coarse pre-pattern for embryonic organ specification. For example, the separation of the *Arabidopsis* embryo proper into apical, central, and basal regions may turn out to yield examples of asymmetric cell division as well (51a).

In summary, one class of asymmetric cell divisions in plants may be regulated by intrinsic mechanisms, and in that case orientation of cell division and partitioning of cell fate determinants should be coupled. Asymmetric divisions of a second class may be regulated entirely by external cues. Nevertheless, oriented cell division can still be an important parallel process, required for the establishment of the regular cell arrangements that accompany the asymmetric cell divisions.

THE FIRST DIVISION OF THE PLANT EMBRYO

The first division of the plant zygote produces two daughter cells with different fates. This asymmetric division has been investigated in lower and in higher plants, which show similarities in their early embryo development despite the large evolutionary distances. Lower plants such as the brown alga *Fucus* have free-living zygotes, and the early steps of embryogenesis can be investigated by direct experimental manipulation. In higher plants, where the developing zygote is buried in maternal tissues, genetic manipulation is an important tool and we focus on genetic analyses in *Arabidopsis*.

Fucus: Vesicle Transport and the Cell Wall

The *Fucus* zygote divides to give rise to two cells with different fates (Figure 6). The apical cell will form the stipe and fronds of the mature plant, whereas the basal cell will give rise to the holdfast, a support structure. Of major significance is the finding that apical and basal cell fates after the two-cell stage are determined by the cell wall, as shown by laser manipulations whereby the progeny of the apical and the basal cell after the first zygotic division were allowed to contact different cell walls (7). While this opens up the possibility that cell fate determinants can be deposited into the cell wall, nuclear factors that execute the differences in cell fate have to be identified. Studies on cell fate at later stages of *Fucus* development indicate that a role of the cell wall in cell fate may be restricted to the zygotic division (9). Hence wall components may only play a role in a small number of asymmetric cell divisions.

How is the orientation of this asymmetric cell division controlled? Prior to the first cell division, the orientation of the apical-basal axis is determined by extrinsic cues such as sperm entry, gravity, and light. This labile polarization is

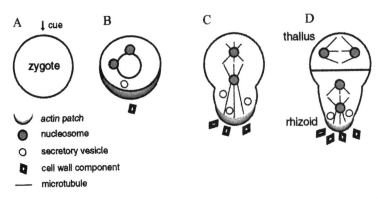

Figure 6   The asymmetric division of the *Fucus* zygote, and the associated acquisition of cellular polarity. The initial cue here is drawn as a light vector, but other cues are possible.

associated with changes in ion currents and deposition of various compounds, among which is filamentous actin, at the presumptive basal ("rhizoid") pole (56). It is not known which of these changes causally relate to the orientation of the apical-basal axis. The axis becomes fixed only just before germination, and this stage is marked by F-granules, Golgi-derived vesicles that accumulate at the basal cortex and are thought to deposit cell wall material (80). Inhibition of actin polymerization and Golgi transport block axis fixation, and also the deposition of basal-specific cell wall material. Thus, vesicle transport, most likely mediated by the actin cytoskeleton, serves to fix the main axis of polarity, possibly through one or several secreted cell wall components.

Two processes occur after axis fixation: Polar growth initiates from the marked pole, and the spindle is aligned with the apical-basal axis by the preferential stabilization of centrosome-derived microtubules at the rhizoid pole, such that the first division occurs perpendicular to the main axis. It is possible that the basally localized actin patch could play a role in this stabilization, but more research is needed to clarify this issue. If actin patches play a role, an interesting analogy emerges with the involvement of actin in spindle rotation in *C. elegans* blastomeres (112). Interestingly, in *Fucus*, randomly oriented cell divisions could occur when a Golgi transport blocker was transiently added, whereas the polarity cue remained capable of instructing polar growth (88). Thus, spindle orientation and cell division can be uncoupled from the initial cue that polarizes the cell. Hence, the Golgi transport process appears to serve two purposes: fixation of initial polar information, and alignment of the spindle. Important questions are which, if any, of the identified factors is required for the asymmetric cell division to occur, which cell wall component(s) specify the apical and basal cell types, and whether these are already laid down in the zygotic cell wall.

Overall, the first asymmetric cell division in *Fucus* appears to adhere to the general principle of intrinsically determined asymmetric division, in that a cortical target site initiates the segregation (or, in this case, deposition) of intrinsic factors as well as spindle rotation. The intriguing possibility that the intrinsic factors are not transcription factors, like yeast Ash1, nor membrane proteins like *Drosophila* Numb, but secreted cell wall molecules, may add another item to the list of cell fate segregating mechanisms.

Arabidopsis: Vesicle Transport Again?

The division of the higher plant zygote produces the apical daughter cell, giving rise to the majority of the later embryo, and the basal daughter that forms the extraembryonic suspensor. Which factors determine cell fate of the two zygotic daughter cells in angiosperms? Genes whose mutant phenotype adheres to the criteria for intrinsic factors involved in cell division (i.e. two daughters of one type upon loss-of-function, and two daughters of opposite type upon gain-of-function) have not been identified so far. Moreover, the analysis of *twin* (*twn*) mutants in *Arabidopsis* has demonstrated that possible fate differences established after the first zygotic division are reversible. In *twn* mutants, a second embryo develops from the basal cell-derived suspensor region (109, 115). However, the *TWN* genes do not encode cell fate–determining factors. For example, *twn2* mutants display altered expression of a valyl-tRNA-synthase, which arrests development of the apical cell progeny. One model states that the apical cell (progeny) normally suppresses the capacity of the basal cell (progeny) to form an embryo, and that the basal cell will only express this potential when this control is absent (Figure 7) (109). An important corollary of this model is that some *twn*-like mutants could be in genes directly involved in this control, but these have not yet been reported.

Even though the determinants that differentiate apical and basal cell fates are still unknown, the histological asymmetry of the zygotic division allows one to ask which factors orient it. The apical-basal axis is aligned with the micropylar-chalazal axis of the female gametophyte (26), suggesting that positional cues from the haploid phase of the life cycle may guide zygotic asymmetry. Furthermore, the oocyte is frequently polarized prior to fertilization (74). Nevertheless, only sparse genetic evidence has surfaced in favor of maternal control of zygotic polarity (e.g. 81), but it should be noted that no systematic screen for maternal-effect mutations has been reported in plants to date. Tissue- and cell-culture studies reveal that plant embryos can develop in the absence of maternal tissue. These studies can be taken to suggest that maternal cues are also not required for the asymmetric division of the zygote, but it is difficult to assess their implications until more is known about the mechanism of in vitro embryogenesis and its relation to zygotic embryogenesis.

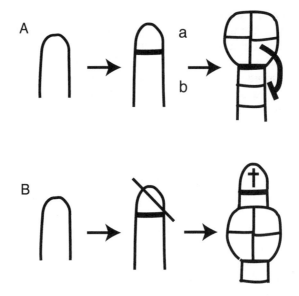

Figure 7   *Arabidopsis* embryogenesis. Zygote, one-cell, and octant stage are shown. *A*. Model for the maintenance/determination of suspensor identity by a negative signal from the embryo proper. *B*. In *twn* mutants, the signal is lacking, allowing the formation of secondary embryos.

Although the cues that orient the asymmetric cell division of the zygote are so far unknown, the *GNOM* gene (also known as *EMB30*) has emerged as a candidate for a role in axis fixation. In *gnom* mutants, the first division of the zygote is frequently skewed or symmetric, and early apical marker genes such as *LTP1* may be expressed with reversed polarity (i.e. in the basal region) later on (66, 109a). Later divisions in *gnom* mutants are also abnormal, indicating that the GNOM protein is a part of a general machinery for correct orientation of cell division. The *GNOM* gene encodes a protein homologous to yeast guanine nucleotide exchange factors involved in vesicle transport to the Golgi network (13, 70, 76, 91). It is tempting to contemplate similarities between a requirement for vesicle transport in axis fixation of higher plants, and the involvement of Golgi-derived vesicles in the same process in algae. More work is needed to assess whether both higher and lower plants do indeed utilize a similar mechanism to stabilize the axis for the asymmetric division of the zygote.

In *Arabidopsis*, the zygotic nucleus is positioned in the subapical region and the subsequent division produces a smaller apical and a larger basal cell. Is this difference in size instrumental in determining cell fate, as it appears to be in the green alga *Volvox* and perhaps also in *Bacillus* (see above)? A survey of the first zygotic cell division in higher plants provides circumstantial evidence against this view. The size difference of the apical and basal cells is absent or opposite in a significant number of angiosperms, indicating that it is not of general importance for the segregation of cell fates (94).

Taken together, much of the control of the first zygotic cell division in higher plants remains unknown. It is not evident whether intrinsic or extrinsic cues are utilized, although pre-existing polarities suggest the presence of intrinsic mechanisms. However, the flexibility in fate of the zygotic daughters in *twn* mutants strongly suggests that a simple scenario with a differentially segregated intrinsic fate determinant is not sufficient to explain the different fate of the two daughters: Subsequent signaling events of unknown nature can modify cell fates later on. Identification of components involved in the zygotic division of lower plants, and extension of these findings to higher plants can be one way to address the many open questions. Another approach may be a rigorous genetic screen to define early steps in zygote development, perhaps aided by novel tools like early marker genes.

RADIAL PATTERNING OF THE ROOT

In the root meristem, all initial cells undergo stem cell–like asymmetric divisions as they regenerate themselves and produce a daughter that differentiates. Some daughter cells go through additional asymmetric divisions when they divide to form the precursors of different cell lineages. In *Arabidopsis*, with its regular cell lineages, these divisions can be followed with ease (25a). An example is the cortex/endodermal initial cell (CEI) that gives rise to the cortex and endodermal lineages. In the *Arabidopsis* root meristem, this cell divides first anticlinally to regenerate an initial with the same stem cell properties and a daughter cell that divides asymmetrically in a periclinal orientation to form cells of two types, cortex and endodermis (Figure 8A).

Candidate genes encoding determinants for the cortical/endodermal fate should, in the ideal case, lead to symmetric cell divisions in the ground tissue. However, mutations in the *SHORT-ROOT* and *SCARECROW* genes, which affect cell fate in the ground tissue, concomitantly result in the loss of the periclinal asymmetric division of the CEI. The fate of the daughter cell differs for the two mutants. In *short-root* (*shr*) the descendants of the resulting cell lack endodermal features, suggesting that SHR is responsible for initiating or maintaining the pathway leading to the endodermal cell fate (6). In *scarecrow* (*scr*), the remaining cell layer has differentiated features of both cortex and endodermis, indicating that SCR is not required for cell fate, per se, but rather plays a role in effecting the asymmetric division (24). Consistent with these interpretations is the result of combining *shr* with *fass* in which there are supernumerary ground tissue layers. The *shr,fass* double mutant contained no

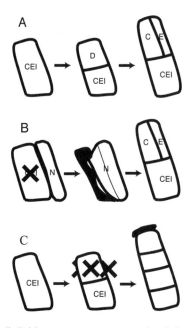

Figure 8 Asymmetric cell division to generate cortex and endodermis in *Arabidopsis*. CEI, cortical-endodermal initial; D, daughter of the initial; E, endodermis; N, neigboring cell. Crosses represent ablations, and black regions are cell corpses. *A.* Formation of the daughter cell from the initial, and asymmetric division to generate cortical and endodermal progenitor cells. *B.* Ablation of the CEI cell leads to replacement from a neighboring cell layer, demonstrating that CEI fate is position-dependent. *C.* Asymmetric division is controlled by more mature D cells.

identifiable endodermal tissue, confirming that SHR is indeed required for the formation of endodermis (85).

Isolation and characterization of the *SCARECROW* gene indicates that it may act as a transcription factor (24). Its expression pattern was particularly revealing. It is expressed in the CEI prior to the first division, then in only one of the two daughter cells resulting from the second division. This would be consistent with its being required for the asymmetric division. Although there is no direct evidence to date as to whether the expression of *SCR* in only the endodermal daughter cell is due to partitioning, indirect evidence argues against it. The *SCR* promoter directs marker gene expression in a pattern identical to that of the endogenous RNA, indicating that transcriptional regulation is sufficient to account for the endodermal daughter-specific expression pattern (24).

How might SHR and SCR function to regulate the asymmetric division? A simple model is that SHR is or produces an endodermal determinant, which, in turn, induces SCR to effect the asymmetric cell division. This model predicts that: (*a*) *shr* should be epistatic to *scr*; if SHR is required to induce SCR activity, then in the absence of SHR it should not matter whether SCR is present; (*b*) SCR activity may be induced by SHR at either a transcriptional or posttranscriptional level. Molecular analysis should enable one to untangle the involvement of *SCR* and *SHR* in determination of cell fate and in cell division.

There is evidence that at least parts of the process that regulates a specific asymmetric division in roots are conserved in shoots. Mutations in *SCR* and *SHR* result in the loss of a normal "starch sheath," a layer occupying an analogous position to the root endodermis, surrounding the vascular tissue in the shoot (34). Thus, these genes appear to be generally required to define ground tissue cell layer(s) around the vascular bundle. Analysis of the ontogeny of shoot-derived ground tissue layers and concomitant expression analysis of *SCR* and of *SHR* should help to determine if similar asymmetric divisions occur in shoot and root.

What orients asymmetric cell divisions in the root meristem? The regularity of cell lineages like those giving rise to cortex and endodermis in the *Arabidopsis* root could be taken as an indication that cell fate is determined by intrinsic factors. Laser ablation of cells within the root was used to test this hypothesis. Contrary to expectations, there was no evidence for cell lineage irreversibly committing a cell to a particular fate. Rather, when any one of the initial cells or their immediate descendants was ablated, a neighboring cell of a different lineage would expand into the empty space, then divide, and one of the daughters would take on the fate of the ablated cell (Figure 8*B*) (107). Ruling out lineage leaves only position as a source of cell fate information. Further ablations revealed a possible source of the information required for at least one type of asymmetric division. When three adjacent cells that were the upper daughters of the first division of the CEI were ablated, the middle initial was unable to make the asymmetric periclinal division but was able to continue the anticlinal divisions (Figure 8*C*). This indicated that information from more mature cells above the initial was required for the cell to perform the asymmetric division to separate the cortical and endodermal fates. Analysis of *SHR* and *SCR* mutants had shown that when these asymmetric divisions were defective in the root tip, corresponding divisions were defective in the complete embryonic axis. Combining the ablation and genetic results led to the hypothesis that an embryonic pre-pattern instructs asymmetric divisions within the root meristem in a "top down" direction (85, 107). As in the EMS/P2 interaction in *C. elegans*, an extrinsic signal may result in the polarization of an internal determinant. *SCR* expression remains on in the entire endodermal lineage (24), which suggests the possibility that SCR could be involved in "top-down" transfer of information. This is not necessarily inconsistent with its role as a transcription factor given the clues for non-cell autonomous behavior of the

KNOTTED homeobox protein (62a) and of the transcription factors GLOBOSA and DEFICIENS (75b). Outstanding questions now are how signaling events of whatever nature can direct the separation of the two cell fates and the orientation of cell division.

STEM CELL DIVISIONS IN THE ROOT MERISTEM

Meristems maintain themselves throughout plant development by retaining a stem cell reservoir, a pool of relatively undifferentiated cells. Cells in the *Arabidopsis* root meristem divide with sufficient regularity to localize the stem cells ("initials") for the different cell lineages, which surround the mitotically almost inactive quiescent center (QC) (25a). Stem cells perform asymmetric cell divisions, as they produce one daughter that will proceed to differentiate, while the other daughter remains a stem cell. The daughters that retain stem cell characteristics are those that are in closest proximity to the QC (Figure 9A). This spatial relationship suggests a function for the QC in the maintenance of stem cells, and thus the influence of QC cells on the surrounding initials was investigated by laser ablation studies (107). When single QC cells were ablated, contacting initial cells of the columella root cap and the cortex/endodermis lost the stem cell status and proceeded with the activity of the daughter cell (Figure 9B). This defect in initial cells contacting eliminated QC cells could not be rescued by intact QC cells, which were at a distance of a few microns. It was concluded that a short-range signal of the QC maintained the stem cell status of cells in its immediate proximity. Phrased in more general terms, an extrinsic cue from the QC biases the fate of the daughters of the asymmetric stem cell divisions. Identical ablation experiments in mutants that lack postembryonic cell division in the root showed that the QC is required in these mutants as well to maintain the initial status. This indicated that cell

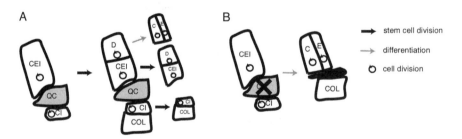

Figure 9   The quiescent center (QC) controls stem cell identity of the cortical-endodermal initial (CEI) and of the columella initial (CI). *A.* Activities of the stem cells contacting the QC. *B.* Result of QC ablation.

differentiation status is controlled directly without being a result of control of cell division.

No determinants are yet identified that specify the stem cell status, and promoter trap or enhancer trap screens for genes expressed specifically in stem cells may be one way to proceed. What signaling pathways can be expected for the maintenance of stem cells? The apparent contact-dependency of the signaling between the QC and the surrounding stem cells reminds one of cell-contact–dependent signaling involving tethered ligands and their transmembrane receptors, such as *Drosophila* Notch and its nematode counterpart GLP1. An important further parallel exists in the shoot apical meristem, where a distinct stem cell population cannot be recognized by anatomical criteria, but where the *CLV* genes are required to regulate the balance between differentiated and undifferentiated cells. *CLV1* encodes a putative transmembrane receptor with an extracellular leucine-rich-repeat domain and an intracellular kinase (19). The *CLV3* gene shows close genetic interactions with *CLV1* and its product is a good candidate to act as a CLV1 ligand (18). Interestingly, a distinct class of *CLV1* homologues is expressed in the root meristem (R Heidstra & B Scheres, unpublished data), and it will be interesting to investigate whether members of this small gene family are candidates to be a part of the QC signaling pathway.

ASYMMETRIC DIVISION DURING POLLEN DEVELOPMENT

During pollen development, each of the products of meiosis undergoes an asymmetric division. The division of the microspore, known as pollen mitosis I (PMI), produces a larger vegetative cell (VC) and a smaller generative cell (GC) that becomes completely engulfed in the VC cell cytoplasm (Figure 10) (reviewed in 104). This process is reminiscent of engulfment of the smaller forespore by the larger mother cell in *B. subtilis*. The GC will later divide to form the two sperm cells, while the VC produces the pollen tube (67).

The two different cell types resulting from PMI of isolated, in vitro cultured microspores are fully functional (29). This provides evidence that the asymmetry of the division is the consequence of the partitioning of intrinsic factors as

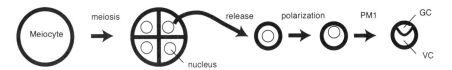

Figure 10   Pollen mitosis is an asymmetric cell division. GC, generative cell; VC, vegetative cell. Where is this called out in the text?

opposed to post-divisional cell-cell contact. The *gemini* (*gem*) mutation results in various division defects, including two-cell pollen in which both cells express VC markers (104). The symmetric division in *gem* does not occur precociously. This may suggest that in *gem* the mutation is in a cell fate determinant. However, a similar phenotype is observed when microspores are cultured in low levels of colchicine, which is expected to perturb the machinery for asymmetric cell division itself. More detailed analysis of the division process in the *gem* mutant will be required to clarify its role in the asymmetric cell division.

What orients the asymmetric cell division in microspores? Evidence for internal polarization of the microspore prior to the asymmetric division comes from histological observations, which reveal an unequal distribution of cytoplasm and organelles, with the lion's share going to the vegetative cell after division. The spindle is also polarized, forming a blunt end at the GC pole and a sharp end at the VC pole (45a). The polarized aspect of the spindle may be a direct result of the localization of the microspore nucleus very close to the wall closest to the future GC. Changing nuclear localization through cold, caffeine treatment, or centrifugation can result in symmetric divisions (99a). Treatment of isolated microspores with the microtubule inhibitor colchicine blocks cell division. The resulting cell has the phenotypic properties of a vegetative cell, indicating that VC differentiation does not require the presence of a generative cell (29). Treatment with lower doses of colchicine results in the occasional occurrence of symmetric divisions in which both daughters express a VC-specific marker. This provides evidence that intact microtubules are required for the asymmetric division. Remarkably, the PMI asymmetric division can still occur in *tetraspore* and in the allelic *stud* mutants that fail to separate the four products of meiosis (95, 49). This suggests that early cues have demarcated multiple domains at the tetrad stage (possibly dependent upon the cytoskeleton), and that these domains are still present in the *tetraspore* mutant (95). The nature of these cues remains to be elucidated.

In addition to *gemini*, there are other mutants in which the asymmetric division is defective. In *sidecar pollen* (*scp*) some pollen grains are formed that contain two VC and one GC (16). Analysis of the division process indicates that the microspore first divides symmetrically to produce two microspores, of which one can follow the normal asymmetric division process while the other differentiates to become a VC. This suggests that the *SCP* gene product may be involved in partitioning of a cytoplasmic determinant such that in the mutant, random distribution occurs in the first division and the daughter cell that happens to get more then goes through the normal asymmetric division process. Another model would have SCP involved in preventing precocious cell division. In the mutant, premature division would result in symmetric fates, but the normal processes that ready the cell for asymmetric division would still take place, allowing for a later asymmetric division in one of the two cells GC (16).

Pollen grains resulting from the *solo* mutation have only a single cell that expresses VC markers (29). The role of the corresponding gene in asymmetric cell division is not yet clear, but it is interesting to note the similarity between *solo* and *shr* mutants: Both block an asymmetric cell division and result in one of the two possible cell types.

Collectively, the genetic and experimental data indicate that VC is the default state (104). There is circumstantial evidence that the difference in cell fate between the vegetative and generative cells is dependent upon an intrinsic factor whose partitioning requires intact microtubules. Whether this unidentified factor is an activator or a repressor of cell fate is unknown, as is the machinery that partitions it.

THE FORMATION OF STOMATA

The formation of stomata, pairs of guard cells that regulate gas exchange, is accompanied by several asymmetric cell divisions that yield cells of unequal size with different fates, as recently reviewed by Larkin et al (57). In monocotyledonous plants, one asymmetric division yields guard mother cells (GMCs) that divide symmetrically to produce guard cell pairs; another asymmetric cell division is subsequently induced in epidermal cells of the adjacent file to give subsidiary cells (Figure 11A). In dicots, a stem cell–like sequence of asymmetric divisions gives rise to stomata. A protodermal cell divides asymetrically to produce a "meristemoid," a stem cell–like stomatal initial cell capable of generating either epidermal pavement cells or GMCs. The GMCs produce guard cell pairs by symmetric divisions (Figure 11B). The orientation of several of these asymmetric divisions is guided by external cues. First, in monocots, the asymmetric divisions are perpendicular to the proximo-distal axis of leaf polarity, with the GMC as the distal daughter (101). Second, in dicots, the first asymmetric division that produces a meristemoid appears to be randomly oriented, but further asymmetric divisions are oriented with respect to developing GMCs (35). Furthermore, both daughters of a scheduled asymmetric division can become epidermal pavement cells to prevent direct contact of GMCs (52). The nature of the proximo-distal cue in monocots and the GMC-related cues in dicots is unknown. The direct accessibility of the epidermis for experimental interference (see 22), and the possibility of monitoring cell division patterns with epidermal peels (52, 113) may provide tools to investigate the nature of the cues involved.

Genetic analysis in *Arabidopsis* may enable one to identify genes involved in the asymmetric divisions that accompany stomatal development. Although candidate genes encoding determinants have not yet been identified, some mutants that affect stomatal patterning have been described. *four lips* mutants, for example, form guard cell pairs by two instead of one symmetric division of

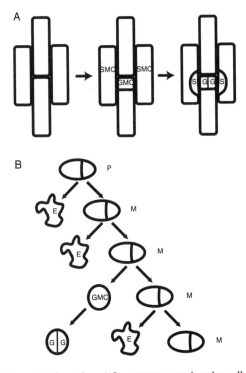

Figure 11 Stomatal complex formation. *A*. In monocots, guard mother cells (GMCs) arise from asymmetric cell divisions oriented along the apical-basal axis, and subsidiary mother cells arise from asymmetric cell divisions that are likely to be induced by the neighboring GMC or its daughters, the guard cells (G). *B*. Repetitive asymmetric divisions retain cells capable of further division, the meristemoids (M). These develop into GMCs that give rise to guard cell pairs or to interstitial epidermal pavement cells (E).

the GMC, and one interpretation is that the corresponding gene is involved in establishment or maintenance of GMC identity (57, 114). While it is uncertain whether this gene will provide an entrance to the mechanisms of asymmetric cell division, promoter/enhancer traps that specifically mark stomatal development should facilitate direct screening for genes involved in asymmetric cell divisions of the epidermis. Analysis of the contribution of intrinsic and extrinsic factors has to await the identification of such genes.

SUMMARY

Although no single asymmetric division has been analyzed with the rigor that has recently been applied to yeast and animal systems, there is mounting evidence that cell fate determination after asymmetric divisions in plants occurs both through the partitioning of cellular determinants and by cellular interactions. The asymmetric division of pollen grains into vegetative and generative cells has all the hallmarks of a strictly intrinsic process. While there is strong evidence that positional information plays an important role in other asymmetric divisions, too little is known to decide to what extent it combines with intrinsic mechanisms. Candidates for intrinsic factors that can be modulated by signaling are the *SCR* and *SHR* gene products, and detailed analysis of their function should provide entrances into at least one asymmetric cell division soon. Overall, it is quite possible that "mix and match" combinations of signaling and intrinsic determinants described in other organisms will also be found in plants.

It is not yet possible to know if most of the signaling will be short distance, cell-to-cell, or longer-distance signaling. To date there is little evidence for a role of the classical plant hormones in regulating specific asymmetric divisions, although this does not rule out that hormonal influences will be found when the different examples of asymmetric cell division are carefully examined. The recent ablation results in *Fucus* indicate a role for factors that travel apoplastically. Alternative signaling pathways might use plasma membrane receptors with tethered diffusible ligands or factors transported through plasmodesmata. The evidence from laser ablation of root meristem cells indicates a role for continuous signaling to maintain the separation of cell fate that is first initiated in the embryo. More generally, the remarkable plasticity of plant cells suggests that once an asymmetric division has occurred the resulting cell fates probably also require active maintenance. In floral development, the gene *CURLY LEAF*, with homology to the polycomb group of cell fate maintenance factors in *Drosophila*, has been shown to provide such a function (39).

To further our understanding of the mechanisms that regulate plant asymmetric divisions, a first step will be to analyze in depth a few of those currently under investigation. We feel that two requirements should be met to face this task.

First, the paradigms that have emerged from yeast and animal systems need to be appreciated by our community of plant scientists. As we have aimed to demonstrate here, these paradigms provide an overall conceptual framework for classification and the interpretation of experimental results, while leaving sufficient space for an open view towards the peculiarities of plant development that will undoubtedly surface.

Second, improvement in techniques for basic cell biological analyses is called for. We need better ways of visualizing plant cell divisions in planta, for example, by using markers that allow the visualization of structures such as the PPB, spindle, and phragmoplast. A step in this direction is the fusion of Green Fluorescence Protein (GFP) to a dynamin-related protein that localizes to the phragmoplast (40). To ascertain the extent to which signaling plays a role in asymmetric divisions, it would be extremely useful to be able to change the location of a plant cell. This may not be as difficult as once imagined, if in vivo cell-specific marking can be successfully combined with micromanipulation and laser microsurgery.

Visit the *Annual Reviews* home page at
http://www.AnnualReviews.org

Literature Cited

1. Arigoni F, Pogliano K, Webb CD, Stragier P, Losick R. 1995. Localization of protein implicated in establishment of cell type to sites of asymmetric division. *Science* 270:637–40

2. Assaad FF, Mayer U, Lukowitz W, Jürgens G. 1997. Cytokinesis in somatic plant cells. *Plant Physiol. Biochem.* 35:177–84

3. Assaad FF, Mayer U, Wanner G, Jurgens G. 1996. The KEULE gene is involved in cytokinesis in Arabidopsis. *Mol. Gen. Genet.* 253:267–77

4. Austin J, Kimble J. 1987. glp-1 is required in the germ line for regulation of the decision between mitosis and meiosis in *C. elegans*. *Cell* 51:589–99

5. Baluska F, Vitha S, Barlow PW, Volkmann D. 1997. Rearrangements of F-actin arrays in growing cells in intact maize root apex tissues: A major developmental switch occurs in the postmitotic transition region. *Eur. J. Cell Biol.* 72:113–21

6. Benfey PN, Linstead PJ, Roberts K, Schiefelbein JW, Hauser M-T, Aeschbacher RA. 1993. Root development in Arabidopsis: four mutants with dramatically altered root morphogenesis. *Development* 119:53–70

7. Berger F, Taylor A, Brownlee C. 1994. Cell fate determination by the cell wall in early *Fucus* development. *Science* 263:1421–23

8. Bobola N, Jansen R-P, Shin TH, Nasmyth K. 1996. Asymmetric accumulation of Ash1p in postanaphase nuclei depends on a myosin and restricts yeast mating-type switching to mother cells. *Cell* 84:699–709

9. Bouget F-Y, Berger F, Brownlee C. 1998. Position dependent control of cell fate in the *Fucus* embryo: role of intercellular communication. *Development* 125:1999–2008

10. Bowerman B. 1995. Determinants of blastomere identity in the early *C. elegans* embryo. *BioEssays* 17:405–14

11. Bowerman B, Tax FE, Thomas JT, Priess JR. 1992. Identification of cell interactions involved in development of the bilaterally symmetrical intestinal valve cells during embryogenesis in *C. elegans*. *Development* 116:1113–22

12. Broadus J, Doe CQ. 1997. Extrinsic cues, intrinsic cues and microfilaments regulate asymmetric protein localization in *Drosophila* neuroblasts. *Curr. Biol.* 7:827–35

13. Bush M, Mayer U, Jürgens G. 1996. Molecular analysis of the *Arabidopsis* pattern formation gene GNOM: gene structure and intragenic complementation. *Mol. Gen. Genet.* 250: 681–91

14. Chant J. 1996. Septin scaffolds and cleavage planes in *Saccharomyces*. *Cell* 84:187–90

15. Chant J, Mischke M, Mitchell E, Herskowitz I, Pringle JR. 1995. Role of Bud3p in producing the axial budding pattern of yeast. *J. Cell Biol.* 129:767–78

16. Chen YC, McCormick S. 1996. Sidecar pollen, an *Arabidopsis thaliana* male gametophytic mutant with aberrant cell divisions during pollen development. *Development* 122:3243–53

17. Chevrier V, Komesli S, Schmit AC, Vantard M, Lambert AM, Job D. 1992. A monoclonal antibody, raised against mammalian centrosomes and screened by recognition of plant microtubule organizing centers, identifies a pericentriolar component in different cell types. *J. Cell Sci.* 101:823–35

18. Clark SE, Running MP, Meyerowitz EM. 1995. CLAVATA3 is a specific regulator of shoot and floral meristem development affecting the same processes as CLAVATA1. *Development* 121:2057–67

19. Clark SE, Williams RW, Meyerowitz EM. 1997. The *CLAVATA1* gene encodes a putative receptor kinase that controls shoot and floral meristem size in *Arabidopsis*. *Cell* 89:575–85
20. Cleary AL, Gunning BES, Wasteneys GO, Hepler PK. 1992. Microtubule and F-actin dynamics at the division site in living *Tradescantia* stamen hair cells. *J. Cell Sci.* 103:977–88
21. Cleary AL, Smith L. 1998. The *Tangled1* gene is required for spatial control of sytoskeletal arrays associated with cell division during maize leaf development. *Plant Cell* 10:1875–88
21a. Colasanti J, Cho S-O, Wick S, Sundaresan V. 1993. Localization of the functional p34cdc2 homolog of maize in root tip and stomatal complex cells: association with predicted division sites. *Plant Cell* 5:1101–11
22. Croxdale J, Smith J, Yandell B, Johnson JB. 1992. Stomatal patterning in *Tradescantia*: an evaluation of the cell lineage theory. *Dev. Biol.* 149:158–67
23. Cyr RJ. 1994. Microtubules in plant morphogenesis: role of the cortical array. *Annu. Rev. Cell Biol.* 10:153–80
24. Di Laurenzio L, Wysocka-Diller J, Malamy JE, Pysh L, Helariutta Y, et al. 1996. The *SCARECROW* gene regulates an asymmetric cell division that is essential for generating the radial organization of the Arabidopsis root. *Cell* 86:423–33
25. Doe CQ, Chu-Lagraff Q, Wright DM, Scott MP. 1991. The *prospero* gene specifies cell fates in the Drosophila central nervous system. *Cell* 65:451–64
25a. Dolan L, Janmaat K, Willemsen V, Linstead P, Poethig S, et al. 1993. Cellular organisation of the Arabidopsis root. *Development* 119:71–84
26. Drews GN, Lee D, Christensen CA. 1998. Genetic analysis of female gametophyte development and function. *Plant Cell* 10:5–17
27. Duncan L, Alper S, Arigoni F, Losick R, Stragier P. 1995. Activation of cell-specific transcription by a serine phosphatase at the site of asymmetric division. *Science* 270:641–44
28. Dunn G, Mandelstam J. 1977. Cell polarity in *Bacillus subtilis*: effect of growth conditions on spore positions in sister cells. *J. Gen. Microbiol.* 103:201–5
29. Eady C, Lindsey K, Twell D. 1995. The significance of microspore division and division symmetry for vegetative cell-specific transcription and generative cell differentiation. *Plant Cell* 7:65–74
30. Falconer MM, Donalson G, Seagull RW. 1988. MTOCs in higher plant cells. An immunofluorescent study of microtubule assembly sites following depolymerization by APM. *Protoplasma* 144:46–55
31. Feucht A, Magnin T, Yudkin MD, Errington J. 1996. Bifunctional protein required for asymmetric cell division and cell-specific transcription in *Bacillus subtilis*. *Genes Dev.* 10:794–803
32. Fowler JE, Quatrano RS. 1997. Plant cell morphogenesis: plasma membrane interactions with the cytoskeleton and cell wall. *Annu. Rev. Cell Dev. Biol.* 13:697–743
33. Frise E, Knoblich JA, Younger-Shepherd S, Jan LY, Jan YN. 1996. The *Drosophila* Numb protein inhibits signaling of the Notch receptor during cell-cell interaction in sensory organ lineage. *Proc. Natl. Acad. Sci. USA* 93:11925–32
34. Fukaki H, Wysocka-Diller J, Kato T, Fujisawa H, Benfey PN, Tasaka M. 1998. Genetic evidence that the endodermis is essential for shoot gravitropism in *Arabidopsis thaliana*. *Plant J.* 14:425–30
35. Galatis B, Mitrakos K. 1979. On the differential divisions and preprophase microtubule bands involved in the development of stomata of *Vigna sinesis* L. *J. Cell Sci.* 37:11–37
36. Gallagher K, Smith LG. 1997. Asymmetric cell division and cell fate in plants. *Curr. Opin. Cell Biol.* 9:842–48
37. Gho M, Schweisguth F. 1998. Frizzled signaling controls orientation of asymmetric sense organ precursor cell divisions in *Drosophila*. *Nature* 393:178–81
38. Goldstein B. 1993. Establishment of gut fate in the E lineage of *C. elegans*: the roles of lineage-dependent mechanisms and cell interactions. *Development* 118:1267–77
39. Goodrich J, Puangsomlee P, Martin M, Long D, Meyerowitz EM, Coupland G. 1997. A Polycomb-group gene regulates homeotic gene expression in Arabidopsis. *Nature* 386:44–51
40. Gu X, Verma DP. 1997. Dynamics of phragmoplastin in living cells during cell plate formation and uncoupling of cell elongation from the plane of cell division. *Plant Cell* 9:157–69
41. Gunning BES, Wick SM. 1985. Preprophase bands, phragmoplasts and spatial control of cytokinesis. *J. Cell Sci. Suppl.* 2:157–79
42. Guo M, Jan LY, Jan YN. 1996. Control

of daughter cell fates during asymmetric division; interaction of Numb and Notch. *Neuron* 17:27–41

43. Guo S, Kemphues KJ. 1996. Molecular genetics of asymmetric cleavage in the early *Caenorhabditis elegans* embryo. *Curr. Opin. Genet. Dev.* 6:408–15

44. Han M. 1997. Gut reaction to Wnt in worms. *Cell* 90:581–84

44a. Hartenstein V, Posakony JW. 1990. A dual function of the *Notch* gene in *Drosophila sensillum* development. *Dev. Biol.* 142:13–30

45. Hemerly A, Engler J de A, Bergounioux C, Van Montagu M, Engler G, et al. 1995. Dominant negative mutants of the Cdc2 kinase uncouple cell division from iterative plant development. *EMBO J.* 14:3925–36

45a. Heslop-Harrison J. 1971. Wall pattern formation in angiosperm microsporogenesis. *Symp. Soc. Exp. Biol.* 25:277–300

46. Hirata J, Nakagoshi H, Nabeshima Y, Matsuzaki F. 1995. Asymmetric segregation of the homeodomain protein Prospero during *Drosophila* development. *Nature* 377:627–30

47. Hitchens AD. 1975. Polarized relationship of bacterial spore loci to the 'old' and 'new' ends of sporangia. *J. Bacteriol.* 121:518–23

48. Horvitz HR, Herskowitz I. 1992. Mechanisms of asymmetric cell division: two Bs or not two Bs, that is the question. *Cell* 68, 237–55

49. Hulskamp M, Parekh NS, Grini P, Schneitz K, Zimmermann I, et al. 1997. The STUD gene is required for male-specific cytokinesis after telophase II of meiosis in *Arabidopsis thaliana. Dev. Biol.* 187:114–41

50. Ikeshima-Kataoka H, Skeath JB, Nabeshima Y, Doe CQ, Matsuzaki F. 1997. Miranda directs Prospero to a daughter cell during *Drosophila* asymmetric divisions. *Nature* 390:625–29

51. Jansen R-P, Dowzer C, Michaelis C, Galova M, Nasmyth K. 1996. Mother cell-specific *HO* expression depends on the unconventional myosin Myo4p and other cytoplasmic proteins. *Cell* 84:687–97

51a. Jürgens G. 1995. Axis formation in plant embryogenesis: cues and clues. *Cell* 81:467–70

51b. Jürgens G, Grebe M, Steinmann T. 1997. Establishment of cell polarity during early plant development. *Curr. Opin. Cell Biol.* 9:849–52

51c. Jürgens G, Mayer U. 1994. Arabidopsis. In *EMBRYOS. Colour Atlas of Devel-*

opment, ed. J Bard, pp. 7–21. London: Wolfe Publ.

52. Kagan ML, Novoplansky N, Sachs T. 1992. Variable cell lineages form the functional pea epidermis. *Ann. Bot.* 69:303–12

53. Kirk MM, Ransic A, McRae SE, Kirk DL. 1993. The relationship between cell size and cell fate in *Volvox carteri. J. Cell Biol.* 123:191–208

53a. Knoblich JA, Jan LY, Jan YN. 1995. Asymmetric segregation of Numb and Prospero during cell division. *Nature* 377:624–26

54. Knoblich JA, Jan LY, Jan YN. 1997. The N terminus of *Drosophila* Numb protein directs membrane association and actin-dependent asymmetric localization. *Proc. Natl. Acad. Sci. USA* 94:13005–10

55. Kraut R, Chia W, Jan LY, Jan YN, Knoblich JA. 1996. Role of *inscuteable* in orienting asymmetric cell divisions in *Drosophila. Nature* 383:50–55

56. Kropf DL. 1997. Induction of polarity in fucoid zygotes. *Plant Cell* 9:1011–20

57. Larkin JC, Marks MD, Nadeau J, Sack F. 1997. Epidermal cell fate and patterning in leaves. *Plant Cell* 9:1109–20

58. Lauber MH, Waizenegger I, Steinmann T, Schwarz H, Mayer U, et al. 1997. The Arabidopsis KNOLLE protein is a cytokinesis-specific syntaxin. *J. Cell Biol.* 139(6):1485–93

59. Lin R, Hill RJ, Priess JR. 1998. POP-1 and anterior-posterior fate decisions in *C. elegans* embryos. *Cell* 92:229–39

60. Liu B, Palevitz BA. 1992. Organization of cortical microfilaments in dividing root cells. *Cell Motil. Cytoskel.* 23:252–64

61. Liu C, Johnson S, Wang TL. 1995. *Cyd*, a mutant of pea that alters embryo morphology is defective in cytokinesis. *Dev. Genet.* 16:321–31

61a. Lloyd CW. 1991. Cytoskeletal elements of the phragmosome establish the division plane in vacuolated higher plant cells. In *The Cytoskeletal Basis of Plant Growth and Form*, ed. CW Lloyd, pp. 245–57. London: Academic

62. Long RM, Singer RH, Meng X, Gonzalez I, Nasmyth K, Jansen R-P. 1997. Mating type switching in yeast controlled by asymmetric localization of *ASH1* RNA. *Science* 277:383–87

62a. Lucas WJ, Bouche-Pillon S, Jackson DP, Nguyen L, Baker L, et al. 1995. Selective trafficking of KNOTTED1 homeodomain protein and its mRNA through plasmodesmata. *Science* 270:1980–83

63. Lukowitz W, Mayer U, Jürgens G. 1996. Cytokinesis in the Arabidopsis embryo involves the syntaxin-related KNOLLE gene product. *Cell* 84:61–71

64. Mango SE, Thorpe CJ, Martin PR, Chamberlain SH, Bowerman B. 1994. Two maternal genes, apx-1 and pie-1, are required to distinguish the fates of equivalent blastomeres in the early *Caenorhabditis elegans* embryo. *Development* 120:2305–15

65. Mansfield SG, Briarty LG. 1991. Early embryogenesis in *Arabidopsis thaliana*. II. The developing embryo. *Can. J. Bot.* 69:461–76

66. Mayer U, Büttner G, Jürgens G. 1993. Apical-basal pattern formation in the *Arabidopsis* embryo: studies on the role of the *gnom* gene. *Development* 117:149–62

67. McCormick S. 1993. Male gametophyte development. *Plant Cell* 5:1265–75

68. Mello CC, Draper B, Priess JR. 1994. The maternal genes apx-1 and glp-1 and establishment of dorsal-ventral polarity in the early *C. elegans* embryo. *Cell* 77:95–106

69. Mickey KM, Mello CC, Montgomery MK, Fire A, Priess JR. 1996. An inductive interaction in 4–cell stage *C. elegans* embryos involves APX-1 expression in the signalling cell. *Development* 122:1791–98

69a. Mineyuki Y, Marc J, Palevitz BA. 1991. Relationship between the preprophase band, nucleus and spindle in dividing Allium cotyledon cells. *J. Plant Physiol.* 138:640–49

70. Mossessova E, Gulbis JM, Goldberg J. 1998. Structure of the guanine nucleotide exchange factor Sec7 domain of human Arno and analysis of the interaction with ARF GTPase. *Cell* 92:415–23

71. Murata T, Wada M. 1991. Effects of centrifugation on preprophase band formation in *Adiantum* protonemata. *Planta* 183:391–98

72. Natesh S, Rau MA. 1984. The embryo. In *Embryology of Angiosperms*, ed. BM Johri. Berlin: Springer-Verlag

73. Nasmyth KA. 1983. Molecular analysis of a cell lineage. *Nature* 302:670–76

74. Nasmyth KA. 1993. Regulating the *HO* endonuclease in yeast. *Curr. Opin. Genet. Dev.* 3:286–94

74a. Oud JL, Nanninga N. 1992. Cell shape, chromosome orientation and the position of the plane of division in *Vicia faba* root cortex cells. *J. Cell Sci.* 103:847–55

74b. Owen HA, Makaroff CA. 1995. Ultrastructure of microsporogenesis and microgametogenesis in *Arabidopsis thaliana* (L.) Heynh. ecotype Wassilewskija (Brassicaceae). *Protoplasma* 185:7–21

75. Palevitz BA. 1986. Division plane determination in guard mother cells of *Allium*: video time-lapse analysis of nuclear movements and phragmoplast rotation in the cortex. *Dev. Biol.* 117:644–54

75a. Palevitz BA, Tiezza A. 1992. Organization, composition and function of the generative cell and sperm cytoskeleton. *Int. Rev. Cytol.* 140:149–85

75b. Perbal MC, Haughn G, Saedler H, Schwarz-Sommer Z. 1996. Non-cell-autonomous function of the Antirrhinum floral homeotic proteins DEFICIENS and GLOBOSA is exerted by their polar cell-to-cell trafficking. *Development* 122:3433–41

76. Peyroche A, Paris S, Jackson CL. 1996. Nucleotide exchange on ARF mediated by yeast Gea1 protein. *Nature* 384:479–84

77. Pickett-Heaps JE, Northcote DH. 1966. Cell division in the formation of the stomatal complex of the young leaves of wheat. *J. Cell Sci.* 1:121–28

77a. Poethig RS. 1987. Clonal analysis of cell lineage patterns in plant development. *Am. J. Bot.* 74:581–94

78. Priess JR, Schnabel H, Schnabel R. 1987. The glp-1 locus and cellular interactions in early *C. elegans* embryos. *Cell* 51:601–11

79. Priess JR, Thomson JN. 1987. Cellular interactions in early *C. elegans* embryos. *Cell* 48:241–50

80. Quatrano RS, Shaw SL. 1997. Role of the cell wall in the determination of cell polarity and the plane of cell division in *Fucus* embryos. *Trends Plant Sci.* 2:15–21

80a. Rappaport R. 1986. Establishment of the mechanism of cytokinesis in animal cells. *Int. Rev. Cytol.* 105:245–81

81. Ray S, Golden T, Ray A. 1996. Maternal effects of the *short integument* mutation on embryo development in *Arabidopsis*. *Dev. Biol.* 180:365–69

82. Rhyu MS, Jan LY, Jan YN. 1994. Asymmetric distribution of Numb protein during division of the sensory organ precursor cell confers distinct fates to daughter cells. *Cell* 76:477–91

83. Rocheleau CE, Downs WD, Lin R, Wittmann C, Bei Y, et al. 1997. Wnt signaling in an APC-related gene specify endoderm in early *C. elegans* embryos. *Cell* 90:707–16

84. Scheres B. 1997. Cell signaling in root development. *Curr. Opin. Genet. Dev.* 7:501–6
85. Scheres B, Di Laurenzio L, Willemsen V, Hauser M-T, Janmaat K, et al. 1995. Mutations affecting the radial organisation of the Arabidopsis root display specific defects throughout the embryonic axis. *Development* 121:53–62
86. Schnabel R, Priess JR. 1997. Specification of cell fates in the early embryo. In *C. elegans* II, ed. DL Riddle, T Blumenthal, BJ Meyer, JR Priess, pp. 361–82. New York: Cold Spring Harbor Lab. Press
87. Shapiro L, Losick R. 1997. Protein localization and cell fate in bacteria. *Science* 276:712–18
88. Shaw SL, Quatrano RS. 1996. The role of targeted secretion in the establishment of cell polarity and the orientation of the division plane in *Fucus* zygotes. *Development* 122:2623–30
89. Shen C-P, Knoblich JA, Chan Y-M, Jiang M-M, Jan LY, Jan YN. 1998. Miranda as a multidomain adapter linking apically localized Inscuteable and basally localized Staufen and Prospero during asymmetric cell division in *Drosophila*. *Genes Dev.* 12:1837–46
90. Shen C-P, Yan LY, Yan YN. 1997. Miranda is required for the asymmetric localization of Prospero during mitosis in Drosophila. *Cell* 90:449–58
91. Shevell D, Leu W-M, Stewart Gillmor S, Xia G, Feldmann K, Chua N-H. 1994. *EMB30* is essential for normal cell division, cell expansion, and cell adhesion in Arabidopsis and encodes a protein that has similarity to Sec7. *Cell* 77:1051–62
92. Shroer TA, Bingham JB, Gill SR. 1996. *Trends Cell Biol.* 6:212–15
93. Sil A, Herskowitz I. 1996. Identification of an asymmetrically localized determinant, Ash1p, required for lineage-specific transcription of the yeast *HO* gene. *Cell* 84:711–22
94. Sivamakrishna D. 1978. Size relationships of apical and basal cell in two-celled embryos in angiosperms. *Can. J. Bot.* 56:1434–39
95. Spielman M, Preuss D, Li FL, Browne WE, Scott RJ, Dickinson HG. 1997. TETRASPORE is required for male meiotic cytokinesis in *Arabidopsis thaliana*. *Development* 124:2645–57
96. Staehelin LA, Hepler PK. 1996. Cytokinesis in higher plants. *Cell* 84:821–24
97. Strathern JN, Herskowitz I. 1979. Asymmetry and directionality in production of new cell types during clonal growth: the switching pattern of homothallic yeast. *Cell* 17:371–81
98. Strathern JN, Klar AJS, Hicks JB, Abraham JA, Ivy JM, et al. 1982. Homothallic switching of yeast mating-type cassettes is initiated by a double strand cut in the MAT locus. *Cell* 31:183–92
99. Strauss E. 1998. How embryos shape up. Meet. Soc. Dev. Biol. *Science* 281:166–67
99a. Terasaka O, Niitsu T. 1987. Unequal cell division and chromatin differentiation in pollen grain cells. I. Centrifugal, cold and caffeine treatments. *Bot. Mag. Tokyo* 100:205–16
100. Thorpe CJ, Schlesinger A, Carter JC, Bowerman B. 1997. Wnt Signaling polarizes an early *C. elegans* blastomere to distinguish endoderm from mesoderm. *Cell* 90:695–705
101. Tomlinson PB. 1974. Development of the stomatal complex as a taxonomic character in the monocotyledons. *Taxon* 23:109–28
102. Torres-Ruiz RA, Jürgens G. 1994. Mutations in the *FASS* gene uncouple pattern formation and morphogenesis in *Arabidopsis* development. *Development* 120:2967–78
103. Traas J, Bellini C, Nacry P, Kronenberger J, Bouchez D, Caboche M. 1995. Normal differentiation patterns in plants lacking microtubular preprophase bands. *Nature* 375:676–77
104. Twell D, Park SK, Lalanne E. 1998. Asymmetric division and cell fate determination in developing pollen. *Trends Genet.* In press
105. Uemura T, Shepherd S, Ackerman L, Jan LY, Jan YN. 1989. *numb*, a gene required in determination of cell fate during sensory organ formation in *Drosophila* embryos. *Cell* 58:349–60
106. Vaessin H, Grell E, Wolff E, Bier E, Jan LY, Jan YN. 1991. *prospero* is expressed in neuronal precursors and encodes a nuclear protein that is involved in the control of axonal outgrowth in *Drosophila*. *Cell* 67:941–53
107. van den Berg C, Willemsen V, Hage W, Weisbeek P, Scheres B. 1995. Cell fate in the Arabidopsis root meristem determined by directional signalling. *Nature* 378:62–65
108. Vantard M, Levilliers N, Hill AM, Adoutte A, Lambert AM. 1990. Incorporation of paramecium axonemal tubulin into higher plant cells reveals functional sites of microtubule assembly.

Proc. Natl. Acad. Sci. USA 87:8825–29

109. Vernon D, Meinke D. 1994. Embryogenic transformation of the suspensor in *twin*, a polyembryonic mutant of *Arabidopsis*. *Dev. Biol.* 165:566–73

109a. Vroeman C, Langeveld S, Mayer U, Ripper G, Jürgens G, et al. 1996. Pattern formation in the Arabidopsis embryo revealed by position-specific lipid transfer protein gene expression. *Plant Cell* 8:783–91

110. Way JC. 1996. The mechanism of bacterial asymmetric cell division. *BioEssays* 18:99–101

111. Weigmann K, Cohen SM, Lehner CF. 1997. Cell cycle progression, growth and patterning in imaginal discs despite inhibition of cell division after inactivation

of Cdc2 kinase. *Development* 124:3555–63

112. White J, Strome S. 1996. Cleavage plane specification in *C. elegans*: How to divide the spoils. *Cell* 84:195–98

113. Williams M, Green P. 1988. Sequential scanning electron microscopy of a growing plant meristem. *Protoplasma* 147:77–79

114. Yang M, Sack FD. 1995. The *too many mouths* and *four lips* mutations affect stomatal production in Arabidopsis. *Plant Cell* 7:2227–39

115. Zhang J, Sommerville CR. 1997. Suspensor-derived polyembryony caused by altered expression of valyl-tRNA-synthase in the *twn2* mutant of *Arabidopsis*. *Proc. Natl. Acad. Sci. USA* 94:7349–55

Annu. Rev. Plant Physiol. Plant Mol. Biol. 1999. 50:539–70

CO$_2$ CONCENTRATING MECHANISMS IN PHOTOSYNTHETIC MICROORGANISMS

Aaron Kaplan and Leonora Reinhold

Department of Plant Sciences, The Hebrew University of Jerusalem, 91904 Jerusalem, Israel; e-mail: Aaronka@vms.huji.ac.il

KEY WORDS: algae, carboxysome, cyanobacteria, photosynthesis, pyrenoid

ABSTRACT

Many microorganisms possess inducible mechanisms that concentrate CO$_2$ at the carboxylation site, compensating for the relatively low affinity of Rubisco for its substrate, and allowing acclimation to a wide range of CO$_2$ concentrations. The organization of the carboxysomes in prokaryotes and of the pyrenoids in eukaryotes, and the presence of membrane mechanisms for inorganic carbon (Ci) transport, are central to the concentrating mechanism. The presence of multiple Ci transporting systems in cyanobacteria has been indicated. Certain genes involved in structural organization, Ci transport and the energization of the latter have been identified. Massive Ci fluxes associated with the CO$_2$-concentrating mechanism have wide-reaching ecological and geochemical implications.

CONTENTS

1040-2519/99/0601-0539$08.00

INTRODUCTION

Rubisco catalyzes the first step along the reductive pathway of CO_2 in photosynthetic organisms. Although the affinity of Rubisco for CO_2 is low relative to ambient CO_2 concentration, many photosynthesizing cells nevertheless achieve high photosynthetic rates by virtue of mechanisms that raise the intracellular CO_2 concentration. Such mechanisms have been recognized in a wide variety of photoautotrophic organisms ranging from bacteria to higher plants (50, 116). The efficiency of an inorganic carbon (Ci) concentrating mechanism (CCM) may be assessed from the ratio between the intrinsic $K_m(CO_2)$ of the particular Rubisco and the apparent photosynthetic K1/2 for extracellular dissolved CO_2. This ratio varies from several thousands in alkalophilic cyanobacteria (6, 54, 84) to close to one in organisms lacking a CCM (99). An inverse correlation exists between the efficiency of the CCM and the inefficiency of Rubisco as deduced from its kinetic characteristics. This observation has led to the suggestion of co-evolution of CCMs and Rubisco in response to changing atmospheric CO_2 (8). The CCM operating in C4 and CAM (crassulacean acid metabolism) plants differs fundamentally from that in microorganisms; the former is based on two sequential carboxylations separated either in space or time, whereas in microorganisms active transport of Ci is involved. Thus another measure of the efficiency of a CCM is the extent of Ci accumulation at a given external Ci concentration (but note that a low accumulation capacity does not necessarily indicate low efficiency). This review focuses on the CCM in aquatic microorganisms (for earlier reviews see 3, 6–8, 15, 25, 55–58, 84, 87, 93, 110, 134, 139, 146).

A few years after the discovery of the CCM, the notion was advanced that in cyanobacteria the elevated internal dissolved CO_2 concentration does not prevail throughout the cell. Rather, it is confined to the carboxysomes (113–115). It was proposed that the generation of CO_2 from HCO_3^- (accumulated by plasmalemma Ci transport mechanisms) is not catalyzed in the cytoplasm, and the Ci species do not reach equilibrium there. The accumulated HCO_3^- penetrates the carboxysomes where formation of CO_2 is catalyzed by carbonic anhydrase (CA) at sites near the carboxylation sites of Rubisco, most of which is located in these bodies (79, 126). This proposal has served as the working hypothesis for much of the current work in the area and experimental evidence has been obtained for some of its principal features. This model may also be applicable to eukaryotic algae where pyrenoids, densely packed with Rubisco

(18, 66, 68, 86, 111, 135) and also containing CA (66), may have the same function as carboxysomes (87, 113, 135, 139).

Elevation of CO$_2$ concentration at the carboxylating site has the following biological significance: First, it activates Rubisco (130); second, it compensates for the relatively low affinity of the enzyme for CO$_2$, and consequently also lowers inhibition of photosynthesis by competing O$_2$ (thus reducing photorespiration and the CO$_2$ compensation point); third, the massive transmembrane Ci fluxes involved may have a role in dissipation of excess light energy and in maintenance of internal pH (148). Any factor that affects the balance between the inward and outward Ci fluxes will significantly alter the level of the internal Ci pool and hence the rate of carboxylation. The photosynthetic rate at limiting CO$_2$ concentration increases during the process of acclimation to low CO$_2$ (20). This observation was explained by the discovery that a fall in ambient CO$_2$ level up-regulates the CCM, and that a syndrome of structural and biochemical changes accompanies this up-regulation (4, 54). The fact that CCM activity is modulated by environmental factors provides an ecological advantage, conferring the ability to acclimate to a wide range of CO$_2$ concentrations (as well as to light and nutrient availability). The degree to which an organism can exploit its potential to concentrate CO$_2$ may influence phytoplankton population dynamics. The CCM also has geochemical significance in that it may strongly affect the stable carbon isotopic composition (δ^{13}C) of the organic matter produced (32, 109, 110).

ACCLIMATION TO CHANGING CO$_2$ LEVEL

The capacity of many photosynthetic microorganisms to acclimate to low CO$_2$ concentrations is attributable to their ability to develop their CCM activity, one of a number of cellular features modulated during this response (55, 104, 139, 144). Many authors have focused on the rise in Ci transport capability as the major parameter determining CCM development. However, a quantitative model that examines changes in isotopic composition (δ^{13}C) during acclimation (O Faber, A Kaplan & L Reinhold, manuscript in preparation) reveals that progressive development of resistance to CO$_2$ leak from the vicinity of the carboxylation site is also essential to this process (see below).

Transfer of cyanobacteria from high- to low CO$_2$ has been thought to involve an increase in the membrane proteins participating in HCO$_3^-$ transfer without appreciably altering their affinity for their substrate. This notion was supported by measurements of initial rate of Ci uptake, which indicated a higher maximal rate in low-CO$_2$–grown cells but only marginally changed affinity (54). Uptake of dissolved CO$_2$ (CO$_{2(aq)}$) was far less affected than that of HCO$_3^-$. Subsequent studies assessing Ci uptake at steady state photosynthesis, in both cyanobacteria and green algae, by means of membrane inlet mass spectrometry (MIMS), led to the contrasting conclusion that the K_m, rather than the V_{max} was affected by the availability of CO_2 during growth (5, 157). However, it should be taken into account that at steady state net Ci influx is equal to the photosynthetic rate; and at saturating Ci concentration the latter is limited by the carboxylation capacity, not by the rate of supply of Ci. Hence, the V_{max} deduced for HCO_3^- transport, which was in fact for net HCO_3^- transport, was underestimated because of the ceiling set by photosynthesis (5). The lack of difference in V_{max} between high- and low-CO_2–grown cells may be due to the minor nature of the difference between their maximal photosynthetic rates. Since V_{max} was underestimated, K_m will also have been underestimated. The proposed effect of acclimation on the kinetic parameters of Ci uptake, and the inference of a high affinity HCO_3^- transporter, thus require further investigation. When the initial rate of CO_2 uptake was measured by MIMS, both K_m and V_{max} appeared to be affected during acclimation (144).

It is interesting that in cultures of *Chlamydomonas* (74), *Chlorella* (89), and *Dunaliella* (142) synchronized by alternating light–dark periods, the cells showed a decline in some of their low-CO_2 characteristics as they approached cell division. Apparently, the cells acclimated to low CO_2 following each cell division, at the beginning of the light period (74, but see 139). A rise in CA activity and its mRNA was observed in synchronized *Chlamydomonas* cultures prior to the onset of light (33, 112), supporting the involvement of a diurnal rhythm. Dependence on the latter may reflect posttranscriptional regulation rather than events at gene level (87). It is not known whether a diurnal rhythm plays an important part in cyanobacterial adaptation. Identification of a DNA region in *Synechococcus* PCC 7942, where mutations in the endogenous rhythm are clustered (63), should enable modulation of the rhythm and hence examination of its potential role.

Regulation of Gene Expression Following Change in Ambient CO_2 Concentration

Analysis of polypeptide patterns in both cyanobacteria and green algae (70, 140) show numerous changes, but the proteins involved have in most cases not been identified. In spite of extensive research, understanding of the regulation of gene expression during acclimation is still poor. One reason is the paucity of mutants impaired in acclimation. The *Chlamydomonas* mutant cia-5 fails to show any of the low CO_2 characteristics (87, 88), and it is likely that the relevant gene is high in the gene hierarchy regulating its acclimation. When the lesion in this mutant is clarified, it may identify a protein participating in signal perception or transduction. Adaptation mutants sensu stricto have not been isolated in cyanobacteria. Phenotypically qualified mutants of *Synechococcus*

PCC 7942 turned out to be impaired in purine biosynthesis under low CO$_2$ (129). Modification of various potentially relevant genes in the *Synechocystis* sp. PCC 6803 genome (53) is likely to produce the desired acclimation mutants.

Progress in identification of genes involved in acclimation is hampered by the multiplicity of the effects of changing CO$_2$ concentration on cell metabolism, leading to modulation of the expression of many genes, only some of which are directly involved in acclimation (55, 58, 87, 94). Many of the changes observed during acclimation are quantitative rather than qualitative, e.g. that on uptake rate and extent of Ci accumulation (25, 57). Also the number of carboxysomes and mitochondria rises during acclimation to low CO$_2$ (64, 79). Furthermore, CO$_2$ stress imposes multiple stresses including carbohydrate starvation, photodynamic damage and oxidative stress.

DNA topology is likely to be involved in cellular response to CO$_2$. Manipulation of *topA*, encoding topoisomerase I, affected growth of *Synechococcus* PCC 7942 following changes in ambient CO$_2$ concentration (45). Specific sigma factors appear following exposure to various environmental stresses and are likely to have a role in regulating the expression of CO$_2$-responsive genes. Inactivation of *sigC* in *Synechococcus* PCC 7002 slowed growth under low CO$_2$; carbon starvation applied by bubbling nitrogen led to a decline in *sigA* but a rise in *sigB* (22, 48).

A change in phosphorylation of PII protein, the gene product of *glnB*, has been observed in a number of cyanobacteria exposed to nitrogen or CO$_2$ stress. It was suggested that PII participates in coordinating carbon and nitrogen metabolism (40, 69). Because saline stress also affects PII phosphorylation, the latter may have a role in cell response to stress in general. When low-CO$_2$-grown *Synechocystis* cells were either exposed to high-CO$_2$ or supplied with glucose, the change in pattern of phosphorylated proteins was similar (16). This suggests that high-CO$_2$ and photoheterotrophic conditions may repress the activity of the CCM via similar routes. *IcfG*, which is possibly involved in coordinating Ci and glucose metabolism (14), may play a role here. A change in phosphorylation of the beta subunit of phycocyanin (N Mann, unpublished data) may be the mechanism whereby exposure to high light or low CO$_2$ treatment modulates the efficiency of energy transfer from phycobilisomes to PSII (49).

CO$_2$-responsive genes have been recognized both in cyanobacteria (55, 104, 127) and in *Chlamydomonas* (33, 87, 139, 150). Certain cyanobacterial genes are transcribed only under low CO$_2$; in the case of others, the level of transcript increases during acclimation (33, 121). Random ligation of small genomic DNA fragments upstream of promoter-less reporter genes led to the identification of CO$_2$-responsive promoters (127). Analysis of the promoter of *cmpA* [encoding a 42-kDa polypeptide located in the cytoplasmic membrane and implicated in HCO$_3^-$ transport (see below)] has led to the identification of positive and negative regulatory elements. Deletion of the latter potentiates transcription under high CO_2 (121). CcmN is essential for expression of *cmpA* in cells grown under low CO_2 (55, 121) but is not required in standing cultures [which, unlike aerated cultures, show Na^+-independent HCO_3^- transport (34)]. Sequence homologies in the promoter regions are reported for various CO_2-responsive genes but their significance remains to be elucidated (55).

A periplasmic alfa type CA activity, encoded by *ecaA*, has been reported in *Anabaena* 7120 and *Synechococcus* PCC 7942 (138). The role of this CA is not readily apparent because it is up-regulated under high CO_2 where CO_2 should not be limiting. It may function in sensing ambient CO_2 level and in maintaining chemical equilibrium between Ci species in the periplasm.

Regulation by ambient CO_2 at the posttranslational level would account for the fast induction observed in *Synechococcus* strains PCC 7942 and PCC 7002, where development of elevated apparent photosynthetic affinity for ambient CO_2 has been observed within several minutes. This process did not seem to require protein synthesis and was sensitive to a protein kinase inhibitor but not to phosphatase inhibitors (144). Inspection of the relationship between HCO_3^- transport and its external concentration [see Sültemeyer et al (144), Figure 3] reveals sigmoidal behavior in the fast induced cells. This recalls positive cooperativity, suggesting a requirement for ligand binding, possibly phosphorylation, for full activation of the transporter. Furthermore, low-CO_2-grown cells exhibited a higher rate of transport at low Ci concentrations than did the fast induced cells. This may be due to synthesis of new transport proteins, e.g. CmpA-D (96), during the longer induction period. The sensitivity of the induction in *Chlorella* to protein synthesis inhibitors (78) supports this suggestion.

What is the Signal that Induces the Response?

The nature of the signal that induces response to ambient CO_2 level is still an enigma. Candidates include total Ci (144), the ratio of $[CO_2]/[O_2]$ (56, 75), and $[CO_{2(aq)}]$ (78). *Synechococcus* PCC 7942 developed low CO_2 characteristics at high pH values when total Ci levels were high and where the $CO_{2(aq)}$ concentration was low (32). It seems likely that the cells respond to the specific Ci species that they are capable of utilizing, and varied conclusions in the literature may stem from this species specificity. It is possible that the signal is perceived as the level of an intermediary metabolite in the carbon metabolic pathways. The observation that photoheterotrophic growth conditions [e.g. supply of acetate or glucose to various cyanobacteria and green algae (16, 38, 76, 125)] suppress low-CO_2 characteristics supports this suggestion.

Other internal signals triggering acclimation might include changes in the reductive state of the photosynthetic electron transport chain, owing to the lower availability of CO_2 as a sink for the photosynthetic electrons. The latter possibility is particularly attractive in view of the role of redox state in transcription regulation in plants in general, and in the expression of CA in *Chlamydomonas* (P Falkowski, unpublished data).

The effect of the $[CO_2]/[O_2]$ ratio on acclimation suggests that photorespiratory metabolites, e.g. phosphoglycolate, might act as an internal signal. Overexpression of *cbbZ* (encoding phosphoglycolate phosphatase in *Ralstonia*), in *Synechococcus* PCC 7942 led to a sustained high level of phosphoglycolate phosphatase activity, in contrast to the transient rise observed in high-CO_2-grown wild-type exposed to low CO_2. Increase in apparent photosynthetic affinity for external Ci, characteristic of acclimation, was retarded in the mutant (55). The pattern of polypeptide synthesis and of polypeptide phosphorylation upon transfer from high to low CO_2 differed between the *cbbZ*-overexpressing mutant and wild-type.

Inhibition of the glycolate pathway by aminooxyacetate significantly retarded acclimation of *Chlamydomonas* and depressed the synthesis of two low-CO_2-induced polypeptides, 21 kDa and 37 kDa. In contrast, synthesis of a third low-CO_2-induced polypeptide, 36 kDa, was not affected by the inhibitor (153) and was inducible in the dark (151). Induction by low CO_2 of *Mca1* and *Mca2*, encoding a mitochondrial CA, occurred in the dark but only when light-dark alternations had produced a diurnal rhythm. Under continuous light, expression of these genes was strongly affected by light intensity, paralleling photosynthesis (150). Clearly, more than one means of triggering by CO_2 concentration may be involved. Furthermore, some of the factors affecting induction may not themselves act as signals but may function in the signal transduction pathway. For example, blue light may act in the signal transduction path when it up-regulates expression of periplasmic CA (19, 29).

THE MECHANISMS THAT BRING ABOUT ELEVATED CO₂ CONCENTRATION AT THE CARBOXYLATION SITE

The CCM in cyanobacteria is believed to involve a number of constituent processes:

1. Membrane transport of Ci across the plasmalemma. Regardless of the Ci species taken up from the medium, HCO_3^- is the species that accumulates in the cytoplasm.

2. Diffusion of the accumulated HCO_3^- into the carboxysomes where CO_2 formation is catalyzed at CA sites, leading to an elevated CO_2 concentration in close proximity to Rubisco. A significant part of the CO_2 is fixed.

3. Outward diffusion of the remaining CO_2, a substantial portion being scavenged by a light energy-dependent process. HCO_3^- also effluxes from the cytoplasm although the net flux is inward.

Although the description above applies to cyanobacteria, in its essential features it may also be valid for green algae, as the ensuing discussion makes clear. Quantitative models for the CCM in microalgae broadly based on the cyanobacterial model have been developed (42) and are reported to account satisfactorily for most experimental data.

Membrane Transport of Inorganic Carbon

Ci SPECIES   The nature of the Ci species utilized from the medium by various phytoplankton has been studied extensively (25, 34, 78, 110, 122, 149, 155). Higher photosynthetic rates than could be accounted for by the theoretical rate of CO_2 supply from the uncatalyzed dehydration of HCO_3^- have been regarded as evidence for direct HCO_3^- utilization (110). Where the estimated CO_2 supply appears sufficient, e.g. under low cell densities, HCO_3^- utilization may nevertheless occur. The net CO_2 efflux observed during net photosynthesis in certain cyanobacteria (67, 124, 148) and eukaryotes (141) provides strong evidence for direct HCO_3^- utilization from the medium; this efflux results from intracellular conversion, catalyzed by CA, of HCO_3^- taken up in a light-dependent process. In other organisms, external CO_2 concentration is maintained in the light, at a steady state, below that expected at CO_2-HCO_3^- equilibrium even in the absence of CO_2 fixation (Figure 1). This indicates that, at steady state, CO_2 is taken up and net HCO_3^- efflux is occurring at a rate equal to the dehydration of ambient HCO_3^-. Most organisms examined to date can utilize CO_2 or HCO_3^- or both, but species-specific preferences are observed. The possibility that CO_3^{2-} serves as the immediate Ci source has also been considered (73).

When the specific transport mechanism for a particular Ci species and the genes involved are under investigation, the experimental conditions must ensure

---→

Figure 1 CO_2 uptake in the presence and absence of net photosynthesis in *Synechococcus* sp. PCC 7942 and a high CO_2-requiring mutant. Wild type in the absence (*A*) and presence (*B*) of iodoacetamide [which inhibits CO_2 fixation (124)]. Uptake of CO_2 lowered the external steady state concentration in the light below that expected at chemical equilibrium with HCO_3^-, even in the absence of net CO_2 fixation, i.e. net O_2 evolution (*B*). Results similar to those in (*B*) were obtained when the high CO_2-requiring mutant was exposed to Ci concentrations below those necessary for net photosynthesis (*C*). These data indicate massive CO_2 uptake even in the absence of net photosynthesis. At the steady state obtained in the light, CO_2 uptake is balanced by CO_2 formation from HCO_3^- in the medium. Because, at this steady state, the cells are not continuously accumulating Ci, massive net HCO_3^- efflux must occur.

The reverse situation, i.e. net CO_2 evolution during net photosynthesis as a result of HCO_3^- uptake, has been observed in certain marine organisms (141, 148).

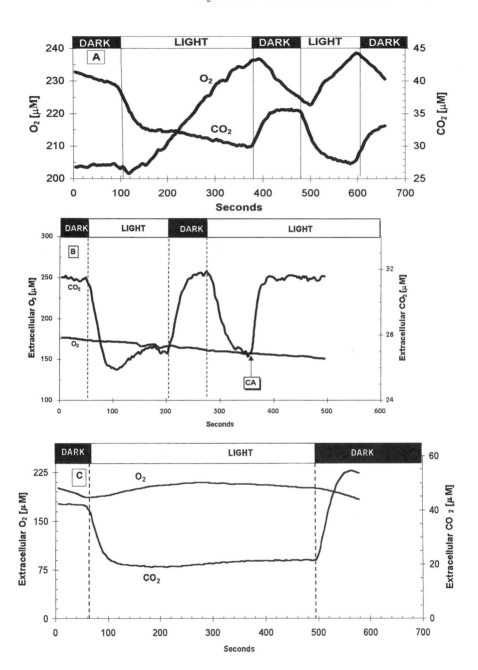

that only the relevant Ci species is taken up. This is achieved in experiments where either CO_2 or HCO_3^- are supplied under disequilibrium conditions; alternatively, where compounds are added that are believed to specifically inhibit CO_2 uptake e.g. COS, Na_2S, and ethoxyzolamide (EZ) (67, 124, 149). Na^+-deprivation, which minimizes HCO_3^- uptake in aerated (as opposed to standing) cultures, has been used to study CO_2 uptake (67).

MEASUREMENTS OF Ci TRANSPORT   Kinetic data based on initial rates of transport are available for Ci uptake in various organisms including cyanobacteria (56), a coccolithophorid (131), a dinoflagellate (11), and diatoms (26). Obtaining such data for *Chlamydomonas* is complicated because steady state internal concentration is attained rapidly (139).

Badger and his colleagues were the first to attempt measurement of the Ci fluxes at photosynthetic steady state, and their work brings to light some novel features of the exchange of Ci between the cells and their environment. The original paper in this series (5) should be consulted for a thoughtful evaluation of some of the errors intrinsic to the method and the assumptions involved. However, additional problems with regard to the method include the following:

1. The assumption that HCO_3^- efflux is negligible is now known to be incorrect (Figure 1) (124, 128). Therefore, net HCO_3^- flux does not adequately represent unidirectional HCO_3^- influx into the cell. The estimated HCO_3^- influx is further in error since the fraction of HCO_3^- uptake that subsequently leaks from the cell in the form of CO_2 [as visualized in the model (41, 113) and recently directly measured (124, 148)] is not taken into account [see Badger et al (5) Equation 5]. This substantial underestimation may have an impact on the conclusion (157) that inactivation of genes such as *psaE*, encoding a functional subunit of PSI, as well as other treatments (144, 145, 157) lead to a considerable change in HCO_3^- but not in CO_2 uptake.

2. Uptake of CO_2 is underestimated by an amount equal to the fraction that subsequently leaves the cell in the form of HCO_3^-. Evidence for substantial cycling of this nature is presented in Figure 1.

CO₂ UPTAKE BY CYANOBACTERIA CO_2 uptake by cyanobacteria shows saturable kinetics (84, 154), leading to the conclusion that CO_2 uptake is mediated by a membrane transport system and, in some cases, to the further inference that it is active (6, 82, 84, 85, 124). The transport system may include a vectorial CA-like moiety converting CO_2 to HCO_3^- during passage through the plasmalemma, since HCO_3^- accumulates in the cytoplasm when CO_2 is supplied (1, 154). The cytoplasmic $CO_{2(aq)}$ concentration is below that expected at chemical equilibrium and, according to the proposed model for the CCM (114), this disequilibrium is related to the fact that CA is located exclusively within the carboxysomes and is absent from the cytoplasm (Figure 2a). Verification of this aspect of the model has been obtained in an elegant experiment where expression of foreign CA in *Synechococcus* PCC 7942 cytoplasm resulted in a mutant requiring high-CO$_2$, presumably because of elevated CO$_2$ efflux (102).

Mediated CO$_2$ transport might be considered improbable across a bilipid membrane with high CO$_2$ permeability. A recent model (41) postulated entry of CO$_2$ into the cell by passive diffusion and subsequent energy-dependent conversion of CO$_2$ to HCO$_3^-$ by a CA-like entity. This model disposes of the need to postulate active CO$_2$ transport through the plasmalemma. Saturation kinetics would be observed for uptake, because maintenance of the inward diffusion gradient would depend on the conversion reaction, and the latter would thus rate-limit CO$_2$ influx. The kinetics would reflect those of the energy-dependent CA-like activity. Energy input might bring about a conformational change in this CA-like entity, resulting in a lower affinity for HCO$_3^-$ and/or a raised affinity for CO$_2$. The hydration reaction would thus be promoted and the dehydration reaction depressed. Alternatively, photosynthetic or respiratory electron transport might raise the pH in the immediate surroundings of the CA (Figure 2b–d). This CA-like activity would not only convert entering CO$_2$ to HCO$_3^-$, it would also scavenge outwardly diffusing CO$_2$.

Although there is strong evidence for the participation of a CA-like moiety, the protein(s) involved and the relevant gene(s) have not been identified. The plasmalemma-located CA activity reported by Bedu et al (10) might result from the CA-like moiety in its nonenergized mode. Alternatively, the CA-like entity might be located in the thylakoid membranes, as has been reported in green algae (106). If there are alkaline "pockets" on the cytoplasmic side (due to the removal of protons by the light-dependeent photosynthetic H$^+$ pump) and if CA is located within them (Figure 2b–d), then HCO$_3^-$ formation due to the alteration in HCO$_3^-$/CO$_2$ equilibrium in these pockets would be accelerated and the immediate dependence on photosynthetic electron transport would be explained. The role of the thylakoid lumen-located CA activity proposed here differs from that suggested (110) for green algae (below).

Studies by Ogawa and colleagues revealed that the phenotype of several PxcA (CotA)-defective mutants of *Synechococcus* PCC 7942 and *Synechocystis* PCC 6803, originally attributed to defective CO$_2$ transport, in fact resulted from impaired sodium-dependent proton extrusion normally observed in cells transferred from dark to light. They concluded that the apparent inhibition of CO$_2$ uptake in these mutants stemmed from impaired ability to regulate the internal pH during Ci utilization (62).

Miller et al (85) questioned the premise that the internal elevated CO$_2$ concentration is confined to the carboxysomes and that the cytoplasmic CO$_2$

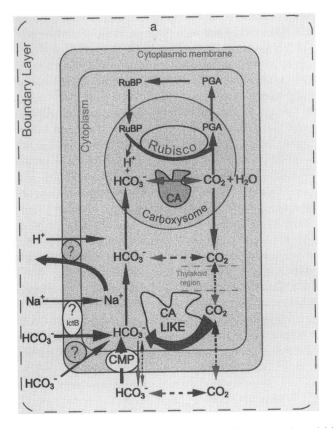

Figure 2 (*a*) Schematic presentation of components involved in the cyanobacterial inorganic carbon concentrating mechanism, showing the boundary layer, the cytoplasmic membrane, and the cytoplasm containing the thylakoid region and a carboxysome. Three possible HCO_3^- transporters are indicated: CMP, a primary HCO_3^- transporter of the ABC type directly fueled by ATP (94; T Omata, unpublished data); IctB, a putative Na^+/HCO_3^- symporter (17) fueled by the Na^+ electrochemical gradient generated either by a primary Na^+ pump (not shown) or a Na^+/H^+ antiporter (shown) secondary to an H^+ pump; and a low affinity HCO_3^- transporter (17) of unknown nature (designated by a question mark). Efflux of HCO_3^- occurs via these transporters (shown only for the case of CMP) and by diffusion. As explained in the text, uptake of CO_2 may occur via diffusion across the cytoplasmic membrane followed by conversion to HCO_3^- by a CA-like moiety. Active CO_2 transport has also been proposed (82). Accumulated HCO_3^- penetrates the carboxysome where CA activity generates CO_2 at high concentration in the vicinity of Rubisco. Part of the CO_2 is fixed, part diffuses outwards. Regeneration of ribulose 1,5-bisphosphate occurs outside the carboxysome.

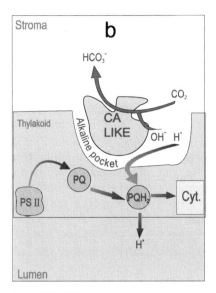

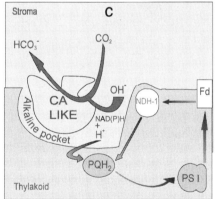

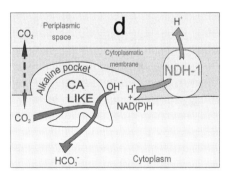

Figures 2   (*b–d*) Suggested mechanisms whereby photosynthetic or respiratory electron transport might lead to the formation of alkaline pockets in thylakoid or cytoplasmic membranes with consequent conversion of CO$_2$ to HCO$_3^-$. Such pockets might result from reduction of PQ (plastoquinone) by either linear (*b*) or cyclic (*c*) photosynthetic electron transport possibly involving NAD(P) dehydrogenase [NDH (94)] or (*d*) from the activity of cytoplasmic membrane-located NDH-1 (104). CA-like activity in the pocket would accelerate HCO$_3^-$ generation consequent on the pH shift. Hydrophobic surfaces would slow diffusion of H$^+$ into the pocket and thus raise the efficiency of CO$_2$ to HCO$_3^-$ conversion.

concentration is low, having recorded substantial net efflux of CO_2 following inhibition of $^{13}CO_2$ uptake by COS, Na_2S, or $^{12}CO_2$. The crucial question is whether the leaking $^{13}CO_2$ they observed existed as such in the cells prior to inhibition of $^{13}CO_2$ uptake. If so, calculations on the basis of their data indicate that the intracellular CO_2 concentration maintained was about 50,000 times higher than the external concentration. This implies an extremely high diffusion resistance for CO_2 across a biological membrane. A similar low diffusivity to O_2 would lead to an internal O_2 concentration of several atmospheres, which does not appear likely.

HCO$_3^-$ UPTAKE IN CYANOBACTERIA   Influx of HCO_3^- is active, i.e. against its elecrochemical potential gradient and therefore dependent on a supply of metabolic energy (56). However, it has not yet been established whether this dependance is direct (primary active) or indirect, via an ion gradient (secondary active). Metabolic energy would in the latter case be required to maintain the ion gradient. It has been proposed that HCO_3^- uptake is fueled by a sodium gradient (56), although it was pointed out that the evidence adduced is equally consistent with a Na^+/H^+ antiport involved in the regulation of internal pH; or with an effect of sodium on the carrier. Evidence regarded as supporting Na^+/HCO_3^- symport, including the effect of the Na^+/H^+ exchanger monensin, has recently been reported (35). The driving force for Na^+/HCO_3^- symport, μNa^+, has been measured in *Synechococcus* PCC 7942 and shown to be adequate to account for the observed HCO_3^- accumulation if the stoichiometry of symport is two or three Na^+ ions per HCO_3^- (117). On the other hand, the measured Na^+ flux was deemed to be too low to support the linked transport processes. Furthermore, it was argued (117) that the proposed 2–3 Na^+/HCO_3^- stoichiometry should depolarize the membrane potential, whereas hyperpolarization was in fact observed on addition of HCO_3^- (56, 117). However, the observed hyperpolarization could reflect stimulation of a primary electrogenic pump. For instance, a primary Na^+ extrusion pump (21, 101) would be stimulated by entering Na^+. The H^+ extrusion pump is an alternative candidate, assuming appropriate stoichiometry. At pH values above 8.0, however, μH^+ would be too small to support HCO_3^- accumulation (56). Mutants impaired in PxcA, defective in light-dependent H^+ extrusion and hence apparently in CO_2 uptake (62), may shed light on the complex interaction between Na^+ ions, H^+ ions, and Ci uptake and thus on the nature of the primary pump.

The possibility that HCO_3^- uptake is mediated by a primary pump, directly fueled by ATP, is suggested by the shared homology of ABC-transporters and CmpA-D in *Synechococcus* PCC 7942 and *Synechocystis* PCC 6803 (94, 95), and by the binding of HCO_3^- by CmpA (T Omata, unpublished data). There are indications that cyanobacteria possess several HCO_3^- transporters (Figure 2a).

Absence of CmpA in a *Synechococcus* mutant scarcely affected CO$_2$ or HCO$_3^-$ uptake (128) and transposon-inactivation of *cmpA* slowed growth of *Synechococcus* and *Synechocystis* only when the mutants were exposed to a level of CO$_2$ lower than 50 ppm (94). Modification of *ictB* in *Synechococcus* PCC 7942 resulted in a mutant impaired in HCO$_3^-$ transport, suggesting that IctB, which shares homology with several transport proteins from various organisms (17), is directly involved in HCO$_3^-$ transport. The fact that HCO$_3^-$ uptake in the IctB-inactivated mutant showed saturable kinetics [with very high K1/2 (HCO$_3^-$)], suggests that *Synechococcus* PCC 7942 contains yet another HCO$_3^-$ carrier. An alternative explanation is that modification of IctB lowered its affinity for its substrate. These findings support the earlier deduction that multiple HCO$_3^-$ transporters operate in cyanobacteria. Pioneering studies by Espie and colleagues (34, 35, 149) demonstrated: (*a*) varying Na$^+$-requirements for HCO$_3^-$ transport under different growth conditions, and (*b*) differing sensitivity of the Na$^+$-dependent and the Na$^+$-independent HCO$_3^-$ transport to EZ. The presence of multiple HCO$_3^-$ transporters would ensure the vital supply of Ci under a wide range of environmental conditions. It would also explain why identifying phenotypes of HCO$_3^-$ carrier–defective mutants has proved difficult.

CO$_2$ UPTAKE BY EUKARYOTIC PHOTOSYNTHETIC MICROORGANISMS   The degree of Ci accumulation in algae is generally lower than in cyanobacteria, a fact possibly connected with the higher affinity of algal Rubisco for CO$_2$. In spite of the relatively low level of accumulation, it has been concluded that uptake is mediated by membrane transport mechanisms (6, 7, 110). In *Chlamydomonas* (87, 139), the membrane concerned may be the chloroplast envelope (Figure 3), which would accord with the proposed origin of the chloroplast as a symbiotic cyanobacterium. Isolated chloroplasts can accumulate Ci to levels lower than those observed in intact cells, but nevertheless higher than could be accounted for by passive diffusion. Furthermore, Ci uptake is higher in chloroplasts isolated from low- than from high-CO$_2$–grown cells. Uptake by whole cells might thus result from two sequential steps, diffusion from the bulk solution via the unstirred layer, and subsequent mediated transfer through the chloroplast envelope. Saturable kinetics for whole cell uptake (2) would be observed if the second step is rate limiting for the overall process.

Interconversion between Ci species in the periplasmic space is catalyzed by a low-CO$_2$–induced CA in many freshwater and marine microorganisms (39, 146). At alkaline pH, inhibition of this CA raises the apparent photosynthetic K$_m$ for external Ci. Dependence of periplasmic CA activity in a diatom on a light-driven electron transport chain in the plasmalemma (N Nimer, unpublished data) is likely to be related to redox control of the enzyme. Therefore, in some organisms where the periplasmic CA is apparently absent, e.g. *Chlorella*

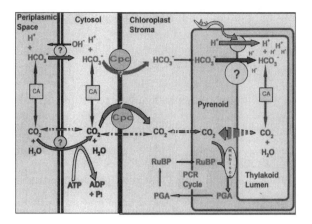

Figure 3   Schematic presentation of components involved in the operation of the inorganic carbon concentrating mechanism in eukaryotic algae (modified from 139). Emphasis is given to the possibility that catalyzed formation of CO_2 in the acidic lumen of pyrenoid-transversing thylakoids raises the concentration of CO_2 near Rubisco (110). It is not known whether the Ci species are at chemical equilibrium in the cytosol and in the stroma. Mediated CO_2 and HCO_3^- uptake are indicated on the plasmalemma and the chloroplast envelope (87, 110, 123). It is possible that Cpc (Lip-36) is involved in Ci transfer into the chloroplast. Direct fueling by ATP has been proposed for CO_2 transport across the plasmalemma (122, 123). Other proteins involved in Ci transport have not been identified (designated by question marks) but it is likely that HCO_3^- uptake is mediated. PCR, pentose carbon reductive cycle; PGA, 3-phosphoglyceric acid; CA, carbonic anhydrase.

11h (146), it may nevertheless exist in a latent form. In other organisms a periplasmic CA is clearly not present because a steady state CO_2 concentration above that expected at CO_2/HCO_3^- equilibrium is maintained in the ambient solution in the light (141).

In *Eremosphaera viridis*, which lacks the periplasmic CA, stimulation of ATPase activity in a microsomal-enriched fraction by CO_2 but not by HCO_3^- was regarded as indicating the possible involvement of a primary pump in CO_2 uptake (123). Inhibition of Ci uptake by the H^+-ATPase inhibitor vanadate, in *Chlamydomonas* cells and isolated chloroplasts, indicates the nature of the primary pump involved, and perhaps also a plasmalemma location for the transporting system (47, 60). Alternative oxidase may play a role in the energization of Ci uptake in *Chlamydomonas* (46) because the activity of this oxidase rises during acclimation to low CO_2 and moreover, Ci uptake is inhibited by salicylhydroxamic acid.

Identification of the lesion in the *Chlamydomonas* mutant impaired in Ci transport, pmp-1, and of the location of the gene product, will help to elucidate whether the Ci transporting mechanism is located on the plasmalemma, the chloroplast envelope, or both (139). Analysis of low-CO_2–induced proteins in *Chlamydomonas* led to the identification of a chloroplast envelope–located polypeptide, Cpc (LIP-36). Cpc is not present in chloroplasts isolated from high-CO$_2$–grown cells, nor in the high CO$_2$–requiring mutants cia-3 and cia-5 (87, 107), indicating it may have a role in CO$_2$ or HCO$_3^-$ transfer into the chloroplast. The deduced homology between Cpc and mitochondrial carrier proteins (24) provides support for this notion.

A *Chlamydomonas* mutant impaired in *ycf10*, encoding CemA, an inner chloroplast envelope protein, has been isolated (118). CemA is similar to PxcA, which has been indirectly implicated in CO$_2$ uptake in cyanobacteria (61, 94). The CemA mutant showed sensitivity to high light intensity, which was attributed to defective CO$_2$ uptake on the basis of MIMS measurements. However, high CO$_2$ concentration neither relieved the light sensitivity nor raised the photosynthetic rate (as would be expected from a mutant defective in Ci transport) and CemA is thus unlikely to be directly involved in CO$_2$ uptake in *Chlamydomonas*.

HCO$_3^-$ UPTAKE BY EUKARYOTIC MICROORGANISMS   As pointed out above, many photosynthetic microorganisms, both freshwater and marine, are capable of direct utilization of HCO$_3^-$ from the medium (90, 110). Net inward passage of HCO$_3^-$ across the plasmalemma (Figure 3) thus occurs despite the fact that, wherever examined, the electrochemical potential gradient for HCO$_3^-$ is outward. Internal Ci levels have frequently been reported that, although below the level in the environment in marine organisms (110), are nevertheless higher than that expected at Nernst equilibrium, taking into account the trans-plasmalemma electrical potential gradient (inside negative). Mediated transport is therefore indicated, although information on the mechanism(s), the proteins involved, and the mode of energization (143) is scarce. A direct link to a primary H$^+$ pump is suggested by the detection, in the plasmalemma of *Ulva*, of a polypeptide that cross-reacted with an antibody raised against band III protein, involved in anion exchange in erythrocytes (30). The trans-plasmalemma OH$^-$ gradient could thus drive HCO$_3^-$ uptake by HCO$_3^-$/OH$^-$ antiport. This antiport is envisaged as electroneutral in erythrocytes, but in algae growing at alkaline pH where the Gibbs free energy for H$^+$ may not be sufficient to drive HCO$_3^-$ across the membrane, electrogenic antiport might be required. Band III inhibitors have been observed to inhibit HCO$_3^-$ uptake in a number of marine microorganisms but not in others (110).

Mechanisms for supplying Ci to the stroma may also be located in the chloroplast envelope. The low-CO$_2$–induced proteins in the envelope, such as Cpc, might in fact be involved in HCO$_3^-$ transport (Figure 3).

Involvement of Photosynthetic Electron Transport

The dependence of Ci transport on light energy indicates a role for the latter in the fueling of the primary pumps involved (143) and for activation of the transport system (56). Acclimation of cyanobacteria to low CO$_2$ is accompanied by a marked rise in the ratio of PSI/PSII activity, in agreement with results of Ogawa and colleagues demonstrating that cyclic electron transport around PSI (photosystem I) is involved in Ci uptake (80, 81, 93). Use of specific electron transport inhibitors and acceptors led to the proposal that transport of CO_2 is supported by cyclic electron flow whereas that of HCO_3^- depends on linear electron flow (67). This conclusion is not yet firmly established; however, if valid, it would have important implications for the nature of the linkage between photosynthetic electron transport and uptake of the two Ci species. Both cyclic and linear electron transport lead to the formation of an overall proton electrochemical gradient across the thylakoid membrane. The proposed importance of a particular segment of the electron transport chain would suggest involvement of a specific electron carrier, for instance NAD(P)H dehydrogenase, which is principally involved in cyclic electron transport (80). A mutant bearing a lesion in *ndh* would thus be expected to show greater loss of CO_2^- than of HCO_3^- uptake capacity. A *Synechocystis* PCC 6803 mutant, Rka, impaired in *ndhB*, was indeed more severely defective in CO_2 uptake than in HCO_3^- uptake (92). Conversion of CO_2 to HCO_3^- in alkaline pockets in the stromal face of the thylakoid membrane [resulting from electron transfer from NAD(P)H dehydrogenase to plastoquinone] or at NAD(P)H dehydrogenase sites in the cytoplasmic membrane (Figure 2c and *d*) would be other possible examples of the importance of local electron transport in Ci uptake.

Confirmation of a role for cyclic electron transport has also emerged from studies of mutants of *Synechocystis* PCC 6803 (91, 92) and *Synechococcus* PCC 7942 (72) defective in *ndhB* and *ndhL* (encoding subunits of NAD(P)H dehydrogenase) and impaired in Ci accumulation. Cyclic electron transport does not operate in these mutants (80). The presence of multiple copies of *ndhI*, *ndhD*, and *ndhF* in *Synechocystis* PCC 6803 (53) raises the possibility of heterogeneity in NAD(P)H dehydrogenase and may explain the failure to isolate a high CO_2–requiring mutant following inactivation of *ndhF* (145). Northern analyses indicated that *ndhD1* and *ndhD4* were constitutively expressed in *Synechocystis* PCC 6803, whereas expression of *ndhD2*, *ndhD3*, and *ndhD5* was induced by low CO_2. Inactivation of only one of the *ndhD* copies, *ndhD3*, significantly slowed growth under low CO_2 (94). It was therefore proposed that there are two functionally different forms of NAD(P)H dehydrogenase (94, 104). Possible modes of NDH-1 involvement in the vectorial, catalyzed conversion of CO_2 to HCO_3^- is presented Figure 2c and 2d.

Studies on *Synechococcus* PCC 7002 mutants where *ndhF* and *psaE* (encoding a subunit of PSI) were inactivated pointed to a key role for PsaE-mediated, and to a lesser extent NdhF-mediated, electron transport in utilization of HCO_3^- by low-CO_2–grown cells. It was further inferred, on the basis of the calculated K1/2 for HCO_3^- transport, that inactivation of *psaE* (145) or *ndhD* (104) resulted in inability to induce high-affinity HCO$_3^-$ transport. However, these experiments did not adequately differentiate between induction of transport and its energization. Furthermore, observed increases in the chlorophyll to phycocyanin ratio, indicating a rise in PSI/PSII (145), demonstrate that the mutants did in fact adapt to the changing CO$_2$ environment. Detection of induced expression of *cmpA* in *ndh*-impaired mutants exposed to low CO$_2$ (T Ogawa, unpublished data) supports the latter conclusion. Moreover, estimated K1/2 for HCO$_3^-$ transport is unreliable because the V$_{max}$ for HCO$_3^-$ transport, as deduced by MIMS, largely reflects maximum photosynthetic rate, and the latter was severely depressed in the mutant.

A *Synechococcus* PCC 7942 mutant impaired in HCO$_3^-$ transport was obtained following insertion of vector sequences in the genomic region of *psaI-L*, encoding PSI subunits. Exposure to low CO$_2$ led to a considerable reduction in the level of the *psaI-L* transcript in the mutant, in contrast to the rise observed in the wild type (119). This result appears to support the participation of PSI in energization of HCO$_3^-$ transport.

Miller and colleagues (83) were the first to note a strong linkage between Ci accumulation and PSII fluorescence quenching. The initial rate of fluorescence quenching was correlated with the initial rate of HCO$_3^-$ transport, and the extent of fluorescence quenching was correlated with the internal Ci concentration (27). Inhibition of CO$_2$ fixation did not alter fluorescence quenching significantly, and the current view holds that dissipation of light energy in the Ci transport process lowers the fluorescence yield. Photoreduction of O$_2$, an indication of linear electron transport through PSI, was stimulated by Ci.

Function of the Carboxysomes

An elevated CO$_2$ concentration in the close vicinity of Rubisco enhances activation of the enzyme, overcomes its relatively low affinity for CO$_2$, and depresses O$_2$ inhibition of photosynthesis. In cyanobacteria, an elevated CO$_2$ level is achieved when Ci accumulated in the cytoplasm, in the form of HCO$_3^-$, penetrates into the carboxysomes (41, 113). The highly organized structure of these bodies brings into close conjunction CA sites (where CO$_2$ is generated from the accumulated HCO$_3^-$) and Rubisco sites where a substantial proportion of the CO$_2$ is fixed. Experimental evidence for the presence of CA in carboxysomes was obtained in fractionation studies (103) and Western blot analysis (137). The closeness of the functional linkage between CA and Rubisco might be assessed from the number of hydration-dehydration cycles the CO$_2$ molecules undergo before being fixed by Rubisco.

The biological advantage of the dense packing of the Rubisco molecules within carboxysomes (and pyrenoids) may well lie in the substantial barrier to diffusion that this arrangement affords (113); it is not known whether any additional barrier is required. Other essential substrates that must diffuse into the carboxysomes from the cytoplasm where they are formed, such as RuBP, reach adequate levels at Rubisco sites owing to the concentrating effect of three-dimensional diffusion from the periphery toward the center of the carboxysome or pyrenoid (113).

Mutants in which carboxysomes are aberrant or absent are frequently found among high CO_2–requiring mutants (6, 31, 43, 57, 58, 71, 93, 94). In all cases examined, their high CO_2–requirement results from a low state of Rubisco activation when exposed to low CO_2 (130). Low intra-carboxysomal CO_2 levels could result from a number of causes, including reduced rate of CO_2 formation from HCO_3^- because of lack of CA, or increased CO_2 leakage due to defective carboxysome organization. It is interesting that internal HCO_3^- levels sometimes even higher than in the wild type have been observed in carboxysome-less mutants. To be consistent with the model, this could only occur if carboxysomal CA is not expressed in these mutants under low CO_2 conditions. Dissipation of the HCO_3^- pool would otherwise occur by CO_2 leakage. The aberrant carboxysomes in various mutants resemble bodies sometimes observed, albeit in a lower frequency, in wild-type cells. It has been proposed that they represent arrested stages in the development of carboxysomes, halted at different points in the various mutants (77, 97).

Replacement of form I Rubisco (L8S8) in *Synechocystis* PCC 6803 by the L2 type (form II, lacking the small subunit) from *Rhodospirillum rubrum* prevented carboxysome formation (100). The same result followed inactivation of *cbbL*, encoding form I in *Thiobacillus neapolitanus*. The latter then expressed *cbbM*, i.e. form II (9). Extension of the C terminus of RbcS resulted in grossly enlarged carboxysomes (130). These results stress the importance of the native Rubisco and its small subunit, in carboxysome assembly. The reported association of the small subunit of Rubisco with the carboxysomal shell (52) may relate to the homology between *ccmM*, encoding a constituent of the shell, and *rbcS*. This association might determine the spatial organization of Rubisco within the carboxysomes. The homology referred to might also help in the retrieval of Rubisco molecules from the cytosol and their incorporation in carboxysomes during biogenesis. Further progress in the identification of carboxysome constituents and their respective genes may enable use of carboxysomes as a model system to study assembly of cellular bodies in prokaryotes.

Shively and colleagues (133, 134) identified the *cso* operon encoding carboxysomal shell proteins in *T. neapolitanus*. It is highly homologous to the *pud* and *eut* operons encoding components of polyhedral bodies in *Salmonella* and to the cyanobacterial *ccm* genes. In *Synechococcus* PCC 7942, six of these genes (*ccmJ* and *ccmK-ccmL-ccmM-ccmN-ccmO*, which were identified with the aid of mutants containing aberrant carboxysomes), are clustered upstream of *rbc* (43, 58, 104, 105, 120). In *Synechocystis* PCC 6803, on the other hand, the *ccm* genes are not located near *rbc* and are clustered in the order *ccmK-ccmK-ccmL-ccmM-ccmN* (94). A *ccmO* has not been recognized in *Synechocystis* but there are five different *ccmK* genes that show high homology to *ccmO*, and a *ccmA* gene (94). It is not established whether the different organization of the *ccm* clusters and their location are of functional significance in cyanobacteria.

Function of the Pyrenoids

Evidence indicating (*a*) that the fraction of Rubisco located in the pyrenoids rises during acclimation to low CO$_2$, reaching a state where nearly all the functional enzyme is concentrated in these organelles (86, 87), and (*b*) that CA is also detectable within these bodies (66, 87) strengthens the proposal that carboxysomes and pyrenoids are functionally analogous (87, 113). A *Chlamydomonas* mutant lacking Rubisco does not possess a pyrenoid (111), which suggests that, as in the case of carboxysomes, Rubisco is essential for the formation of the organelle. The observations that (*a*) a pyrenoid-possessing lichen, *Trebouxia erici*, utilizes low external CO$_2$ concentration more efficiently than *Coccomyxa* which lacks pyrenoids (98); and (*b*) lichens and bryophytes that possess pyrenoids discriminate less against ^{13}CO$_2$ (see next section, 135, 136) further reinforce the notion that pyrenoids play an important role in the CCM. They are surrounded by a starch layer, the thickness of which increases during acclimation to low CO$_2$ (28, 65). The significance, if any, of this sheath in the operation of the CCM remains obscure. Lack of the sheath in starch-less mutants of *Chlorella* (28) and *Chlamydomonas* (152) does not hamper their CCM activity.

In higher plant chloroplasts, a beta-type CA is soluble in the stroma (7). There is no evidence for the presence of CA in the stroma in pyrenoid-containing organisms (87) and it is not known whether the Ci species reach equilibrium there. In *Chlamydomonas*, an alfa-type CA, Cah3, has been detected with a leader sequence targeting it to the thylakoids (59). *Chlamydomonas* mutants where Cah3 is truncated (ca-1) or mistargeted (cia-3, because of a lesion in the leader sequence) were high CO$_2$-requiring (44). This phenotype could result either from lack of CA in the thylakoids (mutant ca-1) or from its expression in the stroma (mutant cia-3). The latter situation would be analogous to the cyanobacterial mutant where human CA was expressed in the cytoplasm (102). Thus the evidence to hand would be consistent with the operation of a CCM in pyrenoid-containing organisms largely analogous to that in those containing carboxysomes.

Catalyzed formation of CO$_2$ from HCO$_3^-$ in an acidic region (110) has been proposed as an alternative mechanism for CO$_2$ supply to the carboxylation site (106). It is envisaged that HCO$_3^-$ is transported into the thylakoid lumen through specific channels in parallel with the H^+ ions driven by the photosynthetic redox pump (110). In the thylakoid lumen, where the pH is substantially lower than the pKa, a local CA generates CO_2 that then diffuses towards Rubisco sites. This notion has been considered quantitatively by Raven (110) who calculated that an adequate HCO_3^- flux need only involve about one twelfth of the thylakoid area, thus avoiding universal uncoupling of photophosphorylation. Thylakoid segments transversing the pyrenoids were likely to comprise the required volume (Figure 3). This mechanism would be useful for those organisms where the capacity for Ci accumulation from the medium is low and where the internal Ci concentration is lower than ambient. The suggested arrangement seems plausible; however, for the system to operate efficiently, not only the HCO_3^- flux but also the location of the thylakoid CA (Cah3) should be confined to the intra-pyrenoid regions.

The thylakoid acid pool hypothesis resembles the carboxysome hypothesis in that they both envisage catalyzed CO_2 release in the close vicinity of tightly packed Rubisco molecules. They differ, however, in that the former requires an additional energy-dependent process to maintain the lumen at a pH that allows CO_2 formation at a rate adequate to support photosynthesis.

ECOPHYSIOLOGY AND ECOPHYSIOLOGICAL IMPLICATIONS

Our understanding of the functioning of the CCM is based almost entirely on laboratory studies and on model organisms. Assessment of its ecophysiological significance in nature requires comprehensive information as to the extent of expression of CCM activity under field conditions. Demonstration of CCM capability in members of the major groups of aquatic primary producers does not necessarily indicate that this potential is in fact realized in nature and data relevant to this problem are scarce (36, 37, 90, 110, 131).

The extent of CCM activity under natural conditions is linked to two major global ecological questions: (*a*) whether CO_2 availability plays a part in limiting primary oceanic productivity; and (*b*) what effect a rising atmospheric CO_2 concentration has on the latter. At first approach, rising CO_2 levels might be expected to boost productivity, particularly where CO_2 is limiting. The question as to whether such limitation in fact occurs in aqueous environments is still controversial (37). Wolf-Gladrow & Riebesell (156) concluded that under normally prevailing cell densities the photosynthetic carbon demand would be fulfilled by the diffusive supply of CO_2 from the bulk medium, even in the case of organisms unable to utilize other Ci species directly. CO_2 limitation would clearly be expected only where the rate of consumption exceeds the rate of diffusive supply, e.g. in massive phytoplankton blooms. Studies on *Peridinium gatunense* showed that, in spite of rising CCM activity in natural populations during the development of the annual bloom (12), the eventual population crash was due to CO$_2$ limitation leading to oxidative stress. On the other hand, CO$_2$-dependent growth even at lower cell densities has been reported for marine phytoplankton in a number of cases (23, 51). Because CO$_2$ scarcity would be expected to induce the development of the CCM, and this would largely over-come CO$_2$ limitation, a rise in CO$_2$ level would not necessarily boost primary productivity. In fact, down-regulation of the CCM by increasing ambient CO$_2$ might lead to declining productivity. Modulation of the CCM by a rising CO$_2$ concentration might also affect the associated Ci fluxes and hence the δ^{13}C of the internal Ci pool (see below).

The magnitude of the biological advantage conferred on CCM possessors will naturally depend on the energy cost entailed. Recently observed massive Ci cycling that increased with rising light intensity would be expected to have a substantial impact on the energy demand for CO$_2$ fixation, as well as on the maintenance and regulation of internal pH (because the excess OH$^-$ ions or pro-tons must be transported or neutralized). The eco-physiological significance of this apparently futile, energy-dependent, Ci circulation is not yet established (141, 148). The Ci fluxes continue to rise at light intensities that saturate pho-tosynthesis, which suggests that they may serve as a means to dissipate excess light energy and hence to minimize photodynamic damage. Moreover, they may also explain the relatively low level of the internal Ci pool observed in marine photosynthetic organisms (49, 110, 141) where net CO$_2$ efflux is partic-ularly prominent. The cycling may be of particular benefit to organisms that markedly modulate their light-energy–absorbing cross section and which may consequently be particularly vulnerable to photodamage by a sudden rise in light intensity.

Biogeochemical Aspects

The massive light-dependent Ci cycling between the cells and their medium also bears biogeochemical implications for the meaning and significance of the fractionation of stable C isotopes (δ^{13}C) observed in organic matter. Frac-tionation in photosynthetic organisms has been thought to result mainly from discrimination by Rubisco against the ^{13}C isotope (13, 110). Rubisco isolated from various organisms exhibits different degrees of discrimination (13). It is clear, however, that for a given value of δ^{13}C for the Rubisco reaction, the δ^{13}C in the organic matter produced will be determined by the isotopic com-position of the internal Ci pool that serves as a substrate for Rubisco. This isotopic composition will in turn depend on the flux balance between Ci influx, efflux, and net carboxylation and on the nature of the Ci species taken from the medium, CO$_2$ or HCO$_3^-$, because at equilibrium the former is approximately

7–9‰ lighter than the latter. Variations in $\delta^{13}C$ under various environmental conditions (13, 110), as well as those associated with different rates of growth, are likely to result from changes in these fluxes.

It has been proposed (132) that operation of the CCM would strongly affect the $\delta^{13}C$ of the organic carbon formed, owing to the tendency of the $^{13}CO_2$ discriminated against by Rubisco to accumulate in the internal pool. Equations that predict the degree of fractionation have been developed on the basis of schemes for the overall fluxes into and out of CCM-possessing cells (147). A fall in discrimination by *Chlamydomonas* cells has in fact been observed during acclimation and was concomitant with a rise in net CO_2 uptake (132). A model for fractionation in cyanobacteria, taking internal compartmentation into account (O Faber, D Tchernov, A Kaplan & L Reinhold, manuscript in preparation) shows that declining fractionation would only be obtained if resistance to CO_2 efflux developed progressively during acclimation. Such resistance would presumably be related to the density of packing of Rubisco molecules in the carboxysomes or pyrenoids (113).

The massive Ci cycling associated with the CCM thus impacts the validity of the present use of $\delta^{13}C$ for various purposes, e.g. as a CO_2 paleobarometer (108). In view of the light dependence of the cycling, the $\delta^{13}C$ of ancient sediments may in fact reflect not CO_2 availability but the light intensity prevailing during their formation. It is essential to understand the interrelations between the Ci fluxes and the $\delta^{13}C$ in order to interpret $\delta^{13}C$ values correctly and to evaluate use of the latter as a probe indicating past and present environmental change.

CONCLUDING REMARKS

The past decade has seen significant advances in this field, notably the recognition that the pyrenoid, long regarded as an enigmatic feature of the algal chloroplast, is in fact the seat of the carboxylation reaction and probably analogous to the carboxysome in prokaryotes. It has also seen the identification of CO_2-responsive genes, and the accumulation of experimental evidence confirming some of the key features of the currently held hypothesis for the working of the CCM. Elucidation of the function of the products of the genes already identified as having a role in the CCM, the mode of their regulation by changing ambient CO_2, and identification of more of the genes involved in its operation will promote our understanding of the CCM. Expression of the Ci transport genes in the chloroplast envelope of C3 plants might contribute towards raising their productivity under the prevailing limiting CO_2 concentration for C3 photosynthesis. The discovery of apparently futile massive Ci fluxes even in cells grown under abundant CO_2 supply raises the intriguing and testable speculation that the Ci transporting machinery may originally have evolved as a means of protection against photodynamic damage during eras where CO_2 was abundant. Subsequently, when ambient CO_2 level declined, the transporting system was incorporated into a system capable of concentrating CO_2 at the carboxylation site.

ACKNOWLEDGMENTS

We thank Drs. M Badger, B Bowien, G Espie, E Laws, J Moroney, N Mann, N Nimer, T Ogawa, T Omata, D Price, U Riebesell, J Shively, and M Spalding for providing results and manuscripts prior to publication. Our research was supported by the USA-Israel Binational Science Foundation (BSF), Jerusalem and the Israel Science foundation.

Visit the *Annual Reviews home page* at
http://www.AnnualReviews.org

Literature Cited

1. Abe T, Tsuzuki M, Miyachi S. 1987. Transport and fixation of inorganic carbon during photosynthesis of *Anabaena* grown under ordinary air. I. Active species of inorganic carbon utilized for photosynthesis. *Plant Cell Physiol.* 28:671–77
2. Amoroso G, Sueltemeyer D, Thyssen C, Fock HP. 1998. Uptake of HCO_3^- and CO_2 in cells and chloroplasts from the microalgae *Chlamydomonas reinhardtii* and *Dunaliella tertiolecta*. *Plant Physiol.* 116:193–201
3. Badger MR, Andrews TJ, Whitney SM, Ludwig M, Yellowlees DC, et al. 1998. The diversity and co-evolution of Rubisco, plastids, pyrenoids and chloroplast-based CCMs in the algae. *Can. J. Bot.* 76:1052–71
4. Badger MR, Kaplan A, Berry JA. 1980. The internal inorganic carbon pool of *Chlamydomonas reinhardtii*: evidence for a CO_2 concentrating mechanism. *Plant Physiol.* 66:407–13
5. Badger MR, Palmqvist K, Yu JW. 1994. Measurement of CO_2 and HCO_3^- fluxes in cyanobacteria and microalgae during steady-state photosynthesis. *Physiol. Plants* 90:529–36
6. Badger MR, Price GD. 1992. The carbon dioxide concentrating mechanism in cyanobacteria and microalgae. *Physiol. Plants* 84:606–15
7. Badger MR, Price GD. 1994. The role of carbonic anhydrase in photosynthesis. *Annu. Rev. Plant Physiol. Plant Mol. Biol.* 45:369–92
8. Badger MR, Spalding MH. 1999. CO_2 *Acquisition, Concentration and Fixation in Cyanobacteria and Algae*. Dordrecht, Netherlands: Kluwer. In press
9. Baker SH, Jin S, Aldrich HC, Howard GT, Shively JM. 1998. Insertion mutation of the form I cbbL of RuBisCO in *Thiobacillus neapolitanus* results in the expression of form II RuBisCO, the loss of carboxysomes, and an increased CO_2 requirement for growth. *J. Bacteriol.* 180:4133–39
10. Bedu S, Beuf L, Joset F. 1992. Membrane and soluble carbonic anhydrase activities in a cyanobacterium, *Synechocystis* PCC6803. See Ref. 88a, pp. 819–22
11. Berman-Frank I, Erez J. 1996. Inorganic carbon pools in the bloom-forming dinoflagellate *Peridinium gatunense*. *Limnol. Oceanogr.* 41:1780–89
12. Berman-Frank I, Kaplan A, Zohary T, Dubinsky Z. 1995. Carbonic anhydrase activity in the bloom-forming dinoflagellate *Peridinium gatunense*. *J. Phycol.* 31:906–13
13. Berry JA. 1994. Studies of mechanisms affecting the fractionation of carbon isotopes in photosynthesis. In *Stable Isotopes in Ecological Researches*, ed. PW Rundel, JR Ehleringer, KA Nagy, pp. 82–94. Berlin: Springer-Verlag
14. Beuf L, Bedu S, Durand MC, Joset F. 1994. A protein involved in co-ordinated regulation of inorganic carbon and glucose metabolism in the falcutative photoautotrophic cyanobacterium *Synechocystis* PCC6803. *Plant Mol. Biol.* 25:855–64
15. Bhaya D, Schwarz R, Grossman AR.

1999. Molecular response to environmental stress. In *Ecology of Cyanobacteria: Their Diversity in Time and Space*, ed. M Potts, BA Whitton. Dordrecht, Netherlands: Kluwer. In press

16. Bloye SA, Silman NJ, Mann NH, Carr NG. 1992. Bicarbonate concentration by *Synechocystis* PCC6803 modulation of protein phosphorylation and inorganic carbon transport by glucose. *Plant Physiol.* 99:601–6

17. Bonfil DJ, Ronen-Tarazi M, Sültemeyer D, Lieman-Hurwitz J, Schatz D, Kaplan A. 1998. A putative HCO_3^- transporter in the cyanobacterium *Synechococcus* sp. strain PCC 7942. *FEBS Lett.* 430:236–40

18. Borkhsenious ON, Mason CB, Moroney JV. 1998. The intracellular localization of ribulose-1,5-bisphosphate carboxylase/oxygenase in *Chlamydomonas reinhardtii. Plant Physiol.* 116:1585–91

19. Borodin V, Garderstrom P, Samuelsson G. 1994. The effect of light quality on the induction of efficient photosynthesis under low CO_2 conditions in *Chlamydomonas reinhardtii* and *Chlorella pyrenoidosa. Physiol. Plants* 92:254–60

20. Briggs GE, Whittingham CP. 1952. Factors affecting the rate of photosynthesis of *Chlorella* at low concentrations of carbon dioxide and in high illumination. *New Phytol.* 51:236–49

21. Brown II, Fadeyev SI, Kirik II, Severina II, Skulachev VP. 1990. Light-dependent delta Na-generation and utilization in the marine cyanobacterium *Oscillatoria brevis. FEBS Lett.* 270:203–6

22. Caslake LF, Gruber TM, Bryant DA. 1997. Expression of two alternative sigma factors of *Synechococcus* sp. strain PCC 7002 is modulated by carbon and nitrogen stress. *Microbiol. Uk.* 12:3807–18

23. Chen CJ, Durbin EG. 1994. Effects of pH on the growth and carbon uptake of marine phytoplankton. *Mar. Ecol. Prog. Ser.* 109:83–94

24. Chen ZY, Lavigne LL, Mason CB, Moroney JV. 1997. Cloning and overexpression of two cDNAs encoding the low-CO_2-inducible chloroplast envelope protein LIP-36 from *Chlamydomonas reinhardtii. Plant Physiol.* 114:265–73

25. Coleman JR. 1991. The molecular and biochemical analyses of CO_2 concentrating mechanisms in cyanobacteria and microalgae. *Plant Cell Environ.* 14:861–67

26. Colman B, Rotatore C. 1988. Uptake and accumulation of inorganic carbon by a freshwater diatom. *J. Exp. Bot.* 39:1025–32

27. Crotty CM, Tyrrell PN, Espie GS. 1994. Quenching of chlorophyll a fluorescence in response to Na^+-dependent HCO_3^--transport-mediated accumulation of inorganic carbon in the cyanobacterium *Synechococcus* UTEX 625. *Plant Physiol.* 104:785–91

28. Del Pino Plumed M, Villarejo A, De Los Rios A, Garcia Reina G, Ramazanov Z. 1996. The CO_2-concentrating mechanism in a starchless mutant of the green unicellular alga *Chlorella pyrenoidosa. Planta* 200:28–31

29. Dionisio ML, Tsuzuki M, Miyachi S. 1989. Blue light induction of carbonic anhydrase activity in *Chlamydomonas reinhardtii. Plant Cell Physiol.* 30:215–19

30. Drechsler Z, Sharkia R, Cabantchik ZI, Beer S. 1993. Bicarbonate uptake in the marine macroalga *Ulva* sp. is inhibited by classical probes of anion exchange by red blood cells. *Planta* 191:34–40

31. English RS, Jin S, Shively JM. 1995. Use of electroporation to generate a *Thiobacillus neapolitanus* carboxysome mutant. *Appl. Env. Microbiol.* 61:3256–60

32. Erez J, Bouevich A, Kaplan A. 1998. Carbon isotope fractionation by the freshwater cyanobacterium *Synechococcus* PCC 7942. *Can. J. Bot.* 76:1109–18

33. Eriksson M, Villand P, Gardestrom P, Samuelsson G. 1998. Induction and regulation of expression of a low-CO_2-induced mitochondrial carbonic anhydrase in *Chlamydomonas reinhardtii. Plant Physiol.* 116:637–41

34. Espie GS, Kandasamy RA. 1992. Na^+-independent HCO_3^- transport and accumulation in the cyanobacterium *Synechococcus* UTEX 625. *Plant Physiol.* 98:560–68

35. Espie GS, Kandasamy RA. 1994. Monensin inhibition of Na^+-dependent HCO_3^- transport distinguishes it from Na^+-independent HCO_3^- transport and provides evidence for Na^+/HCO_3^- symport in the cyanobacterium *Synechococcus* UTEX 625. *Plant Physiol.* 104:1419–28

36. Falkowski PG. 1994. The role of phytoplankton photosynthesis in global biogeochemical cycles. *Photosyn. Res.* 39:235–58

37. Falkowski PG, Raven J. 1997. *Aquatic Photosynthesis.* Oxford, UK: Blackwell Sci.

38. Fett JP, Coleman JR. 1994. Regulation of periplasmic carbonic anhydrase expression in *Chlamydomonas reinhardtii* by acetate and pH. *Plant Physiol.* 106:103–8

39. Fisher M, Gokhman I, Pick U, Zamir A. 1996. A salt-resistant plasma membrane carbonic anhydrase is induced by salt in *Dunaliella salina*. *J. Biol. Chem.* 271:17718–23

40. Forchammer K, Tandue de Marsac N. 1994. The PII protein in the cyanobacterium *Synechococcus* sp. strain PCC 7942 is modified by serine phosphorylation and signals the cellular N status. *J. Bacteriol.* 176:84–91

41. Fridlyand L, Kaplan A, Reinhold L. 1996. Quantitative evaluation of the role of a putative CO₂-scavenging entity in the cyanobacterial CO₂-concentrating mechanism. *Biosystems* 37:229–38

42. Fridlyand LE. 1997. Models of CO₂ concentrating mechanisms in microalgae taking into account cell and chloroplast structure. *Biosystems* 44:41–57

43. Friedberg D, Kaplan A, Ariel R, Kessel M, Seijffers J. 1989. The 5′-flanking region of the gene encoding the large subunit of ribulose-1,5-bisphosphate carboxylase/oxygenase is crucial for growth of the cyanobacterium *Synechococcus* sp. strain PCC 7942 at the level of CO₂ in air. *J. Bacteriol.* 171:6069–76

44. Funke RP, Kovar JL, Weeks DP. 1997. Intracellular carbonic anhydrase is essential to photosynthesis in *Chlamydomonas reinhardtii* at atmospheric levels of CO₂. *Plant Physiol.* 114:237–44

45. Gabay C, Lieman-Hurwitz J, Hassidim M, Ronen-Tarazi M, Kaplan A. 1998. Modification of *topA* in *Synechococcus* sp. PCC 7942 resulted in mutants capable of growing under low but not high concentration of CO₂. *FEMS Microbiol. Lett.* 159:343–47

46. Goyal A, Tolbert NE. 1990. Salicylhydroxamic acid (SHAM) inhibition of the dissolved inorganic carbon concentration process in unicellular green algae. *Plant Physiol.* 92:630–36

47. Goyal A, Tolbert NW. 1989. Uptake of inorganic carbon by isolated chloroplasts from air-adapted *Dunaliella*. *Plant Physiol.* 89:1264–69

48. Gruber TM, Bryant DA. 1998. Characterization of the alternative sigma-factors SigD and SigE in *Synechococcus* sp. strain PCC 7002. SigE is implicated in transcription of post-exponential-phase-specific genes. *Arch. Microbiol.* 169:211–19

49. Hassidim M, Keren N, Ohad I, Reinhold L, Kaplan A. 1997. Acclimation of *Synechococcus* strain WH7803 to ambient CO₂ concentration and to elevated light intensity. *J. Phycol.* 33:811–17

50. Hatch MD. 1992. C₄ Photosynthesis: an unlikely process full of surprises. *Plant Cell Physiol.* 33:333–42

51. Hein M, Sand-Jensen K. 1997. CO₂ increases oceanic primary production. *Nature* 388:526–27

52. Holthuijzen YA, Van Breemen JFL, Konings WN, Van Bruggen EFJ. 1986. Electron microscopic studies of carboxysomes of *Thiobacillus neopolitanus*. *Arch. Microbiol.* 144:258–62

53. Kaneko T, Sato S, Kotani H, Tanaka A, Asamizu E, et al. 1996. Sequence analysis of the genome of the unicellular cyanobacterium *Synechocystis* sp. strain PCC 6803. II. Sequence determination of the entire genome and assignment of potential protein-coding regions. *DNA Res.* 3:109–36

54. Kaplan A, Badger MR, Berry JA. 1980. Photosynthesis and intracellular inorganic carbon pool in the blue-green algae *Anabaena variabilis*: response to external CO₂ concentration. *Planta* 149:219–26

55. Kaplan A, Ronen Tarazi M, Zer H, Schwarz R, Tchernov D, et al. 1998. The inorganic carbon-concentrating mechanism in cyanobacteria: induction and ecological significance. *Can. J. Bot.* 76:917–24

56. Kaplan A, Schwarz R, Ariel R, Reinhold L. 1990. The CO₂ concentrating system in cyanobacteria: perspectives and prospects. In *Regulation of Photosynthetic Processes*, ed. R Kanai, S Katoh, S Miyachi, pp. 53–72. Bot. Mag., spec. issue

57. Kaplan A, Schwarz R, Lieman-Hurwitz J, Reinhold L. 1991. Physiological and molecular aspects of the inorganic carbon-concentrating mechanism in cyanobacteria. *Plant Physiol.* 97:851–55

58. Kaplan A, Schwarz R, Lieman-Hurwitz J, Ronen-Tarazi M, Reinhold L. 1994. Physiological and molecular studies on the response of cyanobacteria to changes in the ambient inorganic carbon concentration. In *The Molecular Biology of the Cyanobacteria*, ed. D. Bryant, pp. 469–85. Dordrecht, Netherlands: Kluwer

59. Karlsson J, Clarke AK, Chen ZY, Hugghins SY, Park YI, et al. 1998. A novel alpha-type carbonic anhydrase associated with the thylakoid membrane in *Chlamydomonas reinhardtii* is required for growth at ambient CO₂. *EMBO J.* 17:1208–16

60. Karlsson J, Ramazanov Z, Hiltonen T, Gardestrom P, Samuelsson G. 1993. Effect of vanadate on photosynthesis and

the ATP/ADP ratio in low-CO_2-adapted *Chlamydomonas reinhardtii* cells. *Planta* 192:46–51

61. Katoh A, Lee KS, Fukuzawa H, Ohyama K, Ogawa T. 1996. CemA homologue essential to CO_2 transport in the cyanobacterium *Synechocystis* PCC6803. *Proc. Natl. Acad. Sci. USA* 93:4006–10

62. Katoh A, Sonoda M, Katoh H, Ogawa T. 1997. Absence of light-induced proton extrusion in *cotA*-less mutant of *Synechocystis* sp. strain PCC6803. *J. Bacteriol.* 178:5452–55

63. Kondo T, Mori T, Lebedeva NV, Aoki S, Ishiura M, Golden SS. 1997. Circadian rhythms in rapidly dividing cyanobacteria. *Science* 275:224–27

64. Kramer D, Findenegg GR. 1978. Variations in the ultrastructure of *Scenedesmus obliquus* during adaptation to low CO_2 level. *Z. Pflanzen.* 89:407–10

65. Kuchitsu K, Tsuzuki M, Miyachi S. 1988. Changes of starch localization within the chloroplast induced by changes in carbon dioxide concentration during growth of *Chlamydomonas reinhardtii*: independent regulation of pyrenoid starch and stroma starch. *Plant Cell Physiol.* 29:1269–78

66. Kuchitsu K, Tsuzuki M, Miyachi S. 1991. Polypeptide composition and enzyme activities of the pyrenoid and its regulation by CO_2 concentration in unicellular green algae. *Can. J. Bot.* 69:1062–69

67. Li Q, Canvin DT. 1998. Energy sources for HCO_3^- and CO_2 transport in air-grown cells of *Synechococcus* UTEX 625. *Plant Physiol.* 116:1125–32

68. Lin S, Carpenter EJ. 1997. Pyrenoid localization of Rubisco in relation to the cell cycle and growth phase of *Dunaliella tertiolecta* (Chlorophyceae). *Phycologia* 36:24–31

69. Liotenberg S, Campbell D, Castets AM, Houmard J, Tandeau De Marsac N. 1996. Modification of the P-II protein in response to carbon and nitrogen availability in filamentous heterocystous cyanobacteria. *FEMS Microbiol. Lett.* 144:185–90

70. Maestri O, Fulda S, Hagemann M, Joset F. 1998. Variations of protein profiles upon shifts in inorganic carbon regime in the cyanobacterium *Synechocystis* PCC6803. *FEMS Lett.* 164:177–85

71. Marco E, Martinez I, Ronem Tarazi M, Orus MI, Kaplan A. 1994. Inactivation of *ccmO* in *Synechococcus* sp. strain PCC 7942 results in a mutant requiring high levels of CO_2. *Appl. Env. Microbiol.* 60:1018–20

72. Marco E, Ohad N, Schwarz R, Lieman-Hurwitz J, Gabay C, Kaplan A. 1993. High CO_2 concentration alleviates the block in photosynthetic electron transport in an *ndhB*-inactivated mutant of *Synechococcus* sp. PCC 7942. *Plant Physiol.* 101:1047–53

73. Marcus Y. 1997. Distribution of inorganic carbon among its component species in cyanobacteria: Do cyanobacteria in fact actively accumulate inorganic carbon? *J. Theor. Biol.* 185:31–45

74. Marcus Y, Schuster G, Michaels A, Kaplan A. 1986. Adaptation to CO_2 level and changes in the phosphorylation of thylakoid proteins during the cell cycle of *Chlamydomonas reinhardtii*. *Plant Physiol.* 80:604–7

75. Marek LF, Spalding MH. 1991. Changes in photorespiratory enzyme activity in response to limiting carbon dioxide in *Chlamydomonas reinhardtii*. *Plant Physiol.* 97:420–25

76. Martinez F, Orus MI. 1991. Interactions between glucose and inorganic carbon metabolism in *Chlorella vulgaris* strain UAM 101. *Plant Physiol.* 95:1150–55

77. Martinez I, Orus MI, Marco E. 1997. Carboxysome structure and function in a mutant of *Synechococcus* that requires high levels of CO_2 for growth. *Plant Physiol. Biochem.* 35:137–46

78. Matsuda Y, Colman B. 1995. Induction of CO_2 and bicarbonate transport in the green Alga *Chlorella ellipsoidea*. II. Evidence for induction in response to external CO_2 concentration. *Plant Physiol.* 108:253–60

79. McKay RML, Gibbs SP, Espie GS. 1993. Effect of dissolved inorganic carbon on the expression of carboxysomes, localization of Rubisco and the mode of inorganic carbon transport in cells of the cyanobacterium *Synechococcus* UTEX 625. *Arch. Microbiol.* 159:21–29

80. Mi H, Endo T, Ogawa T, Asada K. 1995. Thylakoid membrane-bound, NADPH-specific pyridine nucleotide dehydrogenase complex mediated cyclic electron transport in the cyanobacterium *Synechocystis* sp. PCC6803. *Plant Cell Physiol.* 36:661–68

81. Mi H, Endo T, Schreiber U, Ogawa T, Asada K. 1992. Electron donation from cyclic and respiratory flows to the photosynthetic intersystem chain is mediated by pyridine nucleotide dehydrogenase in the cyanobacterium *Synechocystis* PCC 6803. *Plant Cell Physiol.* 33:1233–37

82. Miller AG, Espie GE, Canvin DT. 1991.

Active CO$_2$ transport in cyanobacteria. *Can. J. Bot.* 69:925–35

83. Miller AG, Espie GS, Bruce D. 1996. Characterization of the non-photochemical quenching of chlorophyll fluorescence that occurs during the active accumulation of inorganic carbon in the cyanobacterium *Synechococcus* PCC 7942. *Photosynth. Res.* 49:251–62

84. Miller AG, Espie GS, Canvin DT. 1990. Physiological aspects of CO$_2$ and HCO$_3^-$ transport by cyanobacteria: a review. *Can. J. Bot.* 68:1291–302

85. Miller AG, Salon C, Espie GS, Canvin DT. 1997. Measurement of the amount and isotopic composition of the CO$_2$ released from the cyanobacterium *Synechococcus* UTEX 625 after rapid quenching of the active CO$_2$ transport system. *Can. J. Bot.* 75:981–97

86. Morita E, Kuroiwa H, Kuroiwa T, Nozaki H. 1997. High localization of ribulose-1,5-bisphosphate carboxylase/oxygenase in the pyrenoids of *Chlamydomonas reinhardtii* (Chlorophyta), as revealed by cryofixation and immunogold electron microscopy. *J. Phycol.* 33:68–72

87. Moroney JV, Chen ZY. 1998. The role of the chloroplast in inorganic carbon uptake by eukaryotic algae. *Can. J. Bot.* 76:1025–34

88. Moroney JV, Husic HD, Tolbert NE, Kitayama M, Manuel LJ, Togasaki RK. 1989. Isolation and characterization of a mutant of *Chlamydomonas reinhardtii* deficient in the carbon dioxide concentrating mechanism. *Plant Physiol.* 89:897–903

88a. Murata N, ed. 1992. *Research in Photosynthesis: Proc. Int. Congr. Photosynthesis.* Dordrecht, Netherlands: Kluwer

89. Nara M, Shiraiwa Y, Hirokawa T. 1989. Changes in the carbonic anhydrase activity and the rate of photosynthetic O$_2$ evolution during the cell cycle of *Chlorella ellipsoidea* C-27. *Plant Cell Physiol.* 30:267–76

90. Nimer NA, Iglesias-Rodriguez MD, Merrett MJ. 1997. Bicarbonate utilization by marine phytoplankton species. *J. Phycol.* 33:625–31

91. Ogawa T. 1991. Cloning and inactivation of a gene essential to inorganic carbon transport of *Synechocystis* PCC6803. *Plant Physiol.* 96:280–84

92. Ogawa T. 1991. A gene homologous to the subunit-2 gene of NADH dehydrogenase is essential to inorganic carbon transport of *Synechocystis* PCC6803. *Proc. Natl. Acad. Sci. USA* 88:4275–79

93. Ogawa T. 1993. Molecular analysis of the CO$_2$ concentrating mechanism in cyanobacteria. In *Photosynthetic Responses to the Environment*, ed. H Yamamoto, C Smith, pp. 113–25. Rockville, MD: Am. Soc. Plant Physiol.

94. Ohkawa H, Sonoda M, Katoh H, Ogawa T. 1998. The use of mutants in the analysis of the CCM in cyanobacteria. *Can. J. Bot.* 76:1025–34

95. Omata T. 1992. Characterization of the downstream region of *cmpA*: identification of a gene cluster encoding a putative permease of the cyanobacterium *Synechococcus* PCC7942. See Ref. 88a, pp. 807–10

96. Omata T, Ogawa T. 1986. Biosynthesis of a 42 kD polypeptide in the cytoplasmic membrane of the cyanobacterium *Anacystis nidulans* strain R$_2$ during adaptation to low CO$_2$ concentration. *Plant Physiol.* 80:525–30

97. Orus MI, Martinez F, Rodriguez ML, Marco E. 1993. Ultrastructural study of carboxysomes from high-CO$_2$-requiring mutants of *Synechococcus* PCC 7942. See Ref. 88a, pp. 787–90

98. Palmqvist K, De Los Rios A, Ascaso C, Samuelsson G. 1997. Photosynthetic carbon acquisition in the lichen photobionts *Coccomyxa* and *Trebouxia* (Chlorophyta). *Physiol. Plants* 101:67–76

99. Palmqvist K, Sültemeyer D, Baldet P, Andrews TJ, Badger MR. 1995. Characterisation of inorganic carbon fluxes, carbonic anhydrase(s) and ribulose-1,5-biphosphate carboxylase-oxygenase in the green unicellular alga *Coccomyxa*. *Planta* 197:352–61

100. Pierce J, Carlson TJ, Williams JGK. 1989. A cyanobacterial mutant requiring the expression of ribulose bisphosphate carboxylase from a photosynthetic anaerobe. *Proc. Natl. Acad. Sci. USA* 86:5753–57

101. Popova L, Balnokin Y, Dietz KJ, Gimmler H. 1998. Na$^+$-ATPase from the plasma membrane of the marine alga *Tetraselmis* (Platymonas) *viridis* forms a phosphorylated intermediate. *FEBS Lett.* 426:161–64

102. Price GD, Badger MR. 1989. Expression of human carbonic anhydrase in the cyanobacterium *Synechococcus* PCC7942 creates a high carbon dioxide-requiring phenotype: evidence for a central role for carboxysomes in the carbon dioxide concentrating mechanism. *Plant Physiol.* 91:505–13

103. Price GD, Coleman JR, Badger MR. 1992. Association of carbonic anhydrase activity with carboxysomes isolated

from the cyanobacterium *Synechococcus* PCC7942. *Plant Physiol.* 100:784–93

104. Price GD, Sültemeyer D, Klughammer B, Ludwig M, Badger MR. 1998. The functioning of the CO_2 concentrating mechanism in several cyanobacterial strains: a review of general physiological characteristics, genes, proteins and recent advances. *Can. J. Bot.* 76:973–1002

105. Price GD, Howitt SM, Harrison K, Badger MR. 1993. Analysis of a genomic DNA region from the cyanobacterium *Synechococcus* sp. strain PCC7942 involved in carboxysome assembly and function. *J. Bacteriol.* 175:2871–79

106. Pronina NA, Borodin VV. 1993. CO_2 stress and CO_2 concentration mechanism: investiguation by means of photosystem-deficient and carbonic anhydrase deficient-mutants of *Chlamydomonas reinhardtii*. *Photosynthetica* 28: 515–22

107. Ramazanov Z, Mason CB, Geraghty AM, Spalding MH, Moroney JV. 1993. The low CO_2-inducible 36-kilodalton protein is localized to the chloroplast envelope of *Chlamydomonas reinhardtii. Plant Physiol.* 101:1195–99

108. Rau GH. 1994. Variations in sedimentary organic $\delta^{13}C$ as a proxy for past changes in ocean and atmospheric [CO_2]. In *Carbon Cycling in the Glacial Ocean: Constraints on the Ocean's Role in Global Climate Change*, ed. R Zahn, M Kamiski, LD Labeyrie, TF Pedersen, pp. 307–22. Berlin: Springer

109. Rau GH, Riebesell U, Wolf Gladrow D. 1996. A model of photosynthetic ^{13}C fractionation by marine phytoplankton based on diffusive molecular CO_2 uptake. *Mar. Ecol. Prog. Ser.* 133:275–85

110. Raven JA. 1997. Inorganic carbon acquisition by marine autotrophs. *Adv. Bot. Res.* 27:85–209

111. Rawat M, Henk MC, Lavigne LL, Moroney JV. 1996. *Chlamydomonas reinhardtii* mutants without ribulose-1,5-bisphosphate carboxylase-oxygenase lack a detectable pyrenoid. *Planta* 198: 263–70

112. Rawat M, Moroney JV. 1995. The regulation of carbonic anhydrase and ribulose-1,5-bisphosphate carboxylase/oxygenase activase by light and CO_2 in *Chlamydomonas reinhardtii. Plant Physiol.* 109: 937–44

113. Reinhold L, Kosloff R, Kaplan A. 1991. A model for inorganic carbon fluxes and photosynthesis in cyanobacterial carboxysomes. *Can. J. Bot.* 69:984–88

114. Reinhold L, Zviman M, Kaplan A. 1987.

Inorganic carbon fluxes and photosynthesis in cyanobacteria—a quantitative model. In *Progress in Photosynthesis Research*, ed. J Biggins, 4:289–96. The Hague: Martinus Nijhoff

115. Reinhold L, Zviman M, Kaplan A. 1989. A quantitative model for inorganic carbon fluxes and photosynthesis in cyanobacteria. *Plant Physiol. Biochem.* 27:945–54

116. Reiskind JB, Madsen TV, Van Ginkel LC, Bowes G. 1997. Evidence that inducible C4-type photosynthesis is a chloroplastic CO_2-concentrating mechanism in *Hydrilla*, a submersed monocot. *Plant Cell Environ.* 20:211–20

117. Ritchie RJ, Nadolny C, Larkum AWD. 1996. Driving forces for bicarbonate transport in the cyanobacterium *Synechococcus* R-2 (PCC 7942). *Plant Physiol.* 112:1573–84

118. Rolland N, Dorne AJ, Amoroso G, Sültemeyer DF, Joyard J, Rochaix JD. 1997. Disruption of the plastid ycf10 open reading frame affects uptake of inorganic carbon in the chloroplast of *Chlamydomonas. EMBO J.* 16:6713–26

119. Ronen-Tarazi M, Bonfil DJ, Schatz D, Kaplan A. 1998. Cyanobacterial mutants impaired in bicarbonate uptake isolated with the aid of an inactivation library. *Can. J. Bot.* 76:942–48

120. Ronen-Tarazi M, Lieman-Hurwitz J, Gabay C, Orus MI, Kaplan A. 1995. The genomic region of *rbcLS* in *Synechococcus* sp. strain PCC7942 contains genes involved in the ability to grow under low CO_2 concentration and in chlorophyll biosynthesis. *Plant Physiol.* 108:1461–69

121. Ronen-Tarazi M, Schwarz R, Bouevitch A, Lieman-Hurwitz J, Erez J, Kaplan A. 1995. Response of photosynthetic microorganisms to changing ambient concentration of CO_2. In *Molecular Ecology of Aquatic Microbes*, NATO ASI series, ed. I Joint, pp. 323–34. Berlin: Springer-Verlag

122. Rotatore C, Colman B, Kuzma M. 1995. The active uptake of carbon dioxide by the marine diatoms *Phaeodactylum tricornutum* and *Cyclotella* sp. *Plant Cell Environ.* 18:913–18

123. Rotatore C, Lew RR, Colman B. 1992. Active uptake of carbon dioxide during photosynthesis in the green alga *Eremosphaera viridis* is mediated by a carbon dioxide-ATPase. *Planta* 188:539–45

124. Salon C, Mir NA, Canvin DT. 1996. Influx and efflux of inorganic carbon in *Synechococcus* UTEX 625. *Plant Cell Environ.* 19:247–59

125. Satoh A, Shiraiwa Y. 1996. Two polypeptides inducible by low levels of CO_2 in soluble protein fractions from *Chlorella regularis* grown at low or high pH. *Plant Cell Physiol.* 37:431–37

126. Satoh R, Himeno M, Wadano A. 1997. Carboxysomal diffusion resistance to ribulose 1,5-bisphosphate and 3-phosphoglycerate in the cyanobacterium *Synechococcus* PCC7942. *Plant Cell Physiol.* 38:769–75

127. Scanlan DJ, Bloye SA, Mann NH, Hodgson DA, Carr NG. 1990. Construction of *lacZ* promoter probe vectors for use in *Synechococcus*: application to the identification of carbon dioxide-regulated promoters. *Gene* 90:43–50

128. Schwarz R, Friedberg D, Kaplan A. 1988. Is there a role for the 42 kilodalton polypeptide in inorganic carbon uptake by cyanobacteria? *Plant Physiol.* 88:284–88

129. Schwarz R, Lieman-Hurwitz J, Hassidim M, Kaplan A. 1992. Phenotypic complementation of high-CO_2-requiring mutants of the cyanobacterium *Synechococcus* sp. PCC 7942 by inosine 5′-monophosphate. *Plant Physiol.* 100:1987–93

130. Schwarz R, Reinhold L, Kaplan A. 1995. Low activation state of ribulose 1,5-bisphosphate carboxylase/oxygenase in carboxysome-defective *Synechococcus* mutants. *Plant Physiol.* 108:183–90

131. Sekino K, Shiraiwa Y. 1994. Accumulation and utilization of dissolved inorganic carbon by a marine unicellular coccolithophorid, *Emiliania huxleyi*. *Plant Cell Physiol.* 35:353–61

132. Sharkey TD, Berry JA. 1985. Carbon isotope fractionation in algae as influenced by inducible CO_2-concentrating mechanism. In *Inorganic Carbon Uptake by Aquatic Photosynthetic Organisms*, ed. WJ Lucas, JA Berry, pp. 389–401. Rockville, MD: Am. Soc. Plant Physiol.

133. Shively JM, Bradburne CE, Aldrich HC, Bobik TA, Mehlman JL, et al. 1998. Sequence homologs of the carboxysomal polypeptide CsoS1 of the Thiobacilli are present in cyanobacteria and enteric bacteria that form carboxysomes/polyhedral bodies. *Can. J. Bot.* 76:906–16

134. Shively JM, van Keulen G, Meijer WM. 1998. Something from almost nothing: carbon dioxide fixation in chemoautotrophs. *Annu. Rev. Microbiol.* 52:191–230

135. Smith EC, Griffiths H. 1996. The occurrence of the chloroplast pyrenoid is correlated with the activity of a CO_2-concentrating mechanism and carbon isotope discrimination in lichens and bryophytes. *Planta* 198:6–16

136. Smith EC, Griffiths H. 1996. A pyrenoid-based carbon-concentrating mechanism is present in terrestrial bryophytes of the class *Anthocerotae*. *Planta* 200:203–12

137. So AKC, Espie GS. 1998. Cloning, characterization and expression of carbonic anhydrase from the cyanobacterium *Synechocystis* PCC 6803. *Plant Mol. Biol.* 37:205–15

138. Soltes Rak E, Mulligan ME, Coleman JR. 1997. Identification and characterization of a gene encoding a vertebrate-type carbonic anhydrase in cyanobacteria. *J. Bacteriol.* 179:769–74

139. Spalding MH. 1998. *CO₂ Acquisition: Acclimation to Changing Carbon Availability*. Dordrecht, Netherlands: Kluwer

140. Spalding MH, Jeffrey M. 1989. Membrane-associated polypeptides induced in *Chlamydomonas* by limiting CO_2 concentrations. *Plant Physiol.* 89:133–37

141. Sukenik A, Tchernov D, Huerta E, Lubian LM, Kaplan A, Livne A. 1997. Uptake, efflux and photosynthetic utilization of inorganic carbon by the marine eustigmatophyte *Nannochloropsis* sp. *J. Phycol.* 33:969–74

142. Sültemeyer D. 1997. Changes in the CO_2 concentrating mechanism during the cell cycle in *Dunaliella tertiolecta*. *Bot. Acta* 110:55–71

143. Sültemeyer D, Biehler K, Fock HP. 1993. Evidence for the contribution of pseudocyclic photophosphorylation to the energy requirement of the mechanism for concentrating inorganic carbon in *Chlamydomonas*. *Planta* 189:235–42

144. Sültemeyer D, Klughammer B, Badger MR, Price GD. 1998. Fast induction of high-affinity HCO_3^--transport in cyanobacteria. *Plant Physiol.* 116:183–92

145. Sültemeyer D, Price GD, Bryant DA, Badger MR. 1997. PsaE- and NdhF-mediated electron transport affect bicarbonate transport rather than carbon dioxide uptake in the cyanobacterium *Synechococcus* sp. PCC7002. *Planta* 201:36–42

146. Suzuki E, Shiraiwa Y, Miyachi S. 1994. The cellular and molecular aspects of carbonic anhydrase in photosynthetic microorganisms. *Prog. Phycol. Res.* 10:1–45

147. Takahashi K, Yoshioka T, Wada E, Sakamoto M. 1997. Temporal variations in carbon isotope ratio of phytoplankton in eutrophic lake. *J. Plankton Res.* 12:799–808

148. Tchernov D, Hassidim M, Luz B, Sukenik A, Reinhold L, Kaplan A. 1997. Sustained net CO_2 evolution during photosynthesis by marine microorganisms. *Curr. Biol.* 7:723–28

149. Tyrrell PN, Kandasamy RA, Crotty CM, Espie GS. 1996. Ethoxyzolamide differentially inhibits CO_2 uptake and Na^+-independent and Na^+-dependent HCO_3^- uptake in the cyanobacterium *Synechococcus* sp. UTEX 625. *Plant Physiol.* 112:79–88

150. Villand P, Eriksson M, Samuelsson G. 1997. Carbon dioxide and light regulation of promoters controlling the expression of mitochondrial carbonic anhydrase in *Chlamydomonas reinhardtii*. *Biochem. J.* 327:51–57

151. Villarejo A, Garcia Reina G, Ramazanov Z. 1996. Regulation of the low-CO_2-inducible polypeptides in *Chlamydomonas reinhardtii*. *Planta* 199:481–85

152. Villarejo A, Martinez F, Del Pino Plumed M, Ramazanov Z. 1996. The induction of the CO_2 concentrating mechanism in a starch-less mutant of *Chlamydomonas reinhardtii*. *Physiol. Plants* 98:798–802

153. Villarejo A, Martinez F, Ramazanov Z. 1997. Effect of aminooxyacetate, an inhibitor blocking the glycolate pathway, on the induction of a CO_2-concentrating mechanism and low-CO_2-inducible polypeptides in *Chlamydomonas reinhardtii* (Chlorophyta). *Eur. J. Phycol.* 32:141–45

154. Volokita M, Zenvirth D, Kaplan A, Reinhold L. 1984. Nature of the inorganic carbon species actively taken up by the cyanobacterium *Anabaena variabilis*. *Plant Physiol.* 76:599–602

155. Williams TG, Colman B. 1995. Quantification of the contribution of CO_2, HCO_3^-, and external carbonic anhydrase to photosynthesis at low dissolved inorganic carbon in *Chlorella saccharophila*. *Plant Physiol.* 107:245–51

156. Wolf-Gladrow D, Riebesell U. 1997. Diffusion and reactions in the vicinity of plankton: a refined model for inorganic carbon transport. *Mar. Chem.* 59:17–34

157. Yu JW, Price GD, Badger MR. 1994. Characterisation of CO_2 and HCO_3^- uptake during steady-state photosynthesis in the cyanobacterium *Synechococcus* PCC7942. *Aust. J. Plant Physiol.* 21:185–95

Annu. Rev. Plant Physiol. Plant Mol. Biol. 1999. 50:571–99
Copyright © 1999 by Annual Reviews. All rights reserved

PLANT COLD ACCLIMATION:
Freezing Tolerance Genes and Regulatory Mechanisms

Michael F. Thomashow
Department of Crop and Soil Sciences, Department of Microbiology, Michigan State University, East Lansing, Michigan 48824; e-mail: thomash6@pilot.msu.edu

KEY WORDS: cold acclimation, environmental stress, freezing tolerance, gene regulation, signal transduction

---

ABSTRACT

Many plants increase in freezing tolerance upon exposure to low nonfreezing temperatures, a phenomenon known as cold acclimation. In this review, recent advances in determining the nature and function of genes with roles in freezing tolerance and the mechanisms involved in low temperature gene regulation and signal transduction are described. One of the important conclusions to emerge from these studies is that cold acclimation includes the expression of certain cold-induced genes that function to stabilize membranes against freeze-induced injury. In addition, a family of Arabidopsis transcription factors, the CBF/DREB1 proteins, have been identified that control the expression of a regulon of cold-induced genes that increase plant freezing tolerance. These results along with many of the others summarized here further our understanding of the basic mechanisms that plants have evolved to survive freezing temperatures. In addition, the findings have potential practical applications as freezing temperatures are a major factor limiting the geographical locations suitable for growing crop and horticultural plants and periodically account for significant losses in plant productivity.

CONTENTS

571

1040-2519/99/0601-0571$08.00

INTRODUCTION

Over the course of the year, plants from temperate regions vary dramatically in their ability to survive freezing temperatures. In the warm growing seasons, such plants display little capacity to withstand freezing. However, as the year progresses, many sense changes in the environment that signal the coming winter and exhibit an increase in freezing tolerance. The primary environmental factor responsible for triggering this increase in freezing tolerance is low nonfreezing temperatures, a phenomenon known as cold acclimation. Nonacclimated rye, for instance, is killed by freezing at about $-5°C$, but after a period of exposure to low nonfreezing temperature can survive freezing down to about $-30°C$.

What genes have important roles in cold acclimation? What are their functions? How do plants sense low temperature and activate the cold-acclimation response? These are some of the key questions that investigators working in the field of cold acclimation are actively engaged in answering. Knowledge in these areas is not only important for an overall understanding of how plants sense and respond to changes in the environment, but also has potential practical applications. Freezing temperatures are a major factor determining the geographical locations suitable for growing crop and horticultural plants and periodically account for significant losses in plant productivity. Determining the nature of the genes and mechanisms responsible for freezing tolerance and the sensing and regulatory mechanisms that activate the cold-acclimation response provide the potential for new strategies to improve the freezing tolerance of agronomic plants. Such strategies would be highly significant as traditional plant breeding approaches have had limited success in improving freezing tolerance (100). The freezing tolerance of wheat varieties today, for instance, is only marginally better than those developed in the early part of this century (16).

The primary purpose of this review is to summarize recent developments regarding the identification of freezing tolerance genes and the regulatory and sensing mechanisms involved in controlling the cold acclimation process. To set the stage for these findings, however, I begin by presenting a brief overview of our current understanding of the causes of freezing injury and mechanisms of freezing tolerance. For additional coverage of these and other aspects of cold acclimation, the reader is referred to other reviews (27, 42, 43, 117) and books (67, 68, 99).

FREEZING INJURY AND TOLERANCE MECHANISMS

A wide range of studies indicate that the membrane systems of the cell are the primary site of freezing injury in plants (67, 105). In addition, it is well established that freeze-induced membrane damage results primarily from the severe dehydration associated with freezing (105, 108). As temperatures drop below 0°C, ice formation is generally initiated in the intercellular spaces due, in part, to the extracellular fluid having a higher freezing point (lower solute concentration) than the intracellular fluid. Because the chemical potential of ice is less than that of liquid water at a given temperature, the formation of extracellular ice results in a drop in water potential outside the cell. Consequently, there is movement of unfrozen water down the chemical potential gradient from inside the cell to the intercellular spaces. At −10°C, more than 90% of the osmotically active water typically moves out of the cells, and the osmotic potential of the remaining unfrozen intracellular and intercellular fluid is greater than 5 osmolar.

Multiple forms of membrane damage can occur as a consequence of freeze-induced cellular dehydration including expansion-induced-lysis, lamellar-to-hexagonal-II phase transitions, and fracture jump lesions (107, 117). Thus, a key function of cold acclimation is to stabilize membranes against freezing injury. Indeed, cold acclimation prevents expansion-induced-lysis and the formation of hexagonal II phase lipids in rye and other plants (107, 117). Multiple mechanisms appear to be involved in this stabilization. The best-documented are changes in lipid composition (107, 117). However, the accumulation of sucrose and other simple sugars that typically occurs with cold acclimation also seems likely to contribute to the stabilization of membranes as these molecules can protect membranes against freeze-induced damage in vitro (1, 111). In addition, as discussed below, there is emerging evidence that certain novel hydrophilic and LEA (late embryogenesis abundant) polypeptides also participate in the stabilization of membranes against freeze-induced injury.

Although freezing injury is thought to result primarily from membrane lesions caused by cellular dehydration, additional factors may also contribute to freezing-induced cellular damage. There is evidence that freeze-induced production of reactive oxygen species contributes to membrane damage (77) and that intercellular ice can form adhesions with cell walls and membranes and cause cell rupture (90). In addition, there is evidence that protein denaturation occurs in plants at low temperature (25), which could potentially result in cellular damage. In these cases, the enhancement of antioxidative mechanisms

(77), increased levels of sugars in the apoplastic space (73), and the induction of genes encoding molecular chaperones (32), respectively, could have protective effects.

In sum, we have gained important insights into both the causes of freezing injury and mechanisms involved in freezing tolerance. However, our overall understanding of freezing injury and tolerance mechanisms remains far from complete. We are not yet able to design from a core set of principles new wheat varieties that match the freezing tolerance of their close relative, rye, let alone create freezing-tolerant cucumber or banana plants. The identification and characterization of genes that have roles in freezing tolerance should contribute to the development of such "core principles" and potentially provide tools that can be used to improve plant freezing tolerance.

IDENTIFICATION AND CHARACTERIZATION OF FREEZING TOLERANCE GENES

Classical genetic studies have demonstrated that the ability of plants to cold acclimate is a quantitative trait involving the action of many genes with small additive effects [see (114)]. In recent years, three major approaches have been taken to determine the nature of genes with roles in freezing tolerance: the isolation and characterization of genes induced during cold acclimation; the isolation and characterization of mutants affected in freezing tolerance; and QTL mapping using molecular probes to identify freezing-tolerance loci. Recent studies have provided important advances in these areas.

Role of Cold-Regulated Genes in Cold Acclimation

In 1985, Guy et al (33) established that changes in gene expression occur with cold acclimation. Since then, considerable effort has been directed at determining the nature of cold-inducible genes and establishing whether they have roles in freezing tolerance. This has resulted in the identification of many genes that are induced during cold acclimation (see Supplementary Materials Section, http://www.annualreviews.org; Tables 1, 2, and 3 for examples). A large number of these genes encode proteins with known enzyme activities that potentially contribute to freezing tolerance (see Supplementary Materials Section, http://www.annualreviews.org; Table 1). For instance, the *Arabidopsis FAD8* gene (18) encodes a fatty acid desaturase that might contribute to freezing tolerance by altering lipid composition. Cold-responsive genes encoding molecular chaperones including a spinach *hsp70* gene (2) and a *Brassica napus hsp90* gene (58) might contribute to freezing tolerance by stabilizing proteins against freeze-induced denaturation. Also, cold-responsive genes encoding various signal transduction and regulatory proteins have been identified including a mitogen-activated protein (MAP) kinase (80), a MAP kinase kinase kinase (81), calmodulin-related proteins (95), and 14-3-3 proteins (50). These proteins might contribute to freezing tolerance by controlling the expression of freezing tolerance genes or by regulating the activity of proteins involved in freezing tolerance. Whether these cold-responsive genes actually have important roles in freezing tolerance, however, remains to be determined.

Although many of the genes that are induced during cold acclimation encode proteins with known activities, many do not. Indeed, the largest "class" of cold-induced genes encode polypeptides that are either newly discovered (http://www.annualreviews.org; Table 2) or are homologs of LEA proteins (http://www.annualreviews.org; Table 3), polypeptides that are synthesized late in embryogenesis, just prior to seed desiccation, and in seedlings in response to dehydration stress (9, 13, 44). The polypeptides encoded by the cold-regulated novel and LEA genes fall into a number of families based on amino acid sequence similarities [see (115)]. Interestingly, however, most of these have a set of distinctive properties in common: They are unusually hydrophilic; many remain soluble upon boiling in dilute aqueous buffer; many have relatively simple amino acid compositions, being composed largely of a few amino acids; many are composed largely of repeated amino acid sequence motifs; and many are predicted to contain regions capable of forming amphipathic α-helices. For example:

- *Arabidopsis COR15a* (70) encodes a novel 15-kDa polypeptide that is targeted to the stromal compartment of the chloroplasts ((70); SJ Gilmour & MF Thomashow, unpublished results). The mature 9.4-kDa polypeptide COR15am (Figure 1) is highly hydrophilic; remains soluble upon boiling; is rich is alanine, lysine, glutamic acid, and aspartic acid (they account for 64% of the amino acid residues); and is composed largely of a 13-amino acid sequence that is repeated (imperfectly) four times. Regions of the polypeptide that include the repeated sequences are predicted to form amphipathic α-helices.

- Alfalfa *cas15* (82) encodes a novel 15-kDa polypeptide CAS15 (Figure 1) that is highly hydrophilic; is rich is glutamate, glycine, histidine, and lysine (they account for 68% of the amino acid residues); and nearly a third of the protein is composed of a 10-amino acid sequence that is repeated (imperfectly) four times. Regions of the polypeptide that include the repeated sequences are predicted to form amphipathic α-helices.

- Wheat *wcs120* (41) encodes a 39-kDa polypeptide WCS120 (Figure 1) that is a member of the LEA II group of polypeptides (9). It is highly hydrophilic; remains soluble upon boiling; is rich in glycine, histidine, and threonine

Arabidopsis COR15am (novel hydrophilic)

```
AKGDGNILDDLNEATK KASDFVTDKTKEA LADGE
                 KAKDYVVEKNSET ADTLGKEAE
                 KAAAYVEEKGKEA AN
                 KAAEFAEGKAGEA KDATK
```

Alfalfa CAS15 (novel hydrophilic)

```
MAGIMNKIGDALHGGGDKKEGEH KGEQHGHVGG EHHGEY
                        KGEQHGFVGG HAGDH
                        KGEQHGFVGG HGGDY
                        KGEQHGFGHG DHKEGYHGEEHKEGFADKIKDKIHGEGADGEK
                                   KKKKEKKKHGEGHEHGHDSSSSDSD
```

Wheat WCS120 (LEA II)

```
MENQAHIA GEKKGIMEKIKEKLPGGHGDHKE TAGTHGHPGTATHGAPA TGGAYGQQGHAGTT GTGLHGAHA
         GEKKGVMENIKDKLPGGHQDHQQ                   TGGTYGQQGHTGTA THGTPA
                                                   TGGTYGQQGHTGTA THGTPA
                                                   TGGTYGEQGHTGVT GTGTHGT
         GEKKGVMENIKEKLPGGHGDHQQ                   TGGTYGQQGHTGTA THGTPA
                                                   GGGTYEQHGHTGMT GTGTHGT
         GEKKGVMENIKDKLPGGHGDHQQ                   TGGTYGQQGHTGTA TQGTPA
                                                   GGGTYEQHGHTGMT GAGTHST
         GEKKGVMENIKEKLPGGHSDHQQ                   TGGAYGQQGHTGTR HMAPL
                                                   PAGTYGQHGHAGVI GTETHGTTA
                                                   TGGTHGQHGHTGTT GTGTHGSDGI

         GEKKSLMDKIKDKLPGQH
```

Barley HVA1 (LEA III)

```
MASNQNQGSYHAGETKARTEEKTGQM MGATKQKAGQT
                           TEATKQKAGET
                           AEATKQKTGET
                           AEAAKQKAAEA KDKTAQT
                           AQAAKDKTYET
                           AQAAKERAAQG KDQTGSALGEK
                           TEAAKQKAAET
                           TEAAKQKAAEA
                           TEAAKQKASDT AQYTKESAVAGKDKTGSVLQQAGE
                                       TVVNAVVGAKDAVANTLGMGGDNT
                                       SATKDATTGATVKDTTTTTRNH
```

Figure 1   Examples of novel hydrophilic and LEA proteins encoded by genes induced during cold acclimation. The amino acid sequences are presented with the repeated motifs aligned (in bold) to highlight this attribute of each protein. Additional details about the proteins are presented in the text.

residues (they account for 54% of the amino acid residues); and is composed largely of a lysine-rich sequence, referred to as a K-segment, that is repeated (imperfectly) six times and a glycine-rich sequence, referred to as a ϕ-segment, that is repeated (imperfectly) 11 times. The K-segments are present in all LEA II proteins and are predicted to form amphipathic α-helices (9).

- Barley *HVA1* (37, 38) encodes a 22-kDa polypeptide HVA1 (Figure 1) that is member of the LEA III group of proteins (13). It is highly hydrophilic;

rich in alanine, lysine, and threonine (they account for 53% of the amino acid residues); and is composed largely of an 11-amino acid sequence that is repeated (imperfectly) nine times. The repeated sequences are present in all LEA III proteins and are predicted to form amphipathic α-helices (13).

An intriguing possibility regarding the similar biochemical properties shared by many of the novel hydrophilic and LEA polypeptides is that they may reflect some common underlying mechanism of action. Indeed, a possibility discussed in the following section on *Arabidopsis COR* genes is that the amphipathic α-helical regions predicted to be present in many of the novel and LEA proteins may have roles in stabilizing membranes against freezing damage. Whether the regions predicted to form amphipathic α-helices actually form such structures is uncertain (71). Regardless, based on the expression characteristics of the cold-induced novel and LEA genes, namely their induction in response to conditions associated with water deficit—abscisic acid (ABA), drought, high salt, osmotic stress, and seed desiccation—and the close relationship between freezing and dehydration injury, it has often been suggested that these genes might contribute directly to freezing tolerance by protecting cells against the potentially damaging effects of dehydration associated with freezing. Summarized below are recent results obtained with the *Arabidopsis COR* and spinach *CAP* genes that provide direct support for this hypothesis.

ARABIDOPSIS COR GENES   The *COR* genes—also designated *LTI* (low temperature-induced), *KIN* (cold-inducible), *RD* (responsive to desiccation), and *ERD* (early dehydration-inducible)—comprise four gene families, each of which is composed of two genes that are physically linked in the genome in tandem array (88, 122, 124–126, 130). At least one member of each gene pair is induced in response to low temperature or other conditions associated with water deficit including drought, high salinity, and ABA. The *COR78* (40, 88, 130), *COR15* (70, 126) and *COR6.6* (59, 60, 122) gene pairs encode newly discovered "boiling soluble" hydrophilic polypeptides (see Supplementary Materials Section, http://www.annualreviews.org; Table 2). The *COR47* gene pair encodes hydrophilic boiling soluble polypeptides that belongs to the LEA II protein family (see Supplementary Materials Section, http://www.annualreviews.org; Table 3), also known as dehydrins and LEA D11 proteins (19, 124, 125).

COR15a expression enhances freezing tolerance   Artus et al (4), working with *COR15a*, provided the first direct evidence for a cold-induced gene having a role in freezing tolerance. As mentioned above, *COR15a* encodes a 15-kDa polypeptide that is targeted to the chloroplasts. Upon import into the organelle, COR15a is processed to a 9.4-kDa polypeptide designated COR15am. Artus et al (4) demonstrated that constitutive expression of *COR15a* in nonacclimated transgenic *Arabidopsis* plants increases the freezing tolerance of both chloroplasts frozen in situ and isolated leaf protoplasts frozen in vitro by 1 to 2°C over the temperature range of −4 to −8°C. In these experiments, it appeared that expression of *COR15a* might also have a slight negative effect on freezing tolerance of protoplasts over the temperature range of −2 to −4°C, but subsequent results by Steponkus et al (106) indicate that this is not the case. It was originally assumed that the protoplasts isolated from the leaves of nonacclimated transgenic plants would have the same intracellular osmolality as those isolated from wild-type plants. In fact, the intracellular osmolality of protoplasts isolated from *COR15a* transgenic plants is approximately .413 osm, whereas that of protoplasts isolated from wild-type plants is about .400 osm (106). The reason for this slight difference is not known. However, when it is taken into account, protoplast survival tests indicate that expression of *COR15a* has only a positive effect on freezing tolerance over the temperature range of −4 to −8°C.

Function of COR15a   How does expression of *COR15a* bring about increased freezing tolerance? The results of Artus et al (4) indicated that it involves the stabilization of membranes. This conclusion followed from the fact that protoplast survival was measured using fluorescein diacetate, a vital stain that reports on retention of the semipermeable characteristic of the plasma membrane. The question then became how *COR15a* expression brings about this effect. One possibility was suggested by the work of Steponkus and colleagues (108, 117). These investigators have shown that the formation of hexagonal II phase lipids is a major cause of membrane damage in nonacclimated plants, including *Arabidopsis* (116), over the temperature range of about −4 to −8°C (other forms of membrane damage occur at higher and low temperature due to lesser and greater degrees of dehydration, respectively). Thus, the possibility was that *COR15a* expression might decrease the propensity of membranes to form hexagonal II phase lipids in response to freezing. Indeed, Steponkus et al (106) have found that over the temperature range of −4.5 to −7°C, expression of *COR15a* decreases the incidence of freezing-induced lamellar-to-hexagonal II phase transitions that occur in regions where the plasma membrane is brought into close apposition with the chloroplast envelope as a result of freezing-induced dehydration.

How does COR15am decrease the propensity of membranes to form the hexagonal II phase? More specifically, how can COR15am, which is located in the stromal compartment of the chloroplast, defer to lower temperatures (and thus lower degrees of hydration) the formation of lamellar-to-hexagonal II phase transitions resulting from the interaction of the chloroplast envelope with the plasma membrane? An elegant hypothesis has recently been proposed by Steponkus to explain the phenomenon (106). It is based, in part, on the fact that certain amphipathic α-helices have been shown to have a strong effect on the intrinsic curvature of monolayers and to affect their propensity to form the hexagonal II phase (14). In brief, Steponkus has proposed that COR15am defers freeze-induced formation of the hexagonal II phase to lower temperatures by altering the intrinsic curvature of the inner membrane of the chloroplast envelope. The onset (the freezing temperature) of hexagonal II phase formation is suggested to be determined by the membrane that has the greatest propensity to form the hexagonal II phase. For the chloroplast envelope-plasma membrane ensemble, there is evidence that the "weak link" is the inner membrane of the chloroplast envelope [see (106)]. Thus, COR15am, which is predicted to have regions that form amphipathic α-helices, is envisioned to alter the intrinsic curvature of the monolayer that comprises the inner membrane of the chloroplast envelope such that its propensity to form the hexagonal II phase is deferred to lower temperatures. A sensitive test of whether a polypeptide has an effect on monolayer curvature is to determine whether the polypeptide causes a shift in the lamellar-to-hexagonal II phase transition temperature. Indeed, Steponkus et al (106) have found that the COR15am polypeptide increases the lamellar-to-hexagonal II phase transition temperature of dioleoylphosphatidylethanolamine and promotes formation of the lamellar phase in a lipid mixture composed of the major lipid species that comprise the chloroplast envelope.

The intrinsic curvature hypothesis advanced by Steponkus provides a possible mechanism for how COR15am increases the freezing tolerance of plant cells. In addition, it may also explain how many of the other novel hydrophilic and LEA polypeptides contribute to increased freezing tolerance. As discussed above, many of these proteins are predicted to contain regions that form amphipathic α-helices. Thus, the intriguing possibility raised is that they too might help stabilize membranes against the dehydration associated with freezing—as well as other environmental conditions such as drought and high salinity—by affecting the intrinsic curvature of membrane monolayers. It will be of interest to determine whether any of the other novel hydrophilic and LEA polypeptides cause a shift in the lamellar-to-hexagonal II phase transition temperature of lipid mixtures. It will also be important to determine whether the regions of the novel and LEA proteins predicted to form amphipathic α-helices actually form such structures (71).

Induction of the CRT/DRE-regulon enhances freezing tolerance   Do any other *Arabidopsis COR* genes have roles in freezing tolerance? To address this question, Jaglo-Ottosen et al (49) made transgenic *Arabidopsis* plants that constitutively express the entire battery of *COR* genes and compared the freezing tolerance of these plants to those that expressed *COR15a* alone. Induction of the *COR* genes was accomplished by overexpressing the *Arabidopsis* transcriptional activator CBF1 (CRT/DRE binding factor 1) (109). This factor, which is discussed in greater detail below, binds to the CRT (C-repeat)/DRE (drought responsive element) DNA regulatory element present in the promoters of the *COR* genes (and presumably other as yet unidentified cold-regulated genes) and activates their expression without a low temperature stimulus (49, 109). What Jaglo-Ottosen et al (49) found was that expression of *CBF1* resulted in a greater increase in freezing tolerance than did expressing *COR15a* alone. In one set of experiments, the electrolyte leakage test (112) was used to assess the freezing tolerance of detached leaves from nonacclimated transgenic and wild-type plants. In this test, plant tissues are frozen to various temperatures below zero degrees Celsius, thawed, and cellular damage is estimated by determining the amount of electrolytes that leach out of the cells, a sign that the plasma membrane has lost its semipermeable characteristic. Jaglo-Ottosen et al (49) did not detect a significant enhancement of freezing tolerance by expressing *COR15a* alone, but detected a 3.3°C increase in freezing tolerance in plants that overexpressed CBF1 and, consequently, the CRT/DRE-regulon of genes. In addition, induction of the CRT/DRE-regulon resulted in an increase in whole plant freezing survival whereas expression of *COR15a* alone did not. Taken together, these results implicate additional cold-regulated genes in freezing tolerance. In addition, given the nature of the electrolyte leakage test, the results indicate that a role of these freezing tolerance genes is to protect membranes against freezing injury.

Liu et al (72) have independently demonstrated that induction of the CRT/DRE-regulon increases the freezing tolerance of *Arabidopsis* plants. In their case, they activated gene expression by overexpressing a homolog of *CBF1*, designated *DREB1A*. Significantly, the results of Liu et al (72) indicate that expression of the CRT/DRE-regulon not only increases freezing tolerance, but also increases tolerance to drought. This finding provides strong support for the notion that a fundamental role of cold-inducible genes is to protect plant cells against cellular dehydration. One additional important finding was that overexpression of *DREB1A* resulted in a dwarf phenotype. This phenotype was not observed by Jaglo-Ottosen et al (49) in plants that overexpress *CBF1*. The reason for this difference is not yet known. It could be due to differences in the level of expression of the transcriptional activators in the two studies. Alternatively, it might be due to differences in the activators used in the experiments. Regardless, the results of both studies provide direct evidence that the CRT/DRE-regulon includes genes that have fundamental roles in cold acclimation.

SPINACH CAP GENES   Guy and colleagues have been studying a group of genes, designated *CAP* (cold acclimation protein) (29–31), that are induced during cold acclimation. Two of these, *CAP85* (26, 86) and *CAP160* (26), are also induced in response to dehydration and ABA and encode hydrophilic polypeptides that remain soluble upon boiling. The CAP85 protein belongs to the LEA II group of proteins; the CAP160 polypeptide is novel, but has a low degree of sequence similarity with the *Arabidopsis* COR78 protein. Both CAP85 and CAP160 are soluble and appear to be located primarily in the cytoplasm, though a portion of CAP160 fractionates with mitochondria, apparently due to an association of the protein with the organelle.

To determine whether the *CAP85* and *CAP160* genes might have roles in freezing tolerance, Kaye et al (54) made transgenic tobacco plants that constitutively expressed the spinach CAP85 and CAP160 proteins and assessed the freezing tolerance of the plants using the electrolyte leakage test. The results suggested that production of the CAP85 and CAP160 proteins, either individually or in combination, had no discernible effect on the freezing tolerance of detached leaves; i.e. the EL_{50} values (temperature that resulted in leakage of 50% of electrolytes) were the same for control and transgenic plants. However, additional experimentation indicated that the proteins slowed the rate of freeze-induced cellular damage; the amount of electrolyte leakage with time of freezing at $-2°C$ was less in the transgenic plants. Thus, both proteins have a detectable effect on freezing tolerance. Again, given the nature of the electrolyte leakage test, it appears that both proteins act to stabilize the plasma membrane against freezing injury. In this regard, it is intriguing that CAP85 is a LEA II protein containing K-segments predicted to form amphipathic α-helices (discussed above) and that regions of CAP160 are predicted to form amphipathic α-helices. Whether these regions of the protein are critical for their apparent activity and whether they affect the intrinsic curvature of membrane monolayers would be interesting to determine.

OTHER POTENTIAL FREEZING TOLERANCE PROTEINS   A number of other proteins that accumulate with cold acclimation seem likely to contribute to freezing tolerance (115). These include the cryoprotectin protein of spinach (103), the WSC120 protein family of wheat (101), and the antifreeze proteins that have been described in rye and other cereals (3, 23), bittersweet nightshade (11), and carrot (128). Direct evidence that these proteins have roles in freezing tolerance, however, is not yet available.

Identification of Freezing Tolerance Genes by Mutational Analysis

ARABIDOPSIS SFR GENES   Warren and colleagues (78, 123) have used a mutational approach to identify genes in *Arabidopsis* that have roles in freezing tolerance. They screened M_3 seed pools, derived from 1804 chemically mutagenized M_2 plants, for lines that displayed no adverse effects during the cold acclimation treatment (i.e. did not display a chilling-sensitive phenotype that may have indirectly affected freezing tolerance), but did not attain normal levels of freezing tolerance. These efforts resulted in the identification of five

SFR (sensitivity to freezing) genes that appear to have significant roles in cold acclimation: *SFR1, 2, 4, 5,* and *6.* Whereas wild-type *Arabidopsis* seedlings that are cold-acclimated for 2 weeks at 4°C suffer no obvious damage upon being frozen at −6°C for 24 h (followed by incubation at normal growth temperature), seedlings carrying the *sfr1, 2, 4, 5*-1, *5*-2, and *6* mutant alleles do, the nature of which varies with the mutation. The *sfr1* mutation affects the freezing tolerance of only young leaves; the *sfr6* mutation has its most severe effects on young leaves, but affects all leaves to some extent; and the *sfr2, 4, sfr5*-1, and *sfr5*-2 mutations affect all leaves equally. Significantly, all of the mutations affect the cryostability of the plasma membrane, as indicated by the electrolyte leakage test. With the *sfr1, 4, 5,* and *6* mutations, the severity of the freezing damage observed in the whole plant freezing tests corresponds with the results of the electrolyte leakage test. Thus, the freezing sensitivity caused by these mutations appears to result largely from a decrease in membrane cryostability. In contrast, the *sfr2* mutation results in severe injury in the whole plant freeze test, but only minor damage in the electrolyte leakage test. Thus, the freezing-sensitive lesion caused by this mutation might not have a primary effect on cellular membranes.

Determining the nature and functions of the *SFR* genes should provide significant new insight into our understanding of the cold-acclimation response. Indeed, the *sfr2* and *sfr5* mutations do not have any obvious effects on the alterations in fatty acid composition or the increases in sucrose and anthocyanin levels that normally occur with cold acclimation. Thus, the study of these genes may lead to the discovery of freezing-tolerance mechanisms that have not yet been considered. In contrast, the *sfr4* mutation results in reduced accumulation of sucrose, glucose, and anthocyanin and lowered levels of 18:1 and 18:2 fatty acids. Given the likely role of sugars as cryoprotectants and roles of fatty acid composition in membrane cryostability, it is reasonable to speculate that the effects that the *sfr4* mutation have on sugar and fatty acid composition account, at least in part, for the freezing-sensitive phenotype of these mutants. How the *sfr4* mutation brings about such pleiotropic effects, however, is less clear. An interesting possibility, however, is that the *SFR4* gene may have a role in regulating the activation of cold acclimation.

ARABIDOPSIS ESKIMO1 GENE   Xin & Browse (129) have also used a mutational approach to identify *Arabidopsis* genes with important roles in cold acclimation. In their case, the investigators screened 800,000 chemically mutagenized M_2 seedlings for mutants that displayed "constitutive" freezing tolerance; i.e. mutants that were more freezing tolerant than wild-type plants without cold acclimation. This resulted in the identification of a gene, *eskimo1* (*esk1*), that has a major effect on freezing tolerance. Whereas nonacclimated wild-type plants had an LT_{50} of $-5.5°C$ in a whole-plant freeze test, nonacclimated *esk1* mutant plants had an LT_{50} of $-10.6°C$. Moreover, the *esk1* mutation increased the freezing tolerance of cold-acclimated plants. Wild-type plants that had been cold-acclimated had an LT_{50} of $-12.6°C$, while cold-acclimated *esk1* plants had an LT_{50} of $-14.8°C$.

The molecular basis for the increase in freezing tolerance displayed by the *esk1* mutation is not yet certain. However, the concentration of free proline in the *esk1* mutant was found to be 30-fold higher than in wild-type plants. It seems likely that this dramatic increase in proline contributes to the increased freezing tolerance of the *esk1* plants as proline has been shown to be an effective cryoprotectant in vitro (7, 98). In addition, total sugars are elevated in the *esk1* mutant about twofold and expression of the *RAB18* cold-responsive LEA II gene is elevated about threefold. These alterations may also contribute the increase in freezing tolerance. Significantly, the *esk1* mutation does not appear to affect expression of the *COR* genes; the transcript levels for *COR15a*, *COR6.6*, *COR47*, and *COR78* remained at low levels under normal growth conditions in the *esk1* plants and were greatly induced in response to low temperature. Xin & Browse (129) suggested that these results may mean that there are multiple signaling pathways involved in activating different aspects of the cold-acclimation response and that activation of one pathway may result in considerable freezing tolerance without activation of the other pathways. As discussed above, overexpression of the CBF1 transcription factor induces expression of the CRT/DRE-regulon and results in a significant increase in freezing tolerance. The "CBF1 pathway" might, therefore, control one set of cold-acclimation responses. Similarly, the *ESK1* gene may participate in the control of another set of freezing tolerance responses that includes synthesis of proline, and to a lesser degree, the synthesis of sugars and expression of *RAB18*. The mechanism of *ESK1* action is not known. However, the fact that the two available *esk1* alleles are recessive suggests that *ESK1* may act as a negative regulator (129).

Mapping Freezing Tolerance Genes—The Wheat Vrn1-Fr1 Interval

The ability of plants to cold acclimate is a quantitative trait [see references in (114)]. Indeed, in wheat, there is evidence that nearly all chromosome pairs can contribute to freezing tolerance. Recent studies have identified a locus on chromosome 5A, the *Vrn1-Fr1* interval, that has a major effect on freezing tolerance (17).

The *Vrn1-Fr1* interval contains the *Vrn1* gene, a major determinant of growth habit (6, 96, 104). Winter-type plants, which are sown in autumn, carry recessive *vrn1* alleles. Such plants require a period of vernalization (exposure to low temperature) to promote floral development. The vernalization requirement is thought to have evolved to insure that overwintering plants do not flower before the warm growing season. In contrast, spring-type plants can be sown in spring as they carry dominant *Vrn1* alleles that allow floral development without vernalization. Significantly, winter-type plants carrying *vrn1* alleles are almost exclusively more freezing tolerant than spring-type plants carrying *Vrn1* alleles, which indicates that either *Vrn1* itself is a freezing tolerance gene(s) or that it is tightly linked to a freezing tolerance gene(s). The results of Galiba et al (17) support the latter possibility. These investigators analyzed the progeny from a cross between substitution lines of 'Chinese Spring' (a spring-type wheat) carrying 5A chromosomes from either 'Cheyenne', a freezing-tolerant winter wheat or a freezing-sensitive spring-type accession of *Triticum spelta*. Among the progeny they found a single recombinant line that carried the *vrn1* allele (i.e. was a winter-type) but was freezing sensitive. The freezing tolerance gene linked to *Vrn1* was designated *Fr1*.

Additional important information about the *Vrn1-Fr1* interval has come from a study by Storlie et al (110). These investigators addressed the question of whether differences in freezing tolerance among winter wheat varieties involved differences at the *Vrn1-Fr1* interval. This was accomplished by examining the freezing tolerance of near isogenic lines (NILs) carrying *Vrn1-Fr1* intervals from different varieties. Specifically, NILs were derived from five back-crosses between 'Marfed', a freezing-sensitive (LT_{50}–8.2°C) spring wheat that was used as the recurrent parent, and two winter wheat donor parents that differed in freezing tolerance, 'Suweon 185' (LT_{50}–13.6°C) and 'Chugoku 81' (LT_{50}–12.7°C). An analysis of the progeny indicated that those carrying the winter *vrn1-Fr1* locus were about 4°C more freezing tolerant than those carrying the spring *Vrn1-fr1* locus. Also, progeny carrying the *vrn1-Fr1* locus from 'Suweon 185' were about 0.5°C more freezing tolerant than those carrying the *vrn1-Fr1* locus from 'Chugoku 81'. The *Vrn1-Fr1* interval accounted for 70 to 90% of the difference in the freezing tolerance of the NILs, substantiating the importance of this locus in cold acclimation. In addition, the results indicate that differences in freezing tolerance between winter cultivars can, in at least some cases, result from differences at this locus.

The mechanism whereby the *Vrn1-Fr1* interval affects freezing tolerance remains to be determined. Limin et al (69) have shown that cold-induced expression of the *wcs120* genes, which are located on chromosomes 6A, 6B, and 6C, is higher in a winter-type 'Chinese Spring' ('Cheyenne' 5A) substitution line than in the spring-type parent 'Chinese Spring'. In addition, the freezing tolerance of the 'Chinese Spring' ('Cheyenne' 5A) line is greater than that of 'Chinese Spring'. Thus, the possibility raised is that the *Vrn1-Fr1* interval encodes a protein(s) involved in regulating the expression of cold-inducible genes that have roles in freezing tolerance.

A final point regards conservation of the *Vrn1-Fr1* interval in plants. QTL mapping in barley using a cross between the winter variety 'Dicktoo', which is relatively freezing tolerant, and the spring variety 'Morex', which is relatively freezing sensitive, has resulted in the identification of a 21-cM region on chromosome 7 that has a major role in freezing tolerance (35, 93, 119). This region accounted for 32% of the variance in LT_{50} freezing tolerance values and 39–79% of the variance in winter field survival observed in the population. In addition, the region accounted for 47% of the variation for the winter-spring growth habit. Thus, this region encodes major genes for both freezing tolerance and vernalization. Significantly, these genes are likely to correspond to those included in the *Vrn1-Fr1* interval of wheat. This is suggested by the findings that chromosome 7 of barley is homologous to the group 5 chromosomes of wheat (48); the Xwg644 and Xcdo504 molecular markers that are linked to the *Vrn1-Fr1* locus of wheat are contained within the 21-cM freezing tolerance interval of barley [see discussion in (17)]; and the *Sh2* vernalization locus of barley is linked to the Xwg644 marker (65). Whether the freezing tolerance gene(s) contained within the *Vrn1-Fr1* interval is present outside of cereals remains to be determined.

REGULATION OF THE COLD ACCLIMATION RESPONSE

Current evidence suggests that multiple mechanisms are involved in activating the cold-acclimation response. As discussed above, the *Arabidopsis eskimo1* gene (129) appears to have a role in regulating freezing tolerance that is independent of the mechanism that regulates expression of the *Arabidopsis* freezing tolerance CRT/DRE-regulon. Moreover, cold-regulated gene expression itself involves multiple mechanisms including transcriptional (5, 40, 121, 130) and post-transcriptional (12, 94) processes and both "ABA-dependent" (64, 88) and "ABA-independent" pathways (21, 87, 124, 131). In addition, the changes in lipid composition and accumulation of sugars that are likely to contribute to freezing tolerance do not necessarily rely on changes in gene expression, but may be brought about, at least in part, by alterations in the activities of enzymes involved in their synthesis. A complete understanding of how the cold-acclimation response is activated by low temperature will require considerable effort. However, significant insights have begun to emerge.

The CBF/DREB1 Regulatory Genes

There is direct evidence that the *Arabidopsis COR* genes have roles in cold acclimation (49, 72). Thus, understanding how these genes are activated by low temperature should reveal at least one pathway important in regulating the cold-acclimation response. Toward this end, recent studies have led to the identification of the *CBF/DREB1* family of regulatory genes, "master-switches" (100) involved in *COR* gene induction and cold acclimation.

REGULATION OF COR GENES BY THE CBF/DREB1 TRANSCRIPTIONAL ACTIVATORS Gene fusion studies have demonstrated that the promoters of the *Arabidopsis COR15a* (5), *COR6.6* (121), and *COR78* (40, 130) genes are induced in response to low temperature. The cold-regulatory element that appears to be primarily responsible for this regulation was first identified by Yamaguchi-Shinozaki & Shinozaki (131) in their study of the *RD29A* (*COR78*) promoter. It is a 9-bp element, TACCGACAT, referred to as the DRE (dehydration responsive element). The DRE, which has a 5-bp core sequence of CCGAC designated the CRT (C-repeat) (5), stimulates gene expression in response to low temperature, drought, and high salinity, but not exogenous application of ABA (131). The element is also referred to as the LTRE (low temperature regulatory element) (51, 88).

Stockinger et al (109) isolated the first cDNA for a protein that binds to the CRT/DRE sequence. The protein, designated CBF1 (CRT/DRE binding factor 1), has a mass of 24 kDa, a putative bipartite nuclear localization sequence, and an acidic region that potentially serves as an activation domain. In addition, it has an AP2 domain, a 60-amino acid motif that has been found in a large number of plant proteins including *Arabidopsis* APETALA2 (52), AINTEGUMENTA (55), and TINY (127); the tobacco EREBPs (ethylene response element binding proteins) (89); and numerous other plant proteins of unknown function [see (97)]. Ohme-Takagi & Shinshi (89) have demonstrated that the AP2 domain includes a DNA-binding region. Interestingly, Stockinger et al (109) noted that the tobacco ethylene response element, AGCCGCC, closely resembles CRT/DRE sequences, GGCCGAC and TACCGAC, present in the promoters of *Arabidopsis COR15a* and *COR78*, respectively. Thus, Stockinger et al (109) suggested that CBF1, the EREBPs, and perhaps other AP2 domain proteins may be members of a superfamily of DNA binding proteins that recognize a family of *cis*-acting regulatory elements that have CCG as a common core sequence. Differences in the sequence surrounding the CCG core element were suggested to result in recruitment of distinct AP2 domain proteins that are integrated into signal transduction pathways activated by different environmental, hormonal, and developmental cues.

The CBF1 protein binds to the CRT/DRE sequence and activates expression of reporter genes in yeast carrying the CRT/DRE as an upstream regulatory sequence (109). These results indicated that CBF1 is a transcriptional activator that can activate CRT/DRE-containing genes and, thus, was a probable regulator of *COR* gene expression in *Arabidopsis*. Indeed, as discussed above,

Jaglo-Ottosen et al (49) have shown that constitutive overexpression of *CBF1* in transgenic *Arabidopsis* plants results in expression of CRT/DRE-controlled *COR* genes without a low temperature stimulus. Thus, CBF1 appears to be an important regulator of the cold-acclimation response, controlling the level of *COR* gene expression, which, in turn, promotes freezing tolerance.

The results of Stockinger et al (109) and Jaglo-Ottosen et al (49) have recently been extended by Gilmour et al (22), Liu et al (72), and Shinwari et al (102). These investigators have established that *CBF1* is a member of a small gene family encoding three closely related transcriptional activators. The three genes, referred to as either *CBF1*, *CBF2*, and *CBF3* (22) or *DREB1B*, *DREB1C*, and *DREB1A*, respectively (72, 102), are physically linked in direct repeat on chromosome 4 near molecular markers PG11 and m600 (~71 cM) (22, 102). They are unlinked to their target CRT/DRE-controlled genes, *COR6.6*, *COR15a*, *COR47*, and *COR78*, which are located on chromosomes 5, 2, 1, and 5, respectively (22). Like CBF1, both the CBF2 and CBF3 proteins can activate expression of reporter genes in yeast that contain the CRT/DRE as an upstream activator sequence, indicating that these two family members are also transcriptional activators (22). Indeed, Liu et al (72) have shown that overexpression of *DREB1A/CBF3* in transgenic *Arabidopsis* plants results in constitutive expression of *RD29A* (*COR78*) and, as described above, enhances both the freezing and drought tolerance of the transgenic plants.

LOW TEMPERATURE REGULATION OF THE CBF/DREB1 GENES   The transcript levels for all three *CBF/DREB1* genes increase dramatically within 15 min of transferring plants to low temperature, followed by accumulation of *COR* gene transcripts at about 2 h (22, 72). Thus, Gilmour et al (22) suggested that *COR* gene expression involves a low-temperature signaling cascade in which *CBF* gene expression is an early step. Regulation of the *CBF/DREB1* genes appears to occur, at least in part, at the transcriptional level as hybrid genes containing the *CBF/DREB1* promoters fused to reporter genes are induced at low temperature (102; D Zarka, M Thomashow, unpublished results). As noted by Gilmour et al (22), the fact that *CBF* transcripts begin accumulating within 15 min of plants being exposed to low temperature strongly suggests that the low-temperature "thermometer" and "signal transducer" are present at warm noninducing temperatures. Gilmour et al (22) have, therefore, proposed that there is a transcription factor already present at warm temperature that recognizes the *CBF* promoters. This factor would not appear to be the CBF proteins themselves as the promoters of the *CBF* genes lack the CRT/DRE sequence and overexpression of *CBF1* does not cause accumulation of *CBF3* transcripts (22). Gilmour et al (22) have, therefore, proposed that *COR* gene induction involves a two-step cascade of transcriptional activators in which the first step,

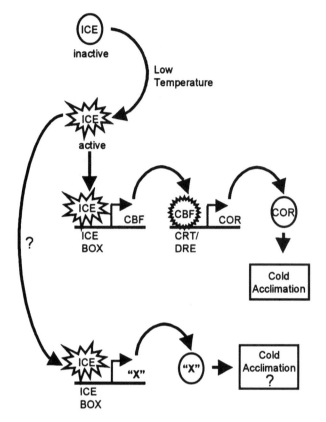

Figure 2   Model for *CBF* regulation of *COR* gene expression (22). See text for details. Reprinted by permission of *The Plant Journal*, 1999.

CBF induction, involves an unknown activator that they tentatively designated "ICE" (inducer of CBF expression) (Figure 2). ICE presumably recognizes a cold-regulatory element, the "ICE Box," present in the promoters of each *CBF* gene. At warm temperature, ICE is suggested to be in an "inactive" state, either because it is sequestered in the cytoplasm by a negative regulatory protein or is in a form that does not bind to DNA or does not activate transcription effectively. Upon exposing a plant to low temperature, however, a signal transduction pathway is suggested to be activated that results in modification of either ICE or an associated protein, which, in turn, allows ICE to induce *CBF* gene expression. As noted by Gilmour et al (22), it is possible that ICE may not only regulate the expression of the *CBF* genes, but might induce expression of other genes ("X") that may also have roles in cold acclimation.

As previously mentioned, the CRT/DRE not only imparts cold-regulated gene expression, but also dehydration-regulated gene expression. The fact that the *CBF* genes are up-regulated in response to low temperature is consistent with the CBF proteins acting at the CRT/DRE to induce expression of *COR* and other cold-regulated CRT/DRE-controlled genes. However, the results of both Gilmour et al (22) and Liu et al (72) indicate that the *CBF* genes are not induced significantly in response to dehydration stress. How then does the CRT/DRE impart dehydration-induced gene expression? Liu et al (72) have provided an answer for this question. They have identified two additional genes, *DREB2A* and *DREB2B*, that are induced in response to dehydration stress and encode proteins that bind to the CRT/DRE. The DREB2 proteins have AP2 domains that are very similar in sequence to those found in the CBF/DREB1 proteins. Outside of the AP2 domains, however, the DREB2 and CBF/DREB1 proteins share little sequence similarity. Liu et al (72) have proposed that induction of the *DREB2* genes leads to the synthesis of DREB2 proteins that bind to the CRT/DRE and activates gene expression. Interestingly, unlike overexpression of the *CBF* genes that results in constitutive expression of the CRT/DRE-controlled *COR* genes, constitutive overexpression of *DREB2A* has only a minor effect on expression of *COR78*. Thus, Liu et al (72) have suggested that the *DREB2* proteins are likely to be activated posttranslationally in response to dehydration stress.

CONSERVATION OF THE CBF/DREB1 REGULATORY PATHWAY   An important question is whether the CBF/DREB1 regulatory pathway is highly conserved in plants. The available data are sketchy on this point, but lend support to the notion that it is. In particular, Singh and colleagues (51) have shown that the promoter of the cold-regulated *Brassica napus BN115* gene (an ortholog of *Arabidopsis COR15a*) is induced in response to low temperature and that this induction is dependent on the action of CRT-containing LTREs present within the promoter. Similarly, Sarhan and colleagues have shown that the cold-regulated *wcs120* gene of wheat has a cold-inducible promoter with two putative CRT/LTREs (120), and have presented results of a deletion analysis that are consistent with the CRT/LTREs having a role in low-temperature induction of the promoter (92). In addition, these investigators have presented results indicating that the *Wcs120* promoter is cold-inducible in the monocotyledonous plants barley, rye, and rice, as well as the dicotyledonous plants alfalfa, *Brassica*, and cucumber (though not in tomato and pepper) (92).

Role of ABA

In 1983, Chen et al (8) reported that ABA levels increase transiently in response to low temperature in *Solanum commersonii*, a plant that cold acclimates, but not in *S. tuberosum*, a plant that does not cold acclimate. Moreover, they found that exogenous application of ABA increases the freezing tolerance of *S. commersonii* plants at warm temperatures. These findings led Chen and colleagues to hypothesize that cold acclimation is activated through the action of ABA; i.e. that low temperature brings about an increase in ABA, which triggers the activation of freezing tolerance mechanisms. Subsequent studies extended the observations of Chen et al (8), establishing that ABA levels increase, at least transiently, in a diverse group of plant species during cold acclimation (28, 62, 63, 74) and that exogenous application of ABA at warm nonacclimating temperature enhances the freezing tolerance of several plant species that cold acclimate (8, 45, 91).

If ABA has a critical role in activating the cold-acclimation response, then one would expect that the freezing tolerance of plants carrying mutations in ABA synthesis or the ability to respond to ABA would be less than that of wild-type plants. This has been shown to be the case in *Arabidopsis*. Seedlings carrying either the *aba1* (formerly *aba-1*) or *abi1* mutations that, respectively, result in impairment of ABA synthesis and insensitivity to ABA [see (57)], are less freezing tolerant than wild-type plants (21, 36, 75); *ABA1* encodes a zeaxanthin epoxidase (76) and *ABI1* encodes a phosphatase 2C (66, 79). These results have led some to conclude that ABA does indeed have a key role in activating cold acclimation. However, the increase in ABA that occurs in response to low temperature is transient in *Arabidopsis*—ABA levels peak at 24 h and return to essentially normal levels by two days (63)—yet freezing tolerance continues to increase for about a week and remains elevated for at least three weeks (20, 61, 116). Moreover, although the *aba1* and *abi1* mutations result in a decrease in *Arabidopsis* freezing tolerance, these mutations also have pleiotropic effects. Plants carrying either the *aba1* or *abi1* mutation display a wilted phenotype and have reduced vigor, for instance. Thus, Gilmour & Thomashow (21) cautioned that the decrease in freezing tolerance caused by the ABA mutations might not be a direct consequence of ABA having a fundamental role in activating cold-acclimation mechanisms, but instead, might be an indirect effect of ABA having important integral roles in plant growth and development. Simply put, "sick" plants might not be able to cold acclimate as well as "healthy" plants.

If ABA does not have a role in activating the cold-acclimation response, why would its exogenous application increase the freezing tolerance of *Arabidopsis* and other plants? A possible explanation is offered by considering the regulation of *Arabidopsis COR* genes. As discussed above, expression of the *Arabidopsis* CRT/DRE-regulon at normal growth temperatures results in an increase in plant freezing tolerance (49, 72). Significantly, at least some of the genes in this regulon, including members of the *COR* gene family—*COR78*,

COR47, *COR15a*, and *COR6.6*—are highly expressed in response to exogenous application of ABA (19, 34, 60, 88, 130). The activation of these genes by exogenous ABA is consistent with observing an increase in freezing tolerance.

Does ABA have an important role in activating expression of CRT/DRE-controlled genes during cold acclimation? There are indications that it does not. It has been shown that cold-induced expression of *COR78*, *COR47*, and *COR6.6* is essentially normal in plants carrying the *aba1* mutation (21, 87). Moreover, the *abi1* mutation essentially abolishes ABA-induced accumulation of transcripts for *COR78*, *COR47*, and *COR6.6*, but has little or no effect on cold-induced accumulation of these transcripts (21, 87). Thus, both Nordin et al (87) and Gilmour & Thomashow (21) proposed that cold-regulated expression of these genes occurs through an ABA-independent pathway. The discovery of the DRE DNA regulatory element by Yamaguchi-Shinozaki & Shinozaki (131) seemingly proved this hypothesis correct. As mentioned above, the element, which is present in the promoters of *COR78*, *COR47*, *COR15a*, and *COR6.6*, imparts cold-regulated gene expression, but does not stimulate transcription in response to exogenous application of ABA. Recently, however, the notion of an ABA-independent pathway regulating CRT/DRE-controlled *COR* genes has been challenged by Zhu and colleagues (47). These investigators have reported the isolation of *Arabidopsis* mutants that "hyper-express" *COR78* in response to both cold and ABA, as well as mutants that are diminished in their expression of *COR78* in response to both cold and ABA. Thus, Zhu and colleagues have proposed that cold and ABA regulatory pathways are not completely independent, but instead have points at which they "cross-talk." Thus, it is not yet certain whether cold-regulated expression of the freezing tolerance CRT/DRE-regulon is completely independent of ABA.

Is cold-regulated expression of any gene dependent on the action of ABA? This appears to be the case for the *Arabidopsis RAB18* and *LTI65* genes; cold-induced accumulation of transcripts for these genes is severely impaired in plants carrying either the *aba1* or *abi1* mutations (64, 88, 124). ABA-regulated expression of these genes is presumably mediated through the action of the putative ABA-responsive elements (ABREs) (24) present in the promoters of these genes (64, 88, 131). As has been shown for other ABA-regulated genes, bZIP transcription factors are likely to bind to these elements in cold-regulated genes and activate gene expression (15). The induction of both *RAB18* and *LTI65* in response to low temperature, however, is very weak. Indeed, one study concluded that *LTI65* (*RD29B*) is not responsive to low temperature at all (131). In contrast, both of these genes are highly responsive to exogenous application of ABA and to dehydration stress (64, 88, 131). The dramatic difference in the relative responses of these genes to drought and low temperature is not surprising, however, as ABA levels are induced to much higher levels in drought-stressed plants than they are in cold-treated plants (63). Whereas low temperature brings about a transient three- to fourfold increase in ABA content, drought stress brings about more than a 20-fold increase in ABA levels (63).

In sum, the issue of whether ABA has a fundamental role in activating the cold-acclimation response beyond its "general" role in plant growth and development is unresolved. The available evidence seems to suggest that ABA has a relatively minor role in inducing the expression of genes in response to low temperature, but firm conclusions even in this regard are probably premature. A final resolution of the role of ABA in activating the cold-acclimation response will require a more detailed understanding of the specific mechanisms that have roles in freezing tolerance and a determination of whether ABA has a critical role in regulating the activity of these mechanisms.

Low Temperature Signal Transduction

ROLE OF CALCIUM   There is mounting evidence that calcium is an important second messenger in a low temperature signal transduction pathway involved in regulating the cold-acclimation response (56, 83, 85, 113). In both *Arabidopsis* (56, 95) and alfalfa (83), cytoplasmic calcium levels increase rapidly in response to low temperature, due largely to an influx of calcium from extracellular stores. Through the use of chemical and pharmacological reagents, it has been shown that this increase in calcium is required for full expression of at least some cold-regulated genes, including the CRT/DRE-controlled *COR6.6* and *KIN1* genes of *Arabidopsis*, and for plants to increase in freezing tolerance (56, 83, 85, 113). For instance, Dhindsa and colleagues (83, 85) have shown that in alfalfa, calcium chelators such as BAPTA (1,2-*bis*(*o*-aminophenoxy)ethane *N*, *N*, *N'*, *N'*-tetraacetic acid) and calcium channel blockers such as La^{3+} inhibit cold-induced influx of calcium and cause both decreased expression of the cold-inducible *cas15* gene and block the ability of alfalfa to cold acclimate. In addition, they have shown that *cas15* expression can be induced at 25°C by treating cells with A23187, a calcium ionophore that causes a rapid influx of calcium.

Two important issues now are to identify the nature of the channels that are presumably responsible for the influx of calcium that occurs with low temperature in *Arabidopsis* and alfalfa and to determine the steps between calcium influx and the activation of gene expression and cold acclimation. In regard to channels, it is of interest that onion has been reported to have a mechanosensitive calcium-selective cation channel that is activated in response to low temperature (10). As for the steps in signal transduction following calcium influx, little is known. However, recent results strongly suggest that protein phosphorylation is involved. In particular, Monroy et al (84) have shown that low temperature induction of alfalfa *cas15* is inhibited by the protein kinase inhibitor staurosporine and is induced at 25°C by the protein phosphatase inhibitor okadaic acid. More-over, they have found that low temperature causes a rapid and dramatic decrease in protein phosphatase 2A (PP2A) activity and that this is dependent on calcium influx. Taken together, these results suggest that low temperature leads to an influx in calcium, which inhibits PP2A activity, and that this, in turn, leads to the phosphorylation of one or more proteins involved in inducing *cas15* expression and activating cold acclimation.

The protein kinase(s) responsible for inducing the expression of cold-regu-lated genes and activating freezing tolerance mechanisms is not known. There are, however, a number of interesting candidates. One is a mitogen-activated protein (MAP) kinase described by Jonak et al (53). This kinase, designated p44^{MMK4}, is activated within 10 min of alfalfa plants being exposed to low temperature. Significantly, two other alfalfa MAP kinases, MMK2 and MMK3, are not activated by low temperature, indicating that there is specificity in cold activation of MAP kinases (53). It is also interesting that the transcript levels for p44^{MMK4} increase rapidly (within 20 min) in response to low temperature (though the amount of p44^{MMK4} protein does not change). Indeed, the transcript levels for a number of protein kinases have been shown to increase in response to low temperature. In *Arabidopsis*, genes encoding a MAP kinase kinase kinase, an S6 ribosomal protein kinase, and a MAP kinase are simultaneously induced in response to low temperature, as well as touch and dehydration stress (81). In addition, there is evidence that the transcript levels for calcium-dependent protein kinases (CDPKs) in *Arabidopsis* (113) and alfalfa (83) accumulate in response to low temperature, as do transcripts for an *Arabidopsis* receptor-like protein kinase (39) and two *Arabidopsis* two-component response regulator-like proteins (118). At present, all of these kinases would seem to be candidates for having roles in cold acclimation.

ROLE OF ARABIDOPSIS HOS1 GENE   As alluded to earlier, Zhu and colleagues (47) have described the isolation of *Arabidopsis* mutants altered in cold-regu-lated gene expression. One mutation identified, *hos1*-1, alters the temperature at which the *RD29A* promoter becomes activated; in wild-type plants, the pro-moter is essentially inactive until the temperature falls below 10°C, whereas low-level induction of the promoter can be detected in the *hos1*-1 mutant even at 19°C (46). Moreover, the *hos1*-1 mutation results in "superinduction" of *RD29A, COR47, COR15a, KIN1*, and *ADH* in response to low temperature (46). Genetic analysis indicates that the *hos1*-1 mutation is recessive. Thus Zhu and colleagues (46) have suggested that the *HOS1* gene encodes a negative regula-tor of low temperature signal transduction. Interestingly, the *hos1*-1 mutation also results in decreased expression of *COR15a, KIN1, RAB18*, and *RD29B* in response to ABA, high salt, and high osmoticum (polyethylene glycol). Thus, the functions of *HOS1* would appear to include a positive role in the induction of at least some genes in response to ABA or osmotic stress (46). Determining the nature of the gene product encoded by *HOS1* should add significantly to an understanding of low temperature gene regulation and, potentially, the interaction of low temperature and ABA signaling pathways.

CONCLUDING REMARKS

Cold acclimation research is in a very exciting phase. Genes and proteins with roles in freezing tolerance are being identified, their mechanisms of action determined, and insights into how the cold-acclimation response is activated in response to low temperature are emerging. In addition, novel strategies for improving plant freezing tolerance are being considered in light of the new results. As discussed by Storlie et al (110), it may be possible to exploit allelic variation at the *Vrn1-Fr1* interval of winter wheat to improve the freezing tolerance of this important crop. Also, as initially alluded to by Artus et al (4), it may be possible to use the *Arabidopsis CBF* genes (or orthologs from other plants) as "master switches" (100) to "manage" activation of freezing tolerance regulons and thereby improve freezing (and possibly drought) tolerance in a broad range of plants. Indeed, it would seem that the next few years promise to bring a burst of fundamental new discoveries regarding the mechanisms and regulation of cold acclimation and efforts to design and evaluate new approaches to improve plant freezing tolerance.

> Visit the *Annual Reviews home page* at
> http://www.AnnualReviews.org

Literature Cited

1. Anchordoguy TJ, Rudolph AS, Carpenter JF, Crowe JH. 1987. Modes of interaction of cryoprotectants with membrane phospholipids during freezing. *Cryobiology* 24:324–31
2. Anderson JV, Li QB, Haskell DW, Guy CL. 1994. Structural organization of the spinach endoplasmic reticulum-luminal 70-kilodalton heat-shock cognate gene and expression of 70-kilodalton heat-shock genes during cold acclimation. *Plant Physiol.* 104:1359–70
3. Antikainen M, Griffith M. 1997. Antifreeze protein accumulation in freezing-tolerant cereals. *Physiol. Plant.* 99:423–32
4. Artus NN, Uemura M, Steponkus PL, Gilmour SJ, Lin CT, Thomashow MF.

1996. Constitutive expression of the cold-regulated *Arabidopsis thaliana COR15a* gene affects both chloroplast and protoplast freezing tolerance. *Proc. Natl. Acad. Sci. USA* 93:13404–9
5. Baker SS, Wilhelm KS, Thomashow MF. 1994. The 5′-region of *Arabidopsis thaliana cor15a* has *cis*-acting elements that confer cold-, drought- and ABA-regulated gene expression. *Plant Mol. Biol.* 24:701–13
6. Brule-Babel AL, Fowler DB. 1988. Genetic control of cold hardiness and vernalization requirement in winter wheat. *Crop Sci.* 28:879–84
7. Carpenter JF, Hand SC, Crowe LM, Crowe JH. 1986. Cryoprotection of phosphofructokinase with organic solutes:

characterization of enhanced protection in the presence of divalent cations. *Arch. Biochem. Biophys.* 250:505–12

8. Chen H-H, Li PH, Brenner ML. 1983. Involvement of abscisic acid in potato cold acclimation. *Plant Physiol.* 71:362–65

9. Close TJ. 1997. Dehydrins: a commonality in the response of plants to dehydration and low temperature. *Physiol. Plant.* 100:291–96

10. Ding JP, Pickard BG. 1993. Modulation of mechanosensitive calcium-selective cation channels by temperature. *Plant J.* 3:713–20

11. Duman JG. 1994. Purification and characterization of a thermal hysteresis protein from a plant, the bittersweet nightshade *Solanum dulcamara*. *Biochim. Biophys. Acta* 1206:129–35

12. Dunn MA, Goddard NJ, Zhang L, Pearce RS, Hughes MA. 1994. Low-temperature-responsive barley genes have different control mechanisms. *Plant Mol. Biol.* 24:879–88

13. Dure L III. 1993. A repeating 11-mer amino acid motif and plant desiccation. *Plant J.* 3:363–69

14. Epand RM, Shai Y, Segrest JP, Anantharamaiah GM. 1995. Mechanisms for the modulation of membrane bilayer properties by amphipathic helical peptides. *Biopolymers* 37:319–38

15. Foster R, Izawa T, Chua NH. 1994. Plant bZIP proteins gather at ACGT elements. *FASEB J.* 8:192–200

16. Fowler DB, Gusta LV. 1979. Selection for winterhardiness in wheat. I. Identification of genotypic variability. *Crop Sci.* 19:769–72

17. Galiba G, Quarrie SA, Sutka J, Morounov A, Snape JW. 1995. RFLP mapping of the vernalization (*Vrn1*) and frost resistance (*Fr1*) genes on chromosome 5A of wheat. *Theor. Appl. Genet.* 90:1174–79

18. Gibson S, Arondel V, Iba K, Somerville C. 1994. Cloning of a temperature-regulated gene encoding a chloroplast omega-3 desaturase from *Arabidopsis thaliana*. *Plant Physiol.* 106:1615–21

19. Gilmour SJ, Artus NN, Thomashow MF. 1992. cDNA sequence analysis and expression of two cold-regulated genes of *Arabidopsis thaliana*. *Plant Mol. Biol.* 18:13–21

20. Gilmour SJ, Hajela RK, Thomashow MF. 1988. Cold acclimation in *Arabidopsis thaliana*. *Plant Physiol.* 87:745–50

21. Gilmour SJ, Thomashow MF. 1991. Cold acclimation and cold-regulated gene expression in ABA mutants of *Arabidopsis thaliana*. *Plant Mol. Biol.* 17:1233–40

22. Gilmour SJ, Zarka DG, Stockinger EJ, Salazar MP, Houghton JM, Thomashow MF. 1998. Low temperature regulation of the *Arabidopsis CBF* family of AP2 transcriptional activators as an early step in cold-induced *COR* gene expression. *Plant J.* 16:433–42

23. Griffith M, Antikainen M, Hon W-C, Pihakaski-Maunsbach K, Yu X-M, et al. 1997. Antifreeze proteins in winter rye. *Physiol. Plant.* 100:327–32

24. Guiltinan MJ, Marcotte WR Jr, Quatrano RS. 1990. A plant leucine zipper protein that recognizes an abscisic acid response element. *Science* 250:267–71

25. Guy C, Haskell D, Li QB. 1998. Association of proteins with the stress 70 molecular chaperones at low temperature: evidence for the existence of cold labile proteins in spinach. *Cryobiology* 36:301–14

26. Guy C, Haskell D, Neven L, Klein P, Smelser C. 1992. Hydration-state-responsive proteins link cold and drought stress in spinach. *Planta* 188:265–70

27. Guy CL. 1990. Cold acclimation and freezing stress tolerance: role of protein metabolism. *Annu. Rev. Plant Physiol. Plant Mol. Biol.* 41:187–223

28. Guy CL, Haskell D. 1988. Detection of polypeptides associated with the cold acclimation process in spinach. *Electrophoresis* 9:787–96

29. Guy CL, Haskell D. 1987. Induction of freezing tolerance in spinach is associated with the synthesis of cold acclimation induced proteins. *Plant Physiol.* 84:872–78

30. Guy CL, Haskell D. 1989. Preliminary characterization of high molecular mass proteins associated with cold acclimation in spinach. *Plant Physiol. Biochem.* 27:777–84

31. Guy CL, Haskell D, Yelenosky G. 1988. Changes in freezing tolerance and polypeptide content of spinach and citrus at 5°C. *Cryobiology* 25:264–71

32. Guy CL, Li QB. 1998. The organization and evolution of the spinach stress 70 molecular chaperone gene family. *Plant Cell* 10:539–56

33. Guy CL, Niemi KJ, Brambl R. 1985. Altered gene expression during cold acclimation of spinach. *Proc. Natl. Acad. Sci. USA* 82:3673–77

34. Hajela RK, Horvath DP, Gilmour SJ, Thomashow MF. 1990. Molecular cloning and expression of *cor* (cold-regulated) genes in *Arabidopsis thaliana*. *Plant Physiol.* 93:1246–52

35. Hayes PM, Blake T, Chen THH, Tragoonrung S, Chen F, et al. 1993. Quantitative

trait loci on barley (*Hordeum vulgare* L.) chromosome 7 associated with components of winterhardiness. *Genome* 36:66–71

36. Heino P, Sandman G, Lång V, Nordin K, Palva ET. 1990. Abscisic acid deficiency prevents development of freezing tolerance in *Arabidopsis thaliana* (L.) Heynh. *Theor. Appl. Genet.* 79:801–6

37. Hong B, Barg R, Ho T-HD. 1992. Developmental and organ-specific expression of an ABA- and stress-induced protein in barley. *Plant Mol. Biol.* 18:663–74

38. Hong B, Uknes S, Ho T-HD. 1988. Cloning and characterization of a cDNA encoding a mRNA rapidly induced by ABA in barley aleurone layers. *Plant Mol. Biol.* 11:495–506

39. Hong SW, Jon JH, Kwak JM, Nam HG. 1997. Identification of a receptor-like protein kinase gene rapidly induced by abscisic acid, dehydration, high salt, and cold treatments in *Arabidopsis thaliana*. *Plant Physiol.* 113:1203–12

40. Horvath DP, McLarney BK, Thomashow MF. 1993. Regulation of *Arabidopsis thaliana* L. (Heyn) *cor78* in response to low temperature. *Plant Physiol.* 103:1047–53

41. Houde M, Dhindsa RS, Sarhan F. 1992. A molecular marker to select for freezing tolerance in Gramineae. *Mol. Gen. Genet.* 234:43–48

42. Hughes MA, Dunn MA. 1996. The molecular biology of plant acclimation to low temperature. *J. Exp. Bot.* 47:291–305

43. Huner NPA, Oquist G, Sarhan F. 1998. Energy balance and acclimation to light and cold. *Trends Plant Sci.* 3:224–30

44. Ingram J, Bartels D. 1996. The molecular basis of dehydration tolerance in plants. *Annu. Rev. Plant Physiol. Plant Mol. Biol.* 47:377–403

45. Ishikawa M, Robertson AJ, Gusta LV. 1990. Effect of temperature, light, nutrients and dehardening on abscisic acid-induced cold hardiness in *Bromus inermis* Leyss suspension cultured cells. *Plant Cell Physiol.* 31:51–59

46. Ishitani M, Xiong L, Lee H, Stevenson B, Zhu JK. 1998. *HOS1*, a genetic locus involved in cold-responsive gene expression in Arabidopsis. *Plant Cell* 10:1151–61

47. Ishitani M, Xiong L, Stevenson B, Zhu JK. 1997. Genetic analysis of osmotic and cold stress signal transduction in Arabidopsis: interactions and convergence of abscisic acid-dependent and abscisic acid-independent pathways. *Plant Cell* 9:1935–49

48. Islam AKMR, Shepherd KW, Sparrow DHB. 1981. Isolation and characterization of wheat-barley chromosome addition lines. *Heredity* 46:161–74

49. Jaglo-Ottosen KR, Gilmour SJ, Zarka DG, Schabenberger O, Thomashow MF. 1998. *Arabidopsis CBF1* overexpression induces *COR* genes and enhances freezing tolerance. *Science* 280:104–6

50. Jarillo JA, Capel J, Leyva A, Martínez-Zapater JM, Salinas J. 1994. Two related low-temperature-inducible genes of *Arabidopsis* encode proteins showing high homology to 14-3-3 proteins, a family of putative kinase regulators. *Plant Mol. Biol.* 25:693–704

51. Jiang C, Iu B, Singh J. 1996. Requirement of a CCGAC cis-acting element for cold induction of the *BN115* gene from winter *Brassica napus*. *Plant Mol. Biol.* 30:679–84

52. Jofuku KD, den Boer BG, Van Montagu M, Okamuro JK. 1994. Control of Arabidopsis flower and seed development by the homeotic gene *APETALA2*. *Plant Cell* 6:1211–25

53. Jonak C, Kiegerl S, Ligterink W, Barker PJ, Huskisson NS, Hirt H. 1996. Stress signaling in plants: A mitogen-activated protein kinase pathway is activated by cold and drought. *Proc. Natl. Acad. Sci. USA* 93:11274–79

54. Kaye C, Neven L, Hofig A, Li QB, Haskell D, Guy C. 1998. Characterization of a gene for spinach CAP160 and expression of two spinach cold-acclimation proteins in tobacco. *Plant Physiol.* 116:1367–77

55. Klucher KM, Chow H, Reiser L, Fischer RL. 1996. The *AINTEGUMENTA* gene of *Arabidopsis* required for ovule and female gametophyte development is related to the floral homeotic gene *APETALA2*. *Plant Cell* 8:137–53

56. Knight H, Trevavas AJ, Knight MR. 1996. Cold calcium signaling in Arabidopsis involves two cellular pools and a change in calcium signature after acclimation. *Plant Cell* 8:489–503

57. Koornneef M, Leon-Kloosterziel KM, Schwartz SH, Zeevaart JAD. 1998. The genetic and molecular dissection of abscisic acid biosynthesis and signal transduction in *Arabidopsis*. *Plant Physiol. Biochem.* 36:83–89

58. Krishna P, Sacco M, Cherutti JF, Hill S. 1995. Cold-induced accumulation of *hsp90* transcripts in *Brassica napus*. *Plant Physiol.* 107:915–23

59. Kurkela S, Borg-Franck M. 1992. Structure and expression of *kin2*, one of two cold- and ABA-induced genes of

Arabidopsis thaliana. Plant Mol. Biol. 19: 689–92

60. Kurkela S, Franck M. 1990. Cloning and characterization of a cold- and ABA-inducible *Arabidopsis* gene. *Plant Mol. Biol.* 15:137–44

61. Kurkela S, Franck M, Heino P, Lång V, Palva ET. 1988. Cold induced gene expression in *Arabidopsis thaliana* L. *Plant Cell Rep.* 7:495–98

62. Lalk I, Dörffling K. 1985. Hardening, abscisic acid, proline, and freezing resistance in two winter wheat varieties. *Physiol. Plant.* 63:287–92

63. Lång V, Mäntylä E, Welin B, Sundberg B, Palva ET. 1994. Alterations in water status, endogenous abscisic acid content, and expression of *rab18* gene during the development of freezing tolerance in *Arabidopsis thaliana. Plant Physiol.* 104:1341–49

64. Lång V, Palva ET. 1992. The expression of a rab-related gene, *rab18*, is induced by abscisic acid during the cold acclimation process of *Arabidopsis thaliana* (L.) Heynh. *Plant Mol. Biol.* 20:951–62

65. Laurie DA, Pratchett N, Bezant J, Snape JW. 1995. RFLP mapping of five major genes and eight quantitative trait loci controlling flowering time in a winter x spring barley (*Hordeum vulgare* L.) cross. *Genome* 38:575–85

66. Leung J, Bouvier Durand M, Morris PC, Guerrier D, Chefdor F, Giraudat J. 1994. *Arabidopsis* ABA response gene ABI1: features of a calcium-modulated protein phosphatase. *Science* 264:1448–52

67. Levitt J. 1980. *Responses of Plants to Environmental Stresses.* New York: Academic. 2nd ed.

68. Li PH, Chen THH. 1997. *Plant Cold Hardiness. Molecular Biology, Biochemistry and Physiology.* New York: Plenum

69. Limin AE, Danyluk J, Chauvin LP, Fowler DB, Sarhan F. 1997. Chromosome mapping in low-temperature induced *Wcs120* family genes and regulation of cold-tolerance expression in wheat. *Mol. Gen. Genet.* 253:720–27

70. Lin C, Thomashow MF. 1992. DNA sequence analysis of a complementary DNA for cold-regulated *Arabidopsis* gene *cor15* and characterization of the COR15 polypeptide. *Plant Physiol.* 99:519–25

71. Lisse T, Bartels D, Kalbitzer HR, Jaenicke R. 1996. The recombinant dehydrin-like desiccation stress protein from the resurrection plant *Craterostigma plantagineum* displays no defined three-dimensional structure in its native state. *Biol. Chem.* 377:555–61

72. Liu Q, Kasuga M, Sakuma Y, Abe H, Miura S, et al. 1998. Two transcription factors, DREB1 and DREB2, with an EREBP/AP2 DNA binding domain separate two cellular signal transduction pathways in drought- and low-temperature-responsive gene expression, respectively, in Arabidopsis. *Plant Cell* 10:1391–406

73. Livingston DP III, Henson CA. 1998. Apoplastic sugars, fructans, fructan exohydrolase, and invertase in winter oat: responses to second-phase cold hardening. *Plant Physiol.* 116:403–8

74. Luo M, Liu J, Mohapatra S, Hill RD, Mohapatra SS. 1992. Characterization of a gene family encoding abscisic acid and environmental stress-inducible proteins of alfalfa. *J. Biol. Chem.* 267:15367–74

75. Mäntylä E, Lång V, Palva ET. 1995. Role of abscisic acid in drought-induced freezing tolerance, cold acclimation, and accumulation of LTI78 and RAB18 proteins in *Arabidopsis thaliana. Plant Physiol.* 107:141–48

76. Marin E, Nussaume L, Quesada A, Gonneau M, Sotta B, et al. 1996. Molecular identification of zeaxanthin epoxidase of *Nicotiana plumbaginifolia*, a gene involved in abscisic acid biosynthesis and corresponding to the ABA locus of *Arabidopsis thaliana. EMBO J.* 15:2331–42

77. McKersie BD, Bowley SR. 1997. Active oxygen and freezing tolerance in transgenic plants. See Ref. 68, pp. 203–14

78. McKown R, Kuroki G, Warren G. 1996. Cold responses of *Arabidopsis* mutants impaired in freezing tolerance. *J. Exp. Bot.* 47:1919–25

79. Meyer K, Leube MP, Grill E. 1994. A protein phosphatase 2C involved in ABA signal transduction in *Arabidopsis thaliana. Science* 264:1452–55

80. Mizoguchi T, Hayashida N, Yamaguchi-Shinozaki K, Kamada H, Shinozaki K. 1993. ATMPKs: a gene family of plant MAP kinases in *Arabidopsis thaliana. FEBS Lett.* 336:440–44

81. Mizoguchi T, Irie K, Hirayama T, Hayashida N, Yamaguchi-Shinozaki K, et al. 1996. A gene encoding a mitogen-activated protein kinase kinase kinase is induced simultaneously with genes for a mitogen-activated protein kinase and an S6 ribosomal protein kinase by touch, cold, and water stress in *Arabidopsis thaliana. Proc. Natl. Acad. Sci. USA* 93: 765–69

82. Monroy AF, Castonguay Y, Laberge S, Sarhan F, Vezina LP, Dhindsa RS. 1993. A new cold-induced alfalfa gene is associated with enhanced hardening at subzero

temperature. *Plant Physiol.* 102:873–79

83. Monroy AF, Dhindsa RS. 1995. Low-temperature signal transduction: induction of cold acclimation-specific genes of alfalfa by calcium at 25°C. *Plant Cell* 7: 321–31

84. Monroy AF, Sangwan V, Dhindsa RS. 1998. Low temperature signal transduction during cold acclimation: protein phosphatase 2A as an early target for cold-inactivation. *Plant J.* 13:653–60

85. Monroy AF, Sarhan F, Dhindsa RS. 1993. Cold-induced changes in freezing tolerance, protein phosphorylation, and gene expression. *Plant Physiol.* 102:1127–35

86. Neven LG, Haskell DW, Hofig A, Li QB, Guy CL. 1993. Characterization of a spinach gene responsive to low temperature and water stress. *Plant Mol. Biol.* 21:291–305

87. Nordin K, Heino P, Palva ET. 1991. Separate signal pathways regulate the expression of a low-temperature-induced gene in *Arabidopsis thaliana* (L.) Heynh. *Plant Mol. Biol.* 16:1061–71

88. Nordin K, Vahala T, Palva ET. 1993. Differential expression of two related, low-temperature-induced genes in *Arabidopsis thaliana* (L.) Heynh. *Plant Mol. Biol.* 21:641–53

89. Ohme-Takagi M, Shinshi H. 1995. Ethylene-inducible DNA binding proteins that interact with an ethylene-responsive element. *Plant Cell* 7:173–82

90. Olien CR, Smith MN. 1977. Ice adhesions in relation to freeze stress. *Plant Physiol.* 60:499–503

91. Orr W, Keller WA, Singh J. 1986. Induction of freezing tolerance in an embryogenic cell suspension culture of *Brassica napus* by abscisic acid at room temperature. *J. Plant Physiol.* 126:23–32

92. Ouellet F, Vazquez-Tello A, Sarhan F. 1998. The wheat *wcs120* promoter is cold-inducible in both monocotyledonous and dicotyledonous species. *FEBS Lett.* 423:324–28

93. Pan A, Hayes PM, Chen F, Chen THH, Blake T, et al. 1994. Genetic analysis of the components of winterhardiness in barley (*Hordeum vulgare* L.). *Theor. Appl. Genet.* 89:900–10

94. Phillips JR, Dunn MA, Hughes MA. 1997. mRNA stability and localisation of the low-temperature-responsive barley gene family *blt14*. *Plant Mol. Biol.* 33: 1013–23

95. Polisensky DH, Braam J. 1996. Cold-shock regulation of the *Arabidopsis TCH* genes and the effects of modulating intracellular calcium levels. *Plant Physiol.* 111:1271–79

96. Pugsley AT. 1971. A genetic analysis of the spring-winter habit in wheat. *Aust. J. Agric. Res.* 22:21–31

97. Riechmann JL, Meyerowitz EM. 1998. The AP2/EREBP family of plant transcription factors. *Biol. Chem.* 379:633–46

98. Rudolph AS, Crowe JH. 1985. Membrane stabilization during freezing: the role of two natural cryoprotectants, trehalose and proline. *Cryobiology* 22:367–77

99. Sakai A, Larcher W. 1987. *Frost Survival of Plants: Responses and Adaptation to Freezing Stress.* Berlin: Springer-Verlag

100. Sarhan F, Danyluk J. 1998. Engineering cold-tolerant crops—throwing the master switch. *Trends Plant Sci.* 3:289–90

101. Sarhan F, Ouellet F, Vazquez-Tello A. 1997. The wheat *wcs120* gene family. A useful model to understand the molecular genetics of freezing tolerance in cereals. *Physiol. Plant.* 101:439–45

102. Shinwari ZK, Nakashima K, Miura S, Kasuga M, Seki M, et al. 1998. An *Arabidopsis* gene family encoding DRE/CRT binding proteins involved in low-temperature-responsive gene expression. *Biochem. Biophys. Res. Commun.* 250: 161–70

103. Sieg F, Schroder W, Schmitt JM, Hincha DK. 1996. Purification and characterization of a cryoprotective protein (cryoprotectin) from the leaves of cold-acclimated cabbage. *Plant Physiol.* 111:215–21

104. Snape JW, Law CN, Worland AJ. 1976. Chromosome variation for loci controlling ear-emergence time on chromosome 5A of wheat. *Heredity* 37:335–40

105. Steponkus PL. 1984. Role of the plasma membrane in freezing injury and cold acclimation. *Annu. Rev. Plant. Physiol.* 35: 543–84

106. Steponkus PL, Uemura M, Joseph RA, Gilmour SJ, Thomashow MF. 1998. Mode of action of the *COR15a* gene on the freezing tolerance of *Arabidopsis thaliana*. *Proc. Natl. Acad. Sci. USA* 95:14570–75

107. Steponkus PL, Uemura M, Webb MS. 1993. A contrast of the cryostability of the plasma membrane of winter rye and spring oat—two species that widely differ in their freezing tolerance and plasma membrane lipid composition. In *Advances in Low-Temperature Biology*, ed. PL Steponkus, 2:211–312. London: JAI Press

108. Steponkus PL, Uemura M, Webb MS. 1993. Membrane destabilization during freeze-induced dehydration. *Curr. Topics Plant Physiol.* 10:37–47

109. Stockinger EJ, Gilmour SJ, Thomashow MF. 1997. *Arabidopsis thaliana CBF1* encodes an AP2 domain-containing transcriptional activator that binds to the C-repeat/DRE, a *cis*-acting DNA regulatory element that stimulates transcription in response to low temperature and water deficit. *Proc. Natl. Acad. Sci. USA* 94: 1035–40

110. Storlie EW, Allan RE, Walker-Simmons MK. 1998. Effect of the *Vrn1–Fr1* interval on cold hardiness levels in near-isogenic wheat lines. *Crop Sci.* 38:483–88

111. Strauss G, Hauser H. 1986. Stabilization of lipid bilayer vesicles by sucrose during freezing. *Proc. Natl. Acad. Sci. USA* 83:2422–26

112. Sukumaran NP, Weiser CJ. 1972. An excised leaflet test for evaluating potato frost tolerance. *HortScience* 7:467–68

113. Tahtiharju S, Sangwan V, Monroy AF, Dhindsa RS, Borg M. 1997. The induction of *kin* genes in cold-acclimating *Arabidopsis thaliana*. Evidence of a role for calcium. *Planta* 203:442–47

114. Thomashow MF. 1990. Molecular genetics of cold acclimation in higher plants. *Adv. Genet.* 28:99–131

115. Thomashow MF. 1998. Role of cold-responsive genes in plant freezing tolerance. *Plant Physiol.* 118:1–7

116. Uemura M, Joseph RA, Steponkus PL. 1995. Cold acclimation of *Arabidopsis thaliana*. Effect on plasma membrane lipid composition and freeze-induced lesions. *Plant Physiol.* 109:15–30

117. Uemura M, Steponkus PL. 1997. Effect of cold acclimation on membrane lipid composition and freeze-induced membrane destabilization. See Ref. 68, pp. 171–79

118. Urao T, Yakubov B, Yamaguchi-Shinozaki K, Shinozaki K. 1998. Stress-responsive expression of genes for two-component response regulator-like proteins in *Arabidopsis thaliana*. *FEBS Lett.* 427:175–78

119. Van Zee K, Chen FQ, Hayes PM, Close TJ, Chen THH. 1995. Cold-specific induction of a dehydrin gene family member in barley. *Plant Physiol.* 108:1233–39

120. Vazquez-Tello A, Ouellet F, Sarhan F. 1998. Low temperature-stimulated phosphorylation regulates the binding of nuclear factors to the promoter of *Wcs120*, a cold-specific gene in wheat. *Mol. Gen. Genet.* 257:157–66

121. Wang H, Datla R, Georges F, Loewen M, Cutler AJ. 1995. Promoters from *kin1*

and *cor6.6*, two homologous *Arabidopsis thaliana* genes: transcriptional regulation and gene expression induced by low temperature, ABA, osmoticum and dehydration. *Plant Mol. Biol.* 28:605–17

122. Wang H, Georges F, Pelcher LE, Saleem M, Cutler AJ. 1994. A 5.3-kilobase genomic fragment from *Arabidopsis thaliana* containing *kin1* and *cor6.6*. *Plant Physiol.* 104:291–92

123. Warren G, McKown R, Marin A, Teutonico R. 1996. Isolation of mutations affecting the development of freezing tolerance in *Arabidopsis thaliana* (L.) Heynh. *Plant Physiol.* 111:1011–19

124. Welin BV, Olson A, Nylander M, Palva ET. 1994. Characterization and differential expression of *dhn/lea/rab*-like genes during cold acclimation and drought stress in *Arabidopsis thaliana*. *Plant Mol. Biol.* 26:131–44

125. Welin BV, Olson A, Palva ET. 1995. Structure and organization of two closely related low-temperature-induced *dhn/lea/rab*-like genes in *Arabidopsis thaliana* L. Heynh. *Plant Mol. Biol.* 29: 391–95

126. Wilhelm KS, Thomashow MF. 1993. *Arabidopsis thaliana cor15b*, an apparent homologue of *cor15a*, is strongly responsive to cold and ABA, but not drought. *Plant Mol. Biol.* 23:1073–77

127. Wilson K, Long D, Swinburne J, Coupland G. 1996. A *Dissociation* insertion causes a semidominant mutation that increases expression of *TINY*, an Arabidopsis gene related to *APETALA2*. *Plant Cell* 8:659–71

128. Worrall D, Elias L, Ashford D, Smallwood MCS, Lillford P, et al. 1998. A carrot leucine-rich-repeat protein that inhibits ice recrystallization. *Science* 282: 115–17

129. Xin Z, Browse J. 1998. *eskimo1* mutants of *Arabidopsis* are constitutively freezing-tolerant. *Proc. Natl. Acad. Sci. USA* 95:7799–804

130. Yamaguchi-Shinozaki K, Shinozaki K. 1993. Characterization of the expression of a desiccation-responsive *rd29* gene of *Arabidopsis thaliana* and analysis of its promoter in transgenic plants. *Mol. Gen. Genet.* 236:331–40

131. Yamaguchi-Shinozaki K, Shinozaki K. 1994. A novel *cis*-acting element in an *Arabidopsis* gene is involved in responsiveness to drought, low-temperature, or high-salt stress. *Plant Cell* 6:251–64

Annu. Rev. Plant Physiol. Plant Mol. Biol. 1999. 50:601–39

THE WATER-WATER CYCLE IN CHLOROPLASTS: Scavenging of Active Oxygens and Dissipation of Excess Photons

Kozi Asada

Department of Biotechnology, Faculty of Engineering, Fukuyama University, Gakuen-cho 1, Fukuyama, 729-0292, Japan; e-mail: asada@bt.fubt.fukuyama-u.ac.jp

KEY WORDS: ascorbate peroxidase (APX), microcompartmentalization of scavenging enzymes, monodehydroascorbate (MDA) reductase, superoxide dismutase (SOD), suppression of photoinhibition

---

ABSTRACT

Photoreduction of dioxygen in photosystem I (PSI) of chloroplasts generates superoxide radicals as the primary product. In intact chloroplasts, the superoxide and the hydrogen peroxide produced via the disproportionation of superoxide are so rapidly scavenged at the site of their generation that the active oxygens do not inactivate the PSI complex, the stromal enzymes, or the scavenging system itself. The overall reaction for scavenging of active oxygens is the photoreduction of dioxygen to water via superoxide and hydrogen peroxide in PSI by the electrons derived from water in PSII, and the water-water cycle is proposed for these sequences. An overview is given of the molecular mechanism of the water-water cycle and microcompartmentalization of the enzymes participating in it. Whenever the water-water cycle operates properly for scavenging of active oxygens in chloroplasts, it also effectively dissipates excess excitation energy under environmental stress. The dual functions of the water-water cycle for protection from photoinihibition are discussed.

CONTENTS

601

INTRODUCTION

The evolution of O_2 in leaves was first documented by Priestley in 1771 in his observations on a candle's continuing combustion when enclosed in a glass chamber with plants. Since then, in oxygenic phototrophs, chloroplasts and thylakoids have been identified to be the photogenerating site of O_2 from water, but no O_2 uptake was considered to occur in them. In 1951, however, Mehler found the photoreduction of O_2 to H_2O_2 in thylakoids as detected by acetaldehyde formation in the presence of ethanol and catalase (oxidation of ethanol by H_2O_2 catalyzed with a peroxidatic activity of catalase, $H_2O_2 + C_2H_5OH \rightarrow CH_3CHO + 2H_2O$), and subsequently by light-dependent uptake of $^{18}O_2$ (120, 121). This was the first indication of the concomitant, light-dependent evolution and reduction of O_2 in thylakoids, and O_2 was established to be a Hill oxidant. Subsequently, the photoreducing site of O_2 was indicated to be PSI, and in 1974, its primary product was identified to be superoxide anion radical (O_2^-) (18). In thylakoids, therefore, H_2O_2 is photoproduced via the spontaneous disproportionation of O_2^-, but not directly through the two-electron reduction of O_2.

In contrast to thylakoids, intact chloroplasts photoreduce $^{18}O_2$ and photoevolve $^{16}O_2$ from water at the same rate in the absence of bicarbonate, and no $H_2^{18}O_2$ accumulates, even in the absence of catalase. These results indicate that chloroplasts have a system to reduce O_2 to water using the electrons derived from water (14, 138). The stoichiometry of the evolution and uptake of O_2 and the absence of catalase in chloroplasts indicate that the H_2O_2 derived from the O_2^- photogenerated in PSI is reduced to water via a peroxidase reaction using the reductant generated in PSI as the electron donor. The occurrences of ascorbate-specific peroxidase in spinach chloroplasts (139) and *Euglena* cells (181), and of dehydroascorbate reductase (49) indicate that the electron donor for the peroxidase is ascorbate (See Figure 1). In this review, the O_2-photoreducing system to water in chloroplasts is referred to as the water-water cycle because the electrons from water in PSII are donated to O_2 for reduction to water in

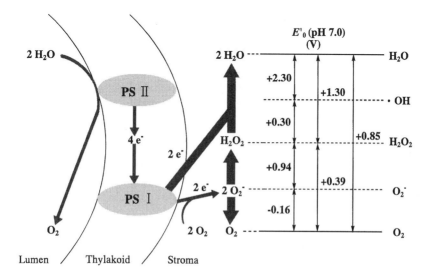

Figure 1   Outline of the water-water cycle in chloroplasts and the redox potentials of the reduced molecular species of oxygen. Photoreduction of $NADP^+$ in PSI, cyclic electron flow around PSI and balance of protons are not included. In PSI two molecules of O_2 are reduced by two electrons ($2\,e^-$) from PSII to superoxide anion radicals (O_2^-). The O_2^- is disproportionated to H_2O_2 and O_2 catalyzed with superoxide dismutase, and the H_2O_2 is reduced to water by ascorbate catalyzed with ascorbate peroxidase. The oxidized ascorbates (monodehydroascorbate radical and its disproportionated product, dehydroascorbate) are reduced back to ascorbate using two electrons ($2\,e^-$) from PSII. One molecule of O_2 is reduced to two molecules of H_2O in PSI by the four electrons ($4\,e^-$) derived from two molecules of H_2O in PSII without any net change in O_2. The thickness of the arrows approximates the reaction rates at each step by the half life; electron transport from water in PSII to the reaction center of PSI, $5\times10^{-3}\,s^{-1}$ at maximum; photoreduction of O_2 at PSI, $4\times10^{-2}\,s$, when the rate is $25\;O_2^-\;P700^{-1}\,s^{-1}$; and the disproportionation of O_2^- and the reduction of H_2O_2 including the regeneration of ascorbate, below $7\times10^{-5}\,s$ on thylakoids and below $2\times10^{-3}\,s$ in the stroma (see Figure 2). Thus, the photoreducing rate of reduced species of oxygen to water is faster by several orders of magnitude than that of O_2 (12). Redox potentials among O_2, O_2^-, H_2O_2, •OH, and H_2O are shown on the right side, calculated by using molar concentration of O_2.

PSI without a net change of O_2. The Mehler-peroxidase reaction has been used for this cycle, but this term does not include the reducing steps of the oxidized ascorbates for regeneration of ascorbate, which is indispensable for the reduction of H_2O_2 and its dual physiological functions. The molecular mechanism of the water-water cycle is the first topic of this review. [For additional references of the production and scavenging of active oxygens in photosynthetic tissues, see (11, 12, 15, 21, 48, 165)].

The second topic of this review is the physiological function of the water-water cycle. The most important function of this cycle is a rapid, immediate scavenging of O_2^- and H_2O_2 at the site of its generation prior to their interaction with the target molecules. The hydroxyl radical ($\cdot$OH) is generated by the interaction of H_2O_2 with reductants such as reduced transition metal ions [Fe(II) and Cu(I)], which are generated via the reduction of Fe(III) and Cu(II) by the O_2^-. The reduced species of oxygen, O_2^-, H_2O_2, and $\cdot$OH, are highly reactive as compared with triplet dioxygen at the ground state ($^3O_2$), their redox potentials being higher than $^3O_2$ (Figure 1). Among them, $\cdot$OH is the most reactive causing oxidative damage to chloroplast components. No specific scavenging enzyme is available for $\cdot$OH, since its reactions with cellular components proceed at diffusion-controlled rates (10^8–10^9 M^{-1} s^{-1}), and its diffusion distance from the generation site is shorter than several amino acid residues in the case of proteins. This is one reason why the cell should be equipped with a perfect scavenging system for O_2^- and H_2O_2 to protect itself from oxidative damage (12, 21). The $\cdot$OH radical is generated by disorder of the water-water cycle, which is unavoidable, especially in aged leaves, and polyols are one of the effective scavengers of $\cdot$OH radicals. Chloroplasts from mannitol-enriched transformants show tolerance to oxidative stress (180). In addition to rapid scavenging of these active oxygens to suppress their interactions with the target molecules, this cycle can safely dissipate excess photon energy. The dual functions of the water-water cycle, therefore protect leaves from photoinhibition under environmental conditions where the photon intensity is in excess of that used for CO_2-assimilation, which is not unusual for plants in nature (109).

PHOTOREDUCTION OF DIOXYGEN TO SUPEROXIDE

Photoreduction of Dioxygen in Thylakoids

The univalent photoreduction of O_2 to O_2^- has been observed in washed, ferredoxin (Fd)-depleted thylakoids. The thylakoid-bound [4Fe-4S] clusters X on psaA and psaB or A/B on psaC of the PSI complex is the most likely electron donor to O_2 (11). At a neutral pH the photoproduction rate of O_2^- is not as high; around 20 μmol mg Chl^{-1} h^{-1} (21), corresponding to only 4 O_2^- $P700^{-1}$ s^{-1} if P700 contains one molecule per 600 molecules of chlorophyll. The apparent

K_m value for O_2 of O_2-photoreduction in thylakoids is in the range of 2 to 10 μM, which is lower than that in chloroplasts (19, 61, 194). The apparently low photoproduction rates as determined using either the O_2^--probes or $^{18}O_2$-uptake are inferred by the O_2^--dependent cyclic electron transfer around PSI (11). The O_2^- generated in the PSI complex has a longer lifetime than that in aqueous medium (196), because protons are not available within the aprotic membranes and neither spontaneous nor SOD-catalyzed disproportionation of O_2^- can proceed. The O_2^- photoproduced within the membranes thus may donate its electron to oxidized cytochrome f (Cyt f) or plastocyanin (PC) (O_2^- + Cyt f or PC $\rightarrow$ O_2 + reduced Cyt f or PC) at around 10^6 M^{-1} s^{-1} (11), and the reduction of the electron donors to PSI by O_2^- allows the O_2^--mediated cyclic electron flow around PSI. This would explain why the photoproduction rate of O_2^- is low as determined by the O_2^--probes outside the membranes such as ferricytochrome c (Cyt c-Fe(III) + O_2^- $\rightarrow$ Cyt c-Fe(II) + O_2) or the exchange rate of oxygen. However, the photoproduction rate of O_2^- by either isolated PSI complex or Triton X-100-treated thylakoids is at least one order of magnitude higher than that by intact thylakoids; 70 O_2^- P700^{-1} s^{-1} (19, 194). Thus, it seems to be essential to keep the supermolecular structure of the PSI complex for the putative O_2^--mediated cyclic electron flow. Photoreduction of O_2 by thylakoids is saturated at low intensities of light (around 10 μmol photons m^{-2} s^{-1}) (61, 131), which also accounts for a low ratio of the diffused O_2^- produced within the membranes to the medium at high intensities of light.

Further evidence for the O_2^--mediated cyclic electron flow is an increase of the photoreduction of O_2 under conditions where protons are supplied to the membranes by ammonium ions, amines (196), or low pH in the medium (68). In either case, the O_2^- photoproduced within the membranes can be protonated and is spontaneously disproportionated to H_2O_2 and O_2 within the membranes. Because of a pK value of superoxide of 4.8 (H^+ + O_2^- $\leftrightarrow$ HO_2), at a neutral pH the protonated superoxide radical (HO_2) is only a minor species, but at a lower pH the ratio of the protonated, neutral form (perhydroxyl radical) increases. The diffusion rate of HO_2 through the membranes would be higher than that of the anionic superoxide (O_2^-), and the disproportionation rate of superoxide between HO_2 and O_2^- at pH 4.8 is maximum. Thus, when protons are available within the membranes, the interaction of the O_2^- with the electron donors of PSI is suppressed. Because the diffusion of H_2O_2 through the membranes is very rapid relative to that of O_2^- (195), the H_2O_2 produced within the membranes could easily diffuse out from the membranes. Under such conditions, the photoreduction of O_2 is apparently increased. In such a way, the putative O_2^--mediated cyclic electron flow suppresses the release of O_2^- from the membranes, and protects chloroplasts from photoinhibition when the reducing system of H_2O_2 does not properly operate.

PSII membranes photoproduce O_2^- under specified conditions via autooxidation of the electron acceptor of the reaction center complex (38), but not in an intact state of the membranes (66). The photoproduction of O_2^- as observed in intact thylakoids is inhibited by either 3-(3,4-dichlorophenyl)-1,1-dimethylurea (DCMU) or dibromothymoquinone (DBMIB) (18, 131), indicating little contribution of either PSII or the plastoquinones to the photoproduction of O_2^-. Further evidence for the participation of PSI is no photoreduction of O_2 in PSI-depleted mutants of *Oenothera* (61) and *Scenedesmus* (164). Another uptake reaction of O_2 in chloroplasts is chlororespiration; electron transfer from NAD(P)H to putative terminal oxidase via the intersystem electron carrier plastoquinones (26, 46). The primary reduced product of O_2 in the reaction of the putative oxidase has not yet been identified, but if the terminal oxidase has properties similar to mitochondrial cytochrome c oxidase, production of neither O_2^- nor H_2O_2 is expected. Furthermore, the electron transfer to the putative oxidase is suppressed when PSI is excited.

Photoreduction of Dioxygen in Chloroplasts

Photoreduction of O_2 in PSI of chloroplasts in leaves and algal cells has been shown by the light-dependent uptake of $^{18}O_2$. Photorespiratory uptake of O_2 by the reactions of ribulose 1,5-bisphosphate oxygenase in chloroplasts and glycolate oxidase in peroxisomes could be suppressed by either low concentrations of O_2 (around 2%) to lower both reactions or high concentrations of CO_2 to lower the oxygenase reaction. Under the conditions whereby the photorespiratory activity is suppressed, uptake of $^{18}O_2$ was observed in intact chloroplasts (14, 52, 116) and also in leaf cells (25) and tissues (28, 32). Uptake of $^{18}O_2$ has also been found in cyanobacterial and eukaryotic algal cells (23, 30, 108, 123, 162, 163), confirming the photoreduction of O_2 by PSI in all oxygenic phototrophs. Mesophyll cells of C_4 plants where no photorespiratory system operates also show $^{18}O_2$ uptake (53). In leaf cells and tissues, the apparent K_m value for O_2 of O_2-photoreduction is about 80 μM, and its value in intact chloroplasts is 60 μM (178, 179). The K_m for O_2 in algae is similar to that in leaf tissues (163). Thus, apparent K_m values for O_2 in chloroplasts is higher than that in thylakoids, indicating that the photoreduction of O_2 in leaf tissues is saturated only in the atmosphere containing over 13% O_2 (160 μM). Furthermore, the photoreduction of O_2 in leaves is saturated at higher intensities of light than that in thylakoids (32). Thus, the photoreduction of O_2 in chloroplasts cannot be accounted for by that in thylakoids. The O_2 concentration in leaf chloroplasts under active photosynthesis is slightly higher than that in the air-saturated water (250 μM). Therefore, the atmospheric concentration of O_2 is not a regulatory factor of the water-water cycle.

Photoreduction of $^{18}O_2$ is enhanced under environmental conditions where CO_2-fixation is suppressed. In *Scenedesmus* cells, the rate of $^{18}O_2$-uptake is the same as that of $^{16}O_2$-evolution from water when CO_2 cannot be fixed owing either to CO_2-deficiency (164) or a latent state of the Calvin cycle enzymes just after illumination (162). These results suggest that in algal cells all the electrons from PSII are used for the photoreduction of O_2 to water when the CO_2-fixation cycle does not operate, stoichiometrically similar to that observed in chloroplasts in the absence of CO_2 (14).

Simultaneous determination of quantum efficiencies of the photosynthetic electron transport through PSII, as estimated by Chl fluorescence, and of the CO_2-fixation allow estimation of the distribution ratio of the electrons from PSII to the CO_2-fixation, photorespiration, and the photoreduction of O_2 to water in PSI. Using this method, high O_2 concentrations have been shown to be required for its photoreduction (105). In C_3 plants, 2% O_2 has been employed to suppress photorespiration, but this is too low to assess the actual photoreduction of O_2 to water in leaves. Estimation of the electron flux through the water-water cycle in 2% O_2 would give lower than actual values. In C_4 plants, a decrease of the electron flux to the CO_2-fixation represents an increase in that to the water-water cycle due to low photorespiratory activity, and the water-water cycle increases under chilling temperatures (50). In combination with $^{18}O_2$ uptake, the electron flux to the photoreduction of O_2 to water has been estimated under various conditions where the excitation energy is in excess of the CO_2-assimilation, such as drought (28, 207), salt stress (34), bright light (153), or senescence (94). In tropical trees, the linear electron flow through the water-water cycle ranges between 10 and 20% of the total flow (110). In a symbiotic alga of coral, zooxanthellae, the water-water cycle appears to protect the cells from photoinhibition at high temperatures where coral bleaching is induced (106, 107b).

The electron flux for the photoreduction of O_2 to water is estimated to be about 250 μmol e^- mg Chl^{-1} h^{-1} in *Hirschfeldia incana* leaf under bright light (32, 153) and in wheat leaf under drought (28), assuming that the chlorophyll content of leaves is 500 mg (m leaf area)$^{-2}$. In both cases, the flux through the water-water cycle is around 30% that of the total flux, and the rate corresponds to 25 e^- $P700^{-1}$ s^{-1} for the production of O_2^- and 50 e^- $P700^{-1}$ s^{-1} for the reduction of O_2 to water.

Enhanced Photoreduction of Dioxygen by Flavodehydrogenases

Where cells are exposed to photons in excess of its utilization capacity, the photoreduction of O_2 in chloroplasts is more than fivefold that in thylakoids. The photoreduction of O_2 in thylakoids is saturated at low intensities of light and low concentrations of O_2, which also is different from that observed in leaf tissues and chloroplasts. Therefore, stromal components are thought to participate in the photoreduction of O_2 in chloroplasts.

Autooxidation of the photoreduced Fd is a possible source of O_2^-, but its rate is very low (redFd + O_2^- → Fd + O_2^-, k_{obs}: 0.081 s^{-1} in air-saturated conditions) (70). Fd is reduced by the [4Fe-4S] cluster A/B in psaC of the PSI complex, and the photoreduced Fd donates electrons to NADP$^+$ catalyzed with the thylakoid-bound Fd-NADP$^+$ reductase (FNR). On addition of over 10 μM Fd to thylakoids, the photoproduction of O_2^- increases several fold when no electron acceptor other than O_2 is available (51). When NADP$^+$ is available for chloroplasts, i.e. the photon-utilizing capacity is high, the reduced Fd has little chance to interact with O_2. Even in the absence of NADP$^+$, autooxidation of the reduced Fd is unlikely to occur because of the competitive reactions for the photoreduced Fd. First, the photoreduced Fd reduces the MDA radical at a rapid rate to regenerate ascorbate from MDA produced in the APX reaction (125). Second, reduced Fd also donates electrons to the plastoquinones for the Fd-dependent cyclic electron flow around PSI (16, 40, 44, 130). Third, there are several Fd-mediated reactions catalyzed with nitrite reductase, thioredoxin reductase, and glutamate synthase.

A possible candidate of the mediator in the stroma for the photoreduction of O_2 is flavodehydrogenase. NAD(P)H-reduced flavodehydrogenases (NAD(P)$^+$-bound charge-transfer complex) including FNR generate O_2^-, but their autooxidation is very slow (118). Flavodehydrogenases including the chloroplastic enzymes FNR, glutathione reductase, and MDA reductase stimulate the photoproduction of O_2^- in thylakoids to around 300 μmol mg Chl^{-1} h^{-1}, which corresponds to 60 e$^-$ P700^{-1} s^{-1}. The flavodehydrogenase-dependent reduction of O_2 is saturated by 1 to 2 μM enzymes, which is at least one order of magnitude lower than Fd required for the photoreduction of O_2. In addition, these reactions require higher intensities of light (200 μmol photons m^{-2} s^{-1} for saturation) and higher concentrations of O_2 (K_m; around 100 μM) than that of the photoreduction of O_2 by thylakoids alone (131), which is similar to the characteristics of the photoreduction of O_2 in leaf and algal cells and chloroplasts. The flavodehydrogenase-stimulated photoproduction of O_2^- is inhibited by DCMU and DBMIB, and the addition of Fd does not affect the rate. Thus, these flavodehydrogenases are directly reduced on the reducing side of PSI, and the reduced enzymes successively donate electrons to O_2, producing two molecules of O_2^-.

FNR binds to the thylakoid membranes in healthy leaves (155), but the Fd-reduced, bound-FNR cannot catalyze the photoreduction of O_2 (131). The bound-FNR is released from the membranes by paraquat treatment of leaves or H_2O_2 (155). Then, the released FNR would stimulate the photoreduction of O_2 (56, 131). The content of glutathione reductase is about one tenth that of MDA reductase (14 μM) in chloroplasts. Therefore, MDA reductase is the most likely mediator for the photoreduction of O_2 at the rates observed in leaves.

NADH-reduced MDA reductase (charge transfer complex) is a poor electron donor to O_2 (73), as are other flavodehydrogenases. Thus, the PSI-reduced form of MDA reductase appears to be different from the NADH-reduced form in respect of the interaction with O_2.

DISPROPORTIONATION OF SUPEROXIDE

$$O_2^- + O_2^- + 2\,H^+ \rightarrow H_2O_2 + O_2$$

Chloroplastic Superoxide Dismutases

The O_2^- photogenerated either directly by the PSI complex or indirectly by the stromal factor-mediated reaction is disproportionated to H_2O_2 and O_2 catalyzed by SOD. No enzymes for the reduction of O_2^- to H_2O_2 or the oxidation of O_2^- to O_2 have been found to date in either plants or other organisms. Spontaneous reactions of O_2^- with the stromal components such as GSH, ascorbate, and Mn^{2+} do not contribute to the scavenging of O_2^- because of a low reactivity of O_2^- with the stromal components (10^5–10^6 M^{-1} s^{-1}) as compared with that of SOD (21).

In almost all plants, chloroplasts contain CuZn-SOD as the major isoform of SOD, and several plants such as tobacco also contain Fe-SOD (102). In Fe-SOD-containing plants, it is exclusively localized in the chloroplast stroma. The amino acid sequence of chloroplastic isoform of CuZn-SOD is different from those of the "cytosolic" isoforms localized in cellular compartments other than chloroplasts (29, 92, 175), and the two isoforms can be immunologically distinguished from each other (91). Plant CuZn-SOD, like that from other organisms, is a homodimer and contains one atom each of Cu and Zn in each subunit (16 kDa). The three-dimensional structure of the chloroplastic CuZn-SOD (95) is characterized by a β-barrel structure, also similar to those from other eukaryotes. Plants have several isoforms of the cytosolic CuZn-SOD, but only one for the chloroplastic isoform. Isoforms of the cytosolic CuZn-SOD would be localized in respective cellular compartments and have been found in nuclei, tonoplasts (146), and peroxisomes (88). Furthermore, the cytosolic CuZn-SOD is also localized in apoplastic compartments (64, 146, 192) and participates in lignin biosynthesis (147).

The catalytic cycle of CuZn-SOD for the disproportionation of O_2^- is shown below:

$$\text{SOD-Cu(II)} + O_2^- \rightarrow \text{SOD-Cu(I)} + O_2$$

$$\text{SOD-Cu(I)} + O_2^- + 2\,H^+ \rightarrow \text{SOD-Cu(II)} + H_2O_2,$$

where SOD-Cu(II) and SOD-Cu(I) represent oxidized and reduced enzymes, respectively. In water, both reaction steps proceed at a diffusion-controlled rate, $2 \times 10^9 \text{ M}^{-1} \text{ s}^{-1}$, in a wide range of pH between 4.8 and 9.7. The diffusion-controlled, rapid reaction of CuZn-SOD with O_2^- is facilitated by the electrostatic guidance of O_2^- with the Arg and other basic residues in the vicinity of the Cu (55), and also by the donation of proton to O_2^- from the protonated His residue of the Cu ligand at the second step. Because of the diffusion-controlled reactions, its rate is lowered in a high-viscosity medium. The viscosity of the stroma would be 69 relative to that of water owing to a high protein content of about 40%. From this value, the SOD-catalyzed disproportionation rate in the stroma is estimated to be $2 \times 10^8 \text{ M}^{-1} \text{ s}^{-1}$, i.e. one order of magnitude lower than that in water (148).

Attachment of CuZn-Superoxide Dismutase on Thylakoid Membranes

The chloroplastic CuZn-SOD is contained at around one molecule P700^{-1} in spinach chloroplasts (12). Both CuZn-SOD and Fe-SOD are soluble proteins, but immunogold labeling of the chloroplastic CuZn-SOD indicates localization of at least 70% of the enzyme in a 5-nm layer on the stroma thylakoids (148). Since the molecular dimension of CuZn-SOD is $16 \times 4.6 \times 8.6$ nm (95), this finding confirms the attachment of CuZn-SOD on the thylakoid membranes where the PSI complex is located, giving a local concentration of the enzyme on the membranes of about 1 mM. Mg^{2+} appears to participate in the attachment of CuZn-SOD on the thylakoid membranes through their ionic interactions (145).

The steady-state concentrations of O_2^- at different distances from the surface of the thylakoid membranes are simulated from the photoproduction rate of O_2^-, concentration of CuZn-SOD, the reaction rate constant of CuZn-SOD with O_2^- correcting the diffusion constants in the stroma due to high concentrations of protein. The simulation shows that if CuZn-SOD is distributed uniformly in the stroma, O_2^- can diffuse from the membrane, at a higher concentration in the nearly 100-nm layer of the stroma than that when CuZn-SOD is attached on the membranes. By the attachment, O_2^- is scavenged within 5 nm from the surface (148). If the diameters of the stromal enzyme molecules are assumed to be 5 to 7 nm, 100 nm corresponds to a layer of 14 to 20 molecules of enzymes. The Calvin-cycle enzyme complex also attaches on the thylakoid membranes (193, 212). Therefore, the active oxygen-sensitive, stromal enzymes (90) are likely located in the 100-nm layer from the membrane, and these enzymes may interact with O_2^- and the H_2O_2 derived from O_2^- if CuZn-SOD is not attached on the membranes. Transformants of tobacco, deficient in chloroplastic CuZn-SOD by the antisense method but with Mn-SOD targeted into chloroplasts, show photobleaching under bright light, and the PSI complex is inactivated (145). Even when Mn-SOD is contained in chloroplasts at higher levels than is CuZn-SOD, the lack of the membrane-attached CuZn-SOD induces the photoinhibition. These results indicate the importance of targeting not only to cell organelles but also of intrachloroplastic targeting of the scavenging enzymes to attain a stress tolerance.

Fe-SOD also is localized in the chloroplast stroma, and the tobacco transformants overproducing Fe-SOD in chloroplasts acquire tolerance against photooxidative stress (5, 210). In cyanobacteria, Fe-SOD is located in the cytosol and Mn-SOD is bound to the thylakoids (149). An Fe-SOD-lacking mutant of *Synechococcus* is sensitive to photooxidative stress, and its PSI complex is inactivated (63, 169, 205), similar to the chloroplastic CuZn-SOD-lacking tobacco. These observations indicate the attachment of Fe-SOD on the thylakoid membranes as well and scavenging of O_2^- at the site where it is generated, prior to its diffusion to the stroma or cytosol.

REDUCTION OF HYDROGEN PEROXIDE TO WATER

$H_2O_2 + 2$ ascorbate $\rightarrow 2\, H_2O + 2$ monodehydroascorbate.

Ascorbate Peroxidases

The H_2O_2 produced via the disproportionation of O_2^- catalyzed with the thylakoid membrane-attached SOD is reduced to water by ascorbate catalyzed with ascorbate-specific peroxidase (APX). H_2O_2-scavenging peroxidases in plants and eukaryotic algae use ascorbate as the electron donor, in contrast to GSH in mammals (seleno enzyme) and cytochrome *c* in yeast (heme enzyme). Chloroplasts contain phospholipid hydroperoxide-scavenging glutathione peroxidase, which is not a selenoenzyme (45, 59, 135). In addition, 2-Cys peroxiredoxin occurs in chloroplasts (24). These enzymes may participate in the reduction of lipid hydroperoxide of thylakoid membranes to its alcohol to suppress the chain oxidation of thylakoid phospholipids because APX cannot reduce lipid hydroperoxides.

APX is a heme peroxidase, and its amino acid sequence indicates that APX belongs to the same superfamily of heme-peroxidase (Class I) as cytochrome *c* peroxidase (CPX). APX occurs also in mammals (211). The three-dimensional structure of APX is similar to that of CPX, and the Trp residue for the Compound I formation of CPX is located on the same proximal His side as that of CPX (158). The ancestor of Class I peroxidase might be a bacterial catalase/peroxidase, which has also been found in cyanobacteria (137) and photosynthetic bacteria (47). Classical guaiacol peroxidases of plants (GPX, Class III), such as horseradish peroxidase, are distinguishable from APX in both amino acid sequence (39, 213) and in physiological function. GPX participates in metabolic reactions such as biosynthesis of lignin, decomposition of IAA, and defense to pathogens, but not in scavenging H_2O_2 (9, 13).

APX, like GPX, is inhibited by cyanide and azide by ligation to the heme. However, APX is also inhibited by thiol-modifying reagents, and by hydroxyurea, *p*-aminophenol (36), and thiols (37) via a suicide mechanism. Since these inhibitors do not affect GPX, APX and GPX can be separately assayed in plant extracts (2). The elicitor salicylic acid does not inhibit APX (103, 129). GPX has the four disulfide bridges of the conserved eight Cys residues and the residues for binding of carbohydrate and Ca^{2+}. APX, however, does not contain carbohydrate (35) and has no corresponding residues for glycosylation and Ca-binding. GPX prefers aromatic electron donors such as guaiacol and pyrogallol, and oxidizes ascorbate usually at a rate of around 1% that of guaiacol. Recently, however, a high ascorbate oxidizing peroxidase with an amino acid sequence similar to GPX has been found in tea (104).

The catalytic cycle of APX is the same as that of GPX, the peroxidase ping-pong mechanism. A two-electron oxidized intermediate of APX, Compound I, is formed, which then oxidizes ascorbate, successively producing two molecules of MDA and is reduced back to the resting ferric state;

$$APX\text{-}Fe(III)\text{-}R + H_2O_2 \rightarrow APX\text{-}Fe(IV) = O\text{-}R^+ + H_2O$$

$$k_1 = 1.2 \times 10^7\,M^{-1}\,s^{-1} \quad \text{Compound I}$$

$$APX\text{-}Fe\text{-}(IV) = O\text{-}R^+ + AsA \rightarrow APX\text{-}Fe(IV) = O\text{-}R + MDA$$

$$k_2 > 10 \times k_3 \quad \text{Compound II}$$

$$APX\text{-}Fe\text{-}(IV) = O\text{-}R + AsA \rightarrow APX\text{-}Fe(III)\text{-}R + MDA + H_2O$$

$$k_3 = 2.1 \times 10^6\,M^{-1}\,s,$$

where R represents porphyrin or the conserved Trp residue. The reaction rates shown are the values for thylakoid-bound APX (127). For Compounds I of CPX and GPX, R^+ is the Trp radical and porphyrin cation radical, respectively. The Trp residue is conserved in both CPX and APX, but R^+ of Compound I of at least cytosolic APX is a porphyrin cation radical (156, 158, 159).

Thylakoid-Bound and Stromal Ascorbate Peroxidases

Chloroplasts contain APX, but not GPX (139). Its content is about 1 molecule $P700^{-1}$ in two isoforms; thylakoid-bound (tAPX) and soluble stromal enzymes (sAPX) (124, 127). At least one half of the chloroplastic APX is tAPX, but the ratio of tAPX/sAPX varies according to plant species and, possibly, leaf age. tAPX binds to the stroma thylakoids where the PSI complex is located (124). sAPX is thought to be localized in the stroma, but its intrastromal distribution has not been determined. The local concentration of tAPX on the thylakoid surface is estimated to be 1 mM, and that of sAPX 40 μM, assuming its uniform distribution in the stroma.

tAPX is about 5.5 kDa larger than sAPX (32.2 kDa), but both the isoforms are monomers and share similar enzymatic properties (127). Chloroplastic APXs are highly specific to ascorbate as the electron donor, different from the cytosolic APX. The additional sequence of tAPX is found in the C terminus and is a hydrophobic domain for binding to the thylakoids (79, 219). Interestingly, the amino acid sequence of tAPX up to the 315th residue from the N terminus is identical to that of sAPX except for the residue of the C terminus of sAPX (79), and the biosynthetic ratio of the two APXs is controlled by alternative splicing (81, 115). Identification of the control mechanism of the biosynthetic ratio of the two chloroplastic isoforms should help elucidate their respective functions.

In addition to chloroplastic APX, plants contain the cytosolic isoforms (cAPX) (13, 35). The amino acid sequence of cAPX is different from those of chloroplastic APXs, and may participate in scavenging of H_2O_2 in compartments other than chloroplasts. Similar to GPX, several isoforms of cAPX occur in plants (78, 87, 99, 172). cAPX is a homodimer and its electron donor is not so specific to ascorbate, unlike tAPX and sAPX. Microbody or peroxisome-bound APX, however, is a monomer and has the hydrophobic additional domain in the C terminus for binding to the membranes (31, 80, 220). MDA reductase is also located accompanying APX in peroxisomes and mitochondria, and a scavenging system similar to that in chloroplasts seems to operate for further removal of H_2O_2, which has escaped from disproportionation by peroxisomal catalase (88).

REGENERATION OF ASCORBATE FROM OXIDIZED ASCORBATES

Reduction of Monodehydroascorbate (MDA) by Reduced Ferredoxin

MDA + redFd → ascorbate + Fd.

MDA radical is directly reduced prior to its spontaneous disproportionation by photoreduced ferredoxin (redFd) in thylakoids at a rate of 10^7 M^{-1} s^{-1}, as determined by EPR (125). The Fd-dependent reduction of MDA is also supported by oxygen exchange (84) and the MDA level in thylakoids (57). The redFd competes between MDA and $NADP^+$ in thylakoids, but the reduction rate of MDA is 34-fold higher than that of $NADP^+$ (125). Thus, the redFd is preferably used to reduce MDA, whenever MDA is generated, rather than $NADP^+$, which suggests that MDA is mainly photoreduced via Fd, but not via NAD(P)H with MDA reductase at least in the thylakoidal scavenging system (Figure 2).

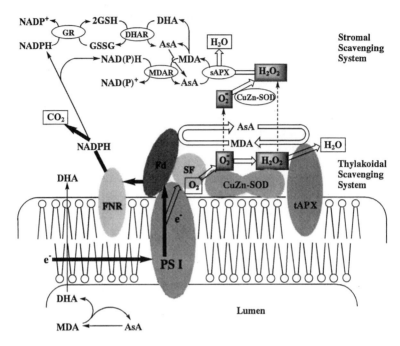

Figure 2 Molecular mechanism of the photoreducing system of dioxygen to water in the water-water cycle and microcompartmentalization of the participating enzymes. O_2^- is photoproduced directly in the PSI complex or indirectly mediated by a stromal factor (SF). A possible candidate of SF is monodehydroascorbate reductase (MDAR). The thylakoidal scavenging system is composed of CuZn-superoxide dismutase (SOD) attached on the thylakoids (in several plants, Fe-SOD), thylakoid-bound APX (t-APX), and ferredoxin (Fd). Fd reduces MDA directly to ascorbate (AsA). The stromal scavenging system is composed of CuZn-SOD localized in the stroma, stromal APX (sAPX), MDAR, dehydroascorbate reductase (DHAR), and glutathione reductase (GR). NAD(P)H for the reduction of either monodehydroascorbate (MDA) or dehydroascorbate (DHA) is photo-generated via ferredoxin-NADP$^+$ oxidoreductase (FNR). MDA is generated also in the lumen in the reaction of violaxanthin deepoxidase, and when AsA donates electrons to PSII or PSI. MDA in the lumen is rapidly disproportionated, and DHA in the lumen penetrates through the thylakoid membranes and is reduced to AsA by the stromal scavenging system.

Reduction of MDA with MDA Reductase

$$2\,\text{MDA} + \text{NAD(P)H} \rightarrow 2\,\text{ascorbate} + \text{NAD(P)}^+.$$

MDA reductase is a FAD enzyme and the first reported enzyme whose substrate is an organic radical (73). Chloroplastic MDA reductase is 8 kDa larger than the cytosolic MDA reductase (47 kDa), although both isoforms share similar enzymatic properties (MA Hossain, C Miyake, M Matsuo, H Aoki & K Asada, unpublished data). Its concentration in spinach chloroplasts is 14 μM, corresponding to 0.2 molecule $P700^{-1}$. A preliminary survey of the intrachloro-plastic distribution of MDA reductase (K Ogawa & K Asada, unpublished data) indicates its location near the thylakoid membranes, similar to CuZn-SOD. This is interesting with respect to the MDA reductase-mediated photoreduc-tion of O_2, as described above, and the microlocation of the stromal scaveng-ing system (Figure 2). The cytosolic isoform occurs in plasma membranes (27), mitochondria (107a), and peroxisomes (88), and is induced by wounding (58).

MDA reductase shows a high specificity to MDA as the electron acceptor and cannot reduce quinones. Its amino acid sequence is specific to this enzyme, and no similar flavodehydrogenase has been found in eukaryotes (136, 170). The enzyme prefers NADH (K_m; 5 μM) rather than NADPH (K_m; 22–200 μM) as the electron donor (171), even the chloroplastic isoform. The catalytic cycle of MDA reductase is represented as follows:

$$E\text{-FAD} + NAD(P)H + H^+ \rightarrow E\text{-FADH}_2\text{-NAD(P)}^+$$

$$E\text{-FADH}_2\text{-NAD(P)}^+ + MDA \rightarrow E\text{-FADH-NAD(P)}^+ + \text{ascorbate} + H^+$$

$$E\text{-FADH-NAD(P)}^+ + MDA + H^+ \rightarrow E\text{-FAD} + \text{ascorbate} + NAD(P)^+.$$

The first step is the reduction of the enzyme-FAD to form a charge transfer complex, and it proceeds at a rate of 1.3×10^8 M^{-1} s^{-1} (171). The reduced enzyme donates electrons successively to MDA, producing two molecules of ascorbate via a semiquinone form (E-FADH-NAD(P)$^+$) at a rate of 2.6×10^8 M^{-1} s^{-1} (97). The diffusion-controlled, rapid reactions at both steps are facilitated by the electrostatic guidance of NAD(P)H and MDA, both anionic forms at a neutral pH, to the enzyme-FAD.

Since redFd more effectively reduces MDA than $NADP^+$, MDA reductase cannot participate in the reduction of MDA in the thylakoidal scavenging sys-tem. Therefore, MDA reductase would function at a site where NAD(P)H is available but redFd is not. In chloroplasts, Fd associates with the basic domains of psaD and psaE of the PSI complex via ionic interaction. It is not known how far Fd can diffuse into the stroma from the PSI complex, but the diffusion of Fd from the PSI complex is likely very limited, given the occurrence of MDA reductase in the vicinity of thylakoid membranes.

Spontaneous Disproportionation of MDA

$$MDA + MDA \rightarrow \text{ascorbate} + DHA.$$

MDA is generated not only by the APX reaction but also by the following re-actions in chloroplasts; the interaction of ascorbate with O_2^-, $\cdot OH$, and thiyl radical when they escape from the scavenging system, and also with organic radicals including tocopherol radical. The tocopherol radical is inevitably produced when thylakoid phospholipids are oxidized by the $^1O_2$ photogenerated, and ascorbate is essential to regenerate tocopherol in the thylakoids.

In addition, MDA is generated in the thylakoid lumen. One is the oxidation of ascorbate to MDA accompanying the de-epoxidation of violaxanthin to zeaxanthin via antheraxanthin with violaxanthin de-epoxidase for the down-regulation of the quantum yield of PSII under the excess photon conditions (69, 141, 142). MDA is also generated by the donation of electrons from ascorbate to PSII when the oxidizing side of PSII is inactivated (114). Further, ascorbate donates an electron to PSI producing MDA when PSII is inactivated (111). In all cases, the MDA produced in the lumen cannot be reduced by either ferredoxin or NAD(P)H. The MDA in the lumen is spontaneously disproportionated to DHA and ascorbate. DHA is a non-dissociated compound and easily penetrates through the thylakoid membranes, whereas anionic MDA cannot (114). DHA preferentially penetrates also through plasma membranes (67).

The rate of spontaneous disproportionation of MDA and the effect of pH on it are very similar to those of superoxide. At pH 3–5, the disproportionation rate is around 10^8 M^{-1} s^{-1}, but decreases to 10^6 M^{-1} s^{-1} at pH 6 and to 5×10^4 M^{-1} s^{-1} at pH 8. Thus, under the conditions where the pH of the lumen is 5–6 and violaxanthin is de-epoxidated, the MDA produced in the lumen is rapidly converted to DHA and ascorbate. In the stroma where the pH is 7–8, the life span of MDA is longer than that in the lumen and is reduced by either redFd or NAD(P)H.

Reduction of Dehydroascorbate (DHA) with DHA Reductase and Regeneration of GSH

$$DHA + 2\, GSH \rightarrow ascorbate + GSSG$$

$$GSSG + NADPH \rightarrow NADP^+ + 2\, GSH.$$

When MDA fails to be directly reduced to ascorbate, DHA is generated via its spontaneous disproportionation. Furthermore, the MDA produced in the lumen is disproportionated, and then the resulting DHA diffuses to the stroma. Thus, the reduction of DHA is indispensable to keeping ascorbate in the reduced state for scavenging of H_2O_2. In a DHA reductase-deficient mutant of *Ficus microcarpa*, chlorophyll is bleached under bright light (221). GSH can nonenzymatically reduce DHA to ascorbate, but its rate is too slow to account for the observed photoreduction of DHA in chloroplasts. In ruptured chloroplasts, DHA serves as the Hill oxidant at a rate of over 130 μmol DHA-reduced mg Chl^{-1} h^{-1}, corresponding to 50 e$^-$ P700^{-1} s^{-1} (139). For the photoreduction of DHA the addition of GSH is required, thus, GSH-dependent DHA reductase is the most likely enzyme for the reduction of DHA in chloroplasts.

DHA reductase using GSH as the electron donor has been purified from leaf (71) and other tissues (43, 93). Its molecular size is 23 kDa and the thiol group participates in the reaction. Trypsin inhibitor in its reduced form and thioredoxins from chloroplasts show DHA-reducing activity (132, 208), but their activities are not high enough to account for the photoreduction of DHA in chloroplasts. Furthermore, thioredoxin reductase (119) and protein disulfide isomerase (1, 214) show a DHA-reducing activity.

Glutathione reductase for the regeneration of GSH is contained at about 1.4 μM (0.02 molecule $P700^{-1}$) in the stroma. [For a recent review of this enzyme, see (134).] Transformants overexpressing glutathione reductase in chloroplasts show tolerance to oxidative stress, as do transformants depleting sensitivity (4). These results indicate the importance of the DHA-glutathione system to protect against photodamage by active oxygens.

THE WATER-WATER CYCLE: PHOTOREDUCTION OF DIOXYGEN TO WATER

The water-water cycle is composed of the following reactions:

[1] $2\,H_2O \rightarrow 4\,[e^-] + 4\,H^+ + O_2$ [photooxidation of water in PSII]

[2] $2\,O_2 + 2\,[e^-] \rightarrow 2\,O_2^-$ [photoreduction of O_2 mediated by SF in PSI]

[3] $2\,O_2^- + 2\,H^+ \rightarrow H_2O_2 + O_2$ [SOD-catalyzed disproportionation of O_2^-]

[4] $H_2O_2 + 2\,ascorbate \rightarrow 2\,H_2O + 2\,MDA$
    [APX-catalyzed reduction of H_2O_2 by ascorbate]

[5] $2\,MDA\,(or\,DHA) + 2\,[e^-] + 2\,H^+ \rightarrow 2\,ascorbate\,(or\,1\,ascorbate)$
    [reduction of oxidized ascorbates]

$$\sum 2\,H_2O + O_2^* \rightarrow O_2 + 2\,H_2O^*.$$

In the water-water cycle, the electrons from water generated in PSII reduce atmospheric O_2 to water in PSI without a net change of O_2. The maximum rate of the linear electron flow in thylakoids is 200 e^- $P700^{-1}$ s^{-1}, with the limiting step at the oxidation of the plastoquinol by cytochrome b/f. The actual photoreduction rate of O_2 in thylakoids, however, is only 4 e^- $P700^{-1}$ s^{-1}, due to the putative O_2^--mediated cyclic electron flow around PSI (11). In leaf tissues, however, the linear electron flow through the water-water cycle is high and cannot be accounted for by the photoreduction of O_2 in thylakoids. A stromal factor (SF), possibly the flavodehydrogenase MDA reductase, enhances the photoreduction of O_2 to O_2^- up to 60 e$^-$ P700^{-1} s^{-1} at maximum (131). Since the limiting step of the water-water cycle is the reduction of O_2, the same reducing equivalent as that for O_2-reduction is transferred to either MDA or DHA. Assuming the MDA reductase-mediated photoreduction of O_2, the maximum flux of the linear electron flow through the water-water cycle is 120 e$^-$ P700^{-1} s^{-1}. Disproportionation of the O_2^- generated in PSI and the reduction of H_2O_2 and of either MDA or DHA proceed at a rapid rate in two compartments: the thylakoidal and stromal scavenging systems (Figure 2).

Thylakoidal and Stromal Scavenging Systems

The enzymes participating in the thylakoidal system are microcompartmented in the vicinity of the PSI complex where O_2^- is photogenerated. CuZn-SOD or Fe-SOD is attached onto, and APX (tAPX) binds to, the membranes of the stroma thylakoids. Fd associates with the PSI complex. Their local concentrations in a 5-nm layer on the thylakoid membranes (1–13 mM) and respective reaction rate constants (SOD + O_2^-, 2×10^9 M^{-1} s^{-1}; H_2O_2 + APX, 10^7 M^{-1} s^{-1}; MDA + redFd, 10^7 M^{-1} s^{-1}) allow estimation of their pseudo-first-order reaction rate constants. The estimated constants of the reactions [3], [4], and [5] are 10^4–2×10^6 s^{-1} (12), which is several orders of magnitude higher than that of reaction [2]. Since reaction [5] requires electrons from PSI, its rate is limited by that of the linear electron flow. The rate of reaction [5] using the electron generated in PSI, however, is faster by several orders of magnitude than that of reaction [2].

When O_2^- and H_2O_2 fail to be scavenged in the thylakoidal system, O_2^- is disproportionated with the CuZn-SOD located in the stroma, and H_2O_2 is then reduced to water with the stromal APX (sAPX). In contrast to the thylakoidal system, in the stromal system, MDA would be reduced by NAD(P)H catalyzed with MDA reductase, and its disproportionation product DHA is reduced by GSH with DHA reductase. Even if the participating enzymes are assumed to be uniformly distributed in the stroma, the pseudo-first-order reaction rate constants of the respective reactions (4×10^2–4×10^4 s^{-1}) are higher than that of the reaction [2] (12). The MDA generated in the lumen is spontaneously disproportionated to ascorbate and DHA. The DHA in the lumen diffuses to the stroma through the thylakoid membranes and is subsequently reduced to ascorbate by the stromal system.

The addition of catalase is recommended to observe CO_2-fixation by intact chloroplasts. Even if a small amount of thylakoids is contaminated in chloroplast preparations, the photosynthetic CO_2-fixation is inhibited by the H_2O_2 derived from O_2^- photogenerated in the thylakoids and protected by the addition of catalase (10). This suggests that the stromal scavenging system is also compartmented in the vicinity of the thylakoid-PSI complex, and is not as effective for scavenging the H_2O_2 that penetrated through the envelope from the medium.

Characteristics of the Water-Water Cycle

The limiting step of the water-water cycle is the photoreduction of O_2, but not the steps of the disproportionation of O_2^- and of the reduction of H_2O_2 and the oxidized ascorbates in both the thylakoidal and stromal scavenging systems. The rapid reactions of [3], [4], and [5] are guaranteed by diffusion-controlled, rapid reactions catalyzed with the respective enzymes and their microcompartmentalization at the generation site of O_2^-. The following observations support the rapid reductions of H_2O_2, MDA, and DHA. The induction of a large photochemical quenching of chlorophyll fluorescence by the addition of H_2O_2 to intact chloroplasts indicates a rapid electron flow in the reducing step of the oxidized ascorbates rather than that in the reduction of O_2 (124, 128, 140). The addition of H_2O_2 to illuminated intact chloroplasts completely suppresses CO_2-fixation, but fixation recovers after the H_2O_2 disappears (138), indicating more rapid reduction of oxidized ascorbates generated by the APX reaction than that of CO_2. Limited accumulation of MDA as observed by EPR in healthy leaves (62, 65, 215) is another line of evidence for the fast reduction of MDA. Further, the evolution of $^{16}O_2$ from water at the same rate of $^{18}O_2$ uptake without accumulation of $H_2^{18}O_2$ in intact chloroplasts (14) is inferred only by the even utilization of the electrons in PSI to reduce $^{18}O_2$ and the oxidized ascorbates.

Rapid rates of the O_2^--disproportionation and reduction of H_2O_2, MDA, and DHA as compared with that of O_2-reduction are the important characteristics of the water-water cycle. The electrons from PSII are divided equally to reactions [2] and [5] by the difference of the rates between the two reactions. Further, there is minimal accumulation of O_2^- and H_2O_2, with very short life spans (the thylakoidal system, 7×10^{-5}–4×10^{-7} s; the stromal system, 2×10^{-3}–2×10^{-5} s), which suppresses their interaction with the target molecules in chloroplasts.

Stability of the Water-Water Cycle

CuZn-SOD is a stable enzyme that is resistant to denaturating stress including increasing the temperature up to 80°C because of its β-barrel structure with a low content of α-helix structure. However, CuZn-SOD is inactivated by H_2O_2 as follows: The reverse reaction of the catalytic cycle of CuZn-SOD causes the enzyme to be reduced with H_2O_2; further reaction of H_2O_2 with the reduced enzyme causes the hydroxyl radical to be generated at the reaction center of the enzyme (92).

$$SOD\text{-}Cu(II) + H_2O_2 \rightarrow SOD\text{-}Cu(I) + O_2^- + 2\,H^+$$

$$SOD\text{-}Cu(I) + H_2O_2 \rightarrow SOD\text{-}Cu(II) + \cdot OH + OH^-.$$

The hydroxyl radical that is generated oxidizes the Cu-ligated His residue to 2-oxo-histidine (209), resulting in inactivation. CuZn-SOD is inactivated by H_2O_2 at an apparent rate constant of 0.8 $M^{-1}\,s^{-1}$ (22); the half time for the inactivation is about 30 min in 0.5 mM H_2O_2 at a neutral pH. CuZn-SOD in chloroplasts is inactivated by H_2O_2 and fragmented by $\cdot OH$ (33). CuZn-SOD is further fragmented (168) and modified by glycosylation (151) by active oxygens. Another stromal SOD, Fe-SOD, also is inactivated by H_2O_2 at a similar rate; 0.6 $M^{-1}\,s^{-1}$ (22). Fe-SOD is not as thermostable as CuZn-SOD, but is stable up to 50°C. Both CuZn-SOD and Fe-SOD are inactivated by H_2O_2, but not Mn-SOD.

A property characteristic of APX, especially of the chloroplastic isoform, is its lability in the absence of its electron donor ascorbate (13). When ascorbate concentration is below 20 μM, both stromal and thylakoid-bound chloroplastic APXs are inactivated with a half-life of only 15–20 s. This may be one reason why APX was not identified in plants until recently, in contrast to classical guaiacol peroxidase, and why its purification is possible only in the presence of ascorbate.

In the presence of ascorbate, the reaction intermediate of APX Compound I successively oxidizes ascorbate to produce two molecules of MDA in the catalytic cycle, as described above. In the absence of ascorbate, however, APX is first inactivated by the interaction of Compound I with H_2O_2 ($7 \times 10^5\,M^{-1}\,s^{-1}$), after which the heme of APX is degraded ($8 \times 10^3\,M^{-1}\,s^{-1}$). APX is inactivated by 10 nanomolar levels of H_2O_2 within 2 min, which is generated via autooxidation of ascorbate at micromolar levels. Since chloroplasts contain ascorbate at 25 mM, APX is not inactivated, in as much as its regeneration system is functional, but it is more labile, with an increase in the H_2O_2/ascorbate ratio (126). Therefore, proper operation of the water-water cycle is indispensable to maintaining APX activity.

DHA reductase is labile in the absence of thiols and inactivated by 0.5 mM H_2O_2 (71). Glutathione reductase is inactivated by NADPH in the absence of GSSG, and labile under photooxidative stress (107, 134).

In illuminated chloroplasts, paraquat (1,1'-dimethyl-4,4'-bipyridylium chloride) traps all the electrons from PSI at a diffusion-controlled rate (197) to form its cation radicals, which generate O_2^- at a rapid rate via autooxidation; ascorbate cannot regenerate from MDA and DHA since Fd and $NADP^+$ cannot be photoreduced. In the presence of paraquat, illuminated chloroplasts start to accumulate H_2O_2 after a lag of several minutes (138). Under such conditions, APX is first inactivated after short illumination (113). Even in the absence of paraquat, the dark addition of H_2O_2 to chloroplasts inactivates APX (72). SOD, DHA reductase, and glutathione reductase are inactivated only after long illumination in the presence of paraquat (82). These results show that, among the scavenging enzymes, APX is the most sensitive, primary target molecule of the water-water cycle.

Molecular assembly of the enzymes of the water-water cycle is essential for effective, rapid reduction of O_2 to water in PSI without releasing reduced species of oxygen. Even if the enzymes retain their activities, disordering of their compartmentalization would lower the scavenging capability, as in the case of CuZn-SOD (145). The MDA radical is hardly detected by EPR in healthy leaves under light, indicating an immediate reduction of MDA generated by the APX reaction. However, MDA is observed in leaves under environmental stresses such as paraquat (62, 215), pollutants (191, 210a), UV-B (65), gas exchange stress, and bright light (62). Thus, the EPR assay of MDA is a sensitive, endogenous probe in leaf tissues for the functional evaluation of the water-water cycle. The fact that disorder occurs in the water-water cycle has been monitored by other methods such as accumulation of H_2O_2 and inactivation of scavenging enzymes, under paraquat (82, 113), SO_2 (17, 201), O_3 (166), and chilling stresses (85, 204, 218).

Ascorbate and Glutathione

In addition to the scavenging enzymes, ascorbate and glutathione are essential mediators in the water-water cycle. In chloroplasts, both compounds are found at similarly high levels of about 25 mM, which is considerably higher than the saturated concentrations for APX and DHA reductase reactions. Under photooxidative stress, their biosynthesis is usually induced to increase their contents, but its regulation mechanism remains obscure. A low ascorbate mutant of *Arabidopsis* is sensitive to oxidative stress such as UV and pollutants (41). The immediate precursor of ascorbate is either sorbosone (167) or galactono-γ-lactone (216), and its dehydrogenase for the synthesis of ascorbate has been characterized (73a, 144, 154). This enzyme is inhibited by an alkaloid, lycorine, but APX, DHA reductase, and MDA reductase are not, which allows the use of this alkaloid to decrease the ascorbate contents in tissues (7). [For further information of the transport and biosynthesis of ascorbate and glutathione, see (143, 184).]

PHYSIOLOGICAL FUNCTIONS OF THE WATER-WATER CYCLE

The water-water cycle allows an appreciable flux of the linear electron flow in oxygenic phototrophs including cyanobacteria and generates the proton gradient across the thylakoid membranes (141, 142, 178, 179). ATP is produced as a result of the proton gradient, but is not consumed in the water-water cycle. This is similar to the case when either nitrite or oxaloacetate is the only electron acceptor. No net production of NADPH and redFd is expected in either the thylakoidal or stromal scavenging systems, but the water-water cycle maintains ascorbate and glutathione at reduced states in chloroplasts. In cyanobacterial cells, the electron flow through the water-water cycle competes with the photoreduction of nitrite (123). During operation of the water-water cycle, so little O_2^- and H_2O_2 are released from the thylakoidal and stromal scavenging complexes that they do not interact with the target molecules around the PSI complex and in the stroma. The physiological functions of the water-water cycle, based on these characteristics, are described as follows.

Protection of Target Molecules from Active Oxygens

If the water-water cycle cannot operate, O_2^- and H_2O_2 diffuse to the stroma and oxidize the target molecules in chloroplasts. When transition metal ions such as Fe and Cu are released from metal proteins, they catalyze the production of the highly reactive ·OH radical. The interaction of H_2O_2 with the photoreduced [4Fe-4S] clusters on the reducing side of PSI also generates ·OH (86). Assuming a photoproduction rate of 25 O_2^- P700^{-1} s^{-1}, a rate observed in intact leaves under bright light, H_2O_2 accumulates in chloroplasts at a rate of 120 μM s^{-1}. Only 2 μM H_2O_2 inactivates APX within several seconds in the absence of ascorbate (126). Ascorbate is consumed by the APX reaction within several minutes when the regeneration system of ascorbate does not operate (138). The CO_2-fixation of chloroplasts is inhibited to a half by 10 μM H_2O_2 (89). Thus, O_2^- and H_2O_2 must be completely scavenged to preserve photosynthetic activity.

Interaction of heme-catalase with O_2^- forms Compound III, which is not a catalytic intermediate, and the activity is lost (98, 183). Little catalase is contained in chloroplasts, so the inactivation of catalase by O_2^- is likely to occur in peroxisomes. Peroxisomal SOD (88) should protect catalase from such inactivation. Under environmental conditions where photorespiration protects leaf cells from photoinhibition, the peroxisomal catalase-deficient tobacco sustains photodamage (217).

At the same levels required to inactivate APX, H_2O_2 inactivates NADP$^+$ glyceraldehyde 3-phosphate dehydrogenase, fructose1,6-bisphosphatase, ribulose 5-phosphate kinase and sedoheptulose 1,7-bisphosphatase in chloroplasts of angiosperms (90, 201). The inhibition of the CO_2-fixation by H_2O_2 is due to the inactivation of these enzymes, and lowers the photon-utilizing capacity, further increasing production of O_2^-. The H_2O_2-sensitive enzymes have the catalytically functional thiol groups and are inactivated by disulfide formation. The oxidized, inactive enzymes can recover their activities by the reduction of the disufide groups with the thioredoxin system in chloroplasts. The corresponding enzymes from eukaryotic algae and cyanobacteria, however, are resistant to H_2O_2 (199, 200). Further, ribulose 1,5-bisphosphate carboxylase/oxygenase is fragmented at the specific site by the ·OH radical generated with H_2O_2 and illuminated thylakoids (76, 77). Glutamate synthase is sensitive to the ·OH radical (190), and the chloroplastic, Fd-dependent glutamate synthase is inactivated under bright light (100). The overexpressing glutamate synthase in chloroplasts in tobacco shows tolerance to bright light, because the Fd-dependent glutamate synthase (42) appears to be the limiting step in the photorespiratory pathway to dissipate excess light energy (100, 198).

The PSI complex also is a target site of active oxygens. Photoinhibition of PSI requires O_2, unlike PSII, and is suppressed by DCMU (75, 173, 174, 185). Thus, the photoinhibition of PSI requires O_2 and the linear electron flow to PSI from PSII, which suggests the inactivation of PSI by the active oxygens photoproduced in PSI. PSII is thought to be the primary target of photoinhibition rather than PSI under the conditions where photon energy is in excess of CO_2-assimilation. In photoinhibition studies, high intensities of light, over 500 μmol m^{-2} s^{-1} have usually been applied. Under such conditions, PSII is photoinactivated and the D1 protein of the PSII complex is then degraded (3, 6), whereas PSI remains active. These observations are also explained by assuming that PSI is apparently protected by the photoinhibition of PSII because no electrons are available to PSI to photoproduce O_2^-. This assumption is supported by remarkable photoinhibition of PSI under moderate light if the water-water cycle is damaged.

PSI is primarily photoinactivated under chilling stress, especially in chilling-sensitive plants, by moderate light prior to the photoinactivation of PSII (60, 185, 203, 206; reviewed in 186, 187). The enzymes of the water-water cycle are inactivated by chilling treatment (85, 218). Cucumber leaves when chilled under light whereby PSI is inactivated, accumulate H_2O_2 due to a low activity of APX (204). PSI is also photoinactivated in *Chlamydomonas* by paraquat even under moderate light prior to that of PSII (117). The chilling-induced photoinhibition of PSI is accelerated when SOD and APX are inhibited by cyanide (206). These observations support the view that PSI is photoinhibited by the active oxygens generated in PSI when the scavenging system is not operating efficiently. The active oxygens photoproduced in PSI, however, do not cause the photoinhibition of PSII, at least, in its initial step. Degradation of D1 protein by active oxygen might be an event in the photoinhibition of PSII at later stages. A chilling-tolerant inbred line of maize contains higher activities of the scavenging enzymes of the water-water cycle than chilling-susceptible line (160). During acclimation to chilling stress, biosynthesis of the scavenging enzymes is induced (50, 150, 161, 176, 218). These observations also support the participation of active oxygen in the preferential photoinhibition of PSI under chilling stress, and protection by the water-water cycle.

The [4Fe-4S]-containing enzymes such as aconitase and 6-phosphogluconate dehydratase are specifically inactivated by O_2^- (54, 202). Similar to these enzymes, when PSI is photoinhibited the [4Fe-4S] clusters are damaged (75, 188, 189, 206). The O_2^- and H_2O_2 photoproduced in these clusters and the ·OH radical generated by the interaction of H_2O_2 with the photoreduced clusters (86) are likely molecules to destroy the clusters.

When PSII is photoinhibited and the D_1 protein is degraded, chloroplasts can rapidly recover PSII activity by incorporating the de novo-synthesized D_1 protein to the D_1-depleted PSII complex (3, 6). However, no such repair or recovery system has been identified to date for PSI; its photoinhibition is irreversible. The water-water cycle protects PSI, but, under severe photon-excess conditions where the water-water cycle is unable to protect PSI, the photoinhibition of PSII seems to function as "an emergency sacrifice" to protect PSI from irreversible inactivation by lowering the electron supply to PSI (15). Even when the emergency sacrifice, D1 protein, is degraded, the PSII complex rapidly recovers its activity.

Adjustment of Photoproduction Ratio of ATP/NADPH

The water-water cycle produces, but does not consume, ATP without net production of NADPH, which has been referred to as pseudocyclic photophosphorylation. The operation of the water-water cycle regulates the photoproduction ratio of ATP/NADPH in chloroplasts. Under anaerobic conditions no CO_2 is fixed in intact chloroplasts because of a low ratio of ATP/NADPH to operate the CO_2-fixation cycle, but the addition of O_2 allows CO_2-fixation to start by increased production of ATP through the water-water cycle (222). The cyclic electron flow around PSI mediated by either ferredoxin or pyridine nucleotide dehydrogenase complex (16, 20, 40, 112, 122, 130, 182) can also supply additional ATP to chloroplasts without producing NADPH, but generally does not operate when there is a substantial electron supply from PSII to PSI. The water-water cycle has no such limitation and can increase the photoproduction ratio of ATP/NADPH depending on the requirement of the reactions in chloroplasts. Thus, the water-water cycle can enhance the photoproduction of ATP so as to increase CO_2-assimilation and protect against photoinhibition that occurs when photon intensity exceeds its utilizing capacity. Higher ATP/NADPH is required to operate photorespiration than for CO_2-fixation (152), and the water-water cycle may supply additional ATP for the photorespiratory pathway. Using a microchlorophyll fluorometer to measure a single cell, it has been demonstrated that, in a guard cell of broad bean, the electron flux is mostly ascribed to the water-water cycle (177). In this case, the water-water cycle may supply the ATP necessary to pump ions for stomata functioning.

Dissipation of Excess Photons

First discovered by Trebst in 1962, photoinhibition under anaerobic conditions has subsequently been shown to occur in intact cells (3, 21, 157) and in intact chloroplasts and thylakoids (101). PSII is the photoinhibiting site, in contrast to the O_2-required photoinhibition of PSI. Anaerobic photoinhibition of PSII in leaf cells is induced by minimal linear electron flux because of inhibited CO_2-assimilation due to a low photoproduction ratio of ATP/NADPH, and because of the suppression of photorespiration. Under such conditions, the unavailability of O_2 as an electron acceptor of PSI prevents linear electron flow, and the intersystem electron carriers are maintained at reduced states, under which PSII is most severely photoinhibited. However, 1–2% O_2 releases the cells from the photoinhibition, not enough to allow recovery of photorespiratory activity but enough to allow the water-water cycle to operate, albeit not at the maximal rate.

If 30% of the flux of the linear electron flow passes through the water-water cycle, its rate is 50 e^- P700^{-1} s^{-1}, about a quarter of the maximum rate of the linear electron flow. Such a high electron flux through the water-water cycle would provide a "safety" way to dissipate excess light energy under environmental stress. Figure 3 shows schematically the fate of photon energy absorbed by chloroplasts depending on the photon intensity. The CO_2-assimilation curve represents the classical Blackman-type response to photon intensity. The CO_2-assimilation response is largely affected by endogenous factors such as sun and shade plants, leaf age, location of chloroplasts in palisade parenchyma and spongy tissues, mineral nutrition, and sink capacity of the assimilation products. Environmental factors such as temperature, water supply, salt stress, gas exchange (CO_2) stress, and pollutants also affect the capacity for CO_2-assimilation. Fluctuations in sunlight intensity in areas other than desert, independent of other environmental factors, frequently expose plants to unfavorable combinations of the factors for photosynthesis such as bright light and chilling temperatures. Together with an increase of photon intensity, the ratio of the photon energy used for the CO_2-assimilation is decreased. In nature, 10, 25, and 80% of absorbed photon energy is used for CO_2-assimilation under full sunlight (about 2000 μmol m^{-2} s^{-1}), 50% and 5% of full sunlight, respectively (109), and the remaining photon energy is dissipated through the water-water-cycle, the down-regulation of PSII triggered by the xanthophyll cycle (69) and photorespiration (8, 100, 152, 198).

Down-regulation of PSII lowers the quantum yield of PSII to at least one half and dissipates the photon energy as heat. This "safety" dissipation of excess

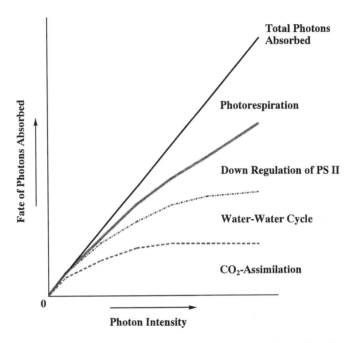

Figure 3   The fate of photon energy absorbed by chlorophylls in chloroplasts under various photon intensities. The photon-utilizing capacity for CO_2-assimilation of leaves is largely affected by environmental stress such as drought, chilling, gas exchange stress, UV, bright light, and mineral stress and also by endogenous stress such as sink stress and senescence. When the photon intensity is in excess of CO_2-assimilation, excess photon energy is dissipated by the water-water cycle, down-regulation of PSII through the generation of proton gradient across the thylakoid membranes and photorespiration. When the physiological electron acceptor, CO_2, is not available to chloroplasts or the CO_2-fixation cycle does not effectively operate, the linear electron flow through the water-water cycle is indispensable to generate the proton gradient, at least initially.

photon energy is induced by a low pH, around 5–6, in the lumen accompanying the de-epoxidation of violaxanthin to zeaxanthin. For this dissipation of excess photon energy, the proton gradient across the thylakoid membrane should be generated (69, 141, 142). Under photon-excess conditions, physiological electron acceptors are usually not available to PSI, and the proton gradient can be generated via either the cyclic electron flow around PSI or the linear electron flow through the water-water cycle. The cyclic electron flow appears to operate preferentially when the electron supply from PSII is limited, for example, by down-regulation of PSII or excitation of PSI only by far-red light, in the PSII-lacking chloroplasts of the bundle sheath cells of C_4 plants or when the PSI complex is functionally separated from the PSII complex, as in cells with a high PSI/PSII ratio. Therefore, in C_3 plants, the cyclic electron flow is unlikely to induce the proton gradient to down-regulate PSII, at least just after exposure to environmental stress. Since the water-water cycle itself does not consume ATP, the proton gradient is effectively generated under any environmental conditions. Thus, the water-water cycle can respond even to a sudden change of environmental factors such as sunflecks. The linear electron flow for the photorespiratory pathway consumes ATP at a higher ratio of ATP/NADPH than does that for the Calvin cycle (152), and would not contribute to generation of the proton gradient. In mesophyll chloroplasts of C_4 plants, where the photorespiratory activity is low, the water-water cycle should generate the proton gradient to down-regulate PSII (83). The proton gradient also regulates the translation elongation of protein synthesis in chloroplasts (133).

The water-water cycle induces and maintains the down-regulation of PSII through generation of the proton gradient. Furthermore, the water-water cycle itself can dissipate excess photons using O_2 as the electron acceptor without releasing O_2^- and H_2O_2 even when the physiological electron acceptors are not available. In a specific case such as in guard cells (177), the water-water cycle functions as the generator of ATP. When CO_2-fixation is suppressed by environmental stress and the water-water cycle only functions to dissipate excess photon energy through the reduction of O_2 to water, ATP is apparently overproduced in chloroplasts. There is a mechanism in illuminated chloroplasts, however, to maintain a constant ratio of ATP/ADP (1.5–3), when either nitrite or oxaloacetate is the electron acceptor that does not consume ATP. Although the molecular mechanism to maintain a constant ratio of ATP/ADP even when ATP is not consumed and to hydrolyze the over-produced ATP is not well characterized, the participation of adenylate kinase and the thiol, light-activated ATPase of thylakoids has been suggested (74, 96). This mechanism would play an important role in maintaining the ADP level for relaxing the proton gradient under conditions whereby ATP is not consumed in chloroplasts.

During environmental stress, which causes lowering of CO_2-assimilation capacity or exposure to high photon intensity, excess photon energy is "safely" dissipated by three pathways (Figure 3). Data indicate that the water-water cycle occupies an appreciable flux of the linear electron flow. The water-water cycle first triggers the down-regulation of PSII and then dissipates excess excitation energy by reducing O_2 to water. Further, the water-water cycle would supply additional ATP required for the photorespiration, which not only dissipates excess photon energy through the electron flow but also supplies the physiological electron acceptor CO_2 to chloroplasts.

CONCLUDING REMARKS

The limiting step of the water-water cycle is the reduction of O_2 to O_2^-; once O_2^- is produced, it rapidly disproportionates to H_2O_2 and O_2, and H_2O_2 is also reduced to water at a rapid rate. Because of this property of the water-water cycle, one half of the electrons generated in PSII from water are donated to O_2 and the other half to oxidized ascorbates, either MDA or DHA, in the thylakoidal and stromal scavenging systems (Figures 1, 2). The diffusion-controlled, rapid rates of the disproportionation of O_2^- and the reductions of H_2O_2 and oxidized ascorbates are guaranteed not only by the enzyme structure, but also by their microcompartmentalization on the stromal side of the PSI complex where O_2^- is photoproduced.

Where the photon intensity exceeds CO_2-assimilation owing to environmental stress or bright light, a considerable flux of the linear electron flow is accounted for by the water-water cycle. The reduction of O_2 through the water-water cycle over 10% of the linear electron flux should be mediated by a stromal factor, for which the flavodehydrogenase MDA reductase is the most likely candidate. Since the turnover of the water-water cycle is determined by the photoreduction rate of O_2, how it is regulated to adjust to various physiological functions is an important topic for future research. Competition of the reducing equivalents in PSI between $NADP^+$ and O_2 is the primary mechanism, but other biochemical mechanisms such as the proton gradient across the thylakoid membranes may participate in regulating the photoreduction of O_2 to meet to the physiological functions of the water-water cycle.

The first physiological function of the water-water cycle is to protect its scavenging enzymes, the stromal enzymes, and the PSI complex from oxidative damage by the O_2^- photoproduced in PSI and other reactive oxygen species derived from O_2^-. To fulfill this function, O_2^- and H_2O_2 should not be released from the thylakoidal and stromal scavenging complexes. As long as the water-water cycle operates correctly, the primary target enzyme of the water-water cycle APX, several H_2O_2-sensitive, stromal enzymes and the [4Fe-4S] clusters of the PSI complex are not damaged. However, when the water-water cycle is disrupted by environmental conditions, CO_2-assimilation is lowered, thereby suppressing the photon-utilizing capacity, and the PSI complex is irreversibly inactivated. In this respect, the water-water cycle is one of the primary target sites of photoinhibition. The second physiological function of the water-water cycle is to reinforce CO_2-assimilation by supplying ATP to operate the CO_2-fixation cycle. The third physiological function is to dissipate excess photon energy (Figure 3). The water-water cycle itself dissipates "safely" the excess photon energy without net change of O_2 and net production of photoreductants. Further, the water-water cycle helps to dissipate excess photon energy via the down-regulation of PSII by forming the proton gradient and via photorespiration by supplying additional ATP. The proper operation of the water-water cycle without releasing O_2^- and H_2O_2 from the scavenging complex is indispensable to the second and third physiological functions. In conclusion, the water-water cycle protects leaf photosynthesis from photoinhibition by environmental stress not only via the scavenging of active oxygens but also the dissipation of excess photon energy.

Anaerobic photosynthetic sulfur bacteria contain Fe-SOD and nonsulfur bacteria Fe-SOD and Mn-SOD (11, 92), indicating that the scavenging enzymes of O_2^- had already been acquired before the appearance of cyanobacteria, most likely to protect the cells from oxidative damage in ultra trace O_2 of the biosphere (below 10^{-4} of the current atmospheric level, 25 nM) over three billion years ago. Cell fusion of sulfur and nonsulfur photosynthetic bacteria and the addition of the water-oxidation system to the PSII-homologous reaction center of the nonsulfur bacteria produced an ancestor of cyanobacteria, and cyanobacteria were the first organisms to acquire the O_2-evolving machinery within cells, despite the very low concentration of O_2 in the environment. Given how dangerous the production of O_2 within cells is, cyanobacteria should be equipped with a perfect system to scavenge O_2^- and H_2O_2. From an evolutionary aspect, while the atmospheric concentration of O_2 was low, the physiological function of the water-water cycle was limited to scavenging active oxygen and protecting the target molecules. At this stage, the stromal factor-mediated photoreduction of O_2 would not be functional as the dissipation system of excess photon energy because over 13% O_2 is required for the stromal factor-mediated photoreduction of O_2 at saturation. Although it is difficult to estimate when the usefulness of the water-water cycle to oxygenic phototrophs became apparent to protect from photoinihibition by dissipating excess light energy, the atmospheric concentration of O_2 would be a key factor in the acquisition of the second and third physiological functions. These functions were likely acquired by oxygenic phototrophs after they emerged on land, 420 MYA, when the atmospheric concentration of O_2 was increased to 2%, because anaerobic photoinhibition can be partially avoided by 1–2% O_2. Since the concentration of O_2 required for the water-water cycle is lower than that for photorespiration, the dissipation system of the excess photon energy via the water-water cycle would be acquired in advance of that via photorespiration.

ACKNOWLEDGMENTS

This review is based largely on research supported by Grants-in-Aid for Scientific Research and International Cooperation Research from the Ministry of Education, Science and Culture, Japan, and also by grants from the Human Frontier Science Program and from the Alexander von Humboldt Award. The author thanks Harumi Ishikawa for preparing the figures.

Visit the *Annual Reviews home page* at
http://www.AnnualReviews.org

Literature Cited

1. Ahn BY, Moss B. 1992. Glutaredoxin homolog encoded by vaccinia virus is a virion-associated enzyme with thiol-transferase and dehydroascorbate reductase activities. *Proc. Natl. Acad. Sci. USA* 89:7060–64

2. Amako K, Chen G-X, Asada K. 1994. Separate assays specific for ascorbate peroxidase and guaiacol peroxidase and for the chloroplastic and cytosolic isozymes of ascorbate peroxidase in plants. *Plant Cell Physiol.* 35:497–504

3. Anderson JM, Park Y-I, Chow WS. 1998. Unifying model for the photoinactivation of photosystem II in vivo under steady-state photosynthesis. *Photosynth. Res.* 56:1–13

4. Aono M, Saji H, Fujiyama K, Sugita M, Kondo N, et al. 1995. Decrease in activity of glutathione reductase enhances paraquat sensitivity in transgenic *Nicotiana tabacum*. *Plant Physiol.* 107:645–48

5. Arisi A-CM, Cornic G, Jouanin L, Foyer CH. 1998. Overexpression of iron superoxide dismutase in transformed poplar modifies the regulation of photosynthesis at low CO_2 partial pressures or following exposure to the prooxidant herbicide methyl viologen. *Plant Physiol.* 117:565–74

6. Aro E-M, Virgin I, Andersson B. 1994. Photoinhibition of photosystem II. Inactivation, protein damage and turnover. *Biochim. Biophys. Acta* 1143:113–34

7. Arrigoni O, De Gara L, Paciolia C, Evidente A, de Pinto MC, et al. 1997. Lycorine: a powerful inhibitor of L-galactono-γ-lactone dehydrogenase activity. *J. Plant Physiol.* 150:362–64

8. Asada K. 1981. Biological carboxylation. In *Organic and Bio-organic Chemistry of Carbon Dioxide*, ed. S Inoue, N Yamazaki, pp. 185–251. Tokyo: Kodansha

9. Asada K. 1992. Ascorbate peroxidase—a hydrogen peroxide-scavenging enzyme in plants. *Physiol. Plant.* 85:235–41

10. Asada K. 1992. Production and scavenging of active oxygen in chloroplasts. In *Molecular Biology of Free Radical Scavenging Systems*, ed. JG Scandalios, pp. 173–92. New York: Cold Spring Harbor Lab. Press

11. Asada K. 1994. Production and action of active oxygen species in photosynthetic tissues. See Ref. 49a, pp. 77–104

12. Asada K. 1996. Radical production and scavenging in the chloroplasts. In *Photosynthesis and the Environments*, ed. NR Baker, pp. 123–50. Dordrecht: Kluwer

13. Asada K. 1997. The role of ascorbate peroxidase and monodehydroascorbate reductase in H_2O_2 scavenging in plants. See Ref. 175, pp. 715–35

14. Asada K, Badger MR. 1984. Photoreduction of $^{18}O_2$ and $H_2^{18}O_2$ with concomitant evolution of $^{16}O_2$ in intact spinach chloroplasts: evidence for scavenging of hydrogen peroxide by peroxidase. *Plant Cell Physiol.* 25:1169–79

15. Asada K, Endo T, Mano J, Miyake C. 1998. Molecular mechanism for relaxation of and protection from light stress. See Ref. 174a, pp. 37–52

16. Asada K, Heber U, Schreiber U. 1993. Electron flow to the intersystem chain from stromal components and cyclic electron flow in maize chloroplasts, as detected in intact leaves by monitoring redox change of P700 and chlorophyll fluorescence. *Plant Cell Physiol.* 34:39–50

17. Asada K, Kiso K. 1973. Initiation of aerobic oxidation of sulfite by illuminated spinach chloroplasts. *Eur. J. Biochem.* 33:253–57

18. Asada K, Kiso K, Yoshikawa K. 1974. Univalent reduction of molecular oxygen by spinach chloroplasts on illumination. *J. Biol. Chem.* 249:2175–81

19. Asada K, Nakano Y. 1978. Affinity for oxygen in photoreduction of molecular oxygen and scavenging of hydrogen peroxide in chloroplasts. *Photochem. Photobiol.* 28:917–20

20. Asada K, Neubauer C, Schreiber U, Heber U. 1992. Methyl viologen-dependent cyclic electron transport in spinach chloroplasts in the absence of oxygen. *Plant Cell Physiol.* 31:557–64

21. Asada K, Takahashi M. 1987. Production and scavenging of active oxygen in photosynthesis. In *Photoinhibition*, ed. DJ Kyle, CB Osmond, CJ Arntzen, pp. 227–87. Amsterdam: Elsevier

22. Asada K, Yoshikawa K, Takahashi M, Maeda Y, Enmanji K. 1975. Superoxide dismutase from a blue-green alga, *Plectonema boryanum*. *J. Biol. Chem.* 250:2801–7

23. Badger MR, Schreiber U. 1993. Effects of inorganic carbon accumulation on photosynthetic oxygen reduction and cyclic electron flow in the cyanobac-

terium *Synechococcus* PCC7942. *Photosynth. Res.* 37:177–91

24. Baier M, Dietz K-J. 1997. The plant 2-Cys peroxiredoxin BAS1 is a nuclear-encoded chloroplast protein: its expressional regulation, phylogenetic origin, and implications for its specific physiological function in plants. *Plant J.* 12:179–90

25. Behrens PW, Marsho TV, Radmer RJ. 1982. Photosynthetic O_2 exchange in isolated soybean cells. *Plant Physiol.* 70: 179–85

26. Bennoun P. 1994. Chlororespiration revisited: mitochondrial-plastid interactions in *Chlamydomonas*. *Biochim. Biophys. Acta* 1186:59–66

27. Berczi A, Moller IM. 1998. NADH-monodehydroascorbate oxidoreductase is one of the redox enzymes in spinach leaf plasma membranes. *Plant Physiol.* 116:1029–36

28. Biehler K, Fock H. 1996. Evidence for the contribution of the Mehler-peroxidase reaction in dissipating excess electrons in drought-stressed wheat. *Plant Physiol.* 112:265–72

29. Bowler C, Van Camp W, Van Montagu M, Inzé D. 1994. Superoxide dismutase in plants. *Crit. Rev. Plant Sci.* 13:199–218

30. Brechignac F, Andre M. 1985. Oxygen uptake and photosynthesis of the red macroalga, *Chondrus crispus*, in sea water. *Plant Physiol.* 78:545–50

31. Bunkelmann JR, Trelease N. 1996. Ascorbate peroxidase; a prominent membrane protein in oilseed glyoxysomes. *Plant Physiol.* 110:589–98

32. Canvin DT, Berry JA, Badger MR, Fock H, Osmond CB. 1980. Oxygen exchange in leaves in the light. *Plant Physiol.* 66: 302–7

33. Casano LM, Gomez LD, Lascano HR, Gonzalez CA, Trippi VS. 1997. Inactivation and degradation of CuZn-SOD by active oxygen species in wheat chloroplasts exposed to photooxidative stress. *Plant Cell Physiol.* 38:433–40

34. Cheeseman JM, Herendeen LB, Cheeseman AT, Clough BF. 1997. Photosynthesis and photoprotection in mangrove under field conditions. *Plant Cell Environ.* 20:597–88

35. Chen G-X, Asada K. 1989. Ascorbate peroxidase in tea leaves: occurrence of two isozymes and their differences in enzymatic and molecular properties. *Plant Cell Physiol.* 30:987–98

36. Chen G-X, Asada K. 1989. Hydroxyurea and *p*-aminophenol are the suicide in-

hibitors of ascorbate peroxidase. *J. Biol. Chem.* 265:2775–81

37. Chen G-X, Asada K. 1992. Inactivation of ascorbate peroxidase by thiols requires hydrogen peroxide. *Plant Cell Physiol.* 33:117–23

38. Chen G-X, Blubaugh DJ, Hormann PH, Golbeck JH, Cheniae GM. 1995. Superoxide contributes to the rapid inactivation of specific secondary donors of the photosystem II reaction center during photodamage of manganese-depleted photosystem II membranes. *Biochemistry* 34:2317–32

39. Chen G-X, Sano S, Asada K. 1992. The amino acid sequence of ascorbate peroxidase from tea has a high degree of homology to that of cytochrome *c* peroxidase from yeast. *Plant Cell Physiol.* 33:109–16

40. Cleland RS, Bendall DS. 1992. Photosystem I cyclic electron transport: measurement of ferredoxin-plastoquinone reductase activity. *Photosynth. Res.* 34: 409–18

41. Conklin PL, Pallanca JE, Last RL, Smirnoff N. 1997. L-Ascorbic acid metabolism in the ascorbate-deficient Arabidopsis mutant *vtc1*. *Plant Physiol.* 115:1277–85

42. Coschigano KT, Melo-Oliveria R, Lim J, Corruzzi GM. 1998. Arabidopsis *gls* mutants and distinct Fd-GOGAT genes: implication for photorespiration and primary nitrogen assimilation. *Plant Cell* 10:741–52

43. Dipierro S, Borraccino G. 1991. Dehydroascorbate reductase from potato tubers. *Phytochemistry* 30:427–29

44. Endo T, Mi H, Shikanai T, Asada K. 1997. Donation of electrons to plastoquinone by NAD(P)H dehydrogenase and by ferredoxin-quinone reductase in spinach chloroplasts. *Plant Cell Physiol.* 38:1272–77

45. Eshdat Y, Holland D, Faltin Z, Ben-Hayyim G. 1997. Plant glutathione peroxidases. *Physiol. Plant.* 100:234–40

46. Feild TS, Nedbal L, Ort DR. 1998. Nonphotochemical reduction of the plastoquinone pool in sunflower leaves originates from chlororespiration. *Plant Physiol.* 116:1209–18

47. Forkl H, Vandekerckhove J, Drews G, Tadros MH. 1993. Molecular cloning, sequence analysis and expression of the gene for catalase-peroxidasae (*cpeA*) from the photosynthetic bacterium *Rhodobacter capsulatus* B10. *Eur. J. Biochem.* 214:251–58

48. Foyer CH. 1997. Oxygen metabolism

and electron transport in photosynthesis. See Ref. 175, pp. 587–621

49. Foyer CH, Halliwell B. 1976. Purification and properties of dehydroascorbate reductase from spinach leaves. *Phytochemistry* 16:1347–50

49a. Foyer CH, Mullineaux PM, eds. 1994. *Causes of Photooxidative Stress and Amelioration of Defense Systems in Plants.* Boca Raton: CRC Press

50. Fryer MJ, Andrews JR, Oxborough K, Blowers DA, Baker NR. 1998. Relationship between CO_2 assimilation, photosynthetic electron transport, and active O_2 metabolism in leaves of maize in the field during periods of low temperature. *Plant Physiol.* 116:571–80

51. Furbank RT, Badger MR. 1983. Oxygen exchange associated with electron transport and photophosphorylation in spinach thylakoids. *Biochim. Biophys. Acta* 723:400–9

52. Furbank RT, Badger MR, Osmond CB. 1982. Oxygen exchange in isolated cells and chloroplasts of C_3 plants. *Plant Physiol.* 70:927–31

53. Furbank RT, Badger MR, Osmond CB. 1983. Photoreduction of oxygen in mesophyll chloroplasts of C_4 plants: a model system for studying an in vivo Mehler reaction. *Plant Physiol.* 73:1038–41

54. Gardner PR, Fridovich I. 1992. Inactivation-reactivation of aconitase in *Escherichia coli.* A sensitive measure of superoxide radical. *J. Biol. Chem.* 267: 8757–63

55. Getzoff ED, Cabelli DE, Fisher CL, Parge HE, Viezzoli MS, Hallewell RA. 1992. Faster superoxide dismutase mutants designed by enhancing electrostatic guidance. *Science* 358:347–51

56. Goetze DC, Carpentier R. 1994. Ferredoxin-NADP$^+$ reductase is the site of oxygen reduction in pseudocyclic electron transport. *Can. J. Bot.* 72:256–60

57. Grace S, Pace R, Wydrzynski T. 1995. Formation and decay of monodehydroascorbate radicals in illuminated thylakoids as determined by EPR spectroscopy. *Biochim. Biophys. Acta* 1229:155–65

58. Grantz AA, Brummell DA, Bennett AB. 1995. Ascorbate free radical reductase mRNA levels are induced by wounding. *Plant Physiol.* 108:411–18

59. Gueta-Dahan Y, Yaniv Z, Zilinskas BA, Ben-Hayyim G. 1997. Salt and oxidative stress; similar and specific responses and their relation to salt tolerance in citrus. *Planta* 203:460–69

60. Havaux M, Davand A. 1994. Photoinhi-bition of photosynthesis in chilled potato leaves is not correlated with a loss of photosystem-II activity: preferential inactivation of photosystem I. *Photosynth. Res.* 40:75–92

61. Heber U, French CS. 1968. Effects of oxygen on the electron transport chain of photosynthesis. *Planta* 79:99–112

62. Heber U, Miyake C, Mano J, Ohno C, Asada K. 1996. Monodehydroascorbate radical detected by electron paramagnetic resonance spectrometry is a sensitive probe of oxidative stress in intact leaves. *Plant Cell Physiol.* 37:1066–72

63. Herbert SK, Samson G, Fork DC, Laudenbach DE. 1992. Characterization of damage to photosystem I and II in a cyanobacterium lacking detectable iron superoxide dismutase. *Proc. Natl. Acad. Sci. USA* 89:8716–20

64. Hérouart D, Van Montagu M, Inzé D. 1994. Developmental and environmental regulation of the *Nicotiana plumbaginifolia* cytosolic Cu/Zn-superoxide dismutase promoter in transgenic tobacco. *Plant Physiol.* 104:873–80

65. Hideg É, Mano J, Ohno C, Asada K. 1997. Increased levels of monodehydroascorbate radicals in UV-B-irradiated broad bean leaves. *Plant Cell Physiol.* 38:684–90

66. Hideg É, Spetea C, Vass I. 1995. Superoxide radicals are not the main promotors of acceptor side induced photoinhibitory damage in spinach thylakoids. *Photosynth. Res.* 46:399–407

67. Horemans N, Asard H, Caubergs RJ. 1997. The ascorbate carrier of higher plant plasma membranes preferentially translocates the fully oxidized (dehydroascorbate) molecule. *Plant Physiol.* 114:1247–53

68. Hormann H, Neubauer C, Asada K, Schreiber U. 1993. Intact chloroplasts display pH 5 optimum of O_2-reduction in the absence of methyl viologen: evidence for a regulatory role of superoxide protonation. *Photosynth. Res.* 37:69–80

69. Horton P, Ruban AV, Walters RG. 1996. Regulation of light harvesting in green plants. *Annu. Rev. Plant Physiol. Plant Mol. Biol.* 47:655–84

70. Hosein B, Palmer G. 1983. The kinetics and mechanism of reaction of reduced ferredoxin by molecular oxygen and its reduced products. *Biochim. Biophys. Acta* 723:383–90

71. Hossain MA, Asada K. 1984. Purification of dehydroascorbate reductase from spinach and its characterization as a thiol enzyme. *Plant Cell Physiol.* 25:85–92

72. Hossain MA, Asada K. 1984. Inactivation of ascorbate peroxidase in spinach chloroplasts on dark addition of hydrogen peroxide: its protection by ascorbate. *Plant Cell Physiol.* 25:1285–95

73. Hossain MA, Asada K. 1985. Monodehydroascorbate reductase from cucumber is a flavin adenine dinucleotide enzyme. *J. Biol. Chem.* 250:12920-26

73a. Imai T, Karita S, Shiratori G, Hattori M, Nunome T, et al. 1998. L-Galactono-γ-lactone dehydrogenase from sweet potato: purification and cDNA sequence analysis. *Plant Cell Physiol.* 39:1350–58

74. Inoue Y, Kobayashi Y, Shibata K, Heber U. 1978. Synthesis and hydrolysis of ATP by intact chloroplasts under flash illumination and in darkness. *Biochim. Biophys. Acta* 504:142–52

75. Inoue K, Sakurai H, Hiyama T. 1986. Photoinactivation of photosystem I in isolated chloroplasts. *Plant Cell Physiol.* 27:961–68

76. Ishida H, Nishimori Y, Sugisawa M, Makino A, Mae T. 1997. The large subunit of ribulose-1,5–bisphosphate into 37-kDa and 16-kDa polypeptides by active oxygen in the lysates of chloroplasts from primary leaves of wheat. *Plant Cell Physiol.* 38:471–79

77. Ishida H, Shimizu S, Makino A, Mae T. 1998. Light-dependent fragmentation of the large subunit of ribulose-1,5-bisphosphate carboxylase/oxygenase in chloroplasts isolated from wheat leaves. *Planta* 204:305–9

78. Ishikawa T, Sakai K, Takeda T, Shigeoka S. 1995. Cloning and expression of cDNA encoding a new type of ascorbate peroxidase from spinach. *FEBS Lett.* 367:28–32

79. Ishikawa T, Sakai K, Yoshimura K, Takeda T, Shigeoka S. 1996. cDNAs encoding spinach stromal and thylakoid-bound ascorbate peroxidase, differing in the presence or absence of their 3′-coding regions. *FEBS Lett.* 384:289–93

80. Ishikawa T, Yoshimura K, Sakai K, Tamoi M, Takeda T, et al. 1998. Molecular characterization and physiological role of a glyoxysome-bound ascorbate peroxidase. *Plant Cell Physiol.* 39:23–34

81. Ishikawa T, Yoshimura K, Tamoi M, Takeda T, Shigeoka S. 1997. Alternative mRNA splicing of 3′-terminal exons generates ascorbate peroxidase isoenzymes in spinach (*Spinacia oleracea*) chloroplasts. *Biochem. J.* 328:795–800

82. Iturbe-Ormaetxe I, Escuredo PR, Arrese-Igor C, Becana M. 1998. Oxidative damage in pea plants exposed to water deficit or paraquat. *Plant Physiol.* 116:173–81

83. Ivanov BN, Edwards GE. 1997. Electron flow accompanying the ascorbate peroxidase cycle in maize mesophyll chloroplasts and its cooperation with linear electron flow to NADP$^+$ and cyclic electron flow in thylakoid membrane energization. *Photosynth. Res.* 52:187–98

84. Ivanov BN, Ignatova LK. 1997. Photoreduction of acceptor generated in an ascorbate peroxidase reaction in pea thylakoids. *Biokhimiya* 62:1082–88 (English ed.)

85. Jahnke LS, Hull MR, Long SP. 1991. Chilling stress and oxygen metabolizing enzymes in *Zea mays* and *Zea diploperenuis*. *Plant Cell Environ.* 14:97–104

86. Jakob B, Heber U. 1996. Photoproduction and detoxification of hydroxyl radicals in chloroplasts and leaves and relation to photoinactivation of photosystems I and II. *Plant Cell Physiol.* 37:629–35

87. Jerpersen HM, Kjoersgard IVH, Ostergaard L, Welinder KG. 1997. From sequence analysis of three novel ascorbate peroxidases from *Arabidopsis thaliana* to structure, function and evolution of seven types of ascorbate peroxidase. *Biochem. J.* 326:305–10

88. Jiménez A, Hernández JA, del Réo LA, Sevilla F. 1997. Evidence for the presence of the ascorbate-glutathione cycle in mitochondria and peroxisomes of pea leaves. *Plant Physiol.* 114:272–84

89. Kaiser WM. 1976. The effect of hydrogen peroxide on CO$_2$-fixation of isolated chloroplasts. *Biochim. Biophys. Acta* 440:476–82

90. Kaiser WM. 1979. Reversible inhibition of the Calvin cycle and activation of oxidative pentose phosphate cycle in isolated intact chloroplasts by hydrogen peroxide. *Planta* 145:377–82

91. Kanematsu K, Asada K. 1990. Characteristic amino acid sequences of chloroplast and cytosol isozymes of CuZn-superoxide dismutase in spinach, rice and horsetail. *Plant Cell Physiol.* 31:99–112

92. Kanematsu S, Asada K. 1994. Superoxide dismutase. In *Molecular Aspects of Enzyme Catalysis*, ed. T Fukui, K Soda, pp. 191–210. Tokyo: Kodansha

93. Kato Y, Urano J, Maki Y, Ushimaru T. 1997. Purification and characterization of dehydroascorbate reductase from rice. *Plant Cell Physiol.* 38:173–78

94. Kingston-Smith AH, Thomas H, Foyer C. 1997. Chlorophyll *a* fluorescence, enzyme and antioxidant analyses provide evidence for the operation of alternative electron sinks during leaf senescence in a *stay-green* mutants of *Festuca pratensis*. *Plant Cell Environ.* 20:1323–37

95. Kitagawa Y, Tanaka Y, Hata M, Kusunoki M, Lee GP, et al. 1991. Three-dimensional structure of Cu Zn-superoxide dismutase from spinach at 2.0 Å resolution. *J. Biochem.* 109:477–85

96. Kobayashi Y, Inoue Y, Furuya F, Shibata K, Heber U. 1979. Regulation of adenylate levels in intact spinach chloroplasts. *Planta* 147:69–75

97. Kobayashi K, Tagawa S, Sano S, Asada K. 1995. A direct demonstration of the catalytic action of monodehydroascorbate reductase by pulse radiolysis. *J. Biol. Chem.* 270:27551–54

98. Kono Y, Fridovich I. 1982. Superoxide inhibits catalase. *J. Biol. Chem.* 257:5751–54

99. Koshiba T. 1993. Cytosolic ascorbate peroxidase in seedlings and leaves of maize (*Zea mays*). *Plant Cell Physiol.* 34:713–21

100. Kozaki A, Takeba G. 1996. Photorespiration protects C_3 plants from photoinhibition. *Nature* 384:557–60

101. Krause GH. 1994. The role of oxygen in photoinhibition of photosynthesis. See Ref. 49a, pp. 43–76

102. Kurepa J, Hérouart D, Van Montagu M, Inzé D. 1997. Differential expression of CuZn- and Fe-superoxide dismutase genes of tobacco during development, oxidative stress and hormonal treatments. *Plant Cell Physiol.* 38:463–70

103. Kvaratskhelia M, Geroge SJ, Thorneley RN. 1997. Salicylic acid is a reducing substrate and not an effective inhibitor of ascorbate peroxidase. *J. Biol. Chem.* 272:20998–1

104. Kvaratskhelia M, Winkel C, Thorneley NF. 1997. Purification and characterization of a novel class III peroxidase isoenzyme from tea leaves. *Plant Physiol.* 114:1237–45

105. Laisk A, Loreto F. 1996. Determining photosynthetic parameters from leaf CO_2 exchange and chlorophyll fluorescence. *Plant Physiol.* 110:903–12

106. Larkum AWD, Jones RJ, Hoegh-Guldberg O. 1998. Temperature-induced bleaching of corals begins with impairment to the carbon dioxide fixation mechanism of zooxanthellae. *Photochem. Photobiol. Biol.* In press

107. Lascano FR, Gomez LD, Casano LM, Trippi VS. 1998. Changes in glutathione reductase activity and protein content in wheat leaves and chloroplasts exposed to photooxidative stress. *Plant Physiol. Biochem* 36:321–29

107a. Leonardis SD, Lorenzo GD, Borraccino G, Dipierro S. 1995. A specific ascorbate free radical reductase isozyme participates in the regeneration of ascorbate for scavenging toxic oxygen species in potato tuber mitochondria. *Plant Physiol.* 109:847–51

107b. Lesser MP. 1997. Oxidative stress causes coral bleaching during exposure to elevated temperatures. *Coral Reefs* 16:187–92

108. Li Q, Canvin DT. 1997. Inorganic carbon accumulation stimulates linear electron flow to artificial electron acceptor of photosystem I in air-grown cells of the cyanobacterium *Synechococcus* UTEX 625. *Plant Physiol.* 114:1273–81

109. Long SL, Humphries S, Falkowski PG. 1994. Photoinhibition of photosynthesis in nature. *Annu. Rev. Plant Physiol. Plant Mol. Biol.* 45:633–62

110. Lovelock CE, Winter K. 1996. Oxygen-dependent electron transport and protection from photoinhibition in leaves of tropical tree species. *Planta* 198:580–87

111. Mano J. 1999. Photooxidation of ascorbate on the donor side of photosystem I in the thylakoid lumen. *Plant Cell Physiol.* 40S: In press

112. Mano J, Miyake C, Schreiber U, Asada K. 1995. Photoactivation of the electron flow from NADPH to plastoquinone in spinach chloroplasts. *Plant Cell Physiol.* 36:1589–98

113. Mano J, Ohno C, Asada K. 1999. Ascorbate peroxidase is the primary target of methylviologen-induced photooxidative stress in spinach leaves: its assessment by in vivo electron spin resonance spectrometry. *Plant Cell Physiol.* 40: In press

114. Mano J, Ushimaru T, Asada K. 1997. Ascorbate in thylakoid lumen as an endogeneous electron donor to photosystem II: Protection of thylakoids from photoinhibition and regeneration of ascorbate in stroma by dehydroascorbate reductase. *Photosynth. Res.* 53:197–204

115. Mano S, Yamaguchi K, Hayashi M, Nishimura M. 1997. Stromal and thylakoid-bound ascorbate peroxidase are produced by alternative splicing in pumpkin. *FEBS Lett.* 413:21–26

116. Marrsho TV, Behrens P, Radmer RJ. 1979. Photosynthetic oxygen reduction

in isolated intact chloroplasts and cells from spinach. *Plant Physiol.* 64:656–59

117. Martin RE, Thomas DJ, Tucker DE, Herbert SK. 1997. The effects of photooxidative stress on photosystem I measured *in vivo* in *Chlamydomonas*. *Plant Cell Environ.* 20:1451–61

118. Massey V, Strickland S, Mayhew SG, Howell LG, Engel PC, et al. 1969. The production of superoxide anion radicals in the reaction of reduced flavins and flavoproteins with molecular oxygen. *Biochem. Biophys. Res. Commun.* 36:891–97

118a. Mathis P, ed. 1995. *Photosynthesis: From Light to Biosphere.* Dordrecht: Kluwer

119. May JM, Mendiratta S, Hill KE, Burk RF. 1997. Reduction of dehydroascorbate to ascorbate by the selenoenzyme thioredoxin reductase. *J. Biol. Chem.* 272:22607–10

120. Mehler AH. 1951. Studies on reactivities of illuminated chloroplasts. I. Mechanism of the reduction of oxygen and other Hill reagents. *Arch. Biochem. Biophys.* 33:65–77

121. Mehler AH, Brown AH. 1952. Studies on reactivities of illuminated chloroplasts. III. Simultaneous photoproduction of and consumption of oxygen studied with oxygen isotopes. *Arch. Biochem. Biophys.* 38:365–70

122. Mi H, Endo T, Ogawa T, Asada K. 1995. Thylakoid membrane-bound, NADPH-specific pyridine nucleotide dehydrogenase complex mediates cyclic electron transport in the cyanobacterium *Synechocystis* sp. PCC 6803. *Plant Cell Physiol.* 36:661–68

123. Mir NA, Salon C, Canvin DT. 1995. Inorganic carbon-stimulated O_2 photoreduction is suppressed by NO_2^--assimilation in air-grown cells of *Synechococcus* UTEX 625. *Plant Physiol.* 109:1295–300

124. Miyake C, Asada K. 1992. Thylakoid-bound ascorbate peroxidase in spinach chloroplasts and photoreduction of its primary product monodehydroascorbate radicals in thylakoids. *Plant Cell Physiol.* 33:541–53

125. Miyake C, Asada K. 1994. Ferredoxin-dependent photoreduction of monodehydroascorbate radical in spinach thylakoids. *Plant Cell Physiol.* 34:539–49

126. Miyake C, Asada K. 1996. Inactivation mechanism of ascorbate peroxidase at low concentrations of ascorbate; hydrogen peroxide decomposes Compound

I of ascorbate peroxidase. *Plant Cell Physiol.* 37:423–30

127. Miyake C, Cao WH, Asada K. 1993. Purification and molecular properties of thylakoid-bound ascorbate peroxidase from spinach chloroplasts. *Plant Cell Physiol.* 34:881–89

128. Miyake C, Michihata F, Asada K. 1991. Scavenging of hydrogen peroxide in prokaryotic and eukaryotic algae: acquisition of ascorbate peroxidase during the evolution of cyanobacteria. *Plant Cell Physiol.* 32:33–43

129. Miyake C, Sano S, Asada K. 1996. A new assay of ascorbate peroxidase using the coupled system with monodehydroascorbate radical reductase. In *Plant Peroxidases, Biochemistry and Physiology*, ed. C Obinger, U Burner, R Ebermann, C Openel, H Greppin, pp. 386–89. Geneva: Univ. Geneva Press

130. Miyake C, Schreiber U, Asada K. 1995. Ferredoxin-dependent and antimycin A-sensitive reduction of cytochrome *b*-559 by far-red light in maize thylakoids: participation of menadiol-reducible cytochrome *b*-559 in cyclic electron flow. *Plant Cell Physiol.* 36:743–48

131. Miyake C, Schreiber U, Hormann H, Sano S, Asada K. 1998. The FAD-enzyme monodehydroascorbate radical reductase mediates photoproduction of superoxide radicals in spinach thylakoid membranes. *Plant Cell Physiol.* 39:821–29

132. Morell S, Fallmann H, De Tullio M, Hägerlein I. 1997. Dehydroascorbate and dehydroascorbate reductase are phantom indicators of oxidative stress in plants. *FEBS Lett.* 414:567–70

133. Muhlbauer SK, Eichacker LA. 1998. Light-dependent formation of the photosynthetic proton gradient regulates translation elongation in chloroplasts. *J. Biol. Chem.* 273:20935–40

134. Mullineaux PM, Creissen GP. 1997. Glutathione reductase: regulation and role in oxidative stress. See Ref. 175, pp. 667–713

135. Mullineaux PM, Karpinski S, Jimenez A, Cleary SP, Robinson C, et al. 1998. Identification of cDNAs encoding plastid-targeted glutathione peroxidase. *Plant J.* 13:375–79

136. Murthy SS, Zilinskas BA. 1994. Molecular cloning and characterization of a cDNA encoding pea monodehydroascorbate reductase. *J. Biol. Chem.* 269:31129–33

137. Mutuda M, Ishikawa T, Takeda T, Shigeoka S. 1996. The catalase-peroxidase

of *Synechococcus* PCC 7942: purification, nucleotide sequence analysis and expression in *Escherichia coli. Biochem. J.* 316:251–57

138. Nakano Y, Asada K. 1980. Spinach chloroplasts scavenge hydrogen peroxide on illumination. *Plant Cell Physiol.* 21:1295–307

139. Nakano Y, Asada K. 1981. Hydrogen peroxide is scavenged by ascorbate-specific peroxidase in spinach chloroplasts. *Plant Cell Physiol.* 22:867–80

140. Neubauer C, Schreiber U. 1989. Photochemical and non-photochemical quenching of chlorophyll fluorescence induced by hydrogen peroxide. *Z. Naturforsch.* 44c:262–70

141. Neubauer C, Yamamoto HY. 1992. Mehler-peroxidase reaction mediates zeaxanthin formation and zeaxathine-related fluorescence quenching in intact chloroplasts. *Plant Physiol.* 99:1354–61

142. Neubauer C, Yamamoto HY. 1994. Membrane barriers and Mehler-peroxidase reaction limit the ascorbate available for violaxanthin de-epoxidase activity in intact chloroplasts. *Photosynth. Res.* 39:139–47

143. Noctor G, Foyer CH. 1998. Ascorbate and glutathione: keeping active oxygen under control. *Annu. Rev. Plant Physiol. Plant Mol. Biol.* 49:249–79

144. Ôba K, Ishikawa S, Nishikawa M, Mizuno H, Yamamoto T. 1995. Purification and properties of L-galactono-γ-lactone dehydrogenase, a key enzyme for ascorbic acid biosynthesis, from sweet potato roots. *J. Biochem.* 117:120–24

145. Ogawa K, Endo T, Kanematsu S, Tanaka R, Ishiguro S, et al. 1997. Tobacco chloroplastic CuZn-superoxide dismutase cannot function without its localization at the site of superoxide generation (PSI). *Plant Cell Physiol.* 38:S35.

146. Ogawa K, Kanematsu S, Asada K. 1996. Intra- and extra-cellular localization of "cytosolic" CuZn-superoxide dismutase in spinach leaf and hypocotyl. *Plant Cell Physiol.* 37:790–99

147. Ogawa K, Kanematsu S, Asada K. 1997. Generation of superoxide anion and localization of CuZn-superoxide dismutase in the vascular tissue of spinach hypocotyls: their association with lignification. *Plant Cell Physiol.* 38:1118–26

148. Ogawa K, Kanematsu S, Takabe K, Asada K. 1995. Attachment of CuZn-superoxide dismutase to thylakoid membranes at the site of superoxide generation (PSI) in spinach chloroplasts: detection by immunogold labeling after rapid freezing and substitution method. *Plant Cell Physiol.* 36:565–73

149. Okada S, Kanematsu S, Asada K. 1979. Intracellular distribution of manganic and ferric superoxide dismutases in blue-green algae. *FEBS Lett.* 103:106–10

150. O'Kane D, Gill V, Boyd P, Burdon R. 1996. Chilling, oxidative stress and antioxidant responses in *Arabidopsis thaliana* callus. *Planta* 198:371–77

151. Ookawara T, Kawamura N, Kitagawa Y, Taniguchi N. 1992. Site-specific and random fragmentation of Cu Zn-superoxide dismutase by glycation reaction. Implication of reactive oxygen species. *J. Biol. Chem.* 267:18505–10

152. Osmond CB. 1981. Photorespiration and photoinhibition: some implications for the energetics of photosynthesis. *Biochim. Biophys. Acta* 639:77–98

153. Osmond CB, Grace SC. 1995. Perspectives on photoinhibition and photorespiration in the field: quintessential inefficiencies of the light and dark reactions of photosynthesis? *J. Exp. Bot.* 46:1351–62

154. Ostergaard J, Persiau G, Davey MW, Bauw G, Van Montagu M. 1997. Isolation of a cDNA coding for L-galactono-γ-lactone dehydrogenase, an enzyme involved in the biosynthesis of ascorbic acid in plants. *J. Biol. Chem.* 272:30009–16

155. Palatnik JF, Valle EM, Carrillo N. 1997. Oxidative stress causes ferredoxin-NADP$^+$ reductase solubilization from the thylakoid membranes in methyl viologen-treated plants. *Plant Physiol.* 115:1721–27

156. Pappa H, Patterson WR, Poulos TL. 1996. The homologous tryptophan critical for cytochrome *c* peroxidase function is not essential for ascorbate peroxidase. *J. Bioinorg. Chem.* 1:61–66

157. Park Y-I, Chow WS, Osmond CB, Anderson JM. 1996. Electron transport to oxygen mitigates against the photoinactivation of photosystem II in vivo. *Photosynth. Res.* 50:23–32

158. Patterson WR, Poulos TL. 1995. Crystal structure of recombinant pea cytosolic ascorbate peroxidase. *Biochemistry* 34:4331–41

159. Patterson WR, Poulos TL, Goodin DB. 1995. Identification of a porphyrin π cation radical in ascorbate peroxidase compound I. *Biochemistry* 34:4342–45

160. Pinhero RG, Rao MV, Paliyath G, Murr DP, Fletcher RA. 1997. Changes in activities of antioxidant enzymes and their relationship to genetic and paclobutrazol-induced chilling tolerance of maize seedlings. *Plant Physiol.* 114:695–704

161. Prasad TK, Anderson MD, Martin BA, Stewart CR. 1994. Evidence for chilling-induced oxidative stress in maize seedlings and a regulatory role for hydrogen peroxide. *Plant Cell* 6:65–74

162. Radmer RJ, Kok B. 1976. Photoreduction of O_2 primes and replaces CO_2 assimilation. *Plant Physiol.* 58:336–40

163. Radmer RJ, Kok B, Ollinger O. 1978. Kinetics and apparent K_m of oxygen cycle under conditions of limiting carbon dioxide fixation. *Plant Physiol.* 61:915–17

164. Radmer RJ, Ollinger O. 1980. Light driven uptake of oxygen, carbon dioxide and bicarbonate by the green alga *Scenedesmus. Plant Physiol.* 65:723–29

165. Robinson JM. 1988. Does O_2 photoreduction occur within chloroplast in vivo? *Physiol. Plant.* 72:666–80

166. Runeckles VC, Vaartnou M. 1997. EPR evidence for superoxide anion formation in leaves during exposure to low levels of ozone. *Plant Cell Environ.* 20:306–14

167. Saito K, Nick JA, Loewus FA. 1990. D-Glucosone and L-sorbosone, putative intermediates of L-ascorbic acid biosynthesis in detached bean and spinach leaves. *Plant Physiol.* 94:1496–500

168. Salo DC, Pacifici RE, Lin SW, Giulivi C, Davies KJA. 1990. Superoxide dismutase undergoes proteolysis and fragmentation following oxidative modification and inactivation. *J. Biol. Chem.* 265:11919–27

169. Samson G, Herbert SK, Fork DC, Laudenbach DE. 1994. Acclimation of the photosynthetic apparatus to growth irradiance in a mutant strain of *Synechococcus* lacking iron superoxide dismutase. *Plant Physiol.* 105:287–94

170. Sano S, Asada K. 1994. cDNA cloning of monodehydroascorbate radical reductase from cucumber: a high degree of homology in terms of amino acid sequence between this enzyme and bacterial flavoenzymes. *Plant Cell Physiol.* 35:425–37

171. Sano S, Miyake C, Mikami B, Asada K. 1995. Molecular characterization of monodehydroascorbate radical reductase from cucumber overproduced in *Escherichia coli. J. Biol. Chem.* 270:21354–61

172. Santos M, Gousseau H, Lister C, Foyer C, Creissen G, et al. 1996. Cytosolic ascorbate peroxidase from *Arabidopsis thaliana* L. is encoded by a small multigene family. *Planta* 198:64–69

173. Satoh K. 1970. Mechanism of photoinactivation in photosynthetic systems. II. The occurrence and properties of two different types of photoinactivation. *Plant Cell Physiol.* 11:29–38

174. Satoh K. 1970. Mechanism of photoinactivation in photosynthetic systems. III. Site and mode of photoinactivation in photosystem I. *Plant Cell Physiol.* 11:187–97

174a. Satoh K, Murata N, eds. 1998. *Stress Responses of Photosynthetic Organisms.* Amsterdam: Elsevier

175. Scandalios JG. 1997. Molecular genetics of superoxide dismutases in plants. In *Oxidative Stress and the Molecular Biology of Antioxidant Defenses*, ed. JG Scandalios, pp. 527–68. NY: Cold Spring Harbor Lab. Press

176. Schöner S, Krause GH. 1990. Protective systems against active oxygen species in spinach: response to cold acclimation in excess light. *Planta* 180:383–89

177. Schreiber U. 1998. Chlorophyll fluorescence: new instruments for special applications. In *Proc. Int. Congr. Photosyn., 11th*. Dordrecht: Kluwer. In press

178. Schreiber U, Hormann H, Asada K, Neubauer C. 1995. O_2-dependent electron flow in intact spinach chloroplasts: properties and possible regulation of the Mehler-ascorbate peroxidase cycle. See Ref. 118a, 2:813–18

179. Schreiber U, Neubauer C. 1990. O_2-dependent electron flow, membrane energization and the mechanism of nonphotochemical quenching. *Photosynth. Res.* 25:279–93

180. Shen BS, Jensen RG, Bohnert HJ. 1997. Increased resistance to oxidative stress in transgenic plants by targeting mannitol biosynthesis to chloroplasts. *Plant Physiol.* 113:1177–83

181. Shigeoka S, Nakano Y, Kitaoka S. 1980. Metabolism of hydrogen peroxide in *Euglena gracilis. Biochem. J.* 186:377–80

182. Shikanai T, Endo T, Hashimoto T, Yamada Y, Asada K, et al. 1998. Directed disruption of the tobacco *ndhB* gene impairs cyclic electron flow around photosystem I. *Proc. Natl. Acad. Sci. USA* 95:9705–9

183. Simizu N, Kobayashi K, Hayashi K. 1984. The reaction of superoxide radical with catalase. *J. Biol. Chem.* 259:4414–18

184. Smirnoff N. 1996. The function and metabolism of ascorbic acid in plants. *Ann. Bot.* 78:661–69

185. Sonoike K. 1995. Selective photoinhibition of photosystem I in isolated thylakoid membranes from cucumber and spinach. *Plant Cell Physiol.* 36:825–30

186. Sonoike K. 1996. Photoinhibition of photosystem I: its physiological significance in the chilling sensitivity of plants. *Plant Cell Physiol.* 37:239–47

187. Sonoike K. 1998. Various aspects of inhibition of photosynthesis under light/chilling stress: "photoinhibition at chilling temperatures" versus "chilling damage in the light". *J. Plant Res.* 111:121–29

188. Sonoike K, Kamo M, Hihara Y, Hiyama T, Enami I. 1997. The mechanism of the degradation of *psa*B gene product, one of the photosynthetic reaction center subunits of photosystem I, upon photoinhibition. *Photosynth. Res.* 43:55–63

189. Sonoike K, Terashima I, Iwaki M, Itoh S. 1995. Destruction of photosystem I iron-sulfur centers in leaves of *Cucumis sativus* L. by weak illumination at chilling temperatures. *FEBS Lett.* 362:235–38

190. Stadtman ER. 1992. Protein oxidation and aging. *Science* 257:1220–24

191. Stegmann HB, Schuler P, Westphal S, Wagner E. 1993. Oxidation stress of crops monitored by EPR. *Z. Naturforsch.* 48c:766–72

192. Streller S, Wingsle G. 1994. *Pinus sylvestris* L. needles contain extracellular CuZn-superoxide dismutase. *Planta* 192:195–201

193. Süss KH, Prokhorenko I, Adler K. 1995. In situ association of Calvin cycle enzymes, ribulose-1,5-bisphosphate carboxylase/oxygenase activase, ferredoxin-NADP$^+$ reductase and nitrite reductase with thylakoid and pyrenoid membranes of *Chlamydomonas reinhardtii* chloroplasts as revealed by immunoelectron microscopy. *Plant Physiol.* 107:1387–97

194. Takahashi M, Asada K. 1982. Dependence of oxygen affinity for Mehler reaction on photochemical activity of chloroplast thylakoids. *Plant Cell Physiol.* 25:1457–61

195. Takahashi M, Asada K. 1983. Superoxide anion permeability of phospholipid membranes and chloroplast thylakoids. *Arch. Biochem. Biophys.* 226:558–66

196. Takahashi M, Asada K. 1988. Superoxide production in aprotic interior of chloroplast thylakoids. *Arch. Biochem. Biophys.* 267:714–22

197. Takahashi Y, Katoh S. 1984. Triplet state in a photosystem I reaction center complex. Inhibition of radical pair recombination by bipyridinium dyes and naphthoquinones. *Plant Cell Physiol.* 25:785–94

198. Takeba G, Kozaki A. 1998. Photorespiration is an essential mechanism for the protection of C$_3$ plants from photooxidation. See Ref. 174a, pp. 37–52

199. Takeda T, Yokota A, Shigeoka S. 1995. Resistance of photosynthesis to hydrogen peroxide in algae. *Plant Cell Physiol.* 36:1089–95

200. Tamoi M, Ishikawa T, Takeda T, Shigeoka S. 1996. Enzymic and molecular characterization of NADP-dependent glyceraldehyde-3-phosphate dehydrogenase from *Synechococcus* PCC 7942: resistance of the enzyme to hydrogen peroxide. *Biochem. J.* 316:685–90

201. Tanaka K, Otsubo T, Kondo N. 1982. Participation of hydrogen peroxide in the inactivation of Calvin-cycle SH enzymes in SO$_2$-fumugated spinach leaves. *Plant Cell Physiol.* 23:1009–18

202. Teixeira HD, Schumacher RI, Meneghini. 1998. Lower intracellular hydrogen peroxide levels in cells overexpressing CuZn-superoxide dismutase. *Proc. Natl. Acad. Sci. USA* 95:7872–75

203. Terashima I, Funayama S, Sonoike K. 1994. The site of photoinhibition in leaves of *Cucumis sativas* L. at low temperatures is photosystem I, not photosystem II. *Planta* 193:300–6

204. Terashima I, Noguchi K, Itoh-Nemoto T, Park Y-M, Kubo A, et al. 1998. The cause of PSI photoinhibition at low temperatures in leaves of *Cucumis sativus*, a chilling sensitive plant. *Physiol. Plant.* 103:295–303

205. Thomas DJ, Avenson TJ, Thomas JB, Herbert SK. 1998. A cyanobacterium lacking iron superoxide dismutase is sensitized to oxidative stress induced with methyl viologen but is not sensitized to oxidative stress induced with norflurazon. *Plant Physiol.* 116:1593–602

206. Tjus SE, Moller BL, Scheller HV. 1998. Photosystem I is an early target of photoinhibition in barley illuminated at chilling temperatures. *Plant Physiol.* 116:755–64

207. Tourneux C, Peltier G. 1995. Effect of water deficit on photosynthetic oxygen exchange measured using $^{18}O_2$ and mass

spectrometry in *Solanum tubersoum* L. leaf disks. *Planta* 195:570–77

208. Trümper A, Follmann H, Häberlein I. 1994. A novel dehydroascorbate reductase from spinach chloroplasts homologous to plant trypsin inhibitor. *FEBS Lett.* 352:159–62

209. Uchida K, Kawakishi S. 1994. Identification of oxidized histidine generated at the active site of Cu,Zn-superoxide dismutase exposed to H_2O_2. *J. Biol. Chem.* 269:2405–10

210. Van Camp W, Capiau K, Van Montagu M, Inzé D, Slooten L. 1996. Enhancement of oxidative stress tolerance in transgenic tobacco plants overproducing Fe-superoxide dismutase in chloroplasts. *Plant Physiol.* 112:1703–14

210a. Veljovic-Jovanovic S, Oniki T, Takahama U. 1998. Detection of monodehydroascorbic acid radical in sulfite-treated leaves and mechanism of its formation. *Plant Cell Physiol.* 39:1203–8

211. Wada N, Kinoshita S, Matsuo M, Amako K, Miyake C, et al. 1998. Purification and molecular properties of ascorbate peroxidase from bovine eye. *Biochem. Biophys. Res. Commun.* 242:256–61

212. Wedel N, Soll J. 1998. Evolutionary conserved light regulation of Calvin cycle activity by NADPH-mediated reversible phosphoribulokinase/CP12/glyceraldehyde-3-phosphate dehydrogenase complex dissociation. *Proc. Natl. Acad. Sci. USA* 95:9699–704

213. Welinder KG. 1992. Superfamily of plant, fungal and bacterial peroxidases. *Curr. Opin. Struct. Biol.* 2:388–93

214. Wells WW, Xu DP, Yang Y, Rocqne PA. 1990. Mammalian thioltransferase (glutaredoxin) and protein disulfide isomerase have dehydroascorbate reductase activity. *J. Biol. Chem.* 265:15361–64

215. Westphal S, Wagner E, Knolmuller M, Loreth W, Schuler P, et al. 1992. Impact of aminotriazole and paraquat on the oxidative defence system of spruce monitored by monodehydroascorbic acid. *Z. Naturforsch.* 47c:1342–46

216. Wheeler GL, Jones MA, Smirnoff N. 1998. The biosynthetic pathway of vitamin C in higher plants. *Nature* 393:365–69

217. Willekens H, Chamnongpol S, Davey M, Schraudner M, Langebartels C, et al. 1997. Catalase is a sink for H_2O_2 and is indispensable for stress defence in C_3 plants. *EMBO J* 16:4806–16

218. Wise RR. 1995. Chilling-enhanced photooxidation: the production, action and study of reactive oxygen species produced during chilling in the light. *Photosynth. Res.* 45:79–97

219. Yamaguchi K, Hayashi M, Nishimura M. 1996. cDNA cloning of thylakoid-bound ascorbate peroxidase in pumpkin and its characterization. *Plant Cell Physiol.* 37:405–7

220. Yamaguchi K, Mori H, Nishimura M. 1995. A novel isoenzymes of ascorbate peroxidase localized on glyoxysomal and leaf peroxisomal membranes in pumpkin. *Plant Cell Physiol.* 36:1157–62

221. Yamasaki H, Heshiki R, Yamasu T, Sakihama Y, Ikehara N. 1995. Physiological significance of the ascorbate regenerating system for the high-light tolerance of chloroplasts. See Ref. 118a, 4:291–94

222. Zium-Hanck U, Heber U. 1980. Oxygen requirement of photosynthetic CO_2 assimilation. *Biochim. Biophys. Acta* 591:166–74

Annu. Rev. Plant Physiol. Plant Mol. Biol. 1999. 50:641–64

SILICON

Emanuel Epstein

Department of Land, Air and Water Resources—Soils and Biogeochemistry,
University of California at Davis, Davis, California 95616-8627;
e-mail: eqepstein@ucdavis.edu

KEY WORDS: essentiality, growth, development, mechanical strength, stress

> "... silicon in life here on earth is little understood as yet. There is a lot of work to be done" RJP Williams (136).

ABSTRACT

Silicon is present in plants in amounts equivalent to those of such macronutrient elements as calcium, magnesium, and phosphorus, and in grasses often at higher levels than any other inorganic constituent. Yet except for certain algae, including prominently the diatoms, and the Equisetaceae (horsetails or scouring rushes), it is not considered an essential element for plants. As a result it is routinely omitted from formulations of culture solutions and considered a nonentity in much of plant physiological research. But silicon-deprived plants grown in conventional nutrient solutions to which silicon has not been added are in many ways experimental artifacts. They are often structurally weaker than silicon-replete plants, abnormal in growth, development, viability, and reproduction, more susceptible to such abiotic stresses as metal toxicities, and easier prey to disease organisms and to herbivores ranging from phytophagous insects to mammals. Many of these same conditions afflict plants in silicon-poor soils—and there are such. Taken together, the evidence is overwhelming that silicon should be included among the elements having a major bearing on plant life.

CONTENTS

1040-2519/99/0601-0641$08.00

INTRODUCTION

Silicon is a quantitatively major inorganic constituent of higher plants but absent or nearly so from even major scientific publications devoted to them. The reason for the astonishing discrepancy between the Si content of plants and the Si content of the enterprise of plant science is the conclusion of the developers of the solution culture technique, in the early 1860s, that Si need not be included in the formulation of nutrient solutions, i.e. that Si is dispensable in the growth of plants. This conclusion subsequently was reinforced by the promulgation of a logically flawed definition of "essentiality," the wide acceptance of which did nothing to dissuade plant physiologists from their disinterest in this element. There is, however, a large body of botanical, agronomic, horticultural, and plant pathological knowledge of Si, along with some plant physiological experimentation, that forcibly drives home the lesson that Si cannot be written off as a plant biological nonentity. In this review evidence is summarized that will drive home the significance to plant life of this ubiquitous element.

"SILICON DEFICIENCY" IN PLANT PHYSIOLOGY

What better place to look for the significance accorded Si than this very series, which has chronicled the advances in our field since 1950? Except for incidental brief references, mainly concerned with clay minerals, the first volume to include a discussion of Si was Volume 5 (1954), in a chapter devoted to the mineral nutrition of phytoplankton (81), in which the well-known requirement of diatoms for Si was discussed. For a decade after that we find nothing but an occasional mention of the element, mainly in connection, again, with algae.

In 1964, Clements (31) sounded a note that will recur in the present review. In a chapter entitled "Interaction of Factors Affecting Yield," Clements, who worked in Hawaii on the physiology of sugarcane, reported a number of instances where applications of silicates to this and other crops gave striking yield responses. He thought that silicate eliminated some toxic agents in those highly weathered soils. This field-oriented line of inquiry was resumed in 1966, Volume 17, where Bollard & Butler (17), in a chapter on mineral nutrition, devoted a section to Si, briefly discussing some aspects of it in relation to soil/plant interactions, Si uptake and transport, and the protection it may afford plants against insects and disease agents.

In 1969, Volume 20, Lewin & Reimann (84) published the first and till now, only chapter with the word silicon in its title: "Silicon and Plant Growth." The authors gave an account of the then current knowledge of Si, in respect to both diatoms and higher plants.

In their discussion of metal toxicity in plants, Foy et al (54) summarized evidence regarding the protection afforded by Si against the damage wrought plants by high concentrations of metal ions in their substrate, such as Al, Mn, and Fe. Metal toxicity is an increasingly acute problem, and we shall have occasion to discuss the role of Si in often mitigating its adverse effects on plant growth.

In the last reference but one to Si in this series, in 1980, Clarkson & Hanson (30) wrote briefly about Si, concluding with the sentence: "When more is known about cell wall biosynthesis in higher plants the need for silicic acid can be examined." But despite the huge body of work on the biosynthesis of cell wall components that has since been done (the preceding volume of this series has no less than three reviews bearing on this topic), the expectation expressed in that sentence has proven illusory. Finally, Carpita (25) recently referred briefly to Si in the walls of grasses.

For another instance of "Si deficiency" in the literature let us turn to a volume specifically devoted to the plant organ directly exposed to Si, the roots (134). This 1002-page tome devoted to plant roots has about four lines devoted to Si. The reasons for this neglect of the element, and further instances of it, are discussed in the next section. These cases are not given in a spirit of finger-pointing but to document our collective, astonishing disregard of an element present in plants in amounts equivalent to those of Ca, Mg, S, and P, and often in excess of them.

HISTORY

Research on the plant physiology of Si depended on the advent of the solution culture technique (47). With some notable exceptions (mainly certain highly weathered tropical soils), soils contain high percentages of Si. In most of them, indeed, Si is second only to oxygen as a soil constituent; the mean values are O, 49% and Si, 31% (125). In the soil solution, the direct source of Si that plant roots draw upon, the element, in the form of H_4SiO_4 over the physiological range of pH values, is present at concentrations normally ranging from 0.1 mM to 0.6 mM (46), roughly two orders of magnitude higher than the concentrations of P in soil solutions, for which Tisdale et al (130) give an average (but highly variable) value of 0.0016 mM. Plants growing in soil therefore are exposed to Si, and control of their Si status is well-nigh impossible.

That difficulty, at the experimental level, changed with the introduction of the solution culture technique around 1860. An account of the then available knowledge of plant nutrition is given by Sachs (110), in a chapter entitled Nutrients (Nährstoffe). He stressed that only with this technique can the investigator control the composition of the inorganic substrate of plants, and determine which elements are indispensable constituents (unentbehrliche Bestandtheile) of plant nutrition (p. 141). He concluded that Si does not belong among the indispensable elements in the same sense as K, P, etc. He did, however, comment on the wide distribution of Si in plants.

The conclusion of Sachs as to the dispensability of Si for plant life was subsequently given further sanction by the publication, in 1939, of a definition of essentiality (4): "an element is not considered essential unless (a) a deficiency of it makes it impossible for the plant to complete the vegetative or reproductive stage of its life cycle; (b) such deficiency is specific to the element in question, and can be prevented or corrected only by supplying this element; and (c) the element is directly involved in the nutrition of the plant quite apart from its possible effects in correcting some unfavorable microbiological or chemical condition of the soil or other culture medium."

The near-universal acceptance of this definition is puzzling in view of its flaws. As for (a), many plants may be quite severely deficient in a nutrient element and yet complete their life cycle; (b) is redundant, (a) already having made reference to the element ("it") under consideration, excluding all others; and (c) presumes that designation of an element as essential has to entail knowledge of its direct involvement in the nutrition of the plant. But at the time when the essentiality of B was established by Katherine Warington, AL Sommer, and CB Lipman in the 1920s (45, 47), nothing was known about its direct involvement in plant nutrition. The evidence simply was that the plants failed unless the element was supplied. The same argument applies to the discovery of the essentiality of several other elements at the time when their essentiality was discovered. At any rate, the promulgation of this definition and its acceptance reinforced the impression that Si could be dispensed with. Thus the classical compilation of the chemical composition of plants by Goodall & Gregory (59) contains not a single value of the Si content of any crop. When, in 1966, Hewitt (63) published the second edition of what still is by far the most comprehensive compilation of the composition of nutrient solutions, his listing of about 140 such formulations includes Si in just 5, meant specifically for research on Si or Si accumulator plants. Even in recent books dealing specifically with plant nutrition (14, 103), Si is given short shrift.

A turning point in research on Si was the publication, in 1967, of a review by Jones & Handreck (78). It remains indispensable reading for anyone interested in Si in soils, plants, and animals. Nearly every topic on Si in plants being pursued today was discussed, or at least touched upon, in that classical paper. A look at its lengthy list of references impresses the reader by the extent to which they come from agricultural and crop science publications rather than from plant physiological ones.

The further history of advances made in our knowledge of Si in plants is recorded in a number of reviews. That by Lewin & Reimann (84) in this series has already been mentioned. Shkolnik (120) has summarized literature not readily available in the West, and Richter & Suntheim (108) have given early references, many to papers in German that have not been cited often. Raven's review (106) deals with transport, primarily, and is required reading. Epstein (46) has discussed the reasons for considering experimental plants grown in conventional nutrient solutions (i.e. without inclusion of Si in their formulation) to be experimental artifacts. Savant et al (116) have recently reviewed silicon management and sustainable rice production. Despite its specialized topic, this is a wide-ranging review drawing on the extensive research published in Pacific and Far East countries, and elsewhere.

Other reviews are mentioned in various contexts, but those referred to will give readers an entrée into the material available concerning this element. So will study of several multi-authored volumes (13, 49, 124). The 1975 book by Voronkov et al (132) provides a large amount of analytical data on the Si content of living things, ranging from bacteria to humans. The authors discuss possible functions and remark on the paucity of what is known about its transport. Vitti et al (131) give a general account of Si in soils and plants, including some references pertaining to Brazil in particular. In the broad sense, the subject of Si in plants is a subset of the inorganic chemistry of life, to which the books by Fraústo da Silva & Williams (55), Kaim & Schwederski (79), and Simkiss & Wilbur (123) are wide-ranging introductions. For trenchant, brief discussions of Si in biology, reference is made to Birchall (15), Exley (50), Perry & Keeling-Tucker (101), and Williams (137).

ESSENTIALITY

At this time only two groups of plants are known to have an absolute and quantitatively major requirement for Si: the diatoms and other members of the yellow-brown or golden algae, the Chrysophyceae (84, 124, 135), and the "scouring rushes," Equisitaceae (27).

As for higher plants in general, however, on which the present review focuses, essentiality of Si in the Arnon-Stout (4) sense has not been proven (5, 126). However, it is difficult to purge solution cultures thoroughly of Si. Woolley (144) has done what still remains the most painstaking experimentation on this subject and reduced the Si content of tomato plant tops to a mere 0.0006% on a dry weight basis, but the growth of the plants was not diminished compared with that of plants supplied with Si. Work in Japan, however, has led to the conclusion, in 1990, that further intensive research "should qualify silicon as an essential element for higher plants in the near future" (126). Werner & Roth (135) are more sanguine and "consider it to be an essential element," without, however, spelling out the criteria for this opinion.

At this point we have to come to grips with the application of the Arnon & Stout (4) definition of essentiality, already criticized above, to Si. This is not the place for a full-blown discussion of this subject, but a few points need to be made nevertheless. They have to do with (a) the limitations of purging elements from experimental cultures; (b) the genotypic variations in nutrient requirements; and (c) quantitative considerations. As for (a), when Woolley (144), in the research already referred to, grew tomato plants in highly purified culture solutions and failed to find differences between their growth and that of plants deliberately supplied with Si, he concluded, correctly, that this did not show Si to be nonessential but only that, if it were essential, it would be required at tissue levels of less than 0.2 μmol $\cdot$ g^{-1} dry weight, or 0.0006%. For demonstrating the essentiality of Ni, it was necessary to purify the medium to the point where tissue levels were reduced to less than 0.00001% (21, 22). Clearly, it has not been possible to reduce tissue levels of Si to comparably low values, and the possibility of a micronutrient function for Si cannot be ruled out. This same point concerning the limitations of purification procedures has been made repeatedly (5, 45, 46). It applies to Si with particular force, because the uncharged H_4SiO_4 passes through the ion exchangers so widely used for preparing laboratory-grade water.

The second reservation about the definition of essentiality is that it conveys an impression of general applicability that is not warranted. Sodium is an example: It is a micronutrient for C_4 plants (23), but not known to be a micronutrient for plants generally. For Si, equal care needs to be taken not to generalize, in view of its essentiality for some groups of plants.

Finally, quantitative considerations. Chlorine is assuredly a micronutrient for higher plants (24), but it may well play a macronutrient role not only in some halophytes (133) but in some genotypes and conditions, for such a common crop as winter wheat (44, 82). For Si, no micronutrient function has been shown; plants unable to grow without it have a quantitatively major requirement for the element.

Thus it is necessary to go beyond mere listings of elements as essential or not known to be so, and to give judicious interpretations of the complexities of the real world. In view of what follows, Si will be considered "quasi-essential" for many of those plants for which its absolute essentiality has not been established. An element is defined as quasi-essential if it is ubiquitous in plants, and if a deficiency of it can be severe enough to result in demonstrable adverse effects or abnormalities in respect to growth, development, reproduction, or viability. That deficient situation prevails for Si in many highly weathered tropical soils, such as Ultisols and Oxisols (53, 115), and, as already pointed out, in conventional, Hoagland-type solution cultures. By extension, these same considerations apply when plants are grown in commercial (or hobby) hydroponics systems (11). The addition of clay to soilless media promoted the growth of greenhouse crops (40); might not that finding be due to Si having become available to the plants from the clay?

SILICON IN PLANTS: THE AMOUNTS

A tabulation of the ranges of the tissue levels of mineral elements found in plants gives the range for Si as 0.1–10% on a dry weight basis, it being understood that both lower and higher values may be encountered (46). Comparison of these values with those for such elements as Ca (0.1–0.6%), S (0.1–1.5%), and others shows Si to be present in amounts equivalent to those of several macronutrient elements, and even exceeding them at the highest levels.

Very comprehensive sets of Si analyses of plant material have been compiled by Japanese workers who grew plants of numerous species in the same soil and then analyzed leaves for a number of elements including Si (98, 127–129), expressing the results in terms of percentages or ppm (mg $\cdot$ kg^{-1}) on a dry weight basis. It is impossible in this small compass to do more than single out some facts. The lowest value for Si recorded in any of these tabulations is 0.01%, for very few species including *Sansevieria trifasciata* Prain (Agavaceae) and *Lycoris radiata* Herb. (Amaryllidaceae) (98). Values below 0.1% were found but rarely. The highest value among these leaf analyses was that for rice, 6.3% (98, 127). The coefficient of variation was higher for Si than for any other element. In the angiosperms, there was a significant negative correlation between values for Si and B, and a positive one for Ca and B.

Figure 1 condenses the information on the Si content of 147 species of angiosperms (98), separately for dicots (left bar), and monocots (right bar). Clearly, quantitatively, the values for Si fall in the same range as those for the inorganic macronutrient elements, but their variability is wide. The observation, often made in earlier publications, that the Si content of monocots is by and large higher than that of dicots is borne out.

As Figure 1 indicates, the Si content of plants varies vastly in different genotypes. Such differences may occur even among genotypes of the same species, as shown, for example, in studies of rice ecotypes (38, 143). Other factors are important for it, as is the case for many elements, but for Si, the differences in Si content are exceptionally large. The availability of Si in the soil or other

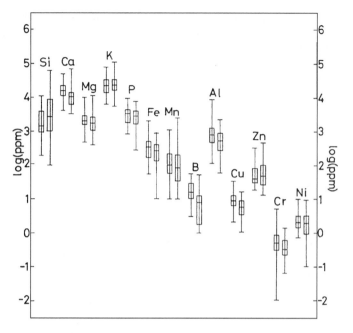

Figure 1 Box and whisker chart showing the statistics of values of maximum, minimum, upper quartile, lower quartile, and median, indicated, respectively, by bars at the top and bottom ends of the vertical lines, upper and lower sides of the boxes, and bars within the boxes. Left symbol and ordinate, dicots; right, monocots. The figure is based on the analysis of 147 species of Angiospermae. From (98), by permission.

substrate does, of course, influence the Si content of the plant. Plants themselves alter the chemical form and availability of soil Si (80; See 14a). Other generalizations emerging from the reviews already cited are that, as a rule, most of the Si absorbed is transferred from the roots to the shoots, and that within the shoots, its distribution is far from even. For a recent (and extreme) instance of the latter phenomenon, Ernst et al (48) analyzed different parts of the fruits (nutlets) of the sedge *Schoenus nigricans* (Cyperaceae) collected at a wet dune slack. The highest values by far were those of the pericarp of the fruit, ranging up to 34%. Vegetative parts of the fruiting shoots had Si values lower by more than an order of magnitude. Such differential, sharply localized distributions of Si have often been noted, in both roots and shoots (68–70, 99, 113, 114).

Customarily, however, it is leaf tissue that is mainly analyzed for Si, as is the case for other mineral elements. Analyses of many plants led Jones & Handreck (78) to propose a rough division of plants into three groups: wetland Gramineae (paddy-grown rice) have the highest values, on the order of 10–15% on a dry weight basis; dryland grasses such as rye and oats are intermediate at

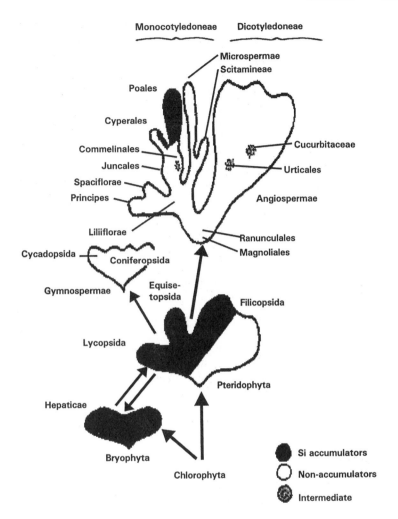

Figure 2 Distribution of Si accumulators in the plant kingdom. Modified after Figure 2 of (129), by permission.

about 1–3%; and most dicots have less than 1%. However, in view of the large variability already referred to (see Figure 1), all such categorizations should be viewed with caution. Finally, attention is drawn to the fact that even relatively "low" values for tissue Si contents, such as 0.1%, are still comparable with the low values found for such macronutrient elements as S, P, and Mg (46) (see also Figure 1).

Figure 2 shows the distribution of Si in the plant kingdom, with marked differences among groups, in that such relatively primitive plants as the Bryophyta are high-Si plants, as are some of the evolutionarily advanced ones, such as the Poales. In the Pteridophytes, both high-Si and low-Si data appear, and there are scattered high-Si groups in the dicots, and in monocots besides the Poales. Evolution seems to have "adopted" Si as a major mineral plant constituent a number of times, and "made do" with more moderate but by no means negligible Si contents for much of the plant kingdom.

The processes of absorption and transport of Si resulting in the eventual Si contents of plants and their parts are unlikely to differ fundamentally from those of other elements, to judge by the findings of Jarvis (75), Stookey (125a), and MM Rafi & E Epstein (unpublished results). However, because it is in the form of an uncharged molecule, H_4SiO_4, and is not redistributed once deposited as "opal" (see below), its translocation within the plant is even more markedly affected by the transpiration stream than is the case for other elements (see below, under Abiotic Stresses). In the marine diatom, *Cylindrotheca fusiformis*, Hildebrand et al (65) recently identified a cDNA that encodes a Si transporter ([68]Ge, a chemical analog of Si, was used as a tracer). The findings represent the first identification of a Si transporter in any organism.

SILICON AND GROWTH AND DEVELOPMENT

The literature on Si in plants is replete with reports that Si promotes the growth of plants. In many instances the growth stimulation was due to the protection that Si afforded plants against the detrimental effects of abiotic and biotic stresses, discussed below. Another factor to keep in mind is that when experimental plants are grown in soil, the interaction between added silicon and soil constituents such as P may produce effects not directly attributable to Si (89, 109, 121). Nevertheless, an appreciable body of evidence supports the conclusion that often Si enhances plant growth and development and does so directly, hence its designation here as "quasi-essential."

Since experiments with solution culture are likely to provide the most un-equivocal evidence in this regard, several such investigations are given first. Adatia & Besford (1) grew cucumber (*Cucumis sativus* cv. Corona) in recir-culating nutrient solutions to which Si had not been deliberately added (less than 0.17 mM Si) and solutions with Si at ten times that concentration. The lower leaves of the high-Si plants were darker green, better disposed for light interception, and were noticeably rougher, especially their petioles, in com-parison with these features of the low-Si plants. Leaf thickness was greater than in the low-Si plants, and the dry weight per unit area of the high-Si plant leaves was also greater. The leaves of the low-Si plants were prone to wilting, and senesced earlier than those of the high-Si plants. Other indices of plant performance were also higher in the high-Si plants. For example, on a unit area basis, the activity of Rubisco was 50% higher in the leaves of the high-Si plants. Fruit yield, on the other hand, was the same in the two treatments.

Other such instances of growth and development being positively affected by Si have been given for such diverse plants as rice, *Oryza sativa* (97), loblolly pine, *Pinus taeda* (43), cotton, *Gossypium hirsutum* (85), annual brome, *Bromus secalinus* (56), and Poinsettia, *Euphorbia pulcherrima* (93), among many, all these reports referring to experiments with solution cultures. Not all such comparisons have yielded similar results, however. Thus Wutcher (145) compared the growth of young orange trees, *Citrus sinensis*, in solution culture without Si addition and with deliberate inclusion of Si; the results indicated "only a limited role of this element in citrus nutrition."

Turning to field-grown plants, or experimental ones in soil in pots, positive responses are not as clear-cut as is the case with plants in solution culture, as already pointed out. Nevertheless, the evidence for effects of Si per se is strong. As early as 1976, Lian (85a) declared that "Si nowadays is generally considered to be an essential element agronomically (though not, in the strict sense, physiologically) for rice growth...today, the application of silicate fertilizers is very common in Japan and Korea. An annual consumption of over one million ton (38) and 400,000 tons (5) has been reported in these countries" (the numbers in parentheses refer to Lian's references by E Takahashi and CS Parks, respectively). The Si material is mostly derived from slags, the refuse of iron ore smelters most prominently. By now, good effects from the application of Si fertilizers to rice have been reported from many locations, indicating "the agronomic essentiality of Si management for increasing and/or sustaining rice yields" (116). Winslow et al (143) have drawn attention to the difference between lowland, Si-rich soils where rice of the indica ecotype is grown in paddy culture, and upland, Si-deficient soils, where the japonica ecotype is cultivated under rainfed conditions. Yamauchi & Winslow (146) and Goto & Goto (60), among others, also stress the importance of Si for the growth of japonica rice in low-Si, upland soils. In many areas where rice is grown, depletion of plant-available Si in soils may be the cause of declining rice yields (115). These authors make clear that this conclusion refers not only to upland, acidic soils but other soils such as calcareous ones as well.

While rice takes pride of place among crops that have been shown to respond positively to Si applications, it is not the only one. As early as 1965 and 1966, Clements (32) and Ayres (6) reported that application of silicate slag stimulated the growth of sugarcane, *Saccharum* spp., on low-Si soils. Other research has borne out that conclusion (3, 41, 42, 121). Silicon is used to good effect as a fertilizer for sugarcane in Brazil, Australia, South Africa, and India (LE Datnoff, personal communication).

As for other species, the depletion of plant-available Si commented on above in connection with the decline of rice yields (116) may well be a factor, but evidence is lacking. Silicate rock (granite) powder, applied to soils of the southwestern part of Western Australia, promoted the growth of wheat (66) and of subterranean clover, *Trifolium subterraneum*, and ryegrass, *Lolium rigidum* (35). The authors attributed the growth responses to K^+ released from the rock powder, but the possible contribution of other elements including Si is under investigation (P Hinsinger, personal communication). Application of rock dust to soils as a means to improve the yield and quality of crops has been touted in the popular press but without documentation and explanation of the putative effects (9).

The quantitatively major "product" of plant growth and development is cell wall, and the physical properties of plants are determined to a large extent by those of its cell walls. Virtually all the reviews mentioned above under the heading, History, refer to the role of Si in minimizing or preventing lodging of plants, especially of grain crops such as rice and wheat. The effect is attributed to the deposition of Si in the form of solid amorphous silica, $SiO_2 \cdot nH_2O$, in the cell walls. The reinforcement of the walls provided by these "opal phytoliths" adds mechanical strength.

Deposition of silica in the surface cell layers, and especially the epidermis, has a bearing as well on physical surface properties. Thus Adatia & Besford (1), in their experiments with cucumber plants growing in circulating nutrient solutions, noted that the basal leaves of the high-Si plants and the petioles were rough, and the cucumber fruit of the Si-supplied plants in the experiments of Samuels et al (112) had a dull appearance, or bloom. The leaf edges and the awns of Si-deprived wheat plants were fairly smooth, but those of plants grown in solutions supplied with Si were exceedingly rough (104). For the awns, the difference between the Si-deprived and Si-replete plants was quantified in terms of the friction force—the force needed to be overcome if the heads of the plants were to slide down an inclined plane. Despite the numerous reports on physical differences between low-Si and high-Si plants, this is so far the only quantification of any such difference. It was due to the deposition of Si in the trichomes of the awns of the Si-replete plants, as determined by X-ray microanalysis, vs its absence, i.e. undetectability, in the Si-deprived ones. These authors (104) provide formulations of +Si and −Si culture solutions.

This Si reinforcement of cell walls, and further effects elicited by Si, protect the plants against the depredations of disease organisms and herbivores including insects, as discussed below under the heading, Biotic Stresses. Indeed, we shall see that it is in relation to abiotic and biotic stresses that the role of Si in plant life is most patent.

ABIOTIC STRESSES

In 1957, Williams & Vlamis (138) made a remarkable discovery, the repercussions of which have recently become more significant than at any time before. When they grew barley, *Hordeum vulgare* cv. Atlas 46, in a conventional nutrient solution with 9.1 μM Mn, the leaves developed dark brown necrotic spots, the formation of which was prevented by adding Si at 357 μM (10 ppm). The yield depression caused by the Mn toxicity was likewise eliminated by the addition of Si. Their findings (138–141) did not indicate that Si diminished the total Mn content of the leaves but rather, that it caused Mn to be more evenly distributed instead of being concentrated in discrete necrotic spots. The interaction between Mn and Si has been studied by several workers, who also give further references (71, 72, 76).

The discovery that Si often confers on plants protection against Mn toxicity in turn led to the finding that this role of Si applies to other metal ions as well, particularly Al. Aluminum (Al^{3+}) is highly toxic at the elevated concentrations in acid soils. For this reason, considerable research has lately been devoted to the effects of this toxic metal on plants, and the role of Si in minimizing or eliminating them. In many of these investigations, no decision could be reached regarding the mechanism of the alleviation of toxicity by Si. Interaction between Si and the metals, reducing the activity of toxic metal ions in the medium, is one possible mechanism (67). The solubility of Si in the medium is another. The concentration of Si in soil solutions is on the order of 0.1–0.6 mM (46), and changes little over the entire range of physiological pH values. When culture solutions with much higher concentrations are used, concerns arise about polymerization and precipitation. The discussion will make clear, however, that not only processes in the substrate, but others occurring within the plant, need to be considered in connection with the interactions between Si and metals such as Mn and Al. There may be considerable differences in the protection Si provides against Al toxicity, as shown, for example, for the indica and japonica ecotypes of rice (105).

Li et al (85) found cotton, *Gossypium hirsutum*, to be highly sensitive to 200 μM Al in solution, the effect being mitigated by additions of Si. Purely external interactions as discussed above may have been responsible. External effects also seem to have predominated in experiments in which Si diminished the toxicity of Al to soybean (10). The precipitation of subcolloidal, inert hydroxyaluminosilicate species seems to have diminished the concentration (activity) of phytotoxic Al in solution. From experiments with corn, *Zea mays*, in which Si afforded protection against the inhibition of root elongation by Al, the authors also concluded that the formation of Al-Si complexes was responsible (88).

This kind of explanation based entirely on extraneous processes in the solution is much less likely, however, in the experiments by Barceló et al with teosinte, a wild form of *Zea mays* that grows in acid soils prone to high Al levels (7). These authors found plants exposed to toxic Al concentrations (60 μM or 120 μM) to be less inhibited in their growth when Si at the low concentration of 4 μM was included in the culture solution. The mechanism of this effect seemed to be mainly a decrease of Al uptake by the plants due to Si. Silicon and Al also changed the concentrations of organic acids in the plants; most significant were the increased concentrations of malic and formic acids in the plants grown in the presence of added Si and Al. The same group, working with an Al-sensitive variety of corn, *Zea mays* var. BR 201 F, compared root elongation of plants that had been pretreated with Si and then exposed to Al with that of plants not pretreated with Si. The Si pretreatment was effective in counteracting the inhibition of root elongation by Al (36). The temporal separation of the Si and Al treatments notwithstanding, the Si pretreatment diminished the absorption of Al by the plants. Since Al and Si were not present in the medium simultaneously a purely external interaction between the two elements is an unlikely explanation of the results. In wheat, too, the mitigation of Al toxicity was shown to be due to internal, physiological processes (33). Hammond et al (61) found Al toxicity in barley to be mitigated by Si. They concluded, surely with ample justification, that the "overall Al/Si effect is likely to be complex," but drew attention to their finding that Si reversed the inhibitory effect of Al on the absorption of Ca by the roots and Ca transport to the shoots.

Readers are referred to the paper by Hodson & Evans (68) and Cocker et al (34) for a review of Al/Si interactions. The authors (68) tabulate investigations of no less than nine species in which Al toxicity as affected by Si has been studied. In most of them, Si did indeed ameliorate various adverse effects of Al. Mechanisms of this amelioration are discussed. In a subsequent paper, Cocker et al (34) have proposed that low-solubility aluminosilicates or hydroxyaluminosilicates (or both) are formed within the root cell wall (apoplastic) space, thereby reducing the concentration of free, toxic Al^{3+} ions. The protection that Si provides against Al toxicity is a lively topic not just in plant physiology but in research on biota generally, including fish (51, 87).

Of the potentially toxic metals Al looms largest, because of its ubiquity in soils and its heightened solubility under acid conditions such as prevail in heavily weathered tropical soils. A second reason for emphasizing Al is that forest decline due to anthropogenic soil acidification (58) has become a grave problem. Barceló et al (8) have called Al phytotoxicity a "challenge for plant scientists," and so it is. These matters give an impetus for research on the mechanisms of Si/Al interactions. Hodson & Evans (68) are surely justified in their injunction: "Workers on Al toxicity must take Si into account." The same comment applies to many contexts besides that of Al toxicity. The frequent failure to heed this admonition represents a triumph of plant physiological "ideology" (the adherence to a flawed definition of essentiality) over the facts of life, and common sense. It has to be kept in mind that plants grow in a medium containing Si, and absorb it: the +Si situation is the norm, or "control." Depriving them of the element amounts often to the gratuitous imposition of an atypical environmental stress.

Salinity is another abiotic stress that has been shown in some investigations to be mitigated by Si. Matoh et al (91) reported such an instance for rice grown in solution culture with additions of either NaCl, seawater, or polyethylene glycol at concentrations up to the equivalent of 20% seawater. Silicate at 0.89 mM reduced the translocation of Na^+ to the shoots and increased dry matter production of the stressed plants over that of the controls. For wheat (2) and barley (86) similar findings have been reported, namely a repression of Na^+ transport in plants growing in salinized solutions supplied with Si, with concomitant improvement in their growth, compared with that of plants in similar solutions to which Si had not been added. The results reported for *Prosopis juliflora*, a small leguminous tree, came from an experiment in which the plants grew in a sand-peat-vermiculite mix (20). At the highest NaCl concentration (260 mM), Si at 0.47 mM had a positive effect on plant growth.

Because salinity has both ionic and osmotic components the question arises as to the role of Si in plant water relations, and vice versa. Because silicon is present over the physiological pH range as uncharged silicic acid, H_4SiO_4, and absorbed as such, the idea has surfaced time and again that it is absorbed passively along with the water and may indeed serve as an index of water use (46, 78, 106), but the idea has lost currency (46, 92). Nevertheless, transpiration favors translocation of Si, and Si often accumulates at the terminal sites of the transpiration stream, at leaf edges, in trichomes, specialized bulliform cells, and other sites of intense transpiration, but exceptions have also been noted (114).

Conversely, effects of Si on water relations have also been reported (91, 116), but the findings so far are contradictory and inconclusive. Nevertheless, the osmotic effect of silicic acid, H_4SiO_4, its proneness to formation of opal, its intimate association with the cell wall matrix (102), and the rigidity that it bestows on cell walls once deposited as opal (46, 104, 106) all make it likely that Si plays a role in plant water relations—a fertile field for investigation. Reference has already been made to the physical strength that Si bestows on cell walls. Some of these effects also fall under the present heading of abiotic stresses, such as resistance to lodging, where the stress is gravity. Roots cope with a stress shoots, except emerging ones, seldom do; roots have to overcome the physical impedance offered by the solid matrix of the soil. It is possible that Si plays a role here, but evidence is lacking.

BIOTIC STRESSES

The interaction between plants and their environment takes place mainly via plant surfaces, both below and above the ground. That statement applies in full force to the encounter between plants and disease and pest organisms. The cell wall with its cuticle is the foremost defense to breach. Bacteria and fungi do so by chemical means, whereas pests including phytophagous insects utilize mechanical means mainly though not exclusively. From early on (78), the assumption has often been made that it is the physical incrustation of cell walls by opal phytoliths that is responsible for the frequently observed protection afforded plants by Si against diseases and pests, and no doubt it plays a role (16). We shall see, however, that Si is involved in very complex, as yet incompletely understood defense mechanisms. When, as is often done for control of the medium, the plants are grown experimentally in conventional solution cultures to which no Si has been added, the plants' own physiological defense mechanisms are set aside: The plants are to some extent "experimental artifacts" (46).

As usual, Jones & Handreck (78) furnish a starting point. They summarized several instances of protection provided plants by Si against fungal and insect attack. A more recent review (46) cites more, so that the present one is confined to just a limited number.

In the experiments by Adatia & Besford (1) on the effects of Si on cucumber plants grown in a recirculating nutrient solution, the plants in the low-Si cultures were attacked by the powdery mildew fungus *Sphaerotheca fuliginea* despite applications of fungicide, whereas the high-Si plants remained nearly free of disease symptoms. Bélanger et al (11, 12) have shown the effectiveness of Si in protecting greenhouse crops against disease, including spectacular pictures (12) of leaves of cucumber grown in solution culture, and the total protection obtained when Si was supplied. For grapevines, *Vitis vinifera*, Si applications proved highly successful in protecting the plants against powdery mildew, a major fungal disease in vineyards (18, 107).

Cucumber roots are subject to attack by fungi of the genus *Pythium*, and their shoots by *Sphaerotheca fulginea* (95, 96). Chérif et al (28, 29) examined the effects of Si added to nutrient solutions at 1.7 mM on the responses of the plants to infection by *Pythium* spp. Silicon in the infected plants elicited resistance to fungal attack and the appearance of precursors of fungitoxic aglycones acting possibly like phytoalexins. Other potentially protective metobolic processes and inhibitory substances were identified in the infected, Si supplied plants. Very recently, Fawe et al (52) identified the active, Si-induced agent protecting cucumber against attacks by powdery mildew as a phytoalexin. These investigations clearly show that Si elicits a veritable cascade of biochemical defense mechanisms in the infected plants. Reviewing these and other findings, these authors (52a) relate Si to one type of induced disease resistance, specifically "systemic acquired resistance (SAR)." It is characterized by accumulation of salicylic acid and pathogenesis-related (PR) proteins. This response is typically elicited by pathogens, but their effect can be mimicked by "activators," of which Si is one (52a, 117). The defense mechanisms mobilized by Si include the accumulation of lignin, and generally, phenolic compounds, as well as chitinases and peroxidases. Fawe et al (52a) make an impressive case for the significance of Si in plant defense mechanisms against disease. Perhaps Si acts as a second messenger.

Monocots are no less immune to diseases than dicots such as cucumber. Indeed, field experience about the role of Si in plant protection against disease is extensive; only a few recent instances can be given here. Winslow (142) found that rice grown on Si-poor, upland soils in West Africa had a shoot Si content much below that recommended for rice. Silicon fertilization gave a large increase in Si content and yield, and significantly reduced the severity of fungal diseases. The responses varied among genotypes.

Rice grown under entirely different conditions, namely, on low-Si organic Histosols in Florida, responded similarly: Incidence of disease was negatively correlated with Si content in comparisons of plots receiving, with those not receiving Si applications (39). Genotypic variations were noted in this investigation as well. For a comprehensive review readers are referred to Savant et al (116), and for a recent investigation of control by Si of rice blast, accompanied by a comprehensive bibliography, to Seebold (118). Latin American instances are discussed by Vitti et al (131). Silicon provides protection not only against disease organisms but herbivores as well, be they phytophagous insects (111) or mammals (57, 94).

As in the dicots discussed above, in monocots, too, there is a complex interplay between Si (or Si deprivation) and biochemistry. Thus in oat, *Avena sativa*, Carver et al (26) found that Si deprivation resulted in greater accumulation of phenolic compounds in epidermal cells under attack by *Blumeria graminis*, a pathogenic fungus. The activity of phenylalanine ammonia-lyase was also enhanced by Si deprivation, even in uninocculated plants. Under Si deprivation, the production of structural phenolic compounds may have substituted for the silicification that normally contributes to the structural strength of cell walls. These response mechanisms are by no means uniform in different species; readers are referred to the paper by Carver et al (26) and its references.

BIOCHEMISTRY

With apologies to Winston Churchill, the biochemistry of Si is a riddle wrapped in a mystery inside an enigma. When Jones & Handreck (78) published their groundbreaking review in 1967, Si biochemistry was one of the few relevant topics they did not include: There was no evidence to discuss. And when Carpita (25) recently referred to the incorporation of Si into cell walls of grasses, he wrote, "Little has been reported on any chemical interaction with other cell wall constituents." The amorphous nature of cell wall silica (90, 147) might support the idea that there is no biochemistry involved.

Various attempts to demonstrate the occurrence of organo-silicon compounds, reported by Peggs & Bowen (100) were unsuccessful, including their own, in which they studied *Equisetum arvense* (horsetail) and *Thuja plicata* (western red cedar, Cupressaceae).

Nevertheless, the evidence discussed above on the role of Si in the response of plants to pathogens indicates that Si is not biochemically inert. Because of its prominent presence in the cell wall, Si is a candidate for association with cell wall components and hence, with polysaccharides (19), and with cell wall–associated proteins.

Inanaga et al (73, 74) have presented evidence that in rice, Si may be involved in Si-aromatic ring associations between lignin and carbohydrate. In *Phalaris canariensis* (canary grass) and *Equisetum telmateia*, low levels of proteins are trapped inside silica bodies, along with monosaccharides (62). In the cell wall of the diatom *Cylindrotheca fusiformis*, a 200-kDa protein is associated with a specific structure of the silica scaffold (83). In this case, trapping of the protein within the cell wall silica was specifically ruled out: The protein is located at the surface of the silica. These workers cloned a family of cDNAs that code for these silica-associated proteins.

Further demonstrating an intimate connection between proteins and Si, Shimizu et al (119) recently identified three proteins that seem to control biosilicification in the sponge (Porifera) *Tethya aurantia*. They called these proteins silicateins (for *silica* pro*teins*); they belong to a single protein family. The most abundant of these proteins, silicatein α, belongs to the cathepsin L class of the papain-like cysteine protease superfamily. The biosilicification in these sponges may therefore be mediated through the enzymatic activity of these silicateins. Cha et al (26a) in subsequent in vitro experiments have extended these exciting findings and strengthened that conclusion.

In the marine diatom *Cylindrotheca fusiformis*, Hildebrand et al (64) isolated Si-responsive cDNA clones derived from Si-responsive mRNAs. One possible role for such cDNAs might be that of encoding Si transport proteins. Indeed, this group, associated with the late "silicophile," BE Volcani, has recently cloned a family of genes that code for Si transport proteins (65a). Perhaps the new type of hydrogen bond recently discussed by Crabtree (37) may be involved in complexes between Si and cell wall components (carbohydrates, proteins) and possibly other Si biochemistry, but that, for now, is speculation.

These recent exciting findings from organisms other than higher plants, and from chemistry, should spur plant physiologists, biochemists, and molecular biologists to probe deeply into the corresponding features, metabolites, and processes in higher plants.

CONCLUSION

This survey has covered a great deal more ground than is customary in annual reviews of very actively pursued topics in plant physiology and plant molecular biology. As shown early on, the ubiquitous and quantitatively prominent element, Si, has not received the attention it requires, in view of the significance of it in plant life. It has thus been necessary to cast the net rather wide and bring together not just strictly plant physiological evidence but to include experience gathered by botanists, agronomists, horticulturists, plant pathologists, and still others. There is no justification for the disregard of Si in so much of the science of plant physiology. The evidence is overwhelming that in the real world of plant life, Si matters.

ACKNOWLEDGMENTS

I thank members of the Davis campus for discussion: RG Burau, WH Casey, RA Dahlgren, RJ Southard, and RJ Zasoski, all of my own department. D Potter of the Department of Pomology assisted in the interpretation of Figure 2, and JE Hill of the Department of Agronomy and Range Science advised me on rice ecotypes. RR Bélanger made available unpublished material and gave encouragement and counsel. LE Datnoff, MJ Hodson, and P Hinsinger did likewise. J Ma directed me to material published in Japanese and helped me with the interpretation of Figure 2, as did M Kobayashi. E Malavolta sent me material from Brazil. None of these colleagues bears any responsibility for shortcomings of this review. The secretarial staff's competence and devotion are deeply appreciated. I thank Peggy for putting up with one more of these distractions. My own research on Si was supported by the Department of Energy. Finally, I express my appreciation to Annual Reviews for inviting me to write this, my third contribution to this series.

> Visit the *Annual Reviews home page* at
> http://www.AnnualReviews.org

Literature Cited

1. Adatia MH, Besford RT. 1986. The effects of silicon on cucumber plants grown in recirculating nutrient solution. *Ann. Bot.* 58:343–51
2. Ahmad R, Zaheer SH, Ismail S. 1992. Role of silicon in salt tolerance of wheat (*Triticum aestivum* L.). *Plant Sci.* 85:43–50
3. Anderson DL, Snyder GH, Martin FG. 1991. Multi-year response of sugarcane to calcium silicate slag on Everglades Histosols. *Agron. J.* 83:870–74
4. Arnon DI, Stout PR. 1939. The essentiality of certain elements in minute quantity for plants with special reference to copper. *Plant Physiol.* 14:371–75
5. Asher CJ. 1991. Beneficial elements, functional nutrients, and possible new essential elements. In *Micronutrients in Agriculture*. ed. JJ Mortvedt, FR Cox, LM Shuman, RM Welch, pp. 703–22. Madison: Soil Sci. Soc. Am. 2nd ed.
6. Ayres A. 1966. Calcium silicate slag as a growth stimulant for sugarcane on

low silicon soils. *Soil Sci.* 101:216–17

7. Barceló J, Guevara P, Poschenrieder C. 1993. Silicon amelioration of aluminum toxicity in teosinte, *Zea mays* L. ssp. *mexicana. Plant Soil* 154:249–55

8. Barceló J, Poschendrieder C, Vásquez MD, Gunsé B. 1996. Aluminium phytotoxicity: a challenge for plant scientists. *Fertil. Res.* 43:217–23

9. Bawden Davis J. 1996. Powder your soil with rock dust. *San Francisco Chron.*, March 6

10. Baylis AD, Gragopoulou C, Davidson KJ, Birchall JD. 1994. Effects of silicon on the toxicity of aluminium to soybean. *Commun. Soil Sci. Plant Anal.* 25:537–46

11. Bélanger RR, Benyagoub M. 1997. Challenges and prospects for integrated control of powdery mildews in the greenhouse. *Can. J. Plant Pathol.* 19:310–14

12. Bélanger RR, Bowen PA, Ehret DL, Menzies JG. 1995. Soluble silicon: its role in crop and disease management of greenhouse crops. *Plant Dis.* 79:329–36

13. Bendz G, Lindquist I, eds. 1977. *Biochemistry of Silicon and Related Problems.* New York: Plenum. 591 pp.

14. Bennett WF, ed. 1993. *Nutrient Deficiencies & Toxicities in Crop Plants.* St. Paul: Am. Phytopathol. Soc. 202 pp.

14a. Bidle KD, Azam F. 1999. Accelerated dissolution of diatom silica by marine bacterial assemblages. *Nature* 397:508–12

15. Birchall JD. 1995. The essentiality of silicon in biology. *Chem. Soc. Rev.* 24:351–57

16. Blaich R, Grundhöfer H. 1998. Silicate incrusts induced by powdery mildew in cell walls of different plant species. *Z. Pflanzenkr. Pflanzenschutz* 105:114–20

17. Bollard EG, Butler GW. 1966. Mineral nutrition of plants. *Annu. Rev. Plant Physiol.* 17:77–112

18. Bowen P, Menzies JG, Ehret DL, Samuels AL, Glass ADM. 1992. Soluble Si sprays inhibit powdery mildew development in grape leaves. *J. Am. Soc. Hortic. Sci.* 117:906–12

19. Boylston EK, Hebert JJ, Hensarling TP, Bradow JM, Thibodeaux DP. 1990. Role of silicon in developing cotton fibers. *J. Plant Nutr.* 13:131–48

20. Bradbury M, Ahmad R. 1990. The effect of silicon on the growth of *Prosopis juliflora* growing in saline soil. *Plant Soil* 125:71–74

21. Brown PH, Welch RM, Cary EE, Checkai RT. 1987. Beneficial effects of nickel on plant growth. *J. Plant Nutr.* 10:2125–35

22. Brown PH, Welch RM, Cary EE. 1987. Nickel: a micronutrient essential for higher plants. *Plant Physiol.* 85:801–3

23. Brownell PL, Crossland CJ. 1972. The requirement for sodium by species having the C4 dicarboxylic phytosynthetic pathway. *Plant Physiol.* 49:794–97

24. Broyer TC, Carlton AB, Johnson CM, Stout PR. 1954. Chlorine—a micronutrient element for higher plants. *Plant Physiol.* 29:526–32

25. Carpita NC. 1996. Structure and biogenesis of the cell walls of grasses. *Annu. Rev. Plant Physiol. Plant Mol. Biol.* 47:445–76

26. Carver TLW, Robbins MP, Thomas BJ, Troth K, Raistrick N, Zeyen RJ. 1998. Silicon deprivation enhances local autofluorescent responses and phenylalanine ammonia-lyase activity in oat attacked by *Blumeria graminis. Physiol. Mol. Plant Pathol.* 52:245–57

26a. Cha JN, Shimizu K, Zhou Y, Christiansen SC, Chmelka BF, et al. 1999. Silicatein filaments and subunits from a marine sponge direct the polymerization of silica and silicones in vitro. *Proc. Natl. Acad. Sci. USA* 96:361–65

27. Chen C-H, Lewin J. 1969. Silicon as a nutrient element for *Equisetum arvense. Can. J. Bot.* 47:125–31

28. Chérif M, Asselin A, Bélanger RR. 1994. Defense responses induced by soluble silicon in cucumber roots infected by *Pythium* spp. *Phytopathology* 84:236–42

29. Chérif M, Benhamou N, Menzies JG, Bélanger RR. 1992. Silicon induced resistance in cucumber plants against *Pythium ultimatum. Physiol. Mol. Plant Pathol.* 41:411–25

30. Clarkson DT, Hanson JB. 1980. The mineral nutrition of higher plants. *Annu. Rev. Plant Physiol.* 31:239–98

31. Clements HF. 1964. Interaction of factors affecting yield. *Annu. Rev. Plant Physiol.* 15:409–42

32. Clements HF. 1965. The roles of calcium silicate slags in sugarcane growth. *Rep. Hawaiian Sugar Technol.,* pp. 103–26

33. Cocker KM, Evans DE, Hodson MJ. 1998. The amelioration of aluminium toxicity by silicon in wheat (*Triticum aestivum* L.): malate exudation as evidence for an *in planta* mechanism. *Planta* 204:318–23

34. Cocker KM, Evans DE, Hodson MJ. 1998. The amelioration of aluminium

toxicity by silicon in higher plants: solution chemistry or an in planta mechanism? *Physiol. Plant.* 104:608–14

35. Coroneos C, Hinsinger P, Gilkes RJ. 1996. Granite powder as a source of potassium for plants: a glasshouse bioassay comparing two pasture species. *Fertil. Res.* 45: 143–52

36. Corrales I, Poschenrieder C, Barceló J. 1997. Influence of silicon pretreatment on aluminium toxicity in maize roots. *Plant Soil* 190:203–9

37. Crabtree RH. 1998. A new type of hydrogen bond. *Science* 282:2000–1

38. Deren CW, Datnoff LE, Snyder GH. 1992. Variable silicon content of rice cultivars grown on Everglades Histosols. *J. Plant Nutr.* 15:2363–68

39. Deren CW, Datnoff LE, Snyder GH, Martin FG. 1994. Silicon concentration, disease response, and yield components of rice genotypes grown on flooded organic Histosols. *Crop Sci.* 34:733–37

40. Ehret DL, Zebarth BJ, Portree J, Garland T. 1998. Clay addition to soilless media promotes growth and yield of greenhouse crops. *HortScience* 33:67–70

41. Elawad SH, Gascho GJ, Street JJ. 1982. Response of sugarcane to silicate source and rate. I. Growth and yield. *Agron. J.* 74:481–84

42. Elawad SH, Street JJ, Gascho GJ. 1982. Response of sugarcane to silicate source and rate. II. Leaf freckling and nutrient content. *Agron. J.* 74:484–87

43. Emadian SF, Newton RJ. 1989. Growth enhancement of loblolly pine (*Pinus taeda* L.) seedlings by silicon. *J. Plant Physiol.* 134:98–103

44. Engel RE, Bruckner PL, Eckhoff J. 1998. Critical tissue concentration and chloride requirements for wheat. *Soil Sci. Soc. Am. J.* 62:401–5

45. Epstein E. 1972. *Mineral Nutrition of Plants: Principles and Perspectives.* New York: Wiley. 412 pp.

46. Epstein E. 1994. The anomaly of silicon in plant biology. *Proc. Natl. Acad. Sci. USA* 91:11–17

47. Epstein E. 1999. The discovery of the essential elements. In *Discoveries in Plant Biology*, Vol. 3, ed. SD Kung, SF Yang. Singapore: World Sci. Publ. In press

48. Ernst WHO, Vis RD, Piccoli F. 1995. Silicon in developing nuts of the sedge *Schoenus nigricans*. *J. Plant Physiol.* 146:481–88

49. Evered D, O'Connor M, eds. 1986. *Silicon Biochemistry*. Ciba Found. Symp. 121. Chichester, UK: Wiley 264 pp.

50. Exley C. 1998. Silicon in life: a bioinorganic solution to bioorganic essentiality. *J. Inorg. Biochem.* 69:139–44

51. Exley C, Birchall JD. 1996. Silicic acid and the biological availability of aluminium. *Eur. J. Soil Sci.* 47:137

52. Fawe A, Abow-Zaid M, Menzies JG, Bélanger RR. 1998. Silicon-mediated accumulation of flavonoid phytoalexins in cucumber. *Phytopathology* 88:396–401

52a. Fawe A, Menzies JG, Bélanger RR. 1999. Soluble silicon: a possible positive modulator of plant disease resistance?

53. Foy CD. 1992. Soil chemical factors limiting plant growth. *Adv. Soil Sci.* 19: 97–149

54. Foy CD, Chaney RL, White MC. 1978. The physiology of metal toxicity in plants. *Annu. Rev. Plant Physiol.* 29: 511–66

55. Fraústo da Silva JJR, Williams RJP. 1991. *The Biological Chemistry of the Elements. The Inorganic Chemistry of Life.* Oxford: Clarendon Press. 561 pp.

56. Gali HV, Smith CC. 1992. Effect of silicon on growth, fertility, and mineral composition of an annual brome, *Bromus secalinus* L. (Gramineae). *Am. J. Bot.* 79:1259–63

57. Gali-Muhtasib HU, Smith CC. 1992. The effect of silica in grasses on the feeding behavior of the prairie vole, *Microtus ochrogaster. Ecology* 73:1724–29

58. Godbold DL, Fritz E. Huttermann A. 1988. Aluminum toxicity and forest decline. *Proc. Natl. Acad. Sci. USA* 85: 3888–92

59. Goodall DW, Gregory FG. 1947. *Chemical Composition of Plants as an Index of their Nutritional Status*. Imp. Bur. Hortic. Plant. Crops Tech. Commun. No. 17. East Malling, U.K. 167 pp.

60. Goto S, Goto I. 1997. Effects of Indonesian electric furnace slag on rice yield and chemical properties of soils. In *Plant Nutrition for Sustainable Food Production and Environment*, ed. T Ando, K Fujita, T Mae, H Matsumoto, S Mori, J Sekiya, pp. 803–4. Dordrecht: Kluwer

61. Hammond KE, Evans DE, Hodson MJ. 1995. Aluminium/silicon interactions in barley (*Hordeum vulgare* L.) seedlings. *Plant Soil* 173:89–95

62. Harrison CC, Lu Y. 1994. *In vivo* and *in vitro* studies of polymer controlled silification. *Bull. Inst. Océan., Monaco*, no spéc. 14.1:151–58

63. Hewitt EJ. 1966. *Sand and Water Culture Methods Used in the Study of Plant*

Nutrition. Tech. Commun. No. 22, Commonw. Bur. Hortic. Plant. Crops, East Malling, UK: CAB. Rev. 2nd ed. 547 pp.

64. Hildebrand M, Higgins DR, Busser K, Volcani BE. 1993. Silicon-responsive cDNA clones isolated from the marine diatom *Cylindrotheca fusiformis. Gene* 132:213–18

65. Hildebrand M. Volcani BE, Gassmann W, Schroeder JI. 1997. A gene family of silicon transporters. *Nature* 385:688–89

65a. Hildebrand M. Dahlin K, Volcani BE. 1998. Characterization of a silicon transporter gene family in *Cylindrotheca fusiformis*: sequences, expression analysis, and identification of homologs in other diatoms. *Mol. Gen. Genet.* 260:480–86

66. Hinsinger P, Bolland MDA, Gilkes RJ. 1996. Silicate rock powder: effect on selected chemical properties of a range of soils from Western Australia and on plant growth as assessed in a glasshouse experiment. *Fertil. Res.* 45:69–79

67. Hiradate S, Taniguchi S, Sakurai K. 1998. Aluminum speciation in aluminum-silica solutions and potassium chloride extracts of acidic soils. *Soil Sci. Soc. Am. J.* 62: 630–36

68. Hodson MJ, Evans DE. 1995. Aluminium/silicon interactions in higher plants. *J. Exp. Bot.* 46:161–71

69. Hodson MJ, Sangster AG. 1984. Observations on the distribution of mineral elements in the leaf of wheat (*Triticum aestivum* L.), with particular reference to silicon. *Ann. Bot.* 62:463–71

70. Hodson MJ, Sangster AG. 1998. Mineral deposition in the needles of white spruce [*Picea glauca* (Moench.) Voss]. *Ann. Bot.* 82:375–85

71. Horiguchi T, Morita S. 1987. Mechanism of manganese toxicity and tolerance of plants. VI. Effect of silicon on alleviation of manganese toxicity of barley. *J. Plant Nutr.* 10:2299–310

72. Horst WJ, Marschner H. 1978. Effect of silicon on manganese tolerance of bean plants (*Phaseolus vulgaris* L.). *Plant Soil* 50:287–304

73. Inanaga S, Okasaka A. 1995. Calcium and silicon binding compounds in cell walls of rice shoots. *Soil Sci. Plant Nutr.* 41:103–10

74. Inanaga S, Okasaka A, Tanaka S. 1995. Does silicon exist in association with organic compounds in rice plants? *Soil Sci. Plant Nutr.* 41:111–17

75. Jarvis SC. 1987. The uptake and transport of silicon by perennial ryegrass and wheat. *Plant Soil* 97:429–37

76. Jarvis SC, Jones LHP. 1987. The absorption and transport of manganese by perennial ryegrass and white clover as affected by silicon. *Plant Soil* 99:231–40

77. Deleted in proof

78. Jones LHP, Handreck KA. 1967. Silica in soils, plants, and animals. *Adv. Agron.* 19:107–49

79. Kaim W, Schwederski B. 1994. *Bioinorganic Chemistry: Inorganic Elements in the Chemistry of Life.* Chichester, UK: Wiley. 401 pp.

80. Kelly EF, Chadwick OA, Hilinski TE. 1998. The effects of plants on mineral weathering. *Biogeochemistry* 42:21–53

81. Ketchum BH. 1954. Mineral nutrition of phytoplankton. *Annu. Rev. Plant Physiol.* 5:55–74

82. Koenig RT, Pan WL. 1996. Chloride enhancement of wheat responses to ammonium nutrition. *Soil Sci. Soc. Am. J.* 60: 498–505

83. Kröger N, Lehmann G, Rachel R, Sumper M. 1997. Characterization of a 200-kDa diatom protein that is specifically associated with a silica-based substructure of the cell wall. *Eur. J. Biochem.* 250:99–105

84. Lewin J, Reimann BEF. 1969. Silicon and plant growth. *Annu. Rev. Plant Physiol.* 20:289–304

85. Li YC, Adva AK, Sumner ME. 1989. Response of cotton cultivars to aluminum in solutions with varying silicon concentrations. *J. Plant Nutr.* 12:881–92

85a. Lian S. 1976. Silica fertilization of rice. In *The Fertility of Paddy Soils and Fertilizer Applications for Rice,* pp. 197–220. Taipei: Food Fertil. Technol. Cent. Asian Pac. Reg.

86. Liang Y, Shen Q, Shen Z, Ma T. 1996. Effects of silicon on salinity tolerance of two barley cultivars. *J. Plant Nutr.* 19:173–83

87. Lumsdon DG, Farmer VC. 1996. Response to: 'Silicic acid and the biological availability of aluminium' by C. Exley and J.D. Birchall. *Eur. J. Soil Sci.* 47:139–40

88. Ma JF, Saski M, Matsumoto H. 1997. Al-induced inhibition of root elongation in corn, *Zea mays* L. is overcome by Si addition. *Plant Soil* 188:171–76

89. Ma J, Takahashi E. 1991. Effect of silicate on phosphate availability for rice in a P-deficient soil. *Plant Soil* 133:151–55

90. Mann S, Perry CC, Williams RJP, Fyfe CA, Gobbi GC, Kennedy CJ. 1983. The characterisation of the nature of silica in

biological systems. *J. Chem. Soc. Chem. Commun.* 4:168–70

91. Matoh T, Kairusmee P, Takahashi E. 1986. Salt-induced damage to rice plants and alleviation effect of silicate. *Soil Sci. Plant Nutr.* 32:295–304

92. Mayland HF, Johnson DA, Asay KH, Read JJ. 1993. Ash, carbon isotope discrimination, and silicon as estimators of transpiration efficiency in crested wheatgrass. *Aust. J. Plant Physiol.* 20:361–69

93. McAvoy RJ, Bible BB. 1996. Silica sprays reduce the incidence and severity of bract necrosis in Poinsettia. *HortScience* 31:1146–49

94. McNaughton SJ, Tarrants JL, McNaughton MM, Davis RH. 1985. Silica as a defense against herbivory and a growth promoter in African grasses. *Ecology* 66:528–35

95. Menzies J, Bowen P, Ehret D, Glass ADM. 1992. Foliar applications of potassium silicate reduce severity of powdery mildew on cucumber, muskmelon, and zucchini squash. *J. Am. Soc. Hortic. Sci.* 117:902–5

96. Menzies JG, Ehret DL, Glass ADM, Samuels AL. 1991. The influence on cytological interactions between *Sphaerotheca fuliginea* and *Cucumis sativus*. *Physiol. Mol. Plant Pathol.* 39:403–14

97. Mitsui S, Takatoh H. 1963. Nutritional study of silicon in graminaceous crops. Part I. *Soil Sci. Plant Nutr.* 9:49–53

98. Nishimura K, Miyaki Y, Takahashi E. 1989. On silicon, aluminium, and zinc accumulators discriminated from 147 species of Angiospermae. *Mem. Coll. Agric. Kyoto Univ.* No. 133:23–43

99. Parry DW, Hodson MJ, Sangster AG. 1984. Some recent advances in studies of silicon in higher plants. *Phil. Trans. R. Soc. London Ser. B* 304:537–49

100. Peggs A, Bowen H. 1984. Inability to detect organo-silicon compounds in *Equisetum* and *Thuja*. *Phytochemistry* 23:1788–89

101. Perry CC, Keeling-Tucker T. 1998. Aspects of the bioinorganic chemistry of silicon in conjunction with the biometals calcium, iron and aluminium. *J. Inorg. Biochem.* 69:181–91

102. Perry CC, Williams RJP, Fry SC. 1987. Cell wall biosynthesis during silification of grass hairs. *J. Plant Physiol.* 126:437–48

103. Porter JR, Lawlor DW, eds. 1991. *Plant Growth: Interaction with Nutrition and Environment*. New York: Cambridge Univ. Press. 284 pp.

104. Rafi MM, Epstein E. 1997. Silicon deprivation causes physical abnormalities in wheat (*Triticum aestivum* L.). *J. Plant Physiol.* 151:497–501. For properly reproduced Figure 1 see *J. Plant Physiol.* 152:592

105. Rahman MT, Kawamura K, Koyama H, Hara T. 1998. Varietal differences in the growth of rice plants in response to aluminum and silicon. *Soil Sci. Plant Nutr.* 44:423–31

106. Raven JA. 1983. The transport and function of silicon in plants. *Biol. Rev.* 58:179–207

107. Reynolds AG, Veto LJ, Sholberg PL, Wardle DA, Haag P. 1996. Use of potassium silicate for the control of powdery mildew [*Uncinular necator* (Schwein) Burrill] in *Vitis vinifera* L. cultivar Bacchus. *Am. J. Enol. Vitic.* 47:421–28

108. Richter W, Suntheim L, Matzel W. 1989. Untersuchungen zur Wirkung von Silizium auf die Ertragsbildung von Sommergerste. *Arch. Acker-Pflanzenbau Bodenkd.* 33:33–40

109. Richter W, Suntheim L. 1986. Zur Bedeutung des Siliciums in der Pflanzenernährung. *Arch. Acker-Pflanzenbau Bodenkd.* 30:737–44

110. Sachs J. 1865. *Handbuch der Experimental-Physiologie der Pflanzen*. Leipzig: Verlag von Wilhelm Engelmann. 514 pp.

111. Salim M, Saxena RC. 1992. Iron, silica, and aluminum stresses and varietal resistance in rice: effects on whitebacked planthopper. *Crop Sci.* 32:212–19

112. Samuels AL, Glass ADM, Ehret DL, Menzies JG. 1993. The effects of silicon supplementation on cucumber fruit: changes in surface characteristics. *Ann. Bot.* 72:433–40

113. Sangster AG, Hodson MJ. 1992. Silica deposition in subterranean organs. In *Phytolith Systematics*, ed. G Rapp Jr, SC Mulholland, pp. 239–51. New York: Plenum

114. Sangster AG, Hodson MJ. 1992. Silica in higher plants. See Ref. 49, pp. 90–107

115. Savant NK, Datnoff LE, Snyder GH. 1997. Depletion of plant-available silicon in soils: a possible cause of declining rice yields. *Commun. Soil Sci. Plant Anal.* 28:1145–52

116. Savant NK, Snyder GH, Datnoff LE. 1997. Silicon management and sustainable rice production. *Adv. Agron.* 58:151–99

117. Schneider S, Ullrich WR. 1994. Differential induction of resistance and enhanced enzyme activities in cucumber and tobacco caused by treatment

with various abiotic and biotic inducers. *Physiol. Mol. Plant Pathol.* 45:291–304

118. Seebold KW Jr. 1998. *The influence of silicon fertilization on the development and control of blast caused by* Magnaporthe grisea *(Hebert) Barr, in upland rice.* PhD thesis, Univ. Fla. 230 pp.

119. Shimizu K, Cha J, Stucky GD, Morse DE. 1998. Silicateine α: cathespin L-like protein in sponge biosilica. *Proc. Natl. Acad. Sci. USA* 95:6234–38

120. Shkolnik MY. 1984. *Trace Elements in Plants*, pp. 267–461. Amsterdam: Elsevier. 463 pp.

121. Silva JA. 1971. Possible mechanisms for crop response to silicate applications. *Proc. Int. Symp. Soil Fertil. Eval.* New Delhi 1:806–14

122. Deleted in proof

123. Simkiss K, Wilbur KM. 1989. *Biomineralization. Cell Biology and Mineral Deposition.* San Diego: Academic. 337 pp.

124. Simpson TL, Volcani BE, eds. 1981. *Silicon and Siliceous Structures in Biological Systems.* New York: Springer-Verlag. 587 pp.

125. Sposito G. 1989. *The Chemistry of Soils.* New York: Oxford Univ. Press. 277 pp.

125a. Stookey MA. 1995. *Silicon uptake in rice and cucumbers.* MS thesis, Univ. B-C.

126. Takahashi E, Ma JF, Miyake Y. 1990. The possibility of silicon as an essential element for higher plants. *Comments Agric. Food Chem.* 2:99–122

127. Takahashi E, Miyake Y. 1976. Distribution of silica accumulator plants in the plant kingdom. (1) Monocotyledons. *J. Sci. Soil Manure* 47:296–300 (In Japanese)

128. Takahashi E, Miyake Y. 1976. Distribution of silica accumulator plants in the plant kingdom. (2) Dicotyledons. *J. Sci. Soil Manure* 47:301–6 (In Japanese)

129. Takahashi E, Tanaka H, Miyake Y. 1981. Distribution of silicon accumulating plants in the plant kingdom. *Jpn. J. Soil Sci. Plant Nutr.* 52:511–15 (In Japanese)

130. Tisdale SL, Nelson WL, Beaton JD, Havlin JL. 1993. *Soil Fertility and Fertilizers.* New York: Macmillan. 5th ed. 634 pp.

131. Vitti GC, de Oliveira FA, Prata F, de Oliveira JA Jr, del C Ferragine M, de A Silveira RLV. 1997. *Silicio No Solo E Na Planta.* Piracicaba: Univ. São Paulo. 90 pp.

132. Voronkov MG, Zelchan GI, Lukevitz. 1975. *Silizium und Leben. Biochemie, Toxikologie und Pharmakologie der*

Verbindungen des Siliziums. Bearb. hrsg. Prof. Dr. Klaus Rühlmann, Dresen. Berlin: Akademie-Verlag

133. Waisel Y. 1972. *Biology of Halophytes.* New York: Academic. 395 pp.

134. Waisel Y, Eshel A, Kafkafi V, eds. 1996. *Plant Roots: The Hidden Half.* New York: Marcel Dekker. 2nd ed. 1002 pp.

135. Werner D, Roth R. 1983. Silica metabolism. In *Inorganic Plant Nutrition, Encyclopedia of Plant Physiology, New Ser.*, ed. A Läuchli, RL Bieleski, 15B:682–94. Berlin: Springer-Verlag

136. Williams RJP. 1977. Summary—Silicon in biological systems. In *Biochemistry of Silicon and Related Problems*, ed. G Bendz, I Lindquist, pp. 561–76. New York: Plenum

137. Williams RJP. 1986. Introduction to silicon chemistry and biochemistry. In *Silicon Biochemistry.* Ciba Found. Symp. 121, ed. D Evered, M O'Connor, pp. 24–39

138. Williams DE, Vlamis J. 1957. Manganese toxicity in standard culture solutions. *Plant Soil* 8:183–93

139. Williams DE, Vlamis J. 1957. Manganese and boron toxicities in standard culture solutions. *Soil Sci. Soc. Am. Proc.* 21: 205–9

140. Williams DE, Vlamis J. 1957. The effect of silicon on yield and manganese-54 uptake and distribution in the leaves of barley plants grown in culture solutions. *Plant Physiol.* 32:404–9

141. Williams DE, Vlamis J. 1967. Manganese and silicon interaction in the Gramineae. *Plant Soil* 27:131–40

142. Winslow MD. 1992. Silicon, disease resistance, and yield of rice genotypes under upland cultural conditions. *Crop Sci.* 32:1208–13

143. Winslow MD, Okada K, Correa-Victoria F. 1997. Silicon deficiency and the adaptation of tropical rice ecotypes. *Plant Soil* 188:239–48

144. Woolley JT. 1957. Sodium and silicon as nutrients for the tomato plant. *Plant Physiol.* 32:317–21

145. Wutscher HK. 1989. Growth and mineral nutrition of young orange trees grown with high levels of silicon. *HortScience* 24:275–77

146. Yamauchi M, Winslow MD. 1989. Effect of silica and magnesium on yield of upland rice in the humid tropics. *Plant Soil* 113:265–69

147. Yoshida S, Ohuishi Y, Kitagishi K. 1962. Chemical forms, mobility and deposition of silicon in rice plant. *Soil Sci. Plant Nutr.* 8:107–13

Annu. Rev. Plant Physiol. Plant Mol. Biol. 1999. 50:665–93

PHOSPHATE ACQUISITION

K. G. Raghothama

Department of Horticulture, Purdue University, West Lafayette, Indiana 47907;
e-mail: Ragu@hort.purdue.edu

KEY WORDS: phosphate starvation, gene expression, homeostasis, phosphate transporters, tissue-specific expression

---

ABSTRACT

Phosphorus is one of the major plant nutrients that is least available in the soil. Consequently, plants have developed numerous morphological, physiological, biochemical, and molecular adaptations to acquire phosphate (Pi). Enhanced ability to acquire Pi and altered gene expression are the hallmarks of plant adaptation to Pi deficiency. The intricate mechanisms involved in maintaining Pi homeostasis reflect the complexity of Pi acquisition and translocation in plants. Recent discoveries of multiple Pi transporters have opened up opportunities to study the molecular basis of Pi acquisition by plants. An increasing number of genes are now known to be activated under Pi starvation. Some of these genes may be involved in Pi acquisition, transfer, and signal transduction during Pi stress. This review provides an overview of plant adaptations leading to enhanced Pi acquisition, with special emphasis on recent developments in the molecular biology of Pi acquisition.

CONTENTS

1040-2519/99/0601-0665$08.00

INTRODUCTION

Phosphate (Pi) is one of the key substrates in energy metabolism and biosynthesis of nucleic acids and membranes. It also plays an important role in photosynthesis, respiration, and regulation of a number of enzymes. Among the many inorganic nutrients required by plants, P is one of the most important elements that significantly affect plant growth and metabolism. Low availability of Pi is a major constraint for crop production in many low-input systems of agriculture worldwide. Phosphate acquisition by plants is one of the more thoroughly studied aspects of plant nutrition, and there is an extensive literature on the biochemical, morphological, and physiological effects of Pi deficiency on plants. There are excellent reviews on the role of Pi in the glycolytic pathway (124, 125), regulation of RNases (61), phosphatases (40), mycorrhizal interactions (see Harrison in this volume, pp. 361–89), root architecture (96), Pi uptake (140), modeling of Pi uptake (3), rhizosphere, and plant nutrition (31). These reviews provide a comprehensive picture of the complex nature of Pi acquisition and utilization by plants. An excellent reference source for an overall understanding of P in soil and plants is *Mineral Nutrition of Higher Plants* (102). This review provides an overview of the physiological and biochemical adaptations of plants to acquire and utilize P, emphasizing recent developments in the molecular biology of Pi acquisition. This is an exciting period for plant nutrition; many genes coding for major nutrient transporters have been cloned and the molecular basis of nutrient acquisition is beginning to unravel. An increasing number of genes induced in response to nutrient stress, especially P, have been identified. Some of these genes may serve as molecular determinants of Pi uptake, use efficiency, and signal transduction.

PHOSPHORUS AVAILABILITY: A GLOBAL PROBLEM

Phosphorus is one of the least available of all essential nutrients in the soil and its concentration is generally below that of many other micronutrients (7). Many soils around the world are deficient in Pi, and even in fertile soils, available Pi seldom exceeds 10 μM (13). In most soils, the concentration ($\sim$2 μM) of available Pi in soil solution is several orders of magnitude lower than that in plant tissues (5–20 mM). Phosphorus availability is of particular concern in the highly weathered and volcanic soils of the humid tropics and subtropics, and in many sandy soils of the semiarid tropics, where crop productivity is severely compromised through lack of available Pi (139). The level of Pi in soil solution is regulated mainly by its interaction with organic or inorganic surfaces in the soil. Aluminum ions, which predominate in acid soils of the world, and iron interact strongly with Pi and render it unavailable to plants. Acid conditions exist in approximately 30% of soils worldwide, and are found in all continents (157a). In addition, a considerable fraction (20–80%) of Pi in soils is found in the organic form (76, 130), which has to be mineralized to the inorganic form before it becomes available to plants. Plants must also compete with microorganisms to obtain Pi under nutrient-limiting conditions. Observations suggest that the low concentration of Pi in the soil solution is a major factor limiting growth in many natural ecosystems.

Phosphorus is considered to be the most limiting nutrient for growth of leguminous crops in tropical and subtropical regions (2). A survey by CIAT (22) indicated that 50% of the bean-growing areas in Latin America are low in Pi and the same may be true for bean-growing areas of Africa. The nonrenewable nature of Pi resources results in continuous depletion of terrestrial Pi in the absence of added fertilizers or organic matter, a very common condition in many developing countries. Because of the unique interaction of Pi with other elements, up to 80% of applied Pi may be fixed in the soil (9, 71), forcing farmers to use up to four times the fertilizer necessary for crop production (58). At the current rate of usage of P fertilizer, readily available sources of phosphate rocks will be depleted over the next 60 to 90 years (132). At present, many tropical regions are faced with excessive mining of nutrients, including P, whereas some temperate regions with intensive, animal-based agricultural systems have, ironically, to deal with excessive soluble P in the soil that is threatening the ecosystem. In many parts of the United States and Europe, where enormous quantities of nutrient-rich manure are spread on the soils, the soluble Pi levels often exceed the crop requirement. Under these conditions, there is a significant potential for Pi movement (144). Excess soil Pi not removed by crops can enter surface water by erosion of Pi-rich soil particles, runoff, and leaching to field drain tiles. Increased Pi concentration in aquatic systems results in eutrophication and degradation of the environment.

RESPONSE OF PLANTS TO Pi STARVATION

In response to persistently low levels of available Pi in the rhizosphere, plants have developed highly specialized physiological and biochemical mechanisms

Table 1   Multiple responses of plants to phosphate deficiency

Morphological responses
Increased root:shoot ratio; changes in root morphology and architecture; increased root hair
    proliferation; root hair elongation; accumulation of anthocyanin pigments; proteoid root
    formation; increased association with mycorrhizal fungi

Physiological
Enhanced Pi uptake; reduced Pi efflux; increased Pi use efficiency; mobilization of Pi from
    the vacuole to cytoplasm; increased translocation of phosphorus within plants; retention of
    more Pi in roots; secretion of organic acids; protons and chelaters; secretion of phosphatases
    and RNases; altered respiration; carbon metabolism; photosynthesis; nitrogen fixation;
    and aromatic enzyme pathways

Biochemical
Activation of enzymes; enhanced production of phosphatases; RNases and organic acids,
    changes in protein phosphorylation; activation of glycolytic bypass pathway

Molecular
Activation of genes (RNases, phosphatases, phosphate transporters, Ca-ATPase, vegetative
    storage proteins, β-glucosidase, PEPCase, novel genes such as *TPSII*, *Mt 4*)

to acquire and utilize Pi from the environment. The ultimate consequences of these modifications are increased Pi availability in the rhizosphere and enhanced uptake. Some of the major adaptive changes by plants in response to limiting Pi in the rhizosphere are listed in Table 1.

Modification of Root Systems

Modification of root growth and architecture is a well-documented response to Pi starvation (96, 97). Increase in the root-shoot ratio under Pi starvation is a hallmark of plant response to Pi deficiency, enhancing the total surface area available for soil exploration and acquisition of nutrients for a particular species of plant. Plants with a more proliferated root system that is efficient in uptake are well suited to exploit soil Pi. In addition to increased root mass, root diameter decreases under Pi stress, while the amount of absorptive surface area relative to root volume increases. The inherent differences in Pi uptake and utilization by plant species have been elegantly demonstrated in a comparative study of Pi efficiencies of seven different species (50). Highly efficient plants had either high influx rates (rape and spinach) or high root-shoot ratios (rye and wheat) compared to species of low efficiency (onion, tomato, and bean), which had low influx rates and low root-shoot ratios. Phosphorus-efficient bean genotypes have a highly branched, actively growing root system compared to those of Pi-inefficient genotypes, which suggests that root architectural traits strongly influence Pi acquisition (98).

A significant difference in Pi uptake is also attributed to the production of more root hairs by Pi-efficient plants in low Pi soil (51). Under Pi deficiency,

90% of the total Pi acquired by plants may occur through root hairs. Direct evidence for the involvement of root hairs in Pi acquisition came from field studies of rye grown in PVC pipes covered with nylon mesh permeable only to root hairs (53). In this study, root hairs contributed up to 63% of total Pi uptake by plants. Root hairs have a smaller diameter and grow perpendicular to the root axis, which allows better exploration of soil due to enhanced absorptive surface area. In addition to increased density of root hairs, Pi deficiency also leads to elongation of root hairs (10). In situ transcript localization experiments have shown that Pi transporter genes are preferentially expressed in the epidermis and root hairs of tomato (30, 90, 127).

Plants also respond to heterogeneously distributed Pi in the soil by enhancing uptake in localized Pi-rich patches (37, 72). This phenotypic plasticity helps to maximize Pi uptake without wasting valuable carbon resources. An 80% increase in uptake by roots growing in enriched soil patches is a clear indication of the physiological plasticity exhibited by plants (72). This physiological flexibility of roots is crucial in compensating for a lack of uptake of Pi by major portions of the root system. Two distinctly different situations, a general unavailability of P, or availability only in certain patches of soil, lead to the similar root proliferation response. Phosphorus deficiency not only enhances root proliferation, root hair production, and elongation, but also leads to modifications in the biochemical processes of roots. Recent molecular analysis of Pi starvation-induced gene expression shows that multiple genes are activated in roots under Pi starvation (127).

There is a high degree of correlation between mycorrhizae formation and Pi status of soil. Mycorrhizae are considered to be an integral part of the Pi absorption and translocation pathway in plants belonging to different genera (159). Mycorrhizae formation involves a complex interaction between fungi and plant roots, and P appears to play a key role in this association. One has to take mycorrhizal association into consideration in studying Pi nutrition of plants. The recent cloning of a Pi transporter gene from mycorrhizal fungi *Glomus versiforme* (65) has provided opportunities to evaluate the contribution of Pi uptake by mycorrhizal fungi to plant nutrition. (See the review article on mycorrhizal fungi by Harrison in this volume for additional information.)

Proteoid Root Formation

Formation of proteoid roots under Pi deficiency is a well-characterized response in white lupins (54, 73, 79, 103). Proteoid roots are composed of bottlebrush-like clusters of rootlets covered with a dense mat of root hairs, specialized in efficient synthesis and secretion of organic acids to the rhizosphere (33, 54, 73). This root adaptation is of widespread occurrence in Proteaceae and other families (33). Internal Pi concentrations influence the formation of proteoid roots in white lupin (79, 103). Organic acids secreted from proteoid roots aid in the release of Pi from Ca, Fe, and Al phosphates, via chelation of the metal. Increased organic acid secretion is correlated with the increased activity of several enzymes involved in organic acid synthesis, including phosphoenolpyruvate carboxylase (PEPC), citrate synthase (CS), and malate dehydrogenase (MDH) (73). Enhanced activities of PEPC are generally associated with increased protein and mRNA levels for PEPC in Pi-deficient proteoid roots (73). In addition to secretion of organic acids, phosphatase production also increased nearly 20-fold in lupins under Pi deficiency (149). Proteoid roots are not only efficient in producing organic acids and phosphatases (56), but they also absorb Pi at a faster rate than non-proteoid roots (158).

Secretion of Organic Acids

In general, dicots, particularly legumes, are more efficient than monocots in producing and excreting organic acids to the rhizosphere to enhance Pi solubilization under Pi deficiency. The root exudates of Pi-deficient plants contain a large number of organic acids (62). Plants such as rape, white lupin, and pigeon pea are very efficient in mobilizing Pi from soil. The roots of rape excrete organic acids into the rhizosphere and solubilize Pi from rock phosphate (70). Organic acid exudation, acidification of rhizosphere, and organic acid release in specific regions of the root are all known to increase concurrently in Pi-starved rape plants. Malic and citric acids are the predominant acids excreted by roots under Pi deficiency. The excretion zone is normally located 1–2 cm behind the root tip, and, interestingly, the zone of excretion is also the region of the root that is directly in contact with rock phosphate (69). The biochemical and spatial flexibility exhibited by these plants in response to localized application of rock phosphate is quite remarkable. This flexibility allows them to target the excretion of valuable carbon compounds in the vicinity of a Pi source. Increased exudation of organic acids in roots is also observed in many other dicots under Pi-limiting conditions. A twofold increase in exudation of citrate was observed under Pi starvation in alfalfa (89). Citric acid secreted by chickpea roots helps them obtain Pi from vertisol soils, where a high proportion of Pi is associated with soil calcium (1). Increased activity of PEP carboxylase in response to Pi starvation was found in many plants including tomato (123), white lupin (73), pea (131), rape (69), and *Brassica nigra* (39). It is presumed that Pi starvation may also lead to activation or synthesis of anion transporters or channels to enhance secretory processes. Measurement of anion currents across membranes of Pi-starved roots by patch clamping should reveal the induction of anion transport system(s) that facilitate organic acid efflux.

Plants also solubilize Pi in the rhizosphere by exudation of compounds that act as chelators. Some of the special biochemical adaptations in plants include production of phenolic compounds such as pisidic acid (p-hydroxybenzyl tartaric acid) by pigeon pea (2), and alfafuran (2-(3′,5′-dihydroxyphenyl)-5,6-dihydroxybenzofuron) by alfalfa (105). Pigeon peas secrete pisidic acid to extract Pi associated with iron; the alcohol, and carboxyl groups of the tartarate appear to be involved in Fe chelation and release of Pi (1, 2). Intercropping of lupin or pigeon pea with cereals enhances the uptake of Pi by the grass species, indicating an increased availability of soluble Pi due to the release of organic compounds.

Biochemical Changes under Pi Deficiency

Changes in biochemical processes occurring in Pi-starved plant cells have been recently reviewed (124, 125). The induction of acid phosphatases in response to Pi starvation is a universal response in higher plants (40). Production of both extracellular and intracellular phosphatases is considered to be an integral part of the plant response to Pi deficiency. Phosphatases are presumed to liberate Pi from organic materials (40, 58, 86, 154). The levels of ATP and all of the nucleotides are significantly reduced during Pi starvation of plant cell cultures and PPi (pyrophosphate), which is maintained at high levels, may function as an autonomous energy donor (124). During Pi starvation, enzymes that do not require nucleotide phosphates or Pi as substrate(s) are activated. These enzymes are considered to be adaptive enzymes of the glycolytic pathway during Pi starvation, and they are involved in "bypass reactions" that circumvent Pi and adenylate-requiring steps in glycolysis, thus permitting carbon metabolism to proceed during Pi starvation (39). Pi limitation also results in activation of an alternate respiratory pathway (133) and decreases in the rate of photosynthesis and stomatal conductance (12, 26, 116, 128). Accumulation of the polyamine putrescine in response to Pi starvation, has also been reported in rice culture (145). Increasing levels of putrescine may be one of the causes for inhibition of cell growth under Pi-starvation conditions.

GENE EXPRESSION AND Pi STARVATION

Activation of specific genes during Pi starvation seems to be a universal phenomenon in all organisms. Differential accumulation of proteins under Pi starvation is indicative of extensive changes in gene expression and/or protein turnover (21, 42, 49, 59, 66, 157). Many proteins, including RNases, phosphatases, Pi transporters and PEPCase, increase under Pi starvation because of de novo synthesis of proteins. Synthesis of a 25-kDa membrane polypeptide and a 65-kDa soluble polypeptide increased in tomato roots under Pi starvation (66). Specific changes in protein profiles are suggestive of altered gene expression during Pi starvation.

Phosphorus Regulated Gene Expression in Plants

The identification of genes expressed in response to Pi deficiency has increased rapidly in recent years (127). The available data indicate that Pi starvation in plants may lead to a coordinated expression of genes, similar to the *PHO*-regulon in yeast (59). Organisms have developed multifaceted response mechanisms to acquire P from Pi-limiting environments. This response mechanism is much more complex in plants than in unicellular organisms such as bacteria and yeast. Over 100 genes may well be involved in adaptation of plants to Pi-deficient conditions. Many of these genes may have specific roles in enabling plants to acquire and utilize Pi efficiently, whereas others may be involved in regulating the expression of multiple Pi starvation–induced genes. Two of the well-characterized senescence-associated RNase genes (*RNS1* and *RNS2*) from *Arabidopsis* were shown to be induced under Pi starvation (8, 61). Similar genes have been cloned and characterized in tomato and tobacco and are also induced by Pi starvation (34, 82). These RNases are presumed to release Pi from RNA molecules in the extracellular matrix including those derived from other organisms in the rhizosphere or present within cells. In the green alga *Chlamydomonas reinhardtii*, the accumulation of specific mRNAs encoding enolase and pyruvate formate-lyase during Pi starvation was reported (41). Aluminum ions, which interfere with Pi nutrition, or Pi starvation may have similar effects on gene expression (48). The vegetative storage protein gene of soybean (*VspB*) is regulated at the level of transcription by P and sucrose (134). A 130-bp region in the *VspB* promoter mediates both the sucrose and Pi responses, which suggests that sugar-responsive genes are activated, in part, by accumulation of sugar-phosphates and a general reduction of cellular Pi levels (134). Several cDNAs for genes with altered expression levels have been isolated from a 7-day Pi-starved *Brassica* suspension culture (100). One Pi starvation–inducible gene (*psr3*) encodes a polypeptide that is homologous to β-glucosidase, which may play a role in the deglycosylation and regulation of acid phosphatases during Pi stress (101). A tomato gene *TPSI1* that is induced during the early stages of Pi starvation has been cloned and characterized (91, 92). This gene is temporally regulated by changes in Pi concentrations. A similar gene (*Mt4*) has been isolated from *Medicago truncatula* (19, 20). Both of these genes have multiple translation initiation codons and are capable of coding for several short peptides. Expression of *Mt4* is suppressed either by supply of P or by formation of a mycorrhizal association. Furthermore, both *TPSI1* and *Mt4* promoter sequences contain *cis* elements similar to those found in many *PHO*-regulon genes of yeast. A purple acid phosphatase gene isolated from *Arabidopsis* is induced during Pi deficiency (122). There is also evidence for increased Ca^{2+} ATPase transcript accumulation in roots of Pi-starved tomato (112). It is presumed that changes in cytosolic Ca^{2+} activities and their modulation through the activity of specific Ca^{2+}-ATPases play a major role in both response and adaptation of plants to Pi starvation.

Do Plants Have a Gene Regulation System Similar to the Yeast PHO-Regulon?

The complex pattern of induction of a number of enzymes and genes in response to Pi starvation suggests there is a highly regulated molecular network in plants to alleviate Pi deficiency. In yeast, expression of structural genes encoding repressible phosphatases and a high-affinity Pi uptake system is controlled at the transcriptional level by an intricate cascade involving both positive and negative regulatory proteins (74, 119, 120, 152). The *PHO*-regulon of yeast, which responds to changes in Pi concentration, includes genes coding for (*a*) structural proteins such as both acidic and alkaline phosphatases (*PHO5, PHO8, PHO10, PHO11*) and a high-affinity Pi transporter (*PHO84*), (*b*) positive regulators (*PHO2, PHO4, PHO81*), and (*c*) negative regulators (*PHO80, PHO85*). The Pho84 protein appears to be associated with the Pho86, Pho87, and Pho88 gene products (18, 160). Furthermore, the synthesis and activity of the Pho84 transporter is modulated by a GTP-binding protein whose gene is located immediately upstream of *PHO84* (17). Recently, another high-affinity Pi transporter gene, *PHO89*, coding for a sodium coupled Pi transporter has been described (104).

The regulation of this regulon is quite complicated. *PHO4* encodes a transcriptional activator protein that binds and activates each of the regulated *PHO* genes. In addition, the protein product of *PHO2* (*BAS2*), a general transcriptional activator, is also involved in the expression of all *PHO* genes except *PHO8* (142). *PHO85* encodes a protein kinase required for Pho80 repressor function (151). The *PHO80* gene product in association with *PHO85* protein, represses *PHO* gene transcription, probably by hyperphosphorylation of the pho4 activator. Evidence also suggests that nuclear localization of Pho4 protein is altered by changes in P levels. *PHO81*, a positive regulatory gene, encodes a protein that inhibits pho80/pho85 function in the absence of Pi. In the presence of Pi, the *PHO81* product is inactivated and its synthesis is reduced. Under such conditions the pho80/pho85 proteins prevent pho4 and pho2 proteins from activating *PHO* gene transcription. In this cascade, Pi or one of its metabolites appear to act as the direct effector molecule, with Pho81 sensing the Pi levels.

There are several similarities in the responses of higher plants and yeast to Pi starvation. Some of the structural genes, such as Pi transporters and phosphatases, are induced in a similar manner in both types of organisms. Phosphate starvation in plants also results in activation of multiple genes of both known and unknown functions. Many of these genes are repressible by the addition of P (90, 127). A growing number of expressed sequence tags (ESTs) and gene sequences from plant DNA sequencing programs around the world have revealed the presence of genes with sequence similarity to the yeast *Saccharomyces cerevisiae* and *Neurospora crassa PHO* regulon genes. Some of the plant Pi starvation–induced gene promoters have *cis* activation sequences that are similar to those found in yeast genes induced by Pi starvation (20, 90). There is also evidence for increased phosphorylation of specific peptides under Pi starvation in *Brassica napus* cell cultures (21). The evidence for an intricate gene regulation system in plants similar to the yeast *PHO* regulon is growing; however, the complexity of plant morphology and biochemistry point to the existence of regulatory mechanisms in addition to those found in microorganisms.

PHOSPHORUS HOMEOSTASIS

The significance of Pi homeostasis in plant metabolism is appropriately stated by Lauer et al (83): "As Pi is itself a key metabolite, a component of every compound in the Calvin cycle, every metabolite in the sucrose synthesis pathway, and a key part of the adenylates and nucleotides, the maintenance of some minimal level of Pi in the cytosol is critical for maintenance of normal plant metabolism." Plants exhibit a strong tendency to maintain constant cytoplasmic concentrations of ions such as N, P, K, and Cl in spite of large fluctuations in the external concentrations of those ions (57). Many lower plants accumulate and store significant quantities of Pi as polyphosphates, whereas higher plants in general store Pi in the vacuole. In higher plants, the cytoplasmic Pi is maintained at a relatively constant level during short-term nutrient deficiency, at the expense of vacuolar P. Numerous reports have suggested the role of the vacuole in Pi homeostasis in the cytoplasm (13, 83, 84, 108, 110, 129, 137, 153). The cytoplasmic Pi concentration (mM) is generally maintained constant under varying levels of Pi supply, whereas the vacuolar Pi levels decreased significantly under Pi deficiency. ^{31}P NMR studies showed that feeding maize roots with Pi resulted in a >80% increase in vacuolar P, whereas the cytoplasmic P levels remained relatively unchanged (153). In P-sufficient plants most cellular Pi (85 to 95%) is typically found in vacuoles (4, 116).

Transport of Pi Across the Tonoplast

Phosphate homeostasis in the cytoplasm is maintained by Pi transport across the tonoplast activated in response to changing concentrations of P. The transport of Pi across the tonoplast requires ATP and is associated with cytoplasmic alkalization (135, 153). ATP-dependent Pi transport activity across tonoplasts has been demonstrated, providing evidence for the role of vacuoles in Pi homeostasis (108, 138). A transient acidification (0.2 pH units) of the cytoplasm was also observed immediately upon Pi supply to cultured cells (137), due to Pi influx via a H^+/Pi symporter. The excess protons entering the cytoplasm during Pi transport may lead to activation of a biochemical pH stat mechanism during and after Pi uptake (136). At present, the transport mechanism for the tonoplast Pi transporter is unclear; however, the possibility of a Pi-specific channel or a symporter operating at the tonoplast cannot be ruled out. The inward and outward flux of Pi across the tonoplast occurs at high concentrations (mM) of Pi in either the vacuole, cytoplasm, or both, which indicates that high V_{max} transport systems are operating at the tonoplast. The tonoplast H-ATPase or pyrophosphatase could provide required energy to maintain an electrochemical gradient across the tonoplast to facilitate Pi transport. Vacuoles apparently play the dual role of sink and source for Pi in plant cells. Regulation of bi-directional movement of Pi across the tonoplast in response to changing concentrations in the cytoplasm should be a promising field of study in the future.

Regulation of Pi Homeostasis

Plants likely have at least two different signaling mechanisms to maintain Pi homeostasis, one operating at the cellular level and another involving multiple organs and probably arising from shoots. At the cellular level, movement of Pi from and to the vacuole, and regulated plasma membrane efflux and influx are the primary mechanisms to maintain Pi homeostasis. Changes in cytosolic or vacuolar Pi could trigger a signal transduction pathway that activates Pi starvation rescue systems similar to those in microorganisms. However, the coordination of the response mechanism at the whole-plant level is much more complex. When the Pi supply to shoots from roots is restricted under Pi deficiency, stored Pi in older leaves is remobilized and transported to younger growing leaves and other active sinks (14, 107). Given the intricate movement of Pi from old to young tissues, and from root to shoot and back to roots, the transition from a relatively weak sink (leaves) to a strong sink (seeds) will likely have an impact on Pi stress signaling and ultimately Pi uptake by roots. Furthermore, during plant development and under Pi stress, different organs change their role from source to sink, and vice versa. Roots, generally considered a source of Pi for other plant parts, become a sink during Pi starvation. This appears to be a deliberate, adaptive response by the plant to promote root proliferation and thereby enhance soil exploration and Pi uptake. A greater proportion of Pi absorbed by plants may also be retained in the roots under Pi deficiency, to enhance root growth and soil exploration. In mildly starved potato plants exposed to Pi, the proportion of the nutrient transferred to the shoot increased significantly, whereas in severely starved plants most of the Pi was retained in the roots (27).

A considerable amount of carbohydrate also moves to the roots during Pi deficiency. Information about the nutrient status of the shoot is perhaps communicated to the root via some of the carbon assimilates or changes in Pi flux in the phloem (29). Plant hormones are less likely to act as signals, because the effects of ABA and cytokinins on ion transport are rather general in nature (23). Expression of the Pi transporter gene in Pi-sufficient tomato cells did not change significantly in response to NAA or ACC, an ethylene precursor, which suggests that hormones may not be directly involved in regulating Pi uptake (80). However, several reports have implicated an indirect role for ethylene in the Pi-starvation response (36, 99). Studies have shown that ethylene functions as a positive regulator of root hair development in plants (150). Although no direct evidence is available, it is possible that ethylene plays a role as a component of the Pi starvation-induced response by altering root architecture and root hair elongation.

P-efflux is another mechanism documented as regulating Pi concentrations in the cytoplasm (14, 28, 45). At higher external Pi concentrations, increased P-efflux almost compensates for the higher P-influx, thus supporting the hypothesis that under non-limiting Pi availability the Pi homeostasis is primarily controlled by P-efflux. The P-efflux increased with increasing external concentration of Pi, from 28% of influx at 50 μM to 90% at 5 mM Pi (28). In a double isotope (^{33}P and ^{32}P) labeling study the rate of efflux, determined by the appearance of ^{32}P in the solution was approximately tenfold higher for the $^{+}$P treatment (44). Anion channels regulated by pH and/or membrane potentials are likely to be involved in Pi efflux.

CHARACTERISTICS OF Pi UPTAKE BY PLANTS

The very low concentration of available Pi in the soil solution and a high demand for Pi in cells poses a problem unique to plants. The P requirement for optimal growth of plants ranges from 0.3 to 0.5% on a dry matter basis during the vegetative stage of growth (102). The concentration of Pi in cytosol is generally in the millimolar range and, therefore, roots have to acquire Pi against a steep concentration gradient (1000-fold or higher) across the plasma membrane. The concentration-dependent uptake of nutrient ions including Pi has been extensively studied and several uptake models have been proposed (11, 46, 68, 81, 117). A dual uptake model for ions is characterized by a high-affinity transporter operating at low (μM) concentration and a low-affinity transporter functioning at high concentration (mM) of ions (46, 81). A number of studies have shown dual uptake patterns for P in plants (13, 141, 143, 155). A multiphasic uptake model has also been proposed, based on statistical analysis of the ion uptake data from several experiments (117). According to this model, the ion uptake is mediated by a single mechanism that changes its characteristics at certain discrete, external solute concentrations. An alternate uptake model has also been proposed, consisting of a linear uptake component in addition to a dual uptake mechanism (11).

Several comprehensive mathematical models have described the uptake of Pi from soil (5, 6, 55, 60, 147). An interesting comparison of different nutrient supply-uptake models is presented in a review by Amijee et al (3). These mathematical models have considered both soil and plant factors in determining Pi uptake by plants. The model proposed by Barber and colleagues (5, 6, 147) describes nutrient uptake by integrating Pi supply to roots, by mass flow and diffusion, changes in root geometry and size, and Pi uptake in relation to Pi concentration at the root surface, and root competition. Kinetic modeling of ion uptake is still a matter of academic discussion. This area needs to be revisited in view of the recent discoveries of multiple transporters and channels for individual nutrient ions. Multiple transporters for the same ion may help resolve some anomalies in the proposed models. However, it is generally accepted that in the micromolar range of Pi, which corresponds to the normal conditions in cultivated soils, Pi uptake is mediated by a low K_m, high affinity Pi transporter.

The transport of inorganic Pi across the plasma membrane has been extensively studied in roots of higher plants and cultured cells (15, 21, 24, 27, 37, 38, 52, 86, 108, 135, 143, 153, 155, 156). As several different forms of Pi exist ($H_2PO_4^-$, $H_2PO_4^{-2}$, PO_4^{3-}), it has been found that the dihydrogen form of the orthophosphate ion ($H_2PO_4^-$) is the Pi species most readily transported into plant cells. An energy-mediated cotransport process, driven by protons generated by a plasma membrane H^+ATPase, has been proposed for Pi uptake in plants (137, 155, 156). Pi uptake results in alkalization of the media, consistent with the hypothesis that Pi uptake is associated with cotransport of more than one proton. Phosphate absorption is accompanied by H^+ influx with a stoichiometry of 2 to 4 H^+/H_2PO_4 transported (135, 155, 156). The number of protons increases as the concentration of Pi in the media decreases; furthermore, the required protons may be supplied by the media or activated proton pumps in the membrane. Also observed are a transient decrease in cytoplasmic pH due to proton:PO_4 cotransport, and transient membrane depolarization (due to uptake of 2 to 4 H^+ with every $H_2PO_4^-$) followed by a repolarization of the membrane by the plasma membrane H^+-ATPase that pumps protons from cytoplasm to maintain cellular pH and to provide the proton motive force to drive continued Pi uptake (16, 155, 156). The Pi pump is presumed to contribute about 35% of total cell PD at a normal Pi uptake rate (16). It should be noted that the metabolic inhibitors that dissipate the proton gradients across the membranes also suppress Pi uptake. The available Pi in soil solution is generally lower than the reported K_m (around 5 μM) for the high affinity transports (43, 75, 143, 146, 155). Furthermore, the cell wall chemistry also has a strong influence on the concentration of Pi and pH in the vicinity of plasma membrane (23, 143). Owing to interactions of the anion with cell walls, the actual concentration of Pi at the plasma membrane may be ten times lower than in the external solution.

Effect of Pi Starvation on Uptake

There is overwhelming evidence supporting enhanced uptake of Pi by roots and cultured cells in response to Pi starvation (26, 38, 85). Both uptake and translocation of Pi to shoots increase significantly following the exposure of Pi-starved plants to P. It is hypothesized that enhanced uptake could be due to increased synthesis of a carrier system under Pi deprivation (38). A distinct increase in Pi uptake in cultured cells following Pi starvation was attributed primarily to the greater number of high-affinity Pi transporters (146). In general, the low-affinity transport system appears to be expressed constitutively in plants, whereas the high-affinity uptake system is regulated by the availability of Pi (52). However, Pi uptake studies in the water fern *Azolla* showed a modest increase in both K_m and V_{max} of the low-affinity transport system (15). In plants, uptake of Pi under nonlimiting concentrations appears to be regulated by the internal levels of P. In some plant species, resupply of Pi following a brief period of Pi starvation resulted in accumulation of toxic levels of Pi in the leaves (26). Toxic effects are mainly due to excess uptake and transfer of Pi in the Pi-starved plants exposed to relatively high concentrations of the nutrient. A similar response was observed when plants adapted to grow in alkaline calcareous soils with low Pi availability were cultured in an acidic medium (115). This is a highly artificial situation. Under natural conditions, plants are seldom exposed to the unrestricted supply of Pi provided under experimental conditions. This response also suggests that the relative insensitivity of Pi uptake in some plants may be due to slow turnover of the Pi transporters (24).

PLANT Pi TRANSPORTERS

Phosphate transport in plants exhibits distinctive uptake kinetic properties at low (μM) and high (mM) concentration of Pi in the external media. High-affinity Pi transporters are membrane-associated proteins translocating Pi from an external media containing μM concentrations to the cytoplasm. At present, the nature and function of the low-affinity (high K_m) Pi transporter is not clearly understood. The availability of *Arabidopsis* ESTs has led to the discovery of several high-affinity Pi transporters (30, 77, 88, 90, 93, 95, 113, 148). The isolated genes have been used to complement a yeast mutant (pho84) deficient in high-affinity Pi uptake (30, 77, 88, 95, 113), and enhance the uptake of Pi by cultured tobacco cells grown in a nutrient-limiting medium (111). Phosphate transporter genes are encoded by a small family of genes in the genome (113). In *Arabidopsis* six high-affinity Pi transporter genes have been isolated

(113, 118, 148). These genes are called by different names and hence, it is necessary to clarify this situation: *AtPT1* = *APT2* = *PHT1*; *APT1* = *PHT2*; *AtPT2* = *PHT4*; *AtPT4* = *PHT3*; *PHT5* and *PHT6* have not been described under any other name. Four genes (*PHT1,2,3,* and *5*) are clustered together in a 25-kb region of chromosome V (118). *PHT4* and *PHT6* are located on chromosome II. The two *Arabidopsis* Pi transporter genes *APT1* and *APT2* (*AtPT1*) have different promoter sequences but nearly identical coding regions (148). None of the cloned Pi transporters is allelic with *pho1* and *pho2* mutants of *Arabidopsis* (118).

Structure of Plant Pi Transporters

A comparison of the deduced amino acid sequences and hydropathy plots of the plant, yeast, *Neurospora*, and the mycorrhizal fungi (*Glomus versiforme*) Pi transporters shows significant structural similarity (113). All the cloned Pi transporters are integral membrane proteins consisting of 12 membrane-spanning regions, separated into two groups of six by a large hydrophilic charged region (Figure 1), a common feature shared by many proteins involved in transport

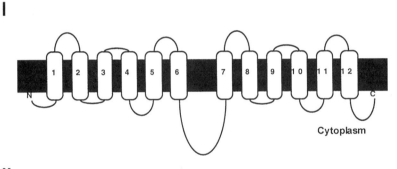

% Identity	LePT1	LePT2	AtPT1	AtPT2	STPT1	STPT2	MtPT1	MtPT2	PIT1
LePT1	100.0	80.5	78.4	82.9	95.5	78.0	76.5	77.3	86.4
LePT2	80.5	100.0	77.2	75.9	80.8	95.0	80.7	81.2	76.8

Figure 1   I. A model of plant Pi transporter containing 12 membrane-spanning domains. II. Amino acid identity (%) among Pi transporters isolated from tomato (*LePT1* and *LePT2*), *Arabidopsis thaliana* (*AtPT1* and *AtPT2*), potato (*StPT1* and *StPT2*), *Medicago truncatula* (*MtPT1* and *MtPT2*), and *Catharanthus roseus* (*PIT 1*).

of sugars, ions, antibiotics, and amino acids (63, 67). Plant Pi transporters are members of subfamily 9 of the Major Facilitator Super family (MFS) (121). The MFS represent one of the largest families of single polypeptide-facilitated carriers capable of transporting small solutes utilizing chemiosmotic ion gradients. This family includes carriers of sugar, nucleosides, Pi, organo phosphate, and other simple molecules. Phosphate transporters, members of MFS, have an interesting phylogenetic origin (121): They may have evolved by a tandem intragenic duplication process. The original structural unit of a six membrane-spanning protein has duplicated to form an active 12 membrane-spanning Pi transporter. The first half of the peptide has several conserved motifs and sequences that are also found in the second half of the protein. Interestingly, the Pi:H symporters have been isolated and characterized only from fungi and plants. The fungal and plant Pi transporters are of similar size (518–587 amino acids) and utilize a H^+ gradient to drive the symport process. Many of the potential phosphorylation and N-glycosylation sites are conserved between plant and fungal polypeptides. The remarkable similarity in structure and function of Pi transporters indicates that these are among the oldest proteins conserved during evolution. Phylogenetically, the Pi transporters isolated from plants and fungi belong to a closely related family, even though the similarity between the plant transporters is significantly higher than that between plant and fungal transporters (113).

Complexity of Pi Transport in Plants

The existence of multiple high-affinity Pi transporters in plants reflects complexity of the transport process. The Pi acquired by roots is rapidly loaded into the xylem, although the nature of the transporter involved in xylem loading of Pi is not clear. The existence of a specific xylem loading transporter has been suggested by a well-characterized *Arabidopsis* mutant (*pho1*) that lacks in the ability to load Pi into the xylem (126). The transfer of Pi to shoots is drastically reduced in this mutant, and leaf Pi content is approximately 5% of that of wild-type plants. The deficiency is caused by a single nuclear recessive mutation at the *pho1* locus of *Arabidopsis*. There is also genetic evidence for the involvement of Pi transporter(s) in phloem loading of the nutrient. The well-characterized *pho2* mutant of *Arabidopsis* shows higher levels of Pi in leaves owing to defective regulation of the amount of Pi loaded into shoots (32). The *pho2* phenotype may be a result of altered function of Pi transporter(s) in shoot or any of the regulatory genes involved in Pi homeostasis in the shoot. Phloem loading and unloading is an important component of the internal movement of Pi in plants. In plants such as soybean, most of the Pi utilized during grain filling originates in the leaf. Significant amounts of Pi also move in a basipetal direction during Pi deficiency. As knowledge about plant genome structure increases, the isolation and characterization of many other tissue-specific Pi transporters should be feasible in the near future.

Functional Characterization of Pi Transporters

Phosphate transporters isolated from *Arabidopsis*, tomato, *Catharanthus roseus*, potato, and *Medicago truncatula* have been used to complement yeast mutants lacking the high-affinity Pi transport mechanism (30, 77, 88, 93, 113). The tomato Pi transporter, *LePT1*, was functionally analyzed by complementing a yeast mutant (PM971) lacking the high-affinity Pi transport mechanism (30). Expression of *LePT1* in the mutant restored growth in medium containing 20 μM Pi, supporting the role of *LePT1* as a Pi transporter. Uptake studies with ^{32}P-orthophosphate showed that Pi uptake in yeast cells complemented with *LePT1* was nearly seven times higher than in the uncomplemented mutant control. The Pi uptake shows a strong dependence on pH, and the uncouplers (DNP and CCCP), which dissipate pH-gradients across membranes, strongly depress Pi uptake. Plant Pi transporters have a pH optimum of 4.5 to 5.0 in the yeast expression system (30, 88), which is comparable to the pH range for optimal Pi uptake in plants (52, 156). This pH optimum indicates preferential uptake of $H_2PO_4^-$ over HPO_4^{2-} by the transporters. Tobacco cell cultures expressing *Arabidopsis* Pi transporters also responded similarly to the addition of protonophores, arsenate, and a plasma membrane H^+-ATPase inhibitor (111). These results support the hypothesis that Pi transport is driven by the pH gradient across the plasma membrane (156). The apparent K_m for *LePT1* in yeast was 30 μM, which is much higher than the K_m reported for the high-affinity Pi transporters in plants. Similar data were also obtained in complementation studies with high-affinity Pi transporters of potato and *Medicago* (88, 93). The higher K_m observed in these experiments is not surprising, considering that complementation of yeast was done with a heterologous transporter. Moreover, the activity of the native Pi transporter pho84 in yeast is regulated by interactions with other proteins in the membrane (18, 160). These proteins may not interact properly with the plant transporter, and the apparent K_m obtained from expression in yeast may not necessarily reflect the K_m of the native transport proteins in plants. This is further supported by the observed K_m value of 3.1 μM in tobacco cells transformed with *Arabidopsis* Pi transporter (111), which is similar to the reported K_m in whole-plant and cell-culture studies. Overexpression of the *Arabidopsis* Pi transporter gene resulted in increased uptake of labeled P and increased fresh weight under low Pi nutrition. These results strongly support the notion that even the constitutive expression of Pi transporters can provide a distinct advantage to cells in obtaining the nutrient under Pi-limiting conditions. The role of Pi transporters was further confirmed by antisense expression of the potato Pi transporters *StPT1* and *StPT2* (G Leggewie, personal communication). Antisense expression of these genes resulted in 40–60% reduction in the message levels and was accompanied by distinct phenotypic changes. The antisense plants showed a loss of apical dominance and bushy growth. Furthermore, early onset of tuberization, a well-characterized symptom of Pi deficiency, was also observed in the antisense plants. The differences between wild-type and transgenic plants were less pronounced under Pi-sufficient conditions. These results provide strong evidence for the role of the cloned Pi transporter genes in the acquisition of this nutrient.

EXPRESSION AND REGULATION OF Pi TRANSPORTERS

Southern blot analysis of genomic DNA isolated from tomato, potato, and *Arabidopsis* indicated that cloned Pi transporters are represented by a small family of genes in the genome (88, 90, 93, 113, 118, 148). The existence of several genes encoding Pi transporters in *Arabidopsis* has been reported (111, 113, 118, 148). Some of these transporters may be temporally and spatially controlled in plants exposed to varying levels of available Pi in the rhizosphere. Phosphate transporters are preferentially expressed in the roots of Pi-starved plants (30, 90, 113). Although six different Pi transporter genes have been isolated from *Arabidopsis thaliana*, RNA transcripts can be detected for only two of the transporters (*AtPT1* and *AtPT2*) on Northern blots (118). The tomato Pi transporters (*LePT1* and *LePT2*) are induced in a temporal and concentration-mediated manner in roots and cell cultures. An increase in Pi transport activity well in advance of any significant effect of Pi deficiency on growth is likely due to an increase in the number of Pi transporters in plants. The induction of Pi transporter mRNA in cell cultures was evident 3 to 6 h after transfer to Pi minus media (80). In plants, both the RNAs and proteins were induced within 24 h of initiation of Pi starvation (90, 114). A rapid activation of the expression of Pi transporters under Pi-limiting conditions is likely the reason for the observed fivefold increase in V_{max} without any apparent changes in K_m (146). Uptake studies with barley (38) and maize (4) provide further evidence for increased V_{max} during Pi starvation.

The results of many Pi uptake experiments confirm that plants adjust their Pi uptake based on their internal Pi status, particularly by increasing I_{max} (maximum influx), whereas changes in K_m and C_{min} are of minor importance in this process (75). Although allosteric regulation of Pi uptake under Pi starvation is suggested (87), it is becoming evident that the capacity for Pi uptake is regulated by increasing the total number of transporter molecules (127). Induction of Pi transporter gene expression is not only a rapid response to Pi starvation but is also reversible upon replenishment of P (90). Many Pi uptake studies showed a similar response (78, 146). Transferring tobacco cells to a Pi-deficient medium led to a gradual increase in the V_{max}, reaching the maximum after 20 h. When these Pi-starved cells were replenished with P, the uptake rate decreased gradually to reach the normal uptake rates after 5 h (146). Similarly, enhanced Pi uptake rates in P-starved tomato plants returned to normal 30 h after Pi replenishment (78), indicating the down-regulation of the uptake process.

The requirement for de novo protein synthesis (146) and the concurrent accumulation of the transcripts and Pi transport protein in the plasma membranes provide strong evidence for the transcriptional regulation of Pi uptake in plants (114). Furthermore, the genes are induced as a specific response to Pi starvation, which supports the earlier observation that stimulation of ion influx is highly selective for the ion previously in short supply (37, 85). Transcriptional activation of Pi transporters in response to Pi starvation seems to be a major mechanism of regulation of Pi uptake. Further studies should reveal the interaction of Pi transporters with other proteins, as observed in regulation of the yeast high-affinity Pi transporter (18, 160).

Split Root Studies

Split root experiments have been used effectively to delineate physiological and biochemical changes in plants grown under different nutrient regimes. Experiments with divided root systems have revealed that shoot Pi levels play a role in regulating its uptake by roots (4, 27, 38). Uptake and transport of nutrients by a single seminal root of barley was stimulated significantly following a brief period of Pi starvation to the remainder of the roots (37, 38). The enhanced uptake was associated with increased transport of Pi to the shoot and to the remainder of the root system. These observations support the hypothesis that internal Pi status of a plant can regulate the rates of Pi uptake by roots and the distribution of Pi within the plant. The rate of uptake may be influenced by the demand created by growth of plant organs and enhanced loading of Pi into the xylem. Phosphorus starvation not only results in increased uptake but also in stimulated radial flow and enhanced efflux from the symplast into xylem vessels (23). In tomato plants where part of the root system was exposed to Pi while another portion was exposed to −Pi solute, the Pi transporter mRNA levels remained comparable to those of Pi-sufficient plants. These data suggest that even if Pi is supplied to a portion of the root system, expression of Pi transporters in other portions of roots exposed to Pi-deficient conditions do not increase as compared to Pi-deficient plants. Once the Pi requirements of plants are satisfied, the level of Pi transporter gene expression is down-regulated in the entire root system. A coordinated regulation of Pi transporter expression in different parts of the roots by internal Pi levels should allow plants to obtain the required amount of the nutrient without leading to toxic effects, and retain valuable energy resources. Similar regulation of Pi uptake in barley roots by internal Pi concentration has also been proposed (38, 87). The divided root studies support the existence of internal signaling mechanisms that lead to an appropriate physiological response. This hypothesis is further supported by uptake studies in the *pho2 Arabidopsis* mutant, a hyperaccumulator of Pi in shoots (35). Pi uptake in *pho2* was about twofold greater than in wild-type plants and much of the acquired Pi was found in shoots. The removal of shoots resulted in normal uptake rates comparable to the rates in roots of wild types (35). Interestingly, the expression levels of two high-affinity Pi transporters were similar in both wild-type and *pho2* mutant (148). These genetic data clearly suggest that signals originating in shoots control Pi uptake by roots. The nature of these signals is not yet clear; however, the genes activated during Pi stress are increasingly being cloned and characterized; it is likely that some of these gene products may play a role in the signal transduction pathway during Pi starvation.

Tissue-Specific Expression of Pi Transporters

Most of cloned Pi transporter genes are preferentially expressed in roots under Pi starvation (88, 90, 113). However, *LePT1* and *StPT1*, representing tomato and potato Pi transporters, respectively, are also expressed in other plant organs including leaves, stem, tubers, and flowers (88, 90). The spatial distribution of the Pi transporter gene expression indicates their function beyond the acquisition of Pi from soil. Some transporters may be involved in the intercellular movement of Pi or scavenging Pi leaked into the apoplast. Tomato Pi transporter genes (*LePT1* and *LePT2*) are expressed predominantly in the root epidermis and root hairs of Pi-deficient plants (30, 90). The *LePT1* message was also observed in the root cortical cells, palisade parenchyma, and vascular bundles in leaves. The epiderm-specific expression of these Pi transporters supports their role in acquiring Pi from the rhizosphere. In a Pi-limiting environment, the epidermis is exposed to a relatively high concentration of Pi compared to other internal cell layers of the root. As discussed earlier, the presence of cell walls drastically affects Pi fluxes and local concentrations of the nutrient in the vicinity of the transporters. Exclusion of anions in general and Pi in particular by cell walls results in a decrease in Pi concentration by one to two orders of magnitude near cell membranes. As cells sense the local Pi concentrations around the membranes (143), the concentration of Pi around the epidermis is relatively higher than that near internal cell layers. Epidermal expression of high-affinity Pi transporters is indicative of specific targeting of the transporters in cell layers exposed to relatively higher concentrations of the nutrient to drive the uptake process. Since the contribution of root hairs to Pi acquisition is significant, particularly in Pi-deficient soils, expression of Pi transporters in root hairs also results in better exploration of the rhizosphere and rapid uptake of P. An interesting observation showed that the root system as a whole retains the potential to transport Pi at an increased rate, despite anatomical and physiological changes associated with aging (38). An increased influx of Pi following a period of starvation is evident in the older root zone close to the shoot, and it spreads along the root toward the apical region (25). Immunological studies with antibodies specific to tomato Pi transporter revealed the accumulation of Pi transporters in all parts of the roots following Pi starvation (US Muchhal & KG Raghothama, unpublished data). Detailed in situ hybridization and immunolocalization studies of different members of the Pi transporter family should help in understanding the complexity of P nutrition in higher plants.

Phosphate Transporter Gene Expression under Natural Conditions

Most gene expression and Pi uptake studies reported in the literature are conducted under controlled experimental conditions, wherein the plants were either uniformly or partially depleted of P. Moreover, the concentrations of Pi used in these studies (250 μM or above) are much higher than the concentrations of available Pi in most soils. These situations may arise in soils immediately following the application of fertilizers or excessive manure, as seen in agriculture in the temperate regions. Given that Pi deficiency is the norm under natural conditions, the level of Pi transporter expression and protein may always be high, and repression of gene expression is likely to be the regulating factor of Pi acquisition. An examination of gene expression in roots of plants grown in soils with differing Pi fixation/buffering capacity is required to confirm this hypothesis. Furthermore, under natural conditions the rate of plant growth is normally adjusted to nutrient availability. A reduced rate of growth leads to redistribution of Pi within the plant, which allows plants to overcome brief periods of Pi starvation without a major effect on growth. The association of plant roots with microorganisms such as Pi solubilizing bacteria and mycorrhizal fungi should help in obtaining an adequate amount of P. Mycorrhizal symbiosis should have a significant effect on the expression of plant Pi transporters. If they are active in acquiring P, then plant Pi transporter gene expression is likely to be suppressed. RNA analyses show that colonization of roots by mycorrhizal fungi and expression of the fungal transporters lead to repression of the plant Pi transporters in *Medicago truncatula* (93). A similar pattern of down-regulation of gene expression was also observed for another Pi starvation–induced gene, Mt4, in *Medicago* (19). These data indicate a highly coordinated interaction between plant roots and mycorrhizal fungi in obtaining and transferring Pi to plants in the natural ecosystem.

CONCLUDING REMARKS

During the 1990s, significant progress has been made in understanding the molecular regulation of processes associated with Pi acquisition and Pi starvation responses. Pi acquisition and homeostasis are complex biological processes that need to be examined in greater detail. In general, Pi nutrition is controlled by two major factors, nutrient availability and the ability of plants to acquire the limiting nutrient. Recently developed molecular tools have allowed researchers to start to dissect ion transport processes. However, understanding how plants adapt to Pi stress remains a challenging task and an open area of research. Elucidating how plants sense and respond to available P in the rhizosphere is also an exciting area that is amenable to more research. This may lead to better understanding of the complex physiological and biochemical responses observed in plants under varying Pi availability. While this is now an exciting period for plant nutrition, the next millennium will be exhilarating. The advent of functional genomics, sequence-tagged reverse genetics, gene-chip technologies, and complete sequencing of the first plant (*Arabidopsis*) genome in the near future will open up research opportunities to dissect the signal transduction pathway leading to transcriptional regulation of Pi acquisition and Pi starvation rescue mechanism in plants. The emerging research area of functional and structural genomics will allow targeted gene expression, serving as molecular determinants of Pi acquisition and adaptation to Pi stress, to specific tissues under the control of regulated promoters. The most promising areas in which manipulation of nutrient absorption could improve the efficiency of crop production in the near future involve the acquisition of P, Fe, and Mn (94). These genetically modified plants will have the potential to respond better to applied fertilizers and perform better under nutrient-limiting conditions. They will also serve as valuable genetic resources for understanding the fundamental mechanisms by which plants adapt to nutrient stress in the environment.

ACKNOWLEDGMENTS

Research in my laboratory is supported by a grant from the United States Department of Agriculture. I thank Drs. Plaxton and Leggewie for providing a manuscript in press and sharing some of the unpublished results. I appreciate the support of the members of my laboratory. My sincere thanks are due to my colleagues Drs. Peter Goldsbrough, Jules Janick, and Robert Schaffert for critical reading of the manuscript. I apologize to the colleagues whose work is not directly cited because of space limitations.

Visit the *Annual Reviews home page* at
http://www.AnnualReviews.org

Literature Cited

1. Ae N, Arihara J, Okada K. 1991. Phosphorus uptake mechanisms of pigeon pea grown in alfisols and vertisols. See Ref. 72a, pp. 91–98

2. Ae N, Arihara J, Okada K, Yoshihara T, Johansen C. 1990. Phosphorus uptake by pigeon pea and its role in cropping systems of the Indian subcontinent. *Science* 248:477–80

3. Amijee F, Barroclough PB, Tinker PB. 1991. Modeling phosphorus uptake and utilization by plants. See Ref. 72a, pp. 62–75

4. Anghinoni I, Barber SA. 1980. Phosphorus influx and growth characteristics of corn roots as influenced by phosphorus supply. *Agron. J.* 172:655–68

5. Barber SA, Cushman J. 1981. Nitrogen uptake model for agronomic crop. In *Modeling Waste Water Renovation-Land Treatment*, ed. IK Iskander, pp. 382–409. New York: Wiley

6. Barber SA, Silverbush M. 1984. Plant root morphology and nutrient uptake. In *Roots, Nutrient and Water Influx and Plant Growth*, pp. 65–88. Madison, WI: ASA Spec. Publ. 49

7. Barber SA, Walker JM, Vasey EH. 1963. Mechanisms for the movement of plant nutrients from the soil and fertilizer to the plant root. *J. Agric. Food Chem.* 11:204–7

8. Bariola PA, Howard CJ, Taylor CP, Verburg MT, Jaglan VD, Green PJ. 1994. The *Arabidopsis* ribonuclease gene *RNS1* is tightly controlled in response to phosphate limitation. *Plant J.* 6:673–85

9. Barrow NJ. 1980. Evaluation and utilization of residual phosphorus in soils. In *The Role of Phosphorus in Soils*, ed. FE Khasawneh, EC Sample, EJ Kamprath, pp 335–55. Madison, WI: Am. Soc. Agron.

10. Bates TR, Lynch JP. 1996. Stimulation of root hair elongation in *Arabidopsis thaliana* by low phosphorous availability. *Plant Cell Environ.* 19:529–38

11. Berstlap AC. 1983. The use of model-fitting in the interpretation of 'dual' uptake isotherms. *Plant Cell Environ.* 6:407–16

12. Biddinger EC, Liu C, Joly RJ, Raghothama KG. 1998. Physiological and molecular responses of aeroponically grown tomato plants to phosphorus deficiency. *J. Am. Soc. Hortic.* 123:330–33

13. Bieleski RL. 1973. Phosphate pools, phosphate transport, and phosphate availability. *Annu. Rev. Plant Physiol.* 24:225–52

14. Bieleski RL, Ferguson IB. 1983. Physiology and metabolism of phosphate and its compound. In *Encyclopedia of Plant Physiology; Inorganic Plant Nutrient*, ed. A Lauchi, RL Bieleski, pp. 422–29. Berlin: Springer-Verlag

15. Bieleski RL, Lauchli A. 1992. Phosphate uptake, efflux and deficiency in the water fern *Azolla. Plant Cell Environ.* 15:665–73

16. Bowling DJF, Dunlop J. 1978. Uptake of phosphate by white clover. 1. Evidence for an electrogenic phosphate pump. *J. Exp. Bot.* 29:1139–46

17. Bun ya M, Harashima S, Oshima Y. 1992. Putative GTP-binding protein, *Gtr1*, associated with the function of the Pho84 inorganic phosphate transporter in *Saccharomyces cerevisiae. Mol. Cell Biol.* 12:2958–66

18. Bun ya M, Shikata K, Nakade S, Yompakdee C, Harashima S, Oshima Y. 1996. Two new genes, *PHO86* and *PHO87*, involved in inorganic phosphate uptake in *Saccharomyces cerevisiae. Curr. Genet.* 29:344–51

19. Burleigh SH, Harrison MJ. 1997. A novel gene whose expression in *Medicago truncatula* roots is suppressed in response to colonization by vesicular-arbuscular mycorrhizal (VAM) fungi and to phosphate nutrition. *Plant Mol. Biol.* 34:199–208

20. Burleigh SH, Harrison MJ. 1998. Characterization of the Mt4 gene from *Medicago truncatula. Gene* 216:47–53

21. Carswell MC, Grant BR, Plaxton WC. 1997. Disruption of the phosphate-starvation response of oilseed rape suspension cells by the fungicide phosphonate. *Planta* 203:67–74

22. CIAT (International Center for Tropical Agriculture). 1987. *CIAT Annual Report 1987.* Cali, Colombia: CIAT

23. Clarkson DT, Grignon C. 1991. The phosphate transport system and its regulation in roots. See Ref. 72a, pp. 49–62

24. Clarkson DT, Hawkesford MJ, Davidian JC, Grignon C. 1992. Contrasting responses of sulfate and phosphate transport in barley roots to protein-modifying reagents and inhibition of protein synthesis. *Planta* 187:306–14

25. Clarkson DT, Sanderson J, Scattergood CB. 1978. Influence of phosphate stress

on phosphate absorption and translocation by various parts of the root system of *Hordeum vulgare* L (barley). *Planta* 139:47–53

26. Clarkson DD, Scattergood CB. 1982. Growth and phosphate transport in barley and tomato plants during development of, and recovery from, phosphate stress. *J. Exp. Bot.* 33:865–75

27. Cogliatti DH, Clarkson DT. 1983. Physiological changes in, and phosphate uptake by potato plants during development of, and recovery from phosphate deficiency. *Physiol. Plant.* 58:287–94

28. Cogliatti DH, Santa Maria GE. 1990. Influx and efflux of phosphorus in roots of wheat plants in non-growth-limiting concentrations of phosphorus. *J. Exp. Bot.* 41:601–7

29. Cooper HD, Clarkson DT. 1989. Cycling of amino nitrogen and other nutrients between shoots and roots in cereals; a possible mechanism integrating shoot and root in the regulation of nutrient uptake. *J. Exp. Bot.* 40:733–62

30. Daram P, Brunner S, Amrhein N, Bucher M. 1998. Functional analysis and cell-specific expression of a phosphate transporter from tomato. *Planta* 206:225–33

31. Darrah PR. 1993. The rhizosphere and plant nutrition: a quantitative approach. In *Plant Nutrition—From Genetic Engineering to Field Practices*, ed. NJ Barrow, pp: 3–22. Dordrecht: Kluwer

32. Delhaize E, Randall PJ. 1995. Characterization of a phosphate-accumulator mutant of *Arabidopsis thaliana*. *Plant Physiol.* 107:207–13

33. Dinkelaker B, Hengeler C, Marschner H. 1995. Distribution and function of proteoid roots and other root clusters. *Bot. Acta* 108:183–200

34. Dodds PN, Clarke AE, Newbigin E. 1996. Molecular characterization of the S-like RNase of *Nicotiana alata* that is induced by phosphate starvation. *Plant Mol. Biol.* 31:227–38

35. Dong B, Rengel Z, Delhaize E. 1998. Uptake and translocation of phosphate by *pho2* mutant and wild-type seedlings of *Arabidopsis thaliana*. *Planta* 205: 251–56

36. Drew MC, He CJ, Morgan PW. 1989. Decreased ethylene biosynthesis, and induction of aerenchyma, by nitrogen- or phosphate-starvation in adventitious roots of *Zea mays* L. *Plant Physiol.* 91:266–71

37. Drew MC, Saker LR. 1978. Nutrient supply and growth of the seminal root

system in barley. III. Compensatory increases in growth of lateral roots and in rates of phosphate uptake in response to a localized supply of phosphate. *J. Exp. Bot.* 29:435–51

38. Drew MC, Saker LR. 1984. Uptake and long-distance transport of phosphate, potassium and chloride in relation to internal ion concentrations in barley: evidence of non-allosteric regulation. *Planta* 160:500–7

39. Duff SMG, Moorhead GBG, Lefebvre DD, Plaxton WC. 1989. Phosphate starvation inducible "bypasses" of adenylate and phosphate dependent glycolytic enzymes in *Brassica nigra* suspension cells. *Plant Physiol.* 90:1275–78

40. Duff SMG, Sarath G, Plaxton WC. 1994. The role of acid phosphatase in plant phosphorus metabolism. *Physiol. Planta.* 90:791–800

41. Dumont F, Joris B, Gumusboga A, Bruyninx M, Loppes R. 1993. Isolation and characterization of cDNA sequences controlled by inorganic phosphate in *Chlamydomonas reinhardtii*. *Plant Sci.* 89:55–67

42. Dumont F, Lopes R, Kremers P. 1990. New polypeptides and *in vitro* translatable mRNAs are produced by phosphate starved cells of the unicellular algae *Chlamydomonas reinhardtii*. *Planta* 182:610–16

43. Dunlop J, Phung HT, Meeking R, White DWR. 1997. The kinetics associated with phosphate absorption by *Arabidopsis* and its regulation by phosphorus status. *Aust. J. Plant Physiol.* 24:623–29

44. Dunlop J, Phung T. 1998. *Efflux and influx as factors determining the relative abilities of ryegrass and white clover to compete for phosphate.* Presented at Int. Symp. Genet. Mol. Biol. Plant Nutr., 6th, Elsinore, Denmark. Abstr.

45. Elliott G, Lynch J, Lauchi A. 1984. Influx and efflux of P in roots of intact maize plants. *Plant Physiol.* 76:336–41

46. Epstein E, Rains DW, Elzam OE. 1963. Resolution of dual mechanisms of potassium absorption by barley roots. *Proc. Natl. Acad. Sci. USA* 49:684–92

47. Deleted in proof

48. Ezaki B, Yamamoto Y, Matsumoto H. 1995. Cloning and sequencing of the cDNAs induced by aluminum treatment and Pi starvation in cultured tobacco cells. *Physiol. Plant.* 93:11–18

49. Fife CA, Newcomb W, Lefebvre DD. 1990. The effect of phosphate deprivation on protein synthesis and fixed

carbon storage reserves in *Brassica nigra* suspension cells. *Can. J. Bot.* 68:1840–47

50. Fohse D, Claassen N, Jungk A. 1988. Phosphorus efficiency of plants I. External and internal P requirement and P uptake efficiency of different plant species. *Plant Soil* 110:101–9

51. Fohse D, Classen N, Jungk A. 1991. Phosphorus efficiency of plants. II. Significance of root radius, root hairs and cation-anion balance for phosphorus influx in seven plant species. *Plant Soil* 132:261–72

52. Furihata T, Suzuki M, Sakurai H. 1992. Kinetic characterization of two phosphate uptake systems with different affinities in suspension-cultured *Catharanthus roseus* protoplasts. *Plant Cell Physiol.* 33:1151–57

53. Gahoonia TS, Nielsen NE. 1998. Direct evidence on participation of root hairs in phosphorus (^{32}P) uptake from soil. *Plant Soil* 198:147–52

54. Gardner WK, Parberry DG, Barber DA. 1982. The acquisition of phosphorus by *Lupinus albus* L. I. Some characteristics of the soil/root interface. *Plant Soil* 68:19–32

55. Geelhoed JS, Mous SLJ, Findenegg GR. 1997. Modeling zero sink nutrient uptake by roots with root hair from soil: comparison of two models. *Soil Sci.* 162:544–53

56. Gilbert GA, Allan DL, Vance CP. 1997. Phosphorus deficiency in white lupin alters root development and metabolism. In *Radical Biology: Advances and Perspectives on the Function of Plant Roots*, ed. HE Flores, JP Lynch, D Eissenstat, pp. 92–103. Rockville, MD Am. Soc. Plant Physiol.

57. Glass ADM, Siddiqi MY. 1984. The control of nutrient uptake rates in relation to the inorganic composition of plants. *Adv. Plant Nutr.* 1:103–47

58. Goldstein AH. 1992. Phosphate starvation inducible enzymes and proteins in higher plants. In *Society for Experimental Biology Seminar Series 49: Inducible Plant Proteins*, ed. JL Wray, pp. 25–44. Cambridge: Cambridge Univ. Press

59. Goldstein AH, Mayfield SP, Danon A, Tibbot BK. 1989. Phosphate starvation inducible metabolism in *Lycopersicon esculentum. Plant Physiol.* 91:175–82

60. Grant RF, Robertson JA. 1997. Phosphorus uptake by root systems: mathematical modeling in *ecosys. Plant Soil* 188:279–97

61. Green PJ. 1994. The ribonucleases of higher plants. *Annu. Rev. Plant Physiol. Plant Mol. Biol.* 45:421–45

62. Grierson PF. 1992. Organic acids in the rhizosphere of *Banksia interifolia* L.F. *Plant Soil* 44:259–65

63. Griffith JK, Baker ME, Rouch DA, Page MGP, Skurray RA, et al. 1992. Membrane transport proteins: implications of sequence comparisons. *Curr. Opin. Cell Biol.* 4:684–95

64. Grinsted MJ, Hedley MJ, White RE, Nye PH. 1982. Plant-induced changes in the rhizosphere of rape (*Brassica napus* var. Emerald) seedlings. I. pH change and the increase in P concentration in the soil solution. *New Phytol.* 91:19–29

65. Harrison MJ, van Buuren ML. 1995. A phosphate transporter from the mycorrhizal fungus *Glomus versiforme. Nature* 378:626–29

66. Hawkesford MJ, Belcher AR. 1991. Differential protein synthesis in response to sulfate and phosphate deprivation: identification of possible components of plasma-membrane transport systems in cultured tomato roots. *Planta* 185:323–29

67. Henderson PJF. 1993. The 12-transmembrane helix transporters. *Curr. Opin. Cell Biol.* 5:708–21

68. Hodges TK. 1973. Ion absorption by plant roots. *Adv. Agron.* 25:163–207

69. Hoffland E, Boogaard RVD, Nelemans J, Findenegg G. 1992. Biosynthesis and root exudation of citric and malic acids in phosphate-starved rape plants. *New Phytol.* 122:675–80

70. Hoffland E, Findenegg GR, Nelemans JA. 1989. Solubilization of rock phosphate by rape. *Plant Soil* 113:155–60

71. Holford ICR. 1997. Soil phosphorus: its measurement, and its uptake by plants. *Aust. J. Soil Res.* 35:227–39

72. Jackson RB, Manwaring JH, Caldwell MM. 1990. Rapid physiological adjustment of roots to localized soil enrichment. *Nature* 344:58–60

72a. Johansen C, Lee KK, Sahrawat KL, eds. 1991. *Phosphorus Nutrition of Grain Legumes in the Semi-Arid Tropics.* India: ICRISAT

73. Johnson JF, Allan DL, Vance CP. 1996. Phosphorus deficiency in *Lupinus albus.* Altered lateral root development and enhanced expression of phosphoenolpyruvate carboxylase. *Plant Physiol.* 112:31–41

74. Johnston M, Carlson M. 1992. Regulation of carbon and phosphate utilization. In *The Molecular and Cellular Biology*

of the Yeast Saccharomyces: Gene Expression, ed. EW Jones, JR Pringle, JR Broach. 2:193–281. Cold Spring Harbor: Cold Spring Harbor Lab. Press

75. Jungk A, Asher CJ, Edwards DG, Meyer D. 1990. Influence of phosphate status on phosphate uptake kinetics of maize and soybean. *Plant Soil* 124:175–82

76. Jungk A, Seeling B, Gerke J. 1993. Mobilization of different phosphate fractions in the rhizosphere. *Plant Soil* 155/156:91–94

77. Kai M, Masuda Y, Kikuchi Y, Osaki M, Tadano T. 1997. Isolation and characterization of a cDNA from *Catharanthus roseus* which is highly homologous with phosphate transporter. *Soil Sci. Plant Nutr.* 43:227–35

78. Katz DB, Gerloff GC, Gabelman WH. 1986. Effects of phosphate stress on the rate of phosphate uptake during resupply to deficient tomato plants. *Physiol. Plant.* 67:23–28

79. Keerthisinghe G, Hocking PJ, Ryan PR, Delhaize E. 1998. Effect of phosphorus supply on the formation and function of proteoid roots of white lupin (*Lupinus albus* L.). *Plant Cell Environ.* 21:467–78

80. Kim DH, Muchhal U, Raghothama KG. 1998. Tomato phosphate transporters respond to altered phosphorus levels in cell cultures. *Plant Physiol.* p. 136. (Abstr.)

81. Kochian LV, Lucas WJ. 1982. A re-evaluation of the carrier-kinetic approach to ion transport in roots of higher plants. *What's New Plant Physiol.* 13:45–48

82. Kock M, Loffler A, Abel S, Glund K. 1995. Structural and regulatory properties of a family of phosphate starvation induced ribonucleases from tomato. *Plant Mol. Biol.* 27:477–85

83. Lauer MJ, Blevins DG, Gracz SH. 1989. ^{32}P-nuclear magnetic resonance determination of phosphate compartmentation in leaves of reproductive soybeans (*Glycine max* L.) as affected by phosphate nutrition. *Plant Physiol.* 89:1331–36

84. Lee RB, Ratcliffe RG. 1993. Nuclear magnetic resonance studies of the location and function of plant nutrients *in vivo*. *Plant Soil* 155/156:45–55

85. Lee RB. 1982. Selectivity and kinetics of ion uptake by barley plants following nutrient deficiency. *Ann. Bot.* 50:429–49

86. Lefebvre DD, Duff SMG, Fife CA, Julien-Inalsingh C, Plaxton WC. 1990. Response to phosphate deprivation in *Brassica nigra* suspension cells. *Plant Physiol.* 93:504–11

87. Lefebvre DD, Glass ADM. 1982. Regulation of phosphate influx in barley roots: effects of phosphate deprivation and reduction in influx with provision of orthophosphate. *Physiol. Plant.* 54:199–206

88. Leggewie G, Willmitzer L, Riesmeier JW. 1997. Two cDNAs from potato are able to complement a phosphate uptake-deficient yeast mutant: identification of phosphate transporters from higher plants. *Plant Cell* 9:381–92

89. Lipton DS, Blancher RW, Blevins DG. 1987. Citrate, malate and succinate concentrations in exudates from P-sufficient and P-starved *Medicago sativa* L. seedlings. *Plant Physiol.* 85:315–17

90. Liu C, Muchhal US, Mukatira U, Kononowicz AK, Raghothama KG. 1998. Tomato phosphate transporter genes are differentially regulated in plant tissues by phosphorus. *Plant Physiol.* 116:91–99

91. Liu C, Muchhal US, Raghothama KG. 1997. Differential expression of TPSI1, a phosphate starvation-induced gene in tomato. *Plant Mol. Biol.* 33:867–74

92. Liu C, Raghothama KG. 1995. A practical method for cloning cDNAs generated in a mRNA differential display. *Biotechniques* 20:576–79

93. Liu H, Trieu AT, Blaylock LA, Harrison MJ. 1998. Cloning and characterization of two phosphate transporters from *Medicago truncatula* roots. Regulation in response to phosphate and to colonization by arbuscular mycorrhizal (AM) fungi. *Mol. Plant Microbe Interact.* 11:14–22

94. Loneragan JF. 1997. Plant nutrition in the 20th and perspectives for the 21st century. *Plant Soil* 196:163–74

95. Lu YP, Zhen RG, Rea PA. 1997. AtPT4: a fourth member of the *Arabidopsis* phosphate transporter gene family (accession no. U97546). *Plant Physiol.* 114:747

96. Lynch JP. 1995. Root architecture and plant productivity. *Plant Physiol.* 109:7–13

97. Lynch JP. 1997. Root architecture and phosphorus acquisition efficiency in common bean. In *Radical Biology: Advances and Perspectives on the Function of Plant Roots*, ed. HE Flores, JP Lynch, D Eissenstat, pp. 81–92. Rockville, MD Am. Soc. Plant Physiol.

98. Lynch JP, Beebe SE. 1995. Adaptation of beans (*Phaseolus vulgaris* L.) to low phosphorus availability. *HortScience* 30:1165–71

99. Lynch JP, Brown KM. 1997. Ethylene

and plant responses to nutritional stress. *Physiol. Plant.* 100:613–19

99a. Lynch JP, Deikman J, eds. 1998. *Phosphorus in Plant Biology, Regulatory Roles in Molecular, Cellular, Organismic, and Ecological Processes.* Penn State Summer Symp. Rockville, MD: Am. Soc. Plant Physiol.

100. Malboobi MA, Lefebvre DD. 1995. Isolation of cDNA clones of genes with altered expression levels in phosphate-starved *Brassica nigra* suspension cells. *Plant Mol. Biol.* 28:859–70

101. Malboobi MA, Lefebvre DD. 1997. A phosphate-starvation inducible *β*-glucosidase gene (*psr 3.2*) isolated from *Arabidopsis thaliana* is a member of a distinct subfamily of the BGA family. *Plant Mol. Biol.* 34:57–68

102. Marschner H. 1995. *Mineral Nutrition in Plants.* San Diego, CA: Academic. 2nd ed.

103. Marschner H, Romheld V, Cakmak I. 1987. Root-induced changes of nutrient availability in the rhizosphere. *J. Plant Nutr.* 10:1175–84

104. Martinez P, Persson BL. 1998. Identification, cloning and characterization of a derepressible Na$^+$-coupled phosphate transporter in *Saccharomyces cerevisiae. Mol. Gen. Genet.* 258:628–38

105. Masaoka YU, Kojima M, Sugihara S, Yoshihara T, Koshino M, Ichihara A. 1993. Dissolution of ferric phosphate by alfalfa (*Medicago sativa* L.) root exudates. *Plant Soil* 155/156:75–78

106. McLachlan KD. 1976. Comparative phosphorus responses in plants to a range of available phosphorus situations. *Aust. J. Agric. Res.* 27:323–41

107. Mimura T. 1995. Homeostasis and transport of inorganic phosphate in plants. *Plant Cell Physiol.* 36:1–7

108. Mimura T, Dietz KJ, Kaiser W, Schramm MJ, Kaiser G, Heber U. 1990. Phosphate transport across biomembranes and cytosolic phosphate homeostasis in barley leaves. *Planta* 180:139–46

109. Mimura T, Reid RJ, Smith FA. 1998. Control of phosphate transport across the plasma membrane of *Chara corallina. J. Exp. Bot.* 49:13–19

110. Mimura T, Sakano K, Shimmen T. 1996. Studies on the distribution, retranslocation and homeostasis of inorganic phosphate in barley leaves. *Plant Cell Environ.* 19:311–20

111. Mitsukawa N, Okumura S, Shirano Y, Sato S, Kato T, et al. 1997. Overexpression of an *Arabidopsis thaliana* high-affinity phosphate transporter gene in tobacco cultured cells enhances cell growth under phosphate-limited conditions. *Proc. Natl. Acad. Sci. USA* 94:7098–102

112. Muchhal US, Liu C, Raghothama KG. 1997. Calcium-ATPase is differentially expressed in phosphate starved roots of tomato. *Physiol. Planta* 101:540–44

113. Muchhal US, Pardo JM, Raghothama KG. 1996. Phosphate transporters from the higher plant *Arabidopsis thaliana. Proc. Natl. Acad. Sci. USA* 93:10519–23

114. Muchhal US, Varadarajan D, Damsz B, Raghothama KG. 1998. Tomato phosphate transporter (LePT1) is localized in plasmamembranes. See Ref. 99a, pp. 365–66

115. Musick HB. 1978. Phosphorus toxicity in seedlings of *Larrea divartica* grown in solution culture. *Bot. Gaz.* 139:108–11

116. Natr L. 1992. Mineral nutrients—a ubiquitous stress factor for photosynthesis. *Photosynthetica* 27:271–94

117. Nissen P. 1971. Uptake of sulfate by roots and leaf slices of barley: mediated by single, multiphasic mechanisms. *Physiol. Planta* 24:315–24

118. Okumura S, Mitsukawa N, Shirano Y, Shibata D. 1998. Phosphate transporter gene family of *Arabidopsis thaliana. DNA Res.* 5:1–9

119. Oshima Y. 1982. Regulatory circuits for gene expression: the metabolism of galactose and phosphate. In *The Molecular Biology of the Yeast Saccharomyces: Metabolism and Gene Expression,* ed. J Strathern, EW Jones, JR Broach, pp. 159–80. Cold Spring Harbor, NY: Cold Spring Harbor Lab. Press

120. Oshima Y, Ogawa N, Harashima S. 1996. Regulation of phosphatase synthesis in *Saccharomyces cerevisiae*—a review. *Gene* 179:171–77

121. Pao SS, Paulsen IT, Saier MH. 1998. Major facilitator superfamily. *Microbiol. Mol. Biol Rev.* 62:1–34

122. Patel K, Lockless S, Thomas B, McKnight TD. 1995. A secreted purple acid phosphatase from *Arabidopsis. Plant Physiol.* (Suppl.) 111:270 (Abstr.)

123. Pilbeam DJ, Cakmak I, Marschner H, Kirkby EA. 1993. Effect of withdrawal of phosphorus on nitrate assimilation and PEP carboxylase activity in tomato. *Plant Soil* 154:111–17

124. Plaxton WC. 1996. The organization and regulation of plant glycolysis. *Annu. Rev. Plant Physiol. Mol. Biol.* 47:185–214

125. Plaxton WC, Carswell MC. 1999. Metabolic aspects of the phosphate starvation response in plants. In *Plant Responses to Environmental Stresses: From Phytohormones to Genome Reorganization*, ed. HR Lerner, pp. 349–72. New York: Dekker

126. Poirier Y, Thoma S, Somerville C, Schiefelbein J. 1991. A mutant of *Arabidopsis* deficient in xylem loading of phosphate. *Plant Physiol.* 97:1087–93

127. Raghothama KG, Muchhal US, Kim DH, Bucher M. 1998. Molecular regulation of plant phosphate transporters. See Ref. 99a, pp. 271–80

128. Rao M, Terry N. 1995. Leaf phosphate status, photosynthesis, and carbon partitioning in sugar beet. IV. Changes with time following increased supply of phosphate to low-phosphate plants. *Plant Physiol.* 107:1313–21

129. Ratcliff RG. 1994. *In vivo* NMR studies of higher plants and algae. *Adv. Bot. Res.* 20:43–112

130. Richardson AE. 1994. Soil microorganisms and phosphorus availability. *Soil Biota* 50–62

131. Rolland RH, Contard P, Betsche T. 1996. Adaptation of pea to elevated atmospheric CO_2: Rubisco, phosphoenolpyruvate carboxylase and chloroplast phosphate translocator at different levels of nitrogen and phosphorus nutrition. *Plant Cell Environ.* 19:109–17

132. Runge-Metzger A. 1995. Closing the cycle: obstacles to efficient P management for improved global food security. In *Phosphorus in the Global Environment: Transfers, Cycles and Management*, ed. H Tiessen, pp. 27–42. New York: Wiley

133. Rychter AM, Misulska M. 1990. The relationship between phosphate status and cyanide-resistant respiration in bean roots. *Physiol. Plant.* 79:663–67

134. Sadka A, DeWald WB, May GD, Park WD, Mullet JE. 1994. Phosphate modulates transcription of soybean VspB and other sugar-inducible genes. *Plant Cell* 6:737–49

135. Sakano K. 1990. Proton/phosphate stoichiometry in uptake of inorganic phosphate by cultured cells of *Catharanthus roseus* (L.) G. Don. *Plant Physiol.* 93:479–83

136. Sakano K, Kiyota S, Yazaki Y. 1998. Degradation of endogenous organic acids induced by Pi uptake in *Catharanthus roseus* cells: involvement of the biochemical pH-stat. *Plant Cell Physiol.* 39:615–19

137. Sakano K, Yazaki Y, Mimura T. 1992. Cytoplasmic acidification induced by inorganic phosphate uptake in suspension cultured *Catharanthus roseus* cells. *Plant Physiol.* 99:672–80

138. Sakano K, Yazaki Y, Okihara K, Mimura T, Kiyota S. 1995. Lack of control in inorganic phosphate uptake by *Catharanthus roseus* (L.) G. Don cells. *Plant Physiol.* 108:295–302

139. Sanchez PA, Shepherd KD, Soule MJ, Place FM, Buresh RJ. 1997. Soil fertility replenishment in Africa. An investment in natural resource capital. In *Replenishing Soil Fertility in Africa*, ed. RJ Buresh, PA Sanchez, F Calhoun, pp. 1–46. Madison, WI: Soil Sci. Soc. Am.

140. Schachtman DP, Reid RJ, Ayling SM. 1998. Phosphorus uptake by plants: from soil to cell. *Plant Physiol.* 116:447–53

141. Schmidt ME, Heim S, Wylegalla C, Helmbrecht C, Wagnor KG. 1992. Characterization of phosphate uptake by suspension cultured *Catharanthus roseus* cells. *J. Plant Physiol.* 140:179–86

142. Sengstag C, Hinnen A. 1988. A 28-bp segment of the *Saccharomyces cerevisiae PHO5* upstream activator sequence confers phosphate control to the CYC1-lacZ gene fusion. *Gene* 67:223–28

143. Sentenac H, Grignon C. 1985. Effect of pH on orthophosphate uptake by corn roots. *Plant Physiol.* 77:136–41

144. Sharpley AN. 1995. Identifying sites vulnerable to phosphorus loss in agricultural runoff. *J. Environ. Qual.* 24:947–51

145. Shih CY, Kao CH. 1996. Growth inhibition in suspension-cultured rice cells under phosphate deprivation is mediated through putrescine accumulation. *Plant Physiol.* 111:721–24

146. Shimogawara K, Usuda H. 1995. Uptake of inorganic phosphate by suspension-cultured tobacco cells: kinetics and regulation by Pi starvation. *Plant Cell Physiol.* 36:341–51

147. Silverbush M, Barber SA. 1983. Sensitivity of simulated phosphorus uptake to parameters used by a mechanistic-mathematical model. *Plant Soil* 74:93–100

148. Smith FW, Ealing PM, Dong B, Delhaize E. 1997. The cloning of two *Arabidopsis* genes belonging to a phosphate transporter family. *Plant J.* 11:83–92

149. Tadano T, Sakai H. 1991. Secretion of acid phosphatase by the roots of several crop species under phosphorus-deficient conditions. *Soil Sci. Plant Nutr.* 37:129–40

150. Tanimoto M, Roberts K, Dolan L. 1995. Ethylene is a positive regulator of root hair development in *Arabidopsis thaliana*. *Plant J.* 8:943–48

151. Toh-e A, Tanaka K, Uesono Y, Wickner RB. 1988. *PHO85*, a negative regulator of the PHO system, is a homologue of the protein kinase gene, *CDC28*, of *Saccharomyces cerevisiae*. *Mol. Gen. Genet.* 214:162–64

152. Toh-e A. 1989. Phosphorus regulation in yeast. In *Yeast Genetic Engineering*, ed. PJ Barr, AJ Brake, P Valenzuela, pp. 41–52. Boston: Butterworths

153. Tu SI, Cananaugh JR, Boswell RT. 1990. Phosphate uptake by excised maize root tips studied by in vivo ^{31}P nuclear magnetic resonance spectroscopy. *Plant Physiol.* 93:778–84

154. Ueki K. 1978. Control of phosphatase release from cultured tobacco cells. *Plant Cell Physiol.* 19:385–92

155. Ullrich-Eberius CI, Novacky A, Fischer E, Lüttge U. 1981. Relationship between energy-dependent phosphate uptake and the electrical membrane potential in *Lemna gibba* G1. *Plant Physiol.* 67:797–801

156. Ullrich-Eberius CI, Novacky A, van Bel AJE. 1984. Phosphate uptake in *Lemna gibba* G1: energetics and kinetics. *Planta* 161:46–52

157. Usuda H, Shimogawara K. 1995. Phosphate deficiency in maize. VI. Changes in the two-dimensional electrophoretic patterns of soluble proteins from second leaf blades associated with induced senescence. *Plant Cell Physiol.* 36:1149–55

157a. Von Vexhull HR Mutert E. 1998. Global extent, development and economic impact of acid soils. In *Plant-Soil Interactions at Low pH: Principles and Management*, ed. RA Date, NJ Grundon, GE Rayment, ME Probert, pp. 5–19. Dordrecht: Kluwer

158. Vorster PW, Jooste JH. 1986. Potassium and phosphate absorption by excised ordinary and proteoid roots of the Proteaceae. *S. Afric. J. Bot.* 52:276–81

159. Wilcox HE. 1991. Mycorrhizae. In *Plant Roots: The Hidden Half*, ed. MY Waisel, A Eshel, U Kafkafi, pp. 731–55. New York: Dekker

160. Yompakdee C, Ogawa N, Harashima S, Oshima Y. 1996. A putative membrane protein, Pho88p, involved in inorganic phosphate transport in *Saccharomyces cerevisiae*. *Mol. Gen. Genet.* 251:580–90

Annu. Rev. Plant Physiol. Plant Mol. Biol. 1999. 50:695–718

ROOTS IN SOIL: Unearthing the Complexities of Roots and Their Rhizospheres

Margaret E. McCully

Department of Biology, Carleton University, Ottawa, Ontario, Canada, K1S 5B6;
e-mail: mmccully@ccs.carleton.ca

KEY WORDS: fine roots, rhizosheaths, root apices, root/microbial interactions, root system function

---

ABSTRACT

The root system of a plant is as complicated as the shoot in its diversity, in its reactions with the matrix of substances, and with the myriad organisms that surround it. Laboratory studies blind us to the complexity found by careful study of roots in soil. This complexity is illustrated in the much-studied corn root system, covering the changes along the framework roots: the surface tissues and their interactions with the soil, the water-conducting xylem, whose gradual elaboration dictates the water status of the root. A conspicuous manifestation of the changes is the rhizosheath, whose microflora differs from that on the mature bare zones. The multitude of fine roots is the most active part of the system in acquiring water and nutrients, with its own multitude of root tips, sites of intense chemical activity, that strongly modify the soil they contact, mobilize reluctant ions, immobilize toxic ions, coat the soil particles with mucilage, and select the microflora.

CONTENTS

1040-2519/99/0601-0695$08.00

INTRODUCTION

The chemical, physical, and biological interactions that occur between roots and the surrounding environment of the soil are easily the most complex experienced by land plants. Recent years have seen tremendous advances in the knowledge of the complexity of, for example, the signal traffic that moves back and forth between roots and soil microbes, roots and soil fauna, roots and other roots, as well as signals relating the nature of local soil chemical and physical properties to nearby roots (e.g. 16, 34, 68, 73, 104, 106, 115, 142).

Advances have also been made in understanding the processes involved in nutrient and water uptake by roots, both under lush conditions (e.g. 19, 101) and in conditions where nutrients like phosphate or iron are in short supply or scarcely available (25, 73). The intricacies of how roots of genetically resistant cultivars respond to pathogens or interact with biocontrol agents (45) or with toxic elements in the soil (22, 73, 112) are also being unraveled.

Much of our knowledge of these processes has been, by necessity, obtained from work with young plants grown under relatively very simple, controlled conditions, remote from the complexities prevailing in the field. Frequently, the plants studied were seedlings, often grown axenically or in the presence of only a single other organism, usually at high initial inoculum in the case of microbial interaction studies, and frequently supplied with nutrients greatly in excess of those available in fertile agricultural soils.

This laboratory-based work has been spectacularly successful in revealing details of physiological parameters and root-microbe interactions as they apply to the specific situations under which the experiments were performed. Extrapolations to field conditions have, however, been hampered, not only by the biological, chemical, and physical complexities of the soil milieu (21, 23, 45, 141) but also by a shortage of accurate and comprehensive information about root systems and how they work throughout the life span of plants in the field (76). There is now recognition of the urgent need to understand the behavior of roots in the field so that laboratory-based findings can be more successfully applied. Some progress is now being made.

This review attempts to draw together some of these studies, with emphasis on the developmental aspects of the form and function of herbaceous root systems in the field. Eshel & Waisel (33) have discussed root multiform and multifunctionality, and readers are directed to this excellent article. Here I expand on some of their major points and extend the consideration of root heterogeneity to include the key role played by fine roots in the acquisition of water and nutrients, and in root adaptation to extreme environments, as well as effects of root heterogeneity on some of the processes that occur at the root surface and in the rhizosphere.

ROOT SYSTEM AND RHIZOSPHERE HETEROGENEITY

Most recent characterizations of root systems of crop plants have been from measurements of pieces washed out from different levels in soil cores, in contrast to the classic excavations of earlier workers (103, 135). The coring method gives equal weight to all roots recovered and obscures the effects of functional heterogeneity among or along the component roots in a system, either inherent or induced by microheterogeneity in the soil milieu.

At a gross level, a mature root system has two major sorts of root: long, usually relatively thick roots that form the framework or major axile components of the system and which define the broad extent of the soil volume occupied; and shorter, fine branch (lateral) roots arising either directly from these framework roots (Figure 1a: See color section at the back of the volume), or indirectly as higher-order branches. Many of these fine roots may be ephemeral, and they often are unnoticed, either because they are in unexpected places [as the hair-like tree feeder roots in the leaf litter in deciduous forests (70)], or they may break off during excavation or washing. A recent study suggests that the fine roots of corn lost during conventional washing procedures may exceed 50% of the total root length (100). I refer to branch roots of diameter <0.8 mm as fine roots.

Fortunately, the random sampling approach to root study is changing; much more emphasis is being placed on the overall configuration or architecture of individual root systems (i.e. the spatial arrangement in situ), particularly of the framework components throughout the course of their development (e.g. 36, 71). This may involve the use of careful excavations as done in the classic studies (e.g. 8, 99, 105, 120, 128) as well as in situ observation devices (rhizotrons), nuclear magnetic resonance, computer-aided tomography, and neutron radiography (5, 10, 14). There is also much interest in using computer models to predict root architecture from data derived from soil cores or rhizotrons. With the incorporation of better knowledge of the placement and functioning of different root types within a root system, it is hoped that these will provide realistic patterns of nutrient and water acquisition. One of the few models that has incorporated experimental data on root heterogeneity (30, 31) is illustrated in Figure 1b (See color section at the back of the volume).

Surprisingly few detailed studies have been made of the structure of the various types of root present in an individual root system. The anatomical and physiological data that are easily available have been obtained largely from the study of axile roots of young seedlings. Yet seedling roots are still drawing from the nutrient store of the seed, and if connected at all to a transpiration stream, this latter connection is weak. They are therefore not typical of most roots in a more mature root system. For example, the work of Dittmer (26, 27) focused attention on the dominance of the fine root component in respect to the total surface area of any root system, and there is usually among root biologists who work with field-grown material the assumption that these roots are playing the dominant role in water and nutrient acquisition. However, until very recently there has been no detailed study of the structure/functional aspects of fine roots, though developmental biologists have been much interested in their initiation and early development in laboratory-grown material.

The importance of the influence of the stage of development and the local functioning of regions along an individual root on the microbiology of the adjacent soil has been emphasized strongly (86, 118). Subsequent studies of root/microbial interactions have, however, often failed to consider these effects, or the possibility that there are different root types with different physiology in a single root system. Thus many of the subtleties of the interactions occurring at the root surface may have been missed.

The root system of corn has been the most studied under field conditions from a structural/functional point of view, and the findings provide an example for work with other species. This species has been a particularly appropriate choice because many of the laboratory-based studies, particularly of root physiology, have been with corn. I use the corn example to illustrate the complexities of the root/soil system, touching on some parallel observations of other grass species, with occasional reference to the very few studies with herbaceous dicotyledons.

CORN AS A MODEL FOR FURTHER INVESTIGATIONS

Framework Roots

Careful excavation of corn roots combined with detailed measurements of structural and physiological parameters have been carried out by a number of research groups on three continents. From these studies, despite differences in cultivars and growth conditions, a remarkably consistent picture is emerging. As in all grasses, the root system consists of two systems of quite different origin and predominance during plant development: the seminal system and the nodal system.

The seminal root system is indispensable for the establishment of the young plant and any weak early development of this system adversely affects the growth of the plant throughout its development (61). The idea still persists that the seminal system of grasses, including corn, is temporary, and that it senesces after the nodal system (often termed the permanent system) is established. This has been many times shown to be incorrect for corn and many other grasses (40, 63, 93, 94, 136). Indeed, the seed-derived root system is not only indispensable for the seedling but it appears to play a continuing important role. The seminal roots of soil-grown corn supply about one fifth of the total water used by the plant during its lifetime. The relative amount of the supplied water decreases after tasseling but, surprisingly, increases again during grain development. The seminal roots are significantly more efficient at supplying water to the shoot on the basis of volume/dry weight or surface area (55, 93, 94). The basis for this efficiency is not known and is surprising because of the relatively fewer large xylem vessels and the greater age of the root system.

The nodal system of corn includes about 40 to 70 axile roots (49, 105, 135), which develop from the stem. The oldest 4 to 5 nodes and internodes of the stem remain below ground for the life of the plant, with very little elongation of these internodes, except the uppermost, which places the node above it approximately at the soil surface. Tiers of nodal roots emerge sequentially at the top of each node; those from the oldest (node 1) emerge when about 3 leaves have unfolded. Roots of successively younger tiers emerge at regular intervals approximately coincident with the appearance of every 2 new leaves (105). Shortly before flowering, roots emerge from the next younger nodes above the ground, and usually two tiers of these (frequently termed prop roots) grow into the soil. The last two tiers (6 and 7) include about half of the total number of nodal roots of the system and 3 times more large xylem conduits than the total for the nodal roots from the lower tiers and the primary root. Measurements of actual water flow through isolated mature regions of these roots showed, for example, that a root of the youngest tier transported 8 times the volume of water transported by those from the oldest tier of nodal roots and 21 times the volume of the primary root (69).

Fine Roots

A surprisingly consistent finding (15, 40, 56, 99, 125) has been that most of the first-order branch roots of corn are short (mode ≤ 3 cm), and only about 2% exceed 10 cm (99). The short roots reach their final length in <2.5 day (15, 40). They have a normal apical meristem and small root cap while extending, but the meristem then grows out, the cap is lost, and tissues differentiate right to the tip, and the surface cells around the tip often develop root hairs (126). These determinate roots persist for the life of the plant but gradually shorten as the distal

ends slowly die back (40, 131). The density of these fine roots along the proximal regions of the framework roots is also remarkably consistent (7 to 12/cm)
from measurements of different cultivars growing in soil in different countries
(40, 56, 89, 99, 125, 135). This density falls to about 4/cm by depths of 60 cm
(99). Only about one third of the first-order fine roots produce further branches
and these are very short and sparse. The fine roots comprise up to 30 times the
total length of the framework roots (99).

The fine roots have a normal component of tissues, epidermis, cortex, and a
narrow stele (132). Much of the epidermis remains alive in moist soils, even in
very old roots. The cortex, which also remains intact in old roots, includes the
two specialized layers, hypodermis and endodermis, both with Casparian strips
and suberized secondary walls. In the narrow secondary branches, the cortex is
often reduced to the two specialized layers (Figure 2d), suggesting that they and
the epidermis are necessary to maintain normal functioning of these little roots.

The calculated axial water-conducting capacity of the fine roots ranges over
five orders of magnitude, depending on the number and diameter of the major
xylem conduits. These vary from three or four tiny vessels (diameter <6 μm)
in the narrowest roots (through which little flow would be expected) to as many
as four large central vessels (diameter about 60 μm) in the widest ones, with
vessel diameters increasing in direct proportion to root diameter (125).

The development of fine roots in corn is affected quite differently from that of
framework roots by differences in soil type and aeration (39). Fine root numbers
and growth are both stimulated by local enrichment of nitrate, whereas the main
axis is not affected (42).

Despite their insignificant appearance, the fine roots of corn are the major
sites of water uptake into mature root systems (124). These roots only become
active in water uptake at about 20 to 30 cm from the main axis tips, at the point
where the main root large xylem conduits become mature. At this point the
fine roots have stopped elongating and many have lost their tips. They provide
about eight times the surface area of the mother root, and collect about eight

$\longrightarrow$

Figure 2   a. Outer surface of the rhizosheath on an immature region of a field-grown corn nodal
root. The root was excavated, shaken free of loose soil, and observed in a cryo-SEM. b. A transverse
face through a similar root and rhizosheath to that in 1a, showing thickness of the rhizosheath (R).
The root has developed extensive aerenchyma (A). Arrowheads indicate large xylem conduits.
Observed in a cryo-SEM. c. A living hair root of the Ericaceous species Lysinema ciliatum mounted
in water. The 50-μm wide root is surrounded by expanded mucilage, which was produced by cells
of the tiny root cap. While in soil the mucilage was contracted and held soil (arrows) tightly
against the root surface. d. A tiny fine root of corn, 70 μm in diameter. The living epidermis
(E) surrounds a minimal cortex comprising a hypodermis and endodermis. Xylem conduits are
minute. Transverse face observed by cryo-SEM.

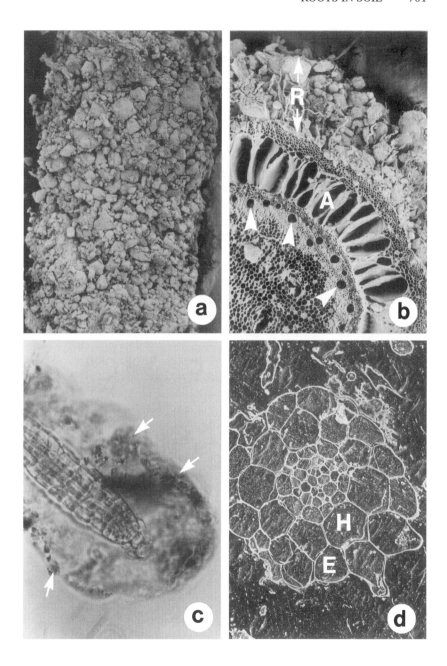

times as much water. The amount of water taken up was maximal at 30 to 60 cm from the main root tip and decreased to about 25% of this maximum in older regions (124). Mature fine roots in soil have the lowest relative water content of the components of the root system, typically dropping to about 50% during transpiration, compared with 60% in the subtending mature regions of axile roots (130).

In soil-grown corn plants, even the oldest fine roots are active in phosphate uptake; some phosphate is transported directly to the main roots, while some is initially accumulated in the fine root cortex, then mobilized during shoot maturation (40, 64). Preferential uptake of nitrate into fine roots compared to that of the main seminal root has also been demonstrated in seedlings (67), but lacks confirmation for soil-grown plants. Substantial nitrate uptake was found in basal regions of axile roots of mature plants in soil (108), and this may have been via the fine branch roots. Fine roots may also have been responsible for the six times greater uptake of water by basal regions of hydroponically grown corn axile roots compared to the tip regions (72).

There is both anatomical and physiological evidence that the fine roots are particularly well supplied with assimilates, further suggesting a very active role in root system function. There is much more phloem in these roots than would be expected by their small tissue mass (132), and the connections of the phloem strands with those of the main root are very extensive (83). Fine roots on mature regions of field-grown axile roots accumulated more ^{14}C-labeled assimilate per unit weight than did subtending main roots, even when they were determinate and old (79).

A dramatic difference in metabolic activity between fine roots and the subtending parent roots in soil has been demonstrated in corn, where mature fine roots lowered the pH at their root/soil interface 1.5 units below that of the bulk soil, while the surface of the subtending axile roots became more alkaline by about 1.5 units (74). Further evidence for fine root metabolic specialization is shown in the differential distribution of glucosinolates (GSLs) in the roots of *Brassica* species (60). The aromatic and aliphatic GLSs, which release pathogen-controlling isothiocyanates, predominated in the coarse roots, whereas indolic-GLSs, which do not release these components, were predominant in the fine roots, possibly leaving them vulnerable, or, alternatively, allowing interaction with beneficial soil bacteria.

Developmental Changes In Individual Roots: Xylem

Traditional methods for preparing plant tissues for microscopy are destructive of the delicate, large cells that become elements of xylem vessels (119). This destruction during trusted procedures, and misplaced confidence in the presence of lignin as a sign of maturity, led to the conclusion, still prevalent in texts, that xylem maturation is complete within at most a few cm from the tip of a root. There were suggestions, and indeed some good physiological data (13, 47a), that maturation of the xylem in corn roots was much further from the tip, but these data were largely ignored. The use of less destructive anatomical methods (119) and cryo-microanalysis (82) produced indisputable evidence that the maturation of the large vessels (Figure 2a, See color section at the back of the volume) in actively growing corn roots are not open for conduction until at least 15 cm, and sometimes more than 40 cm behind the tips. The much narrower small vessels in these roots begin to mature from about 4 to 9 cm from the tip, and the very narrow protoxylem at about 1 to 2 cm (13; ME McCully, unpublished). Burley et al (13) found that some of the small vessels may remain nonconducting as far back from the root tip as 16 cm. Immature xylem does not conduct the transpiration stream but accumulates ions, particularly potassium, to the highest concentrations found in the roots (75, 82). Note that when roots stop elongating, maturation of all tissues, including the xylem, moves closer to the tip. This process proceeds quickly in detached roots, so that any prolonged experimentation with such roots should include an anatomical evaluation of tissue maturation.

Individual plant variation and differences in ages of the individual roots on a plant both must be considered in assessing the axial hydraulic conductivity of any root system. Root length is not necessarily a good indicator of the length of xylem that is open for conduction. Wenzel et al (137) found, for example, that there were no open large vessels in corn roots until leaf area exceeded about 50 cm^2. Comparable delay in xylem development has been observed in the axile roots of several other C4 grasses (137), in barley (53, 113), and in wheat (52). In wheat the distance from the root apex at which both the small and large vessels were mature was strongly temperature dependent.

The large xylem vessels in the fine roots of corn are also slow to mature (131) and in those roots that are still indeterminate (with an intact apical meristem and root cap), all the large elements (these are up to 2 mm long) may be alive, thus nonconducting. The most open vessels observed in these roots was 45% of the total length. In the younger determinate roots, up to 60% of the vessel length is open, with the distal elements still alive.

When the large elements of a typical axile root (with, for example, 12 such elements) mature, the axial conducting capacity of the root portion is increased 1800 times (80). The root is then mature and from there to its base is linked strongly to the transpiration stream. Comparisons of relative water content (RWC) and water potential of the regions of roots with immature or mature large vessels showed the dramatic effect of the extent of the connection to the transpiration stream on the water relations of the two portions of the same root. While transpiration was intense, the RWC of the mature regions dropped from about 63 to 55%, whereas those of the immature regions remained steady at >80%. The determinate branches on the mature regions of these roots were as low as 50% RWC even early in the day. Thus the heterogeneity in vascular maturation produces heterogeneity in the water status of different regions of the root, and the effects extend to the root surface and beyond it into the rhizosphere (see below).

Large xylem conduits are delayed in their maturation in soybean primary roots (59), but the generality of this development in dicotyledons, or its effects on root water relations and the adjacent rhizosphere, are not known.

The point of transition in the grass root from closed to open large xylem vessels is of major significance. It is the point at which the relative water content drops sharply, the soil rhizosheath disintegrates, the epidermis dies, and the microflora change abruptly (see sections below).

Developmental Changes: Surface Features

Framework roots, gently excavated in the field and shaken but not washed, show clearly the developmental heterogeneity along their length. There is a shiny white tip (up to about 5 cm) and a long zone covered with soil (the rhizosheath, 15 to 30+ cm long), except in very wet conditions. Beyond this, the root is largely free of soil, except at the most basal few cm where some soil sheath may persist in the upper nodal roots (80). Fine roots developed within the sheath region are themselves sheathed; older fine roots are bare. Surface changes along framework roots have been extensively investigated (1, 2, 43, 44, 79, 85, 128).

MUCILAGE AND ROOT CAP CELLS   The outer two layers of cells along the flanks and at the tip of the root cap synthesize and secrete mucilage in a process that has been well characterized in seedling roots growing under laboratory conditions by many workers (see review 111), and the composition of the polysaccharide mucilage is well known (6, 88). The mucilage has been credited with several functions. The commonest is that it is a lubricant that smoothes the passage of the root through the soil, and that retains water, slowing root desiccation. These putative functions arose from the appearance of expanded mucilage around the tips of washed roots, or roots mounted in water, or of embedded sections in which the mucilage has been swollen and dispersed during aqueous fixation. The mucilage swells markedly when in contact with water (84). Root tips observed with a hand lens immediately after they are removed from soil, even at moisture content close to field capacity, or examined frozen, do not have expanded mucilage (84). Water potential measurements of the mucilage show that it has no capacity to swell or retain water at potentials significantly lower than 0 (43, 78). This failure to retain water against a low water potential is shared by other biological mucilages except if they are within a solute-rich compartment (78).

The mucilage has other, better-documented functions. Alternate wetting and drying of the mucilage that is left along the surface of the root produces important adhesive properties (133, 134), resulting in soil aggregation in the rhizosphere (87, 98), and adherence of the aggregates to each other and to the roots (see below). Clay plates are dispersed when wet mucilage is added, and strongly ordered when dried (29, 37). These properties, together with its viscosity (107) and acidic residues, are almost certainly of great importance for nutrient acquisition by diffusive exchange (73, 91, 92, 122, 123), and, possibly, for amelioration of the toxic affects of elements like Al^{3+} and Cd^{2+} by binding them outside the vulnerable root tips (50, 88). The mucilage also induces attachment and development of pathogenic fungal zoospores (48, 54).

The peripheral cap cells accumulate mucilage, often in large, dense deposits in the periplasmic space between their walls and plasma membranes. This mucilage is somehow extruded through the wall, and in roots in soil remains as a dense, unexpanded coating over the cap that binds soil particles (84). It does not expand until the root is wetted. Cryo-microscopy has shown that the initial expansion is a sharply defined phase shift at the edge of the condensed mucilage (84), with increasing centrifugal expansion beyond this interface.

The peripheral root-cap cells abscise in an orderly manner, separated in part by polygalacturonase activity (47) and by the expansion of the released mucilage, presumably when the tip is in contact with free water. The loosening of cells appears to reduce the friction between the advancing root tip and the soil (7). Unless the root has been strongly desiccated, the root cap cells are alive when they are released and remain actively streaming and capable of uptake of vital dyes for some time along the edge of the root (44, 46, 80, 128).

As the root tip is pushed through the soil by the elongating cells behind it, the released mucilage and cap cells remain behind in the soil, and then lie over increasingly mature root surface. This gives the appearance that the mucilage is also secreted by the root epidermis and the root hairs, but there is no evidence for this, nor for the suggestion (38, 123) that the detached cap cells continue to secrete mucilage. The release of root-cap mucilage and living cap cells into the soil is characteristic of all grasses and of many dicotyledons that have been examined (46, 85).

EPIDERMAL SURFACE AND ROOT HAIRS   The surface of the young, white tip region has a thick, complex helicoidal wall that thins gradually as the cells elongate and accommodates the protrusion of the root hairs. This thick wall has confusingly been termed epidermal mucilage (85), but in fact it is a stiff, discrete pellicle that can be peeled off the root intact (2, 81), and may stiffen the extending root tip (81). It has a very thin outer layer, thought to be proteinaceous, which obscures cell outlines (1, 81) to which root-cap mucilage and soil do not adhere. When this layer disappears, root hairs emerge, and soil and mucilage then stick to the epidermal surface, forming the rhizosheath (see below). The now elongated epidermal cells bulge outward, leaving deep grooves over their longitudinal anticlinal walls (1).

Initially, the root hairs grow relatively straight, but where the soil begins to stick to them they are distorted, curled, and frequently branched. They, and the epidermis that produces them, persist as far as the region where the large vessels mature. Here, on the drier, mature root tissues, the epidermis disintegrates, and the living interface with the soil is now the newly exposed hypodermis.

It is not known whether the root hairs are functional all along the length of the immature root. As with root tissues in general, particularly those in soil, vital stains often give false negative results because they fail to penetrate incrusted walls (138).

LOSS OF SURFACE LAYERS   The epidermis is the only tissue that is shed by maturing corn roots. Under wet or very dry field conditions, the cortex develops extensive aerenchyma, but the remaining cells are alive, often developing thick lignified walls in the framework roots (80). Cortical loss has been reported in primary roots of corn under severe nutrient and water stress (117) but this has not been observed in the field, in either corn or related teosinte species. In contrast, old roots of many other grasses shed variable amounts of their cortical tissues, and this process appears to be hastened by drought (57). When cortical tissue is lost, the living surface in contact with the soil becomes either the endodermis or, in some cases, inner cortical cells that are strongly lignified.

Water Efflux from the Root Surface

Cryo-microscopy of the root/soil interface of corn has shown that droplets of water are exuded during the night from regions where the epidermis is still intact (77) by a process related to the generation of root pressure (17), and that this water disappears when the plants transpire. Similar exudation was observed in other grasses and has been confirmed in corn by measurements of changes in dielectric constant at the root surface (121). Water efflux from the roots of pearl millet into damp soil has been detected with microtensiometers (129).

Root Developmental Changes are Reflected in Rhizosheath Development and Loss

Roots exert a profound influence on the soil with which they are in contact (e.g. 9, 21, 29, 32a, 35, 62, 73, 74, 114, 123, 134, 141). This influence is partly direct, by the exertion of physical pressure, localized drying and rewetting, change of pH and redox potential, mineralization, mineralogical changes, nutrient depletion, and the addition of a wide variety of organic compounds (including root-cap mucilage and surfactants). Roots also affect the soil indirectly through the activities of the specific microbial communities that are established in the rhizosphere (9, 21, 29, 134). Until recently, root effects on the rhizosphere have not been closely linked to stages of root development.

Enhancement and stabilization of soil aggregation is perhaps the most recognized overall effect of roots, particularly grass roots on the adjacent soil. In many grasses there is an unusual additional effect in which a coherent sheath of aggregated soil develops around and adheres tightly to the immature roots (Figures 2a,b). These rhizosheaths, first thought to be special features of desert grasses (139), also form on many mesophytic grasses, including all soil-grown cereals, corn, and sorghum, and their development in relation to root development has been studied most extensively in the mesophytic grasses. They are present only over those portions of the axile roots in which the large vessels are immature, and are shed when these vessels mature and the epidermis disintegrates (80, 119, 137). Work with a model system (133) has shown that sheath aggregation and cohesion in corn are produced by mucilage, synthesized either by corn root-cap cells, or by bacteria indigenous to the corn rhizosphere. The chemistry of the binding of the two mucilages to soil is different, but both require a cycle of wetting, when mucilage is expanded around the soil particles, followed by drying, when mucilage contraction pulls the particles tightly together. Cohesion is unaffected by further wetting. It has been proposed that the initial sheath formation in the region of the developing root hairs results from an expansion of the residual root-cap mucilage due to the water exuded from the root at night, and a subsequent drying when water is taken up by the growing root during the following day (77, 134). Activity by mucilage-producing bacteria and continuing diurnal wetting and drying of the surface soil will further stabilize the sheath.

When sheathed roots are excavated, the rhizosheath separates cleanly from the surrounding bulk soil, retaining a firm outer surface, in which clay particles tend to be oriented parallel to the root surface (Figure 2a). The volume and coherence of the rhizosheaths depend on the moisture status of the adjacent soil; sheaths in locally dry regions are much better developed than in adjacent wet regions (134).

Rhizosheaths are permeated by root hairs, with the thickness of the sheath roughly the same as the length of the hairs. The volume of the sheath in corn is about two and a half times the volume of the root (Figure 2b). About two thirds of the bulk of the sheath can be removed by sonication, but the rest remains attached to anchor points at distorted regions of the root hairs, and at the root surface.

Rhizosheaths will form beyond the root hairs on the outer surface of fine nylon mesh between the root and the soil (Figure 1c, See color section at the back of the volume) (M Watt & ME McCully, unpublished). As the root tip grows along the inner surface of the mesh, root-cap mucilage is released, which when wetted by the root expands through the mesh (Figure 1*d*, See color section at the back of the volume), contacting the soil particles. These are aggregated into coherent rhizosheaths when water is drawn back into the root. These rhizosheaths overlie the immature regions of framework roots and their fine branches (Figure 1*c*, See color section at the back of the volume).

As mentioned above, the water status of those regions of the root that underlie the rhizosheath is higher than that of mature root regions, and the rhizosheaths are not subjected to the full draw of the transpiration stream. The measured higher moisture content of rhizosheath soil (140) thus derives most likely from the state of the underlying root, water exudation from the root surface, and also the higher organic matter content of the sheath (62) and the smaller gap size between the soil particles.

The importance of the rhizosheath, with its tightly adhering soil and increased water content, for nutrient acquisition is especially well illustrated by the ability of roots to take up zinc from localized regions of very dry soil within which an extensive rhizosheath has developed (91, 92).

Rhizosheaths form on immature regions of the roots of some cacti, and these remain at a higher water potential than the surrounding dry soil, and they restrict water loss from the subtending roots (51, 97). Rhizosheaths are also found on the roots of many desert non-grass monocotyledons (28), on fine roots of some legumes, and on a few dicotyledonous crop plants (ME McCully, unpublished), but these have not been investigated further.

Root Developmental Changes Are Reflected in the Bacterial Flora

Much work over nearly a century has demonstrated the "rhizosphere effect" on soil bacteria (and their predators), i.e. the enhancement of numbers relative to those of the bulk soil, and differential distribution of species in the two locations (9, 21). Soil chemical and physical properties, plant species, and to a limited extent, plant age, are important controllers of indigenous microbial populations in the rhizosphere. (All these factors, including the indigenous microorganisms, also affect root colonization by introduced microbes.) Root heterogeneity within an individual root system in respect to rhizosphere microbes has, however, received little attention, although the importance of root developmental changes was suggested much earlier (86, 118). Mitchell (86) has stated this most clearly: "...to understand the activity of...pathogenic or any other microorganism in relation to the roots of plants, the dynamic character of the rhizosphere must be appreciated...on this scale of the microhabitat of a particular segment of root. ...Just as the character of any specific point on the root changes, so will the composition of the microbial community change as the nature of the substrate changes due to host tissue maturation...."

Any field-based study of rhizosphere microbiology that attempts to link the microflora with the different root regions is very labor intensive, but the common assessment of microbial rhizosphere populations from bulk root samples, or from samples recovered from soil cores, obscures any effects of root heterogeneity. Results from the few studies that have attempted to investigate the linkage suggest that this may be a serious omission. For example, there is a dramatic difference in the population of bacteria culturable from the soil sheath, and from soil adjacent to the mature sheath-free regions of corn framework roots (41). Each population yielded about the same number of bacteria viable on various media. However, many more spore-formers capable of growth on nitrogen-poor media, fluorescent pseudomonads, and mucilage-producing types were cultured from the rhizosheath soil than from soil adjacent to bare roots. Actinomycetes were absent from the rhizosheaths but plentiful in the bareroot rhizospheres. Markedly larger populations of bacteria and fungi were isolated from the rhizospheres of seminal compared with nodal roots of the same age in wheat (116), though no data were given on the actual developmental stages examined. Almost certainly, the nodal roots were much younger. Other workers have recently attempted to determine the effect of corn root developmental stage on the associated rhizosphere bacteria (18, 24, 90). Each study has shown differences in the total culturable bacteria, relative abundance of different bacteria, or speed of growth in culture, which have been attributed to root developmental changes, but the collection methods have not allowed a precise separation of root type or developmental stage. There appear to have been no general studies of bacterial populations specifically associated with fine roots despite the predominance of these roots and their apparent functional specializations.

FINE ROOTS: THE KEY TO ROOT SYSTEM FUNCTION

General Features

Since fine roots comprise most of the total root length of any root system, and (also because of their tiny diameter) are in intimate contact with much the largest volume of soil per unit root volume, they are of major importance in normal nutrient and water acquisition. Fine roots also provide most of the root tips in any root system and are thus the major source of root-cap mucilage and shed cap cells, and in this way exert a major influence on rhizosphere development. These tips also play a crucial role in the release of available forms of phosphate and iron by the secretion of high concentrations of organic acids, protons, or reducing or chelating compounds, the synthesis of which is enhanced in the tips by deficiencies of these elements (25, 35, 73, 74, 110). In aluminum-resistant varieties, normally toxic concentrations of Al^{3+} induce secretion of organic acids from root tips, and these form nontoxic aluminum complexes. Al^{3+} also induces enhanced proton uptake into the tips, raising rhizosphere pH and reducing the availability of toxic aluminum (22, 66, 112, 143). [For a particularly dramatic example of an induced reaction localized to root tips (in this case of proton extrusion enhanced by iron deficiency) see Figure 1E of (110).]

Exploitation of Soil Heterogeneity

Patchy distribution of nutrients and water is characteristic of natural soils, and root systems of many plants exploit this heterogeneity directly (16, 109). A classic example is the local proliferation of fine roots into a pocket of nitrogen- or phosphate-rich soil by plants starved of these nutrients. Previously, such responses were thought to result from metabolically driven redistribution of photosynthate. Now a new paradigm is suggested by the exciting finding (68, 142) that fine roots develop from primordia in direct response to a metabolically independent signaling system in the meristem that detects local NO_3^-. The first step in this detection is the rapid specific activation of a gene with homologies to the MADS-box transcription factor. It remains to be discovered how general such local signal response is to root system adjustment to soil environment.

Drought-Resistant Roots and Meristems

Fine roots play important roles in drought resistance and recovery in several widely different families. In the Brassicaceae, short (<2 mm long) fine roots with radially swollen bases, precociously matured xylem (in some species with markedly enlarged vessels), and inactive meristems develop and persist during drought. When watered, these little roots rapidly develop numerous root hairs and resume elongation (20, 127). Drought induces the development of fine root primordia in species of other families (11, 32, 57, 96), often in response to death or exhaustion of the parent root apical meristem. These primordia are drought resistant, and when watered, rapidly develop into fine roots, which resume water and nutrient uptake.

Proteoid Roots

These specialized fine roots, also commonly called cluster roots or proteoid-like roots, were first described in species of the family Proteaceae, but are now known to occur in a number of other species from diverse families, including one species of agricultural importance, *Lupinus albus* (25). They have been thought to develop only when phosphate and/or iron is limited, but, while proteoid roots are not developed under the relatively high phosphate concentrations that have previously been used to inhibit them in culture, those of lupin at least will develop at P levels found in fertile agricultural soils (58).

Proteoid roots are branch roots of varying orders that develop compact (usually 0.5 to about 1 cm long), often ellipsoid-shaped clusters of rootlets at intervals along their length. In seedlings these clusters first form on the early roots developing from the hypocotyl. Each cluster includes numerous (frequently hundreds) short, determinate rootlets, and many clusters, each separated by normally branched regions, can develop on an individual proteoid root. The result, in all proteoid roots, is a marked increase in the surface area of young, glistening white apices. This immature surface is ephemeral and soon darkens, often within a few days, and the clusters, or even the whole proteoid roots eventually senesce. Continuing formation of new proteoid roots maintains the augmentation of young tips.

Proteoid roots aid phosphate uptake by exuding high concentrations of organic acids, principally citric acid or citrates, and in some cases reducing compounds such as phenols (25). This release is localized to the young root tip and may occur over very short times (95; M Watt, personal communication). These roots are also thought to be particularly suited to the uptake of phosphate and iron, and in some species there is also evidence for enhanced uptake of Mn and nitrogen (25). Proteoid roots exude water from their surface at night (65, 102) and also bind soil strongly.

Short, clustered fine roots that bind soil strongly are also characteristic of species in the families Cyperaceae and Restionaceae (25, 65). To date, their function is unknown.

Hair Roots

Characteristic of many Ericaceous plants, these are the narrowest fine roots known, most ranging in diameter from about 70 μm to as small as 20 μm, and <10 mm long. In the epacrid species in which they have been most extensively studied (3, 4), they develop as first-order branches on normal axile roots, or as second- or higher-order branches from other hair roots. They show perhaps the ultimate reduction possible in a functional, terrestrial root, consisting of an epidermis, a cortex reduced to a hypodermis, and endodermis (each with Casparian strips), and a tiny stele with xylem conduits frequently as narrow as 2.4 μm. These reduced conduits, as well as the suberization that develops around the hypodermal cells, and the observation that this latter tissue completely surrounds the tip of a mature root, would seem to make hair roots unlikely candidates for efficient water uptake. The short lengths and very high numbers of these roots may compensate for this inefficiency (3). When first excavated, these roots have a coating of tightly bound soil particles. When wetted they are enclosed in a thick swollen mucilage, which retains the soil particles on its outer edge (Figure 2c). As in corn, this mucilage is produced only by the root-cap cells (12).

SOME KEY QUESTIONS TO BE ADDRESSED

1. How general are the effects of delayed xylem maturation on water efflux from roots to soil, root surface and rhizosphere development, and microbial/root interactions?

2. What restricts the development of rhizosheaths to certain plant species?

3. Is fine root specialization more widespread than presently recognized?

4. What is the significance of the extensive loss of older fine root tips of corn, and is this specific only to this species?

5. How widespread is the control of root development through transduction of local signals from the rhizosphere?

ACKNOWLEDGMENTS

I thank Anne Ashford for Figure 2c, Claude Doussan for Figure 1b, and Michelle Watt for Figures 1c and d, the Natural Sciences and Engineering Research Council of Canada for an operating grant, the Librarians of the Hancock Library, Australian National University, and the Black Mountain Library of the CSIRO, Canberra, for their help, and Martin Canny for discussions.

Visit the *Annual Reviews home page* at
http://www.AnnualReviews.org

Literature Cited

1. Abeysekera RM, McCully ME. 1993. The epidermal surface of the maize root tip. I. Development in normal roots. *New Phytol.* 125:413–29
2. Abeysekera RM, McCully ME. 1994. The epidermal surface of the maize root tip. III. Isolation of the surface and characterization of some of its structural and mechanical properties. *New Phytol.* 127:321–33
3. Allaway WG, Ashford AE. 1996. Structure of hair roots in *Lysinema ciliatum* R. Br. and its implications for their water relations. *Ann. Bot.* 77:383–88
4. Ashford AE, Allaway WG, Reed ML.

1996. A possible role for the thick-walled epidermal cells in the mycorrhizal hair roots of *Lysinema ciliatum* R. Br. and other Epacridaceae. *Ann. Bot.* 77:375–81
5. Aylmore LAG. 1993. Use of computer-assisted tomography in studying water movement around plant roots. *Adv. Agron.* 49:1–54
6. Bacic A, Moody SF, Clarke AE. 1986. Structural analysis of secreted root slime from maize (*Zea mays* L.). *Plant Physiol.* 80:771–76
7. Bengough AG, McKenzie BM. 1997. Sloughing of root cap cells decreases the

frictional resistance to maize (*Zea mays* L.) root growth. *J. Exp. Bot.* 48:885–93

8. Böhm W. 1979. *Methods of Studying Root Systems*. Berlin: Springer. 188 pp.

9. Bowen GD, Rovira AD. 1991. The rhizosphere. The hidden half of the hidden half. In *Plant Roots. The Hidden Half*, ed. Y Waisel, A Eshel, U Kafkaf, pp. 641–69. New York: Marcel Dekker. 948 pp. 1st ed.

10. Box JE Jr. 1996. Modern methods for root investigations. See Ref. 129a, pp. 193–237

11. Brady DJ, Wenzel CL, Fillery IRP, Gregory PJ. 1995. Root growth and nitrate uptake by wheat (*Triticum aestivum* L.) following wetting of dry surface soil. *J. Exp. Bot.* 46:557–64

12. Burgeff H. 1961. *Mikrobiologie des Hochmoores*. Stuttgart: Gustav Fischer. 178 pp.

13. Burley JWA, Nwoke FIO, Leister GL, Popham RA. 1970. The relationship of xylem maturation to the absorption and translocation of ^{32}P. *Am. J. Bot.* 57:504–11

14. Bushamuka VN, Zobel RW. 1998. Differential genotypic and root type penetration of compacted soil layers. *Crop Sci.* 38:776–81

15. Cahn MD, Zobel RW, Bouldin DR. 1989. Relationship between root elongation rate and diameter and duration of growth of lateral roots of maize. *Plant Soil* 119:271–79

16. Caldwell MM. 1994. Exploiting nutrients in fertile soil microsites. In *Exploitation of Environmental Heterogeneity by Plants. Ecophysiological Processes Above- and Belowground*, ed. MM Caldwell, RW Pearcy, pp. 325–47. New York: Academic. 429 pp.

17. Canny MJ. 1998. Applications of the compensating pressure theory of water transport. *Am. J. Bot.* 85:897–901

18. Chiarini L, Bevivino A, Dalmastri C, Nacamulli C, Tabacchioni S. 1998. Influence of plant development, cultivar and soil type on microbial colonization of maize roots. *Appl. Soil Ecol.* 8:11–18

19. Clarkson DT. 1996. Root structure and sites of ion uptake. See Ref. 129a, pp. 483–510

20. Couot-Gastelier J, Vartanian N. 1995. Drought-induced short roots in *Arabidopsis thaliana*: structural characteristics. *Bot. Acta* 108:407–13

21. Curl EA, Truelove B. 1986. *The Rhizosphere*. Berlin: Springer. 288 pp.

22. Degenhardt J, Larsen PB, Howell SH, Kochian L. 1998. Aluminum resistance in the *Arabidopsis* mutant *alr-104* is caused by an aluminum-induced increase in rhizosphere pH. *Plant Physiol.* 117:19–27

23. de Weger LA, van der Bij AJ, Dekkers LC, Simons M, Wijffelman CA, et al. 1995. Colonization of the rhizosphere of crop plants by plant-beneficial pseudomonada. *FEMS Microbiol. Ecol.* 17: 221–28

24. Di Cello F, Bevivino A, Chiarini L, Fani R, Paffetti D, et al. 1997. Biodiversity of a *Burkholderia cepacia* population isolated from the maize rhizosphere at different plant growth stages. *Appl. Environ. Microbiol.* 63:4485–93

25. Dinkelaker B, Hengeler C, Marschner H. 1995. Distribution and function of proteoid roots and other root clusters. *Bot. Acta* 108:183–200

26. Dittmer HJ. 1937. A quantitative study of the roots and root hairs of a winter rye plant (*Secale cereale*). *Am. J. Bot.* 24:417–20

27. Dittmer HJ. 1938. A quantitative study of the subterranean members of three field grasses. *Am. J. Bot.* 25:654–57

28. Dodd J, Heddle EM, Pate JS, Dixon KW. 1984. Root patterns of sandplain plants and their functional significance. In *Kwongan Plant Life of the Sandplain*, ed. JS Pate, JS Beard, pp. 146–77. Nedlands, WA: Univ. West. Australia Press. 284 pp.

29. Dorioz JM, Robert M, Chenu C. 1993. The role of roots, fungi and bacteria on clay particle organization. An experimental approach. *Geoderma* 56:179–94

30. Doussan C, Pagès L, Vercambre G. 1998. Modelling of the hydraulic architecture of root systems: an integrated approach to water absorption. Model description. *Ann. Bot.* 81:213–23

31. Doussan C, Vercambre G, Pagès L. 1998. Modelling of the hydraulic architecture of root systems: an integrated approach to water absorption. Distribution of axial and radial conductances in maize. *Ann. Bot.* 81:225–32

32. Dubrovsky JG, North GB, Nobel PS. 1998. Root growth, developmental changes in the apex, and hydraulic conductivity for *Opuntia ficus-indica* during drought. *New Phytol.* 138:75–82

32a. Errede LA. 1983. Correlation of water uptake and root exudation. *Ann. Bot.* 52:373–80

33. Eshel A, Waisel Y. 1996. Multiform and multifunction of various constituents of one root system. See Ref. 129a, pp. 175–92

34. Estabrook EM, Yoder JL. 1998. Plant-plant communications: rhizosphere signaling between parasitic angiosperms and their hosts. *Plant Physiol.* 116:1–7

35. Fischer WR, Flessa H, Schaller G. 1989. pH values and redox potentials in microsites of the rhizosphere. *Z. Pflanzenernahr. Bodenk.* 152:191–95

36. Fitter A. 1996. Characteristics and function of root systems. See Ref. 129a, pp. 1–20

37. Floyd RA, Ohlrogge AJ. 1971. Gel formation on nodal root surfaces. Some observations relevant to understanding its action at the root-soil interface. *Plant Soil* 34:595–606

38. Foster RC, Rovira AD, Cock TW. 1983. *Ultrastructure of the Root-Soil Interface.* St. Paul: Am. Phytopathol. Soc. 157 pp.

39. Fusseder A. 1984. Der Einfluss von Bodenart, Durchlüftung des Boden, N-Ernährung und Rhizosphärenflora auf die Morphologie des seminalen Wurzelsystems von Mais. *Z. Pflanzenernaehr. Bodenk.* 147:553–64

40. Fusseder A. 1987. The longevity and activity of the primary root of maize. *Plant Soil* 101:257–65

41. Gochnauer MB, McCully ME, Labbé H. 1989. Different populations of bacteria associated with sheathed and bare regions of roots of field-grown maize. *Plant Soil* 114:107–20

42. Granato TC, Raper CD Jr. 1989. Proliferaton of maize (*Zea mays* L.) roots in response to localized supply of nitrate. *J. Exp. Bot.* 40:267–75

43. Guinel FC, McCully ME. 1986. Some water-related properties of maize root-cap mucilage. *Plant Cell Environ.* 9:657–66

44. Guinel FC, McCully ME. 1987. The cells shed by the root cap of *Zea*: their origin and some structural and physiological properties. *Plant Cell Environ.* 10:565–78

45. Handelsman J, Stabb EV. 1996. Biocontrol of soil-borne pathogens. *Plant Cell* 8:1855–69

46. Hawes MC. 1990. Living plant cells released from the root cap: a regulator of microbial populations in the rhizosphere? *Plant Soil:* 129:19–27

47. Hawes MC, Lin HJ. 1990. Correlation of pectolytic enzyme activity with the programmed release of cells from root caps of pea (*Pisum sativum*). *Plant Physiol.* 94:1855–59

47a. Higginbotham N, Davis RF, Mertz SM, Shumway LK. 1972. Some evidence that radial transport in maize roots is into living vessels. In *Ion Transport in Plants*, ed. WP Anderson, pp. 493–506. London: Academic. 630 pp.

48. Hinch JM, Clarke AE. 1980. Adhesion of fungal zoospores to root surfaces is mediated by carbohydrate determinants of the root slime. *Physiol. Plant Pathol.* 16:303–407

49. Hoppe DC, McCully ME, Wenzel CL. 1986. The nodal roots of *Zea*: their development in relation to structural features of the stem. *Can. J. Bot.* 64:2524–37

50. Horst WJ, Wagner A, Marschner H. 1982. Mucilage protects root meristems from aluminium injury. *Z. Pflanzenphysiol.* 105:435–44

51. Huang B, North GB, Nobel PS. 1993. Soil sheaths, photosynthate distribution to roots, and rhizosphere water relations for *Opuntia ficus-indica*. *Int. J. Plant Sci.* 154:425–31

52. Huang B, Taylor HM, McMichael BL. 1991. Effects of temperature on the development of metaxylem in primary wheat roots and its hydraulic consequences. *Ann. Bot.* 67:163–66

53. Huang CX, Van Steveninck RFM. 1988. Effect of moderate salinity on patterns of potassium, sodium and chloride accumulation in cells near the root tip of barley. Role of differentiating metaxylem vessels. *Physiol. Plant.* 73:525–33

54. Irving HR, Grant BR. 1984. The effects of pectin and plant root surface carbohydrates on encystment and development of *Phytophthora cinnamomi* zoospores. *J. Gen. Microbiol.* 130:1015–18

55. Ješko J, Navara J, Dekánková. 1997. Root growth and water uptake by flowering maize plants under drought conditions. In *Biology of Root Formation and Development*, ed. A Altman, Y Waisel, pp. 270–71. New York: Plenum. 376 pp.

56. Jordan M-O, Harada J, Bruchou C, Yamazaki K. 1993. Maize nodal root ramification: absence of dormant primordia, root classification using histological parameters and consequences on sap conduction. *Plant Soil* 153:125–43

57. Jupp AP, Newman EI. 1987. Morphological and anatomical effects of severe drought on the roots of *Lolium perenne* L. *New Phytol.* 105:393–402

58. Keerthisinghe G, Hocking PJ, Ryan PR, Delhaize E. 1998. Effect of phosphorus supply on the formation and function of proteoid roots of white lupin (*Lupinus albus* L.). *Plant Cell Environ.* 21:467–78

59. Kevekordes KG, McCully ME, Canny MJ. 1988. Late maturation of large metaxylem vessels in soybean roots: significance for water and nutrient supply to the shoot. *Ann. Bot.* 62:105–17

60. Kirkegaard JA, Sarwar M. 1998. Glucosinolate profiles of Australian canola (*Brassica napus annua*) and Indian mustard (*Brassica jubcea*) cultivars—implications for biofumigation. *Aust. J. Agric. Sci.* In press

61. Klepper B. 1987. Origin, branching and distribution of root systrems. In *Root Development and Function*, ed. PJ Gregory, JV Lake, DA Rose, pp. 103–24. Cambridge: Cambridge Univ. Press. 206 pp.

62. Kodama H, Nelson S, Yang AF, Kohyama N. 1994. Mineralogy of rhizospheric and non-rhizospheric soils in corn fields. *Clay Miner.* 42:755–63

63. Krassovsky I. 1926. Physiological activity of the seminal and nodal roots of crop plants. *Soil Sci.* 21:307–25

64. Kraus M, Fusseder A, Beck E. 1987. Development and replenishent of the P-depletion zone around the primary root of maize during the vegetative period. *Plant Soil* 101:247–55

65. Lamont B. 1982. Mechanisms for enhancing nutrient uptake in plants with particular reference to Mediterranean South Africa and Western Australia. *Bot. Rev.* 48:597–689

66. Larsen PB, Degenhardt J, Tai CY, Stenzler LM, Howell SH, Kochian LV. 1998. Aluminum resistant *Arabidopsis* mutants that exhibit altered patterns of aluminum accumulation and organic acid release from roots. *Plant Physiol.* 117:9–18

67. Lazof DB, Rufty TW Jr, Redinbaugh MG. 1992. Localization of nitrate absorption and translocation within morphological regions of the corn root. *Plant Physiol.* 100:1251–58

68. Leyser O, Fitter A. 1998. Roots are branching out in patches. *Trends Plant Sci.* 3:203–4

69. Luxovà M, Kozinka V. 1970. Structure and conductivity of the corn root system. *Biol. Plant. (Praha)* 12:47–57

70. Lyford WH. 1975. Rhizography of non-woody roots of trees in the forest floor. In *The Development and Function of Roots*, ed. JG Torrey, DT Clarkson, pp. 179–96. New York: Academic. 618 pp.

71. Lynch J. 1995. Root architecture and plant productivity. *Plant Physiol.* 109:7–13

72. Maertens C. 1971. Etude expérimentale de l'alimentation minérale et hydrique du mais. Capacité d'absorption des parties basales et apicales de *Zea mays*. *C. R. Acad. Sci. (Paris)* 273:730–32

73. Marschner H. 1995. *Mineral Nutrition of Higher Plants*. New York: Academic. 889 pp. 2nd ed.

74. Marschner H, Römheld V, Horst WJ, Martin P. 1986. Root-induced changes in the rhizosphere: importance for the mineral nutrition of plants. *Z. Pflanzenernähr. Bodenk.* 149:441–56

75. McCully ME. 1994. Accumulation of high levels of potassium in the developing xylem elements in roots of soybean and some other dicotyledons. *Protoplasma* 183:116–25

76. McCully ME. 1995. How do real roots work? Some new views of root structure. *Plant Physiol.* 109:1–6

77. McCully ME. 1995. Water efflux from the surface of field-grown grass roots. Observations by cryo-scanning electron microscopy. *Physiol. Plant.* 95:217–24

78. McCully ME, Boyer JS. 1997. The expansion of maize root-cap mucilage during hydration. 3. Changes in water potential and water content. *Physiol. Plant.* 99:169–77

79. McCully ME, Canny MJ. 1985. Location of translocated ^{14}C in roots and root exudates of field-grown maize. *Physiol. Plant.* 65:380–92

80. McCully ME, Canny MJ. 1988. Pathways and processes of water and nutrient uptake in roots. *Plant Soil* 111:159–70

81. McCully ME, Canny MJ. 1994. Contributions of the surface of the root tip to the growth of *Zea* roots in soil. *Plant Soil* 165:315–321

82. McCully ME, Canny MJ, Van Steveninck RFM. 1987. Accumulation of potassium by differentiating metaxylem elements of maize roots. *Physiol. Plant.* 69:73–80

83. McCully ME, Mallett J. 1993. The branch roots of *Zea*. III. Vascular connections and bridges for nutrient recycling. *Ann. Bot.* 71:327–41

84. McCully ME, Sealey LJ. 1996. The expansion of maize root-cap mucilage during hydration. 2. Observations on soil-grown roots by cryo-scanning electron microscopy. *Physiol. Plant.* 97:454–62

85. Miki NK, Clark KJ, McCully ME. 1980. Histological and histochemical comparison of the mucilages on the root tips of several grasses. *Can. J. Bot.* 58:2581–93

86. Mitchell JE. 1976. The effect of roots on the activity of soil-borne plant pathogens. In *Encyclopedia of Plant Physiology*, New Ser. 4, ed. R. Heitefuss, PH

Williams, pp. 104–28. Berlin: Springer. 890 pp.

87. Morel JL, Habib L, Plantureaux S, Guckert A. 1991. Influence of maize root mucilage on soil aggregate stability. *Plant Soil* 136:111–19

88. Morel JL, Mench M, Guckert A. 1986. Measurement of Pb^{2+}, Cu^{2+} and Cd^{2+} binding with mucilage exudates from maize (*Zea mays* L.) roots. *Biol. Fertil. Soils* 2:29–34

89. Morita S, Thongpae S, Abe J, Nakamoto T, Yamazaki K. 1992. Root branching in maize. 1. "Branching index" and methods for measuring root length. *Jpn. J. Crop Sci.* 61:101–6

90. Nacamulli C, Bevivino A, Dalmastri C, Tabacchioni S, Chiarini L. 1997. Perturbation of maize rhizosphere microflora following seed bacterization with *Burkholderia cepacia* MCI7. *FEMS Microbiol. Ecol.* 23:183–93

91. Nambiar EKS. 1976. The uptake of zinc-65 by roots in relation to soil water content and root growth. *Aust. J. Soil Res.* 14:67–74

92. Nambiar EKS. 1976. Uptake of Zn^{65} from dry soil by plants. *Plant Soil* 44:267–71

93. Navara L, Ješko T, Duchoslav S. 1994. Participation of seminal roots in water uptake by maize root system. *Biologia (Bratislava)* 49:91–95

94. Navara J, Ješko T, Ziegler W, Duchoslav S. 1993. Water uptake by maize (*Zea mays* L.) root system. *Biologia (Bratislava)* 48:113–17

95. Neumann G, Dinkelaker B, Marschner H. 1996. Kurzzeitige Abgabe organischer Säuren aus Proteoidwurzeln von *Hakea undulata* (Proteaceae). In *Pflanzliche Stoffaufnahme und Microbielle Wechselwirkungen in der Rhizosphäre*, ed. W Merbach, pp. 128–36 Stuttgart: Teubner

96. North GB, Huang B, Nobel PS. 1993. Changes in structure and hydraulic conductivity for root junctions of desert succulents as soil water status varies. *Bot. Acta* 106:126–35

97. North GB, Nobel PS. 1997. Drought-induced changes in soil contact and hydraulic conductivity for roots of *Opuntia ficus-indica* with and without rhizosheaths. *Plant Soil* 191:249–58

98. Oades JM. 1978. Mucilages at the root surface. *J. Soil Sci.* 29:1–16

99. Pagès L, Pellerin S. 1994. Evaluation of parameters describing the root system architecture of field grown maize plants (*Zea mays* L.). II. Density, length, and branching of first-order lateral roots. *Plant Soil* 164:169–76

100. Pallant E, Holmgren RA, Schuler GE, McCraken KL, Drbal B. 1993. Using a fine root extraction device to quantify small diameter corn roots (≥ 0.025 mm) in field soils. *Plant Soil* 153:273–79

101. Passioura JB. 1988. Water transport in and to roots. *Annu. Rev. Plant Physiol. Plant Mol. Biol.* 39:245–65

102. Pate JS, Dawson TE. 1998. Novel techniques for assessing the performance of woody plants in uptake and utilization of carbon, water and nutrients: implications for designing agricultural mimic systems. In *Agriculture as a Mimic of Natural Ecosystems*, ed. EC Lefroy, RJ Hobbs, MH O'Connor, JS Pate. Dordrecht: Kluwer. In press

103. Pavlychenko TK. 1937. Quantitative study of the entire root systems of weed and crop plants under field conditions. *Ecology* 18:62–79

104. Phillips DA. 1992. Flavonoids: plant signals to soil microbes. *Recent Adv. Phytochem.* 26:201–31

105. Picard D, Jordan M-O, Trendel R. 1985. Rythme d'apparition des racines primaires du maïs (*Zea mays* L.) I. Etude détaillée pour une variété en un lieu donné. *Agronomie* 5:667–76

106. Pueppke SG, Bolaños-Vásquez MC, Werner D, Bec-Ferté M-P, Promé J-C. et al. 1998. Release of flavonoids by the soybean cultivars McCall and Peking and their perception as signals by the nitrogen-fixing symbiont *Sinorhizobium fredii*. *Plant Physiol.* 117:599–608

107. Read DB, Gregory PJ. 1997. Surface tension and viscosity of axenic maize and lupin mucilages. *New Phytol.* 13:623–28

108. Reidenbach G, Horst WJ. 1997. Nitrate-uptake capacity of different root zones of *Zea mays* (L.) in vitro and in situ. *Plant Soil* 196:295–300

109. Robinson D. 1996. Resource capture by localized root proliferation: Why do plants bother? *Ann. Bot.* 77:179–85

110. Römheld V, Müller Ch, Marschner H. 1984. Localization and capacity of proton pumps in roots of intact sunflower plants. *Plant Physiol.* 76:603–6

111. Rougier M, Chaboud A. 1985. Mucilages secreted by roots and their biological function. *Isr. J. Bot.* 34:129–46

112. Ryan PR, Ditomaso JM, Kochian L. 1993. Aluminum toxicity in roots: an investigation of spatial sensitivity and the role of the root cap. *J. Exp. Bot.* 44:437–46

113. Sanderson J, Whitbread JC, Clarkson DT. 1988. Resistant xylem cross-walls reduce the axial hydraulic conductivity in the apical 20 cm of barley seminal root axes: implications for the driving force for water movement. *Plant Cell Environ.* 11:247–56

114. Schaller G, Fischer WR. 1985. pH-Änderungen in der Rhizosphäre von Mais- und Erdnusswurzeln. *Z. Pflanzenernaehr. Bodenk.* 148:306–20

115. Siqueira JO, Nair MG, Hammerschmidt R, Safir GR. 1991. Significance of phenolic compounds in plant-soil-microbial systems. *Crit. Rev. Plant Sci.* 10:63–123

116. Sivasithamparam K, Parker CA. 1979. Rhizosphere microorganisms of seminal and nodal roots of wheat grown in pots. *Soil Biol. Biochem.* 11:155–60

117. Stasovski E, Peterson CA. 1991. The effects of drought and subsequent rehydration on the structure and vitality of *Zea mays* seedling roots. *Can. J. Bot.* 69:1170–78

118. Starkey RL. 1929. Some influences of the development of higher plants upon the microorganisms in the soil. II. Influence of the stage of plant growth upon abundance of organisms. *Soil Sci.* 27:355–78

119. St. Aubin G, Canny MJ, McCully ME. 1986. Living vessel elements in the late metaxylem of sheathed maize roots. *Ann. Bot.* 58:577–88

120. Tardieu F, Pellerin S. 1990. Trajectory of the nodal roots of maize in fields with low mechanical constraint. *Plant Soil* 124:39–45

121. Topp GC, Watt M, Hayhoe HN. 1996. Point specific measurement and monitoring of soil water content with emphasis on TDR. *Can. J. Soil Sci.* 76:307–16

122. Uren NC. 1993. Mucilage secretion and its interaction with soil, and contact reduction. *Plant Soil* 155/156:79–82

123. Uren NC, Reisenauer HM. 1988. The role of root exudates in nutrient acquisition. *Adv. Plant Nutr.* 3:79–114

124. Varney G, Canny MJ. 1993. Rates of water uptake into the mature root system of maize plants. *New Phytol.* 123:775–89

125. Varney GT, Canny MJ, Wang XL, McCully ME. 1991. The branch roots of *Zea*. I. First order branches, their number, sizes and division into classes. *Ann. Bot.* 67:357–64

126. Varney GT, McCully ME. 1991. The branch roots of *Zea*. II. Developmental loss of the apical meristem in field-grown roots. *New Phytol.* 118:535–46

127. Vartanien N. 1997. The drought rhizogenesis. See Ref. 129a, pp. 471–82

128. Vermeer J, McCully ME. 1982. The rhizosphere in *Zea*: new insight into its structure and development. *Planta* 156:45–61

129. Vetterlein D, Marschner H. 1993. Use of a microtensiometer technique to study hydraulic lift in a sandy soil planted with pearl millet (*Pennisetum americanum* L. Leeke). *Plant Soil* 149:275–82

129a. Waisel Y, Eshel A, Kafkafi U, eds. 1991. *Plant Roots. The Hidden Half.* New York: Marcel Dekker. 1002 pp. 2nd ed.

130. Wang XL, Canny MJ, McCully ME. 1991. The water status of the roots of soil-grown maize in relation to the maturity of their xylem. *Physiol. Plant.* 82:157–62

131. Wang XL, McCully ME, Canny MJ. 1994. The branch roots of *Zea*. IV. The maturation and openness of xylem conduits in first-order branches of soil-grown roots. *New Phytol.* 126:21–29

132. Wang XL, McCully ME, Canny MJ. 1995. The branch roots of *Zea*. V. Structural features that may influence water and nutrient uptake. *Bot. Acta* 108:209–19

133. Watt M, McCully ME, Jeffree CE. 1993. Plant and bacterial mucilages of the maize rhizosphere: comparison of their soil binding properties and histochemistry in a model system. *Plant Soil* 151:151–65

134. Watt M, McCully ME, Canny MJ. 1994. Formation and stabilization of rhizosheaths in *Zea mays* L. Effect of soil water content. *Plant Physiol.* 106:179–86

135. Weaver JE. 1926. *Root Development of Field Crops.* New York: McGraw-Hill. 291 pp.

136. Weaver JE, Zink E. 1945. Extent and longevity of the seminal root of certain grasses. *Plant Physiol.* 20:359–79

137. Wenzel CL, McCully ME, Canny MJ. 1989. Development of water conducting capacity in the root systems of young plants of corn and some other C4 grasses. *Plant Physiol.* 89:1094–101

138. Wenzel CL, McCully ME. 1991. Early senescence of cortical cells in roots of cereals. How good is the evidence? *Am. J. Bot.* 78:1528–41

139. Wullstein LH, Pratt SA. 1981. Scanning electron microscopy of rhizosheaths of *Oryzosis hymenoides. Am. J. Bot.* 68:408–19

140. Young IM. 1995. Variations in moisture

contents between bulk soil and the rhizo-sheath of wheat (*Triticum aestivum* L. cv Wembley). *New Phytol.* 130:135–39

141. Young IM. 1998. Biophysical interactions at the root-soil interface: a review. *J. Agric. Sci.* 130:1–7

142. Zhang H, Forde BG. 1998. An *Ara-bidopsis* MADS-box gene that controls nutrient-induced changes in root architecture. *Science* 279:407–9

143. Zheng SJ, Ma JF, Matsumoto H. 1998. High aluminum resistance in buckwheat. 1. Al-induced specific secretion of oxalic acid from root tips. *Plant Physiol.* 117:745–51

SUBJECT INDEX

M

CUMULATIVE INDEXES

CONTRIBUTING AUTHORS, VOLUMES 40–50

CHAPTER TITLES, VOLUMES 40–50

751

ACCLIMATION AND ADAPTATION

Economic Botany

Physiological Ecology

Plant Genetics/Evolution

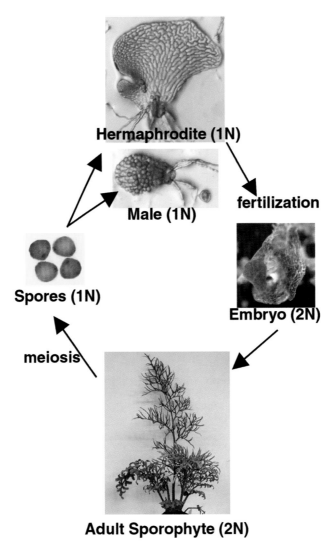

Hermaphrodite (1N)

Male (1N)

fertilization

Spores (1N)

Embryo (2N)

meiosis

Adult Sporophyte (2N)

Figure 1 The life cycle of *Ceratopteris richardii,* a homosporous fern.

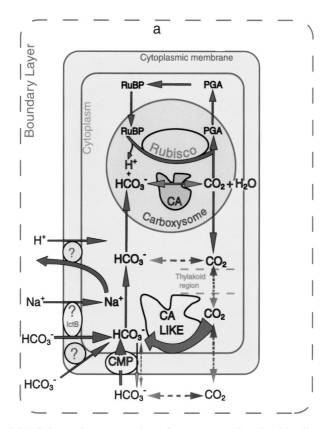

Figure 2(a) Schematic presentation of components involved in the cyanobacterial inorganic carbon concentrating mechanism, showing the boundary layer, the cytoplasmic membrane, and the cytoplasm containing the thylakoid region and a carboxysome. Three possible HCO_3^- transporters are indicated: CMP, a primary HCO_3^- transporter of the ABC type directly fueled by ATP (94; T Omata, unpublished data); IctB, a putative Na^+/HCO_3^- symporter (17) fueled by the $Na+$ electrochemical gradient generated either by a primary Na^+ pump (not shown) or a Na^+/H^+ antiporter (shown) secondary to an H^+ pump; and a low affinity HCO_3^- transporter (17) of unknown nature (designated by a question mark). Efflux of HCO_3^- occurs via these transporters (shown only for the case of CMP) and by diffusion. As explained in the text, uptake of CO_2 may occur via diffusion across the cytoplasmic membrane followed by conversion to HCO_3^- by a CA-like moiety. Active CO_2 transport has also been proposed (82). Accumulated HCO_3^- penetrates the carboxysome where CA activity generates CO_2 at high concentration in the vicinity of Rubisco. Part of the CO_2 is fixed, part diffuses outwards. Regeneration of ribulose 1,5-bisphosphate occurs outside the carboxysome.

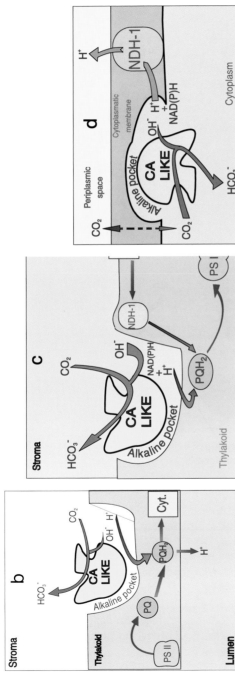

Figure 2 (b–d) Suggested mechanisms whereby photosynthetic or respiratory electron transport might lead to the formation of alkaline pockets in thylakoid or cytoplasmic membranes with consequent conversion of CO_2 to HCO_3^-. Such pockets might result from reduction of PQ (plastoquinone) by either linear (*b*) or cyclic (*c*) photosynthetic electron transport possibly involving NAD(P) dehydrogenase [NDH (94)] or (*d*) from the activity of cytoplasmic membrane-located NDH-1 (104). CA-like activity in the pocket would accelerate HCO_3^- generation consequent on the pH shift. Hydrophobic surfaces would slow diffusion of H^+ into the pocket and thus raise the efficiency of CO_2 to HCO_3^- conversion.

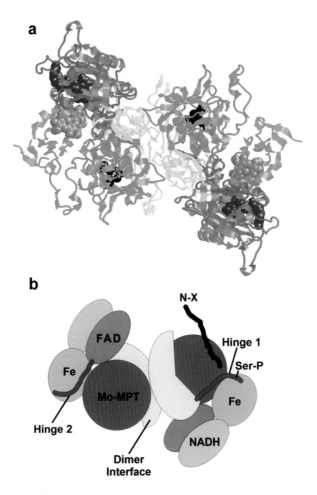

Figure 3 "Working model" for the three-dimensional structure of the holo-nitrate reductase homodimer. (*a*) Ribbon-model rendering of dimeric nitrate reductase with coordinates derived from docking two cytochrome c reductase fragments (57) on an atom replacement model of the dimer of *Arabidopsis* NIA2 (residues 91 to 490) generated from sulfite oxidase A and B chains [51; 1SOX in PDB (Protein Data Base)]. (*b*) Schematic model of nitrate reductase dimer. From the blocked N terminus *(N-X)*, the order of domains and hinge regions are: N-terminal region *(black tube)**; Mo-molybdopterin *(Mo-MPT)* domain *(dark green with Mo-MPT black)*; interface domain *(yellow)*; Hinge 1 *(gray tube with phosphorylated Ser534)**; cytochrome b domain *(light green with heme-Fe purple)*; Hinge 2 *(gray tube)**; cytochrome b reductase fragment [in (*a*) *red* with FAD *blue*; in (*b*) FAD domain *red* and *NADH domain pink*]. Asterisked regions are not included in (*a*).

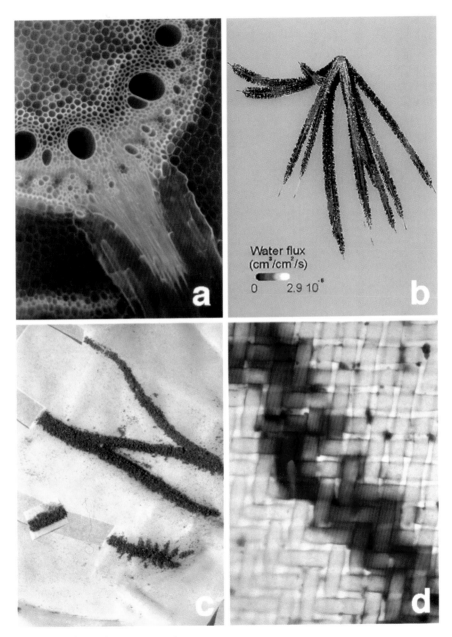

Figure 2a legend appears on the next page.

Figure 2 a Transverse section of a corn nodal root at the point of emergence of a fine branch root. *b* Model of a corn root system in soil incorporating measured values for water flux into the roots. The deepest root extends ~1 m into the soil; the horizontal axis is also ~1 m broad. Reproduced from (31) with permission from *Annals of Botany. c* Rhizosheaths formed by corn roots that were growing in soil along the underside of this nylon mesh (5 μm openings), which was surrounded by soil. Root hairs did not penetrate the mesh. Coherent rhizosheaths formed on the upper side of the mesh and remained firmly attached to the mesh when it was removed from the soil and gently shaken. *d* Root-cap mucilage left behind as a corn root tip extended along the underside of this piece of nylon mesh. Water efflux from the root expanded the mucilage through the mesh, here stained to show the acidic polysaccharide component. Such expanded mucilage in soil binds rhizosheaths as shown in Figure 2c.

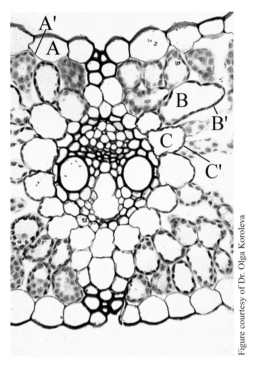

Figure courtesy of Dr. Olga Koroleva

Figure 2 Mapping of the water relations parameters of the apoplast of a barley leaf. The pressure probe and SiCSA is used to measure both P_{cell} and π_{cell} of protoplasts in situ, e.g. *A* (epidermis), *B* (mesophyll) and *C* (bundle sheath). Since the membranes approximate to ideal semipermeable membranes, these cells behave as in situ osmometers and they respond to the water potential of the apoplast (*A'*, *B'*, and *C'*) surrounding each individual cell. If P_{wall} can be raised to atmospheric a second measurement of P_{cell} and π_{cell} will allow an estimate of the components of apoplast water potential (P_{wall} and π_{wall}) individually. ($P_{cell}-\pi_{cell} = \Psi_{cell} = \Psi_{wall} = P_{wall}-\pi_{wall}$). We have found sizeable gradients of both P_{wall} and π_{wall} between *A'*, *B'*, and *C'*. In addition, SiCSA techniques have been used to map individual solute concentrations in cell types *A, B,* and *C* (see text).

Annual Reviews
THE INTELLIGENT SYNTHESIS OF SCIENTIFIC LITERATURE

ANNUAL REVIEW OF:	INDIVIDUALS U.S.	Other countries	INSTITUTIONS U.S.	Other countries
ANTHROPOLOGY				
Vol. 28 (avail. Oct. 1999)	$55	$60	$110	$120
Vol. 27 (1998)	$55	$60	$110	$120
ASTRONOMY & ASTROPHYSICS				
Vol. 37 (avail. Sept. 1999)	$70	$75	$140	$150
Vol. 36 (1998)	$70	$75	$140	$150
BIOCHEMISTRY				
Vol. 68 (avail. July 1999)	$68	$74	$136	$148
Vol. 67 (1998)	$68	$74	$136	$148
BIOMEDICAL ENGINEERING New Series!				
Vol. 1 (avail. August 1999)	$62	$67	$124	$134
BIOPHYSICS & BIOMOLECULAR STRUCTURE				
Vol. 28 (avail. June 1999)	$70	$75	$140	$150
Vol. 27 (1998)	$70	75	$140	$150
CELL & DEVELOPMENTAL BIOLOGY				
Vol. 15 (avail. Nov. 1999)	$64	$69	$128	$138
Vol. 14 (1998)	$64	$69	$128	$138
COMPUTER SCIENCE (suspended)				
Call Customer Service or see our Web site for pricing.				
EARTH & PLANETARY SCIENCES				
Vol. 27 (avail. May 1999)	$70	$75	$140	$150
Vol. 26 (1998)	$70	$75	$140	$150
ECOLOGY & SYSTEMATICS				
Vol. 30 (avail Nov. 1999)	$60	$65	$120	$130
Vol. 29 (1998)	$60	$65	$120	$130
ENERGY & THE ENVIRONMENT				
Vol. 24 (avail. Oct. 1999)	$76	$81	$152	$162
Vol. 23 (1998)	$76	$81	$152	$162
ENTOMOLOGY				
Vol. 44 (avail. Jan. 1999)	$60	$65	$120	$130
Vol. 43 (1998)	$60	$65	$120	$130

BACK VOLUMES ARE AVAILABLE Visit http://AnnualReviews.org for information

ANNUAL REVIEW OF:	INDIVIDUALS U.S.	Other countries	INSTITUTIONS U.S.	Other countries
FLUID MECHANICS				
Vol. 31 (avail. Jan. 1999)	$60	$65	$120	$130
Vol. 30 (1998)	$60	$65	$120	$130
GENETICS				
Vol. 33 (avail. Dec. 1999)	$60	$65	$120	$130
Vol. 32 (1998)	$60	$65	$120	$130
IMMUNOLOGY				
Vol. 17 (avail. April 1999)	$64	$69	$128	$138
Vol. 16 (1998)	$64	$69	$128	$138
MATERIALS SCIENCE				
Vol. 29 (avail. Aug. 1999)	$80	$85	$160	$170
Vol. 28 (1998)	$80	$85	$160	$170
MEDICINE				
Vol. 50 (avail. Feb. 1999)	$60	$65	$120	$130
Vol. 49 (1998)	$60	$65	$120	$130
MICROBIOLOGY				
Vol. 53 (avail. Oct. 1999)	$60	$65	$120	$130
Vol. 52 (1998)	$60	$65	$120	$130
NEUROSCIENCE				
Vol. 22 (avail. March 1999)	$60	$65	$120	$130
Vol. 21 (1998)	$60	$65	$120	$130
NUCLEAR & PARTICLE SCIENCE				
Vol. 49 (avail. Dec. 1999)	$70	$75	$140	$150
Vol. 48 (1998)	$70	$75	$140	$150
NUTRITION				
Vol. 19 (avail. July 1999)	$60	$65	$120	$130
Vol. 18 (1998)	$60	$65	$120	$130
PHARMACOLOGY & TOXICOLOGY				
Vol. 39 (avail. April 1999)	$60	$65	$120	$130
Vol. 38 (1998)	$60	$65	$120	$130
PHYSICAL CHEMISTRY				
Vol. 50 (avail. Oct. 1999)	$64	$69	$128	$138
Vol. 49 (1998)	$64	$69	$128	$138
PHYSIOLOGY				
Vol. 61 (avail. March 1999)	$62	$67	$124	$134
Vol. 60 (1998)	$62	$67	$124	$134

ANNUAL REVIEW OF:	INDIVIDUALS U.S.	Other countries	INSTITUTIONS U.S.	Other countries
PHYTOPATHOLOGY				
Vol. 37 (avail. Sept. 1999)	$62	$67	$124	$134
Vol. 36 (1998)	$62	$67	$124	$134
PLANT PHYSIOLOGY & PLANT MOLECULAR BIOLOGY				
Vol. 50 (avail. June 1999)	$60	$65	$120	$130
Vol. 49 (1998)	$60	$65	$120	$130
POLITICAL SCIENCE New Series!				
Vol. 2 (avail. June 1999)	$60	$65	$120	$130
Vol. 1 (1998)	$60	$65	$120	$130
PSYCHOLOGY				
Vol. 50 (avail. Feb. 1999)	$55	$60	$110	$120
Vol. 49 (1998)	$55	$60	$110	$120
PUBLIC HEALTH				
Vol. 20 (avail. May 1999)	$64	$69	$128	$138
Vol. 19 (1998)	$64	$69	$128	$138
SOCIOLOGY				
Vol. 25 (avail. Aug. 1999)	$60	$65	$120	$130
Vol. 24 (1998)	$60	$65	$120	$130

Also Available From Annual Reviews:

The Excitement & Fascination Of Science	INDIVIDUALS U.S.	Other countries	INSTITUTIONS U.S.	Other countries
Vol. 4 (1995)	$50	$55	$50	$55
Vol. 3 (1990) 2-part set, sold as set only	$90	$95	$90	$95
Vol. 2 (1978)	$25	$29	$25	$29
Vol. 1 (1965)	$25	$29	$25	$29
Intelligence and Affectivity				
by Jean Piaget (1981)	$8	$9	$8	$9
Paperback Collections				
The Cytoskeleton	$21	$21	$21	$21
Genetic Flow	$21	$21	$21	$21
AIDS	$15	$18	$15	$18
Origins of Planets and Life	$15	$20	$15	$20
Hydrologic Processes from Catchment to Continental Scales	$15	$20	$15	$20

Annual Reviews

A nonprofit scientific publisher

4139 El Camino Way • P.O. Box 10139
Palo Alto, CA 94303-0139 USA

BB99

STEP 1 : ENTER YOUR NAME & ADDRESS

NAME

ADDRESS

CITY   STATE/PROVINCE   COUNTRY   POSTAL CODE

TODAY'S DATE   DAYTIME PHONE

E-MAIL ADDRESS   FAX NUMBER

☎ Phone **800-523-8635** (U.S. or Canada)
Orders **650-493-4400 ext. 1** (worldwide)

8 a.m. to 4 p.m. Pacific Time, Monday-Friday

FAX **650-424-0910**
Orders 24 hours a day

Mention
priority code
BB99
when placing
phone orders

STEP 4 : CHOOSE YOUR PAYMENT METHOD

☐ Check or Money Order Enclosed (US funds, made payable to "Annual Reviews")

☐ Bill Credit Card ☐ AmEx ☐ MasterCard ☐ VISA

Account No. _____

Signature _____

Exp. Date **MO/YR** Name _____
(print name exactly as it appears on credit card)

STEP 2 : ENTER YOUR ORDER

QTY	ANNUAL REVIEW OF:	VOL.	Place on Standing Order? SAVE 10% NOW WITH PAYMENT	PRICE	TOTAL
		#	☐ Yes, save 10% ☐ No	$	$
		#	☐ Yes, save 10% ☐ No	$	$
		#	☐ Yes, save 10% ☐ No	$	$
		#	☐ Yes, save 10% ☐ No	$	$
		#	☐ Yes, save 10% ☐ No	$	$

30% STUDENT/RECENT GRADUATE DISCOUNT (past 3 years) Not for standing orders. Include proof of status.

CALIFORNIA CUSTOMERS: Add applicable California sales tax for your location. $

CANADIAN CUSTOMERS: Add 7% GST (Registration # 121149029 RT). $

STEP 3 : CALCULATE YOUR SHIPPING & HANDLING

HANDLING CHARGE (Add $3 per volume, up to $9 max. per location). **Applies to all orders.** $

SHIPPING OPTIONS:
(No UPS to P.O. boxes)

U.S. Mail 4th Class Book Rate (surface). Standard option. FREE. $ N/C

UPS Ground Service ($3/volume. 48 contiguous U.S. states.) $

Please note expedited
shipping preference:

☐ UPS Next Day Air ☐ UPS Second Day Air ☐ US Airmail
☐ UPS Worldwide Express ☐ UPS Worldwide Expedited

Note option at left. We will calculate
amount and add to your total

Abstracts and content lists available on the World Wide Web at
http://AnnualReviews.org. **E-mail orders: service@annurev.org**

TOTAL $

Orders may also be placed through booksellers or subscription agents or through our Authorized Stockists. From Europe, the UK, the Middle East and Africa, contact: Gazelle Book Service Ltd, Lancaster, England, Fax (0) 1524-63232. From India, Pakistan, Bangladesh or Sri Lanka, contact: SARAS Books, New Dehli, India, Fax 91-11-941111.